Fourth Edition

Handbook of
ENVIRONMENTAL
HEALTH

Volume 2

Pollutant Interactions in
Air, Water, and Soil

Herman Koren • Michael Bisesi

NEHA

Co-published with the
National Environmental Health Association

CRC Press
Taylor & Francis Group
Boca Raton London New York

CRC Press is an imprint of the
Taylor & Francis Group, an **informa** business

CRC Press
Taylor & Francis Group
6000 Broken Sound Parkway NW, Suite 300
Boca Raton, FL 33487-2742

First issued in paperback 2017

ISBN-13: 978-1-56670-547-9 (hbk)
ISBN-13: 978-0-8153-8097-9 (pbk)

Library of Congress Cataloging-in-Publication Data

Koren, Herman.
    Handbook of environmental health / by Herman Koren and Michael Bisesi.
        p.   cm.
    Includes bibliographical references and index.
    Contents: v. 1. Biological, chemical, and physical agents of environmental related disease — v. 2. Pollution interactions in air, water, and soil.
    ISBN 1-56670-547-9 (alk. paper)
    1. Environmental health — Handbooks, manuals, etc. 2. Environmental Engineering — Handbooks, manuals, etc. I. Title: Environmental health. II. Bisesi, Michael S. III. Title.

RA565 .K67 2002
363.7—dc21   2002016110

Visit the Taylor & Francis Web site at
http://www.taylorandfrancis.com

and the CRC Press Web site at
http://www.crcpress.com

Fourth Edition

Handbook of
ENVIRONMENTAL
HEALTH Volume 2

*Pollutant Interactions in
Air, Water, and Soil*

Dedication

To Donna Lee Koren and Christine Bisesi, our wives, dearest and very best friends, for all they have done to enhance our lives and encourage us to teach our students the true significance of improving the environment for all people.

Foreword

I have spent a career in this field. My years of experience have taught me that environmental health might best be described as a colorful, complex, and diverse spectrum of interrelated topics. These topics range from the individual to the compounded effects of pollution; to the impacts of contaminated air, water, land, food, and indoor and outdoor environments; and to biological, physical, chemical, and radiological hazards. Because this field is so broad, to protect both our environment and our people properly, it is important that one possesses a credible understanding of basic science, laws and regulations, governmental and private programs, disease and injury identification, and prevention and control. Moreover, because so many environmental concerns impact on other elements of the environment, it is important that even the specialists within our field appreciate and understand the implications of how their issues can impact other environmental concerns. A reading and comprehension of the material in these books can help greatly in building that overarching understanding that professionals in this field need to have.

The *Handbook of Environmental Health* has been, for the past 23 years, an excellent source for gaining that needed understanding of interrelated and current environmental topics. The presentations offered by this publication are particularly helpful inasmuch as they are comprehensive while concise and basic. The two volumes cover basic and applied chemistry relative to both toxicity to humans and the fate of natural and anthropogenic contaminants in the environment; basic and applied microbiology relative to both pathogenicity to humans and the fate of natural and anthropogenic contaminants in the environment; current status of each environmental problem area; discussion of the problem; potential for intervention; resources available for use, standards, practices, and techniques utilized to resolve the problem; surveillance and evaluation techniques; appropriate controls, laws, and regulations; future research needs; large number of current references; state-of-the-art graphics, and major environmental, including industrial hygiene, sampling and analytic instruments.

The fourth edition of Volume I, Chapters 1 and 2, provides a significant understanding of basic new environmental issues, energy, emerging infectious diseases, recent laws, emerging microorganisms, toxicology, epidemiology, human physiology, and the effects of the environment on humans. The remainder of the chapters discuss a variety of indoor environmental issues, including food safety, food technology, insect and rodent control, pesticides, indoor environment, institutional environment, recreational environment, occupational environment, and instrumentation. Some of the new and significantly expanded and updated sections include: new food codes and programs; emerging and reemerging insectborne disease; insecticide resistance; pesticides and water quality; indoor air pollution; asthma; monitoring environmental disease; homelessness and disease; emerging zoonoses; bacteria in hospitals and nursing homes; fungal and viral agents in laboratories; prions; guidelines for infection control; principles of biosafety; a variety of risk assessment techniques; and an updated overview of the occupational environment.

The fourth edition of Volume II discusses a variety of outdoor environmental issues including changes to the Clean Air Act; PM 2.5; toxic air pollutants; risk assessment and air, water, solid, and hazardous waste and the interrelationship between these areas; methyl bromide; air quality index; air, water, and solid waste programs; technology transfer; methyl tertiary butyl ether; toxic releases from waste; technical tracking systems; biological processes and solid and hazardous wastes; storm water runoff; ocean dumping; waste to energy; toxics release inventory; brownfields; contaminated governmental facilities; Superfund update; geographic information systems; pollution prevention programs; environmental justice; new laws; waterborne disease update; national mapping; EnviroMapper; safe drinking water standards; maximum contaminant level; crossconnections, backflow, and back-siphonage; sewage pretreatment technologies; leaching field chambers; constructed wetlands; drip and spray irrigation systems; peat bed filters; zone of initial dilution; new laws; wetlands; nonpoint source pollution; national water quality assessments; dredging waste; trihalomethanes; environmental studies; bioterrorism; and the Federal Emergency Management Agency. There are chapters on air quality management, solid and hazardous waste management, private and public water supplies, swimming areas, plumbing, private and public sewage disposal and soils, water pollution water quality controls, terrorism and environmental health emergencies, and instrumentation.

Two well-regarded environmental health professionals have written these two books. They have each conducted extensive research and are extremely knowledgeable about all areas of the environmental, industrial hygiene, and health related fields.

Dr. Herman Koren is a founding director of the environmental health science and internship program at Indiana State University and is a professor emeritus there. He has gained respect as a researcher, teacher, consultant, and practitioner in the environmental health, hospital, and medical care fields as well as in management areas related to these fields; nursing homes; water and wastewater treatment plants; and other environmental and safety industries for the past 47 years. Since his retirement from Indiana State University in 1995, he has continued his lifelong quest to gain and interpret the latest knowledge possible and to share it with students and other professionals. He also continues to give numerous presentations and workshops and has written and rewritten several books.

Dr. Michael Bisesi, professor and chairperson, department of public health, and associate dean, graduate allied health programs, Medical College of Ohio, School of Allied Health, has been in the environmental and occupational health fields for 20 years. He, too, is respected as a researcher, teacher, consultant, practitioner, and administrator. In addition to his environmental science and industrial hygiene accreditations, he is an expert in human exposure assessment and environmental toxicology. He also holds appointments in the School of Pharmacology and School of Medicine, has written numerous scientific and technical articles and chapters in scientific books, and is the author or co-author of several additional books.

The books are user-friendly to a variety of individuals including generalist professionals as well as specialists, industrial hygiene personnel, health and medical personnel, managers, and students. These publications can be used to look up specific information or to gain deeper knowledge about an existing problem area. The section on surveillance techniques helps the individual decide the extent and nature of a problem. The appropriate and applicable standards, rules, and regulations help the reader resolve a problem. Further information and assistance can be gained through the resource area and through the review of many of the updated bibliographical references. Except for Chapters 1, 2, 11, 12, in Volume I, and Chapters 8 and 9 in Volume II, all chapters follow the same format, thereby making the books relatively easy to use. The extensive index for both volumes in each book is also very useful.

Thank you, Professors Koren and Bisesi, for providing environmental health professionals, new or seasoned, generalist or specialist, with such a helpful resource.

Nelson Fabian
Executive Director
National Environmental Health Association
February 28, 2002

Preface

This handbook, in two volumes, is designed to provide a comprehensive but concise discussion of each of the important environmental health areas, including energy, ecology and people, environmental epidemiology, risk assessment and risk management, environmental law, air quality management, food protection, insect control, rodent control, pesticides, chemical environment, environmental economics, human disease and injury, occupational health and safety, noise, radiation, recreational environment, indoor environments, medical care institutions, schools and universities, prisons, solid and hazardous waste management, water supply, plumbing, swimming areas, sewage disposal, soils, water pollution control, environmental health emergencies, and nuisance complaints.

Sufficient background material is introduced throughout these texts to provide students, practitioners, and other interested readers with an understanding of the areas under discussion. Common problems and potential solutions are described; graphs, computerized drawings, inspection sheets, and flowcharts are utilized as needed to consolidate or clarify textual material. All facts and data come from the most recent federal government documents, many of which date from the late 1990s and early 2000s. Rules and regulations specified will continue to be in effect into the early 2000s. For rapidly changing areas in which the existing material used is likely to become dated, the reader is referred to the appropriate sources under resources and in the bibliography to update a given environmental health area or portion of an area as needed. This enhances the value of the text by providing basic and current materials that will always be needed and secondary sources that will enable the reader to keep up to date.

These books are neither engineering texts nor comprehensive texts in each area of study. Their purpose is to provide a solid working knowledge of each environmental health area with sufficient detail for practitioners and students. The text can be used in basic courses in environmental health, environmental pollution, ecology, and environment and people that are offered at all universities and colleges in the United States and abroad. These courses are generally taught in departments of life science, geology, science education, environmental health, and health and safety. For general areas of study, the instructor can omit specific details, such as resources, standards, practices and techniques, and modes of surveillance and evaluation. This same approach may be used by schools of medicine, nursing, and allied health sciences for their students. These texts are also suitable for basic introductory courses in schools of public health, environmental health, and sanitary science, as well as junior colleges offering 2-year degree programs in sanitary science and environmental science.

Practitioners in a variety of environmental health and occupational health and safety fields will find these volumes handy references for resolving current problems and for obtaining a better understanding of unfamiliar areas. Practitioners and administrators in other areas, such as food processing, water-quality control, occupational health and safety, and solid and hazardous waste management, will also find these reference books useful.

High school teachers often must introduce environmental health topics in their classes and yet have no specific background in this area. These books could serve as a text in graduate education courses for high school teachers as well as a reference source.

Public interest groups and users of high school and community libraries will obtain an overall view of environmental problems by reading Chapter 1; Chapter 2; and the sections in each chapter titled "Background and Status, Problems, Potential for Intervention, Resources, and Control." This volume also supplies a concise reference for administrators in developing nations because it explains tested controls and provides a better understanding of environmental problems; various standards, practices, and techniques; and a variety of available resources.

The material divides easily into two separate courses. Course I would correspond to the content of Volume I and would include Chapter 1, Environment and Humans; Chapter 2, Environmental Problems and Human Health; Chapter 3, Food Protection; Chapter 4, Food Technology; Chapter 5, Insect Control; Chapter 6, Rodent Control; Chapter 7, Pesticides; Chapter 8, Indoor Environment; Chapter 9, Institutional Environment; Chapter 10, Recreational Environment; Chapter 11, Occupational Environment; and Chapter 12, Major Instrumentation for Environmental Evaluation of Occupational, Residential, and Public Indoor Settings.

Course II, corresponding to the content of the Volume II, would include Chapter 1, Air Quality Management; Chapter 2, Solid and Hazardous Waste Management; Chapter 3, Private and Public Water Supplies; Chapter 4, Swimming Areas; Chapter 5, Plumbing; Chapter 6, Private and Public Sewage Disposal and Soils; Chapter 7, Water Pollution and Water Quality Controls; Chapter 8, Terrorism and Environmental Health Emergencies; and Chapter 9, Major Instrumentation for Environmental Evaluation of Ambient Air, Water, and Soil.

Because the problems of the environment are so interrelated, certain materials must be presented at given points to give clarity and cohesiveness to the subject matter. As a result, the reader may encounter some duplication of materials throughout the text.

With the exception of Volume I, Chapters 1, 2, 11, and 12, and Volume II, Chapters 8 and 9, all the chapters have a consistent style and organization, facilitating retrieval. The introductory nature of Volume I (Chapters 1 and 2) as well as the unusual nature of Volume II (Chapter 8) do not lend themselves to the standard format. Volume I (Chapter 12) and Volume II (Chapter 9) discuss instrumentation for the specific areas of each volume and therefore do not follow standard format.

In Volume I (Chapter 1), the reader is introduced to the underlying problems, basic concerns, and basic philosophy of environmental health. The ecological, economic, and energy bases provided help individuals understand their relationship to the ecosystem and to the real world of economic and energy concerns. It also provides an understanding of the role of government and the environmental health practitioner in helping to resolve environmental and ecological dilemmas created by humans. Chapter 2 on human health helps the reader understand the relationship between biological, physical, and chemical agents, and disease and injury causation.

In Volume II, Chapter 8, the many varied facets of terrorism and environmental emergencies, nuisances, and special problems are discussed. Students may refer to other chapters of the text to obtain a complete idea of each of the problems and the potential solutions.

The general format of Volume I, Chapters 3 to 11, and Volume II, Chapters 1 to 7, is as follows:

STANDARD CHAPTER OUTLINE

1. Background and status (brief)
2. Scientific, technological, and general information
3. Problem
   a. Types
   b. Sources of exposure
   c. Impact on other problems
   d. Disease potential
   e. Injury potential
   f. Other sources of exposure contributing to problems
   g. Economics
4. Potential for intervention
   a. General
   b. Specific
5. Resources
   a. Scientific and technical; industry, labor, university; research groups
   b. Civic
   c. Governmental
6. Standards, practices, and techniques
7. Modes of surveillance and evaluation
   a. Inspections and surveys
   b. Sampling and laboratory analysis
   c. Plans review
8. Control
   a. Scientific and technological
   b. Governmental programs
   c. Other programs
   d. Education
9. Summary
10. Research needs
    - The background and status section of each chapter presents a brief introduction to, and the current status of, each problem area. An attempt has been made in each case to present the current status of the problem.
    - The problem section is subdivided into several important areas to give the reader a better grasp of the total concerns. To avoid disruption in continuity of the standard outline, the precise subtitles listed may not be found in each chapter. However, the content of the subtitles will be present. The subtitle, impact on other problems, is given as a constant reminder that one impact on the environment may precipitate numerous other problems.
    - The potential for intervention section is designed to succinctly illustrate whether a given problem can be controlled, the degree of control possible, and some techniques of control. The reader should refer to the controls section for additional information.

- Resources is a unique section providing a listing of scientific, technical, civic, and governmental resources available at all levels to assist the student and practitioner.
- The standards, practices, and techniques section is specifically geared to the reader who requires an understanding of some of the specifics related to surveys, environmental studies, operation, and control of a variety of program areas.
- The modes of surveillance and evaluation section explains many of the techniques available to determine the extent and significance of environmental problems.
- The control section presents existing scientific, technological, governmental, educational, legal, and civic controls. The reader may refer to the standards, practices, and techniques section in some instances to get a better understanding of controls.
- The summary presents the highlights of the chapter.
- Research needs is another unique section intended to increase reader awareness to the constantly changing nature of the environment and of the need for continued reading or in-service education on the future concerns of our society.
- The reference section is extensive and as current as possible. It appears as the last area in each volume and provides the reader with sources for further research and names of individuals and organizations involved in current research.

Acknowledgments

I extend thanks to Boris Osheroff, my friend, teacher, and colleague, for opening the numerous doors needed to obtain the most current information in the environmental health field, and for contributing his many fine suggestions and ideas before and after reading the manuscript; to Ed O'Rourke for intensively reviewing the manuscript and recommending revisions and improvements; to Karol Wisniewiski for giving his review and comments on the manuscript; to Dr. John Hanlon for offering his encouragement and for helping a young teacher realize his potential; to all the environmental health administrators, supervisors, practitioners, and students for sharing their experiences and problems with me and for giving me the opportunity to test many of the practical approaches used in the book; to the National Institute of Occupational Safety and Health, National Institutes of Health, National Institute of Environmental Health Science, U.S. Environmental Protection Agency, U.S. Food and Drug Administration, Cunningham Memorial Library, Indiana State University, Indiana University Library, Purdue University Library, and the many other libraries and resources for providing the material that was used in developing the manuscript; to my wife, Donna Koren, and my student, Evelyn Hutton, for typing substantial portions of the manuscript; to my daughter, Debbie Koren, for helping me organize the materials and for working along with me throughout the night at the time of deadlines to complete the work.

In the second edition, Kim Malone typed portions of the new manuscript, retyped the entire manuscript, and was of great value to me. Pat Ensor, librarian, Indiana State University, was of considerable value in helping gather large numbers of references in all areas of the book. A very special thanks goes to my sister-in-law, Betty Gardner, for typing a substantial portion of the new manuscript, despite recurring severe illness. Her cheerfulness during my low periods helped me complete my work. Finally, thanks go to my wife Donna for putting up with my thousands of hours of seclusion in the den, while I was working, and for encouraging me throughout the project and my life with her. She has truly been my best friend.

In the third edition, Alma Mary Anderson, C.S.C., and her assistants, Carlos Gonzalez and Brian Flynn, redid the existing illustrations and added new ones to enhance the manuscript. Professor Anderson directed the production of all the new artwork. In addition, I thank Bill Farms for his assistance with the original computer-assisted drawings for the chapters on instrumentation in both volumes. My wife Donna typed much new material, and the previously mentioned libraries, Centers for Disease Control, and the University of South Florida were most helpful in my research efforts. I would like to recognize Dr. Michael Bisesi, friend and colleague, who became my co-author.

In the fourth edition, Professor Anderson, advisor of the graphic design area in the department of art, Indiana State University, provided many excellent graphics and endless hours of work inserting new material and correcting previous material in the manuscripts. Without her help this new edition would not have been possible. Thanks also go to my daughter and son-in-law, Debbie and Kenny Hardas, who dragged me into the computer age by purchasing my first computer and by teaching me to use it.

Dr. Michael Bisesi acknowledges his appreciation of his wife Christine Bisesi, M.S., C.I.H., C.H.M.M.; his two sons Antonio (Nino) and Nicolas (Nico); and his parents Anthony (deceased) and Maria Bisesi for their love and support. In addition, he wants to acknowledge his mentors Rev. Francis Young; George Berkowitz, Ph.D.; Raymond Manganelli, Ph.D.; Barry Schlegel, M.S., C.I.H.; John Hochstrasser, Ph.D., C.I.H.; Richard Spear, H.S.D.; Christopher Bork, Ph.D.; Keith Schlender, Ph.D.; and Roy Hartenstein, Ph.D. for sharing their knowledge, wisdom, and encouragement at various phases of his academic and professional career.

About the Authors

Herman Koren, R.E.H.S., M.P.H., H.S.D., is professor emeritus and former director of the environmental health science program, and director of the supervision and management program I and II at Indiana State University at Terre Haute. He has been an outstanding researcher, teacher, consultant, and practitioner in the environmental health field, and in the occupational health, hospital, medical care, and safety fields, as well as in management areas of these areas and in nursing homes, water and wastewater treatment plants, and other environmental and safety industries for the past 47 years. In addition to numerous publications and presentations at national meetings, he is the author of six books, titled *Environmental Health and Safety*, Pergamon Press, 1974; *Handbook of Environmental Health and Safety*, Volumes I and II, Pergamon Press, 1980 (now published in updated and vastly expanded format by Lewis Publishers, CRC Press, as a fourth edition); *Basic Supervision and Basic Management*, Parts I and II, Kendall Hunt Publishing, 1987 (now published in updated and vastly expanded format as *Management and Supervision for Working Professionals*, Volumes I and II, third edition, by Lewis Publishers, CRC Press); *Illustrated Dictionary of Environmental Health and Occupational Safety*, Lewis Publishers, CRC Press, 1995, second edition due in 2004. He has served as a district environmental health practitioner and supervisor at the local and state level. He was an administrator at a 2000-bed hospital. Dr. Koren was on the editorial board of the *Journal of Environmental Health* and the former *Journal of Food Protection*. He is a founder diplomate of the Intersociety Academy for Certification of Sanitarians, a Fellow of the American Public Health Association, a 46-year member of the National Environmental Health Association, founder of the Student National Environmental Health Association, and the founder and advisor of the Indiana State University Student National Environmental Health Association (Alpha chapter). Dr. Koren developed the modern internship concept in environmental health science. He has been a consultant to the U.S. Environmental Protection Agency, the National Institute of Environmental Health Science, and numerous health departments and hospitals; and has served as the keynote speaker and major lecturer for the Canadian Institute of Public Health Inspectors. He is the recipient of the Blue Key Honor Society Award for outstanding teaching and the Alumni and Student plaque and citations for outstanding teaching, research, and service. The National Environmental Health Association has twice honored Dr. Koren with presidential citations for "Distinguished Services, Leadership and Devotion to the Environmental Health Field" and "Excellent Research and Publications."

Michael S. Bisesi, Ph.D., R.E.H.S., C.I.H., is an environmental and occupational health scientist and board certified industrial hygienist working full-time as professor and chairman of the department of public health in the School of Allied Health at the Medical College of Ohio (MCO). He also has a joint appointment in the department of pharmacology in the School of Medicine, serves as the associate dean of allied health programs, and is director of the Northwest Ohio Consortium for Public Health. At MCO, he is responsible for research, teaching, service, and administration. He teaches a variety of graduate level and continuing education courses, including toxicology, environmental health, monitoring and analytical methods, and hazardous materials and waste. His major laboratory and field interests are environmental  toxicology involving biotic and abiotic transformation of organics; fate of pathogenic agents in various matrices; and industrial hygiene evaluation of airborne biological and chemical agents relative to human exposure assessment. He also periodically provides applicable consulting services via Enviro-Health, Inc., Holland, OH.

Dr. Bisesi earned a B.S. and an M.S. in environmental science from Rutgers University and a Ph.D. in environmental science from the SUNY College of Environmental Science and Forestry in association with Syracuse University. He continues to complete additional postgraduate course work, including an MCO faculty development leave at Harvard and Tufts Universities in the summer of 1998. He also completed a fellowship at MCO in teaching and learning health and medical sciences and earned a graduate certificate.

Dr. Bisesi has published several scientific articles and chapters, including two chapters in the *Occupational Environment: Its Evaluation and Control* and three chapters in fourth and fifth editions of *Patty's Industrial Hygiene and Toxicology*. In addition, he is first author of the textbook *Industrial Hygiene Evaluation Methods* and second author of two other textbooks, the *Handbook of Environmental Health and Safety*, Volumes I and II. He is a member of the American Industrial Hygiene Association (fellow), American Conference of Governmental Industrial Hygienists, American Public Health Association, National Environmental Health Association, and Society of Environmental Toxicology and Chemistry.

Contents

Volume II

Contents

Volume I

Air Quality Management

BACKGROUND AND STATUS

Clean air is necessary for safe and healthy living. We as individuals and as a nation must further reduce or control present air pollution and eliminate the potential for future air pollutants. Increases in population, industrialization, urbanization, mobility, and consumption of products must be offset by further effective research, application of air pollution control devices, and increased citizen concern to provide a healthier and cleaner environment for ourselves and our children. Air pollution impairs health by affecting otherwise healthy people, as well as individuals with chronic diseases such as emphysema, bronchitis, and asthma; and by possibly leading to lung cancer. Air pollution irritates the throat, makes eyes water, and causes headaches. Air pollution reduces visibility, spoils scenic areas, makes driving hazardous, soils clothing, deteriorates house paints and metals, dirties monuments and public buildings, and costs billions of dollars per year. Although many large particulates have been brought under control in the United States, gases and smaller particulates are still of concern.

The necessity for a national approach to air pollution control was recognized by the Congress when it passed the Clean Air Act of 1963. This act enabled the U.S. Public Health Service to study air pollution and provided grants and training to state and local agencies to control the condition. The act was strengthened by the Clean Air Act of 1970, which made the Environmental Protection Agency (EPA) the focal point of the national effort. A partnership was created between state and federal governments, which gave these authorities the primary responsibility for preventing and controlling air pollution. The EPA became a support group that conducted research, developed programs, set national standards and regulations, provided technical and financial assistance to the states, and supplemented the state programs where necessary.

In 1970, as a result of the Clean Air Act, the EPA set National Ambient Air Quality Standards for the pollutants that were commonly found throughout the country and that were the greatest threat to air quality. These pollutants, which were called *criteria pollutants*, included ozone, carbon monoxide, airborne particulates, sulfur dioxide, lead, and nitrogen oxides. Primary standards were set to protect human health and secondary standards to protect "welfare," which meant primarily crops, livestock, vegetation, buildings, and visibility. With some pollutants, a single national ambient standard was established. The Clean Air Act also required the EPA to set national emission standards for hazardous pollutants (NESHAPs). Hazardous pollutants were defined as those that could contribute to an increase in mortality or serious illness. Standards were established for asbestos, beryllium, mercury, vinyl chloride, arsenic, radionuclides, benzene, and coke oven emissions; however, numerous other potential hazardous air pollutants exist.

Clean Air Act of 1990

The Clean Air Act of 1990 is a federal law covering the entire United States, which allows various states to have stronger pollution controls than the rest of the country, and requires that the states do much of the work to carry out the act. In no case, can a state have fewer rules. Each state has to develop state implementation plans, which are a collection of the regulations it uses to clean up polluted areas. The public is involved in the development of each of these plans. The plans have to be approved by the EPA. Provisions are made for interstate commissions on air pollution control to develop regional strategies. International air pollution control is also included and Mexico and Canada have formed separate commissions with the United States.

A permit program for larger sources that release pollutants into the year is a major change in the Clean Air Act. The permits are issued by the states, or when the state fails to carry out its responsibility, by the EPA. The permit includes information on which pollutants are released, how much may be released, what kind of steps the owner or operator is taking to reduce pollution, and plans to monitor the pollution.

The EPA, has been given the enforcement powers to penalize the company for violating the Clean Air Act. Deadlines are established for the EPA, states, local governments, and businesses to reduce air pollution. Public participation is an important part of the 1990 Clean Air Act. Market or market-based approaches to the cleanup of air pollution, in an inexpensive, flexible, and efficient manner is part of the law. Economic incentives for cleaning up pollution include credits for gasoline refiners who produce cleaner gasoline. The law includes sections on criteria air pollutants, smog, volatile organic compounds (VOCs), hazardous air pollutants (air toxics), accidental releases of highly toxic chemicals, mobile sources (such as cars, trucks, and buses), cleaner fuels, acid rain, ozone-destroying chemicals, home woodstoves, National Ambient Air Quality Standards, nonattainment requirements, etc.

1997 Changes to the Clean Air Act

Based on new scientific evidence, the EPA has made revisions to the air quality standards for ground-level ozone (smog) and particulate matter. This was the first update in 20 years for the smog standard, and the first update in 10 years for particulate matter. The new standards should protect 125 million Americans including 35 million children, prevent approximately 15,000 premature deaths every year, avoid 1 million cases of significant lung function disease in children, and help reduce 350,000 cases of aggravated asthma. The EPA replaces the previous 1-hour ozone standard with a new 8-hour standard. The new standard is set at a concentration of 0.08 ppm, with the measurement period of 8 hours. The EPA added a new fine particle standard of $PM_{2.5}$ set at a concentration of 15 μg/m³ and a new 24-hour $PM_{2.5}$ standard set at 65 μg/m³. ($PM_{2.5}$ is the particulate matter 2.5 corrected by a sampler that indicates the particles under 2.5 μm in diameter. These particles are considered respirable and have the greatest potential for entering deeper into the respiratory system and adversely affecting an individual's health.) The EPA will complete its next review of particulate matter by the year 2002. By then EPA will have determined whether to revise or maintain this new standard.

Toxic Air Pollutants

Toxic air pollutants are poisonous substances in the air that come from natural sources such as radon gas or from anthropogenic sources such as chemicals from factories. These substances may cause health problems or environmental problems. Title III of the 1990 Clean Air Act lists 189 toxic air pollutants to be considered for emission controls. The EPA must identify categories of major sources that emit any of these pollutants. A major source is one that emits more than 10 tons/year of a single air toxic or 25 tons/year of any combination of air toxics. The EPA is required to develop technology-based emissions standards for these source categories that reflect the "maximum achievable control technology" for all the toxic pollutants emitted by them. The standards reflect some of the best control technologies that had been demonstrated and must be at least the average of the best performing 12%. Data on air toxics have been collected since 1987 in the Toxics Release Inventory, which covers 300 chemicals and is submitted annually by manufacturing facilities with 10 or more employees. This inventory does not provide a complete picture of air toxics emissions, because it does not include all toxic pollutants, small companies, mobile, commercial, and residential areas.

Risk assessment consisting of hazard identification, exposure assessment, dose–response assessment, and risk characterization are important factors in determining the nature and ultimate effect of toxic air pollutants. Health risks are a measure of the probability that a person will experience health problems. Breathing air toxics that cause cancer can increase the risk of cancer in individuals. Breathing air toxics can also increase the risk of noncancer type diseases such as emphysema, respiratory irritation, nervous system problems, and birth defects.

Hazard identification is based on animal or human studies. One good animal study might indicate a possibility for cancer from inhaling the toxic air pollutant. Two or more good animal studies or some evidence in human studies might indicate a probability for cancer. Good evidence in human studies may indicate that the toxic air pollutant is a known cause of cancer. Case reports, unusual number of cases of a specific illness, and epidemiological studies of special groups and their exposure to chemicals are examples of human studies.

Exposure assessment is based on the amount of pollutant individuals inhale during a specific time period. Dose–response assessments are determined by the types of health problems at different exposures, how the toxic air pollutants enter the body, movement and change of the chemical in the body, and how it interferes with normal body functions. For cancer-causing chemicals the EPA assumes that no exposures have a zero risk of disease. For non-cancer-causing chemicals there may be a dose below the minimum health effect level for which no adverse effect occurs. Risk characterization indicates the extra risk to health of individuals and the overall population. It is based on the maximum lifetime exposure to a chemical times the dose–response relationship.

TYPES OF POLLUTANTS

Pollutants that are listed as the criteria group and the hazardous group come from mobile and stationary sources. Mobile sources include cars, trucks, buses, trains, motorcycles, boats, and planes. Stationary sources range from iron and steel plants to oil refineries to dry cleaners and gas stations.

More than half of the country's air pollution comes from mobile sources. The exhaust from these sources contain carbon monoxide, VOCs, nitrogen oxides, particulates, and lead. The VOCs, along with the nitrogen oxides, are the major contributors to the ground-level or tropospheric ozone formation known as *smog*.

Stationary sources generate air pollutants usually by burning fuel for energy and as a by-product of industrial processes. The electric utilities, factories, and residential and commercial buildings burn coal, oil, natural gas, wood, and other fuels; and therefore are the principal sources of sulfur dioxide, nitrogen oxides, carbon monoxide, particulates, VOCs, and lead. Fuel oils contaminated with toxic chemicals, hazardous waste disposal facilities, municipal incinerators, landfills, and electric utilities are some of the sources of toxic air pollutants. In addition, specialized industrial and manufacturing processes may produce toxic air pollutants.

Although the amount of pollutants in the air has been reduced from the period of 1980 through 1997, considerable problems of air quality still need to be resolved. During this time period, PM_{10} levels have decreased by 12%, mainly due to the sharp decrease in residential wood burning. Sulfur dioxide levels have decreased 39% because of a cutback in emissions, due to pollution controls at coal-fired power plants. Sulfur dioxide levels, since 1970, have decreased by about two thirds. Nitrogen oxide levels have decreased slightly in the last 10 years. Ozone levels have decreased 33% since 1970, due to the Federal Motor Vehicle Control Program and the control of stationary sources. However, many urban areas do not meet the standards established by the EPA. Carbon monoxide levels have decreased an additional 25% in the last 10 years because of the reductions brought about by the Federal Motor Vehicle Control Program. Currently, 66% of the SO_x emissions come from burning of fossil fuels in electric power plants, whereas 36% of NO_x emissions come from power plants and 31% come from highway vehicles.

There has been a 98% reduction of lead in the air, although lead is still a very serious problem. The reduction is due to the removal of lead from gasoline used in many vehicles. The airborne toxics problem, the acid rain or acid deposition problem, the potential global atmospheric warming problem, and the indoor air pollution problem, including the radon gas concerns, are additional issues for the present and the future.

Ozone pollution is a persistent problem in the United States, with 70 million people living in areas not meeting ozone standards. Of these areas, 5 have severe problems, 11 have serious problems, and 6 have moderate problems. Of the urban areas, 7 have serious nonattainment where carbon monoxide has a highly localized impact, and 7 additional areas have moderate nonattainment, with 12 million people living in these areas. SO_2 nonattainment includes 31 areas; lead nonattainment includes 7 areas. The establishment of the PM_{10} standard in lieu of the total solid particulates (TSP) standard for particulates has brought about a new study of air quality in all states. This standard has been incorporated into the State Implementation Plans. Of the urban areas, 6 have serious nonattainment problems and 70 others have moderate problems and currently do not meet the PM_{10} standards; a total of 24 million people living in these areas.

It is true that long before the internal combustion engine was built and the Industrial Revolution started, volcanoes, dust storms, and fires were polluting the skies. However, nature had time to make adjustments to these pollutants whereas today the atmosphere is misused by people and is an area of deposit for waste material.

SCIENTIFIC, TECHNOLOGICAL, AND GENERAL INFORMATION

Atmosphere

Air consists of a number of gases, including nitrogen (~78%); oxygen (~21%); and carbon dioxide (~0.03%); and less than 1% of argon, neon, helium, krypton, and xenon. Oxygen oxidizes other substances by serving as an electron acceptor and bonding with them. When oxidation is slow, it is called rust or decay. When it is fast, it is called burning. The air also contains varying amounts of water vapor and a variety of natural and anthropogenic (human-made) pollutants, including greenhouse gases, such as carbon dioxide, methane, and nitrous oxide; these gases help trap heat in the atmosphere. Methane concentrations have more than doubled and nitrous oxide concentrations have risen about 15%.

Global Warming

The concentration of carbon dioxide in the atmosphere has increased about 30% since the Industrial Revolution. Although there is no direct toxic effect on human beings from the increase in carbon dioxide, provided the available oxygen is not greatly reduced, there is an indirect effect on the atmosphere, resulting in a corresponding increase in atmospheric temperature. However, the implications of this change are hard to assess because of the interactions of chemical, geochemical, and biological processes in the disposal of carbon dioxide. The ocean is a natural reservoir and contains 60 times more CO_2 than the atmosphere.

Data from 1979 through 1988 by the *TIROS-N* series of weather satellites proved that Earth's temperature can be measured accurately by instruments probing the atmosphere from space. Earth's atmosphere goes through fairly large year-to-year changes in temperature, and over that 10-year period there were no long-term warming or cooling trends. Significant temperature swings occurred during the 10-year study, but the thermal changes tended to even out on a global basis.

The Northern Hemisphere went up slightly during the 10 years and the Southern Hemisphere went down slightly. The net effect for Earth was basically zero. However, 7 of the 10 warmest years since temperatures have been recorded occurred in the 1990s, with 1998 the warmest year, followed by 1997. Nearly every month in the last 10 years has been above the mean of 1880 to 1998. Global mean temperatures have increased 0.6 to 1.2°F since the late 19th Century. It is predicted that temperatures will rise 1.6 to 6.3°F by 2100, with significant regional variation.

Rising global temperatures are expected to raise sea levels, changing precipitation and other local climate conditions, which could alter forests, crop yield, and water supplies. Lower river flows and lake flows could impair navigation, hydroelectric power generation, and water quality; and add significant cost to all types of products that use or are transported by water. Recreational areas and facilities would be broadly affected. Climate change is a global problem and must be addressed on a global scale, as in the Kyoto Protocol.

The *TIROS* satellite data were the first to add global temperature data to the scientific debate about the greenhouse effect. Most other studies of temperature trends, some extending over more than a century, have come from the records of ground-based thermometers. These readings do not reflect the global temperature, because very few temperature measurements are available for vast areas of Earth's oceans.

Although the controversy over global warming goes on, it is good for one to understand potential health effects that may occur as a result of the increased temperatures. Beside the obvious health effect of increased and skin sunburn cancers, other impacts such as cataracts, depressed immune systems, respiratory effects, heat stress, vectorborne diseases, and malnourishment due to reduced food supply could occur.

Increased temperatures at ground level tend to increase ground ozone pollution, which can irritate and inflame the respiratory tract. When ozone is combined with other pollutants such as acids, pesticides, metals, aerosols, and chlorofluorocarbons, more respiratory problems may occur. Heat stress, although a minor problem for healthy individuals, may become very serious for the very young, the very old, and people with heart or respiratory problems. In July 1995, a heat wave in Chicago killed 700 people.

Vectorborne disease potential increases with higher temperatures and more rain. Because insects become more prevalent, a greater possibility of disease exists if the organisms that cause disease are present in the insects.

Food supply may decrease with increasing temperatures. Higher yields of food are found in temperate or colder climates. Climate change may also reduce the variety of crops.

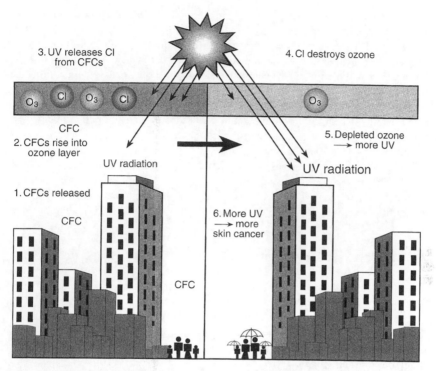

3. UV releases Cl from CFCs

4. Cl destroys ozone

CFC

2. CFCs rise into ozone layer

5. Depleted ozone → more UV

UV radiation

UV radiation

1. CFCs released

CFC

6. More UV → more skin cancer

CFC

Figure 1.1 The ozone depletion process. (From www.epa.gov.)

Ozone Depletion

Sunlight contains ultraviolet-A (UV-A), ultraviolet-B (UV-B), and ultraviolet-C (UV-C) radiation, but only UV-A and UV-B reach Earth; UV-B levels are primarily affected by stratospheric ozone, which helps absorb the radiation before it reaches the Earth. The incidence of typically nonlethal basal cell and squamous skin cancers is directly correlated to the amount of exposure to UV-B radiation. A 2% decrease in stratospheric ozone produces a 4% increase in nonmelanoma skin cancer. A 5% decrease in ozone should increase skin cancer rates sharply. UV-B radiation also affects the eyes by aging lenses of the eyes and causing cataracts.

Ozone depletion may affect the immune system. In animals where herpes simplex is present, UV-B may trigger skin lesions. Also in animals, exposure to UV-B can decrease resistance to mycobacterium causing tuberculosis and leprosy (Figure 1.1).

Atmospheric Regions

The atmosphere, which is above Earth's surface, is divided into a number of regions. Conditions in the atmosphere vary with the height from the ground. The greatest differences are in pressure, temperature, and radiation. That portion of the atmosphere closest to Earth's surface, and the one in which we live, is the tropo-sphere. Above this layer of air we find the stratosphere, the ionosphere, and the exosphere. The troposphere varies from 5 miles in thickness at the North and South Poles to 10 miles in thickness at the equator. The heat energy in warm air expands the layer near the equator, whereas the cold of the poles packs the air molecules tightly together.

The troposphere includes about 80% of all the air molecules around Earth. It also contains beneath it all the mountain peaks and other solid parts of Earth. The troposphere is the layer of air responsible for weather on Earth's surface. As we rise from the surface of Earth through the troposphere, the winds increase in speed and the temperature steadily drops. At the top of the layer, the temperature is about −70°F. The pressure of the air is about one quarter of the pressure at sea level. The percentage of water in the air is highest close to the ground and drops steadily as the altitude increases. We dump thousands of tons of pollutants into the troposphere every day.

Air Currents and Movement

The air around Earth is always in motion. It moves along the surface of Earth as well as up and down. Wind is simply air in motion above Earth's surface. Some air movements are due to the heat effects of the Sun. As the atmosphere is warmed, the air rises higher above Earth's surface. As it is cooled, the air lowers toward the earth. This heating is accomplished through radiation, conduction, and convection related to the Sun (Figure 1.2).

Radiation is the transfer of heat or light by waves of energy. It contributes very little to the heating of the air, because our atmosphere is incapable of absorbing very much of the Sun's energy directly. An interesting phenomenon, known as the green-house effect, occurs when Earth absorbs the waves of light energy from the Sun and reradiates them as heat energy. The water vapor in the air, the carbon dioxide, and the ozone help the atmosphere absorb the heat energy. This greenhouse effect is similar to the glass in the greenhouse holding back the reradiated heat but allowing the light to pass through.

Conduction is the transfer of heat by physical contact between the molecules of the air and of the warm Earth. Convection is the transfer of heat by the movement of air masses. Air movement is caused by expansion due to heat. Lighter air rises and cooler air above sinks. Convection is the key force in the heating of the atmo-sphere. Air movement is essential for life, because without it the atmosphere would become unbearably hot and polluted. Air movement is also due to the effect of the rotation of Earth. The Earth spins from west to east, or counterclockwise. The speed of rotation is about 1000 miles/hr at the equator. At the pole there is no surface speed. Air that moves from the equator is traveling faster to the east than the ground below. Air moving toward the equator has a slower speed than the ground below.

These general patterns of air motion are changed in local areas by differences in temperature, air pressure, and topography. Mountains, deserts, oceans, and other large bodies of water may alter the direction of the winds from day to day. The

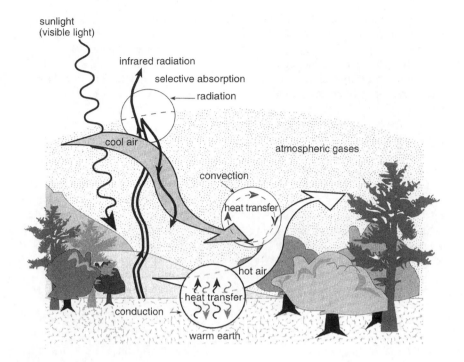

Figure 1.2   Air currents and air motion; radiation, conduction, and convection.

prevailing winds in an area greatly influences the day-to-day weather. The prevailing winds over the United States are westerly. They bring moisture from the Pacific Ocean across the western coast of the United States. They then pass over the Rocky Mountains, causing the air to rise and grow cooler. Because the air loses most of its moisture crossing the mountains, the area east of the mountains is very dry. The Midwest and Eastern Seaboard are affected by the movement of cold or cooled Canadian air masses and a warm, moist flow of tropical air from the Gulf of Mexico. The jet stream has a profound effect on weather conditions.

Dispersal of air pollutants is based on stability of the air, thermal and mechanical turbulence, mixing depths, inversion, wind direction, wind speed, time of day, season, weather, topography of the land, and local obstructions causing cross currents.

Air Temperature

The temperature of the air decreases with an increase in altitude at a given rate, called the adiabatic lapse rate. That rate is theoretically 5.4°F for every 1000 ft in height in dry air. Therefore, if air is permitted to rise without any heat exchange due to the environment the air expands and cools as the pressure decreases. The reverse is also true. However, the actual temperature of the air often changes with altitude at a higher rate than the theoretical rate. When such a situation exists, the air is said to be unstable and warm surface air rises rapidly. This instability is good because it creates an up-and-down mixing of air currents that dilutes air pollutants.

Thermoturbulence is the rapid mixing of hot and cold air. The mixing depth of the air usually extends several thousand feet during the daylight hours of the summer and a few hundred feet during the winter when the sun gives less heat. At night the air close to Earth is cooled by contact with it, whereas the air higher up stays relatively warm. This causes a minimal level of mixing, thereby concentrating the pollutants close to the surface of Earth.

Winds at surface and higher elevations are important factors in determining the rate at which concentrations of pollutants can be dispersed. The upward dispersion of pollutants is more effective than horizontal dispersion for two reasons: upward dispersion immediately removes pollutants from the ground, where their nuisance effect is greatest; and wind velocity normally increases rapidly with height, so that greater possibilities exist for rapid horizontal dispersion at higher levels. For vertical air movement to take place freely without resistance in the atmosphere, the temperature must be cooler by about 5.4°F for each 1000 ft of elevation. This is called the *lapse rate*. If upper strata of air become warmer than the air at the surface, an *inversion* is said to have occurred and upward dispersion ceases.

Inversions

When the temperature increases instead of decreases with altitude, a temperature inversion exists (Figure 1.3). When temperatures near the ground are colder than the air a few hundred feet higher, a *surface inversion* exists. Sometimes the atmosphere is arranged in several layers and an *upper air inversion* is formed. No matter where it is located a temperature inversion creates a *lid*, which stops vertical mixing.

At equal pressure, the same volume of warm air is lighter than cool air, and when air near the ground is heated by the sun during the daytime, it becomes warmer and lighter than the air immediately above it. A vertical exchange of small parcels of air develops. The warmer air near the ground is constantly replaced by the slightly cooler air just above it.

During the night a surface temperature inversion generally occurs with cool (heavier) air located below warm (lighter) air. The atmosphere is very stable and pollutants entering the air near the ground are forced to stay in that layer near the ground. Temperature inversions near Earth's surface and up to 1000 ft are very important in areas subject to heavy air pollution, because they limit the amount of air through which the burden of pollution can be mixed.

The nighttime or radiation inversion is a normal occurrence. This inversion should break up when the morning sun heats the ground and causes the warm air to rise and mix. An inversion is affected by topography. In a valley the cool air flows downward at night and is trapped at the bottom by the warm air that has risen. As the concentration of pollutants increases in the air of the valley that is sealed and the pollutants continue to flow into the air, disastrous effects may occur. The same kind of problem exists in large cities where the skyscrapers create valleys. In the

Figure 1.3 Temperature inversion.

city, the buildings tend to retard or break up the air masses, causing the air to stagnate and become more polluted. The intensity and duration of an inversion directly affects the health of humans and their environment.

Urban and Topographic Effects

Metropolitan areas impose their own effects on the atmosphere. The city acts as a heat island or reservoir by storing heat through absorption from the sun's rays during the day and releasing the stored heat at night. The heat island has the following two direct effects: (1) the nighttime temperatures remain higher than in the surrounding countryside; and (2) the nighttime inversion, which generally is located at the ground, rises to some distance above the ground.

Although the heat islands have been demonstrated, their dimensions and total effects have not been defined. The vertical height has been estimated at about three times the average height of the buildings. Low wind speeds and clear skies intensify the release of heat, because radiation from the countryside allows rapid cooling at night, whereas radiation from building walls reflects back and forth to be eventually absorbed by the air. The warm air rises and carries with it the pollutants; then it expands, flows toward the edges of the city, cools, and sinks. The cooler air, which is loaded with pollution, flows from the edges of the city into the center to replace the rising warm air. In effect, a self-contained circulatory system is set up that perpetuates the dust or haze cover. Only a strong wind can then interrupt this continuous pattern.

Topography also has an effect on weather and pollution conditions. Low-lying areas constantly have lower minimum temperatures than the surrounding hills. As the air cools and becomes dense, it settles into the low places. It carries with it many of the pollutants from the metropolitan areas and can cause high air pollution counts in nonindustrialized parts of the community.

Weather

Weather has an effect on the degree of pollution in the air and the interaction of pollutants. Rain or snow precipitates pollutants, and therefore, makes the air cleaner at the expense of causing surface pollution. Fog, the condensation of water vapors in the air, contains aerosols, the tiny suspended solid or liquid particles found in the air. As the aerosols cool, the moisture in the air adheres to them. Fog can convert harmful gases, such as sulfur dioxide into even more harmful acids, such as sulfuric acid. It is believed that the presence of sulfur dioxide in fog has been a major factor in many of the serious air pollution disasters.

Smog is a combination of smoke and fog generally found in areas where smoke is produced in the operation of homes and factories. Photochemical smog is a combination of fog and chemicals that come from automobile and factory emissions and is acted on by the action of the sun. Nitrogen dioxide, in the presence of the sun and some hydrocarbons, is turned into nitric oxide and atomic oxygen. The atomic oxygen reacts with the oxygen molecules and other constituents of automobile exhaust emissions to form a variety of products including ozone. The ozone is harmful in itself and is also implicated in a highly complex series of continuing reactions. As long as ozone or nitrogen dioxide and sunlight are present, other undesirable reactions can occur. One related chemical product is peroxyacetyl nitrate (PAN).

Physical Properties of Gases

An understanding of the gas laws is necessary from a theoretical and practical point of view to understand the manner in which a gaseous pollutant is released to the atmosphere. This discussion is of necessity brief and incomplete. For more detailed information, see basic textbooks in chemistry, physical chemistry, and physics.

Law for Ideal Gases

Boyle's law states that when the temperature is held constant, the volume of a given mass of a perfect gas with a given composition varies inversely as the absolute pressure, shown as follows:

$$\frac{P_1}{V_1} = \frac{P_2}{V_2} \tag{1.1}$$

Charles' law states that when the pressure is held constant, the volume of a given mass of a perfect gas with a given composition varies directly as the absolute temperature, as follows:

$$\frac{V_1}{T_1} = \frac{V_2}{T_2} \tag{1.2}$$

Both Boyle's and Charles' laws are satisfied in the following equation:

$$PV = \frac{mRT}{M} \tag{1.3}$$

where
 M = molecular weight of a gas (mass per mole)
 m = mass of a gas
 P = absolute pressure
 R = universal gas constant
 T = absolute temperature
 V = volume of a gas

The law states that if a perfect gas has mass (m) and molecular weight (M), it occupies volume (V) at absolute pressure (P) and absolute temperature (T).

Absolute pressure is commonly expressed as 760 mmHg, or 1 atm. Absolute temperature is expressed as °C + 273.15 = degrees Kelvin (K), or °F + 459.6 = degrees Rankine (R).

Dalton's Law of Partial Pressure

Partial pressure is the pressure that one gas or vapor component of a gas or vapor mixture exerts. The total pressure of the gas mixture is the sum of the partial pressures. The mathematical expression of Dalton's law is:

$$P_T = P_1 + P_2 + \cdots + P_n \tag{1.4}$$

where
 P_n = partial pressure of a gas component
 P_T = total pressure

Gas Pressures

Barometric, gauge, and absolute pressures are equal in all directions at a point within a volume of fluid and act perpendicular to a surface. Barometric pressure and atmospheric pressure are synonymous. Such pressure is measured with a barometer and usually is expressed as inches or millimeters of mercury. Standard barometric pressure is the average atmospheric pressure at sea level, 45° north latitude, at 35°F. It is equivalent to a pressure of 14.696 lb/in.^{2} exerted at the base of a column of mercury 29.921 in., or 760 mmHg, high. Barometric pressure varies with weather and altitude.

Measurements of pressure by ordinary gauges are indications of the difference in pressure above or below that of the atmosphere surrounding the gauge. Gauge pressure is ordinarily the pressure indicated by the gauge itself. If the pressure in the system is greater than the pressure prevailing in the atmosphere, the gauge pressure is expressed as positive; if smaller, the gauge pressure is expressed as negative. The term *vacuum* designates a negative gauge pressure.

Because gauge pressure, which may be either positive or negative, is the pressure relative to the prevailing atmospheric pressure, the gauge pressure added algebraically to the prevailing atmospheric pressure (which is always positive) provides a value that has a datum of *absolute zero pressure*. A pressure calculated in this manner is called absolute pressure.

Density of a Perfect Gas

$$p = \frac{PM}{RT} \tag{1.5}$$

where
- p = density (mass per unit volume) of a gas
- P = absolute pressure
- M = molecular weight of a gas (mass per mole)
- R = universal gas constant
- T = absolute temperature

In short, density is the mass of a given volume of gas, expressed in units of grams per liter (g/L).

Density of a Gas Component

$$p_x = \frac{P_x M_x}{RT_{mix}} \tag{1.6}$$

where
- p_x = density of a gas component (mass of gas component per unit volume gas mixture)
- P_x = partial pressure of a gas component

M_x = molecular weight of a gas component (mass per mole)
R = universal gas constant
T_{mix} = absolute temperature of a gas mixture

Gas Temperatures

Fahrenheit and Celsius Scales

The range of units on the Fahrenheit scale between freezing and boiling is 180 (32 to 212°F); on the Celsius scale, the range is 100 (0 to 100°C). The following relationships are used to convert one scale to the other:

$$°F = \frac{9}{5}°C + 32$$

$$°C = \frac{5}{9}\left(°F - 32\right)$$

(1.7)

Absolute Temperature

Experiments with a perfect gas have shown that, under constant pressure, for each change in degree Fahrenheit below 32 the volume of a gas changes (1/491.6). Similarly, for each Celsius degree, the volume change is 1/273. Therefore, if this change in volume per degree is constant, the volume of gas would, theoretically, become zero at 491.6 Fahrenheit degrees below 32, or at a reading of –459.6°F. On the Celsius scale, this condition would occur 273 Celsius degrees below 0, or at a temperature of –273°C.

Absolute temperatures determined by using Fahrenheit units are expressed as degrees Rankine (R); those determined by using Celsius units are expressed as degrees Kelvin (K). The following relationships convert one scale to the other:

$$R = °F + 459.6$$

$$K = °C + 273.15$$

(1.8)

Psychrometric Properties of Air

Because air and gases usually contain water vapor, it is necessary to understand how to determine the amount of moisture present. This is accomplished using the wet-dry bulb thermometer assembly. The water vapor pressure existing in the air passing the assembly when the thermometer or air is in motion is calculated. The amount of water vapor can then be determined from charts of saturation water pressures.

Relative humidity is the relationship of water vapor pressure present in the air to water vapor pressure possible at the same temperature under saturated conditions. The dew point is the temperature at which air can exist saturated with water vapor. Below the dew point condensation occurs. The dew point can be determined by water saturation tables and psychometric charts.

Particulate Matter

Particulate or aerosol pollutants can exist as solid matter or liquid droplets the sizes of which have a specific effect on pollution. Coarse dust particles larger than 10 μm in diameter and fly ash composed of the impurities remaining after coal is burned settle out of the air quickly. They are, therefore, most troublesome near their source. Fume, dust, and smoke particles ranging in size from less than 1 to 10 μm, travel farther. Particles less than 1 μm in diameter move as easily and as far as gases, depending on the wind or air currents.

Polluting particles are composed of a variety of artificial and natural substances. Because their size and, to a lesser degree their state, influence their behavior, they are often identified as smoke, fumes, mist, and particles.

Smoke describes both solid and liquid particles under 1 μm in diameter usually less than 0.05 μm in diameter. Smoke is produced during all forms of incomplete combustion and in such other processes as distillation, which is the removal of impurities from liquids by heating them to the boiling point and then condensing the vapors to form solid particles.

Fumes are the solid particles under 1 μm in diameter that are formed as vapors condense or as chemical reactions take place. Fumes are emitted by many industrial processes, including smelting and refining, which generate metallic oxide fumes.

When solid particles are more than 1 μm in size, they are generally referred to as dust. Dust may be formed from solid organic or inorganic matter by natural attrition or in innumerable industrial and agricultural processes. The parent material is reduced in size through some mechanical process such as crushing, drilling, grinding, or friction.

Mist is made up of liquid particles up to 100 μm in diameter. It is released industrially in such operations as spraying, splashing, foaming, and impregnating; or it is formed by the condensation of vapor in the atmosphere or by the effect of sunlight on automobile exhausts. As mist evaporates, a more concentrated liquid aerosol or mist is formed.

Particles are complex in their chemical composition. The inorganic fraction of collected samples of airborne particles usually contains a few dozen metallic elements combined with the organic carbon or soot and other tarred organic material. The most frequently found metallic elements are silicon, calcium, aluminum, and iron. Relatively high quantities of magnesium, lead, copper, zinc, sodium, and manganese are also found. The organic fraction of collected particle samples is usually even more complex, containing a large number of aliphatic and aromatic hydrocarbons and related substituted compounds.

TSPs refer to minute bits of matter carried in the air. TSPs are thought to be one of the most critical contaminants during most episodes of high pollution. The suspended particulates are of concern, because they affect health, and are in the PM_{10} and $PM_{2.5}$ range. The new $PM_{2.5}$ standard on an annual basis is 15 μg/m^3 and on a 24-hour basis is 65 μg/m^3. (At present no monitoring stations are available.) The standards will not require local controls until 2005, and no compliance determinations will be required until 2008, with possible extensions after that.

Settling Velocity of Particles

The settling velocity of particles is based on Stokes law, Cunningham's corrections to Stokes law, Newton's law of turbulent flow, the intermediate law of transitional flow, and the drag coefficient. Excellent discussions on these topics are found in physics textbooks.

Properties of Particulates

When a liquid or solid substance is emitted into the air as particulate matter, its properties and its effects may be changed. As a substance is broken into smaller particles, more of its surface area is exposed to the air. The substance, whatever its chemical composition, seems to become more attractive to other particulates or gases. The following detrimental results may occur:

1. Very small aerosols (from 0.001 to 0.1 μm in diameter) act as nuclei on which vapor condenses relatively easily. Fogs, ground mists, and rain are thus increased and prolonged.
2. Particles less than 2 or 3 μm in size (about half, by weight, of the particles suspended in urban air are estimated to be that small) reach deep into the part of the lung that is unprotected by mucus, and attract and carry with them such harmful chemicals as sulfur dioxide.
3. Particulates act as catalysts. For example, sulfur dioxide is changed to sulfuric acid using iron oxides as the catalysts.

Particle Retention and Deposition in Lungs

The amount of particles in inspired air reaching the lungs depends on the:

1. Size, shape, and density of the particles
2. Size and shape of the airway
3. Pattern of breathing, that is, breathing through the nose vs. the mouth

The particulate matter is inhaled and larger particulates (5 to 30 μm) impact on the upper nasal airways through inertia. Smaller particles may be deposited in the lungs. Through diffusion or Brownian movement, particles less than 1 μm and probably less than 0.1 μm in size may accumulate in the alveolar region. The particles may cause a localized irritation plus a lung disease in reaction to the particle and what adheres to it, or it may absorb into the circulatory system, ultimately causing a systemic effect.

Alteration and Transportation of Particulate Matter

When a liquid or solid is dispersed to form a cloud or smoke, many of its properties are altered. Its surface area is greatly increased. The nature of the surface influences the behavior of the material in many ways. Particles may absorb reacting gases and thereby catalyze their reaction.

Dusts and other aerosols speed up the rate of oxidation of gases. For example, the oxidation of sulfur dioxide is normally very slow. However, aqueous droplets, which are present in great quantities in contaminated atmospheres, dissolve sulfur dioxide; and the sulfurous acid formed is quickly oxidized to sulfuric acid when heavy metal salts are in the solution. Organic particles, such as those from heated fats and oils, produce similar reactions in organic substances. Many natural particles are complex combinations of organic and aqueous droplets surrounded by an organic surface layer that is relatively insoluble in water.

The effects of thermal radiation become much more pronounced if suspended solid or liquid particles are present in the air. Such particles absorb radiation and conduct heat rapidly to the surrounding gas molecules.

BUBBLE CONCEPT

Typically the control of stationary sources of air pollution had been implemented on a stack-by-stack and equipment-by-equipment basis. A single plant that contained several stacks or pieces of equipment emitting air pollutants was generally subjected to numerous individual rules for emission limitations. The cost of controlling pollutants varied with the process or equipment, volume of exhaust gas, pollutant concentration, and composition and temperature. At some plants it was cheaper and easier to reduce a ton of pollutant at one stack or piece of equipment than at another one at the same plant. In some plants, the control of VOCs can almost be negligible in cost whereas the cost of other plants would run into hundreds of dollars per ton.

The bubble concept was devised to achieve an optimum way of getting control of potential pollutants and still make it financially feasible for the company. Thus, instead of being concerned with the emissions from each stack, the company could control those problems that were cheapest to control as long as the total emissions were not increased and the pollutants were decreased.

AIR POLLUTION PROBLEMS

Regulated Air Pollutants

The regulated air pollutants are made up of two groups: the criteria pollutants and the hazardous air pollutants. The criteria pollutants include particulate matter, nitrogen oxides, sulfur oxides, carbon monoxide, ozone, and lead. The hazardous air pollutants include asbestos, beryllium, mercury, vinyl chloride, arsenic, radionuclides, benzene, coke oven emissions, and many others.

Particulates

Particulates in the air, also called aerosols, such as dust, smoke, and mists, may have a short- and long-term health and environmental effect. These effects range from irritating the eyes and throat, to reducing resistance to infection, to causing chronic respiratory diseases. Fine particulates, about the size of cigarette smoke particles, can cause temporary or permanent damage when they are inhaled deeply and lodged in the lungs. Particulates from diesel engines are suspected of causing cancer. Windblown soil dust can carry toxic substances such as polychlorinated biphenyls (PCBs) and pesticides. These particulates can also damage vegetation and buildings, and reduce visibility. Major sources of particulates include steel mills, power plants, cotton gins, cement plants, smelters, grain storage elevators, construction and demolition sites, wood-burning stoves and fireplaces, dusty areas, and diesel engines.

Carbon particulate matter, a product of incomplete combustion, is known as soot. One of the most prevalent of solid pollutants, it is made up of very finely divided carbon particles clustered together into long chains. Because they are so fine, the particles have an exceptionally broad surface in proportion to their weight, and a very active surface that attracts a great variety of chemicals from the air. This ability to attract chemicals makes soot dangerous. Certain hydrocarbons that are known to produce cancer in animals are adsorbed on the surface of soot particles and ride with them from the scene of combustion into the atmosphere, and then deep into the human lung. The largest single source of outdoor fine particles ($PM_{2.5}$) is from wood burning in fireplaces or woodstoves, except in areas where forest fires are burning.

Gaseous Air Pollutants

Nitrogen Oxides

Nitrogen is a colorless, tasteless, odorless gas that constitutes 78% of our atmosphere. Although usually harmless, it combines with oxygen to form a number of oxides of nitrogen. Of these, nitric oxide and nitrogen dioxide are pollutants. Nitric oxide is formed primarily in automobile cylinders, electric power plants, and other very large energy-conversion processes. In most cities, the automobile is the largest single source of this compound. Nitrogen dioxide is formed readily by photochemical actions. It is also a product or by-product of a number of industries, including fertilizer and explosives manufacturing. Nitrogen dioxide reacts with raindrops or water vapor in the air to produce nitric acid (HNO_3), which, even in small concentrations, can corrode metal surfaces, injure vegetation, or fade fabrics in the immediate vicinity of the source. From 1950 to 1980, nitrogen oxides rose from 10.09 to 24.87 million tons a year. In 1997, emissions were reduced slightly to 23.58 million tons. The principal sources of NO_x were electrical utilities and industrial processes, 10.72 million tons; on-road vehicles, 7.04 million tons; and construction, farm, aircraft, marine vessels, and railroads, 4.56 million tons.

Sulfur Oxides

Sulfur oxides are a major source of pollution for many cities. Sulfur is a non-metallic element found in nature, either free or combined with other elements. This element is almost invariably found as an impurity in coal and fuel oil. When fuels containing sulfur are burned, the sulfur joins with the oxygen of the air and gaseous oxides of sulfur are produced as by-products. These by-products are also produced in chemical plants and in processing metals and burning trash.

The major oxide of sulfur produced in combustion is sulfur dioxide (SO_2), which combines easily with water vapor to become sulfurous acid (H_2SO_3). Sulfurous acid joins easily with the oxygen in the air to become the even more corrosive, sulfuric acid (H_2SO_4). Sulfuric acid is also produced by oxidation of sulfur dioxide. SO_2 + O_2 becomes sulfur trioxide (SO_3). Iron oxides that form on boiler tubes and walls act as catalysts.

As sulfur dioxide leaves the smoke stack, it is usually diffused rapidly, so that oxidation to sulfur trioxide takes place rather slowly. However, with time, sulfur trioxide can build up substantially, and it reacts very quickly with water vapor to become sulfuric acid.

Hydrogen sulfide gas (H_2S) carries the foul smell of rotten eggs. A group of gases called mercaptans are by-products of petroleum refining, kraft pulping for paper production, and various chemical processes.

Several areas of the country exceed the ambient standards for sulfur dioxide. This may present serious health and environmental problems. Excessive levels of sulfur dioxide are associated with a sharp increase in acute and chronic respiratory diseases. Sulfur dioxide may be transported for long distances in the atmosphere, because it binds to particulates of dust, smoke, or aerosols. Sulfur dioxide is released to the air primarily through the burning of coal and fuel oils. About two thirds of all national sulfur dioxide emissions come from electric power plants. Coal-fired plants account for 95% of all power plant emissions. Other sources of sulfur dioxide include refineries, pulp and paper mills, smelters, steel and chemical plants, and energy facilities related to oil shale, synthetic fuels, and oil and gas production. In the residential neighborhood, coal-burning stoves and home furnaces are the primary sources of sulfur dioxide.

Carbon Monoxide and Carbon Dioxide

Carbon (C) is a nonmetallic element found either in its pure state or as a constituent of coal, petroleum, limestone, and other organic and inorganic compounds. Carbon, which is widely used as a fuel, gives off a great deal of unburned or partly burned particles (e.g., smoke), carbon monoxide or carbon dioxide (or both) during the combustion process.

A major source of carbon monoxide is motor vehicles that are burning fuel inefficiently during start up, idling, or moving slowly in congested traffic. Other sources, such as coal, oil or woodstoves and ovens, kerosene or propane furnaces, gas water heaters, gasoline generators, fires, barbecue grills, hibachis, burn fossil fuels and generate carbon monoxide. A decrease in carbon monoxide has occurred from 1978 (116.08 million tons) to 1997 (87.45 million tons). More than half of the emissions are from on-road vehicles.

Carbon dioxide (CO_2) is formed in nature by the decomposition of organic substances and is absorbed from the air by plants during photosynthesis. Carbon dioxide can convert to carbonic acid in the presence of moisture and erode stone. It is partially responsible for the corrosion of magnesium, and perhaps of other structural metals as well.

Hydrocarbons

Hydrocarbons are a class of compounds containing carbon and hydrogen in various combinations. They are found especially in petroleum, natural gas, and coal, and are in the gaseous, liquid, or solid state. There are over a thousand hydrocarbon compounds. Most of these compounds are harmful only in very high concentrations (Figure 1.4). The two groups of hydrocarbon compounds that seem to be of the greatest importance in pollution are the olefin or ethylene series and the aromatic, benzenoid, or benzene series.

Polycyclic aromatic hydrocarbons (PAHs) may be found in combustion emissions. The numerous sources and types of PAHs make them very difficult to be monitored. Some PAHs are carcinogenic and some are mutagenic. High concentrations of PAHs may be present in some urban areas, and possibly a relationship exists between the high incidents of cancer in these urban areas in high risk groups to the PAH concentrations. PAH toxicity factors include adsorption capabilities of the compound, number of carbon atoms in the molecule (carcinogenic potency is low for groups containing less than four benzene rings), metabolites, polycyclic hydrocarbons, epoxides and their effect on toxicity (epoxides produce cancer), and introduction of methyl groups.

PAHs are produced by the combustion of organic materials. PAHs have been used as solvents for gasoline and diesel products, as well as in other industrial uses. PAHs are high in molecular weight. Some of the PAHs include fluoranthene, pyrene,

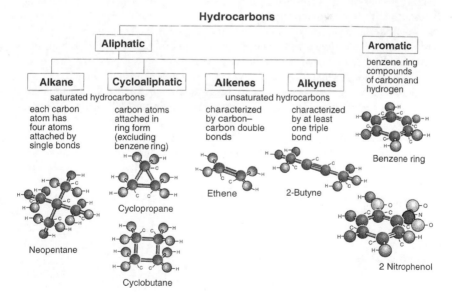

Figure 1.4 Hydrocarbons.

benzo(*a*)anthracene, benzo(*b*)fluoranthene, benzo(*a*)pyrene, chrysene, anthracene, and fluorene. Benzo(*a*)pyrene is a known carcinogen.

The two major categories of sources of PAH are stationary sources and mobile sources. Stationary sources consist of residential heating, industry, power heat generation, and incineration. In residential heating, wood and coal fuel stoves, as well as wood-burning fireplaces produce PAH through the incomplete combustion of fuel during low-temperature burning. In industry, coke production, as well as petroleum catalytic cracking, produces PAH as well as other organic pollutants. Coke ovens are a major source of PAH emissions. The leaks in the oven doors permit the emissions to enter the atmosphere. Carbon black is widely used as a pigment in rubber tires. This is another industrial source of PAH. Several types of coal-fired power plants, when improperly operated, have high emissions of PAH. Biomass or combustion of wood and peat produce relatively large amounts of PAH. Open burning of tires, burning of coal and refuse, forest fires, and agriculture burning all emit large quantities of PAH.

Mobile sources include gasoline fueled automobiles; diesel engines; two stroke engines, such as motorcycles and lawn mowers; rubber tire wear; and aircraft and ship fuel usage.

Photochemical Oxidants

Photochemical oxidants are several different pollutants (notably ozone and the peroxyacyl nitrates [PAN] group of chemicals) that can come from several sources. All these pollutants share three properties: (1) they are all formed by the chemical reaction of other pollutants; (2) the reactions forming them proceed much more rapidly in areas with intense sunlight; and (3) they are extremely reactive chemical substances that act as oxidizing agents.

Among the most effective combinations for producing this class of pollutants are the oxides of nitrogen and reactive hydrocarbon (organic) vapors. Los Angeles, with its sunny climate and large number of cars, offers extremely good conditions for the production of photochemical oxidants, which comprise the main part of that city's smog. Smog is not confined to Los Angeles, because the constituents of photochemical smog are detected in many metropolitan areas. They decrease effective visibility.

Ground-level ozone is an early and continuing product of the photochemical smog reaction, and the presence of ozone in the air assures the continuation of the oxidizing process. For these reasons, ozone is the chemical whose presence is used to measure the oxidant level of the atmosphere at any time.

Still another group of photochemical smog products caused by the automobile are the aldehydes such as acrolein. They result from the union of some hydrocarbon compounds with oxygen.

Organic Gaseous Discharges

The majority of organic discharges into the air occur from natural sources, transportation, and stationary sources. These organic gaseous hydrocarbons include oxygenated hydrocarbons, halogenated hydrocarbons, olefins, and aromatics. The oxygenated hydrocarbons include aldehydes, ketones, alcohols, and acids. In quantity, they produce eye irritation, reduce visibility in the atmosphere, and form additional pollutants along with other components in the atmosphere. Formaldehyde is one of the more common oxygenated hydrocarbons found in industrial discharges. It can cause eye irritation in concentrations as low as 0.25 ppm.

The halogenated hydrocarbons, such as carbon tetrachloride and perchloroethylene, may generate odors and affect the clarity of the air. The olefins are a group of unsaturated hydrocarbons compounds that readily react with other chemicals. In the unsaturated compound, carbon atoms are not at their maximum valence capability. Adjacent carbon atoms share double bonds, whereas in the saturated compounds adjacent carbon atoms share only one bond. Most olefins have no direct effect on animal life in small concentrations. However, they have been found to cause a general reduction in plant growth. Olefins take part in photochemical reactions in the presence of nitrogen oxides and other pollutants.

Aromatics, including benzene, toluene, and xylene, have been found to be carcinogenic. One of the most toxic PAH substances (carcinogenic) in the industrial processes is benzo(a)pyrene. It is used as a standard of emissions to determine the level of hydrocarbons, because it is relatively simple to detect. Organic discharges are important in the formation of additional chemical smogs.

Simplified Reactions in the Atmosphere

The following formulas are a set of simplified reactions that generate the various atmospheric problems related to air pollution:

$$NO_2 + sunlight \rightarrow NO + O$$

$$O + O_2 \leftrightarrow O_3$$

$$O_3 + NO \rightarrow NO_2 + O_2$$

$$HO + RH\left(+ O_2\right) \rightarrow RO_2 + H_2O$$

$$RO_2 + NO \rightarrow RO + NO_2 \tag{1.9}$$

$$RO + O_2 \rightarrow HO_2 + aldehydes, \ etc.$$

$$HO_2 + NO \rightarrow HO + NO_2$$

$$O + SO_2 \rightarrow SO_3$$

$$SO_3 + H_2O \rightarrow H_2SO_4$$

The HO is formed when water is in the presence of certain hydrocarbons. The symbol RO represents a hydrocarbon radical generated from an organic emission, such as formaldehyde or an aromatic derivative. A series of aldehydes plus ketones and other hydrocarbons may be generated from RO. These hydrocarbons then become available for generating nitrogen dioxide from nitrogen oxide. If this occurs, the process of conversion to nitrogen dioxide is more rapid than when nitrogen and oxygen form nitrogen dioxide in combustion processes.

Ozone

Ground Level Ozone

Ozone, one of the most difficult environmental problems to deal with, is widespread. VOCs and nitrogen oxides are precursors to ground-level ozone formation. Ozone, which is the major component of smog, can cause serious respiratory problems such as breathing difficulties, asthma, loss of lung function, and reduced resistance to infection. Chemically, ozone is very reactive and combines practically with every material with which it comes into contact. Therefore, it tends to break down biological tissues and cells.

The national ambient air quality standards for ozone was 0.12 ppm, which was not to be exceeded more than 1 day/year on the average. Ozone was measured at the highest hourly average for the day at a daily peak of 1-hour concentration, but now is 0.08 parts per million (ppm) measured over an 8-hour period. However, the 1-hour ozone standard will stay in place until it is achieved. Existing agreements with communities and businesses will be honored. The new standard will not require local control until 2004, and the no-compliance determination will not be made until 2007, with possible extensions afterward. On smoggy days the peak ozone concentrations occur in the afternoon. The Clean Air Act requires that the EPA set the ambient standards at a level to protect human health with an adequate margin of safety. The standard of 0.15 ppm is where reversible physiological lung function changes and symptomatic effects may occur. Ozone affects crops and vegetation at levels below 0.12 ppm.

At much lower levels in the atmosphere, the unwanted ozone is produced when sunlight triggers a chemical reaction between naturally occurring atmospheric gases and pollutants such as VOCs and nitrogen oxides. VOCs are released into the air through the combustion, handling, and processing of petroleum products. Nitrogen oxides are also produced by combustion sources. Ozone levels are highest during the day, usually after heavy morning traffic has released large amount of VOCs and nitrogen oxides. Drought brings about ozone levels higher than those usually observed. There is also a question about the relationship of ozone to elevated temperatures. Apparently, in abnormally hot years the ozone levels rise.

Upper Level Ozone

Ozone in the upper atmosphere is needed to protect people from UV radiation. The ozone is destroyed by chlorofluorocarbons (CFCs). These chemicals consist of chlorine, fluorine, and carbon atoms; and are part of the halogenated alkane group with a typical formula of 1,1,2-trichloro-1,2,2-trifluoroethane. CFCs are nontoxic and inert; however, acute exposure may cause heart and eye problems. These chemicals vaporize at low temperatures, making them excellent for use as coolants in refrigerators, as propellant gases for spray cans, and as solvents. CFCs are good insulators, and therefore are part of plastic foam materials, such as styrofoam. These chemicals are simple to make and therefore are cheap to manufacture.

When the CFCs are released into the environment the molecule is far more efficient at trapping heat than a molecule of carbon dioxide. This CFC molecule would, therefore, increase the greenhouse effect at a higher rate than carbon dioxide would. The chlorine is released when the molecule breaks up and destroys ozone molecules. This ozone layer is located in the stratosphere between 10 and 30 miles above the surface of Earth. Because CFCs do not degrade easily in the troposphere, they may rise into the stratosphere, where they are broken down by UV light. Each chlorine atom can destroy as many as 10,000 ozone molecules before it is returned to the troposphere.

In 1985, atmospheric scientists of the British Antarctica Survey published their findings that there was an *ozone hole* in the atmosphere over Antarctica. They found that springtime levels of ozone in the stratosphere over Halley Bay, Antarctica had decreased by more than 40% between 1977 and 1984. Measurements taken from space by the *Nimbus-7* satellite showed that the loss was occurring above an area greater than the size of the entire Antarctic continent. This study provided the first evidence that the stratospheric layer of ozone surrounding Earth might be destroyed by CFCs.

Further studies in 1987 concluded that the ozone hole was the largest ever that was caused by CFCs. These holes in the atmosphere continue to be a potential problem. UV light splits the chlorine atom off the CFC molecule. The chlorine atom then attaches to one of the oxygen atoms in the ozone molecule, and a free oxygen atom then breaks up the chlorine monoxide, thereby two oxygen atoms are produced and the chlorine is released to once again combine with an oxygen atom from an ozone molecule.

Other ozone-destroying chemicals include:

1. Halons were used in fire extinguishers, with U.S. production ending in 1994.
2. Carbon tetrachloride is used as a solvent and in chemical manufacture. It causes cancer in animals. U.S. production ended January 1, 1996.
3. Methyl chloroform (1,1,1-trichloroethylene) is widely used as a solvent. U.S. production ended January 1, 1996.
4. HCFCs, which are hydro-CFCs, are CFC substitutes. U.S. production for the most severe ozone-destroying HCFCs ended January 1, 2002; and for lesser problem HCFCs, U.S. production will end January 1, 2030.

Methyl Bromide

Methyl bromide is one of the substances that depletes the ozone layer and therefore must be regulated under the Clean Air Act. Methyl bromide is a broad-spectrum pesticide used in the control of insects, nematodes, weeds, pathogens, and rodents. In the United States, about 42 million lb of methyl bromide are used each year in agriculture, primarily for soil fumigation (85%), commodity and quarantine treatment (10%), and structural fumigation (5%). Methyl bromide is a colorless and odorless gas at normal temperatures and pressures, but can be liquefied under moderate pressure. It is readily soluble in lower alcohols, ethers, esters, etc. When used as a soil fumigant, the chemical is usually injected into the soil before the crop is planted. About 50 to 95% of the methyl bromide can eventually enter the atmosphere.

Exposure to methyl bromide affects not only the target pests but also any organisms. Human exposure to high concentrations can result in central nervous system and respiratory system failure as well as specific and severe deleterious actions on the lungs, eyes, and skin. Exposure of pregnant women may result in fetal defects. Methyl bromide contributes significantly to the destruction of Earth's stratospheric ozone layer. The science of atmospheric methyl bromide is complex and not well understood. With 1991 levels as a baseline, the EPA has ordered a reduction of 25% in 1999, 50% in 2001, 70% in 2003, and 100% in 2005. This chemical will be allowed for critical agricultural uses, or emergency uses.

Other Air Pollutants

Lead

Lead, which occurs naturally and is found in soils, rocks, water, and food, contributes only a small fraction to air pollution. Lead is emitted into the atmosphere from smelters that recover primary lead from ores and secondary lead from scrap. Emissions of lead also result from the combustion of certain fuels, such as coal, which may contain as much as 54 ppm. Dusts and sprays that contain lead arsenate also contribute to levels of atmospheric lead; the attrition of lead paint on buildings may also contribute. Although the combustion of products that contain large amounts of lead, such as storage battery dividers and heavily painted wood scrap, may produce high amounts of atmospheric lead locally, such combustion is not likely to be a widespread problem.

Industrial processes are now responsible for all violations of the lead standard and for 74% of total lead emissions in the United States. On-road vehicles now account for only 0.5% of total lead emissions, due to the phaseout of leaded gasoline. Annual emissions of lead have declined from the 1970 level of 221,000 tons to the 1997 level of 4000 tons.

When lead is taken into the body, it is circulated in the blood and deposited in bone and other tissues. As with mercury, lead absorption in the body can vary with fasting and with low intake of calcium and phosphorus.

Asbestos

Asbestos is a fibrous mineral dust, which, because of its small size, may be inhaled into the lungs if it becomes airborne. Exposure to high levels of asbestos fibers over a period of years causes a serious chronic lung condition called asbestosis. This condition is caused by the accumulation of relatively large amounts of fibrous asbestos that cause fibrotic changes or "scarring" that can physically obstruct the lung's air passages and in other ways damage its ability to function. Asbestos may also cause lung cancer, including a rare form called mesothelioma. People living in a town with an asbestos mine or processing facility are exposed to elevated levels of asbestos dust.

Beryllium

Beryllium is among the most toxic and hazardous of the nonradioactive substances used in industry. Beryllium and its derivatives are not a major item of commerce, although they are very important components in certain metal alloys and rocket fuels.

The chronic disease berylliosis may not develop until after months or years of exposure. It involves the lungs and many other tissues, because beryllium has the ability to interfere with basic biochemical processes in many cells of the body. Lung damage is usually serious and permanent, and may lead to cancer.

Residents of communities with facilities that use beryllium have developed diseases traceable to beryllium exposure. Airborne beryllium apparently was present at high enough levels to make them seriously ill. It seems that the presence of beryllium in any community's air carries with it the clear risk that serious, permanent, and possibly fatal disease will be contracted by residents.

Cadmium

Cadmium is used primarily in electroplating other metals on alloys to protect them against corrosion. It is used extensively in producing alloys, solders, low cadmium copper, and stabilizers for plastics. Cadmium is usually ingested by cigarette smoking. It travels through the bloodstream as other heavy metals do, and concentrates in the liver and kidneys.

Mercury

Mercury, in any of its chemical forms, is very toxic. Exposure to high levels of this pollutant results in very serious damage to many organs of the body, especially to the brain and the kidneys. It can result in death. Exposure to lower levels of mercury can also have serious effects, especially on the brain.

Some recent findings indicate that airborne mercury may be a significant source of mercury contamination of other parts of the environment, such as water. Inorganic or metallic mercury and organic or methylmercury are found in water. The methylmercury, when ingested, enters the bloodstream and is distributed throughout the body. Certain organs, particularly the liver and kidneys, accumulate more of it than other organs. Mercury is mobile in the environment and once released may cycle between air, land, and water for long periods of time. Many activities release mercury into the air. Two common fuels, coal and oil, contain small amounts of naturally occurring mercury. When these fuels are burned, much of this mercury is vented into the air. Some paper is treated with mercury during manufacture. When the paper is burned after its use, mercury is released. These sources supplement the major industrial sources of mercury, such as the processing of mercury-containing ore and chloralkali plants producing mercury cells.

Fluorides

Fluoride compounds occur in vegetation and in most ores, clays, and soils. Fluorides may be either gaseous or solid emissions of such industrial processes as the manufacture of fertilizer and aluminum, iron ore smelting, and ceramics production.

Airborne fluorides damage vegetation. Furthermore, some plants concentrate and accumulate these fluorides, and then livestock eating the plants as forage become ill. Industrial sources of fluoride pollution cause eye and skin irritation, inflammation of the respiratory tract, and breathing difficulty.

Odors

Odors may be a nuisance or they may indicate a release of substances harmful to humans, animals, plants, and property. Odor-producing activities include the following:

1. Animal odors from meat packing and rendering plants include fish oil odors and fish meal from manufacturing plants and food-processing plants, poultry ranches and poultry processing, and leather manufacturing.
2. Odors from combustion processes include gasoline and diesel engine exhaust from cars, buses, trucks, and jet planes; coke oven and coal gas odors; and poorly operated heating systems.
3. General industrial odors are caused by burning rubber from smelting and debonding, odors from dry-cleaning shops, fertilizer plants, asphalt odors (roofing and street paving), paint and solvent manufacturing, plastic manufacturing, kraft pulping, and paper manufacturing.
4. Odors from combustible waste including home incinerators and backyard trash fires, city incinerators burning garbage, open-dump fires, and compost piles.
5. Refinery odors include mercaptans, crude oil and gasoline odors, sulfur, sulfur dioxide, and hydrogen sulfide.
6. Sewage odors include city sewers carrying industrial waste, sewage treatment plants, and overflowing individual sewage systems.

Economic Poisons as Air Pollutants

Economic poisons, which are necessary in agriculture, forestry, and public health, are all those materials that are toxic to humans but are nonetheless used to control pests in the human environment. The term refers to pesticides including insecticides, fungicides, herbicides, and rodenticides. Economic poisons also include other agriculturally important chemicals such as regulators of plant growth, defoliants, and desiccants. Kerosene, copper sulfate, and cyanides are common commercial items that are considered economic poisons when used as pesticides. The hazard involves length of exposure, concentration, individual sensitivity, and toxicity of the material.

Spraying, dusting, and fumigating with pesticides involve especially the citizens of small communities located in farm areas. Because the pesticides become airborne, the individuals are subjected to a constantly varying load of pollutants, depending on the amount of economic poison that becomes airborne. In all aerial applications, part of the material ejected does not reach the target because of weather conditions and possible pilot error. This increases the hazard to humans. Further, additional hazards are created, because about 20% of economic poisons are used in the home and garden, and by public authorities in buildings, parks, and street plantings. Commercial establishments and industry use an additional 5%. Whereas target efficiency for industry and commercial establishments is probably high, with most of this use within confined areas, public authorities generally use insecticides in the open air; and this is accompanied by all the uncertainties of crop and forest treatment.

Radioactive Pollution

Radioactivity in the atmosphere comes from both natural sources and human activities. Natural radioactivity is either of terrestrial origin, consisting mainly of radioactive gases released from soils and rocks, such as radon and thoron, or anthropogenic (human-made) pollution that comes from weapons testing. A potential problem also exists from nuclear energy plants and waste disposal.

Sources of Air Pollution

Natural Sources

Natural sources of air pollution include fog, dust due to desert sandstorms, and soil becoming airborne; salt spray; gases from springs, fissures, and cracks in the earth; volcanoes producing ash, gas, fumes, and smoke; smoke from forest fires; and pollens from various plants and weeds. Important natural air pollutants are pollens, because they cause hay fever and allergies. Airborne organisms, such as bacteria and molds, also pose risk to health.

Multiple Sources

Multiple sources of air pollution include combustion of fuel, attrition, evaporation, and incineration. Combustion is the process of burning in which certain chemicals are combined with oxygen and result in the production of energy and by-products. In complete perfect combustion, which can only theoretically be accomplished, a pure hydrocarbon combines with the right amount of oxygen at the proper temperatures for the right amount of time to produce energy, water vapor, and carbon dioxide. However, because fossil fuels contain contaminants, especially sulfur, an imperfect combustion usually takes place, where the waste materials — unburned or partially burned fuels, oxides of contaminants, fly ash, and carbon monoxide — are emitted to the atmosphere. The high temperatures in near perfect combustion cause oxygen and nitrogen in the air to form nitric oxide and nitrogen dioxide.

Incinerator Air Pollutants

Particulate and gaseous emissions from municipal solid waste incineration depend on the waste stream composition, incinerator engineering, and operation.

These emissions typically include heavy metals such as mercury, lead, arsenic, cadmium, and chromium. They also include polychlorinated dibenzodioxins, dibenzofurans, and polycyclic aromatic hydrocarbons, in addition to acid gases, oxides of nitrogen, sulfur, and carbon. Chlorine and fluorine are emitted as acid gases and contribute to the formation of halogenated aromatic organic compounds.

The U.S. EPA has found measurable quantities of at least 55 hazardous chemicals in the emissions from incinerators. The most common were benzene, toluene, carbon tetrachloride, chloroform, methylene chloride, trichloroethylene, chlorobenzene, naphthalene, phenol, and *bis*(2-ethylhexy)phthalate, in addition to the previously mentioned chemicals. Apparently when the furnace is cold during the start-up of the plant or a large load of plastics or solvents is involved, the amount of pollutants increase substantially. The presence of many of these compounds indicates that the incinerators are operated under conditions that cause incomplete combustion.

Cement Kiln Air Pollutants

Cement kilns are major sources of toxic air emissions. Currently, 120 cement plants operate in the United States, each emitting many millions of pounds of pollution into the air each year; 18 of these plants are burning commercial hazardous waste for fuel in one or more kilns. The plants emit sub-micron particles of cement kiln dust, sulfur dioxide, metals, polycyclic aromatic hydrocarbons, and other pollutants. In addition, a vast number of toxic air pollutants are emitted, depending on the type of waste used as a fuel. About 70% of all liquid and solid hazardous wastes commercially incinerated in the United States are burned in cement kilns. Inadequate combustion also produces dioxins and furans.

Backyard Burning Air Pollutants

Solid wastes contain plastics, heavy metals, and various other synthetic materials. When these materials are burned at inadequate temperatures, they provide a variety of particulates as well as numerous toxic substances. Exposure to smoke may result in burning eyes, headaches, nausea, fatigue, and dizziness; and damage to lungs, nervous system, kidneys, and liver.

External Combustion

External combustion takes place in furnaces. These furnaces include steam–electric generating plants; industrial boilers; commercial and institutional boilers; and commercial and residential combustion units used for homes, hotels, motels, hospitals, universities, schools, government buildings, and other service industries.

The furnace is a combustion chamber in which air and fuel are mixed and heat is applied to release large quantities of stored energy from the fuel. Incomplete combustion results from too much or too little air, too low a temperature, and inadequate time allowed for the burning process.

Industry, smelting, power generation, process heating, and space heating produce pollutants containing metal oxides, sulfur oxides, nitrogen oxides, and particulate emissions. The fossil fuels usually used in external combustion are coal, fuel oil, and natural gas. Other fuels used in relatively small quantities are liquid petroleum gas, wood, coke, refinery gas, blast furnace gas, and other waste or by-product fuels. Coal, oil, and natural gas supply about 95% of the total thermoenergy used in the United States. Coal, which is the most abundant fossil fuel in the United States is burned in a wide variety of furnaces to produce heat and steam. Although it is mostly carbon, coal contains many other compounds, including sulfur, silica, and iron oxides.

Fuel oil is classified as either residual or distillate. The primary difference is that residual oil, which is basically used for industry, contains a higher ash and sulfur content, is more viscous, and therefore is harder to burn properly.

Natural gas, although considered a relatively clean fuel, emits quantities of carbon monoxide, hydrocarbons, and sulfur oxides. Nitrogen oxides are produced in the combustion chamber. Liquefied petroleum gas, such as butane and propane, are considered clean fuels. However, they still produce carbon monoxide, hydrocarbons, and nitrogen oxides.

Internal Combustion

The internal combustion engine is a machine that produces mechanical energy by first producing heat energy through combustion of fuel and then converting it to a force. In the typical automobile engine, a mixture of fuel and air is fed into a cylinder by the carburetor, compressed, and ignited by a spark from the spark plugs. The explosive energy and the burning mixture pushes the pistons, and the motion of the pistons is transmitted to the crank shaft that drives the car. The burned mixture, unburned fuel, and by-products pass out through the tail pipe. Inadequate quantities of air in the combustion process result in emission of carbon monoxide instead of carbon dioxide, unburned gasoline, and assorted hydrocarbons. The extremely hot air in the engine converts the nitrogen and oxygen of the air into the nitrogen oxide that helps produce photochemical smog.

In addition, other emissions such as benzene, 1,3-butadiene, and polycyclic organic material are toxic to humans. The crankcase, containing the crankshaft, contributes heavily to pollution, because air flowing through the crankcase eliminates any gas–air mixture that has blown past the pistons, as well as any evaporated lubricating oil and any escaped exhaust products. Crankcase emissions are known as blowby gases. A small amount of air pollution is due to evaporation from the fuel tank and from the carburetor after the heated engine is turned off. Internal combustion engines are responsible for approximately 73% of the carbon monoxide, 56% of the hydrocarbons, and 50% of the nitrogen oxides emitted in the United States. Proper maintenance of vehicles and catalytic converters helps reduce these pollutants.

Diesel engines, although malodorous, should produce less pollutants than gasoline engines, because the diesel operates at a higher air-to-fuel ratio than the gasoline engine and fuel is injected directly into the combustion chamber. Further, there is no spark ignition system. The air is heated by compression in the engine cylinder until the temperature is raised to about 1000°F. The fuel is then ignited as it is injected into the cylinder. The diesel engine can become, however, a worse source of pollution than the gasoline engine if the vehicle is operated improperly and maintenance is poor. Further, the excessively heated air causes the diesel to form greater quantities of nitrogen oxides.

The chemical composition and magnitude of vehicle exhaust emissions can be directly related to the composition of the gasoline used. Even though a strong relationship exists, there is a deficiency in understanding the chemistry of the combustion process, the effect of engine design, engine operating conditions, control system strategy, and their interrelationship.

Attrition

Attrition is wearing or grinding down by friction. Attrition occurs from the process of sanding, grinding, demolishing, drilling, spraying, or simply walking or driving. In any situation where friction is present, some portion of the surface in contact breaks off in minute particles into the atmosphere.

Evaporation

Evaporation is the change of state of a liquid into a gas at any temperature below its boiling point. This is a physical change only. Evaporation, which contributes to most of the odor-type pollutants, includes the emission of organic solvents from dry-cleaning plants, surface-coating operations, as well as volatilization of petroleum products. Examples of surface coating include the brushing, rolling, and spraying of paint, varnish, lacquer, or paint primer for decorative or protective purposes. Petroleum product evaporation results from petroleum storage and the pumping of gasoline, for example.

Incineration

The Resource, Conservation and Recovery Act (RCRA) defines solid waste as garbage, refuse, sludges, and other discarded solid materials. These are both combustible and noncombustible, resulting from industrial, commercial, agricultural, and community activities. Incinerators are used to reduce the volume of this material. Incineration is a thermal treatment method based on a combustion process. Hazardous organic components serve as a fuel or reducing agent and are mixed with oxygen, which serves an oxidizing agent. Under the conditions of high-oxygen combustion at temperatures ranging from 425 to 1650°C the fuel is thermally broken down and the elemental components are oxidized to yield various combustion products. Combustion efficiency is dependent on duration of incineration (time), mixing and oxygenation (turbulence), and thermal conditions (temperature). The combustion products include only carbon dioxide, water, and heat if the fuel is a pure hydrocarbon and is combusted (thermally oxidized) under conditions of 100% efficiency. Hydrocarbons incinerated at less than 100% combustion efficiency, which is the normal case, would also generate carbon monoxide and smoke in addition. The products of the combustion process are directly influenced by the composition of the incinerated material or fuel. Whereas pure hydrocarbons would yield oxidized forms of carbon and oxygen, chlorinated fuels would yield chlorinated products as well.

Accordingly, the composition products or residues generated from combustion processes typically include carbon dioxide, water, carbon monoxide, sulfur dioxide, nitrogen oxides, hydrochloric acid, particulates (including metals), and ash. Some of these products are typically collected from the emission source and subjected to additional treatment or disposal. Because the solid waste is a conglomeration of many materials, the inefficient incinerator produces a variety of emissions, including fly ash, smoke, gases, and odors. Whereas fly ash and odors are primarily a nuisance, smoke and gases contribute to the overall air pollution problem through reduction in visibility and as part of the smog-forming photochemical reactions taking place in the air. Wet solid waste lowers the temperatures of incinerators. Glass and plastics can form a coating on the grates, thereby lowering incinerator temperatures. The result of these problems is partially burned solid waste that must be hauled away to a landfill and increased levels of air pollutants.

Automobile bodies may be disposed of through incineration in an uncontrolled burning process. Emissions from this process would include particulates, carbon monoxide, hydrocarbons, nitrogen oxide, aldehydes, and organic acids.

Open burning causes the emission of particulates, carbon monoxide, hydrocarbons, sulfur oxides, and many other odorous materials. The kinds and quantities of emissions depend on wind, temperature, composition, and moisture content of the material burned and the compactness of the pile.

Specific Sources

Each industry is a specific source of air pollution. The polluted effluents are the result of the manufacturing process, which includes raw materials, fuels, processing technique, efficiency of the process, and air pollution control measures utilized. Although all industries contribute to the pollution problem, only the major industrial polluters are discussed. They are pulp and paper mills, iron and steel mills, petroleum refineries, metallurgical industries, chemical manufacturers, power plants, and food and agricultural industries. The kinds of pollutants in the air are based on the kinds of industries within the metropolitan area. An area having primarily one category of industry has air pollution problems related to that industry. An area having a diversified economy has a greater variety of air pollution problems.

Pulp and Paper Mills

Wood pulping involves the production of cellulose from wood by dissolving the lignin that binds the cellulose fibers together. This is accomplished chemically by the use of the kraft or sulfate process, the sulfite process, and the neutral sulfite semichemical process. The kraft process involves the cooking of wood chips under pressure in the presence of a cooking liquor composed of an aqueous solution of sodium sulfide and sodium hydroxide. The kraft process produces annoying odors due to the release of sulfur dioxide, hydrogen sulfide, dimethyl sulfide, or methyl mercaptan. In addition, carbon monoxide may be released into the atmosphere. The other wood pulping processes are not discussed due to a lack of valid emission data.

Iron and Steel Mills

Air contaminants are emitted in the open-hearth furnace throughout the 8- to 10-hour process. These contaminants are due to combustion of grease, oil, or other material found on steel scrap or due to furnace fuel. Iron oxides, other metal oxides, and small amounts of fluorides may be produced and emitted. The fluorides are found in some iron ores and are particularly dangerous, because they affect plants, which may in turn cause chronic poisoning of animals.

The basic oxygen furnace produces iron oxides and carbon monoxide as unwanted by-products. The electric arc furnace, which is primarily used to produce special steel alloys or to melt large amounts of scrap for reuse, produces particulates containing oxides of iron, manganese, aluminum, and silicon, as well as moderate amounts of carbon monoxide.

Petroleum Refineries

The modern petroleum refinery is a complex system of many processes that can be divided into four major steps: separating, converting, treating, and blending. The crude oil is separated into selected fractions that include gasoline, kerosene, fuel oil, and lubricating oil. Heavy naphtha, which is of less value than the previously mentioned products, can be converted into gasoline by splitting, uniting, or rearranging the original molecules.

Crude oil, which is a mixture of many different hydrocarbons, contains a variety of impurities. The most significant of these impurities are sulfur, oxygen, and nitrogen. Emissions to the atmosphere include light hydrocarbons, hydrogen sulfide, waste gases, smoke, carbon monoxide, nitrogen oxide, aerosols, sulfur dioxide, and other odorous gases. Sources of emission from oil refineries include storage tanks, catalyst regeneration units, pipeline valves and flanges, pressure relief valves, pumps and compressors, compression engines, cooling towers, loading facilities, wastewater separators, blow-down systems, boilers, process heaters, and air-blowing devices.

Metallurgical Industries

The metallurgical industries can be divided into primary and secondary metal production operations. In primary production, the metal is removed from the ore. These industries include aluminum production, copper smelters, lead smelters, zinc smelters, iron and steel mills, iron alloys, and metallurgical coke. A considerable amount of sulfur oxide, particulate matter, gaseous hydrogen fluoride, hydrocarbons, carbon monoxide, and nitrogen oxide are emitted to the air.

The secondary metal industries include the recovery of metal from scrap and production of alloys. These industries include brass and bronze, gray iron foundries, lead smelting, magnesium smelting, steel foundries, and zinc processing. The major air contaminants are particulates of the various metals and impurities released as fumes, smoke, and dust.

Chemical Manufacturers

The chemical industries include the manufacture of thousands of different chemicals and chemical by-products. These products include petrochemicals; and heavy or industrial chemicals, such as sulfuric acid, soda ash, caustic soda, chlorine, ammonia, phosphoric acid, synthetic detergents, pharmaceuticals, plastics, cosmetics, synthetic fibers, chloralkali, explosives, paint, and varnish. The emissions from these industries are due to any step in the process and are too numerous to name. For further information, see government documents on each of these chemical processes.

Power Plants

Power plants generate electricity by burning coal and oil that may contain sulfur as an impurity. The emissions from the power plants include carbon monoxide, sulfur dioxide, nitrogen dioxide, hydrocarbons, and particulate matter. The quantities of the pollutants help make the power plant a major source of air pollution, particularly in the area of sulfur dioxide.

Food and Agricultural Industries

Food and agricultural products must be processed before consumption. This processing includes refinement, product improvement, preservation, storage, handling, packaging, and shipping. The food and agricultural industries include alfalfa dehydration, coffee roasting, cotton production, feed and grain mill storage, fermentation, fish processing, meat processing, nitrate and phosphate fertilizer utilization, starch manufacturing, and sugarcane processing. The major emissions are dust, particulate matter, and odors. In the utilization of fertilizer, emissions include ammonia, nitric oxides, nitrates, silicon, tetrafluoride, and sulfur oxides.

Air Toxics and Hazardous Air Pollutants

Toxic air pollutants are poisonous substances in the air that come from natural sources, such as radon gas coming from the ground, or from anthropogenic sources, such as chemical compounds coming from factory smokestacks. These pollutants can harm the environment or the individual's health. Breathing the toxic air pollutants can increase your chances of health problems. The pollutants create health risks, directly or indirectly. The 1990 Clean Air Act amendments list various categories: major sources (10 tons/year of any listed hazardous air pollutant, or a combination of listed hazardous air pollutants of 25 tons or more); lesser quantity major sources (pollutants that are highly toxic to human health or the environment, which are present in less than 10 or 25 ton quantities); and area sources (source emissions that are small, but whose collective volume can be hazardous in densely developed areas where large numbers of such facilities are packed together into urban neighborhoods and industrial areas).

Air toxics come from routine emissions from stationary sources, mobile sources, volcanoes, accidental releases, and forest fires. Once released toxic pollutants can be carried by the wind to other locations. These pollutants may remain airborne and contribute to pollution problems far from the original source. The pollutants may be deposited on bodies of land or water. Some of the toxic air pollutants are of special concern such as the metals, because they degrade very slowly or not at all. Chemicals of concern may be found on the Unified Air Toxics Web site of the EPA.

Community sources of specific air toxics are important to know, to develop appropriate control programs, and to determine the potential for human health effects. The following chemicals are used as examples:

1. Benzene comes from pumping gasoline, tobacco exposure, lawn mowers and other small household combustion engines, household solvents and cleaners, art and hobby supplies, glue, and industrial sources.
2. 1,3-Butadiene comes from industrial sites and gasoline vapors.
3. Formaldehyde comes from combustion of hydrocarbons from incinerators, vehicle exhaust, industrial processes, and products.
4. Tetrachloroethylene (perchloroethylene) comes from waste disposal sites and dry-cleaning establishments.
5. Polychlorinated dibenzodioxins and polychlorinated dibenzofurans come from industrial sources (chemical synthesis and processing, manufacture of pesticides), combustion sources (municipal waste, hazardous waste, sewage sludge), and other sources (automobile exhaust, chemical fires, waste fires, and dry cleaning).
6. Polycyclic aromatic compounds come from biomass combustion, wildfires, fireplaces, internal combustion engines, municipal solid waste incineration, cigarette smoke, and industrial sources.

Air toxics or hazardous air pollutants are a very serious problem. They are categorized as organic vapor, organic particulate, inorganic vapor, and inorganic particulate. The hazardous air pollution emission source may be a process point source, a process fugitive source, or an area fugitive source. A point source may be a reactor, distillation column, condenser, furnace, or boiler. A process fugitive source may be pump, valve, compressor, access port, and feed or discharge openings. An area fugitive source may be a waste treatment lagoon, new material storage pile, or a road. Once the emission source is determined, it is necessary to find out the type or mixture of hazardous air pollutants, concentration, temperature, flow rate, heat content, and particle size. Toxic substances are found in a variety of different situations and come from many different processes.

Although the concentrations of the toxic chemicals may be below the level considered to have an effect on humans, the short-term or long-term hazard must be accounted for. Industrial and manufacturing processes, solvent use, sewage treatment plants, hazardous waste handling and disposal sites, municipal waste sites, incinerators, and motor vehicles are all sources of toxic chemicals. Cadmium, lead, arsenic, chromium, mercury, and beryllium are released by smelters, metal refiners, manufacturing processes, and stationary fuel combustion sources. Toxic organic chemicals, such as vinyl chloride and benzene, are released to the atmosphere by a variety of plastic and chemical manufacturing plants. Chlorinated dioxins are emitted by some chemical processes and by the high-temperature burning of plastics in incinerators.

People are exposed to the toxic chemicals when they are emitted to the air from industry or automobiles. The most common route of exposure is through inhalation. Indirect exposure may occur after the airborne particles fall onto the earth and are taken up by crops, animals, or fish. These particles may also contaminate the water supply. Toxics may accumulate over time and become highly concentrated in human fatty tissue or breast milk.

Mercury sources include mercury smelters, chloralkali plants, sewage sludge, and incinerators. Beryllium sources include extraction plants, foundries, ceramics, and rockets. Asbestos sources include asbestos mills, roadways, and manufacturing processes. Vinyl chloride is emitted from ethylene diochloride producers, vinyl chloride producers, and vinyl chloride polymers. Benzene sources include fugitive emissions and coke plant by-products. Radionuclide sources include U.S. Department of Energy facilities, Nuclear Regulatory Commission sites or facilities, and elemental phosphorus plants. Inorganic arsenic sources include high arsenic copper smelters and low arsenic copper smelters.

Major Source Categories

The major source categories for hazardous air pollutants include solvent usage operations, metallurgical industries, synthetic organic chemical manufacturing industries, inorganic chemical manufacturing industries, chemical product industries, mineral product industries, wood product industries, petroleum-related industries, combustion sources, surface coating processes, and food and agriculture processes.

The solvent-usage industries manufacture processes that use solvents such as surface coating operations, dry cleaning, solvent degreasing, waste solvent reclaiming, and graphic arts. They use such compounds as toluene, xylene, ethylene glycol, acetone, and alcohols.

The metallurgical industries are divided into primary, secondary, and miscellaneous metal production operations. Primary refers to production of metal from ore. Secondary refers to recovery of metals from scraps and salvage. Miscellaneous refers to industries that use or produce metals as final products. These industries may potentially produce such pollutants as acetaldehyde, ammonia, antimony, arsenic, benzene, beryllium, chlorides, cyanides, fluorides, hydrogen sulfide, lead, manganese, mercury, nickel, phenol, polycyclic organic compounds, toluene, and xylene.

The synthetic organic chemical manufacturing industry produces organic vapors and some particulates as pollutants. Potential emissions come from:

1. Storage and handling of raw and finished materials
2. Reaction process that alters molecular structure
3. Separation process that divides chemicals into distinct products
4. Fugitive sources from storage and handling
5. Secondhand sources such as waste treatment

The pollutants vary with the raw products used, the intermediates created, and the final products produced.

The inorganic chemical manufacturing industry includes the manufacture of the basic inorganic chemicals before they are used in manufacturing other chemical products. Although the potential for hazardous air pollutants is high because of economics, the actual pollutants are reduced. They include ammonia, arsenic, arsenic compounds, beryllium, chlorine, fluorine, acids, bases, lead, manganese, mercury, phosphorus, etc.

The chemical products industry manufactures carbon black, synthetic fibers, synthetic rubber and plastics, pharmaceuticals, detergents, pesticides, paints, explosives, etc. The potential pollutants produced include aluminum compounds, arsenic compounds, ammonia, halides and their compounds, lead compounds, mercury compounds, phosphorous compounds, etc.

The mineral products industry involves the processing or production of nonmetallic minerals such as cement, coal conversion, glass and glass fibers, and phosphate. The potential pollutants include arsenic, beryllium, benzene, cresole, cadmium, carbon tetrachloride, formaldehyde, mercury, perchloroethylene, nitrosamines, terpenes, toluene, xylene, ammonia, vinyl chloride, hydrogen cyanide, sulfuric acid, lead, etc.

The wood products industry includes industrial processes that convert logs to pulp, pulpboard, hardwood, plywood, particleboard, or related wood products and wood preserving by use of various chemicals. The kraft sulfite and neutral sulfite processes are used in chemical pulping. Plywood is produced by bonding veneers with an adhesive. Wood preserving is accomplished by injecting chemicals with fungicidal, insecticidal, and fire-resistant properties. The potential hazardous air pollutants include arsenic, asbestos, mercury, polycyclic organic compounds, chlorine, chlorobenzene, formaldehyde, methyl mercaptan, dioxin, hydrogen sulfide, phenol, cresols, etc.

The petroleum-related industries include oil and gas production, petroleum refining, and basic petrochemical industries. The potential hazardous air pollutants include paraffins; aromatics such as benzene, toluene, and xylene; sulfur compounds; ammonia; coke; and catalyst fines.

The fuel combustion industry includes units used to produce electricity, hot water, and steam. The combustion units burn coal, fuel oil, natural gas, wood, and various wastes. The potential hazardous air pollutants include fluorides, arsenic, cyanides, polycyclical organic compounds, lead, etc.

Impact on Other Problems

Air contaminants that have been removed by a variety of air pollution control devices create land and water pollution problems. The solid pollutants removed from the air are generally disposed of in landfills. Leaching occurs from the landfill, and the pollutants eventually contaminate underground and surface water supplies. The pollutants, depending on the content of the materials, create hazards to plant and animal life. Where water is used in the air-cleaning process, the wastes plus the water must be disposed of again. The contaminated water may be pretreated on the site of the plant or it may go directly into a receiving stream. Depending on the type of pollutants present, another hazard may be created to plant and animal life.

Disease and Injury Potential

Human exposure to urban air pollution can be estimated by determining the concentration of pollutants including hazardous air pollutants, by collecting "personal" air samples, and by using data from blood samples. Determining human exposure is very complex because information must be evaluated ranging from what and how much is emitted, to seasonal and weather conditions, to how it may be transported and changed in the air, to how much a person inhales, to the individuals personal susceptibility to the pollutants. Humans are exposed to non-genotoxic, as well as a high number of physical or chemical genotoxic agents, some of which take years to induce disease. These agents may act alone or synergistically. Genotoxic agents may be mutagenic, carcinogenic, or both. Nitro aromatic hydrocarbons in particulate pollutants are examples of substances mutagenic to bacterial and mammalian cells. Their behavior is related to the combustion of fossil fuels.

Also, the respiratory-cancer risk is closely related to the content of polycyclic aromatic hydrocarbons found in urban air and industrial air. Once the particles and bound chemicals previously mentioned enter the respiratory tract, they may become metabolized. Metabolism may result in products that are relatively nontoxic and cleared rapidly from the body. On the other hand, the chemicals may be metabolized to a more chemically reactive species capable of reacting with cellular macromolecules such as DNA. Exposure is the contact of a chemical with the skin, nose, or mouth. Potential dose is the amount of chemical available. Applied dose is the amount of chemical coming in contact with the person, and available for absorption. The internal absorbed dose is the amount of chemical penetration in a body barrier. The delivered dose, which is of greatest significance, is the amount of chemical available for interaction with the body.

Respiratory System

The structure and function of the respiratory system may be seriously altered by air pollutants. Certain irritants, gaseous or particulate, can slow down or stop the action of the cilia. The function of the cilia is to sweep and propel the mucus, laden with microorganisms and dirt, out of the respiratory tract. If these cells are inactivated, the microorganisms invade the underlying cells. The irritants can increase or thicken the mucus, constrict the air passages, and also cause swelling of the cells. Breathing may become more difficult because of a mechanical blockage or because of an ineffective exchange of oxygen and carbon dioxide in the alveoli.

Sulfur dioxide, nitrogen dioxide, ozone, carbon monoxide, organic gases, and particulate matter are probably the substances that cause the greatest respiratory problems. A high enough concentration of sulfur dioxide irritates the upper respiratory tract because of its high solubility in body fluids. It also has an effect on breathing and respiratory illness, alters pulmonary defenses, and aggravates existing cardiovascular disease. In low concentrations, sulfur dioxide makes breathing more difficult by constricting the fine air tubes of the lungs. Sulfur dioxide especially affects children, the elderly, and people with asthma, cardiovascular disease, or chronic lung disease. Nitrogen dioxide acts as an acute irritating substance. It is absorbed in both large and small airways. Very high concentrations cause lung injury, fatal pulmonary edema, and bronchial pneumonia. Lower concentrations cause impaired mucociliary clearance, particle transport, macrophage function, and local immunity.

Ozone is a highly irritating, oxidizing gas. A few parts per million may produce congestion, edema, and hemorrhage of the lungs. Ozone also causes upper respiratory irritation, cough, shortness of breath, wheezing, decreased tidal volume, nausea, malaise, and headache. Pulmonary functions may be impaired. People at the greatest risk are those with asthma, chronic lung disease, and heart disease; and those who spend considerable time out-of-doors.

Carbon monoxide combines readily with hemoglobin to produce carboxyhemoglobin, because hemoglobin has an affinity for carbon monoxide, which is 210 times as great as its affinity for oxygen. This causes oxygen deficiency in the body tissues, leading to dizziness, headaches, disorientation, unconsciousness, and eventually death. Carbon monoxide may be related to the development of atherosclerosis, which is a progressive disease characterized by deposits of fat, cholesterol cells, and connective tissue in blood vessels. Formaldehyde, which is an organic gas, is highly irritating to mucous membranes and can cause constriction of the bronchial tubes. Gases and vapors either directly interact with tissue at the site of contact or are absorbed into the circulatory system. This inhalation of gases and vapors is probably the primary mode of exposure to chemicals.

Light and noncyclic hydrocarbons affect human health. Benzene is an established human carcinogen that induces DNA strand breaks. Chloroform and carbon tetrachloride are carcinogens and cause liver and kidney toxicity. Methylene chloride, trichloroethylene, and tetrachloroethylene are strongly suspected carcinogens. Many halogenated alkanes, alkenes, and alkynes are cytotoxic, carcinogenic, or both. Toluene induces DNA strand breaks and can induce central nervous system symptoms, including brain damage, with chronic or acute use. Mixtures of solvents used industrially have been implicated in inducing fatigue, loss of appetite, loss of memory, etc.

Polycyclic aromatic hydrocarbons affect human health. Many of the hundreds of congeners of PCBs, dioxins, and furans appear to create human cancer and toxicity by interacting with cell DNA to cause gene alterations. These compounds may cause neurological changes including altering dopamine concentrations in the central nervous system, bronchial and liver damage, and immune changes.

Diesel exhaust contributes to ambient sulfur oxides, ozone precursors, and aerosols that contribute to chronic respiratory morbidity and mortality, as well as potential increase in cancer risk. Diesel fuel may be a cocarcinogen. Heavy exposure to diesel exhaust is clearly associated with pulmonary inflammation and reduced levels of pulmonary function.

Small particles, 10 μm or less, penetrate deeply into the respiratory system. They tend to concentrate various chemicals or microorganisms on their surfaces, and therefore can produce a locally high concentration of an otherwise very dilute substance. Particles inhaled at the same time as gases or microorganisms may modify the body's response to these foreign substances, thereby causing a greater possibility of damage or disease. Chronic bronchitis, emphysema, and other lung diseases are aggravated by small particles. The particles most closely associated with mortality are 2.5 μm or smaller in diameter. They come from auto exhausts, factory and power plant smokestacks, and other processes that burn oil, gas, and coal. The white haze seen on hot, humid days is evidence of this pollution.

Particulate airway distribution and apparently health effects are dependent on the size of the particles and on the structural and functional characteristics of the airways. Nearly all particles larger than PM_{10} are trapped in the upper airways where they tend to be cleared by mucociliary mechanisms. People with obstructive pulmonary disease have a greater problem becoming cleared. Acute symptoms of particulate inhalation include restricted activity, respiratory illnesses, exacerbations of asthma, and chronic obstructive pulmonary disease. The annual estimated mortality in the United States to particulate inhalation is thought to be in the tens of thousands.

Metals and heavy metals from incineration are of serious concern because 50 to 60% of the inputted metals may be released in the stack gases. Cadmium and mercury are a serious concern because both of them displace zinc, a crucial metal for the living processes including enzyme DNA transcription, immune system activation, and membrane stabilization, therefore reducing the available zinc in the body.

Wood smoke particulates contain over 200 chemicals and causes high ground-level pollution. They may cause pulmonary arterial hypertension, disrupt cellular membranes, depress macrophage activity, and destroy ciliated and secretory respiratory epithelial cells.

Chronic lung diseases effected by air pollution include lung cancer, chronic bronchitis, asthma, and emphysema. Strong evidence exists to suggest a link between air pollution and lung cancer including the following: mortality rates for lung cancer are higher in urban polluted areas than in rural areas, even when cigarette smoking is taken into account; substances producing cancer in experimental animals are found in polluted areas; carcinogens are relatively stable in the atmosphere; carcinogens extracted from air samples produce cancers in experimental animals; and atmospheric irritants affect the protective action of the cilia and the normal flow of mucus. Smoking accentuates all the diseases mentioned, with smoking responsible for the majority of lung cancers.

Chronic bronchitis, which is characterized by excessive mucus secretion in the bronchial tree, is manifested by chronic or recurrent coughs (Figure 1.5). The mortality rate of this disease is associated with the amount of pollution to which the individual is exposed. This has been especially evident in the air pollution disasters in London.

Asthma is an inflammatory disease with episodic symptoms of airflow obstruction, at least partial reversability of airway obstruction, diagnosed after reasonable exclusion of alternative diagnoses. Specific sensitization is associated with asthma and may result from a predisposition combined with a sufficient exposure over time to either biological materials such as mold, pollens, plant materials, insect and animal

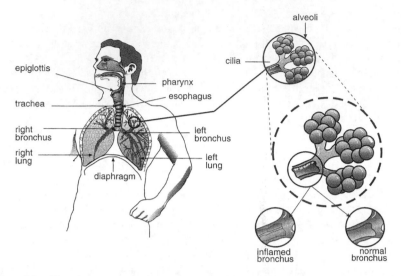

Figure 1.5   Chronic lung diseases affected by air pollution: bronchitis.

materials; or nonbiological sensitizers such as isocyanates, acrylates, amino alcohols, metals (chromium, aluminum, platinum, nickel, cobalt), or pharmaceuticals.

Asthma is characterized by an increased responsiveness of the trachea and bronchi to various stimuli and is manifested by the widespread narrowing of the airways or bronchioles (Figure 1.6). The narrowing may be caused by a muscle spasm, a swelling of the mucous membrane, or the thickening and increase in mucus secretions.

Emphysema is an alteration of the anatomy of the lung characterized by an abnormal enlargement of the air spaces distal to the terminal nonrespiratory bronchial, leading to a destructive change in the alveolar walls (Figure 1.7). In the United States, the disease is more common in polluted urban areas than in rural areas. The important factors relating to air pollution in the development of the disease are chronic recurrent irritations, chronic or recurrent infections, recurrent temporary or permanent paralysis of the cilia, overproduction and increased viscosity of mucus, abnormal prolonged retention of active or inert air pollutants, and injuries from violent coughing.

Acute nonspecific upper respiratory symptoms occurred in individuals after subjection to high levels of sulfur dioxide and particulate matter in the air during the major air pollution disasters in Donora, Meuse Valley, and London. The residents who were least affected still displayed symptoms similar to the common cold.

Of the U.S. population 80% live in large urban areas. These individuals are at greater risk of having lung disease than those living in rural areas. Estimates of mortality due to air pollution vary from 0.1 to 10%. Lack of substantial data on mortality related to air pollutants accounts for the wide range of variance. The air pollutants of greatest concern have already been mentioned in this chapter. The amount of air pollution present is important. Also, other pollutants within the air

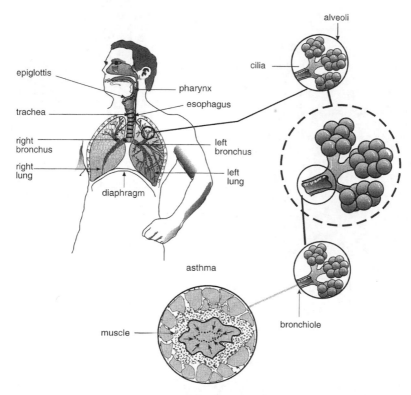

Figure 1.6   Chronic lung diseases affected by air pollution: bronchial asthma.

may interact with the initial pollutant or act as a means of carrying the pollutant into the respiratory system. Most of the epidemiological data available are of very recent origin and therefore it is difficult to have a good picture of what has occurred in the past to individuals in the general population. Laboratory research on animals has been used to study the long-term effects of pollutants, to test various hypotheses, and to study environmental carcinogens and the interaction between pollutants and the age or nutritional status of the subject.

It is essential to recognize that although the specific data are not available, it is known that the pollutants entering the air are hazardous and probably over long periods of time can create substantial difficulties for individuals living within the polluted air.

The Eye

The eye is sensitive to certain air pollutants (Figure 1.8). Because the external coat, conjunctiva, and cornea come into direct contact with the atmosphere, eye irritation occurs if the atmosphere contains irritating gases or particles that have absorbed gases. The resultant tears are a mechanism to flush the foreign material

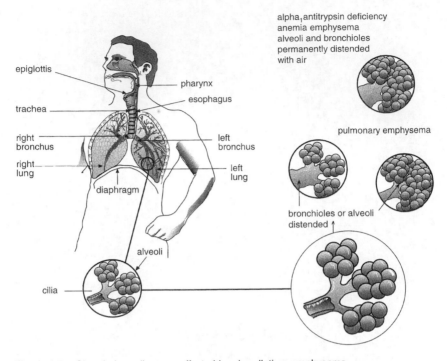

Figure 1.7   Chronic lung diseases affected by air pollution: emphysema.

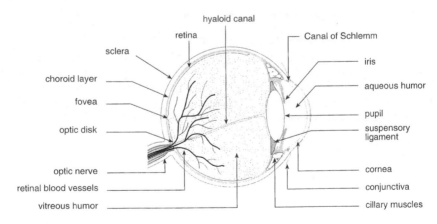

Figure 1.8   The eye.

from the eyes. Photochemical air pollution causes eye irritation and decreased visibility. It is possible that the irritant is a mixture of formaldehyde, acrolein, peroxides of various types, and free radicals. Peroxyacetyl nitrate is a powerful, highly irritating substance at 1 ppm. It is also possible that the particles in the air associated with the photochemical reaction may contribute to the high irritation.

Systemic Impact

Because the respiratory and circulatory system are so closely linked, burdens placed on the respiratory system eventually affect the circulatory system. Where individuals are exposed to unusually severe air pollution conditions, those with heart and respiratory problems are most affected. Individuals who already have cardio-respiratory diseases may be affected by the carbon monoxide levels reached in polluted air. In addition, particles, gases, and vapors absorbed into the bloodstream can potentially damage blood cells, liver, kidneys, nervous system, and other organs.

Digestive System

Air toxics can contaminate food and water. They can bioaccumulate in a variety of foods, including human breast milk. They can accumulate in body tissues after ingestion of contaminated food products such as fish, meat, milk, eggs, fruits, and vegetables. Ingestion of contaminated water, or in the case of children, contaminated soil, adds to the body burden. Exposure to these toxic air pollutants increases the risk of cancer, possibly damaging the immune system, and the neurological, repro-ductive, and developmental systems.

Economic Effects

Effects on Animals

Serious air pollution is a hazard to animals and humans. Canaries, ducks, cattle, sheep, horses, and pigs grow sick and die. The animals breathe air pollutants or they eat feed or grass that has been contaminated by the residues of the various pollutants. Arsenic, produced during the copper-smelting process, has killed numerous sheep, cattle, and horses. The arsenic, which is released into the air as arsenic trioxide, lands on forage and is consumed by animals. Animals grazing within the vicinity of lead and zinc foundries have been sickened and died as a result of lead or zinc poisoning. The lead and zinc dust coats the forage consumed by the animals. Molybdenum is produced as a pollutant during the steel-making process. Again, the pollutant is dispersed through the air on the forage, and the animals, after consuming the contaminated feed, become sick and may die.

The most serious problem for livestock is the ingestion of fluoride. Certain plants used for fodder have the ability to store and concentrate fluoride. When the animals consume this forage, they ingest an enormous overdose of fluoride. The animals' teeth become mottled; the animals lose weight, give less milk, grow spurs on their bones, and become so crippled that they must be sacrificed. Pollutants also cause chickens to yield fewer eggs, cows to produce fewer young or less milk, and sheep to grow a thinner coat of wool.

EFFECTS ON VEGETATION

Vegetation is affected very quickly by the presence of pollution. Crops are destroyed, and shrubbery and ornamental plants become spotted and are eventually devastated.

Sulfur dioxide enters the leaf through the stoma and causes injuries on the blade. Leaves are marked with an ivory or brown to reddish-brown color. Although brief exposure at a low concentration only temporarily injures the plant, sustained exposure causes death of plant cells.

Hydrogen fluoride is toxic to some plants at extremely low concentrations. Fluoride particles can harm a plant when they are absorbed through the surface of the leaf. Leaves exposed to fluoride generally have burned, dried-out edges.

Photochemical smogs are a serious hazard to plants, because they produce both visible and invisible injuries. The stomata or leaf openings, through which the plant draws carbon dioxide, are closed up. Severe smogs kill plants, whereas chronic exposure to lesser smogs retards their growth. The harmful components of the smog are ozone and PAN. Ethylene, a gaseous hydrocarbon, is a product of automotive exhausts. It is part of the photochemical smog process and also part of the destruction of vegetation.

Effects on Materials

Property is damaged by polluted air. The effects on these materials are the result of chemical or physical reactions. Although a certain amount of soiling and deterioration occurs in clean, unpolluted areas, the presence of pollutants accelerate this process. Abrasion, which is a physical erosion of the surface of materials by particulate pollutants, wears away surfaces, pits them, and may eventually destroy them. Soil due to pollutants is deposited on buildings. For aesthetic reasons, sandblasting is used to clean off these structures. Chemical attack on materials may be due to the presence of gaseous or particulate pollutants. Steel, iron, zinc, brass, copper, nickel, lead, tin, and aluminum all corrode faster in urban, industrial areas. The degree of deterioration is roughly proportional to the amount of pollution present. Stone is damaged by a combination of carbon dioxide and water, forming carbonic acid. This acid reacts with the calcium carbonate to form a soluble bicarbonate that is dissolved in water in the air or in rainwater.

Hydrogen sulfide reacts with lead-based paints to form a blackish-red sulfide. The exposed painted surfaces become splotchy and vary in color from gray-brown to black. Repainting is usually needed to correct this condition.

Pollutants affect fabrics. Clothing and draperies must be cleaned more frequently, resulting in excessive wear. Fabrics can actually be damaged by the abrasive particles found in air pollution. Minute particles of sulfuric acid can cause nylon hose to run. Dyes on fabrics are faded and discolored by polluted air. Blue dye becomes lighter or discolored to a reddish tone. Oxides of nitrogen, ozone, and sulfur dioxide also affect fabric dyes.

Smog can cause rubber tires to crack because of oxidation due to ozone and other oxidants present in the smog. Atmospheric sulfur dioxide causes paper to become brittle and crack. Unless preventive measures are taken, many important documents can be destroyed. Leather becomes brittle and cracks when exposed to sulfur dioxide.

The factors that influence the deterioration of materials include the concentration of the pollutant, the reactions of various pollutants mixed together, the critical influence of moisture, the effect of temperature, the amount of sunlight present, and the degree of air movement, including wind speed and wind direction.

Effects on Visibility

Impaired visibility is due to the scattering of sunlight by particulates suspended in the atmosphere. Aerosols increase the frequency and density of fog. This occurs because the hygroscopic particulates collect water from the atmosphere and form a mist. Carbon, tar, and opaque metal particles further contribute to reduced visibility. Nitrogen dioxide is also a problem because it is a yellow-brown gas, which in substantial concentration, reduces light. Visibility is reduced by dense smoke coming from the smoke stacks of industrial plants, burning open dumps, or smoldering culm piles that consist of unused coal waste.

It is obvious from the previous discussions of the effects of air pollutants on people, animals, materials, vegetation, and visibility, that there is a considerable economic loss to society. The annual cost of air pollution is estimated to be in the billions of dollars. Costs include higher medical expenses due to sickness and accidents, loss of livestock, increased maintenance, higher cleaning bills, greater absenteeism, higher food bills, lowered real estate values, reduced or destroyed crops, costlier equipment, and higher taxes.

Acid Deposition

Acid deposition is a serious environmental and economic concern in many parts of the United States as well as many parts of the world (Figure 1.9). Acid deposition starts with the emission of sulfur dioxide, primarily from coal-burning power plants, and nitrogen oxides, primarily from motor vehicles and coal-burning power plants. These pollutants interact with sunlight and water vapor in the upper atmosphere to form acidic substances. The acidic substances may fall to Earth as acid rain or snow, or may join dust and become a dry acid deposition. Over 80% of sulfur dioxide emissions in the United States originate in the 31 states east of or bordering on the Mississippi River. Most of the emissions come from the Ohio River Valley. The prevailing winds transport these emissions hundreds of miles to the northeast across state and national borders. Acid rain causes damage in sensitive areas. In most areas, where there are acid-neutralizing substances in the soil, the deposition may occur for many years without any problems. Where thin soils exist in mountainous or glaciated areas of the Northeast, very little acid-buffering capacity exists; therefore, these areas are vulnerable to damage from the acid deposition. Acid deposition can adversely impact the soil, groundwater and surface water, leaching nutrients in the ground, killing nitrogen-fixing microorganisms that nourish plants, killing fish, and releasing toxic metals.

sulfur dioxide and
nitrogen oxide
interact with sunlight,
chemical oxidants,
and water vapor

industrial wastes
and vehicle exhaust

dry deposits,
acid-rain or
snow

Figure 1.9 Acid deposition.

Acidity attacks this population in several ways. As the pH of the lake decreases, it alters the delicately balanced working of the internal system of the fish. One result is a depletion of calcium from the bone tissue and skeleton. Due to this, a lot of the fish that are found in the lakes are deformed, humpbacked, or dwarfed. The muscles stay as strong as before but the skeleton does not. Therefore, the pulling of the muscles against the skeleton pulls the skeleton into different shapes. Eventually, the fish are unable to function and may die. Aluminum is released from the soil when acid deposition occurs. The aluminum flows into the water and is easily taken into the gills of the fish. The aluminum causes a very slow death to fish.

When the pH drops to 5.5, most species do not lay eggs at normal levels. By a pH drop to 4.5, most species are endangered because of reproductive failures. When the fish and plant populations die, algae, moss, and fungi cover the bottom of the lake and the condition of eutrophication occurs.

Acid shock happens when acidic deposition occurs during the winter and becomes part of snow. When the snow melts, a substantial amount of acid is present and is released in a matter of a short period of time into the water.

Acid can kill amphibians, especially salamanders and frogs. The reason is that most amphibians lay their eggs in the meltwater pools during early spring. These pools become saturated with acid from the buildup over the winter. The salamander is an important feed source for birds and mammals, and therefore the killing off of salamanders also affects these birds and mammals.

Acid rain affects the forest by depriving the roots of plants of essential nutrients that are necessary for the growth of these plants.

Global Atmospheric Changes

The greenhouse effect is a natural phenomenon, largely caused by carbon dioxide (Figure 1.10). Visible light passes through the atmosphere to Earth's surface. Earth then radiates the heat as infrared rays. Some heat escapes, but carbon dioxide and other gases in the troposphere trap the heat, thereby warming Earth. Without the greenhouse effect, Earth would be a frozen planet like Mars. An increasing amount of carbon dioxide is present in the atmosphere from the burning of fossil fuels, such as coal, oil, and natural gas. A warming trend could possibly occur that may raise

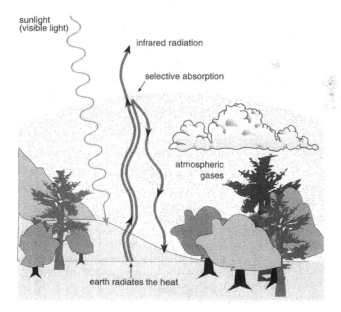

Figure 1.10 Greenhouse effect.

global temperatures. The clearing of rain forests also contributes carbon dioxide and other greenhouse gases to the atmosphere when the wood is burned. Also, the plants that were in the forest helped remove carbon dioxide from the air. There is a concern that other greenhouse gases such as methane, ground-level ozone, nitrous oxide, and CFCs are in such concentrations that they interfere with Earth's normal balanced process of releasing heat into space. For instance, a CFC molecule has 10,000 to 20,000 times the warming impact of a molecule of carbon dioxide.

Refrigerants and other chemicals classified as CFCs are commonly released into the environment. When airborne and dispersed approximately 10 km above Earth's surface and into the stratosphere, CFCs may contribute to the depletion of the ozone layer. The higher intensity UV light at this altitude can photochemically break the intramolecular bonds in CFC molecules and release chlorine atoms or radicals. The highly unstable, yet very reactive, chlorine radicals readily combine with and destroy the triatomic ozone (O_3) molecules while forming various oxidized products of chlorine. If the depletion of the ozone exceeds the replenishment, then there is concern that the ozone layer as a whole will be depleted. As a result, the ozone layer's normal attenuation of harmful and relatively penetrating UV-B rays from the sun will be compromised and more of this form of nonionizing radiation will reach Earth's surface.

POTENTIAL FOR INTERVENTION

Intervention occurs through the use of isolation, substitution, shielding, treatment, or prevention techniques. In the isolation technique, the industrial plant is moved to a rural area, where the air quality is excellent. Substitution includes the use of different fuels and the changing of the process. Shielding is not an appropriate technique. Treatment is by far the technique most frequently used in air pollution control. Whether it is treatment of the various pollutants from stationary sources or from automobiles, the removal of the pollutants reduces air problems. Prevention is best achieved through proper planning, proper maintenance of equipment, and adequate location of various industrial sources.

In general, air pollution is reduced by utilizing many of the techniques known today. Additional research is needed in developing inexpensive means of removing contaminants from fossil fuels before the fuels are used as an energy source. Specific controls are generally available for most industries.

The potential for intervention in air pollution problems is excellent. However, the cost of the intervention must be balanced with the tremendous problems relating to loss of jobs and loss of industrial potential.

RESOURCES

Scientific and technical resources include the National Environmental Health Association, American Chemical Society, American Public Health Association,

American Medical Association, University of Michigan School of Public Health, University of California School of Public Health, Harvard Medical School, University of North Carolina School of Public Health, and others.

Civic associations include the Air Pollution Control Association, Tuberculosis Association, Environmental Defense Fund, and Sierra Club.

Governmental resources include the U.S. Public Health Service and its various branches, such as the Centers for Disease Control and Prevention and National Institutes of Health. Other governmental agencies interested in air pollution include the EPA, Department of Defense, National Oceanic and Atmospheric Administration, Department of Energy, U.S. Geological Service, etc.

Other specific resources include the Air Quality, Planning and Standards Office of the EPA at Research Triangle Park, NC; various regional offices of the EPA; health, agricultural, and air pollution control boards of the 50 states, and many of the large cities. Further other professional associations include the Air Pollution Control Association, Box 2861, Pittsburgh, PA; the American Association for Aerosol Research Indoor Environment Program, 1 Cyclotron Road, Berkeley, CA; Automotive Exhaust Symptoms Manufacturers Council, 300 Sylvan Avenue, Box 1638, Englewood Cliffs, NJ; Council of Pollution Control Financing Agencies, 1225 23rd Street, N.W., Washington, D.C.; National Council of the Paper Industry for Air and Stream Improvement, 260 Madison Avenue, New York, NY; State and Territorial Air Pollution Program Administrators, 444 North Capitol Streets, N.W., Suite 306, Washington, D.C. A good source of current information pertaining to hazardous air pollutants is the National Air Toxics Information Clearinghouse operated by the EPA Office of Air Quality Planning and Standards. The clearinghouse provides an on-line database containing all toxic-related information from state and local agencies and supporting bibliographical citations. The EPA Air Pollution Hotline is 202-260-2080.

STANDARDS, PRACTICES, AND TECHNIQUES

The EPA, under the provision of the Clean Air Act, established standards for sulfur dioxide, PM_{10}, $PM_{2.5}$, carbon monoxide, ozone, nitrogen dioxide, and lead. These standards, which are listed in Table 1.1, describe the limits that a given pollutant may reach annually and on a 24-hour basis. The air quality standard defines the limit for pollutants.

The air pollution levels in America's cities change from day to day, usually as a result of meteorologic conditions and some variation in the emissions from local and stationary sources. The daily levels of air pollution in excess of health standards are important to individuals who suffer from diseases that are aggravated or caused by air pollution. The public is informed of the air pollution problem on a daily basis by means of the Air Pollution Index Value (Equation 1.10). The Air Quality Index, formerly the Pollutant Standard Index, was updated August 4, 1999. It makes the air pollution index values readily understandable. Table 1.2 gives these index values; the air quality level associated with them; the specific pollutant levels in terms of total solid particles, PM_{10}, $PM_{2.5}$, sulfur dioxide, carbon monoxide, ozone, and

Table 1.1   National Ambient Air Quality Standards (NAAQS)

Pollutant	Averaging Time	Primary Standards	Secondary Standards
Particular			
(PM$_{10}$)	Annual (arithmetic mean)	50 µg/m^3	50 µg/m^3
	24 hr	150 µg/m^3	150 µg/m^3
(PM$_{2.5}$)	Annual (arithmetic mean)	15 µg/m^3 [a]	15 µg/m^3 [a]
	24 hr	65 µg/m^3 [a]	65 µg/m^3 [a]
Sulfur oxides	Annual (arithmetic mean)	80 µg/m^3 [a] (0.03 ppm)	
	24 hr	365 µg/m^3 (0.14 ppm)	—
	3 hr	—	1300 µg/m^3 (0.5 ppm)
Carbon monoxide	8 hr	10 mg/m^3 (9 ppm)	
	1 hr	40 mg/m^3 (35 ppm)	
Nitrogen dioxide	Annual (arithmetic mean)	100 µg/m^3 (0.053 ppm)	100 µg/m^3 (0.053 ppm)
Ozone	Maximum daily 1-hr average	235 µg/m^3 (0.12 ppm)	235 µg/m^3 (0.12 ppm)
	Maximum daily 8-hr average	157 µg/m^3 [a] (0.08 ppm)	157 µg/m^3 [a] (0.08 ppm)
Lead	Maximum quarterly average	1.5 µg/m^3	1.5 µg/m^3

[a] Not enforced yet.

From U.S. Environmental Protection Agency, Office of Air Quality and Standards, Updated Air Quality Standards Web site, July 10, 2000.

nitrogen dioxide; the potential health effects; the general health effects; and the precautions to be taken. The formula used to determine the index value for a pollutant is as follows:

$$I_p = \frac{I_{Hi} - I_{Lo}}{BP_{Hi} - BP_{Lo}}\left(C_p - BP_{Lo}\right) + I_{Lo} \tag{1.10}$$

where

I_p = index value for pollutant$_p$
C_p = truncated concentration of pollutant$_p$
BP_{Hi} = breakpoint that is greater than or equal to C_p
BP_{Lo} = breakpoint that is less than or equal to C_p
I_{Hi} = AQI value corresponding to BP_{Hi}
I_{Lo} = AQI value corresponding to BP_{Lo}

Updating Clean Air Standards

The clean air standards are to be updated periodically, as mandated by the Clean Air Act. Public health protection from potential disease is an important consideration in the review and updating of standards. As an example of the process the ozone and particulate matter standards have been upgraded in the last few years. To do

Table 1.2 Air Quality Index (AQI) Values

AQI	Pollutant Level Breakpoints							Health Effect Descriptor	General Health Effects	Cautionary Statements	Color Code
	PM_{10} μg/m³	$PM_{2.5}$ μg/m³	SO_2 ppm	Co ppm	O_3 (1 hr) ppm	O_3 (8 hr) ppm	NO_2 ppm				
401–550	505–604	350.5–500.4	0.805–1.004	40.5–50.4	0.505–0.604	Use 1 hr values	1.65–2.04	Hazardous	Premature death of ill and elderly; healthy people experience adverse symptoms that affect their normal activity	All persons should remain indoors, keeping windows and doors closed; all persons should minimize physical exertion and avoid traffic	Maroon
301–400	425–504	250.5–350.4	0.605–0.804	30.5–40.4	0.405–0.504	Use 1 hr values	1.25–1.64	Hazardous	Premature onset of certain diseases in addition to significant aggravation of symptoms and decreased exercise tolerance in healthy person	Active children, elderly adults, and persons with existing diseases should stay indoors and avoid physical exertion; general population should avoid outdoor activity	Maroon
201–300	355–424	150.5–250.4	0.305–0.604	15.5–30.4	0.205–0.404	0.125–0.374	0.65–1.24	Very unhealthy	Significant aggravation of symptoms and decreased exercise tolerance in persons with heart or lung disease, with widespread symptoms in the healthy population	Active children, elderly adults, and persons with heart or lung disease should avoid all outdoor exertion	Purple

Table 1.2 (continued) Air Quality Index (AQI) Values

| | Pollutant Level Breakpoints | | | | | | | | | | |
AQI	PM_{10} $\mu g/m^3$	$PM_{2.5}$ $\mu g/m^3$	SO_2 ppm	Co ppm	O_3 (1 hr) ppm	O_3 (8 hr) ppm	NO_2 ppm	Health Effect Descriptor	General Health Effects	Cautionary Statements	Color Code
150–200	255–354	65.5–150.4	0.225–0.304	12.5–15.4	0.165–0.204	0.105–0.124	—	Unhealthy	Mild aggravation of symptoms in susceptible persons, with irritation symptoms in the healthy population	Active children, adults and persons with existing heart or respiratory ailments should avoid prolonged physical exertion and outdoor activity	Red
101–150	155–254	40.5–65.4	0.145–0.224	9.5–12.4	0.125–0.164	0.085–0.104	—	Unhealthy for sensitive groups		Active children and adults and people with respiratory diseases should limit activities	Orange
51–100	55–154	15.5–40.4	0.035–0.144	4.5–9.4	—	0.065–0.084	—	Moderate		Unusually sensitive people should consider limiting outdoor exertion	Yellow
0–50	0–54	0.0–15.4	0.000–0.034	0.0–4.4	—	0.000–0.064	—	Good		None	Green

Note: Formerly known as Pollutant Standard Index.
From Rules and regulations, Fed. Regist., 64(149), 42548–42549, August 4, 1999.

this work, scientifically, effectively, and efficiently with appropriate cost controls, the EPA is required to use the federal advisory committee established by the Federal Advisory Committee Act enacted in 1972. This has been updated through the Clean Air Act Advisory Committee associated with the Clean Air Act of 1990. The specific responsibilities of this committee include advising the EPA on the following:

1. Approaches for new and expanded programs
2. Potential health, environmental, and economic effects of programs required by the amendments and how the potential impacts affect the public, regulated community, state, and local governments
3. Policy and technical content of proposed major EPA rule making and guidance
4. Integration of existing policies, regulations, standards, guidelines, and procedures into programs for implementing new requirements

The subcommittees are as follows:

1. Mobile source technical review
2. Transportation, energy, and air quality concerns
3. Economic incentives and regulatory innovations
4. Permits, new source review, toxics
5. Ozone, particulate matter, and regional haze implementation program
6. Climate change
7. Accident prevention

The scientific process involves the review of thousands of studies by the science advisers; an extensive period of public comment involving many thousands of individuals, small businesses, associations, and other citizens; and an intensive interagency review. For the proposed ozone standards, some 3000 reputable scientific studies were reviewed. For the proposed particulate matter standards, some 2000 reputable scientific studies were reviewed. These reviews were conducted primarily by non-EPA researchers from prestigious academic institutions. EPA then prepared a criteria document that summarized the state of scientific knowledge for each of the pollutants. A staff paper was prepared assessing the policy implications of the new science. The criteria document and the staff paper were presented in a series of public meetings to the Clean Air Scientific Advisory Committee, and then revised based on the comments from this committee and the public. The committee then recommended to the EPA that the current standards needed to be revised and what the revisions should entail.

The health concerns behind the updated standards were significant, real-world, quality of life issues for the American public. The scientific review determined that the previous standards did not adequately protect public health. For ozone, longer term exposures, at lower levels than under the previous standard, cause significant health effects, such as asthma attacks, breathing and respiratory problems, loss of lung function, possible long-term lung damage, and lowered disease immunity. For particulate matter, exposure to smaller sized particles such as $PM_{2.5}$ at lower concentrations than EPA previously regulated causes premature deaths and respiratory problems.

The new ozone standard will prevent approximately 1 million cases per year of significant decreases in children's lung functions and child activity, reducing hospital admissions and emergency room visits. The new particulate matter standard will prevent approximately 15,000 premature deaths each year. This standard will prevent hundreds of thousand of cases of aggravated asthma in children and adults, and reduce the risk of thousands of hospital admissions and emergency room visits from the elderly and those with heart and lung disease.

Part of the process of implementing the new air standards is to use a common-sense approach that is flexible and cost effective for communities and businesses. The timeline established for the standards recognizes the fact that a period of transition is needed to avoid significant disruptions. However, this sensible approach set forth in the Clean Air Act still allows for significant public health standards.

Once a standard is set, the law lays out a series of milestones for federal, state, and local agencies including:

1. Gathering data about the air quality in the specific area
2. Designing areas based on their air quality
3. Developing and implementing plans to reduce pollution
4. Meeting the new standards

The EPA implementation strategy focuses on the largest sources that provide the most cost-effective pollution reductions to meet the new standards. State and local governments have the flexibility to reduce local emissions in whatever way they think best. The Clean Air Act prevents EPA from considering cost when setting public health standards. However, costs and benefits are assessed and the best techniques are utilized.

Air Toxics

Section 112 of the Clean Air Act of 1990 refers to the establishment of standards, permits, schedules for compliance, and prevention of accidental releases of hazardous air pollutants (air toxics). The law contains a list of 189 hazardous air pollutants and provisions for the EPA to add or delete pollutants as needed. The act also provides for a listing of all categories and subcategories of major and area sources that emit hazardous air pollutants. The EPA must develop the National Emissions Standards for Hazardous Air Pollutants (NESHAP) program. They must develop the necessary standards, or maximum achievable control technology (MACT) standards, as well as several related programs to enhance and support the NESHAP program. The law requires the prevention of accidental releases of regulated hazardous air pollutants and other extremely hazardous substances. The EPA developed a list of over 100 substances, which in cases of an accidental release are known to cause or may reasonably be anticipated to cause death, injury, or serious adverse effects to human health or the environment.

The Technology Transfer Network of the EPA has a Unified Air Toxics Web site that lists chemical fact sheets providing a hazard summary for each of the hazardous air pollutants. The fact sheets were developed based on available human or animal data. It is extremely useful in assessing various hazardous chemicals.

The top 20 chemicals released into the air, either as nonpoint air emissions or stack or point source air emissions are as follows: methanol, zinc compounds, ammonia, nitrate compounds, toluene, xylene, carbon disulfide, *n*-hexane, manganese compounds, chlorine, hydrochloric acid, phosphoric acid, methyl ethyl ketone, copper compounds, dichloromethane, styrene, glycol ethers, chromium compounds, ethylene, and lead compounds. This information comes from the Toxics Release Inventory, which is also discussed in Chapter 2 on solid and hazardous wastes.

Methanol accounts for approximately 10% of all toxic releases, amounting to over 240 million lb/year. Zinc compounds amount to approximately 207 million lb released each year. Ammonia amounts to 193 million lb released each year. Nitrate compounds amount to 163 million lb released each year. Toluene amounts to 127 million lb released each year.

Methanol is used as a solvent, raw material in the synthesis of organic chemicals, fuel, deicing agent, and denaturing agent. Methanol is readily absorbed from the gastrointestinal tract and the respiratory tract and is toxic to humans in moderate to high doses. In the body, methanol converts into formaldehyde and formic acid. At high dose levels it causes central nervous system damage and blindness. It may affect the liver and blood.

Zinc is used for coating iron and steel and in making brass metal alloys. It is used in industry to make paint, rubber, dye, and wood preservatives. Zinc is a group D carcinogen, which means that evidence is insufficient at present to classify its cancer-causing potential. It can damage the fetus as shown in animal studies.

Ammonia is used in the manufacture of nitrogen compounds, including chemicals used as fertilizers and plastics. It is also used in refrigeration, paper pulp production, explosives, cleaners, and metal-treating operations. Anhydrous ammonia is a corrosive and severely irritating gas with a pungent odor. It irritates the skin, eyes, nose, throat, and upper respiratory system.

Nitrate compounds have many uses, such as in fertilizers, as oxidizing agents, in explosives and in the glass, enamel, metallurgy, and alloy industries. Nitrate compounds that are soluble in water release nitrate ions that can develop a condition that reduces the blood's ability to carry oxygen.

Toluene is a flammable liquid used in the manufacture of organic chemicals and is a solvent for paint, gums, and resins. It is also part of gasoline. It can cause headaches, confusion, weakness, memory loss, and may affect kidney and liver function. Toluene may contribute to the formation of ozone in the lower atmosphere and therefore affects sensitive individuals with asthma or those with allergies.

MODES OF SURVEILLANCE AND EVALUATION

Surveillance and evaluation generally consist of two basic types of programs. The first is a field enforcement service, where the practitioner conducts a regular inspection, evaluates and studies a complaint, goes out on field surveys to determine the gross polluters by use of the Ringlemann chart (a means of measuring blackness of a smoke discharge, the Ringlemann scale ranges from 0 for white to 5 for black), or is involved in a specific emergency situation. The second is a carefully planned evaluation of the air pollution problems in a given area by means of air sampling and evaluation (refer to Chapter 9). The second technique determines level of air quality, level of pollutants, and adherence to the air quality standards established by the EPA. It is important when conducting the detailed survey to ensure that the site selection for sampling is made in such a way that it reflects the entire network design and is a representative sample consistent with objectives, and that the sampler considers the meteorologic and topographical restraints, as well as the sampling schedules. Sampling must be done accurately by skilled individuals who will handle the sample without contaminating it. Analysis and evaluation of the sample must be accurate, and the data handling and evaluation systems must also be accurate and consistent. The initial major step in surveillance and evaluation is proper planning. Planning is the advance thinking and organizing of a sequence of actions needed to accomplish the proposed objectives and to communicate the information to other individuals. The planning of the survey must include selection of the site for sampling, sampling equipment to be used, actual sample collection, sample analysis, data processing, data evaluation, and comprehensive report writing.

The major objectives for monitoring air pollution are to:

1. Provide an early warning system for potential health effects
2. Assess air quality against standards
3. Track air pollution trends and specific polluters
4. Make major changes to alter pollutant levels
5. Make major changes to lower air toxics

AIR POLLUTION CONTROLS

Air Pollution Control Devices

Air pollutants are controlled at the source or diluted after emission into the atmosphere. Source pollution control is accomplished by preventing the pollutant from forming. This can be done by changing existing industrial operations through modification or replacement of raw materials, fuels, equipment, or production methods; developing new products or processes that minimize air pollution problems; developing equipment that destroys, alters, or traps the pollutant; and destroying, masking, or counteracting odorous materials.

The pollutant that is reduced in quantity but not eliminated before leaving the stack can be diluted in the outside atmosphere if meteorologic conditions are appropriate.

Control of Particulates

Particulate matter leaves the industrial process in the gases or smoke that is emitted through the smokestack. A considerable quantity of these particles can be removed from this gas stream in the following ways:

1. The velocity of the gas is reduced sufficiently to allow it to settle by gravity, for example, a settling chamber.
2. The direction of the gas flow is changed suddenly, causing the particles to flow straight ahead because of inertia, for example, a cyclone or louver collector.

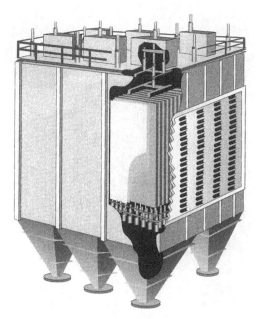

Figure 1.11   Electrostatic precipitator, Western Precipitation model. (Reprinted with permission of Allen-Sherman-Hoff, Malvern, PA.)

3. The dust-laden gas is filtered, for example, a baghouse collector.
4. Electrostatic attraction causes the charged particles to be attracted to objects with the opposite charge, for example, an electrostatic precipitator.
5. Water is used to make the particles heavier than the surrounding gas, and therefore to aid in the separation of the pollutant; examples include type D scrubber, electrostatic precipitator, reverse air baghouse, Multiclone® mechanical dust collector, and type D scrubber (Figures 1.11 to 1.14).

Particulate Matter Collection Systems

The four major particulate matter collection systems consist of mechanical collectors, electrostatic precipitators (ESPs), fabric filters, and wet scrubbers. The particulate matter emissions vary in particle size from less than 1 μm in diameter to more than 200 μm in diameter. The larger particles that comprise the greatest portion of the emissions are the easiest to collect. The smaller inhalable particles (less than 15 μm) and the respirable particles (less than 3 μm) are more difficult and costly to collect. These four systems remove at least 90% of the particulates in the gas stream leaving the source.

Mechanical collectors, such as cyclones and multitube cyclones have been used to control particulate matter emissions but their efficiencies range from 50 to 90%. The mechanical collectors are often used for preliminary treatment when other secondary treatment systems are available. Multitube cyclones are more efficient than the single tube cyclones and may be utilized for final particulate collection.

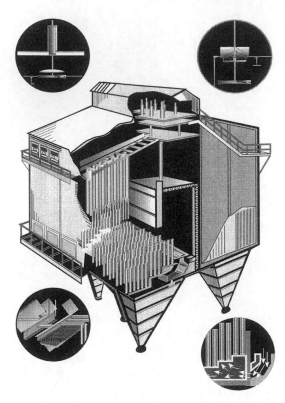

Figure 1.12   Reverse air baghouse. (Reprinted with permission of Allen-Sherman-Hoff, Malvern, PA.)

The cyclones use the principle of centrifugal separation to collect the particulate matter in a dry form. The spinvanes in each vertical tube of the multitube cyclone cause a high rational velocity to be present when the gas stream enters the tube. As the gas stream spirals downward through the tubes, centrifugal forces push the suspended particles toward the walls of the tubes. The particles then fall from the open bottoms of the tubes into collection hoppers, when the gas flow makes a sharp change in direction. The dust particles are then transferred from the collection hoppers to storage, and finally to disposal.

The efficiency of the multitube cyclones depends mainly on the velocity of the gas coming in, the diameter and length of the individual tubes, and the range of particle size in the gas stream. Higher inlet gas velocities, smaller tube diameters, and longer tube lengths increase particle removal efficiency and also increase the resistance to gas flow. These systems can be used for any removal of fly ash from coal-fired boilers and can be used for dust control from the processing of minerals.

ESPs are used to remove particulate matter from waste gases in a variety of different industries. The aluminum industry produces particulates in baking ovens, reduction cells, carbon plants, bauxite dryers, alumina calciners, and remelt furnaces.

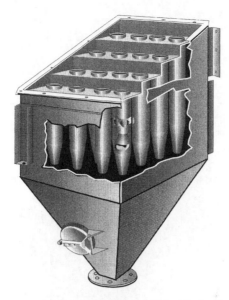

Figure 1.13 Multiclone, mechanical dust collector. (Reprinted with permission of Allen-
Sherman-Hoff, Malvern, PA.)

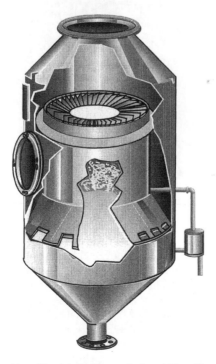

Figure 1.14 Type D scrubber. (Reprinted with permission of Allen-Sherman-Hoff, Malvern, PA.)

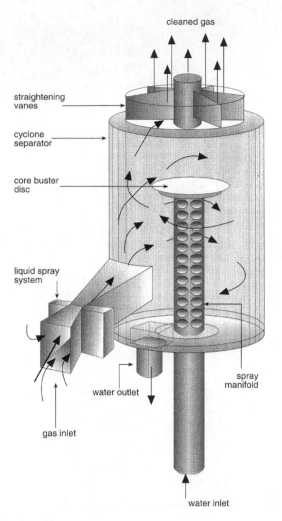

Figure 1.15 Venturi scrubber.

ESPs are also used in removing particles from carbon black production, sulfuric acid recovery, asphalt blowing stills, phosphoric acid production, tar and oil recovery from waste or fuel or gases, fluid-bed catalytic crackers, coal-fired boilers, pulp and paper industry, cement industry, gypsum industry, iron and steel industry, municipal waste incinerators, glass manufacturing, phosphate rock crushing, lime industry, and copper smelters. Particle removal efficiencies of more than 99.9% have been achieved in some of these processes. The collection efficiency of an electrostatic precipitator is dependent on the characteristics of the particulates, that is, its size and electrical resistivity and the amount of collection electrode plates surface area used. In an

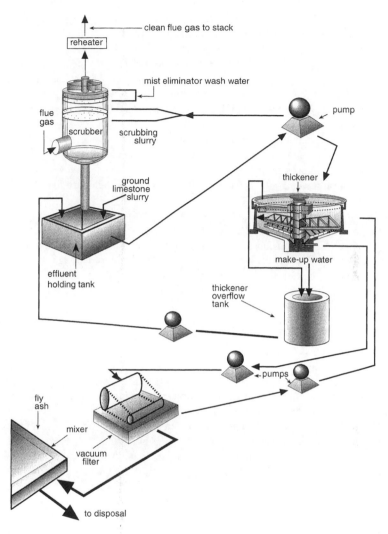

Figure 1.16 Flue gas desulfurization (lime–limestone process).

electrostatic precipitator installation, the four major components include the electrostatic precipitator, the duct, the fan, and the collected solids removal system.

Particulate matter removal occurs in three stages: (1) suspended particles are given an electrical charge, (2) charged particles migrate to a collecting electrode of opposite electrical charge, and (3) collected particles are dislodged from the collecting electrodes by mechanical banging of the electrodes. This causes the particles to fall into hoppers beneath the ESP. The dust removal system from the ESP depends on the size of the system. In the small ESP systems, dust is usually discharged directly from hoppers to dumpsters and taken to a landfill. For a large ESP system, a pneumatic dust removal system transmits the dust to a storage silo before disposal. Wetting the ash may be a method of keeping it from causing a dust problem in removal.

Fabric filters are used to control dry particulate matter emissions in a variety of industries whenever dry bulk solids are processed, particularly metals, minerals, and grains. Fabric filters are also used on coal-fired boilers. Fabric filtration is the physical straining or sieving of particulate matter from gas streams. The gas stream passes through a fabric filter, usually in the shape of a bag. The particulate matter from the inlet gas deposits mainly on the surface of the filter bag where a dust layer accumulates. The collection efficiency and the pressure drop across the bag's surface, as the dust layer builds up. The bags are cleaned to remove the collective solids when an excessive drop in pressure occurs. The collection efficiency of fabric filters is dependent on the characters of the fabrics used, the particle size distribution, and the porosity of the dust cake.

The filters are usually cleaned in one of three ways. In shaker cleaning, the filter bags are oscillated by a small electric motor. This dislodges varying amounts of dust into a hopper, depending on the shaking frequency and amplitude. In reverse air cleaning a reverse airflow is used to collapse the bags and to dislodge the dust cake. To accomplish either of these two techniques the baghouse has to be in sections so that one section can be cleaned at a time. The pulse jet cleaning technique is a short pulse of compressed air shot through nozzles at the bag exit and is directed toward the bottom of the bag. The primary pulse of air entrains a pulse of secondary airflow as it passes through the nozzle. The pressure produces a shock that expands the bag and dislodges the surface dust layer.

Small fabric filter systems typically discharge collected solids directly into bag-house hoppers to dumpsters and are taken to landfills. Large fabric filter systems need pneumatic dust removal systems, which transfer the dust to a storage silo. The dust may be wetted down to keep it from getting into the air.

At times the inert gas is pretreated or precleaned before it comes to the fabric filter. Because the fabric filter system cannot handle gases hotter than 550°F, and in fact some fabrics cannot handle gases hotter than 175°F, the gas must be cooled before it can come into the fabric filters. However, if water is used to reduce the temperature, the temperatures must be high enough to prevent moisture condensation within the baghouse or in the filter area. Moisture would cause the collected particles to plug the bags.

Particulate matter removal efficiencies of greater than 99.9% are achieved in a variety of applications of the fabric filters known as high-efficiency particulate air (HEPA) filters, which are HEPA filters. They are effective down to the submicron range. HEPA filters are usually used for particulate sizes below 10 µm and have excellent removal efficiencies for particle sizes less than 1 µm. They can remove over 99.97% of particulates of 0.3 µm and greater.

A Venturi wet scrubber is a collection device where an aqueous stream or slurry is used to remove particulate matter or gaseous pollutants. Scrubbers are classified by energy consumption in terms of gas-phase pressure drop. Low-energy scrubbers, including spray chambers and towers, have pressure drops of less than 5 inches of water. Medium-energy scrubbers such as impingement scrubbers have pressure drops of 5 to 15 inches of water. Higher energy scrubbers such as high-pressure drop Venturi scrubbers, have pressure drops exceeding 15 inches of water. The most common scrubbers used for the moderate removal of particulate matter are medium-energy impingement and Venturi scrubbers. Greater removal of particulate matter can be achieved with high-energy venturi scrubbers. The collection efficiency of scrubbers is dependent on the characteristics of the particulate matter, such as particle size and energy input to the scrubber, as measured in a pressure drop.

Venturi scrubbers are used for the collection of hygroscopic and corrosive sub-micron particles (Figure 1.15). In a typical Venturi scrubber, the particle-laden gas first contacts the liquor stream in the core and throat of the Venturi section. The gas and liquor streams then pass through the annular orifice formed by the core and throat, atomizing the liquor into droplets that are impacted by particles in the gas stream. The impaction is caused by the high differential velocity between the gas stream and the atomized droplets. The particulate-laden droplets then are removed from the gas stream by centrifugal action in a cyclone separator or a mist elimination section. Venturi scrubbers are used for particulate matter control in the coal-cleaning industry, in the phosphate fertilizer industry, in the lime and asphalt plant industry, and in the metal industry. When the particulate matter accumulates in the circulating scrubber liquid, a portion of the liquid is removed and sent to disposal or reused in the treatment. Fresh scrubber liquid is then added to the circulating stream. The pH of the scrubber circulating liquid can be altered by the incidental absorption of acidic gases, such as sulfur dioxide, or by the collection of alkaline materials, such as lime dust. It may be necessary to alter or to change the pH on removal of the liquor. The performance of typical wet scrubbers are affected by the gas velocity, the liquid-to-gas ratio, the particle size distribution, and the inlet gas particulate matter concentration. The gas-phase pressure drop is usually the major factor affecting removal of the particulates.

Control of Gases

Gases are controlled by process modification, which is the same as the process modification for control of particulates; combustion control through proper use of temperature, oxygen, time, turbulence, and catalysis; absorption of gases by use of water or other liquids; adsorption of gases on activated carbon, silica gel, lithium chloride, activated alumina, and activated bauxite (desorption or removal of the pollutant from the adsorbing bed is done by raising the temperature of the bed above the boiling temperature of the adsorbed gas); and dilution of gases in the atmosphere in the same manner as particulates.

Industrial Waste

Industrial waste often contains halogens or sulfides. Therefore, when the gases are passed through water, the process is more efficient if the water contains sodium hydroxide. The sodium hydroxide reacts with the hydrochloric acid to form water and sodium chloride. The sodium hydroxide reacts with sulfuric acid to form sodium sulfate and water. The neutralization of these acids are a function of the concentration of the acids in the gas stream, the temperature of the gas stream, the concentration of sodium hydroxide in the water, the presence of other compounds in the water, the water temperature, and the contact time of gas and water flow.

Flue Gas Desulfurization

Sulfur oxide emissions, mostly sulfur dioxide, come from fuel combustion and also from industrial processes. Flue gas desulfurization scrubbing systems are used to absorb sulfur dioxide gas from combustion gases. The two major systems of flue gas desulfurization are wet scrubbing and spray drying systems. The wet flue gas desulfurization scrubbing process includes lime–limestone, nonregenerable sodium alkali scrubbing, dual alkali, magnesium oxide, and Wellman–Lord.

In the lime–limestone process, a slurry containing calcium hydroxide or calcium carbonate removes sulfur dioxide from flue gas in the wet scrubber. These are nonregenerable processes and produce a large amount of solid waste for disposal. Particulate matter is usually removed upstream from the flue gas desulfurization system. After the particulate matter removal, the flue gas enters the scrubber, where the sulfur dioxide is absorbed by contact with the slurry of lime or limestone and water. The sulfur dioxide chemically reacts with the lime or limestone to form calcium sulfite and calcium sulfate. Both of these substances are only slightly soluble in water and precipitate from the solution. The slurry passes through the scrubber into a holding tank, which allows enough time for solids precipitation to occur. Crushed lime or limestone is then added to the tank. Most of the slurry is recirculated to the scrubber (Figure 1.16). The remainder is removed on a continuous basis from the holding tank to solid–liquid separation by ponding or clarification–vacuum filtration processes. In ponding, the pond serves as the final solid disposal method, as well as a liquid clarifier. Solids from filtration are usually discarded in a landfill. The clarified liquid is returned to the scrubber system for reuse. The lime–limestone process is the most extensively used system for utility boilers, which represent more than 80% of the systems either under operation or under construction.

In the sodium alkali scrubbing system, a scrubbing solution of sodium hydroxide, sodium carbonate, or sodium bicarbonate absorbs sulfur dioxide from flue gases. The sulfur dioxide chemically reacts to form sodium sulfite and sodium bisulfite, which remains dissolved in solution. Part of the sulfite in solution reacts with oxygen and the flue gas to form sodium sulfate. The sodium sulfite in sulfate salts are removed from the system in solution as a liquid waste. Sodium carbonate or sodium hydroxide are generally selected as the makeup sodium alkali added to the circulating scrubber solution to compensate for the quantity that reacts with the sulfur dioxide. The sodium alkali scrubbing differs from the limestone and dual alkali system in that no solid waste is formed. However, large quantities of liquid waste containing sodium sulfite, sodium bisulfite, and sodium sulfate must be discarded. Disposal may consist of holding ponds for evaporation of liquids or deep-well injection. This system is the simplest of the flue gas desulfurization systems. However, it is costly and therefore not used as frequently as the others.

The dual-alkali process uses some features of the previous two processes. The scrubbing liquid is a solution of soluble sodium-based alkali-containing sodium carbonate, sodium bicarbonate, sodium sulfite, and sodium hydroxide. Calcium-based solids, such as the ones used in the lime–limestone process may be produced by adding lime to the stream of spent scrubbing liquor. After removal of the particulate matter, the flue gas enters the scrubber where the sulfur dioxide is absorbed. The sulfur dioxide reacts chemically with sodium alkali to form sodium sulfite and sodium bisulfite. Some of the sodium sulfite reacts in solution with oxygen from the flue gas to form sodium sulfate. The acid stream of scrubbing solution is then reacted with slaked lime to form calcium sulfite and calcium sulfate, which precipitates the solution. This helps in regeneration of the sodium-based alkali for recycling through the scrubber. The sidestream slurry that contains the calcium sulfite and sulfate solids is sent to a thickener, where the solids are concentrated by sedimentation. This forms a sludge and is further thickened, and the solids are then finally discarded.

The magnesium oxide or magnesia slurry absorption process is a regenerable process that uses magnesium hydroxide to absorb sulfur dioxide in a wet scrubber. After particulate removal, the flue gas may be pretreated for chloride removal to prevent an excess concentration of chloride in the scrubbing liquid. Then the flue gas is put in contact with the magnesia slurry in the scrubber to remove the sulfur dioxide. The sulfur dioxide reacts with the magnesium oxide to form magnesium sulfite and magnesium bisulfite. Some of the sulfite reacts with oxygen to form magnesium sulfate. The spent slurry is sent to a holding tank, where adequate time is allowed for the solids to precipitate. The holding tank effluent is split into two streams. The first stream is relatively clear liquid, which is combined with fresh magnesium oxide recycled slurry. This stream is then recycled to the scrubber. The second stream of liquid and settled solids is sent to a thickener for solids concentration and finally to a centrifuge where the magnesium sulfite or magnesium sulfate hydrated crystals are regenerated to produce fresh magnesium oxide. The next generation occurs in two steps. First the water is driven off in an oil-fired rotary kiln and the dried magnesium sulfite and magnesium sulfate crystals are calcinated. During this process magnesium sulfite and magnesium sulfate are converted to magnesium oxide and the sulfur dioxide gas comes off. The gas can then be captured.

The Wellman–Lord scrubbing system is a regenerable process that uses sodium sulfite solution to absorb sulfur dioxide. It produces a concentrated stream of sulfur dioxide that can be processed into sulfur, sulfuric acid or liquid sulfur dioxide. The particulate matter is first removed from the flue gas and also chlorides are removed. The humidified gas from the prescrubber enters the absorption tower, where it comes in contact with the scrubbing liquor. Sodium sulfite reacts in solution with oxygen from the air to form sodium sulfate. The sodium sulfate must be removed from the solution for the scrubbing liquid to maintain its ability to absorb sulfur dioxide. A portion of the spent scrubbing liquid is sent to treatment where a heated sulfate crystallizer or a refrigerated sulfate crystallizer are used. Either process produces a slurry of sodium sulfate solid crystals, which are centrifuged and turned into a cake of solids, dried, and disposed of as solid waste. The rest of the liquid is regenerated by converting sodium bisulfate to sodium sulfite. This is done by heating the liquid in a series of double-effect forced-circulation evaporators. Sodium sulfite crystals formed during the regeneration are dissolved in water and recycled to a sulfur dioxide absorber. The concentrated sulfur dioxide may be used for the production of elemental sulfur, sulfuric acid, or liquid sulfur dioxide.

Argonne National Laboratory has developed new processes and new additives for flue gas, to reduce NO_x and SO_2 emissions. They include the use of ethylenedi-aminetetraacetate in a wet flue-gas-desulfurization system, a spray dryer system working on high-sulfur coal, and spray dryer and fabric filter system.

Air Pollution Control Devices for Automobiles

Automobile emissions are controlled by the use of positive crankcase ventilation (PCV), where the PCV valve regulates the flow rate of unburned crankcase vapors back to the combustion chamber; an air injection system, where a controlled amount of air is added to the exhaust ports of the combustion chambers, thereby burning carbon monoxide and hydrocarbons; evaporation emission control systems, where a charcoal canister is used to collect fuel vapors from the fuel tank; exhaust gas recirculation, where the exhaust gases are fed to the combustion chamber as inert dilutants to minimize combustion temperatures and to reduce nitrogen oxide emissions; and retarded spark, where the spark is delayed briefly for a cooler burn to be achieved, thereby reducing carbon monoxide, hydrocarbons, and nitrogen oxides.

Experiments are under way on engines other than internal combustion engines to reduce air pollution. These engines include the Rankine cycle, rotary, and Stirling cycle engines.

Innovative Air Pollution Control Devices

Six innovative air pollution control devices are discussed next.

TiO$_2$ Photocatalytic Air Treatment

TiO_2 photocatalytic air treatment was developed by Matrix Photocatalytic Inc. The device is a titanium dioxide (TiO_2) photocatalytic air treatment technology used to remove and destroy VOC and semivolatile organic compounds from airstreams. The technology is an ambient temperature solid-state process in which contaminated air flows through a fixed TiO_2 catalyst bed activated by light. Typically, destruction of organic contaminants occurs in fractions of a second.

The TiO_2 photocatalytic air treatment technology can effectively treat dry or moist air. The technology has been demonstrated to purify steam directly, thus eliminating the need to condense. Systems of 100 ft^3/min have been successfully tested on vapor extraction operations, air stripper emissions, steam from desorption processes, and VOC emissions from manufacturing facilities.

Vaporsep™ Membrane Process

The Vaporsep™ membrane process was developed by Membrane Technology and Research Inc. This device uses synthetic polymer membranes to remove organic vapors from contaminated airstreams. The process generates a clean airstream and a liquid organic stream for reuse or disposal.

Air laden with organic vapor contacts one side of a membrane that is 10 or 100 times more permeable to the organic compound than to the air. The membrane separates the gas into two streams: a permeate stream containing most of the organic vapor and a clean residual airstream. The organic vapor is condensed and removed as a liquid; the purified airstream may be vented or recycled.

It can treat most airstreams containing flammable or nonflammable halogenated and nonhalogenated organic compounds, including chlorinated hydrocarbons, CFCs, and fuel hydrocarbons.

Methanotrophic Biofilm Reactor

The methanotrophic biofilm reactor was developed by Remediation Technologies Inc. The device uses methanotrophic organisms in fixed-film biological reactors to treat chlorinated VOC. Treatment occurs while the VOCs are in a gas phase. Gases enter the bottom of the reactor and flow up through a medium that has a high surface area and favorable porosity for gas distribution.

Methane must be supplied to the biofilm reactor to maximize biomass. In the reactor, the methanotrophic organisms oxidize the methane as an energy source. Volatile chlorinated hydrocarbons are co-metabolized into various acids and chlorides that are subsequently degraded to carbon dioxide and chloride by other heterotrophic bacteria.

It is used to treat chlorinated volatile hydrocarbons in gaseous streams, such as those produced from air stripping or *in situ* vacuum extraction operations.

Acoustic Barrier Particulate Separation

Acoustic barrier particulate separation was developed by General Atomics, Nuclear Remediation Technologies Division. The device separates particulates in a high temperature gas flow. The separator produces an acoustic waveform directed against the gas flow, causing particulates to move opposite the flow. Eventually, the particulates drift to the wall of the separator, where they aggregate with other particulates and precipitate into a collection hopper. The acoustic barrier separator differs from other separators in that it combines both high efficiency and high temperature capability.

It can treat off-gas streams from thermal desorption, pyrolysis, and incineration of soil, sediment, sludges, other solid wastes, and liquid wastes. It is a high temperature, high throughput process with a high removal efficiency for fine dust and fly ash.

Bioscrubber

The bioscrubber device, developed by Aluminum Company of America, uses a bioscrubber to digest hazardous organic emissions from soil, water, and air decontamination processes. The bioscrubber consists of a filter with an activated carbon medium that supports microbial growth. This unique medium, with increased microbial population and enhanced bioactivity, converts diluted organics into carbon dioxide, water, and other nonhazardous compounds. The filter provides biomass removal, nutrient supplement, and moisture addition.

The device is used to remove organic contaminants in airstreams from soil, water, or air decontamination processes. The technology is especially suited to treat streams containing aromatic solvents, such as benzene, toluene, xylene, alcohols, ketones, hydrocarbons, and others.

Reactor and Filter System

The reactor and filter system device, developed by Energy and Environmental Research Corporation, is designed to treat gaseous and entrained particulate matter emissions from the primary thermal treatment of sludges, soils, and sediments.

The device is used to remove entrained particulates, volatile toxic metals, and condensed-phase organics present in high temperature (800 to 1000°C) gas streams generated from the thermal treatment of contaminated soils, sludges, and sediments.

New Control Devices

The number of new methods of controlling particulates and gases are increasing sharply. It is not within the purview of this book to describe all these techniques. However, the individual may search out the project summaries, and if wanted, the reports of the EPA Environmental Research Laboratory in Research Triangle Park, NC, to find out more about current control measures.

Hazardous Air Pollution Control Devices

Vapor Emissions Controls

Point source control of organic vapor emissions is accomplished by combustion and recovery. Combustion devices include thermal incinerators, catalytic incinerators, flares, and process heaters. Recovery devices include condensers, absorbers, and adsorbers. Combustion devices destroy the hazardous air pollutant, whereas recovery devices trap the substances that then have to be discarded.

Thermal incinerators control a variety of VOCs that are in a continuous stream with a destruction efficiency of 99%+. These devices are not well suited to highly variable flow rates. Thermal incineration is a widely used air pollution control technique where organic vapors are oxidized at high temperatures. Virtually any gaseous organic stream can be safely and cleanly incinerated if properly designed, engineered, installed, operated, and maintained. The incinerator's time of combustion, combustion temperature, and amount of oxygen present determine its efficiency.

Catalytic incinerators are similar to thermal, but contain a catalyst that allows the reaction to take place at a lower temperature. These incinerators are used primarily to oxidize VOCs. Catalytic units are not used as often as thermal because their performance can be affected by the pollutant characteristics. Phosphorus, bismuth, lead, arsenic, antimony, mercury, iron oxide, tin, zinc, sulfur, and the halogens can poison the catalyst and severely reduce its activities. Typically, 95% of hazardous air pollutant removal can occur.

A catalyst is a substance that accelerates the rate of a reaction at a given temperature without undergoing change during the reaction. Platinum and palladium are used for VOCs, whereas metal oxides are used for chlorinated compounds. The catalytic incinerator is affected by the operating temperature, velocity of movement of the VOCs, composition and concentration of VOCs, properties of the catalyst, poisons or inhibitors in the emissions.

Flares can be used to control hazardous air pollutants. A 98% destruction efficiency occurs. Flares are open flames used for disposing of waste gases during normal operations and emergencies. Flare performance depends on exit gas velocity, gas heating value, time in the combustion zone, flame temperature, and waste gas–oxygen mix.

Boilers or process heaters can control emissions and burn the chemical for its heating value with a 98% efficiency. Carbon or other adsorbers can remove 95 to 99% of VOCs. Absorbers or scrubbers are 99%+ efficient in removing concentrated VOCs. Condenser efficiency in controlling high levels of VOCs varies from 50 to 90%.

Inorganic vapors are a small part of total hazardous air pollutants. They include such gases as ammonia, hydrogen sulfide, carbon disulfide, and metals with hydride and carbonyl complexes, chloride, oxychloride, and cyanide. These vapors are controlled by absorbers, adsorbers, and condensers. Absorption is over 99% efficient, whereas adsorption is over 95% efficient. Activated carbons are the most commonly used adsorbants.

Process fugitive emissions come from a process or piece of equipment away from the main vent or stack. The fumes escape from valves, pumps, compressors, access ports, and feed or discharge access ports to a process. The emissions may be organic or inorganic in nature. Control techniques include leak detection and repair programs to seal the leaks. Further, the use of sophisticated pumps and pump seals, as well as valve and valve seals, can reduce emissions.

Control techniques for organic or inorganic vapor emissions from area fugitive sources as lagoons and ponds is very difficult. The best approach is to reduce sharply the hazardous pollutant before it goes to the lagoon or pond and becomes a hazardous air pollutant.

Particulate Emissions Control

Particulate emissions control from point sources include fabric filters (baghouses), electrostatic precipitators, and venturi scrubbers. These have been previously discussed. Fugitive sources include process sources and area sources. Process sources can be controlled by use of hoods and capture of process emissions. Loss of materials from conveyors at the feeding, and transfers and discharge points due to spillage and mechanical agitation cause area fugitive emissions. These emissions can be controlled by wet suppression methods using water, chemicals, and foam. Other area fugitive emissions come from loading and unloading dust paved and unpaved roads, storage piles, and waste disposal sites. Work should be performed in low wind conditions when possible, wind screens should be used if feasible, and wetting agents or foams should be used where applicable.

Air Pollution Control Programs

The management of our air resources, to protect the health and welfare of our people, must be carried out through joint efforts of local, state, and federal agencies; industry; and the population at large.

A good program must include using legal authority to institute and carry out an air pollution control program, using a continuous air quality monitoring program; establishing an emission-source inventory and a continuous updating of this inventory; developing air quality goals and standards based on air quality criteria; developing a thorough understanding of local meteorologic conditions and their relationship to the movement of air pollutants; making land-use planning decisions based on air quality control and other environmental factors; developing a good public information and educational program; training of available personnel in the use of monitoring equipment, samplers, and laboratory analysis of pollutants; implementing air-use plans for existing industries; approving plans for new industries; and identifying polluters and using enforcement techniques when needed to achieve compliance.

Components of the Federal Program

Prior to 1955, the U.S. Public Health Service conducted and supported limited air pollution research. In July 1955, as a result of serious air pollution episodes in Donora, PA, and Los Angeles, CA, Congress authorized a federal air pollution program in which the federal government would conduct air pollution research and provide technical assistance to state and local governments. In 1960, the federal government started to study motor vehicle pollution. The Clean Air Act of 1963 authorized the federal government to give funds to state and local agencies to help them develop, establish, or improve control programs. The federal authority also took action to abate interstate air pollution problems. This act authorized accelerated research, training, and technical assistance for motor vehicle and sulfur oxide pollution from the burning of coal and fuel oil. In 1967, new legislation was approved requiring the establishment of national emission standards for industries and the creation of regional commissions to establish air quality standards.

The Clean Air Act Amendment of 1970 established new legal techniques of pollution control accompanied by strict federal pollution control standards. The polluters must now prove that their actions are not harmful to the health and welfare of the people. The EPA, which enforces the Clean Air Act, is responsible for setting standards, enforcement, conducting research, and providing funds and technical assistance to states and communities to meet the requirements of the Clean Air Act. The EPA established national ambient air quality standards, standards of performance for new stationary sources, national emission standards for hazardous air pollutants, and motor vehicle and aircraft emission standards. EPA has designated 247 air quality control regions in the United States, crossing state political boundaries.

The EPA was authorized under the Clean Air Act of 1970, Public Law 91-604, to take whatever action is necessary, including court orders, to shut down polluters whenever air pollution poses an imminent and substantial health hazard. This emergency power deals with air pollution episodes occurring when periods of stagnant air occur, allowing pollutants to reach abnormally high concentrations. The pollutants measured under emergency standards include sulfur dioxide, particulates, carbon monoxide, photochemical oxidants, and nitrogen oxide.

Air Quality Standards

The national standards are in two parts. The primary standard protects public health by setting a limit on the amount of pollutant in ambient air that is safe for humans. The secondary standard protects public welfare by setting a standard (usually more stringent than the primary standard) that sets limits on the amount of pollutants considered safe for clothing, buildings, metals, vegetation, crops, and animals. However, today primary and secondary standards tend to be the same.

Maximum concentrations of pollutants used in national air quality standards are based on scientific evidence of the pollutant's effect on public health and welfare. These effects are listed in criteria documents that must be published by the EPA. The agency also publishes information on the known techniques and methods of controlling each of the pollutants stated in the National Air Quality Standard. Technical information includes the cost of emission control, the availability of controlled technology, and the alternate methods available. The air quality standards, when established, are met by the state when the state enforces limits on emissions of the varying pollutants from their sources. Whereas the air quality standard is the limit of the amount of pollutant permitted in the air, the emission standard is the maximum amount of the pollutant that can be discharged from a specific source. In establishing the National Emissions Standards, the EPA identifies the hazardous air pollutants, issues proposed national emissions standards, holds hearings, and enforces emissions standards for hazardous pollutants. The pollutant is not permitted to be released into the air anywhere in the United States. The EPA has the right to allow a 2-year period for implementing this measure if human health is protected against immediate danger. All new plants must meet federal air pollution performance standards for fossil fuel and other hazards. Motor vehicle emission standards required sharp reductions in carbon monoxide and hydrocarbon emissions.

The EPA has the right to request a fine, after a 30-day notice, of $25,000 for each day of violation of hazardous emission standards, plus 1 year in prison for each day of violation. Subsequent convictions can bring fines of up to $50,000 a day per violation and 2 years imprisonment. Where automobiles are in violation of EPA laws, the individual can be fined $10,000 for each car or engine that is operating illegally, and those who handle motor vehicle fuel can be fined up to $10,000 a day for violation of the motor vehicle fuel standards. Although on the surface, it sounds good to be able to fine individuals so stringently for these violations, in practice it is really difficult to achieve this level of punitive action. It is always better in the field of environmental health to utilize educational and consultation approaches as far as possible before court action is taken.

Clean Air Act of 1990

The Clean Air Act of 1990, enacted after 13 years of discussion, includes tighter controls on tailpipe exhaust, starting with 1994 cars, which results in a 60% reduction in nitrogen oxide and a 40% reduction in hydrocarbons; reducing acid rain by cutting back 50% of sulfur dioxide emission and 2 million tons a year of nitrogen oxide by the year 2000; industry reduction by 90% of the use of 189 chemicals that are known carcinogens over the next 10 years; and protection of the ozone layer by phasing out CFCs and carbon tetrachloride by the year 2000, methyl chloroform by 2002, and hydrochlorofluorocarbons by 2030.

Title I — Title I specifies that areas be classified according to their air quality. The EPA has helped state governments to reclassify and establish boundaries for areas that have not yet met the national ambient air quality standards. This is the first step in developing the state implementation plans (SIPs).

Title II — Title II uses a multifaceted approach to reducing motor vehicle emissions. This includes cleaner vehicles, cleaner fuels, and motor vehicle inspection and maintenance programs. Starting with 1994 vehicles, stricter standards apply to all vehicular pollutants including hydrocarbons, carbon monoxide, nitrogen oxides, and diesel particulates. Fuel volatility and the tendency of gasoline to evaporate and pollute the air are to be reduced. Reformulated, cleaner burning gasoline including oxygenated fuels must be sold in the worst smog areas of the United States by 1995. VOC emissions should be cut almost 300 million lb/year in the 9 dirtiest cities in the United States. There is to be a 2.7% oxygen content in the gasoline. Some people pumping the new gasoline have complained of dizziness or headaches. Motor vehicle inspection and maintenance programs are required in the worst smog and carbon monoxide areas. Moderate ozone areas now are also required to implement these programs.

Title III — Title III provides new regulations for the routine and accidental release of toxic air pollutants that are known or suspected of causing or contributing to various serious health problems such as cancer or reproductive disorders. Over a 10-year period EPA will issue regulations for all major sources of toxic air pollution.

Title IV — Title IV is the acid rain control program, which includes a market-based system of sulfur dioxide allowances, a flexible permitting system, a continuous emissions monitoring system, and an emissions cap for utility emissions of sulfur dioxide.

Title V — Title V requires federal permits for all major sources of air pollution to be allowed to operate.

Title VI — Title VI establishes a comprehensive program to phase out most ozone-depleting substances by the year 2000.

Title VII — Title VII increases the range of civil and criminal sanctions available in enforcement.

Titles VIII and IX — Titles VIII and IX authorize extensive research to measure, monitor, and analyze air pollutants and investigate alternate energy sources.

Title X — Title X deals with disadvantaged businesses.

Title XI — Title XI provides for financial assistance, training, and employment services for workers who are laid off or fired because of the implementation of the Clean Air Act.

1997 Changes to the Clean Air Act of 1990

Based on new scientific evidence, revisions have been made in the air quality standards for ground-level ozones (smog) and particulate matter ($PM_{2.5}$). The court of appeals for the D.C. Circuit said that the EPA had too much authority. It did not challenge any of the health studies, but instead the level at which the standards were set. This can be appealed.

Climate Change Action Plan

On October 19, 1993, the President released a plan to help curb global warming titled, Climate Change Action Plan. It is designed to reduce greenhouse gas emissions to their 1990 levels by the year 2000 by means of a volunteer industry effort. The plan was created as a partial response to the Climate Change Treaty at the Earth Summit held in Rio de Janeiro in 1992. It contains 44 action items needed to reduce harmful emissions and establishes three partnerships. They include the Climate Challenge, a partnership between the Department of Energy and major electric utilities that have agreed to reduce their greenhouse emissions; Climate Wise, a joint effort among the Department of Energy, EPA, and industries that have pledged to set strict emission targets that can be met in a cost-effective way; and Motor Challenge, a joint effort of the Department of Energy, motor system manufacturers, industrial motor users, and utilities to install energy-efficient motor systems in industrial applications.

The 150 governments signing the Framework Convention on Climate Change at the Earth Summit in Rio de Janeiro in 1992 stipulated (and was accepted) that the developed countries would attempt to return to the 1990 levels of greenhouse emissions by the year 2000. By 1995, the Intergovernmental Panel on Climate Change under the supervision of the United Nations and the World Meteorological Organization published its first assessment of the current emissions trends. The committee stated that under current conditions the mean global temperature would rise 0.3°C per decade. In 1995, this committee released its Second Assessment Report, which stated that human activity had a discernible influence on the global climate. The largely voluntary targets created by the 1992 convention were insufficient. By the end of 1997, carbon dioxide had increased in all but a few developed nations and the chances for meeting the year 2000 target were very poor. In December 1997, at the third conference in Kyoto, Japan, more than 160 countries produced a protocol that stated the industrialized nations agreed to reduce their aggregate emissions of a basket of 6 greenhouse gases by at least 5% below 1990 levels in the period from 2008 to 2012.

From 1986 to 1995, world emissions of carbon dioxide from fossil fuel burning and gas flaring (excluding emissions from cement manufacturing) rose 15% to 6.06 billion tons of carbon. The highest amount of carbon dioxide emissions came from the Far East, where a combination of population growth, rapid industrialization, and use of coal as the primary fuel occurred. Industrialized nations account for about 20% of the world population, but consume about two thirds of the world's energy. In 2001, the United States decided not to join the Kyoto proposals.

Components of the State Program

Each state develops an implementation plan, including the establishment of legal authority for air pollution control (an air pollution board must be established), control plans, schedules of compliance, rules for controlling air pollution episodes, air quality monitoring systems, existing emission inventories, air quality goals, and liaisons with other agencies.

Results of the Federal Program

State and local governments, with the support and guidance of the EPA, have established programs for the operation and maintenance of vehicle inspection. The programs are used to test automobile emission levels. They also determine if cars built for unleaded gas are altered or their emission control devices are removed.

The EPA closely monitors the compliance status of about 30,000 stationary air pollution sources that are regulated by the states. The agency reviews whether the states are meeting the ambient air standards for individual criteria pollutants. If these standards are not met, the states are required to develop new plans to do so. As a result of the federal and state actions, ambient levels of all criteria pollutants have decreased. A good example of this is the sharp reduction in lead emissions and ambient air levels of lead due to the reduction of lead in gasoline. Between 1970 and 1997 ambient levels of lead in the air have declined by 98.2%.

Because of EPA and state monitoring of criteria pollutants in the ambient air, especially in cities, a sharp drop in criteria pollutants has occurred. Particulate levels of pollutants have decreased because of the installation of pollution control equipment and a reduction in industrial activity. Sulfur dioxide levels have decreased because of a cutback of emissions from coal-fired power plants. Nitrogen dioxide levels have decreased because of pollution controls. Ozone levels have decreased primarily because of the Federal Motor Vehicle Control Program as well as control efforts on stationary sources. Carbon monoxide levels have decreased because of the Federal Motor Vehicle Control Program.

Radiation is addressed in four primary areas: radiation from nuclear accidents, radon emissions, land disposal of radioactive waste, and radiation in groundwater and drinking water. The EPA has set standards for radioactive emissions under the Clean Air Act, Atomic Energy Act, and Uranium Mill Tailings Radiation Control Act.

Emission of VOCs from large stationary sources, such as chemical plants, refineries, and industrial processes, have been substantially controlled through EPA, state, and local regulatory and enforcement efforts. Further studies are under way on paint manufacturers, dry cleaners, and gas stations.

To control new and existing sources of VOCs, emission standards for a variety of new sources based on the best control technology have been proposed. Cars will be required to have equipment to control gasoline vapors while refueling. The EPA will assist state and local governments in promoting the use of alternative fuels, such as methanol, ethanol, and compressed natural gas.

In 1971, the EPA issued a National Ambient Air Quality Standard for total suspended particulates covering all kinds and sizes. In July 1987, EPA published new standards based on particulate matter smaller than 10 µm in size (PM_{10}). In 1997, it proposed $PM_{2.5}$ standards. The most serious health threat comes from the smaller inhalable particulates, because they tend to become lodged in the lungs and remain in the body for a long period of time. Although some particulates may be controlled by conventional means, the EPA and states have tried to meet the particulate standard by limiting emissions from industrial facilities and other sources. Industries have installed pollution controls, such as electrically charged plates and huge filters, to meet these standards.

The EPA has also set emission standards for diesel automobiles. Better techniques of paving streets, better street cleaning, and limits on agriculture and forest-burning practices, as well as a ban on backyard burning in urban areas, have helped to reduce particulate concentrations. Standards have been set for the emission of particulates for diesel trucks and buses. These standards took effect for diesel vehicles within the 1988 model year and became more stringent in the 1991 to 1994 model years. Because of the PM_{10} standards, the EPA has classified all counties in the United States into three groups based on the probability of not meeting the new standards. For high probability of nonattainment, the EPA requires the states to revise their implementation plans. For areas with moderate probabilities of non-attainment, the states must carefully monitor the air quality. For areas with low probabilities, the current control strategies are considered to be adequate.

The EPA is continuously issuing NESHAPs under the Clean Air Act. The first eight hazardous air pollutants listed were asbestos, beryllium, mercury, vinyl chloride, benzene, arsenic, radionuclides, and coke oven emissions. The EPA works with state and local agencies to identify other types of potentially hazardous air pollutants. They are assessing the health effects of many additional chemicals, including ammonia, chlorine, formaldehyde, chromium, chloroform, carbon tetrachloride, and other suspected carcinogens.

Because the air toxic problem is far more complex than thought by the writers of the Clean Air Act of 1970, in 1987 the EPA developed a 5-year plan and strategy for controlling air toxics. This plan includes establishing federal programs to identify and regulate air toxics from stationary and mobile sources; helping states evaluate and decide on the regulations of high-risk point sources that pose significant local risk; enhancing state and local programs by assisting and planning, finance, and technical support; enforcing NESHAPs and mobile source regulations; and extending and improving long-term air toxics monitoring programs. In the 1990 Clean Air Act, 189 toxic air pollutants were identified.

The EPA has identified various categories of sources of toxic air pollutants, such as gasoline service stations, electrical repair shops, coal-burning power plants, chemical plants, incinerators, and cement kilns. It has been determined which sources are major or large and which sources are areal or small. The EPA has issued regulations, and in some cases has specified exactly how to reduce pollutant releases. Wherever possible companies are allowed to be flexible and use MACT to reduce the pollutant releases. The early reduction program allows the company extra time to finish cleaning up the remaining 10% of pollutants, if it cleans up the initial 90% before the EPA regulates the chemical.

Prior to 1990, the Clean Air Act directed the EPA to regulate toxic air pollutants based on the risks each pollutant caused to human health. Because this approach had severe limitations, Congress adopted the new MACT strategy in 1990, which mandates a more practical approach to reducing emissions of toxic air pollutants. The standards of this approach are based on the emissions levels already achieved by the best performing similar facilities. This performance-based approach produces standards that are both reasonable and effective in reducing toxic emissions. For existing sources the MACT floor or baseline must equal the average emissions limitations currently achieved by the best performing 12% of sources in that source category, if there are 30 or more existing sources. If there are fewer than 30 existing sources, it is for the best performing 5 sources in the category. For new sources the baseline must equal the level of emissions control currently achieved by the best controlled similar source.

As of January 1998, the EPA had issued 23 air toxics MACT standards affecting 48 categories of major industrial sources, such as chemical plants, oil refineries, aerospace manufacturers, and steel mills, as well as 8 categories of small resources, such as dry cleaners, commercial sterilizers, secondary lead smelters, and chromium electroplating facilities. EPA had also proposed a number of rules covering 22 source categories, such as polyurethane foam production, wool fiberglass operations, and phosphoric acid and phosphate fertilizer production. Additional standards will be developed over the next 10 years.

The decline in sulfur dioxide emissions is largely due to the use of fuels that lower the average sulfur content, the introduction of scrubbers, which remove sulfur oxides from flue gases, and controls on industrial processes. The decrease in sulfur dioxide levels in residential and commercial areas are due to a combination of energy conservation measures and the use of cleaner fuels.

Other Air Quality Trends

Acid deposition has been decreased because of the decrease of sulfur dioxide emissions and nitrogen oxide emissions. In addition, the National Acid Precipitation Assessment Program, which was established under the provisions of the Clean Air Act, is a major research effort to resolve the critical problems and uncertainties about the effects of acid rain. This effort has been renewed by Congress.

The Clean Air Act of 1990 took a new nationwide approach to the acid rain problem. The law set up a market-based system designed to lower sulfur dioxide pollution levels. The reduction of sulfur dioxide results in a major reduction in acid rain. Phase I of the acid rain reduction program went into effect in 1995. Big coal-burning boilers in 110 power plants in 21 Midwest, Appalachian, southeastern, and northeastern states had to reduce releases of sulfur dioxide. In 2000, phase two of the acid rain program began, further reducing the sulfur dioxide releases from the country's power plants and other smaller polluters. The reductions in sulfur dioxide releases are essential in the national program of emission (release) allowances.

The EPA issues allowances to power plants covered by the acid rain program, with each allowance worth 1 ton of sulfur dioxide released from the smokestack. Plants may only release as much sulfur dioxide as they have allowances. To obtain reductions in sulfur dioxide pollution, allowances are set below the current level of sulfur dioxide releases. If a plant expects to release more sulfur dioxide than it has allowances, it has to get more allowances, by buying them from another power plant that has reduced its sulfur dioxide releases below its number of allowances and therefore can sell the surplus. The acid rain program provides bonus allowances to power plants for installing clean coal technology that reduces sulfur dioxide releases, or by using renewable energy sources.

The NOX program follows many of the same principles as the sulfur dioxide trading program in design. NOX is results oriented and flexible in the method to achieve emission reductions, and it reaches program goals through measurement of the emissions. However, it does not cap NOX emissions and it does not use an allowance trading system. The use of the program allows flexibility because the facility can either comply with the individual emission rate for a boiler or average the emission rates over multiple boilers.

In 1978, the EPA and other federal agencies banned the nonessential use of CFCs as propellants. On an international basis, because the amount of propellants have continued to grow, 31 nations representing the majority of the CFC-producing countries signed the Montreal Protocol of 1987. The protocol requires developed nations to freeze consumption of CFCs at 1986 levels by mid-1990 and to use only half as much by 1999. In the United States CFCs were to be eliminated by the year 2000.

The national power plant cleanup campaign is designed to force policy change at the federal level, and at the state level to dramatically reduce harmful air emissions from the country's electric power system where appropriate. Targeted pollutants include sulfur dioxide, nitrogen oxides, toxic emissions, and carbon dioxide. The major focus of the campaign is to close the 30-year-old loophole that allows more than two thirds of the nation's coal and oil fire generating units to emit key pollutants at 3 to 4 times the rate of new coal plants. The electric power industry continues to be one of the most polluting industries in the world. Many plants that were built in the 1940s to 1960s contribute to air pollution. The cleanup effort is currently pursued by 75 funded state and regional organizations in 27 different states.

Visibility impairment is measured by the amount of haze during summer months at 280 monitoring stations located at airports across the country. Haze increased greatly between 1970 and 1980, and decreased slightly between 1980 and 1990. These trends follow overall trends in emissions of sulfur oxides during these periods. From 1990 to 1997, the haziest visibility days appeared to have stayed the same in the East, whereas visibility improvement has occurred in the West.

The Air Quality Index is derived from pollutant concentrations and is reported daily in all metropolitan areas of the United States with populations exceeding 200,000. It is reported as a value between 0 and 500 and is associated with the general descriptions of air quality. An Air Quality Index >100 indicates that at least one criteria pollutant, with the exception of nitrogen dioxide, exceeded the level of the national air quality standards and therefore indicates unhealthy air. Analysis of trends of the country's 94 largest metropolitan areas from 1988 through 1997 shows a 56% decrease in Air Quality Index values >100 in southern California and a 66% decrease in the remaining major cities across the country. Although progress has been made and air quality trends are improving nationally, both urban and rural areas still have concentrations above the level of the national standard, and some areas even have worsening trends. These areas are nonattainment areas and are subject to the EPA rule-making process. As of September 1998, a total of 130 nonattainment areas existed with 113 million people residing there.

The mobile source control program is a result of the 1990 Clean Air Act amendments, which impose stricter controls on mobile sources of air pollution, including cars, trucks, buses, and other nonroad engines. Further, the Clean Air Act changes fuel requirements so that cleaner fuels are required in the most polluted areas. In some cases, alternative fuels such as natural gas or electricity must be used to power vehicles. In other cases, gasoline must be reformulated so that it has at least a 2% oxygen content, which leads to more complete combustion. Reformulated gasoline must also have lower emissions of toxics such as benzene. Fuel sulfur is a major problem causing air pollution in mobile sources.

Fuel sulfur creates immediate and long-lasting impacts in the air. It increases exhaust volatile organic chemicals by 40% and nitrogen oxide emissions by 150% under special test conditions. It also increases carbon monoxide emissions. It does this by decreasing the efficiency of the three-way catalyst used in current and advanced emission control systems. Sulfur sensitivity is variable and depends on both the catalyst formulation and the vehicle operating conditions. Sulfur sensitivity is temperature dependent. Sulfur adheres to the catalyst surface more thoroughly at lower catalyst temperature than at higher temperatures. Sulfur therefore takes up space on the catalyst so that the other chemicals cannot react to form water, nitrogen, oxygen, and carbon dioxide. It therefore becomes essential to drastically reduce fuel sulfur as part of the mobile source control program.

From 1970 to 1975, the first standards related to tailpipe emissions were enacted through the 1970 Clean Air Act. From 1977 to 1988, the standards were tightened. From 1990 to 1994, tier 1 emissions standards were passed calling for a nitrogen oxides standard of 0.6 ppm for cars, effective in 1994. This standard was a 40% reduction from the 1981 standard. In 1998, the administration reached a voluntary agreement for cleaner cars with the auto industry and the Northeast states. These cars operate with a nitrogen oxide standard of 0.3 ppm. Also, in 1998, the EPA issued a tier 2 report lowering tailpipe emissions in the year 2004 for new cars and light trucks to 0.07 ppm. The new standard also reduces vehicle emissions of nonmethane organic gases consisting primarily of hydrocarbons and VOCs. The reduction in nitrogen oxides helps reduce the formation of ozone and particulate matter. Further the new standards reduces the average sulfur content to 30 ppm with a cap of 80 ppm.

Reformulated gasoline is required to be sold in the nine worst ozone nonattainment areas in the country, including Los Angeles, San Diego, Hartford, New York, Philadelphia, Baltimore, Houston, Milwaukee, and Chicago. The chemical used to increase the oxygen content of the gasoline is methyl-*tert*-butyl ether. Health complaints have occurred when individuals have been exposed to this chemical. The symptoms include headaches, dizziness, nausea, flulike symptoms, eye irritation, coughing, and burning of the nose or throat. The chemical has been found both in the air and in the water in the communities where this reformulated gasoline is used.

In July 1999, the EPA proposed that Congress should no longer require oil companies to add this chemical to gasoline, because it pollutes water.

Air Pollution Episodes

The Air Pollution Control Board, in addition to its other functions of establishing standards, reviewing applications, and ruling on legal problems related to polluters, establishes the rules for controlling air pollution episodes.

It is within the best interest of the public to prevent high levels of air pollution or episodes that may cause acute, harmful health problems. A warning and control system is instituted when unusually high levels of air pollution persist for many hours. This usually occurs when a high pressure air mass stagnates. The system has been developed on the basis of evidence that a synergistic action occurs between particulates and other pollutants, such as sulfur dioxide. The air pollution alert levels include the watch-and-alert level, the yellow alert, and the red alert.

When a watch-and-alert level occurs, the technical staff provides 24-hour coverage for monitoring equipment, and all agencies are advised of the impending problem. When a yellow-alert level occurs, the weather bureau has forecast at least a 36-hour high air pollution potential, and yellow-alert values of pollutants are equaled or exceeded for 4 hours consecutively in the same geographic area. In the event of a yellow alert, the general public is informed of the hazard, plants burning high-pollution fuels are requested to convert to emergency fuels, large incinerators are requested to stop operation, sources that emit large quantities of air contaminants are requested to reduce emissions, and the public is requested to reduce nonessential driving.

When a red-alert level occurs, the weather bureau has forecast stagnation for 12 hours after 72 hours of prior stagnation. The public is informed that conditions have worsened and that precautions should be taken for individuals with high lung disease potential and also elimination of all nonessential driving. Commercial and industrial operations are requested to cease the emission of contributing air contaminants.

SUMMARY

Significant progress has been made in improving air quality. However, nitrogen oxides and ozone are still serious problems in many parts of the country. Particulate matter has to be removed from the air. Airborne toxics are serious problems.

One of the variables affecting all cities is the weather. When the weather is stagnant, the pollutants found in the ambient air sharply increase. The level of pollution has not been increased by the sources, but instead by concentration of the pollutants as a result of weather conditions. For instance, the TSP standard in Chicago, at one station, exceeded the standard on 20 days. At a second sampling station, the TSP exceeded the standard on 10 days. In Pittsburgh, in one station the TSP exceeded the standard on 22 days, and at a second station the TSP exceeded the standard on 18 days. In Cleveland, the TSP exceeded the standard on 52 days at one station and on 35 days at a second station, and so forth. The same types of patterns can be shown for all the criteria pollutants.

Air pollution is here to stay. However, the amount can be controlled if the overall problems of health are placed in their proper perspective, at a high point in the human value system. Only humans can limit pollutants to a point where they are not harmful to humans or to the ecosystem.

RESEARCH NEEDS

Research is needed on controlled exposures of animals to varying levels of pollutants in varying ranges of temperatures and relative humidities that are found in community and industrial settings. The actual dose that the individual gets, instead of the amount that is present in the air, must be studied to determine the effect of the dose on the individual. This can be accomplished by setting up research studies where personal monitoring devices are utilized, either in the community at large or within the industrial setting. A need exists for the establishment of specific research centers concerned with clinical environmental research in the area of air pollutants as they relate to cardiopulmonary and immunologic disorders.

In future epidemiological studies, not only past data but also current exposures of the individual and the population based on personal sampling devices and other samplers, should be used to determine a broad range of pollutants. Studies should be made to understand the correlation between various human activities and the amount of pollutants taken into the human organism. For the specific major pollutants, specialized studies are needed. Sulfur oxides should be researched in depth. Studies should be made to evaluate the possibility that the pH of an aerosol plays an independent role in determining its biological toxicity. Studies are needed to evaluate the size of the aerosol related to its effects. It is necessary to determine the effects of particles of varying sizes contained in the TSP count and how these particles interact with sulfur dioxide present in the air. Studies are needed to measure the acute, as well as the chronic, response to all the major pollutants.

Long-term studies are needed for the evaluation of individuals living in communities subjected to exposures of varying pollutants. There is a need to determine the effects of low levels of carboxyhemoglobin that are produced in the blood when carbon monoxide is present. What effect does carboxyhemoglobin have on the psychomotor performance of older people? Also, what effect does it have in relation to smoking and diet? Photochemical oxidants, particularly ozone, need to be studied to determine the chronic effects on the individual related to age, diet, and physical condition. What occurs when photochemical oxidants are mixed with other pollutants in the atmosphere? Nitrogen oxides, which are extremely irritating, must be evaluated for their effect on the individual. How does the nitrogen oxide level vary from the source of production, through the interaction with other substances in the air, to the point where the individual inhales it, and what effect does it have on different individuals?

Techniques are needed to reduce ozone and carbon monoxide levels. Quality data are needed for determining the amounts of small particles that may be formed in the atmosphere as the result of various chemical and physical processes. These

PM_{10} particles are also produced from a variety of sources. Control measures need to be developed to further reduce the PM_{10} levels, and to reduce the $PM_{2.5}$ levels.

Scientific assessment of the risk of toxic air pollutants needs to be conducted to determine how to manage these risks properly. The toxic air pollutants problem is a very complex and a controversial scientific and policy issue.

Further reduction of sulfur dioxide and nitrogen oxide emissions is necessary to reduce the problem of acid rain. Continued research is necessary to determine the problems related to acid deposition and the most efficient and economical means of controlling these emissions. Better understanding of the effects of stratospheric ozone depletion and global warming on human health, agriculture, and natural ecosystems is needed. More scientific investigation should be conducted about the Antarctic ozone hole and its implications for the region and for the entire world.

CHAPTER 2

Solid and Hazardous Waste Management

BACKGROUND AND STATUS

Although Americans account for only 7% of the world's population, they consume almost one half of Earth's industrial raw materials. As a result they produce not only many fine products but also an enormous amount of solid waste. Until recently little attention has been paid to this problem.

During the course of the growth of the United States, the natural resources seemed limitless. Little attempt was made to reuse materials, because it was cheaper to obtain the virgin materials. However, in our society today, raw materials are becoming more scarce. Various international cartels, such as the oil cartel, are attempting to control prices again. The United States, along with many other industrialized nations, are now in the process of attempting to find a better solution to the removal, disposal, and reclamation of solid waste materials, where possible. As of 2000, the United States produced more than 6 billion tons of solid waste a year. Of this total 830 million tons were household, municipal, and industrial wastes; 3.0 billion tons were agricultural wastes; and 2.34 billion tons were mineral wastes. Municipal solid waste before recycling was 210 million tons, of which 38.1% was paper; 13.4%, yard waste; 10.4%, food waste; 9.4%, plastics; 7.7%, metals; 5.9%, glass; 5.2%, wood; and 9.9%, other. By the year 2010, this is projected to grow to 262 million tons/year. It is estimated that the average amount of solid waste will increase from the year 2000 level of 4.3 lb/person/day. The cost of removal of solid waste is enormous. Currently, over $10 billion each year is used for storage, collection, processing, transportation, and final disposal of urban waste. It is further estimated that this figure may increase by at least 15% each year for the next 5 years. Not only do individuals pay for the previously mentioned costs, but also they pay for the hidden costs attributed to rodent and insect infestation, depreciation of land values, odors, and a sharp increase in pollution (air, land, and water).

The variety of materials that Americans use and discard is increasing at a very rapid rate. Today we have multiple packaging, no deposit and no return containers, and planned-obsolescence products. Approximately 2000 new products are created each year. These new products sharply increase the level and amount of solid waste.

The annual throw-away includes approximately 52 billion cans, 30 billion bottles and jars, 4.5 million tons of plastic, 8 million television sets, 7 million cars and trucks, and about 33 million tons of paper. By the year 2004, 315 million computers will become obsolete. This could add 8.5 million tons of waste to landfills. Personal computers contain lead, mercury, cadmium, and other persistent toxic compounds that bioaccumulate.

Contemporary industrial society characteristically generates large quantities of waste material. To reduce these quantities, many states and localities have investigated waste disposal systems that are useful for the recovery of energy and the recycling of valuable resources. Landfill sites, high costs of disposal, and rising energy and material prices are forcing these communities to seek even better resource recovery technology.

About 70 to 80% of residential and commercial solid waste is combustible. If all of the solid waste in the United States were to be converted into energy, it would equal more than 206 million barrels of oil per year. With the increasing population and increasing per capita waste generation, this figure will continue to climb. Currently, a number of communities have resource-recovery facilities and others have feasibility studies under way.

Extraction and production techniques are associated with waste, industrial chemicals, spent process water, refuse, abandoned cars, building materials, animal manure, fuel residues, newspapers, and other solid waste. Although the process of recycling is important, it also constitutes potentially serious hazards. An exposure may occur to humans through the food chain or through other environmental factors. It is known that polychlorinated biphenyls (PCBs) are found in recycled paper, hexachlorobenzene (HCB) has been found in land disposals, and organic mercury compounds have been found in process waste.

Even if resource recovery is an excellent process, our concerns for the future will grow, because it is estimated that waste products will increase by 23%. An evaluation of the potential hazards associated with recycling vs. the benefits of recycling will have to be made and then appropriate decisions made concerning the techniques to be utilized.

A large number of existing solid waste landfills do not meet the current federal and state standards for safe design and operation. This includes monitoring groundwater for possible pollution; having natural or artificial liners; collecting the polluted leachate or liquid waste that is produced when liquids percolate through the refuse; treating the leachate; providing controls for water pollution that results from the runoff of rainwater from their sites; and imposing restrictions on the receipt of bulk liquid waste. Most of the landfills were developed many years ago and have received wastes that are now considered to be hazardous. Thousands of closed landfills are being investigated for potentially hazardous waste problems. In 1978, 20,000 landfills existed; by 1991, 6000; and by 1997, 2200. However, landfill capacity has increased in most states.

When communities have tried to develop new waste facilities and waste-processing technology, people within the community have started enormous controversies over the location of the landfill and the potential environmental problem resulting from the landfill. Individuals are saying that they do not want this in their backyard.

SCIENTIFIC, TECHNOLOGICAL, AND GENERAL INFORMATION

Definition of Solid Waste

The term *solid waste* is very broad. It includes not only traditional nonhazardous solid waste, such as municipal garbage, but also hazardous solid waste. The Resource Conservation and Recovery Act (RCRA) defines solid waste as garbage, refuse, sludges, and other discarded materials. The garbage comes from human activity. The refuse includes a list of items discussed under residential waste. The sludges come from waste treatment plants, water-supply treatment plants, or pollution control facilities. Other discarded materials include solid, semisolid, liquid, or contained gaseous materials resulting from industrial, commercial, mining, agricultural, and community activities. Under the RCRA definition of solid waste, it is essential to recognize that all solid waste is not solid. Many solid wastes are liquid, semisolid, or gaseous. Although the definition of solid waste includes hazardous waste, the Environmental Protection Agency (EPA) subtitle D program is concerned primarily with nonhazardous waste. Exceptions to the definition of solid waste under RCRA include:

1. Domestic sewage, which is defined as untreated sanitary wastes that pass through a sewer system
2. Industrial wastewater discharges, which are regulated under the Clean Water Act
3. Irrigation return flows
4. Nuclear material, or by-products, as defined by the Atomic Energy Act of 1954
5. Mining materials that are not removed from the ground during the extraction process

Definition of Hazardous Waste

Under RCRA, *hazardous waste* is defined as a solid waste or a combination of solid wastes that because of their quantity, concentration, or physical, chemical, or infectious characteristics may:

1. Cause, or significantly contribute to, an increase in mortality or an increase in serious irreversible, or incapacitating reversible, illness
2. Pose a substantial present or potential hazard to human health or the environment when improperly treated, stored, transported or disposed of, or otherwise managed

RCRA defines hazardous waste in terms of properties of a solid waste. Therefore, if a waste is not a solid waste as previously defined, it cannot be a hazardous waste. To make these definitions clearer, the EPA was permitted by Congress to list or to specify that a solid waste was hazardous if it met one of four conditions. These included:

1. Exhibits, on analysis, any of the characteristics of a hazardous waste
2. Has been named as a hazardous waste and listed
3. Is a mixture containing a listed hazardous waste and a nonhazardous solid waste
4. Is not excluded from regulation as a hazardous waste

The by-products of the treatment of any hazardous waste are also considered hazardous unless specifically excluded. The EPA has identified four characteristics for hazardous waste. Any solid waste that exhibits one or more of them is classified as hazardous under RCRA. The characteristics are ignitability, corrosivity, reactivity, or extraction procedure (EP) toxicity (the latter is designed to identify waste likely to leach hazardous concentrations of particular toxic constituents into the groundwater if they are not properly managed). Ignitability refers to a solid waste that is a liquid except for solutions containing less than 24% alcohol, which has a flash point less than 140°F; a nonliquid capable under normal conditions of spontaneous combustion; and an ignitable compressed gas as listed in the Department of Transportation regulations. The reason for this listing for ignitability was to identify waste that could cause fires during transport, storage, or disposal. Corrosivity was defined as a solid waste that is an igneous material with pH ≤ 2 or ≥ 12.5 or a liquid that corrodes steel at a rate greater than ¼ in./year at a temperature of 130°F. The pH was chosen as an indicator of corrosivity because waste with high or low pH can react dangerously with other waste or cause toxic contaminants to migrate from certain waste. Steel corrosion may occur as a result of these wastes. Reactivity is defined as a solid waste that exhibits any of the following properties due to it reactivity:

1. Normally unstable and reacts violently without detonating
2. Reacts violently with water
3. Forms an explosive mixture with water
4. Generates toxic gases, vapors, or fumes when mixed with water
5. Contains cyanide or sulfide, and generates toxic gases, vapors, or fumes at a pH between 2 and 12.5
6. Capable of detonation if heated under confinement or subjected to strong initiating sources
7. Capable of detonation at standard temperature and pressure
8. Listed by the Department of Transportation as a class A or B explosive

The activity was chosen as a characteristic to identify unstable waste and waste that may pose a problem at any stage of the waste management cycle. Examples of the active waste include water from trinitrotoluene (TNT) operations and used cyanide solvents. EP toxicity was chosen as a means of identifying waste that can leach from landfills.

Definition of Code of Federal Regulations

The Code of Federal Regulations (CFR) is a document containing all finalized regulations. Federal rules and regulations utilized not only in the control of solid and hazardous waste but also in many other areas of the environment, are typically referred to as CFR, with the number in front and then the parts. This would refer you to a specific place in the CFR for example, 40 CFR, Parts 15 to 22.

Types of Solid Waste

Introduction

The EPA has stated that paper, plastics, and aluminum have been growing as percentages of the total municipal solid waste stream, whereas glass and steel have been declining. In the home, paperboard cartons and plastic containers have replaced glass milk bottles. Aluminum beverage cans have replaced steel, and plastic bottles have replaced glass soft drinks and food jars. Plastic bags in the supermarket are now replacing paper.

Residential Waste

Residential waste is composed of garbage, refuse, ashes, and bulky materials. The garbage comes from food preparation, cooking, serving, handling, and storage. The refuse consists of paper, cartons, boxes, barrels, disposable diapers, clothing, wood, tree branches, yard trimmings, metals, tin cans, dirt, glass, crockery, and minerals. The ashes are residues of fuel and the combustion of solid waste. The bulky wastes consist of such items as wood furniture, bedding, metal furniture, refrigerators, ranges, rubber tires, lead acid batteries, carpets and rugs, and abandoned automobiles. The composition of the waste varies across communities and areas of a city, depending on household practices, use of home incinerators, backyard burning, salvaging, trash smashers, and garbage grinders. The inner-city areas generally produce larger quantities of bulky waste. In the wealthier and suburban areas, one would expect to find more packaging materials and greater quantities of putrescent materials. Generally, the storage of domestic waste is a greater problem in the inner city than in other areas of the community.

Commercial Waste

Commercial waste comes from wholesale and retail trade, hotels, restaurants, offices, and institutions. (The special problems of institutional and research laboratories are discussed separately.) The commercial waste is usually collected by private waste collectors, instead of municipal collectors, although in major cities municipal services may be available. The type of wastes include garbage, refuse, ashes, demolition wastes, urban renewal wastes, wastes from building expressways, construction wastes, and remodeling waste. The refuse is similar to residential waste, but in different quantities. The demolition and construction wastes consist of lumber, pipes, brick, masonry, asphalt, other construction materials, scrap lumber, concrete, and large quantities of soil. (Hazardous wastes are discussed separately.) These wastes also include plaster, plastics, and insulation. Many of the materials should be recycled.

Municipal Waste

Municipal waste consists of street litter; discarded auto bodies; power plant and incinerator ashes and residue; sludge from sewage; dead animals, such as cats, dogs, horses, and fish; and abandoned cars and trucks. This waste also includes fly ash, incinerator residue, boiler slag, metal scraps, shavings, minerals, solids from screening and grit chambers, organic materials, charcoal, and plastic residues.

Institutional and Research Laboratory Waste

Institutions, especially hospitals, have a special problem, because all their solid waste, including garbage, refuse, disposables, dressings, and so forth must be considered contaminated; therefore, all materials are handled in a special manner. Further, the hospital has contaminated waste from patients with known infections or communicable diseases. These wastes receive special treatment to prevent the spread of infection. Hospitals and research laboratories produce a variety of substances, such as tissue from research animals, which are potentially hazardous. Hospitals and research laboratories also produce a considerable volume of hazardous gases and chemicals, and must dispose of the containers in which the chemicals have been stored. Again, special processing techniques are needed. Institutions produce the types of solid waste found in the domestic, commercial, and municipal segments plus specialized contaminated and research waste.

Industrial Waste

Because it is impossible to thoroughly and completely discuss the wastes of each industry, a representative sample has been chosen, and further detailed. The selected industries contribute a substantial quantity of important waste to the pollution problem. Approximately 415 million tons of industrial solid waste are produced annually. They are currently increasing by 3% annually.

Food Waste

Americans throw away about 13 million tons of food waste from their plates each year into garbage cans. The amount of food waste in the waste stream has decreased from 14.9% of the waste stream in 1960 to 10.4% of the waste stream in 2000. The reason for this is that we throw away a lower portion of the edibles than was done several generations ago. Modern packaging, processing, and refrigeration has accounted for this decrease in the waste stream.

Most food waste, like refuse, ends up in landfills. These landfills contribute to the formation of leachate and methane gas. Food waste may be burned in incinerators but their heat content is so low and their moisture content so high that they affect the incinerator's performance.

Yard Waste

Yard waste includes grass clippings, leaves, and tree prunings. These wastes are an important part of the municipal solid waste stream. In 2000, municipal systems collected and disposed of 27 million tons of yard waste. This was more than 13% of the total municipal solid waste for the year. The quantities and types of solid waste vary from community to community and from year to year. In any case, it is material that can be reutilized. More than 50% of the United States now has laws banning yard trimmings from landfills. Over 3300 yard trimming compost facilities are now in operation.

Food Processing

The growing, harvesting, processing, and packaging of fruits, vegetables, and other food crops generate large quantities of solid waste. As the technology of harvesting improves and crops are rapidly harvested, considerably more waste has passed along with the crops to urban areas. This waste must be disposed of within the urban complex. Food product wastes consist of meats, fats, oils, bones, vegetables, fruits, shells, cereals, and assorted packaging and shipping containers.

Metal Waste

Industrial scrap metal is produced at a rate of about 20 million tons a year. It comes from municipal incinerators, refuse dumps, sanitary landfills, abandoned automobiles, warehouses, and various industrial plants. Other industries that produce scrap metal include plumbing; heating and air-conditioning; metal furniture plants; primary metal industries involved in casting, forging, rolling, and forming; fabricated metal plants; electrical and transportation plants; and large varieties of miscellaneous manufacturing plants. Steel can and appliance recycling has increased about 250% since 1988.

Paper

Paper and paperboard are the largest contributors to the waste stream. Paper waste is generated by furniture, metal, paper and allied products, printing and publishing, machinery, transportation equipment, and miscellaneous manufacturing plants.

In 2000, Americans recycled about 42% of the paper produced. Most of our waste paper that is recycled becomes corrugated cardboard. The United States provides more than one half of the world's recycled paper.

Paper products burn readily and account for about 50% of the heat content of an average ton of municipal solid waste. A major, although not completely understood, problem with incineration and recycling of paper is the presence of toxic chemicals found in inks and in other processing chemicals. Certain papermaking processes use chlorine, which might result in the production of small quantities of dioxin. Whereas paper is biodegradable and disappears readily from the environment, if it is compressed in a landfill and kept away from oxygen and water, it degrades very slowly.

Plastics

Plastics are used more frequently in packaging and consumer products. They are also increasing in the municipal solid waste streams and cause potential environmental problems. Once in the waste stream, the plastics are incinerated, buried, or recycled. Plastics can generate almost as much energy as fuel oil. Potential hazardous emissions from incinerated plastics include hydrogen chloride, dioxin, cadmium, and fine particulate matter. Plasticizers may be hazardous substances and have been found in a number of leachate samples at various concentrations. Plastic waste that constitutes about 10% by weight and about 20% by volume of the municipal waste stream is generally nondegradable, therefore occupying a disproportionate amount of landfill space. Plastics recycling prior to putting these materials into the municipal solid waste stream is quite effective. However, plastics recycling within the waste stream is difficult and costly. Plastics may be used in beverage bottles, pipes, cables, film, trash bags, coatings, auto battery cases, screw on caps, food tubs, housewares, electronics, fast-food packaging, food utensils, sporting goods, luggage, auto parts, packaging, etc. Plastics have many advantages, some of which follow:

1. They are lightweight and therefore reduce transport costs substantially.
2. Trucks can carry up to 63% more load with fuel savings.
3. Plastics use less energy to produce than paper, glass, or metal containers.
4. Other packaging materials would require more weight, and more energy.
5. Most plastics can be recycled.
6. Plastic products used in hospitals help reduce rates of infection.

Plastics contribute to litter in community settings. They also contribute to a global waste crisis and are using up scarce petroleum reserves, because they are made from petroleum products. Biodegradable plastics are a recent innovation that seems to be promising. They can be made out of products obtained from raw materials such as starch and cellulose. Plastics made from corn decompose 20 times faster than other plastics. Studies indicate that biodegradable trash bags containing 6% cornstarch and 94% synthetic polymers decompose within 3 to 5 years. Grocery and utility bags made of biodegradable plastics are currently in use in Europe.

Glass

The two economic approaches for managing waste glass are reuse and recycling, and landfilling. Glass is detrimental to the waste incineration process, because it does not burn, is abrasive to equipment, and tends to form slags on the grates. More than 90% of the glass in the municipal solid waste stream is container glass from beverages and foods.

Formerly, refillable glass containers were used, such as milk bottles. These containers were all suited to decentralized bottling industries. Today, because the number of bottlers have dropped from about 3000 to 300, it is more economical to have disposable containers for our liquids. Glass in the solid waste stream is recovered at a rate of 24%, with glass containers recovered at a rate of 37%.

Wood

Wood accounts for about 7% of the waste stream. It comes from furniture and wood packing materials. About 75% of the wood is recovered.

Aluminum

Aluminum is recycled at a rate of about 34%. Aluminum cans alone are recycled at a 65% rate.

Chemical Waste

The chemical industry is one of the most complex. Its output has increased to many thousands of commercially available chemicals. In addition, this industry produces plastics that are both valuable and problematic for disposal. The chemical industry generates hazardous unusable or unsalable by-products that must also be discarded. The chemical waste is highly heterogeneous. It is estimated that over 22,700 facilities released 2.25 billion lb of listed toxic chemicals into the environment, with 69% in air emissions, 15% into injection wells, 13% in landfills and other land disposal, and 3% in water. For recycling, energy recovery, treatment, and disposal, 3.8 billion lb of toxic chemicals were sent off-site. The wastes include organic and inorganic chemicals, metals, plastics, rubber, glass, oils, paints, solvents, pigments, drugs, explosives, asphalt, tars, asbestos, paper, cloth, and fiber.

Rubber

Rubber tires are discarded by Americans at a rate of one tire per person per year. The rate of retreading has dropped from about 20% in 1960s to about 5% currently. Tires are a serious problem for waste managers. Vulcanized rubber cannot be melted down and recast. All tires are difficult to bury in landfills, because they work their way to the top of the piles. Currently, recovery of rubber from tires is about 15% of production.

Tires are ideal breeding grounds for mosquitoes and other disease vectors. Tires hold water and protect the insects from pesticide sprays. The Centers for Disease Control and Prevention (CDC) determined that the Asian tiger mosquito, which carries dengue fever, has entered the United States in used tires and is rapidly spreading. The lacrosse and culex mosquitoes, which carry encephalitis, are also moving across the United States in used tires. Used tires end up in tire dumps, which not only pose a health hazard from mosquitoes or other insects and rodents but also pose a significant potential fire hazard, perhaps resulting in high levels of air pollution.

Radioactive Waste

Two basic types of radioactive waste are low level and high level. Adequate treatment and disposal of these materials are essential in preventing contamination of the environment with radioactivity. These materials are in gaseous, liquid, and solid waste forms. They come from reactors, other nuclear facilities, by-products of fuel production, and radionuclide waste used in healthcare and industry. Low-level solid wastes are packaged and then shipped for storage to area sites owned by or licensed by the Energy and Research Development Authority of the United States. Approximately 8 million m^3 of low-level waste is stockpiled in the United States.

High-level waste is produced after the used fuel is removed from reactors. In the past these wastes were packaged and shipped in solid form to reprocessing plants. After reprocessing, they were concentrated and stored in tanks. The liquid wastes, an estimated 80 million gal, were solidified to reduce the amount of necessary storage space. Today the disposal of radioactive wastes is a unique and necessary part of the nuclear fuel cycle. Due to the toxicity of these wastes, they must be isolated. High-level radioactive wastes generate significant amounts of heat and can be extremely troublesome when stored. It has been proposed that this heat be utilized as an energy source, and that some of the waste be converted to radioactive isotopes for medical purposes. Various disposal methods have been suggested, including storage in underground salt formations, further chemical separation of the waste to reduce necessary surveillance time, extraterrestrial disposal, burial in Antarctic rocks, and burial in the interim near the surface storage in mausoleum-like structures. The last proposal was widely criticized. Additional research on the disposal of radioactive waste must be conducted to prevent our generation and future generations from becoming contaminated.

Mining Waste

During the past 30 years, the mining industries have produced over 25 billion tons of mineral solid waste. The task of handling this huge quantity of waste is increasing sharply. The mining industries are estimating an additional 2 billion tons of solid waste will be produced annually. If ocean and oil shale mining become major industries, approximately 4 billion tons of waste will be generated annually. Although about 80 mineral industries generate wastes, 8 are responsible for about 80% of the total. The copper industry is the largest contributor, followed by the iron, steel, bituminous coal, phosphate rock, lead, zinc, alumina, and anthracite industries. Most of the remaining 20% result from smelting; nonmetallic mineral mining, including sand and gravel; gold dredging; stone and clay; and chemical processing of ores and products. Mining wastes are largely rocks and submarginal grade ore from open pit or surface mining. These huge waste accumulations, often hundreds of feet high, cover extensive land areas. They degrade the natural environment and create serious hazards. Culm banks and coal waste piles are residues of processing coal. They contain high percentages of carbonaceous matter. Rock and mud slides from these towering piles threaten the life and property of towns within the vicinity of the mining areas.

Agricultural Waste

The primary agricultural and forestry wastes produced in the United States are animal manures, vineyard and orchard prunings, crop harvesting residues, animal carcasses, greenhouse wastes, and pesticide containers. Gross estimates of agricultural wastes suggest that over a billion tons per year are produced. In addition, domestic animals produce over 1.5 billion tons of fecal material yearly and over 600 million tons of liquid waste. Used bedding, dead carcasses, and so forth add an additional 800 to 900 million tons of waste annually. Today agricultural waste production is five to six times greater than domestic and industrial waste combined. The problems of storage and disposal are enormous. In the past, most agricultural waste materials were recycled by spreading them on the land. This simple solution, however, is no longer available because the continued economic squeeze on farmers has resulted in special-ization and concentration of production units; the increase in human population requires more food and fiber; the continuous urbanization of previous agricultural areas has reduced the supply of arable land; and the availability of inexpensive chemical fertilizers has decreased the demand for plant and animal wastes as fertilizers.

Many of the plant residues are recycled by spreading and working back into the land. Some are used in livestock feed for roughage and carbohydrate-rich nutrients. However, the burning of stalks, stems, and stubble causes air pollution, and the decomposing of fruits and vegetables causes tremendous fly and odor problems. Animal feces and wastes are fly attractants, cause considerable odor problems, and may also be rodent attractants. A serious problem is created when these wastes are fed into sewage systems, because they raise the biological oxygen demand (BOD) to an unusually high level.

The production wastes of cattle consist of manure and runoff, which are primary and secondary wastes. The primary wastes include solids, nutrients, organics, odor, and dust. The secondary wastes include pesticides, dead animals, afterbirths, litter and bedding, spilled feed, and wash water.

Recreational Waste

Recreational solid waste is a growing concern in the United States. The once unspoiled beauty of state and national parks is systematically obliterated by many millions of individuals who would not think of littering their own living rooms but think nothing of littering recreational areas. Compounding this problem is a sharp increase in the number of visitors to recreational areas. The overall demand for outdoor recreation will increase sharply in the years to come. At present, hundreds of millions of individuals spend some time in state or national parks, and other outdoor recreational facilities. Numerous other problems exist, beside the sheer numbers of careless people. These include the distance from recreational solid waste storage areas to disposal sites, the techniques of disposal, the amount of available facilities, and the varying uses of the facilities made by visitors. The average rate of solid waste generated at campgrounds exceeds 1 lb/day for each visitor. The types of waste vary with visitors. Campers burn or partially burn food waste, and are, of course, more prone to use disposable than reusable dishes and silverware. Animals are attracted to, and scatter, the wastes when the wastes are not removed immediately.

Abandoned Automobiles

As an integral part of society, the automobile contributes not only to our advance-ment but also to a vast number of problems, ranging from crime to environmental pollutants. More than 15 million vehicles are sold annually in the United States. As the economy improves and our population grows, this number may increase accord-ingly. Unfortunately, when the automobile no longer functions, it is discarded. The discarded portion may be partially reclaimed, but the remains become a solid waste problem. Frequently, instead of paying to have a car towed to a junkyard or to have it disposed of properly, individuals simply remove the license plates and abandon the cars. Usable parts, such as tires and radios, are removed and the carcass is left behind. Several industries are involved in the removal and recycling of automobiles. These include the dismantler, scrap processor, and steel mill. Unfortunately, these industries are unrelated, and as a result recycling is not a smooth efficient operation.

The use of steel in cars is static or declining, whereas the use of nonferrous materials, primarily in options, is increasing. This material consists of plastic and rubberlike materials. A typical automobile weighing 3574 lb contains 1309 lb of light steel, 1222 lb of heavy steel, 511 lb of cast iron, 31 lb of copper, 54 lb of zinc, 50 lb of aluminum, 20 lb of lead, 145 lb of rubber, 87 lb of glass, 127 lb of other combustibles, and 15 lb of other noncombustibles. With the onset of the use of fiberglass bodies, there has been a sharp reduction in the use of reusable steel and a sharp increase in nonreusable fiberglass.

The typical dismantler removes the easily marketed parts (such as the radiator, battery, motor, and related parts), and sells the remaining hulk. Unfortunately, the remaining hulk still contains other contaminants, including copper. These other contaminants pose problems for use of the hulk in steel production. If the reusers or the scrap purchasers are unable to utilize the hulks in the steelmaking process, as time goes on, the number of hulks piling up in junkyards across the country may increase sharply.

No good nationwide calculations are available for the number of abandoned automobiles in our country; however, it is estimated that millions of vehicles are discarded, with a significant number abandoned. It is difficult to prove the validity of this estimate. Much depends on the economy, the nature of the cars, the individuals involved, and the various dealers and insurance companies. The four basic types of automobile disposal are dismantlers, processors, solid waste disposal sites, and (unacceptable) public or private property.

Packaging Materials

Packaging materials are extremely visible to the American public, because they are discarded everywhere. In fact, they comprise much of the litter found in our country. A small amount of packaging materials is recycled; the rest becomes solid waste. The amount of packaging material is increasing rapidly. The many reasons for this great increase in packaging material include increased use of self-service merchandising in food stores and drug stores. Packaging also helps sell products; as a result, the manufacturers and wholesalers are continually trying to improve the type and amount of packaging appeal. The basic materials used in packaging include paper, glass, metals, wood, plastics, and textiles. Paper and paperboard accounted for about 50% of the packaging materials. Plastic accounted for an enormous amount of packaging by volume, if not as much by weight. The problems of removal are sharply intensified by the increased volume. The number of collections has sharply increased to accommodate the additional volume, and space must be provided for disposal of this material. Beverage containers are a problem within the packaging area. Although some states have passed bills requiring the return of beverage containers, the problem has decreased, but has not been alleviated. Containers are especially used for beer and soft drinks.

Refuse-Derived Fuels

Refuse-derived fuel (RDF) systems are practical commercially and are a proven alternative to mass burning of municipal waste. These systems use the fuel properties of the material to aid in the burning of material and in the production of heat, which can be turned into steam or hot water. The RDF varies considerably with the type of waste. In any case, it is necessary to separate out the types of materials that are not readily combustible such as metal, glass, other trapped inorganics, and oversized bulky waste. The RDF can then be processed in several different ways and utilized to generate additional energy.

Heavy Metals

Heavy metals tend to be toxic to living organisms in relatively low concentrations. Toxic heavy metals include mercury, cadmium, lead, and arsenic, among others. Trace metals are circulating through the biosphere at increasing rates because of human activity. Incineration and landfilling of refuse is an important source of cadmium, mercury, lead, vanadium, copper, and zinc. The potential for circulation of toxic metals from the soil, water, and air through the food chain is an important environmental issue.

Because lead pollution from gasoline has decreased very sharply, one of the major contributors of lead to the environment is in the solid waste stream. Automobile lead–acid storage batteries are the biggest source of lead in the municipal waste stream. About 94% of the 90 million plus batteries are recycled each year. The rest are in temporary storage or in the waste stream. The market for reclaimed lead is extremely volatile. Once again, economics and the environment interact to determine whether substantial amounts of a pollutant are released into the environment.

Toxic waste regulations have reduced the recycling of refrigerators, stoves, and other appliances, as well as engines developed before 1980. These appliances and engines frequently contain PCBs and other chemicals classified as toxic.

Increasing amounts of cadmium are found in the waste stream because of the disposal of nickel–cadmium rechargeable batteries. These batteries are used in portable consumer products such as rechargeable flashlights, hand vacuums, food mixers, and screw drivers. When the battery finally fails, the entire device, plastic case, battery, and other parts, are frequently tossed into the trash.

Toxic Releases

A toxic release is a discharge of a toxic chemical into the environment. Releases include emissions to the air, discharges to bodies of water, releases at the facility onto the land, releases at other facilities or during transportation, and underground injection wells. Air emissions are either from a point source or fugitive sources. Point source emissions, or stack emissions come from confined airstreams such as stacks, vents, ducts, or pipes. Fugitive emissions are all releases to the air not from a confined airstream, such as equipment leaks, evaporated losses from surface impoundments and spills, and releases from building ventilation systems. Releases to water include discharges to streams, rivers, lakes, oceans, or other bodies of water. The releases come from industrial process outflow pipes, trenches, runoff, including storm water.

On-site land releases include disposal of toxic chemicals in landfills in which the wastes are buried; and land treatment and application farming, in which a waste containing the chemical is incorporated into the soil, spills, or leaks. Transfers to off-site disposal can result in releases during the transportation, onto the land, or during the injection well process.

Underground injection is the subsurface placement of fluids by means of wells. These chemicals come from manufacturing, petroleum industry, mining, commercial and service industries, and federal and municipal government activities. They must not endanger underground sources of drinking water, public health, or the environment. The different types of authorized injection wells are as follows:

1. Class 1 includes industrial, municipal and manufacturing wells that inject fluids into deep, confined and isolated formations below potable water supplies.
2. Class 2 includes oil and gas related wells that reinject produced fluids for disposal, enhanced recovery of oil, or hydrocarbon storage.
3. Class 3 includes wells associated with the solution mining of minerals.
4. Class 4 includes wells that may inject hazardous or radioactive fluids directly or indirectly into underground sources of drinking water, if the injection is part of an authorized Comprehensive Environmental Response, Compensation, and Liability Act (CERCLA) and RCRA cleanup operation.
5. Class 5 includes wells that are generally shallow drainage wells, such as floor drains connected to dry wells or drain fields. Injection can only occur if it does not endanger underground sources of drinking water, public health, or the environment.

Hazardous waste generators are individuals that produce hazardous waste, usually from an industrial process. Transporters are individuals or companies that move the hazardous waste. In either case they may be responsible for toxic wastes releases. They must follow the Uniform Hazardous Waste Manifest, which is a form used to track the movement of hazardous waste from the point of generation to the ultimate point of disposal.

A special type of toxic release occurs during wars. Terrorists, various factions, and people involved in sabotage use chemicals found in industrial nations to create improvised explosives, incendiaries, and chemical agents. The Agency for Toxic Substances and Disease Registry has developed a ten-step procedure to analyze, mitigate, and prevent public health hazards caused by terrorism involving the use of industrial chemicals.

Residues from waste management facilities may include a variety of contaminants, which may be potentially hazardous. These residues come from incinerators, material processing facilities, and compost facilities. Typically they are placed in a landfill.

Pollution Prevention and Waste Minimization

Pollution prevention and waste minimization is part of the four major steps of waste management. Waste minimization includes source reduction and recycling. Pollution prevention is a combination of source reduction and protection of natural resources through conservation or increased efficiency in the use of energy, water, or other materials. It focuses on source reduction instead of treatment and disposal. Source reduction is the reduction of the amount of waste at the source through

changes in industrial processes such as cost-effective changes in production, operation, and raw materials use. Recycling is the reuse and recycling of waste for the
original or some other purpose, such as materials recovery or energy production.
Waste management includes incineration treatment, which is the destruction and
detoxification, as well as neutralization of waste into less harmful substances. The
fourth type of waste management is secure land disposal, where deposits of waste
are placed in land using volume reduction; encapsulation; leachate containment;
monitoring; and control of air, surface, and subsurface waste releases.

Waste minimization has been successful for many organizations. These organizations save money by reducing waste treatment and disposal costs, raw material
purchases, and other operating costs. These groups meet state and national waste
minimization policy goals and reduce potential environmental liabilities. They protect public health and worker health and safety, as well as the environment.

Waste minimization is a policy specifically mandated by the U.S. Congress in
the 1984 Hazardous and Solid Wastes Amendments to the RCRA. This mandate,
along with other RCRA provisions, has led to an unprecedented increase in the cost
of waste management, which has caused industries to consider ways of minimizing
waste and controlling their costs as well as profits. Further, these industries wanted
to reduce their environmental impairment liabilities under the provisions of CERCLA,
or Superfund.

The 1990 Pollution Prevention Act has enhanced the 1984 law. This minimization
is the reduction, to the extent feasible, of hazardous waste as generated or subsequently treated, stored, or discarded. This policy also includes any source reduction
or recycling activity undertaken by a generator of waste that results in the reduction
of total volume or quantity of hazardous waste, the reduction of toxicity of the
hazardous waste, or both, as long as the reduction is consistent with the goal of
minimizing the present and future threats to human health and the environment.

Source reduction is the activity that reduces or eliminates the generation of
hazardous waste at the source, usually within a process. Recycling occurs when a
material is used, reused, or reclaimed. A material is used or reused if it is either
employed as an ingredient to make a product or employed in a particular function
as an effective substitute for a commercial product. A material is reclaimed if it is
processed to recover a useful product or if it is regenerated. An example of reclaiming
would be the recovery of lead from spent batteries and the regeneration of spent
solvents.

Source reduction consists of several steps. First would be good housekeeping
techniques including waste stream segregation, inventory control, employee control,
spill and leak prevention, and scheduling improvement. Then make product changes,
which include product substitution, product conservation, and change in product
composition or package design. Redesigning products or packages reduces the
quantity of materials or the toxicity of the material, and also helps reduce energy
costs. Source control consists of changing the input material, changing the technology, and permitting good operating practices. Changes in input material include
material purification and material substitution. Changes in technology include process changes, equipment, piping or layout changes, additional automation, and
changes in operational settings. Good operating practices include procedural measures,

loss prevention, good management practices, waste stream segregation, material-handling improvements, production scheduling, and reuse of products and packages. Recycling both on-site and off-site includes the use and reuse of materials and reclamation. The use and reuse of materials include returning to the original process and using a raw material substitute for another process. Reclamation includes processes established for resource recovery and processes used to utilize the by-product. Glass bottles and wood pallets are good examples of reusable materials.

To establish a waste minimization program, it is necessary to develop planning and organization, assessment, and feasibility analysis phases, as well as implementation and evaluation. The planning and organizing phase includes the establishment of the goals, objectives, procedures, techniques of program operation, and the actual operation and evaluation of the program. The assessment phase includes the collecting of process and facility data, the prioritizing and selecting of assessment targets, the selecting of people for assessment teams, the review of data and inspecting of the site, the generation of options, and the screening and selection of options for further study. The feasibility analysis phase includes the technical and economic evaluations, and the evaluation of select options for implementation. The implementation includes justification of projects and obtaining funding, installation of equipment, implementation procedure, and evaluation procedure. The evaluation includes all the previous parts of the process mentioned, whether there has actually been a minimization of waste and how much.

The assessment phase of waste minimization is of great importance. It is necessary to collect and compile data. To do this, the following information must be gathered:

1. What are the waste streams generated from the plant, and how much?
2. Which processes or operations do these waste streams come from?
3. Which wastes are classified as hazardous and which are not? What makes them hazardous?
4. What are the input materials used that generate the waste streams of a particular process or plant area?
5. How much of a particular input material enters each waste stream?
6. How much of a raw material can be accounted for through fugitive losses?
7. How efficient is the process?
8. Is unnecessary waste generated by mixing otherwise recyclable hazardous waste with other processed waste?
9. What types of housekeeping practices are used to limit the quantity of waste generated?
10. What types of process controls are used to improve process efficiency?

The gathering of waste management assessment information for a facility is an excellent example of how the environmental health practitioner should approach all types of complex environmental situations and issues. It is necessary to gather design information, environmental information, raw material and production information, economic information, information on tracking of waste, information on prioritizing waste streams, and any other pertinent information. Design information includes process flow diagrams; material and heat balances, production processes, or pollution control processes; operating manuals and process description; equipment lists; equipment specifications and data sheets; piping and instrument diagrams; plot and elevation plans; equipment layouts and work flow diagrams. Environmental information includes hazardous waste manifests; emission inventories; biennial hazardous waste reports; waste analyses; environmental audit reports; and permits or permanent applications.

Raw material and production information includes product composition in batch sheets; material application diagrams; material safety data sheets; product and raw material inventory records; operator data logs; operating procedures; and production schedules. Economic information includes waste treatment and disposal costs; product, utility, and raw material costs; operating and maintenance cost; and departmental cost of accounting reports. The tracking of waste information includes the measurement of waste mass flows and compositions. Seasonal variations and waste flows for single, large waste streams need to be distinguished from the continual and constant flows. The prioritizing of waste streams and of operations helps determine which parts of the operation are the most important problems and which ones can be handled the most readily. The other information includes company environmental policy statements, standards and procedures, and organizational charts.

Waste minimization can be brought about by the use of good equipment-cleaning procedures, good operating practices, appropriate training, material and waste tracking and inventory control, loss prevention programs, and reduction of spills and leaks. The cleaning operations should be established for each of the materials and the pieces of equipment that need to be serviced. Information on how to do this in an appropriate manner may be available through the EPA or through the company's professional organization. Operating practices include waste minimization assessments, environmental audits and review, loss prevention programs, waste segregation, and preventive maintenance programs. Waste minimization assessments are carried out by a team of qualified individuals who could determine the problems and potential techniques for resolving the problems. These individuals assemble pertinent documents, conduct environmental process reviews, and carry out instructions as part of the environmental audits and reviews.

Loss prevention programs need to be established to prepare plans for spill prevention control and countermeasures when the spill occurs. The program should also include hazard assessment in the design and operating phases of the industry. Waste segregation includes the prevention of the mixing of hazardous waste with nonhazardous waste, the isolation of hazardous waste by containment, and the isolation of liquid waste from solid waste. A preventive maintenance program would include a master preventive maintenance schedule and reporting system, as well as necessary manuals and lists of individuals or companies that service the equipment.

Training is essential for continuation of minimizing waste as a result of good operating practices. Proper training is needed for safe operation of the equipment, proper materials handling, understanding the economic and environmental ramifications of hazardous waste generation and disposal, detection of releases of hazardous material, establishment of emergency procedures, and use of safety gear. Effective supervision is the key to the operation of all programs in the environmental health field. Effective supervision can be taught and geared toward any given situation.

Indiana State University, through its Continuing Education and Special Services Department (phone no. 800-234-1639), provides courses in supervision and management for individuals in a variety of program areas. These programs are tailor-made to the group and are presented through workshops and through home study modules. Employee participation in programs is essential, because employees and their supervisors are most readily able to determine where waste occurs and how to reduce it. Production scheduling and planning is part of the supervisory process.

To reduce materials and waste, it is essential to provide adequate inventory control and waste tracking. Waste reduction and the saving of substantial sums of money can be brought about by the following: avoid overpurchasing, accept raw materials only after inspection, ensure that inventory does not go to waste, ensure that no containers stay in inventory longer than a specified period, review material procurement specifications, return expired material to suppliers, validate shelf-life expiration dates, test outdated materials for effectiveness, conduct frequent inventory checks, use computer-assisted plant inventory systems, conduct periodic materials checks, determine whether containers are properly labeled, and set up controls on the dispensing of chemicals and collecting of waste.

Loss prevention programs can be very effective if the following techniques are used:

1. Use properly designed tanks and vessels only for their intended purposes.
2. Install overflow alarms for all tanks and vessels.
3. Maintain the physical integrity of all tanks and vessels.
4. Set up written procedures for all loading and unloading as well as transfer operations.
5. Set up secondary containment areas.
6. Forbid operators to bypass interlocks, alarms, or significantly alter preestablished points in the system without authorization.
7. Isolate equipment or process lines that leak or are not in service.
8. Use seal-less pumps.
9. Use bellows-seal valves.
10. Document all spillage.
11. Perform overall material balances and estimate the quantity and dollar value of all losses.
12. Use floating-roof tanks for VOC control.
13. Use conservation vents on fixed roof tanks.
14. Use vapor recovery systems.

Finally, spills and leaks are essential parts of waste and, therefore, essential parts of waste reduction. To reduce spills and leaks, the following should be completed:

1. Store containers in such a way as to allow for visual inspection for corrosion and leaks.
2. Stack containers in a way to minimize the chance of tipping or breaking.
3. Prevent concrete sweating by raising the drum off the storage floor.
4. Maintain material safety data sheets for use in correctly handling spill situations.
5. Provide adequate lighting in the storage area.
6. Maintain a clean, even surface in transportation areas.
7. Keep aisles clear of obstruction.

8. Maintain a distance between incompatible chemicals.
9. Maintain a distance between different types of chemicals to prevent cross contamination.
10. Avoid stacking containers against process equipment.
11. Follow manufacturer's suggestions on the storage and handling of all raw materials.
12. Check the insulation and inspect the electrical circuitry for corrosion and potential sparking.

Waste reduction can also be carried out through the development and use of products requiring less material per unit of product. An example of this would be to develop smaller automobiles and also to use thinner walled containers. It may be feasible to develop and use products with longer lifetimes. Substitute reusable products for single-use disposal products.

Waste reduction can also be brought about by substituting less hazardous chemicals for more hazardous chemicals. Use product substitution with less hazardous products made. Consider changes in production in chemical processes that result in less chemical use or safer chemical use. Improve production management, including the reduction of spills, leaks, vaporization, reduced mixing errors, and bad production runs. Reclaim, recycle, and reuse chemicals. Treat and detoxify chemicals at the point of generation. Improve waste management through better storage and treatment processes.

Waste minimization has been shown to include the provision of proper information, management, maintenance, design, and planning. The waste minimization techniques included source reduction, recycling, and treatment. The treatment phase includes neutralization, precipitation, filtration, evaporation, and incineration.

Municipal Solid Waste Reduction

Yard trimmings make up about 28% of municipal solid wastes. Grass clippings can be left on lawns through the use of lawn mulchers, or as compost. Recyclables can be brought to recycling centers or picked up at the curb after sorting. Buyback centers purchase scrap metal, aluminum cans, paper, etc. Used materials, appliances, clothing, and other items can be sold at garage sales and reused by others. Paper and paperboard products make up 39% of total waste, much of which can be recycled to make other products, or be used.

Hazardous Waste and Resource Conservation Recovery Act

Emissions from municipal solid waste incinerators must meet the requirements of the Municipal Waste Combustion Standards, Title 40, part 60.3 subparts and part 62. Municipal solid waste is defined as refuse that has more than 50% of municipal-type waste, such as a mixture of paper, wood, yard waste, plastics, leather, rubber, and other combustibles; and noncombustibles, such as glass and rock. The air emissions from the incinerator must not exceed a rate of 0.012 g of particulate per dry standard cubic foot of exhaust gas. Municipal incinerators are also covered under the subpart of the standard. This portion establishes the emissions from incinerators burning sewage sludge generated by sewage treatment facilities. Particulate emission must not exceed 1.3 lb per dry ton of sludge charged. Visible emissions must not exceed 10% opacity.

The RCRA defines a waste as a hazardous waste according to its ignitability, corrosivity, reactivity, extraction procedure toxicity, acute hazardous waste, or toxicity. The RCRA is the comprehensive authority for managing all aspects of hazardous waste, including generation, transportation, and facilities for treatment, storage, or disposal.

Ignitability occurs when a liquid has a flash point less than 140°F or is a substance other than a liquid that can cause a fire through friction, adsorption of moisture, spontaneous chemical changes under standard temperature, and pressures of certain flammable solids and explosive gases. Corrosivity occurs when an aqueous waste has a pH of ≤2.0 or ≥12.5. A liquid that corrodes carbon steel at a rate greater than 0.250 in./year is corrosive. A substance that is normally unstable and undergoes violent physical or chemical change without detonating is a reactive substance. Reactivity also means a substance reacts violently with water and can generate harmful gases, vapors, or fumes when mixed with water. This also is a substance that is readily capable of detonation at standard temperatures and pressures. Extraction procedure toxicity refers to the extract or leachate from a representative sample of the waste. If this sample contains contamination in excess of ten times that which is allowed in drinking water, the contaminant is considered to represent a toxic level. An acute hazardous waste is a substance that has been found to be fatal to humans in low doses or may be fatal to laboratory animals in corresponding human concentrations. Toxicity refers to waste that has been found to be carcinogenic, mutagenic, or tetratogenic through laboratory studies.

The hazardous waste regulations require that the principle organic hazardous constituent (POHC) of a waste be defined. Where waste contains one or more of the hazardous constituents, it is necessary for the EPA to determine which of these components is the principal one. After the POHC is identified, the incinerator must be designed and operated to achieve a destruction and removal efficiency of at least 99.99%. If the hazardous waste generates an excess of 4 lb/hr of hydrogen chloride in the exhaust gas stream of the incinerators, 99% of the HCL produced must be removed prior to discharge into the atmosphere. Waste containing PCBs in a concentration greater than 50 ppm by weight is covered under the Toxic Substances Control Act (TSCA).

This act establishes standards for incinerators that include:

1. The incinerator must operate at 99.90% combustion efficiency or better.
2. Scrubbers are required to remove hydrochloric acid from the furnace exit gases.
3. Continuous monitoring of furnace temperature, stack oxygen, and carbon monoxide, as well as the polychlorinated biphenyl (PCB) feed rate must be done.
4. An automatic shutoff of feed supply is needed in case monitoring equipment fails.
5. Incinerators burning solid PCB waste must have a 99.999% destruction and removal efficiency. The TSCA regulations state that a furnace must operate at 2200°F. The gas must be present in the furnace for at least 2 sec and the gas composition must include at least 3% oxygen.

Chlorine may be a problem in municipal solid waste and definitely is a problem in hazardous waste. Chlorine contributes to the formation of acid gases (HCl) and chlorinated species of the dioxin and furan classes of compounds. An analysis of the components of municipal solid waste and the total amount of urban waste shows quantities of chlorine present. The chlorine may come from both paper and plastics in the waste stream of the community.

Dioxins have a molecular framework consisting of two benzene rings connected by two oxygen bridges. The dioxin molecular framework is also the basic framework for polychlorinated dibenzofurans and PCBs. The dioxins may be produced in the incineration process. Further furans may be produced by the incomplete combustion of PCBs and dioxins. Furans are thought to be potentially carcinogenic.

Heavy metals are contained in all fractions of the waste stream. Lead may enter the waste stream in the pigments used to color paper and plastics. Pigments, inks, enamels, paints, and lacquers may be responsible for over 50 to 70% of the chromium, 50% of the cadmium, 60% of the cobalt, 20% of the nickel, and a significant portion of the zinc and copper found in the waste stream.

Household refuse contains measurable quantities of toxic and hazardous waste, such as caustics, paint thinners, solvents, turpentine, oil-based paints, copper cleaners, silver polish, oven cleaners, as well as a wide range of solvents, pesticides, and other cleaning materials.

Emission Standards for Hazardous Air Pollutants — 1999

Section 112 of the Clean Air Act requires emissions standards for hazardous air pollutants to be based on the performance of the maximum achievable control technology (MACT). Hazardous waste combustors contain these three categories: hazardous waste burning incinerators, hazardous waste burning cement kilns, and hazardous waste burning lightweight aggregate kilns. Emission standards have been set for chlorinated dioxins and furans; particulate matter, to include antimony, cobalt, manganese, nickel, and selenium; semivolatile metals such as lead and cadmium; low volatile metals such as arsenic, beryllium, and chromium; hydrogen chloride and chlorine gas combined; carbon monoxide and hydrocarbons; and nondioxin and furan organic hazardous air pollutants. For dioxins, a carcinogenic slope factor of natural log 1.56e-4 per picogram per kilogram of body weight each day is still in use. For mercury, as methylmercury the standard is 0.1 μm/kg of body weight per day. For lead, the standard is below 10 μg/dl of blood. For other metals, the standards for occupational exposure are the only ones that are available this moment. For hydrogen chloride, the standard is 0.02 mg/m^3 of air. For chlorine gas, the standard is 0.001 mg/m^3 of air.

Solid Waste Storage Systems

Solid waste storage techniques are necessary to protect the health and safety of the public. Without them, insects and rodents, odors, fires, and accidents would sharply increase. Proper storage is needed to expedite the collection of the solid waste and to reduce cost. The responsibility for storage belongs to everyone — individuals, communities, agriculturalists, and industries. Individuals must decide that they want to live in a litter-free, disease-free, and accident-free environment.

On-Site Volume Reduction Systems

In the on-site volume reduction system, mechanical compacting, incineration, pulping, or composting is utilized to reduce the volume of the solid waste before it is collected and finally discarded. On-site reduction eliminates or reduces food and harborage for a variety of insects and rodents; it decreases the quantity of refuse for storage and removal, and also reduces the chance of accidents and fires. Economically, it lowers the cost of refuse handling during the collection and disposal phases of solid waste management.

Composting has considerable advantages, because the waste is reduced to about one quarter of its original volume. Compost is also used as soil conditioner. Garbage grinders are a very effective technique for reducing the volume of the solid waste to about 10% of the original volume. These grinders are convenient, reduce the length of storage time, and also eliminate home garbage cans. On-site incinerators reduce the volume of the solid waste and the number of storage facilities needed.

In pulping, paper wastes are ground in water and squeezed semidry. The advantages of this system are a fast means of reducing and handling wastes. Units capable of 80% volume reduction are suitable for high-rise developments, and more than one unit can be installed within a large building.

Material may be compacted on-site mechanically. In this case, the paper sacks or containers are compressed in a piece of mechanical equipment. The advantages of this system include reduction of waste by about 75%, reduction of space needed for waste storage, elimination of trash barrels, quieter operation, elimination of on-site incineration, reduction of personnel in the movement of solid waste, and reduction of handling and hauling costs.

In any event, on-site volume reduction should definitely be considered when new high-rise buildings are planned. Adequate research should be conducted to determine which of the available on-site reduction systems is most efficient, least costly, and best able to be utilized for reduction of the solid waste pollution problem.

Solid Waste Collection

The collection process accounts for about 75% of the better than $10 billion annual cost of solid waste management. About 170,000 workers in public and private enterprise collect and dispose of solid wastes. This is a large business, which if improperly run creates tremendous hazards to the population of the country. When the collection system is operated by government, it comes out of tax revenues. When it is operated by private contractors, it is paid for in the same way as any other service. The major cost is labor. Unfortunately, labor is generally unskilled and the collection techniques are still the same as they have been for many years. Injury rates are high, working conditions are poor, and opportunity for advancement is slim. As new and more dangerous solid wastes are produced and as the quantities increase, there will be a definite need for better evaluation of existing technology, financing, workforce utilization, equipment selection and maintenance, and routing to provide the best possible service in the safest way for the least number of dollars.

Individual and Community

To establish a proper solid waste collection system and to understand the workings of such a system, one must recognize the fixed and variable factors affecting the system. The fixed factors include climate, topography, physical layout of areas, population density, and types of solid waste produced. Routing of the vehicles must be based on topography, and equipment may be affected by the nature of the land. The physical layout of the area includes alleys, one-way streets, and dead-end streets. Population density affects solid waste collection. In low-density areas, the distance between pickups may be great, whereas in high-density areas, the pickup vehicle can only do a certain amount of collection before it must leave for the disposal site. Time must be provided for the transportation of the collected material from the area to the disposal site. The types of solid waste produced include yard and garden waste, bulky waste, hazardous waste, packing materials, and garbage.

The variable factors in collection include type of storage, salvaging of materials, on-site disposal, frequency of collection, crew organization, collection equipment, and political factors. Solid waste may be stored in 30-gal cans, paper sacks, plastic sacks, or large metal containers. The weight of the containers, and the number, size, shape, type, location, and accessibility all affect the collection procedure. Where on-site disposal occurs, breakdown of equipment, inadequate incineration, or backyard burning must be taken into account. The frequency of collection may be once a week, twice a week, or more often. The crew may consist of one, two, or more individuals. Collection equipment varies in size and type. The political factors involved in collection vary with special interest groups, nature of the neighborhoods, specialized programs, and whether the operation is a government or private enterprise. Additional variable factors include who is responsible for disposal, disposal techniques, who is responsible for the disposal sites, type of material handled and collection frequency, mix of municipal contracts and private collection agencies, and location of the various disposal sites.

Municipal departments of sanitation or solid waste collection are operated under the auspices of the city government. The employees are on the public payroll and the equipment is under the direction of a department official. The advantages of this type of system are proper sanitation and direct accounting and responsibility to the public. An additional advantage is that in those areas where snow is extensive, these workers can assist in snow removal. The disadvantages of the municipal department include the potential for political influence, frequent turnover of crew and supervisory staff, operation of a special service by untrained officials, and emphasis on reduced expenses instead of efficiency.

The contract method of collecting solid waste is utilized by cities who find contracting less expensive or are unequipped or unwilling to handle their own solid waste problem. Solid waste collection is made on a bid basis. The lowest bidder gets the contract. The advantages of a contract are that the service is run as a business, potential political influence is reduced, governmental part of the service is reduced substantially, cost is established in advance, contractors must raise the capital for equipment, and governmental units can set and enforce proper collection standards. The disadvantages are that the profit motive instead of cleanliness may be emphasized, contracting allows for less flexibility, contracts may be broken and an alternate service may not be available immediately, contractors are not necessarily assured of the contract renewal, and in many cases contracts are given out on a yearly basis; a tremendous potential exists for lack of continuity and increased political influence.

Private collection is performed by individuals or companies who make contractual agreements with individuals, businesses, or industry. The advantage is that service is offered where no service is ordinarily available and more frequent service can be contracted for. The disadvantages are an expensive overlapping of routes and too much competition, which reduces the efficiency of service.

Under any conditions, solid waste removal is basically a function of government, and therefore government must prepare and enforce all necessary rules, regulations, and codes to protect the health and safety of the public. In the municipally run operation, the key individuals are the supervisor and the environmental health practitioner. In the private operation, the key individuals are the private owners of the companies and the environmental health practitioner.

The selection of the equipment to be used for collecting solid waste and the collection system followed are important for proper operation and low-cost maintenance. Varying types of equipment and systems of collection are obtainable. Several are discussed to provide some brief examples of techniques available.

Manually loaded compacting bodies are utilized in many areas. The vehicle has the advantage of carrying a considerable load, because the material is compacted and can be loaded readily, and the loads are easily emptied at the disposal site. They are leakproof, easily cleaned, and have many safety advantages, such as traffic signals, mirrors, handholds, steps, and emergency stop bars. The crew for a 16 to 24 yd³ truck consists of a driver and one or two helpers. Occasionally incorporated into this system is the satellite vehicle system that is discussed a little further on.

The mechanically loaded bodies are almost all detachable container systems. They are a partially mechanized service because the customer and collector still have to do some of the work. The customer puts the refuse in the container. The container is emptied at the storage site into a compactor, such as a large metal compactor unit, or is carried to the disposal site and emptied there, such as a large metal mobile unit. Solid waste is compacted at the point of origin by a stationary packer. Detachable container systems include 3 to 15 yd³ capacity containers that are lifted and carried, 1 to 3 yd³ containers that are side loaders, containers that are rear loaders, front-end loaders that lift 1 to 10 yd³ by inserting arms through metal fittings in the containers and carrying them to the disposal site, and any combination of the others. Vacuum systems are used for leaf collection trucks and also for disposal of material from litter barrels. Open-body collection vehicles with special dump mechanisms are used for hauling bulky items such as furniture, appliances, wrecked cars, logs, and brush.

Equipment maintenance including preventive maintenance is an important facet of the entire collection system. Unless the equipment is maintained properly, it will break down. Preventive maintenance includes inspections of all parts of the vehicle every 1000 miles or every 30 days, established maintenance schedule utilized by competent mechanics, accumulation of operating and maintenance data to determine the types of breakdowns, and other service problems.

The collection methods vary in the same way as the equipment varies because of different laws, regulations, habits of the citizens, climate, and other local conditions. The various types of pickup service include:

1. Curb or alley service, where the loading is from the curbs or margins of alleys. The advantage to this is speed, less expense, and no trespassing by the worker. The disadvantages are less service, containers are on the street for a long time, and rigid routing and scheduling must be used.
2. Set-out service, where the collector carries the full containers from the householder's premises to the curb or alley a few minutes before the collection vehicle arrives. The advantages are that the full containers are at the street for a short period of time, with little cost involved and flexible routing possible. The disadvantages are that the workers must trespass, the householder still participates, and the empty containers are left at the street.
3. Set-out or set-back services are the same as mentioned except that the empty containers are brought back to the premises. This creates an additional disadvantage in cost but gives an advantage in that the householder is not involved.
4. Backyard carry service, where the collector empties the solid waste from the place of storage on the premises into large carrying containers and takes it out to the trucks. The advantages are maximum service, containers are never in the street, and flexible routing and scheduling are possible. The disadvantages are the cost, trespassing, and inefficient use of equipment.

The method of organizing the work includes:

1. Establishing a daily route where the community is divided into a day's work and the crew is finished when the work is done
2. A single-load method where an area is divided to provide a full load of solid waste under normal conditions, and where the individual crew has several routes, with the earliest one farthest from the disposal site
3. The large route method, where the work area is a week's activity for a single crew and must be completed within the working week
4. The definite working day, where the individual crew works 8 hours and when the route is completed, the process begins again
5. A relay method where a substitute truck is used while the initial load is taken to the disposal site

Various other ways to utilize crews include using swing crews, changing the size of the crew, providing relief where necessary, and then having a special reservoir route for all crews to pick up material when they have finished their other work.

As can be seen, collection routing is a very difficult task. It is impractical and time consuming to develop exact methods or to establish complex mathematical models or computer programs for the removal of solid waste. The best technique is common sense in developing routes by district, subdistrict, and sub-sub-district; and by assigning specific routes for the crews to follow. These routes are readjusted when they are improperly balanced, overlapped, or fragmented. It is essential that supervisory personnel in radio cars determine where the trucks are and what is occurring. In larger cities, it would be well from time to time to use helicopters to determine whether the routes are working properly. A change is needed if one or more of the following factors change: frequency of collection, point of collection, crew size, truck size or equipment, location of processing or disposal site, type of storage, separation of types of waste, number of services given per week, and collection methods.

The collection methods include:

1. The shuttle system is where the collectors are shuttled to help another crew while their driver is traveling to and from the processing or disposal site.
2. The reservoir system is where all crews work a large area after they have completed their respective routes.
3. Collection on one side of the street is used when the streets are wide, heavily traveled, and in densely populated areas.
4. Collection of both sides of the streets is when the streets are narrow, lightly traveled, or on one-way streets.

To initially determine the routing for solid waste collection, and subsequently, to determine whether rerouting is necessary, the following information by day and by week is needed: the average time on the route, the average round trip time to the disposal site, the number of trips to the disposal site, the average weight of solid waste per truck, the average on-route mileage, the average off-route mileage, the number of services on each route, and the average weight of solid waste per full load. In all cases natural boundaries such as railroads, expressways, and rivers must be considered in routing.

The satellite vehicle waste collection system is utilized to collect waste at alleys or curbs, and to remove it by means of a three- or four-wheeled vehicle to a storage point or packer truck. These vehicles weigh between 1200 and 2600 lb and hold 1 to 3 yd^3 of waste. They are equipped with hydraulic lifts for unloading into the packer trucks. They work best in areas of low- to medium-density housing where single-family homes are most prominent.

Transfer stations are utilized as halfway points between the collection service and the ultimate disposal site. Transfer stations are most valuable in rural and recreational areas. The large transfer station usually consists of 1 to 3 acres of land where the solid waste material is brought for eventual loading and removal to solid waste disposal areas. The vehicle used for transfer depends on the type of road on which the trailers must travel. Small transfer stations are placed on approximately ½ acre of land, in any flood-free location that is accessible to main roads. Usually an 8 yd^3 container can handle the uncompacted waste from 50 families for a 3- to 4-day period. Many transfer stations are located on a 50- to 100-ft lot. The station may have two stationary compactors and two tractor trailers. The material is compacted and then transferred onto the trailers. The trailers are removed when needed to the disposal site.

Another technique that is in the experimental stage is pipe transport. The waste materials are converted into a slurry and piped to an area from where they are hauled or discarded. The difficulty involved in pipe transport is loss of pressure.

Another system in use is the automated vacuum collection device, where wastes are dropped into gravity chutes and stored temporarily. A lateral vacuum pipe then carries the waste to a storage silo that may be as much as a mile away from the source of the waste. The material moves at a speed of about 25 to 30 mi/hr. The air velocity within the pipe ranges from 50 to 60 mi/hr. The speeds plus the abrasive nature of the refuse may damage the pipe walls. Although this system is used widely in Europe, it is still experimental in the United States.

Industrial Waste

Industrial waste may be transferred by truck, train, or barge. The waste should be removed on a 7 day/week basis to central stations where industrial incinerators or burial sites are located. Any of the techniques mentioned may be used. However, the least expensive method is barge transport. The major considerations in any movement of industrial solid waste are the hauling distance, the amount of material to be hauled, the nature of the material, and the hazards associated with the materials. In the removal and disposal of industrial solid waste, it should be determined whether the waste can be recycled, which can reduce the volume and also regain valuable natural resources.

Agricultural Waste

Agricultural solid waste collection and disposal are mutually inclusive. There-fore, in this section waste disposal is discussed in depth. Agricultural production has changed from the concept of open ranges for cattle, small pens for hogs, and chicken coops near farmhouses into a major industry where thousands of animals are housed on feed lots, and where thousands of birds are housed in small, compact areas. As a result of this production line technique, a tremendous problem has developed in the disposal of solid wastes. For instance, more than 1.5 billion tons of raw manure must be disposed of annually.

A variety of systems have been designed for collection, scraping, and hauling of this material from the collection source to the final point of disposal. Sloped lots, paved lots, underdrains, interceptor ditches, grassy waterways, and other measures have reduced labor and equipment expenses. However, because of the uncertainty of meteorologic conditions, including temperature, precipitation, and solar radiation, new problems have been created by the concentration of the material and runoff discharge that contaminate water supplies and recreational areas, or ruin aquatic life. It is necessary to develop an effective and safe system of disposal.

Some of the techniques utilized for this system follow (Figure 2.1 illustrates potential disease hazards):

1. Spreading of manure on the land is an effective means of disposal. However, manure does not have the same nutrient value as a chemical fertilizer and the cost may become excessive.
2. Aerobic lagoons, or oxidation ponds, have the advantage of operating without bad odors. They should be shallow, not more than 3 to 4 ft in depth. They function best where warmth and sunshine are present and algae can grow. A large amount

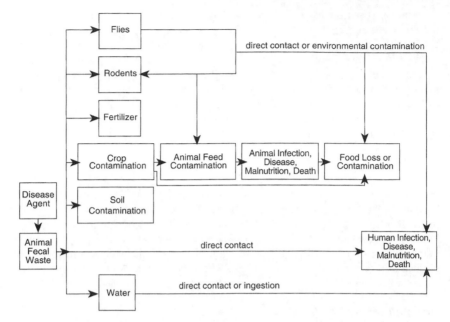

Figure 2.1   Animal fecal waste and disease relationships (postulated). (From Hanks, T.G.,
Solid Waste/Disease Relationships: A Literature Survey, Bureau of Disease Prevention and Environmental Control, Cincinnati, OH, 1967, p. 78.)

of surface area is required for the oxidation pond to work properly. Overloading
of the system causes it to become anaerobic, creating many problems.

3. Aerated lagoons are equipped with mechanical aerators to supply additional oxygen when needed. These lagoons may be 15 to 20 ft deep. The mechanically
injected air allows for proper reduction of the solid waste.

4. An anaerobic lagoon breaks down organic matter. Its purpose is the destruction
and stabilization of organic matter, not water purification. The overflow effluent
must be further treated before it can be released into a body of water. The solids
settle out and the effluent is sent to an aerobic unit. The disadvantage of the
anaerobic lagoon is that it creates bad odors and needs higher temperatures for
optimum operation. Further problems occur when the temperature drops, because
the colder the temperature is, the less the digestion. During the winter months the
system may be totally inoperative.

5. The oxidation ditch is a mechanical method of aerobically treating the liquid
waste. It is essentially odor free and produces an effluent that can be released to
water bodies. The mechanical part of the system must be constantly operating to
keep the waste circulating. In this way an adequate supply of air to the waste for
decomposition is available.

6. Dehydration is a system of seemingly little value. It is expensive to dry manure,
and the waste must still be hauled away. Incineration and composting also are of
little value.

7. Other agricultural solid waste such as corn husks and other plant materials, are
fed to animals or are spread on the fields to aid in fertilization.

Plastic Waste

Plastics are the fastest growing part of the solid waste stream. In 1960, we discarded about 0.8 billion lb of plastics. This represented about 0.5% of the total waste stream by weight. By 1986, we discarded more than 20 billion lb or 7.3% of the total waste stream. In 1994, an estimated 26 billion lb or 8.3% of the total waste stream was plastic. By the year 2000, 9.4% of the waste was plastic, or approximately 39.4 billion lb. Since 1960, plastic production has grown at two to three times the rate of the gross national product. More than half of all discarded plastic is packaging.

Because plastic is very light, it takes up considerably more space in a landfill than is indicated by its weight. A cubic yard of mixed municipal waste weighs from 800 to 1200 lb. The increase in use of plastic is a significant factor in filling our landfills.

A pound of plastic has almost as much heat value as a pound of residual fuel oil. However, there are questions concerning the emissions from incinerators created by the burning of plastics, whether toxins are included in the ash, and whether incinerating the resin produces dioxins and furans. Litter is one of the special problems of plastics. Plastics typically are not usually biodegradable.

Polyethylene, which is a common type of plastic, consists of molecules of ethylene joined together into long polymer chains. If the chains are straight, the polymer molecule is symmetrical and crystalline. If the main chain has branches, the molecule is less dense and less crystalline. The crystallinity varies from 55 to 99%. The greater the crystallinity, the less susceptible the molecule is to chemical attack. The number of double bonds of polyethylene also influences its susceptibility to attack, because oxygen adds to the bond breaking of the polymer chain. The hydrogen atoms on the double-bonded carbons are very reactive. Polyethylene degrades through a chain reaction. As energy is added to the polymer, a hydrogen atom is removed and a free radical is formed. This process self-perpetuates. Eventually, the polymers break into small or large fragments. When burning under ideal conditions, the end product is carbon dioxide and water. However, because ideal conditions do not occur, a variety of polymers of different sizes are produced, leading to a variety of problems.

Oxygen and air do not degrade polyethylene at a significant rate. Ozone will accelerates the oxidative degradation of plastics. Nitrogen tetroxide and nitric acid are also utilized for this purpose. In thermal degradation, the heating breaks the polymers into large or small fragments. The larger fragments decrease in size as the temperature increases. At lower temperatures an increase occurs in the size of the larger fragments but fewer of them are formed. The importance of this occurrence is that larger fragments are harder to burn completely and are more likely to pollute the air, whereas smaller fragments burn easily and release more heat, but the extra heat may damage the incinerators.

In mechanical degradation shearing forces tear the bonds apart. The susceptibility of the molecules to this stress varies considerably with the technique. Ultrasonic degradation is utilized as a shearing technique to tear the material into small particles. Although it is effective, it is also very costly. Solar radiation is also a degrading agent. However, only certain types of radiant energy penetrate the polymer molecule. Effective radioactive materials, such as cobalt-60, may be dangerous to humans. Biodegradation is an extremely slow and ineffective technique.

From the preceding discussion, it is clear that research is needed in the disposal and degradation of plastics. As the quantity and types of plastics increase, the disposal problems may increase accordingly.

Rubber

Tires have a high heat value, and when properly shredded, burn well in municipal incinerators.

Infectious and Medical Waste

Collection and disposal of infectious and medical waste are best discussed simultaneously. All infectious waste should be placed in color-coded opaque plastic trash can liners, which are then placed within a second liner and transported by collection carts to the site of disposal. It is imperative that these carts are not used for any other purpose. Infectious waste should never be thrown down a trash chute, because the possibility of bag rupture and spewing of organisms throughout an institution is a serious potential hazard. Infectious wastes must be placed within an incinerator operating at a high enough temperature to effectively destroy all the material present within the sacks. The incinerator should have an automatic charging unit, ash removal, and energy recovery system. It is poor practice to put infectious waste into landfills, because the workers and the environment may become contaminated. Highly infectious wastes should be sterilized prior to removal to the incinerator area.

Volume Reduction — Central Systems

Even though on-site volume reduction is increasing, it is still necessary to have a centralized system of volume reduction to properly dispose of, or recycle, solid waste material. Techniques used include grinding, baling, compaction, and liquid waste pulping. Reduction may be a mechanical, thermal, biological, or chemical operation.

Principles of Size Reduction

Size reduction is an important operation in municipal solid waste processing, in that it facilitates efficient separation, recovery processes, and material bulk handling. In the past, hand sorting was utilized and size reduction was only applied to such materials as oversized burnable wastes and material that could be composted. Now greater waste generation and increased cost of hand sorting have brought about a mechanization, making the operation cheaper.

Size reduction is the mechanical separation of solid waste material into smaller pieces. Tension, compression, and shear forces are utilized. The hammer mill striking a mass and breaking off a section is an example of a tension force. The energy involved in creating the smaller pieces is the force applied multiplied by the distance through which it moves.

Compression forces fracture the material by causing internal tension and shearing. The energy applied, again, is the force multiplied by the distance. Strong brittle materials, such as glass may require high forces for fracture. Shearing forces are usually accompanied by both tension and compression. The energy required to shear a body into smaller pieces is similar to the other two methods. The shearing of flat material requires a minimum of work and may be the most economical means of municipal solid waste size reduction.

Various pieces of equipment are used in size reduction including crushers, cage disintegrators, shears, shredders, cutters and chippers, rasp mills, drum pulverizers, disc mills, pit pulpers, and hammer mills. Crushers are relatively slow, efficient applicators of compression to material. Cage disintegrators are high-speed machines that produce a contrarotating action that impacts on material. The heavy cage bars help break such materials as glass and rocks. Shears are used to cut bulky material, such as wood timbers and metal automobile bodies. Shredders, cutters, and chippers use the tension and shear to pierce, tear, and cut. They are generally used on fibrous or ductile materials such as those found in the paper and box industry.

Rasp mills are massive cylindrical machines, 20 ft or more in diameter, that use tension, shear, and compression. A large variety of solid waste can be put into the sizable opening in the machine. An internal rotor traveling at five to six revolutions per minute swings heavy arms that push the waste around and over rasping pins and down through holes to produce 2-in. size material. At some point, bulky items incapable of reduction must be rejected. This equipment is best used in a composting process. Drum pulverizers are similar to rasp mills.

Disk mills are precise machines with high-speed single or contrarotating discs that tear material to pieces. They work best with those materials that can be converted to pulp such as paper, boxes, and grain. Wet pulpers are similar to single disc mills, with the exception that they have vertical axes and a slurry of about 90% water is produced. This equipment is best used for fibrous materials. Hammer mills use the rapid application of tension, compression, and shear forces to reduce the size of durable bulky items. This unit is most frequently used for the reduction of community solid waste.

Grinding Refuse

Grinding is used primarily in conjunction with composting. Recently, grinding has been considered a technique to be used along with baling, sanitary landfills, and incineration. Grinding will bring about some volume reduction, eliminate empty space in the solid waste, allow for easier handling and compaction of the material, homogenize the material, and improve the handling and burning characteristics of solid waste in some incinerators. However, some materials cannot be ground. Grinding is helpful to the baling process because it increases the bale density, creates more uniform moisture distribution, creates less presorting, and speeds up the rate of decomposition.

Baling and Compacting

Baling is utilized with materials ranging from paper to metal. The advantages of baling include volume reduction, which results in easier handling and better densities in sanitary landfills; increased loads after transfer; better dust and odor control; and possible use of the bales. The disadvantages include the existence of nonbalable items; too much moisture, which can cause liquid to exude from the bale; and downtime, which may occur.

The use of high-pressure compaction of baled materials can produce compact units of 3 ft^3, which can then be placed on railroad cars and hauled up to 700 miles from the point of origin. This technique increases the distance the disposal site can be from the site of production of the solid waste and is of considerable importance to such areas as the East and West Coasts of the United States, where the greatest percentage of the population resides. By using railroad cars for transportation of compacted solid waste, the cost of disposal is kept at a reasonable level. The amount of moisture created by varying weather conditions is a factor in the baling and compacting of waste. The more moisture, the less actual solid waste is removed. Furthermore, after pressure is released from the baled waste, it expands. The compacted waste at 3000 to 3500 lb/in.^{2} has an average density of 61 to 65.5 lb/ft^3 after expansion. This indicates that compaction is worthwhile and that it is a technique of considerable value to municipalities.

Liquid Waste Pulping

Pulping equipment is primarily used for paper. Water is added to paper and cardboard, and then the paper is ground. The pulped material is forced through a plate. A junk chute carries off the nongrindable materials automatically. Grit, such as ground glass and heavy particles, is separated in a liquid cyclone. The pulp is then dewatered and the process water is recycled. The pulp can then be transported to a landfill. The purpose of pulping is to decrease the quantity of material to be disposed of and to increase efficiency.

Pipeline Transport

Pipeline transport, at present, is still in the research stages. Many questions must be resolved concerning the movement of a vast variety of solid waste along pipes to areas of disposal. The material needs initial treatment to reduce its size. Then techniques must be developed to move the material, with as little damage as possible to the pipes, to a distant location.

Solid Waste Disposal

The final disposal point or treatment is an open dump, a sanitary landfill, an incinerator, composting, biological treatment, disposal at sea (not acceptable), pyrolysis, reclamation and recycling, energy source, rendering plant, part of hog feeding, wet oxidation, biofractionalization, anaerobic digestion, and chemical processing

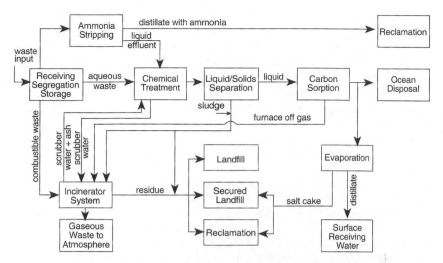

Figure 2.2   Conceptual modular flow diagram for disposal of solid waste. (From U.S. Environmental Protection Agency, Report to Congress. *Disposal of Hazardous Wastes,* U.S. Government Printing Office, Washington, D.C., 1974, p. 18.)

(Figure 2.2). The solid waste collected in the United States goes to over 200,000 facilities for treatment, disposal, or recycling.

Sanitary Landfills

Landfills historically have been the least expensive way to dispose of solid waste. About 55% of municipal solid waste is deposited in landfills. There has been a significant diversion of yard trimmings. Municipal solid waste going to landfills contain approximately 72% household waste, 19% commercial waste, 6% construction and demolition waste, and 3% sewage sludge, industrial waste, and other waste.

The landfill capacity in the United States is rapidly declining. In 1997, about 2200 municipal landfills were operating in the United States; however, many have grown in capacity. Many of these landfills are in the process of closing. The declining numbers of landfills are due to the fact that many of them are becoming full and can no longer handle the waste that is produced by the areas close to them. Currently, a number of landfills are proposed or listed on the National Priorities List (NPL) for Superfund cleanup.

As a result of the declining disposal capacity, many communities and solid waste haulers are transporting municipal waste great distances for disposal. This has resulted in increased disposal costs for many communities. It has also brought about a difficult political situation where waste is shipped from the East Coast to the Midwest and is buried in the Midwest. Many of the individuals in midwestern states are extremely resentful of this waste coming to their area and, in fact, this waste is taking up the space that they will need for their own waste over the coming years.

A sanitary landfill is an engineered means of disposing of solid waste on land by spreading the waste into thin layers, compacting it into the smallest practical volume, and then covering it with soil at the end of each working day. Burning of solid waste is not permitted at the sanitary landfill. The use of submarginal land makes this an acceptable and economical means of solid waste disposal. It is important to choose the site carefully, and then to design and operate the landfill utilizing sound engineering principles. It is necessary to understand the process of decomposition of solid waste and how the rate of decomposition is affected by many variables. The physical stability of the fill and the movement of gases generated, plus the potential of leachate seeping into water, must all be considered during the selection of the site and during the operation and final closing of the landfill.

Solid wastes deposited in the sanitary landfill degrade chemically and biologically, producing solid, liquid, and gaseous products. Ferrous and other metals are oxidized. Organic and inorganic wastes are utilized by microorganisms, either in an aerobic or an anaerobic process. Liquid waste products brought about by microbial degradation, such as organic acids, increase the chemical activity within the fill. Food wastes or garbage degrade quickly, whereas plastics, rubber, glass, and wastes from demolition and construction degrade very slowly. Factors affecting decomposition include the mix of the waste; its physical, chemical, and biological properties; amount of oxygen and moisture within the fill; temperature, type, and size of microbial populations; and type of synthesis. Biological activity usually follows a specific pattern. Solid waste at first decomposes aerobically and then eventually, as the oxygen supply is decreased, it decomposes anaerobically. During the anaerobic phase, methane gas is produced. During the aerobic phase, carbon dioxide, water, and nitrates are produced.

Leachate is a combination of groundwater or surface water that has infiltrated the solid waste and suspended solid matter, microbial waste products, and chemicals. Leachate comes to the surface or percolates down to the groundwater. The types of chemicals or biological agents found in leachate are based on the solid waste and a variety of other factors.

Leachate must be engineered out of the landfill before the landfill is utilized. This can be done by selecting proper land and soil, by proper grading of the landfill, and by ensuring that the landfill is an adequate distance from underground water tables and surface bodies of water. Natural purification of the leachate by means of ion exchange, filtration, adsorption, precipitation, and biodegradation helps reduce some of the hazards of the leachate. However, the best solution is through the use of proper engineering techniques (Figure 2.3).

Gas is produced naturally when solid waste decomposes. The quantity of gas in the landfill and its composition depend on the types of decomposing solid waste. Microbial life causes the production of the gas. The types of gas are basically methane and carbon dioxide. However, hydrogen sulfide may also be formed. Methane, by far, is the most dangerous gas because it is explosive. Carbon dioxide becomes carbonic acid and may cause a mineralization of groundwater. Hydrogen sulfide is very odorous. The permeability of the soil affects the degree to which the gases penetrate the soil (Figure 2.4).

The life of a sanitary landfill is determined by the present composition of the solid waste and what is projected. This information is also important in determining the decomposition rates within the landfill and the equipment that will be needed.

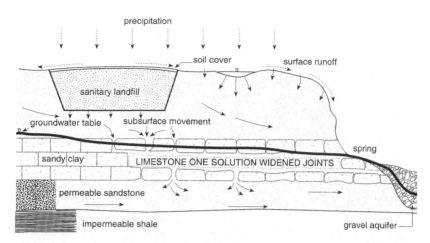

Figure 2.3 Leachate and infiltration movements are affected by the characteristics of the soil and bedrock. (From Bruner, D.R. and Keller, D., *Sanitary Landfill Design and Operation,* U.S. Environmental Protection Agency, U.S. Government Printing Office, Washington, D.C., 1972, p. 13.)

When the landfill has been completed and finally covered, it is used as grassland for pastures or feed production, in agriculture for growing such crops as wheat and corn, for recreational purposes (such as parks, golf courses, and playgrounds), and as a site for light buildings. However, heavy construction should only be allowed if proper engineering studies have been completed. If heavy construction is feasible, pilings will have to be sunk through the landfill to the bedrock or deeply into the original ground. Adequate maintenance of the landfill area is necessary to avoid erosion of the ground. Because portions of the landfill sink as decomposition occurs, it is necessary to regularly check the ground and use additional fill wherever necessary. The final slope of the site should be such that surface water readily runs off.

Landfills had been associated with a series of potential environmental health problems. Because almost 50% of the U.S. population uses groundwater drinking water supplies; and because groundwater is used extensively for agriculture, industry, and recreation, it is essential to keep landfill contaminants from creating health and environmental problems in this water supply. Landfills had been found to contaminate the water supply with arsenic, barium, benzene, cadmium, carbon tetrachloride, chromium, 2,4-dichlorophenoxy acetic acid, 1,4-dichlorobenzene, 1,2-dichloroethane, 1,1-dichloroethylene, endrin, fluoride, lindane, lead, mercury, methoxychlor, nitrate, selenium, silver, toxaphene, 1,1,1-trichloromethane, 2,4,5-trichlorophenoxy acetic acid, trichloroethylene, vinyl chloride, etc.

Groundwater monitoring systems must be set up at all landfills. Enough ground water monitoring wells need to be established to accurately assess the quality of the uppermost aquifer beneath the landfill before it has passed the landfill boundary, and at a relevant point away from the landfill to determine potential flow of contamination. Where problems exist appropriate technical solutions must be set in place to prevent the contamination or treat the water.

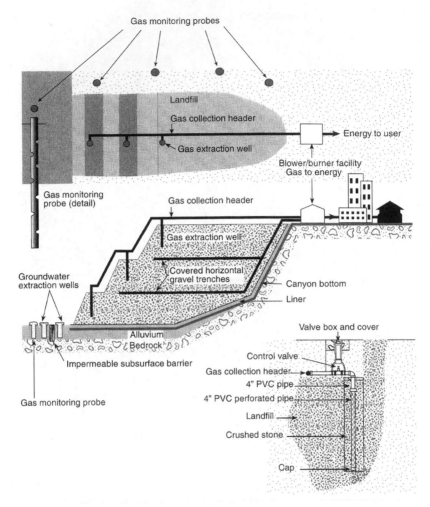

Figure 2.4   Landfill gas collection and conversion to usable energy. (Adapted from Capturing and Converting Landfill Gas. U.S. Department of Energy, DOE/GO 10094-015.)

EPA Groundwater Information Tracking System and Statistics Software (GRITS/STAT) is used at waste management facilities implementing groundwater monitoring requirements under the RCRA Subtitle C (hazardous waste) and RCRA subtitle D (municipal waste) programs. It can be used at sites conducting groundwater monitoring to store, analyze, and report groundwater data, and to conduct statistical tests to determine whether regulatory thresholds have been met or exceeded.

The six location restrictions that apply to municipal landfills are as follows:

1. Municipal landfills located within 10,000 ft of an airport runway used by turbojet aircraft or within 5000 ft of a runway used only by piston-type aircraft must demonstrate that the unit does not pose a bird hazard.
2. Landfills located in a 100-year floodplain cannot restrict the flow of the flood, reduce the temporary water storage capacity of the floodplain, or allow the washing out of solid waste.
3. Landfills cannot be built or expanded in wetlands.
4. New landfills or lateral expansions cannot be built within 200 ft of fault areas.
5. When a new or laterally expanding landfill is located in a seismic impact zone, the landfill must be designed to resist the effects of ground motion due to earthquakes.
6. In unstable areas, it must be demonstrated that destabilizing events such as heavy rainfall, sinkholes, and rock falls cannot destabilize the landfill.

In addition to the preceding points, the owners or operators must provide for the appropriate cover material, landfill liners, and surface water protection. They must not accept, unless specifically approved, hazardous waste, explosive gases, bulk liquid waste, and containers that may contain the residues of hazardous materials. They have to protect air quality and storm runoff, and prevent vectors from breeding and transmitting disease to people.

Geosynthetic clay liners are a relatively thin layer of processed clay (typically bentonite) either bonded to a geomembrane or fixed between two sheets of geotextile. A geomembrane is a polymeric sheet material that is impervious to liquid as long as it maintains its integrity. A geotextile is a woven or nonwoven sheet material less impervious to liquid than a geomembrane, but more resistant to penetration damage.

Plastic landfill liners may fail, because incinerator ash placed in the liner typically contains toxic metals, which may cause holes to form in the plastic. The concerns are how long the liner will last, and whether the toxic metals seep into the ground and eventually into the ground water supply.

Bentonite is an extremely absorbent, granular clay formed from volcanic ash. Bentonite attracts positively charged water particles, thereby becoming rapidly hydrated. As the clay hydrates it swells, giving it the ability to seal all holes in the liner. Shear strength (the maximum stress a material can withstand without losing structural integrity) is important in designing a barrier system to keep water and leachate from leaving the landfill.

Landfill reclamation is used to expand municipal solid waste landfill capacity and avoid the high cost of acquiring additional land. The reclamation costs are often offset by the sale or use of recoverable materials, such as recyclables, and waste that can be burned as fuel. The problem is that there is a potential release of methane and other gases during the process. It is also possible that hazardous materials may be discovered. The excavation work involved in reclamation may cause adjacent landfill areas to sink or collapse. The landfill reclamation process is as follows:

1. An excavator removes the contents of the landfill cell. A front-end loader then organizes the excavated materials into manageable stockpiles and separates out bulky material, such as appliances and other steel or metals.
2. A trommell (a revolving cylinder sieve), or a vibrating screen separates the soil including the cover material from the excavated material.

3. The reclaimed or recyclable material, such as metal, plastics, glass, and paper, is then removed for processing.
4. The other materials are placed back in the cell, allowing more room for additional waste disposal.

Open Dump Closing

Mission 5000 was created by the EPA in 1972, as a citizens' solid waste management project. The project offered to citizens and citizen groups the chance to make a direct and lasting improvement in the environments of their own communities. It helped produce a way to reduce air and water pollution, eliminate breeding grounds for disease, and restore land to usefulness and beauty. The project's goal was to eliminate open dumps.

For years, as our population grew, we neglected to develop a technique for disposing of unwanted solid waste. Instead, citizens and communities threw their solid waste on various parcels of land that became open dumps that disfigured the community and contributed heavily to odors and unsightly messes. It is known that nearly half of the open dumps contributed to water pollution. Nearly three fourths of the open dumps polluted the air.

A dump is an unplanned and unstructured solid waste operation. The closing of the dump should include the following four major elements: (1) advising the public and others involved in refuse handling of necessary changes and enlist their support in making the changes; (2) obtaining acceptable disposal facilities to replace those closed; (3) providing support facilities including better collection techniques, financing, and so forth to ensure a continuing viable operation; and (4) eliminating the existing problems, such as insects and rodents at the closed dump, and restoring the site to an acceptable condition.

One of the most important steps in dump closing is to win the consent and support of the public. The public must understand why it is necessary to halt dumping operations and how it should be done. It is important that the citizens are well aware of the situation and that they are willing to make or lend support to the necessary corrections. Good press relations with newspapers, radio, and television can be invaluable in this situation.

Because the open dump is a necessary part of the community, some substitute is needed. This substitute, which is a sanitary landfill, an incinerator, recycling plus the use of landfill and incinerator, or composting, has to be within the financial means of the community. The financial aspects of solid waste management become a serious concern. Money must be provided for land, equipment, and structures. Operating costs include labor, maintenance and repair, utilities, and overhead. In addition, depreciation of equipment and facilities must be considered. The new solid waste program can be supported either by the use of general tax revenues or by users' fees. In either case, the initial outlay probably comes from either general obligation bonds, which are backed by real estate tax revenues, or revenue bonds, which are backed by the anticipated revenue from the solid waste system.

Along with the disposal site, an adequate collection system must be provided because it is difficult or even impossible for residents to take solid waste to a few central locations. An adequate collection system must include frequent collection (a minimum of once a week, although twice a week is recommended), and collection of all materials including normal household waste and bulky materials. Responsibility for solid waste collection belongs to a governmental agency or a private concern. In any case, supervision of the solid waste program is to be carried out by public health personnel.

Thousands of closed landfills are under investigation for potential threats to the environment. Leachate had not been collected properly and treated before running off into surface bodies of water or into the groundwater supply. In addition, other problems continue to be recorded from landfills.

To minimize the potential problems of leachate in the future, the EPA has established rules requiring that the landfill has a plan for closure at the time of licensing and that the landfill owners provide assurance that money will be available to cap off, monitor, and maintain the landfill for as long as needed to prevent water pollution or gas explosions. Open dumps now are only of an illegal nature.

Types of Properties of Solid Waste

Municipal and commercial solid waste is a combination of mixed garbage refuse, construction and demolition waste, street-cleaning residues, and sewage treatment plant residues. The mixed garbage contains organic material from animals and vegetables along with a small portion of paper or packaging. The density is approximately 600 to 1000 lb/yd^3. The material is highly putrescent, and a nuisance. Refuse is composed of combustible and noncombustible materials other than garbage, such as paper, plastic, cans, wood, ceramics, and glass. The density varies from 200 to 700 lb/yd^3. The material is generally slow to decompose. Construction and demolition waste is composed of building materials, rubble, and other materials from remodeling, repairing, or razing of structures. The density is approximately 2000 lb/yd^3. With the exception of wood, the material is relatively inert. Street-cleaning residue includes dirt, leaves, litter, and small dead animals. The density is approximately 1000 lb/yd^3.

A second type of solid waste is industrial solid waste. This material has greater salvage value and is more homogeneous. Food products industry waste is similar to garbage. The metal industry has varied wastes including metals that can be recycled. Wood and woodworking materials are recycled or compacted and discarded. Plastics are a special problem and are discussed separately. Industry also produces special waste, including refinery sludges, hospital waste, radioactive materials, metal plating, and cleaning waste.

Agricultural waste includes crop residues and livestock waste. It is important to estimate in a given area the amount of waste that comes from residential use, community use, industrial use, and agricultural use to develop an adequate collection and disposal system.

Hydrology and Climatology

A major concern in selecting the landfill site is the hydrology of the area. Will the formation of leachate produce a water pollution problem? Because solid wastes

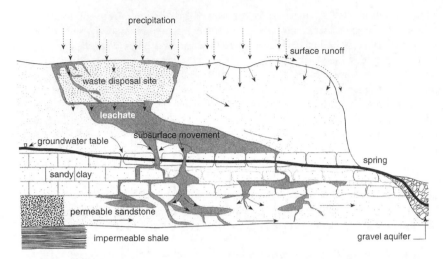

Figure 2.5   Groundwater contamination. Modification of leachate and infiltration movements.
(From Bruner, D.R. and Keller, D., *Sanitary Landfill Design and Operation*, U.S.
Environmental Protection Agency, U.S. Government Printing Office, Washington,
D.C., 1972, p. 13.)

in a sanitary landfill vary tremendously in moisture content, it is important to
understand not only the hydrology but also the type of waste deposited. In general,
the mixed solid waste coming from the community has a moisture content of 20 to
30% by weight. This moisture fluctuates with the kinds of climatic conditions
occurring during storage and collection. Generally, in this range of moisture content,
the leachate problem should not occur if the solid waste is fairly well mixed and
well compacted, because the amount of moisture that is produced during the degrading of the organic materials can be absorbed into the other dry materials. Leachate
is not produced until the entire sanitary landfill is saturated with water or a sizable
portion is saturated with water that comes from the outside. Precipitation, surface
runoff, evapotranspiration, and location and movement of groundwater in relationship to the solid waste are important factors.

Surface water that penetrates the cover soil and enters the solid waste can increase
the rate of waste decomposition and eventually create water pollution or groundwater
contamination problems due to leachate leaving the solid waste (Figure 2.5). The
permeability of the soil, which is a measure of the amount of water that can pass
through it, is affected by the texture, gradation, and structure of the soil, and the
degree of compactness of the soil. Course grain soils are usually more permeable
than fine grain soils, such as silts and clays. To determine the specific permeability
of soils, it is wise to contact the Soil Conservation Service of the U.S. Department
of Agriculture. This agency can assist in soil analysis and can determine whether
the soils are usable for landfill purposes. The amount of water that permeates the
soil cover also depends on the length of time that surface water stays on top of the
cover soil. It is therefore important to grade and slope the daily and final cover to
divert waters that might flow across the disposal site and to decrease the permeability

of the cover material. Cracks, fissures, improperly compacted material as cover material, and poor maintenance of the cover material cause considerable variation in the potential for effluent leaving the solid waste site.

Groundwater is contained within the zone of saturation of soil or rock. The level of groundwater varies tremendously from one area of the country to another and from one part of an area of a community to another. The water table is the water surface that stands in wells at atmospheric pressure. The water table is the top of the zone of saturation. However, in fine grained formations of soil, capillary action causes the water to rise above the zone. There are several zones of saturation as one digs deeper into the ground. Because the conditions affecting groundwater are complex, it is important that a qualified geologist or engineer, who understands groundwater hydrology investigates the landfill site and makes recommendations as to whether the land can be used. Because leachate from a landfill can contaminate the groundwater, it is necessary to determine in advance the level of contamination of the groundwater prior to the landfill use. The aquifer's flow rate and direction should also be determined. This can be done by using fluorescent dye.

Wind, rain, and temperature directly affect the design and operation of the sanitary landfill. All sanitary landfills should have some form of litter fence. However, in windy sites it is essential that the litter fence be high enough and at an appropriate distance from the fill to prevent litter from blowing across the fence onto highways, private property, or forest lands. These areas need to be policed daily. Because wind and sun help dry soil, a dust condition may exist. It is necessary to utilize water-sprinkling techniques to reduce dust on the landfill site and on the various roads leading to the landfill from the main road. Trees planted along the perimeter of the landfill also help reduce dust and litter. Rain can become a problem: if the soil is infiltrated by large quantities of water, the landfill site becomes a pollution hazard. Further, wet soils, especially those that are best for soil cover, such as clay, become sticky and clinging, and are difficult to spread and compact. Rain and snow have a decided effect on the movement of vehicles into the landfill site. Freezing temperatures and ice create additional problems, because the stockpiled cover soil may be difficult to break up for utilization at the end of the working day.

Soils and Geology

The soils and geologic conditions of the area for a proposed sanitary landfill need to be studied to determine the necessary limitations of the area. The study should identify and describe the soils present, the variation and distribution, the physical and chemical properties of bedrock, and how they relate to the movement of water and gas. Rock materials are usually classified as sedimentary, igneous, or metamorphic. Sedimentary rocks are formed from the products of erosion of older rocks and deposits of organic materials and chemical precipitates. Igneous rocks come from the molten mass in the depths of Earth. Metamorphic rock is formed from both igneous and sedimentary rocks, and is altered chemically by intense heat or pressure. Sand, gravel, and clay are sedimentary in origin. Sedimentary rocks are sometimes called aqueous rocks and are often very permeable. Groundwater and leachate flow easily through these rocks. The greatest amount of leachate will pass through sedimentary formations. Leachate also travels through other types of sedimentary rocks, such as limestone, sandstone, and conglomerates.

Sedimentary rock formations frequently have fractures and joints. This contributes further to the porosity of the rock, allowing water and leachate to flow more readily. Igneous and metamorphic rock, such as marble and granite, have a low level of permeability. If these rocks are fractured and jointed, they serve as aquifers. The geology of a proposed site may be obtained from the U.S. Geological Service, U.S. Corps of Engineers, State Geological and Soil Agencies, and various geology departments of universities.

From an aesthetic viewpoint, the greatest difference between the dump and sanitary landfill is the soil cover on top of the compacted waste. For the cover to perform the functions of rodent prevention, fly prevention, minimizing of moisture and gas leakage, and growth of vegetation, it must consist of suitable materials. However, the type of soil that is suitable for one environmental purpose may be poor for another. For instance, clean sand is excellent for providing a pleasing appearance and controlling blowing paper. However, it is poor for keeping flies from emerging and for keeping moisture from entering the fill.

The soil cover serves as a roadbed for the collection vehicles. Therefore, it must be strong enough and compacted sufficiently to support the weight of the vehicles. The cover must also be capable of growing vegetation, given adequate nutrients. The best kind of soil for all purposes is one that is well drained, has low permeability for standing water, prevents fire from spreading, and does not allow gas to be vented through the final cover. If the soil is highly permeable for venting of gases, then a leachate collection and treatment facility is needed to prevent water pollution. The soils should be tested at the excavation sites and classified so that the best soil is selected.

Clay soils are very fine in texture, even though they do contain a small amount of silt and sand. They vary according to particle size, type of clay minerals, and water content present. Dry clay can be hard and tough as rock, and when wet, it becomes soft, sticky, and slippery. It swells when wet, and cracks and shrinks when dry. Clay is probably least desirable as a final or top soil for cover material. The suitability of gravel and sand for cover material depends on the size and shape of the grains and the amount of clay and silt present. A gravel layer of 6 in. probably discourages rats and other rodents from burrowing into the fill, and provides good litter control. When a small amount, even less than 3%, of fines are present in gravel, the compaction characteristics are better and the cover soil is better. The only soils that must be completely ruled out as cover material are peat and highly organic soils, because they are hard to compact and are extremely porous.

A landfill can be constructed on almost any kind of terrain. However, the best type is either flat or gently rolling land that is not subject to flooding. Depressions, such as canyons, ravines, or gullies, can be used. However, it is important to keep surface waters from flowing into them. In some areas, worked-out stone and clay quarries, open pit mines, and sand and gravel pits that are useless and dangerous eyesores are utilized for sanitary landfills. Care must be taken that these artificial pits are not in the water table and that leachate does not find its way into some aquifer. (See Table 2.1 for types of soils suitable for sanitary landfills.)

Planning and Design

The planning of a sanitary landfill must take into consideration not only good sanitary landfill design but also public concerns; public relations; political acceptance; satisfaction of state and local health and nuisance laws; and availability of sewers, roads, water for fire control, potable drinking water, telephones, and electricity. The design of the landfill must consider the volume of material to be compacted and the amount of cover material, which ranges from 4:1 to 3:1, to be utilized.

As part of the design, studies are made of the property description and location, topographical description, soil types and suitability, bedrock elevations and rock types, estimated groundwater table elevation, surface water location, prevailing winds, annual rainfall predictions, frost penetration, temperature variations, and distance from existing collection sites. The plan must include the clearing of shrubs, trees, and bushes; avoidance of erosion and scarring of the land; natural windbreaks; green belts; permanent and temporary roads; and types of buildings to be erected and fencing to be utilized. The most desirable fence is 6 ft in height with three strands of barbed wire projecting at a 45° angle. Litter fences at the actual work area range from 6 to 10 ft in height.

Control of surface water is an important part of the plan. Therefore, the plan must include the techniques to be used to divert surface water. These techniques include pipes, gullies, ravines, canyons, drainage ditches, and sump pumps. Mineral pollutants travel much greater distances than organic pollutants. At times it is possible to permanently or temporarily lower the groundwater in free draining gravel or sandy soil. This is done by using drains, canals, and ditches.

The plan must also account for gas movement control. Methane, which is a colorless, odorless gas that is highly explosive in concentrations of 5 to 15% in the presence of oxygen, is produced in the landfill. The plan must indicate the method that is to be used to get rid of the methane. The method may be a permeable technique where vent pipes are inserted to some outside area to vent the gas, or the lateral gas movement can be prevented by using a material that is more permeable than the surrounding soil. Impermeable methods can be utilized by using a top cover that blocks the flow of gas. The most practical means is to use compacted clay, but it must be kept moist to prevent shrinking and cracking.

Site Selection

In selecting a site, it is necessary to determine the land requirements, zoning restrictions, accessibility, hauling distance, cover material, geology, climate, fire control facilities, anticipated community growth, legal aspects involved, political considerations, and the ultimate land use. The volume of space required depends on the character and quantity of solid waste that will be placed within the landfill, the efficiency of compaction, the depth of the fill, and the desired length of time that the fill is to be left open. Obviously, in-depth studies must be made to determine how much space is required for the community in question. It would be reasonable to expect a rate of 4.5 lb of solid waste per person per day now and in the future

Table 2.1 Unified Soil Classification System and Characteristics Pertinent to Sanitary Landfills

Major Divisions		SYMBOL Letter Matching Color		Name	Potential Frost Action	Drainage Characteristics
coarse-grained soils	gravel and gravelly soils	GW	RED	Well-graded gravels or gravel–sand mixtures, little or no fines	None to very slight	Excellent
		GP		Poorly graded gravels or gravel–sand mixtures, little or no fines	None to very slight	Excellent
		GM	YELLOW	Silty gravels, gravel–sand–silt mixtures	Slight to medium	Fair to poor / Poor to practically impervious
		GC		Clayey gravels, gravel–sand–clay mixtures	Slight to medium	Poor to practically impervious
	sand and sandy soils	SW	RED	Well-graded sands or gravelly sands, little or no fines	None to very slight	Excellent
		SP		Poorly graded sands or gravelly sands, little or no fines	None to very slight	Excellent
		SM	YELLOW	Silty sands, sand–silt mixtures	Slight to high	Fair to poor / Poor to practically impervious
		SC		Clayey sands, sand–clay mixtures	Slight to high	Poor to practically impervious
fine-grained soils	silts and clays LL is less than 50	ML	GREEN	Inorganic silts and very fine sands, rock flour, silty or clayey fine sands or clayey silts with slight plasticity	Medium to very high	Fair to poor
		CL		Inorganic clays of low to medium plasticity, gravelly clays, sandy clays, silty clays, lean clays	Medium to high	Practically impervious
		OL		Organic silts and organic silt–clays of low plasticity	Medium to high	Poor
	silts and clays LL is greater than 50	MH	BLUE	Inorganic silts, micaceous or diatomaceous, fine sandy or silty soils, elastic silts	Medium to very high	Fair to poor
		CH		Inorganic clays of high plasticity, fat clays	Medium	Practically impervious
		OH		Organic clays of medium to high plasticity, organic silts	Medium	Practically impervious
Highly Organic Soils		Pt	ORANGE	Peat and other highly organic soils		

up to 8 lb of solid waste per person per day with a density of 1000 lb/yd^3 and 1 part earth cover to 4 parts of waste.

In addition to the actual size of the community, it is also important to consider the anticipated community growth in both size and geographic direction. Will commercial or industrial development or redevelopment take place within a given area? Zoning is the community's means of controlling its growth and development.

Table 2.1   (Continued)

Value for Embankments	Permeability cm per sec	Compaction Characteristics[†]	Std AASHO Max Unit Dry Weight lb per cu ft[‡]	Requirements for Seepage Control
Very stable, pervious shells of dikes and dams	k>10-2	Good, tractor, rubber-tired steel-wheeled roller	125–135	Positive cutoff
Reasonably stable, pervious shells of dikes and dams	k>10-2	Good, tractor, rubber-tired steel-wheeled roller	115–125	Positive cutoff
Reasonably stable, not particularly suited to shells, but may be used for impervious cores or blankets	k=10-3 to 10-6	Good, with close control, rubber-tired, sheepsfoot roller	120–135	Toe trench to none
Fairly stable, may be used for impervious core	k=10-6 to 10-8	Fair, rubber-tired, sheepsfoot roller	115–130	None
Very stable, pervious sections, slope protection required	k>10-3	Good, tractor	110–130	Upstream blanket & toe drainage or wells
Reasonably stable, may be used in dike section with flat slopes	k>10-3	Good, tractor	100–120	Upstream blanket & toe drainage or wells
Fairly stable, not particularly suited to shells, but may be used for impervious cores or dikes	k=10-3 to 10-6	Good, with close control, rubber-tired, sheepsfoot roller	110–125	Upstream blanket & toe drainage or wells
Fairly stable, use for impervious core for flood control structures	k=10-6 to 10-8	Fair, sheepsfoot roller, rubber-tired	105–125	None
Poor stability, may be used for embankments with proper control	k=10-3 to 10-6	Good to poor, close control essential, rubber-tired, sheepsfoot roller	95–120	Toe trench to none
Stable, impervious cores and blankets	k=10-6 to 10-8	Fair to good, sheepsfoot, roller rubber-tired	95–120	None
Not suitable for embankments	k=10-4 to 10-6	Fair to poor, sheepsfoot roller	80–100	None
Poor stability, core of hydraulic dam, not desirable in rolled fill construction	k=10-4 to 10-6	Poor to very poor, sheepsfoot roller	70–95	None
Fair stability with flat slopes, thin cores, blankets and dike sections	k=10-6 to 10-8	Fair to poor, sheepsfoot roller	75–105	None
Not suitable for embankments	k=10-6 to 10-8	Poor to very poor, sheepsfoot roller	65–100	None

NOT RECOMMENDED FOR SANITARY LANDFILL CONSTRUCTION

[*] Values are for guidance only; design should be based on test results.
[†] The equipment listed will usually produce the desired densities after a reasonable number of passes when moisture conditions and thickness of lift are properly controlled.
[‡] Compacted soil at optimum moisture content for Standard AASHO (Standard Proctor) compactive effort.

From Brunner, D.R. and Keller, D.J., *Sanitary Landfill Design and Operation,* U.S. Environmental Agency, U.S. Government Printing Office, Washington, D.C., 1972, p. 17.

Advance planning is necessary to zone those areas needed for sanitary landfills. In addition, the legal problems of jurisdiction by the state, county, or local governments must be considered; and the effect on public opinion, political powers, and public relations must be analyzed prior to the selection of a given site.

The site should be easily accessible to highways or other major streets, where trucks can normally travel without causing complaints. The roads to the landfill site should be wide enough and constructed of all-weather materials to handle the necessary truck traffic. Consideration should be given to the size of the trucks and the weight of the trucks when fully loaded. This is an especially serious concern during adverse weather conditions. The on-site roads must be laid out to eliminate crossing of traffic resulting in tie-ups. Waiting spaces should be provided for trucks. Additional space must be provided for the employees' automobiles and for extra equipment. It is desirable to have several access routes in the event that one becomes unusable.

The hauling distance from the collection of the solid waste to the disposal site is an extremely important economic factor. The more waste that can be handled in shorter periods of time and the cheaper the operation, the greater is the likelihood of an effective operation. The cover material's availability is another important economic factor. If the material must be hauled to the landfill site, it causes additional cost, making the site less desirable. The soil that makes up the cover material has to be workable. Occasionally it is necessary to bring in additional cover material and mix it with existing materials to adjust the soil cover. To determine if this is necessary, soil analysis should be made and evaluated prior to actual site selection. Geology, climate, and hydrology have already been discussed.

Another important facet of site selection is to determine the techniques to be utilized in the event of fire. Usually, simply smothering the fire with soil will be satisfactory. However, it is preferable to have water available in the event that this is an unsatisfactory fire-control technique. The ultimate use of the land should be determined prior to site selection. Because completed landfills can be used for parks and playgrounds, industrial sites, or farmland, the land can assume a value making the sanitary landfill a positive part of community planning.

Equipment

A wide variety of equipment is used on a sanitary landfill. The size, type, and amount of equipment depend on the landfill, its method of operation, and the experience and preference of the operators and landfill designers. The most common equipment includes the crawler or rubber tire tractor (Figure 2.6), which can be utilized with a dozer blade, trash blade, or front-end loader (Figure 2.7). It usually performs all the operations, including spreading, compacting, covering, trenching, and hauling of cover material. The crawler-type tractor is utilized in certain situations. Other equipment utilized include scrapers, compactors, drag lines, and graders. On small landfill sites, a 5- to 15-ton tractor may be used. On large landfill sites, a 15- to 30-ton tractor may be used.

The actual selection of equipment is usually based on the type of refuse to be handled, the compaction requirements, and the versatility of the equipment. Track-type tractors are usually best suited for area fills and ramp operations, because they

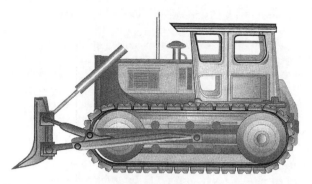

Figure 2.6   Crawler tractor with dozer blade.

Figure 2.7   Front-end loader.

are effective in digging cover material and bulldozing up to 300 ft. They are also versatile in site preparation, site finishing, building of access roads, and hauling of cover material. Track-type loaders are usually best suited for trench operations, because they carry cover material short distances, compact material close to the sides of the trench, and can also be used in small towns on a part-time basis. Wheel loaders serve the dual purpose of maintenance and snow removal, and also carry cover material short distances. Steel-wheeled compactors increase the density of the solid waste.

Auxiliary equipment used on the landfill site includes water trucks to keep down the dust, special rollers, dump trucks, refuse shedders, and possibly drag lines.

Other Types of Land Disposal

Beside using landfills, in which each type of waste is isolated in discrete cells or trenches for hazardous waste disposal, other types of land disposal are currently utilized. Surface impoundments are natural or depressions made by people in which the waste is treated, stored, or discarded. Surface impoundments may be of any shape or of any size. They are often referred to as pits, ponds, lagoons, and basins.

Underground injection wells are steel and concrete encased shafts placed deep into the earth into which hazardous waste are deposited by force and under pressure. Typically, liquid hazardous wastes are disposed of in underground injection wells (Figure 2.8).

In a landfill situation, hazardous waste is often covered with soil before a special cover system is installed. The EPA RCRA regulations require stringent landfill design and construction features such as double liners and leachate collection systems.

Waste piles are not contained. Accumulations of solid but nonflowing hazardous waste are found in these piles. Most waste piles are used for temporary storage until the waste can be transferred to a final disposal site.

Land treatment is a disposal process in which hazardous waste is applied on top of, or incorporated into, the soil surface. Natural microbes in the soil break down or immobilize the hazardous constituents. Land treatment facilities are also called land application or land farming facilities.

Landfill Gas into Electricity

Greenhouse gases can be reduced if the methane gas produced from the landfill can be utilized as a source of energy instead of just burning it off. In this way the greenhouse gases from the additional energy never are produced. Source reduction and recycling also reduce the level of greenhouse gases, because the material is never burned or put into the landfill to decompose.

Methane, the most predominant gas in a landfill, has roughly 21 times the global warming effects of carbon dioxide. Methane is also highly explosive and has been responsible for many landfill fires and explosions, which have resulted in injury and death. Methane contributes to local smog and can cause unpleasant odors. The EPA has designated municipal solid waste landfill emissions as a pollutant because the volatile organic compounds in landfill gas interact with nitrous oxides to form ozone. Instead of flaring the gas, landfill operators can burn it in gas engines or turbines to generate electricity or heat. The technology for recovering landfill gas is already on-line in a number of communities. Landfill gas, with minimal cleaning, can be used directly in boilers for industrial uses. In Raleigh, NC, a boiler fueled by landfill gas generates steam at an average rate of 24,000 lb/hr to meet the needs of a pharmaceutical plant.

Real Cost of Landfill Disposal

There has been a sharp increase in landfill disposal costs due to more stringent environmental standards. Nationwide tipping fees have risen over 400% since 1985. Costs, which continue to rise sharply, are based on planning costs; long-term and incremental development costs; operational costs including labor, insurance, benefits, office supplies and materials, equipment purchases, fuel and maintenance, and erosion control and utilities; closure and postclosure costs; groundwater monitoring costs; and costs related to impact of fill dirt, impact of compaction rate, and impact of tonnage received.

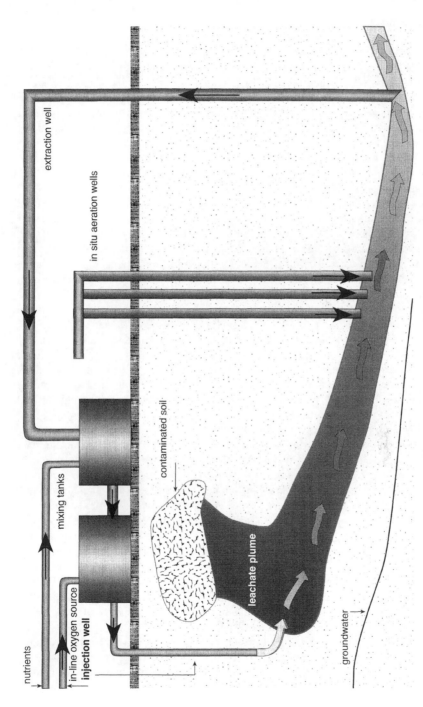

Figure 2.8 Underground injection well used in bioreclamation of soil and groundwater.

Incineration

Municipal waste incinerators are used to burn solid waste. Most of these incinerators are resource-recovery or waste-to-energy facilities that convert the heat generated from the burning waste to electricity or steam. Approximately 156 incinerators are operational. Of these facilities, 80% have energy recovery from the burning waste. The volume of municipal solid waste that is incinerated is about 15%. Incineration reduces the volume of solid waste by up to 90%. Disposal of the remainder of the material, as well as the water used to cool down the waste, must be handled. Approximately 30 million tons of the municipal solid waste generated each year cannot be incinerated. Such materials include construction waste, concrete, old appliances, and bricks.

Incineration is the destruction of waste materials by the controlled application of heat. Incineration varies from improper open burning at homes, dumps, and salvage operations to proper burning within incinerators. Incineration is used as a means of disposal, because it allows for shorter hauling distances and collection points; is usable where landfill sites are not available; and may be a more effective technique for a community, regardless of the initial cost. The purpose of incineration is to provide as complete combustion as possible to reduce the volume of the solid waste; to minimize pollution of the air, water, or land; and to remove as many combustible and putrescent materials as possible from the solid waste. The residue should contain a maximum of 5% combustibles and 1% putrescences by total dry weight.

The burning of the solid waste, which may be semisolid, gaseous, or combustible, should proceed to the point where an inoffensive gas is produced and the sterile residue may be removed to a landfill. The incineration process includes weighing and unloading of refuse; mixing and charging of refuse; and drying, ignition, combustion, and handling of the effluent, which is composed of the residues and flue gases.

To understand the incineration process, it is necessary to understand certain terminology. The furnace is the primary chamber; the combustion area is the secondary chamber. The underfire air is the primary air; the overfire air is the secondary or combustion air. Theoretical air is the amount of air needed for complete combustion of the refuse under ideal conditions. Additional air is the amount purposely introduced or flowing into the incinerator; it is needed over and above the theoretical air requirements for adequate burning of the refuse and prevention of slag formation on the walls of the incinerator. The residue is the inert material left at the end of the incinerator process. Fly ash is material emitted from the smokestack that becomes a particulate form of air pollution. The grate is the area where the material is burned. The settling chamber is the area in which the residue collects. The heating value, which is measured in British thermal units (Btus), is the amount of heat that is produced by the solid waste when it is burned.

It is essential to have an understanding of the composition of the refuse to know how best to incinerate it. The quantity, composition, and sources of solid waste are important to know for the establishment of a proper burning operation. This information can be determined by looking at previous data through comprehensive area surveys or through physical analysis of combustibility, moisture content, and density, and chemical analysis of heating value and reactions of the solid waste.

Solid waste is classified into seven major types as follows:

1. Type 0 is composed of highly combustible refuse, such as paper, cardboard, wooden boxes, and combustible floor sweepings from commercial and industrial activities. The moisture content is 10%, incombustible solids are 5%, and heating value per pound as fired is 8500 Btu.
2. Type 1 refuse is a mixture of combustible waste such as paper, cardboard cartons, wood scraps, foliage, and combustible floor sweepings from domestic, commercial, and industrial activities. The moisture content is 25%, incombustible solids are 10%, and heating value as fired is 6500 Btu/lb.
3. Type 2 refuse consists of approximately an even mixture of refuse and garbage by weight. This usually comes from apartments and residences. The moisture content is 50%, incombustible solids are 7%, heating value as fired is 4300 Btu/lb.
4. Type 3 is garbage consisting of animal and vegetable waste from restaurants, cafeterias, hotels, hospitals, and markets. The moisture content is 70%, incombustible solids are 5%, and heating value as fired is 2500 Btu/lb.
5. Type 4 is human and animal remains, which consists of carcasses, organs, and solid organic waste from hospitals, laboratories, slaughterhouses, and animal pounds. The moisture content is 85%, incombustible solids are 5%, and heating value as fired is 1000 Btu/lb.
6. Type 5 is a combination of by-product waste, which consists of gaseous, liquid, or semiliquid materials, such as tars, paints, solvents, sludge, and fumes from a variety of industrial operations. The moisture and incombustible contents vary with the waste and must be determined on an individual basis. The heating value also varies with the waste.
7. Type 6 comprises solid by-product waste such as rubber, plastics, and wood wastes, from a variety of industrial operations. The moisture, incombustible contents, and heating values vary with the waste and must be determined accordingly.

Combustors are classified in the following variety of ways:

Class 1 combustor is a portable, packaged, completely assembled, direct-feed combustor with 5 ft³ of storage capacity. It operates with a burning rate of 25 lb/hr and is used for type-two wastes.

Class 1A is a portable, packaged, completely assembled direct-feed combustor with a 5 to 15 ft³ primary chamber, a 25- to 100-lb burning rate for type-zero, type-one, or type-two waste; and a 25- to 75-lb/hr burning rate for type-three waste.

Class 2 is a flue-fed single-chamber combustor with more than 2 ft² of burning area and is suitable for type-two waste. It has one vertical flue that functions as a charging chute and a chimney. It is used in apartment houses or other multiple dwellings that do not exceed five stories.

Class 2A is a chute-fed, multiple-chamber combustor with more than 2 ft² of burning area. It is suitable for type-one or type-two waste, with the exception of industrial waste. It has two flues, one for charging refuse and the other for functioning as a chimney.

Class 3 is a direct-feed combustor that has a burning rate of 100 lb/hr or more. It is suitable for type-zero, type-one, or type-two waste.

Class 4 is a direct-feed combustor with a burning rate of 75 lb/hr and more. It is suitable for type-three wastes.

Class 5 is a municipal combustor with a capacity rated in tons per hour or tons per 24 hr. It is suitable for burning type-zero, type-one, type-two, type-three wastes, or a combination of all wastes.

Class 6 is a crematory and pathological combustor with varied capacity. It is suitable for type-four wastes.

Class 7 is a combustor designed for specific by-product wastes, and is used for type-five or type-six waste products.

Current History

In the United States, 15% of the municipal solid wastes are burned in combustion facilities. Before 1976, most incinerators were used to reduce the weight and volume of waste by converting it to smoke and ash. After 1976, energy recovery, usually electric generation by steam or steam heat, became an important purpose for incineration of waste materials.

About 156 municipal trash-burning facilities operate in the United States. The three categories of facilities include mass burn, modular, and RDF. A mass burn plant consumes the entire waste stream, except for such large items as refrigerators. In a typical mass burn combustor, refuse is placed on a grate that moves through the combustor, and appropriate underfire air and overfire air are supplied. They typically burn between 50 and 1000 tons/day. A modern mass burn plant is equipped with a water wall furnace for energy recovery. Some facilities have two or more combustion trains to have backup and to accommodate waste streams of 3000 tons/day or more. Modular combustors are small, approximately 5 to 100 tons/day, two-chamber factory-built facilities. They take the refuse without any preprocessing.

RDF plants are designed to incinerate only the burnable portion of the waste stream. The incoming solid waste must be processed through a series of shredders, screens, and density classifiers, which separate out metals, glass, and other nonburnable constituents. These materials are either recycled or sent to landfills. The burnable residue, usually paper and plastics, is shredded or finely divided and burned in industrial boilers or cofired along with pulverized coal or oil in utility furnaces. Many controversies remain about establishing sites, regional agreements, and pollution control, along with uncertainties about how to manage the ash produced by mass burn plants.

Many combustion facilities are owned and operated by private companies who contract with municipal governments. The municipalities typically guarantee a certain amount of trash, for which they are charged a tipping fee based on tonnage. If less trash is delivered, the municipality still has to pay. Public utilities are obliged by federal law to take the energy produced by the mass burn facility and pay a price for it based on the avoidable cost of building a new generator. This price is often greater than the market value of the energy. This tends to cause disputes between the electric utility companies and the resource-recovery companies about the quantity of energy accepted into the grid and the price paid.

Single-chamber combustors are used for the destruction of solid waste. The charge is placed on the grate and is fired. Air is caused to flow within the chamber through the underfire and overfire air ports. The load is charged, fired, and then allowed to completely burn. Single-chamber combustors with automatic charging and auxiliary fuel are also available. If an afterburner is present, it helps destroy the unburned combustibles within the gases.

Teepee burners are combustors that are extremely popular for the destruction of refuse, wood, and agricultural waste. Waste is usually dropped from the entrance near the top of the cone to a series of cones elevated from the teepee floor. Strict air pollution control regulations make this type of incineration a difficult one to utilize.

Multiple chamber combustors provide a complete burnout of combustion products and decrease the airborne particulates in the flue gas. The primary chamber is used for combustion of solid waste. The secondary chamber provides the time and fuel for combustion of the unburned gaseous products and particulates. The two basic types of multiple chamber combustors are the retort combustor, which is a compact cube-type combustor with multiple internal baffles; and the in-line combustor, which is much larger and allows the flow of combustion gases to go straight through the combustor.

The central disposal combustor system is utilized by municipalities for disposal of household waste and is also used by many industries for their waste products. Because of the size of the unit, heat recovery equipment may be present, which produces hot water or steam. The traveling grate is the most common grate system used in the United States for the large waste combustors.

Pyrolysis is a process that is related to incineration. Whereas incineration requires that an adequate amount of oxygen be provided to combust the waste, primarily to carbon dioxide and water vapor, in pyrolysis a destructive distillation of the solid material takes place in the presence of heat and in the absence of air or oxygen. Pyrolysis is an endothermic reaction, which means that heat must be provided for the reaction to occur. An exothermic reaction, such as incineration, is one in which heat is produced by the action. Pyrolysis usually results in methane, carbon monoxide, and moisture. The carbon monoxide and methane are combustible and provide a positive heating value. The carbon residual contains metals, metallic oxide, and other minerals. The only commercial pyrolysis system used takes care of industrial semiliquid and sludge wastes.

The controlled air combustor, also known as the modular combustion unit or starved air combustion unit, consists of two combustion chambers. The waste is charged into the primary chamber and the amount of air is injected at a controlled rate. Because of very little air, true combustion does not occur; however, it is not a pyrolysis reaction, because some air is present. Enough air is injected to start the waste material burning until sufficient heat is present for the maintenance of the process. The temperature in the primary chamber is usually in the range of 1400 to 1600°F. Air is injected into the secondary chamber for combustion of the off-gas which contains organic material released from the primary chamber. Temperatures in the secondary combustion chamber may range from 1600 to 2200°F.

The rotary kiln combustor is the most flexible and versatile of the combustors used today. It is used for the destruction of industrial waste and also for the incineration of municipal solid waste and sludges. It is a horizontal, refractory lined structure that rotates about its horizontal axis. When the ash falls into a hopper for removal and disposal, an afterburner burns out the off-gases.

Sludge combustion is carried out in the multiple hearth combustor and the fluid-bed combustor. The multiple combustor is a vertical cylindrical structure with refractory brick and a series of refractory hearths positioned one beneath the other. The center shaft is hollow and allows the passage of clean air through it. It rotates within the incinerator at approximately one revolution per minute sweeping the sludge across the hearths. Sludge is dried on the top hearths and starts to burn toward the center of the furnace. Its burnout ash accumulates at the bottom of the furnace. The temperatures within the combustion zone are 1400 to 1800°F and the temperatures at the gas outlets are about 600°F less.

The fluid-bed combustor is a steel refractory lined chamber with a supporting plate separating the windbox from the reactor. Air is heated and injected into the sand bed at a temperature of 600°F. The air provides adequate turbulence to produce fluidization. The sand appears to act as a fluid. When the sludge is injected, it comes in contact with the turbulent sand and air bed. It dries rapidly and oxidation begins. Liquid wastes are burned in a liquid waste incinerator. Air or steam is usually injected into the waste nozzle helping to atomize the waste stream to increase burning efficiency. Gaseous wastes are burned through thermal and catalytic combustion. Catalytic combustors are limited, because they cannot handle all gases.

Thermal or direct flame combustion occurs when waste gas is brought into contact with a high enough temperature and held at that temperature for an adequate time to cause the gaseous components to oxidize or convert into innocuous compounds. Organic gases usually require a temperature of 1400°F for at least 0.5 sec.

Catalytic combustion helps organic components become destructed by allowing oxidation to occur at lower temperatures than direct combustion. Catalytic combustion can save a substantial amount of the energy needed to combust chemicals.

Air Emissions

Depending on the characteristics of the municipal solid waste and the combustion conditions in the municipal waste combustor the following pollutants can be emitted: particulate matter, metals, acid gases, carbon monoxide, nitrogen oxides, and organic compounds. Under normal conditions, solid fly ash particulates formed from inorganic, noncombustible substances are released into the flue gas. The particulate matter can vary greatly in size, diameter, and constituency. Metals and toxic organics may adsorb onto particulates that are less than 10 μm in diameter, those that have the greatest potential for inhalation and destructive effects in the body. The metals present vary with the types of materials in the waste. Paper, newsprint, yard waste, wood, batteries, metal cans, etc. potentially produce a variety of chemical hazards. The chief acid gases from combustion of municipal solid waste are hydrochloric acid and sulfur dioxide. Hydrogen fluoride, hydrogen bromide, and sulfur trioxide are also found. As the waste burns in a fuel bed it releases carbon monoxide, hydrogen gas, and unburned hydrocarbons. Carbon monoxide concentration is a good indicator of combustion efficiency. Nitrogen oxides are products of all fuel and air combustion processes. Nitrogen oxides are also formed during combustion through oxidation of nitrogen in the waste material. A variety of organic compounds including those mentioned in the air pollution chapter are found in these air emissions. Mercury is a special concern in municipal waste-to-energy plants and hazardous waste combustors. Emissions of mercury also come from burning coal.

General Design

The general design and site selection for a municipal combustor is the same as for a sanitary landfill, with the following additional items:

1. The basic data must include the proposed furnace capacity for the present and for a projected 20-year period.
2. The grate area must be adequate.
3. The physical composition of the solid waste, by percentage, by weight of food waste, garden waste, paper products, plastics, rubber and leather, textiles, wood, metals, glass and ceramics, ash, dirt, and rocks must be determined.
4. The characteristics of the solid waste by moisture content and by heat value must be determined.
5. Potential noise problems due to mechanical equipment should be modified or eliminated.
6. The potential problem of visible high stacks causing an air traffic obstruction must be avoided.
7. The site topography is important, because the natural flow of refuse by gravity and removal of residue by gravity makes the process much cheaper.
8. The prevailing wind carrying the stack gases must be considered.
9. The disposal of fly ash, residue, residue quench waters, other process wastewaters, and sanitary wastewaters must be considered.

For each combustor, the design capacity, rate of capacity, and actual capacity must be identified. The design capacity is the quantity of solid waste in tons per 24-hour period that the plant is expected to handle. The rate of capacity is the quantity of solid waste in tons per 24-hour period that the incinerator can process while meeting all the environmental standards. The actual capacity is the quantity of solid waste in tons per 24-hour period that the operator actually believes the incinerator can process. (Figures 2.9 and 2.10 illustrate basic types and designs of incinerators.)

Residue Analysis and Disposal

The incinerator residue includes all solid materials that remain after burning. The physical and chemical characteristics of the residue depends on the composition of the material initially, as well as the design and operation of the incinerator. The residue normally contains tin cans, glass, rock, cinders, clinkers, and fly ash. Fly ash may be found in the gas stream leaving the incinerator as an air pollutant or may be trapped and collected to become part of the residue to be hauled away to a landfill. The amount of fly ash in a municipal incinerator varies from 10 to 60 lb/ton of refuse burned. The grate siftings and residues should be removed on a continuous basis. Fly ash is generally removed through one of the air pollution control devices.

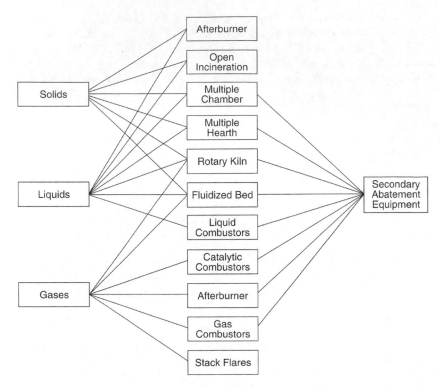

Figure 2.9   Types of incinerators and their applications. (From U.S. Environmental Protection
Agency, Report to Congress. *Disposal of Hazardous Wastes,* U.S. Government
Printing Office, Washington, D.C., 1974, p. 63.)

The residue may be hazardous and on landfilling may contribute to leachate problems. The residue must be deposited in double-lined landfills; and should be checked
for excessive amounts of lead, cadmium, and chromium.

Incinerator Process Water

All incinerators use water for residue quenching. Water is also used in wet bottom
expansion chambers, cooling charging chutes, fly ash sluicing, residue conveying,
and air pollution control. Water requirements vary from 500 to 2000 gal/ton of refuse
processed at a rate of 50 to 1000 gal/min. The wastewater must be treated before it
can go into a receiving stream. Before treatment is possible, it is necessary to
determine the physical and chemical characteristics of the water.

In the air pollution control system, the water is used for dust removal, gas cooling
and absorption, and chemical reactions. Water use is at a rate of 100 to 6000 gal/min
and the effluent is approximately 500 gal/min. The effluent generally is turbid, acidic,
hot, and loaded with gases. It also contains chlorides, phenols, cyanides, sulfates,
heavy metals, and fluorides. Where the process water is put into sewers, it must not
exceed a maximum temperature of 150°F, and it must have a pH of 5.5 to 9.5. The

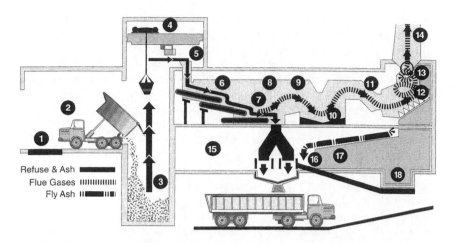

Refuse & Ash ▬▬▬
Flue Gases ||||||||||||||
Fly Ash ||▬▬||▬|

1. Scales	7. Burning Grates	13. Induced Draft Fan
2. Tipping Floor	8. Primary Combustion Chamber	14. Stack
3. Storage Bin (Pit)	9. Secondary Combustion Chamber	15. Garage - Storage
4. Bridge Crane	10. Spray Chamber	16. Ash Conveyors
5. Charging Hopper	11. Breeching	17. Forced Draft Fan
6. Drying Grates	12. Cyclone Dust Collector	18. Fly Ash Settling Chamber

Figure 2.10   Basic incinerator design.

materials carried by the water must be biodegradeable, noncorrosive, and nontoxic. A minimal amount of solids must be present in the water. Water should at least be neutralized, clarified, and filtered prior to discharge. It should also be chemically — chemical oxygen demand (COD) — and biologically — biological oxygen demand (BOD) — oxidized. The water must also be treated for removal of specific pollutants.

Special Waste Handling

Special wastes include oversized combustible waste, industrial and commercial waste, hospital waste, construction and demolition waste, sewage sludge, and dead animals. In most cases, incinerators can handle the special waste if the amounts received are small. When the volume is large, it becomes a hazard, not only to the operators of the incinerators but also to the actual operation of the incinerator. Special wastes should be treated at their source or separated sending some portion to sanitary landfills or, depending on the kind of waste, to recycling processes.

Oversized combustible wastes are items that are too big to be collected and passed through the charging chute, or too big to be picked up during collection. These wastes are too dense to be completely burned within the refuse-retention time within the furnace. This category of waste includes rubber tires, brush and branches, tree stumps and trunks, stuffed and wooden furniture, large boxes, crates, pallets, and mattresses. Some techniques that may be utilized are size reduction through shredding or impaction, or use of special incinerators that have been fed with an auxiliary fuel.

Most hazardous industrial and commercial wastes create a very serious problem in incineration, because the material may consist of corrosive, volatile, toxic, or explosive materials. Hazardous materials may be burned in special incinerators, chemically treated, or disposed of in specially lined sanitary landfills. Other industrial wastes, such as food processing, synthetic textile, plastics, and leather, create special types of problems, depending on quantity. Commercial waste, such as waste oil from gas stations and cleaning fluids from dry-cleaning stores, should not be handled in regular incinerators.

Hospital wastes generally have a very high moisture content and are very putrescent. They contain stool specimens, sputum cups, surgical wastes, pathological wastes, parts of human bodies, research animals, and normal cafeteria wastes. Except for the normal food waste, the rest should be incinerated in special high-temperature incinerators, preferably on-site.

Construction and demolition waste is made up of building materials, such as wood, plaster, steel, bricks, and concrete. This waste is usually very bulky and low in fuel value. It is preferable that this material be used as solid fill instead of disposed of through incineration.

Sewage sludge, if properly conditioned, can be incinerated. The difficulties are the cost of conditioning the sludge, transportation, and scheduling with incoming solid waste for complete mixing.

Dead animals are very difficult to eliminate in an incinerator when processed with refuse because of the high moisture content. Generally, they should be incinerated in special high-temperature crematories. Cold storage rooms are needed for storage of the animals prior to cremation.

Radioactive waste should not be taken to municipal incinerators. Agricultural solid waste, generally, is a problem because of its size and weight. It is advisable not to mix this waste with regular waste in an incinerator.

Composting and Biological Treatment

Composting is the biochemical degrading of organic materials to a humuslike substance. This process is constantly taking place in nature. For many centuries, farmers and gardeners have composted by placing their vegetable matter and animal manures in piles or into pits for the material to decompose. The first municipal composting took place in 1925 in India. Municipal refuse has been composted for many years in Europe and the United States. European composting included considerable research and the technology is well advanced. In the United States, composting plants were established in various communities during the last 20 years.

Physical and Chemical Processes

Composting is a solid waste recovery technique that requires proper temperature, particle size, oxygen, pH, carbon and energy source, nutrients, and moisture (Figure 2.11). Compost can be used as a soil conditioner to improve soil structure, increase the moisture-holding capacity of the soil, reduce leaching of soluble inorganic nitrogen, increase the phosphorus available for growing plants, and increase

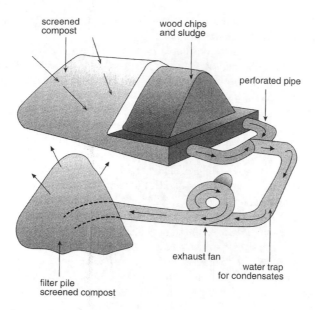

screened
compost

wood chips
and sludge

perforated pipe

exhaust fan

water trap
for condensates

filter pile
screened compost

Figure 2.11   Composting system.

the buffering capacity of the soil. Compost is not fertilizer. Compost results from the aerobic, biological stabilization, or digestion of biodegradable organic materials in solid waste. Because separation of other material may be difficult or costly, the nonbiodegradable material usually passes through the digestion process without changing. However, materials may be removed from the solid waste and recycled, prior to composting.

All microorganisms have an optimum temperature range. For composting this range is between 32 and 60°C. Temperatures below or above this range affect the metabolism of the cells. Composting can occur at a range of temperatures, but the optimum temperature range for thermophilic microorganisms is preferred, to promote rapid composting and to destroy pathogens and weed seeds. Larger compost piles buildup and conserve heat better than smaller piles. Ambient air temperatures have little effect on the composting process, providing the mass of the material composted can retain the heat generated by the microorganisms. Pathogen destruction occurs at a temperature of greater than 55°C for at least 3 days.

The particle size of the material composted is critical. Smaller particles usually have more surface per unit of weight, which facilitates more microbial activity on their surfaces. Grinding up particles may cause them to pack tightly. This must be avoided, because spaces are necessary for air to circulate.

Composting is an aerobic process. Anaerobic conditions can produce annoying odors. Decomposition occurs under both aerobic and anaerobic conditions, but occurs at a much faster rate when the pile is aerobic. In some composting operations, air may be mechanically forced into the piles to maintain adequate oxygen levels.

A 10 to 15% oxygen concentration is considered adequate, although a concentration as low as 5% may be sufficient for leaves.

A pH between 6 and 8 is considered optimum. The pH affects the amount of nutrients available to the microorganisms, the solubility of heavy metals, and the overall metabolic activity of the microorganisms. The adjustment of the pH is rarely needed. Organic materials are naturally well-buffered with respect to pH changes. Microorganisms in the compost process have about the same nutritional needs (nitrogen, phosphorus, and other trace solvents) as large plants.

The one important exception is that the compost microorganisms rely on the carbon in organic material as their carbon and energy source instead of carbon dioxide and sunlight, which is used by higher plants. The carbon in natural or anthropogenic organic materials may or may not be biodegradable. The ease with which a material is biodegraded depends on the genetic makeup of the microorganisms present and the makeup of the organic molecules that the organisms decompose. Carbon is typically not a limiting factor in the composting process, because most municipal and agriculture organics contain adequate quantities of biodegradable forms of carbon.

Nitrogen is of greatest concern if lacking in some of the materials. Phosphorus and potassium are typically not problems. The ratio of carbon to nitrogen is critical in determining the rate of decomposition. The ratio must be established on the basis of available carbon and not total carbon. An initial ratio of 30:1, carbon to nitrogen, is considered ideal. Higher ratios tend to retard the process of decomposition, whereas ratios below 25:1 may cause odors. To lower the carbon to nitrogen ratio, nitrogen-rich materials such as yard trimmings, animal manure, or biosolids may be added. The use of commercial fertilizers to raise the nitrogen level creates other problems.

Most materials have a lower than ideal water content for the composting process and therefore they slow the desired outcome. A moisture content of 50 to 60% of total weight is considered ideal. However, the moisture content should not be such that it causes excessive free flow of water. Excessive moisture and flowing water form leachate, which in itself may become a serious problem.

Biological Processes

To have a good compost process the microorganisms require not only specific chemical and physical conditions but also specific biological conditions throughout all stages of composting. Microorganisms such as bacteria, fungi, and actinomycetes are involved actively in decomposing the organic materials. Larger organisms such as insects and earthworms are also involved in the composting process, but are less significant compared to the microorganisms. As the microorganisms begin to decompose the organic material, the carbon is converted into carbon dioxide, water, and humus, releasing heat. Some of the carbon is used to produce additional microorganisms. If the organic molecules they require are not present, the microorganisms may become dormant or die. The humic end products from metabolic activity of one group of microorganisms may become the food or energy source for another generation or type of microorganisms. The organic material remaining is called compost. This compost is made up largely of microbial cells, microbial skeletons, by-products of microbial decomposition, and undecomposed particles of organic and inorganic material. The composting process takes place at a temperature of about 150 to 170°F. It is essential that the compost heap remains aerobic to minimize odor problems and to ensure a rapid process.

Composting Systems

Composting systems are identified by the name of their inventors or their proprietary names and vary accordingly. In most systems the refuse is prepared for digesting by grinding or, in some methods, by breaking into small pieces. Then digestion is carried out in windrows, pits, trenches, cells, tanks, drums, bins, or multistoried towers or buildings. A windrow is a pile, triangular in cross section, whose length exceeds its width and height, with the width usually twice the height. For most materials the ideal height is 4 to 8 feet with a width of 14 to 16 feet.

The five basic steps involved in composting include preparation, digestion, curing, finishing or upgrading, and storing. In the preparation process, the sorting, separating, grinding, and adding of sewage sludge occurs. The moisture content of the ground refuse ranges from 45 to 65%. The digestion or decomposition phase is carried out in open windrows or enclosures. In most modern plants, an aerobic process is used for decomposition. This is accomplished by supplying oxygen or air to the mass. A heat of 140 to 160°F or higher is generated during the process. The heat destroys pathogenic organisms, weed seeds, and fly ova. If the decomposing mass is not aerated adequately, the microflora change. Anaerobes take over and the process becomes anaerobic.

The anaerobic process is much slower. It usually takes 4 to 6 months vs. the aerobic process, which takes about 6 weeks. The anaerobic process proceeds at a much lower temperature of 100 to 130°F, produces odors, and provides a medium where pathogens may survive. However, specially designed anaerobic processes have been used extensively for biologically stabilizing biosolids from municipal sewage treatment plants. In anaerobic processes, facultative bacteria break down organic materials in the absence of oxygen and produce methane and carbon dioxide.

Curing time, which occurs immediately after stabilization, is the time needed to allow the compost to stabilize further. If the compost is used in unplanted fields or gardens, the compost has time to cure. If planting takes place immediately, the compost must be stabilized to avoid removing nitrogen from the soil. In the windrowing system, the compost must be frequently turned to get adequate aeration throughout the material. The process of curing varies with the system.

In finishing, screening, grinding, or similar techniques are used to remove plastics, glass, and other materials that have not decomposed. If the compost is used for erosion control in isolated areas, it is not necessary to finish the product. If it is used, however, in gardening, then it is important to upgrade or finish the material.

Storage is an essential step, because the compost is produced on a continuous basis but the demand for compost is greatest during the spring and fall. Therefore, storage capacity must be available for a period of at least 6 months.

Composting occurs in a trench that is 2 to 3 ft deep, where the material is placed in alternate layers with soil. The usual retention time is 120 to 180 days. In other systems, the compost may be compressed into blocks and stacked from 30 to 40 days and then aerated; or the compost may be placed into rotating drums, silos, circular tanks, open tanks, or conveyor belts.

Innovative Uses of Compost

Compost may be used in the following manner:

- As a soil amendment
- As a component in manufactured topsoil
- As a mulch or top dressing
- As a means of erosion control
- In the process of hydroseeding
- In wetlands mitigation
- In filtration berms
- In bioremediation and pollution prevention
- In disease control for plants and animals
- In soils contaminated by explosives
- In reforestation, wetlands restoration, and habitat revitalization

Compost has been known to be a valuable soil amendment for centuries. It helps improve plant growth. It also reduces erosion and nutrient runoff, alleviates soil compaction, and helps control disease, pest infestation, and plants. Compost also reduces the use of chemical fertilizers and conserves natural resources.

In the poultry industry, composting has become a cost-effective method of mortality management. It destroys disease organisms and creates a nutrient-rich product that can be used or sold.

Compost can be used on land that has been contaminated with explosives. The compost can effectively be used to remediate the soil at munitions sites. The contaminated soil is excavated, mixed with other feedstocks, and composted. The final product is a contaminant-free humus that is used to enhance landscaping. This process is considerably cheaper than soil excavation and incineration.

Compost can help restore original wetland plants. It provides for tree seedlings added vigor for survival and growth. These native plants within a habitat are more than just beautiful, for they produce food and grasses, and become humus to support the growth of other living plants.

Compost bioremediation refers to the use of a biological system of microorganisms in a mature, cured compost to sequester or break down contaminants in water or soil. Microorganisms consume contaminants in soils, in ground- and surface waters, and from the air. The contaminants are digested, metabolized, and transformed into humus, carbon dioxide, water, and salts. Compost bioremediation can degrade or alter many types of contaminants, such as chlorinated and nonchlorinated hydrocarbons, wood-preserving chemicals, solvents, heavy metals, pesticides, petroleum products, and explosives. The remediation helps turn the land into its prior condition, through revegetation and soil stabilization with compost soil conditioning and soil nutrients.

Compost bioremediation has been used to effectively treat lead and other heavy metal contaminated soils in Pennsylvania, Maryland, and Virginia, in both urban and rural areas. The lead found at high levels in soils adjacent to houses painted with lead-based paints had its bioavailability effectively reduced by the compost.

Soil at the Seymour Johnson Air Force Base near Goldsboro, NC, is contaminated as a result of frequent jet fuel spills and excavation of underground oil storage tanks (USTs). The contaminants include gasoline, kerosene, fuel oil, jet fuel, hydraulic fluid, and motor oil. The base has implemented a bioremediation process that uses compost made from yard trimmings and turkey manure, instead of excavating and high-temperature incineration. The remediation process includes spreading compost on a 50- by 200-ft. unused asphalt runway; and applying the contaminated material, another layer of compost, and then a layer of turkey manure. Fungi in the compost produce a substance that breaks down petroleum hydrocarbons, allowing bacteria in the compost to metabolize them. A typical compost pile has 75% contaminated soil, 20% compost, and 5% turkey manure. A mechanical compost turner mixes the layers to keep the piles aerated. A vinyl-coated nylon tarp covers the piles to protect them from wind and rain, and to maintain the proper moisture and temperature for optimal microbial growth.

Storm water runoff, which is excess water not absorbed by soil after heavy rains, carries a large number of potentially harmful environmental chemicals, such as metals, oil and grease, pesticides, and fertilizers. A compost storm water filter, which is a large cement box with three baffles to allow water to flow inside, is designed to remove floating debris, surface scum, chemical contaminants, and sediment from storm water by allowing it to pass through layers of specially prepared compost. The porous structure of the compost filters traps the physical debris while it degrades the chemical contaminants. Scum baffles along the side of the unit trap large floating debris and surface films. The small volume of specially prepared compost removes over 90% of all solids, 85% of oil, and between 82 and 98% of heavy metals from storm water runoff. It typically has a low operating and maintenance cost and has the ability to treat large volumes of water, up to 8 ft^3/sec. When the compost is no longer effective, it can be removed, tested, recomposted to remove contaminants, and used as daily landfill cover, because the metals are bound by the compost.

Compost bioremediation can be used to remove volatile organic compounds (VOCs) and reduce odors. The microorganisms in the compost digest the organic, odor-causing compounds. Billions of aerosol cans are produced and used each year in the United States. When they are disposed of, they still carry remnants of the paints, lubricants, solvents, cleaners, and other products containing VOCs. Activated carbon has been traditionally used to treat these cans prior to disposal. However, vapor-phase biofilters using compost are now becoming an alternative technology for treating aerosol cans. The biofilters break down the contaminants into harmless products, instead of trapping them, as is done in activated carbon filters.

Ocean Dumping

Approximately 60 million yd^3 of dredged sediment from U.S. waterways is disposed of in the ocean at designated sites. The U.S. EPA and the U.S. Army Corps of Engineers regulate the disposal of dredge material in the ocean and inland waters. Before sediments can be permitted to be dumped in the ocean they are evaluated to ensure that the dumping will not cause significant harmful effects to human health or the marine environment. The EPA also encourages the reuse of appropriate dredge material to restore or create wetlands or other habitat where feasible. The 1972 Marine Protection, Research, and Sanctuaries Act prohibits the dumping of material into the ocean that would unreasonably degrade or endanger human health or the marine environment. This act was amended in 1988 to ban ocean dumping of industrial waste and sewage sludge. Nonpoint sources of pollution are also addressed.

The EPA environmental criteria under this act basically provide that no ocean dumping is to be allowed if the dumping causes significant harmful effects; or if it cannot be determined whether the dumped material can cause significant harmful effects.

The act requires the following activities:

1. The EPA develops regulations detailing the environmental criteria to be used to evaluate proposals for ocean dumping.
2. The EPA designates specific locations where dumping may occur subject to a permit and manages the sites.
3. The Army Corps of Engineers issues permits for ocean dumping of individual dredging projects, and the EPA reviews these individual permits for potential problems in the dredged material and must concur in granting of the permit.
4. The Clean Water Act requires a similar method for the discharge of dredged material into inland waters.

Evaluation of the dredged material to be dumped helps protect human health and marine environment. The sediments can be contaminated by chemical pollutants, which may accumulate in marine organisms and cause death to the organism. It may also bioaccumulate and become a hazard to people who consume the food.

Oceans may also become polluted from ships. Thus, the Oil Pollution Act of 1990 revised vessel design and operation requirements and extended a system of planning for, and responding to, oil spills. This act required all tankers in U.S. waters to have double hulls, be responsible financially for all damages, and contribute to a trust fund maintained by a 5 cent per barrel fee on oil.

Pyrolysis

Pyrolysis is the chemical decomposition of a material by heat in the absence of oxygen. Pyrolysis is often called destructive distillation because organic material is heated to temperatures of 1000 to 2000°F. These high temperatures cause a chemical breakdown into three main components: a gas consisting primarily of hydrogen, methane, carbon monoxide, and carbon dioxide; a tar or oil that is liquid at room temperatures and includes such organic chemicals as acetic acid, acetone, and methanol; and a char consisting of almost pure carbon, plus other inert materials, such as glass, metals, and rock. Because pyrolysis is an advanced waste treatment technique, its application in solid waste disposal has been limited.

Laboratory and pilot plant pyrolysis units have proved the technical feasibility of pyrolysis of municipal refuse. Studies have been conducted by the University of California at Berkeley, Bureau of Mines, Rensselaer Polytechnical Institute, and

New York University. Salable synthetic heating fuels, glass, and magnetic metals have been removed from municipal refuse by the pyrolysis technique.

Heating gas on a sample basis or on a pilot plant basis has been produced from the solid char that is the product of pyrolysis in the upper portion of a reactor. The char is oxidized in the bottom part of a reactor by a mixture of oxygen and steam. The hot reaction product gases continue upward and release their heat to cause charring of the inner refuse. Finally, the residual heat in the gases evaporates moisture from entering refuse at the top of the reactor. This not only helps to prepare the solid waste for pyrolysis but also to leave a group of gases that contain hydrogen, oxides of carbon, water vapor, and mixture of hydrocarbons. These gases can then be burned in a secondary burner for heat.

Pyrolysis Oils

Pyrolysis oils increasingly are becoming an extremely valuable by-product of the pyrolysis process. Fast pyrolysis is a high-temperature process in which the biomass is rapidly heated in the absence of oxygen. As a result it tends to generate mostly vapors and aerosols and some charcoal. After cooling and condensing, a dark brown mobile liquid is formed that has a heating value about half that of conventional fuel. Pyrolysis liquid is called pyrolysis oil, bio-oil, bio-crude-oil, bio-fuel-oil, etc. It has an acrid smoky smell. Its physical properties include moisture content of 15 to 30%; pH of 2.5; specific gravity of 1.20. Its elemental analysis on a dry basis consists of carbon at 56.4%; H at 6.2%; oxygen at 37.3%; N at 0.1%; and ash at 0.1%. The liquid fuel can be substituted for conventional fuels, but it is not as stable as fossil fuels. Pyrolysis oils can also be used for production of other chemicals, which may become a valuable commodity.

Carbon Black from Pyrolysis Oils

The disposal of 280 million tires generated each year in the United States is a serious problem. It is estimated that 2 to 3 billion tires are already stored in tire piles in this country. The tires take up large amounts of valuable landfill space and also may cause fires, be the habitat for mosquitoes, and cause other health hazards. The tires may be processed by a pyrolysis system. Oils derived from waste tire pyrolysis may be turned into carbon black, which is a valuable feedstock for the manufacture of other tires, other rubber products, paints, pigments, ink, powder coating, toner, etc. The carbon black produced from tire oils contains the well-defined grapelike structures that are characteristic of carbon black. No contamination from char residue is observed. This makes the pyrolysis-produced carbon black an excellent substitute for new oil, thereby conserving a valuable natural resource. This process can be used, if all the new research shows that it is successful, to recycle a huge stream of solid waste, which mostly ends up in landfills.

Reclamation and Recycling

Solid waste can be a resource instead of an expensive nuisance. Municipal solid waste contains by weight approximately 38.1% paper and paperboard, 13.4% yard wastes, 10.4% food wastes, 5.9% glass, 7.7.% metals, 5.2% wood, 9.4% plastics, and 9.9% other.

The major question is not how valuable the material in solid waste is, but how expensive it is to get the recyclable material out of the solid waste. It is the demand for recycled material that counts and not the availability of it. Cost is an overriding factor when it comes to the use of virgin materials vs. recycled materials. Unfortunately, in many instances, individuals receive depletion allowances for utilizing raw materials. This makes raw material usage a good payoff instead of making it a more costly form of obtaining materials for processes. Considerable discussion will be needed in future years concerning the best technique to make recycled material less expensive, and as a result, more usable. Obviously, future techniques of recycling will have to be improved. No longer is it feasible to hand pick recyclable materials out of solid waste. Other techniques have been developed and must be used.

By 2000, the yearly recycling programs were deemed successful, because the cost of the program was less than the sum of revenues from sales of recycled materials and the cost of landfilling or incineration. In the United States, thousands of successful recycling programs were in use. Successful community recycling programs were custom designed by local people to meet the needs of their community. Good public education programs, use of volunteer citizen leaders, specially designed and well-marked separation containers, and appropriate collection times for the recycled material aided in the success of these programs. Typically, a program had as its goal the separation of glass, paper materials, and metals, and in some cases, other types of disposables. A new special problem is the recycling of old computers. The National Safety Council estimates that over 315 million computers will become obsolete by the year 2004, and this may be low because reliable numbers were not available between 1980 and 1992. Trashing all these computers would add an estimated 8.5 million tons of waste to U.S. landfills. These units contain lead, mercury, cadmium, and other bioaccumulative and toxic compounds.

Separation

Size reduction is the first step in reclamation and recycling. Techniques of size reduction in composting are also applicable to all other types of reclamation and recycling. (Refer to previous sections for detailed information in this area.) Following size reduction, separation is the next critical step.

Because typical solid waste is heterogeneous, and because the diversity of recovery operations are substantial, the efficiency of the operations from an economic standpoint must be considered when determining which operation to utilize and which to discard. Hand picking and sorting from conveyors is still used, although it tends to be an expensive operation due to the cost of personnel. Clean newsprint, cardboard, rags, metal, glass, and plastic can be removed this way on a small-scale basis. For large recovery it is limited, because salvage prices are low.

Many mechanical means are available for separating materials for recycling purposes. Magnetic separation removes ferrous materials from industrial and municipal solid waste. The recovery of scrap in metal fabrication plants and automobile salvage operations is valuable. In municipalities, this technique is now used in compost plants in Houston, TX; Gainesville, FL; and Johnson City, TN. In Atlanta, GA and Chicago, IL, magnetic separators remove tin cans for salvage. Generally two types of separators are adaptable to the removal of ferrous materials from mixed refuse. These are classified as suspended-type and pulley-type separators. The suspended-type separator makes the first removal of ferrous material from the refuse. The pulley type acts as a second-stage cleanup to remove additional ferrous material. Efficient magnetic separation is affected mostly by the amount of size reduction and how much the ferrous material has been physically separated from other materials.

The separation of nonmagnetic conductive materials, such as copper, aluminum, and zinc, is based on the eddy current phenomenon. This phenomenon occurs when an electric current is passed through a coil surrounding a core made of a conducting material. The magnetic flux that develops changes with the amount of electricity supplied. The varying flux helps in the separation of the materials. This technique needs considerable more research.

Vibrating screens are used to separate materials based on particle size. The screen may be wet or dry. Screens are currently used in two compost plants in Houston, TX, and one in Altoona, PA, to remove glass and other unwanted materials.

Spiral classifiers are pieces of equipment in which minerals may be removed. Fine solids are removed from coarse solids of approximately the same specific gravity. By controlling the design and operation of the equipment, a separation of materials may occur.

Flotation is basically a gravity separation process. Dissimilar solids are separated by creating an aqueous slurry or pulp and then inserting air. The heavier solids sink to the bottom and the lighter materials remain at the surface. Dense media separation is a form of sink float gravity separation where materials with a greater specific gravity sink, whereas those with a lighter specific gravity float.

Separating tables are utilized to separate the materials into low-specific gravity, intermediate-specific gravity, and high-specific gravity. This process tends to separate the metals from the other types of waste materials. Mineral jigs are mechanical devices that separate presized materials of different specific gravities by the pulsation of a stream of liquid through the bed of material. The heavy material works its way to the bottom of the bed and the lighter material rises to the top.

Optical sorting machines use surface light with reflectional properties to separate a variety of materials. Inertial separation is now in the research phase. The inertial separators segregate materials according to the densities of the substances. The conveyor depends on both the densities and elastic properties of the substances. A myriad of techniques attempt to separate various substances found in solid waste. In time, with adequate research, a sound economical technique is likely to be determined.

Plastics

The use of plastics exceeds 40 million tons per year. Plastics average about 19.9% of municipal solid waste by volume. Little of the plastic scrap is recycled other than that reused within the manufacturing plant. The two types of plastics are thermaplastics and thermosetting plastics. The thermosets, which comprise about

20% of plastics used, cannot be softened and reshaped through heating and therefore cannot be recycled. Also, plastics used for coating or adhesives are impossible to recycle. However, roughly 75% of the plastics used can be recycled. The packaging industry uses 20% of the plastics produced and the construction industry uses 25%. By weight, plastics from packaging account for about 60% of the plastics found in solid waste material. Plastics are used in such items as water hoses, weatherstripping, toys, cheap housewares, and pipes. They also serve as a fuel supplement in energy generation because they have a value of 11,500 Btu/lb vs. a value of 8000 Btu/lb for paper and 12,000 Btu/lb for coal.

Unfortunately, because plastics originate as pure material and become progressively contaminated, the technology of purifying plastics is undeveloped. One of the principal difficulties in recycling is the different polymers used. Polyethylene and polyvinyl chloride are incompatible, making it difficult to separate these different plastics. Separation of plastics mixed into other wastes is also difficult and therefore expensive. Probably the best means of recycling plastic cheaply is reuse of plastic bottles. However, before this can be done, considerable research has to be conducted on the types of contaminants found in plastic containers and the techniques needed to sterilize the containers before reuse. The cost of sterilization may be greater than the cost of the original production of the plastic container.

Less than 1% of postconsumer plastic is recycled. Almost all such plastic comes from bottles made from polyethylene terephthalate (PET) and high-density polyethylene (HDPE). PET containers are used most widely for large soft drink bottles and HDPE is used in milk and detergent containers.

Many problems are related to recycling of plastic. First, most uses of recycled plastic require that it be in the pure form of the plastic. However, many products are made up of layers of different resins laminated together or affixed rigidly. Mixed plastic waste can be made into a woodlike construction material, which can be cut, sawed, and nailed to make fencing, boardwalks, docks, outdoor furniture, road signs, and other items that are able to resist weathering. The second problem is the fact that a scrap market has not been developed for plastic other than PET and HDPE. States that require mandatory soft drink deposits provide most of the PET currently recycled.

PVC pipe industry could easily absorb the 560 to 570 million lb of PVC currently in use for packaging. However, no mechanism is available to identify, collect, and transport this material to the industries. A third problem is the low density of the plastic; the bulk of the plastic is such that it is expensive to transport. The fourth problem is that the consumers are not aware of the recycling potential of plastic as they are of paper, aluminum, and glass and, therefore, tend to throw the plastic away.

Paper

Paper is one of the major manufactured materials consumed in the United States and is the largest single component of municipal solid waste. It comprises 38.1% by weight of all municipal waste. Since 1969, in the United States use of paper has almost doubled. The three major categories of paper include paper, paperboard, and construction paper. Since 1969, the amount of paper products recycled has about tripled. Paper can be classified in four major grades: mixed, which accounts for 27.4%; newspaper, which accounts for 19.8%; corrugated, which accounts for 32.6%; and high grade, which accounts for 20.2%. Waste paper comes from residential, commercial, and other sources. Waste paper is generated in conversion operations where paper and board are used in consumer products. This material is almost all recovered. It is easily accessible, generally uncontaminated, and almost half of such wastes are composed of desirable high grade. However, paper wastes from residential sources are widely dispersed, and are highly contaminated with adhesives and coating, and other materials. It becomes costly and difficult to remove them. Therefore, almost none of this material is recycled. The only paper that is available in significant quantities from residential wastes and is of value, is that used in newspapers. This material can be recovered and utilized.

Commercial wastes consist largely of business papers, mail, packing materials, and specially corrugated boxes. These can be recycled. The question is not whether newsprint or paper can be recycled, but instead whether it is cheaper to recycle or to use new wood pulp in the paper production. Many interrelated factors discourage the use of waste paper when virgin pulp may be utilized. These include the following:

1. Paper is collected from diverse sources and must be transported several times, whereas pulp travels from a single source to processing.
2. Contaminants are present in waste paper and affect recycling.
3. Waste paper prices fluctuate widely.
4. Improvements in wood pulping technology allow for cheaper use of the virgin raw materials.
5. Most paper mills have their own forests and also own most of the paper equipment installed since 1945. This equipment is wood pulp based and located close to the virgin raw materials.
6. Tax breaks through cost depletion allowances and capital gains allowances make it more economical to use the virgin wood instead of the recycled waste paper.

The paper industry contaminates the pulp during the course of preparation. Two examples include the sulfite or acid pulp process and the sulfate or kraft pulp process. In the sulfite process, the wood is cooked with sulfurous acid and a base. The sulfite goes into most printing grades of paper, such as business paper. In the sulfate process, materials are produced by chemical methods, using an alkaline solution of caustic soda and sodium sulfide. This produces paper used for paperboard, coarse papers, unbleached grades, packaging materials, and printing grades. In either case, when the papers are destroyed, sulfite or sulfate is part of the material and may therefore produce sulfur dioxide or hydrogen sulfide during a burning process.

Glass

Glass comprises 2% by volume of all municipal solid waste. Almost no recovery of glass takes place from mixed waste. A small amount is recycled through voluntary collection centers and from cullet dealers. This amount seems to be increasing. Recycled glass is utilized in asphalt, which is a road-surfacing material, or glass cement blocks of cullet–terrazzo. Recycled glass should be readily usable, because it can be substituted for the virgin raw materials in glass furnaces. The two problems that occur are costs and the difficulty of recovering large amounts of cullet, or crushed waste glass, from municipal waste.

Cullet is the key to glass reuse in the United States today. The use of cullet by glass manufacturing industries has grown from 24,000 tons a year in 1970 to more than a million tons a year today. Almost any glass factory can use a certain amount of cullet as feedstock. Cullet is a highly desirable component of the mix going into the glass furnace, because it melts at a lower temperature than the other raw materials used in glass, such as silica. The advantages of mixing in cullet include savings in energy cost, reduction in air pollutants, and extension of the life of the furnace refractory linings because the temperature needed for production is lower. The problems of using cullet are reliability, quality, and transportation of the material.

"Furnace ready" cullet used by container manufacturers has to meet fairly detailed specifications. Cullet sorting and purity (this means keeping, paper, plastic, metals, dirt, and ceramics out of the glass) are critical factors for bottle and jar makers. If the cullet is used to make fiberglass, insulation, or "glasphalt," the specifications for the cullet are much less stringent. Quality can be achieved by communities and their residents separating cullet glass from other glass and making sure the glass recycling container is kept clean and free of other materials.

Although most glass furnaces can accept cullet, they react poorly to rapid changes in the mix of cullet and other raw materials. This creates a supply problem. The communities have to have a fairly constant mix of glass to resolve this problem.

Because glass factories tend to locate near deposits of high-quality sand and other raw materials, the cost of transportation may be considerable. These raw materials frequently are not close to the greatest population concentrations where the largest amount of glass would be produced. Inexpensive means of transportation of this glass is important for recycling and reuse.

Ferrous Materials

Ferrous solid waste, which comes primarily from food and beverage containers plus discarded appliances, comprises about 7.7% of collected municipal solid waste. Steel cans and appliance metal recycling has increased about 250%. Iron and steel products contain iron ores and obsolete scrap metals. Part of the recyclables circulate throughout the industry. They are consumed in one area, produced, and finally recycled to start the process again. The industry uses between 30 and 40% of available obsolete scrap each year. The absolute amount varies with the economy, as does the amount of steel produced. Unfortunately, over the years the percentage of recyclable ferrous metals has not increased, even though the need for recycling these metals has sharply increased. The type of steelmaking has a direct bearing on scrap usage. Three types of furnaces are used, including the open hearth furnace, which uses about 45% scrap charge; the basic oxygen furnace, which uses approximately 30% scrap charge; and the electrical furnace, which uses 100% scrap. Basic oxygen furnaces have replaced open hearth furnaces. In the foundry industry, scrap accounts for about 85% of the metallic input. Scrap steel is purchased by Japan and recycled through their steel-making process. Copper precipitation is the major market for the use of steel can scrap.

Because of rigid steel production specifications, it is necessary to remove contaminants and impurities. Only 0.6% nonferrous metals, excluding manganese, is permitted in iron and steel scrap. Manganese may be at a 1.65% level. The metallic impurities, called residual alloys, are nickel, which is limited to 0.045%; chromium, which is limited to 0.20%; and molybdenum, which is limited to 0.10%. The various steels and irons found in solid waste contain higher percentages of these metals and therefore are not usable. Steel cans generally contain aluminum, which fortunately is not a critical contaminant in steel, and lead at 0.5% by weight. Lead is difficult to form as an alloy with steel, and as a result it becomes harmful to furnace bottoms and refractories. Copper, found in copper wire, is in the municipal waste stream; it tends to plate ferrous metals, making removal from the incinerators difficult. The copper level in steel is estimated at 0.1%. However, even at this level it is undesirable because it weakens the steel. One of the purposes of steelmaking is to remove the sulfur from the original ores. The sulfur, which is introduced in scrap, causes difficulty in the steelmaking process.

The major problems in the use of scrap ferrous metals are the quality of the scrap, the changing technology of iron and steelmaking, the low growth rate of consuming industries, the cost of the scrap, and finally the tax advantage in the use of the raw iron ore.

Steel recycling has improved in one area as a result of minimills, which produce steel products almost exclusively from scrap. Minimills increasingly account for more steel made in the United States.

Nonferrous Metals

Nonferrous metals are divided into three basic categories: (1) common metals, such as aluminum, copper, zinc, and lead; (2) exotic metals, such as nickel, cobalt, chromium, titanium, zirconium, molybdenum, tungsten, columbium, tantalum, and beryllium; and (3) precious metals, including gold, silver, platinum, palladium, rhodium, and iridium. Although magnesium and tin are categorized as common nonferrous metals, they are consumed at a rate of only about 10% of these other metals. The common nonferrous metals are of greatest concern in solid waste recycling. Because the value of such elements as zinc and copper are so much higher than iron and steel, the incentive to recycle these metals is much greater. However, they appear in small quantities and are often combined with other metals. They require processing before they are of value for recycling.

Aluminum is used today in cans, foils, trays, and other kinds of containers. The integrated primary aluminum industry converts bauxite to fabricated aluminum products. The nonintegrated aluminum industry uses scrap, and primary and secondary aluminum ingots. The secondary smelters convert scrap aluminum into secondary ingots. The major consumers of scrap are the secondary smelters, followed by the integrated producers. Much of this scrap is internal, produced during the process of making various aluminum products.

In addition, aluminum scrap is obtained from fabrication and conversion, and from obsolete products. The new scrap is pure and greater in value. The old scrap, which comes from junked airplanes, dismantled cars, power cables, household products, and so forth, is of value in the secondary aluminum industry. Because economics is the key to all recycling, the fact that secondary alloys sell cheaper than primary alloys encourages the use of aluminum scrap in certain types of products. Collection and removal of aluminum cans to central places where they can be utilized by aluminum processors, makes aluminum reclamation a useful program. However, if personnel costs exceed the value of separating the aluminum from the rest of the solid waste, once again this waste becomes a burden on our environment.

The other nonferrous materials occur in small quantities in the mixed municipal waste. Metals are found in major appliances, bulky waste, brass and bronze fixtures, automobile grills, addressographs, and so on. With the exception of lead, many of these materials are extremely difficult to recover and utilize. Lead is recoverable from obsolete batteries at a very high rate, and therefore can be recycled. The recycled copper usually comes from construction sites and the recycled zinc from new scrap. Although nonferrous materials are valuable, the difficulty of separating them from other waste limits recycling.

The production of aluminum in the past had been frequently criticized because of the large use of energy and the environmental cost associated with its manufacturing and disposal. Today producing a new aluminum can from a recycled old one uses only 5 or 10% as much energy; 65% of the aluminum cans produced in the United States are recycled.

The aluminum industry has undergone a considerable structural change. Although the domestic production of the metal from the ore has remained fairly flat or declined in recent years, the production from scrap has moved rapidly ahead. Because of technological developments, there has been a 20% reduction in the weight of the average aluminum can.

Textiles

Textiles comprise about 1.9% of municipal solid waste in the United States. In the production of textiles, reusable wastes are created for use within the industry in the production of paper or board products; or in textile products, stuffings, fillings, and packings. Reliable information on textile waste generation and consumption appears to be limited. One of the major problems in recycling textiles today is that one of the major markets, rag paper, is affected by the amount of polyesters introduced into clothing. In the past, old clothing was used in rag paper. At present, rag paper can only have a 1 to 2% maximum of contaminants in the cotton that is utilized. Another major problem in recycling is that wool, which contains impurities, must be listed on the garment label. People are more apt to buy a garment containing virgin wool instead of secondary processed wool. Used textiles go into the wiping cloth cycle. It takes 6 to 12 tons of used clothing collected from households to make 1 ton of wiping material. Obviously, if personnel cost is involved, the cost of recycling becomes very expensive. It appears that, as time goes on, less textiles will be recycled than in the past.

Rubber

Rubber is reclaimed from old tires, retreads, and tire parts. Then it is processed and reused. The reclaimed rubber is competing with natural and synthetic rubber and other materials. With the exception of tires, rubber in mixed municipal waste is expensive to remove and becomes an impractical recyclable material.

Organic Waste

Organic waste includes slaughtering, food processing, agricultural, and wood wastes. Slaughtering waste comes from slaughtering plants, stores, restaurants, and collection of dead animals. The rendering plant produces grease and tallow, meat meal, bonemeal, and feather meal from this material. Animal fats are made into soap, fatty acids, animal and poultry feeds, and lubricants.

The food processing industry disposes of meat, poultry, and various other food wastes. Generally, the wastes are uneconomical or violate environmental health laws, and therefore must be disposed of in ways other than recycling. Some food plants utilize their waste for hog feeding. Food waste can be processed into animal feeds, where high protein or oil content is needed. Agricultural wastes are generally recycled as fertilizer.

Wood wastes come from logging operations and other tree-trimming or tree-cutting activities, as well as from lumber mills. Wood wastes are used as agricultural mulches, in building board mills and paper mills, as fuel, and in chemical applications (such as distilleries and vinegar manufacturing).

Inorganic Waste

Inorganic waste consists of building rubble and ashes. The inorganic portion of building rubble is composed of concrete, bricks, rock, masonry, plaster, clay, glass, nonferrous materials, and steel. These materials are produced when roads and streets are torn up and when buildings and other structures are razed. Almost all this material goes to the landfill. Because the material is inorganic, it provides a good substructure for subsequent construction if it is packed properly.

Ashes originate in incinerators. If the incinerator works properly, the ashes are an inert residue. They may be used for roadbed maintenance for railroads and as fill. Fly ash is produced by electrical utilities in the combustion of coal. The ash is used in concrete, road construction as a soil stabilizer, asphalt paving filler, and production of lightweight aggregate. It is also added to foundry sands or masonry mortar.

Used Oil

Millions of gallons of used oil are sent to municipal landfills, disposed of in backyards, or poured down sewers, thereby contaminating the environment and wasting a nonrenewable resource. In California alone, over 20,000,000 gal of used oil are illegally discarded. The recycling of used oil is of great importance to our society. Vehicle maintenance produces about 700 million gal of used oil each year, of which about 380 million gal are recycled. Recycling of the used oil reduces potential pollution of the surface water, groundwater, soil, and air. To recycle used oil it is necessary to:

1. Collect used oil from the vehicle maintenance area.
2. Store the used oil in nonleaking drums or tanks.
3. Avoid mixing degreasing solvents or other substances into the used oil.
4. Have the used oil picked up by a reputable collector and taken to a used-oil processor.

Yard Waste

The separation of yard waste from other solid waste is easy. Most of the yard waste comes from suburban lawns and gardens. Yard waste can be converted into compost or other soil conditioners or mulches that can be sold to the public, to garden and plant shops, and to farms; or used in public parks, in gardens, and on roadways.

Energy Conservation

European countries have pioneered in heat recovery from the incineration of various municipal wastes. These incinerators have been working in France, Germany, and Switzerland for years. The heat produces steam, which is used for either heating or generation of electrical power. The two types of furnaces used are water wall and refractory wall furnaces. The usual refractory wall furnaces operate at a 150 to 200% excess air. The water wall furnaces, although somewhat more expensive at the outset and in operation, need only 50 to 60% excess air and operate at considerably higher temperatures. This increases the efficiency and reduces the amount of air pollution control equipment needed to control pollutants.

Most of the municipal solid waste combustion in the United States incorporates recovery of energy as steam or electricity. The total U.S. municipal solid waste combustion referred to as waste to energy, has a design capacity of over 32 million tons/day. More than 100 waste-to-energy facilities are in operation, with the pre-ponderant number in the northeastern and southern regions. These facilities generate enough power to meet the needs of 2.4 million homes. More than 39 million people in 31 states use the source of energy. Waste to energy generates about 20% of total biomass generation of power. Beside combusting mixed municipal solid waste, there is a growing amount of combustion of source-separated municipal solid waste. Rubber tires are used as fuel in special facilities, or as fuel in cement kilns. Wood waste, paper, and plastic waste are also burned in special boilers. An average waste-to-energy unit with a capacity of 1000 tons/day can burn 310,000 tons of trash per year, recovering 2 million Btu of energy, which is enough to light, heat, and cool 60,000 homes a year. Since 1980, the National Aeronautics and Space Administration (NASA) at the Langley Research Center in Hampton, VA has saved over 50 million gal of fuel oil by converting the agency waste to energy.

Hydrolysis

Hydrolysis is a means of using agricultural residues and wood waste. Glucose is obtained from hydrolysis of the cellulose portions of the plants. The lignin remaining after recovery of hydrolysis products is a potentially valuable chemical raw material. Lignin is also hydrolyzed into alcohols and acids.

Other Conservation Methods

Extraction of various materials from solid waste by solvents is possible. However, it is quite costly. Carbonization of carbonaceous waste has been employed to obtain activated carbon. This is done on a commercial scale.

SOLID AND HAZARDOUS WASTE PROBLEMS

Magnitude and Composition of Solid Waste Materials

Solid waste includes nonhazardous solid waste, as well as hazardous solid waste (see RCRA definition). Refuse is an older term for solid waste. It includes all the putrescent and nonputrescent solid waste, except body waste. Garbage is putrescent animal and vegetable waste resulting from handling, preparation, and consumption of foods. Rubbish is all nonputrescent solid waste, except ashes. Ashes are the residue from the burning of wood, coal, coke, or other solid combustible waste.

The management of solid waste is not only a problem of sharply increasing quantities of many new kinds of products and packaging materials, and various pesticides and other hazardous materials, but also benign neglect. In the past it was easier to discard unwanted material in open dumps. Saving, which was one of the original ethics of our country, has been discarded much as a throw-away can is discarded. We have become a society of affluent individuals who have lost their foresight by living only for today. The collection and disposal problems are enormous, and it is usually difficult for communities to cope. As a result, even though much of the solid waste discussed in this text is collected, we still have residential, commercial, and institutional wastes scattered or dumped illegally, which are contributing to a variety of environmental pollutants. Further, we have developed slag heaps, culm piles, and mill tailings from our mining operations that are not only unsightly but also potentially dangerous.

Public indifference is a part of the overall problem of solid waste. People seem to lack concern about the enormous quantity of solid waste that they produce. Reuse of materials is too infrequently the concern of citizens, because it is easier and in some cases less expensive to buy a new product than to repair an old one. The cost of labor for repairing television sets, washing machines, and other products has increased the problems of solid waste.

Urban wastes are extremely diverse. Even though they comprise a small part of the total amount of solid waste, they are most offensive because they are in areas where large numbers of people congregate. Urban wastes lead to odors, insects, and rodents, and of course become serious accident hazards. Urban residential waste differs from city to city, day to day, month to month, and season to season. During the summer, urban waste contains a substantial quantity of garbage, is much more prone to attract insects and rodents, and is much more difficult to incinerate. On Sundays, the newspapers add a substantial amount of the actual volume and weight of solid waste. During Christmas, there is a sharp increase in packaging materials; of course after Christmas, the enormous problem of disposing of millions of Christmas trees adds a further burden to the solid waste problem.

Packaging is very important in the market today, because the package in many cases sells or protects the product. It is estimated that the consumer pays for 75% of all packaging costs. The consumer wants containers that cannot burn, break, crush, degrade or dissolve. The environment needs containers that can burn, break, crush, and degrade or dissolve. In the United States, about 90% of packaging materials are discarded. The packaging industry produces all types, sizes, shapes, and materials. These materials include paper, metals, glass, wood, plastics, and textiles. Containers are important in the food industry for frozen, pressurized, freeze-dried, instant, and canned foods. They are also important in all other kinds of situations where the citizen wants convenience. It is easier to buy a package for four bulbs than one. It is easier to throw away the jelly jar than to reuse it as a drinking glass. It is easier to dispose of a plastic butter tub than to use it for storage of leftovers.

Land, Air, and Water Pollution

Land Pollution

The land and soil are blighted and polluted because of improper storage and disposal techniques. Abandoned automobiles, paper, bottles, and cans are found everywhere, from the city streets to the national parks. A properly operating sanitary landfill is quite effective because the solid waste is buried and the land reused; however, an improperly operated or improperly located landfill causes considerable soil, air, and water pollution. Some of the difficulties encountered include the production of methane gas in the ground, the contamination of various water bodies, and the lack of suitable and acceptable landfill sites.

Deep mines should not be used as dumps for solid waste because they may eventually contribute to water pollution, or the solid waste may start to increase in temperature and cause underground fires in the mines, leading to a deterioration of the site, as well as a constant air pollution problem. Soils are also contaminated with a vast number of chemicals that are dumped into the soil. These chemicals are supposedly placed into satisfactory landfill sites, but eventually they create a hazard if they do not degrade. Therefore, additional thought must be given to disposal. Further, many nondegradable plastics, very slowly degradable metals, and other types of containers are dumped and remain in the landfill for long periods of time. These constitute a continuous hazard to the soil, because they release their contaminants for extended periods.

Incinerator Residue

The residue from normal incineration is about 20 to 25% by weight and 10 to 15% by volume of solid waste charged. The residue must be quenched after leaving the grate and placed on a transport vehicle. The residue is corrosive and abrasive, and causes considerable wear on all equipment. It is taken to landfills and buried.

The air pollutants including particulates, metals, and toxic substances that would normally have been released to the air from the incineration process, are now, for the most part controlled by the use of electrostatic precipitators, fabric filters known as baghouses, and scrubbers. Some facilities have all three anti-pollutant devices.

These strict air pollution controls produce another even more difficult environmental problem, the disposal of ash. Combustion facilities yield two kinds of ash: bottom ash and fly ash. Those facilities that are equipped with scrubbers cause the fly ash to combine with chemicals added to the flue gas, usually alkaline reagents such as lime. The combination of ash and lime equal 10% of the waste stream by weight. The mass burn plants reduce the weight of trash by 75%. The rest must be disposed of, usually in a landfill. Most of the ash is bottom ash or slag. About 10% of the residue is fly ash, the material that would become airborne if not captured by precipitators and baghouses.

Incinerators in the United States produce about 3 million tons/year of ash. It is believed that this amount may double or triple as new units come on-line and as air pollution control standards are tightened. The ash contains many of the chemicals found in the waste plus combinations generated by the combustion. Usually, the fly ash is much more hazardous, in terms of toxic metals and dangerous organic chemicals, such as dioxins and furans, than the bottom ash. However, the bottom ash may also contain significant quantities of lead and other hazardous materials. The two types of ash are mixed before landfilling to lower the concentration of hazardous materials. The behavior of the hazardous constituents of incinerator ash in landfills is uncertain. Although the dioxins that are present in the fly ash may be firmly bound to the ash particles, contact with solvents in the landfills can cause the dioxins to leach into the groundwater.

Leachate

All mixed refuse contains some pollution potential of either a chemical or a microbiological nature. These pollutants are produced either by the chemicals found within the solid waste that leach into the soil or by the decomposition of organic material that contaminates the soil or underlying water strata. Leachate from a landfill contains a series of contaminants. The pH generally is acidic, with a mean pH of 5.5. Apparently the greater the flow rate, the more acidic the leachate. Acidic leachate contributes to potential pollution, because the low pH tends to reduce the exchange capacities of the renovating soils at the time when the quantities are high.

Iron production seems to be higher when leachate production is high. The leachate iron may be in excess of 1600 mg/l. Zinc concentrations are usually between 15 and 30 mg/l with highs of 120 to 135 mg/l. The significant zinc levels are not shown until about 430 days have elapsed. This indicates that zinc ions are not released until certain types of refuse start to break down.

High phosphate concentrations occur shortly after the leachate begins to flow. Generally, however, the levels do not exceed 30 mg/l. At times, phosphates are not found at all. Sulfates are found as long as leachate is released from the landfill. Generally, sulfate concentrations increase as the leachate increases. Chlorides are present during the entire period that the leachate is present. Concentrations range from about 200 to 300 mg/l, although higher levels may be obtained. Sodium is present during the entire leachate process. Usually the concentrations vary from 200 to 300 mg/l, although much higher concentrations may be reached.

Nitrogen, generally is available at about 8 mg/l in the leachate, although higher concentrations may appear. Hardness in the form of calcium carbonate ranges from 2250 to 2750 mg/l. Peak periods may be much higher. Suspended solids average about 200 mg/l, although higher levels may he reached. Total solids range between 10,000 and 28,000 mg/l with higher peaks. Nickel may average between 0.2 and 0.3 mg/l, although higher levels may be found. Generally, nickel is not detected in the leachate until 150 days have elapsed. Copper concentrations are erratic. Generally, they are less than 0.1 mg/l although as much as 5 to 7 mg/l may be found.

Gas samples indicate the presence of carbon monoxide, hydrogen sulfide, nitrogen, carbon dioxide, oxygen, and methane in the leachate. Generally, the percentage of oxygen decreases with the depth of the leachate in the soil and with increasing time, whereas the percentage of carbon dioxide and methane increases. Chemicals found in the leachate more than likely will find their way down into a water stratum if the soil is sufficiently permeable. The biological contaminants found in the solid waste generally aid in the decomposition and stabilization of organic material. However, the by-products of decomposition may become a water pollutant. These by-products include hydrogen sulfide, methane, and ammonia.

The effect that the sanitary landfill wastes have on the environment is based on meteorologic factors, such as precipitation and evaporation–transpiration; geologic factors, such as type and sequence of material at the landfill; mineral and organic composition and concentration; permeability of the soil; attenuation characteristics of minerals through ionic exchange, dilution, dispersion, complexing, and filtration. Other geologic factors that control permeability rates of groundwater flow include translocation and removal of contaminants, depending on the type of soil and the size of the openings in the soil.

Hydrologic factors include the position of groundwater saturation, the characteristics and direction of water movement, and the zones of aeration and saturation. Surface water factors are controlled by the influent stream, the effluent stream, and the rising and falling of the stream due to climatic conditions and floods.

The leachate, which is, in effect, contaminated liquid, (Figure 2.12) can be produced by percolation of water vertically downward, movement of water horizontally through the side of the fill, water entering through the bottom of the fill, water present in the fill site initially, moisture within the refuse itself, and decomposition of the refuse. The entire landfill must not necessarily be saturated for leachate to be produced.

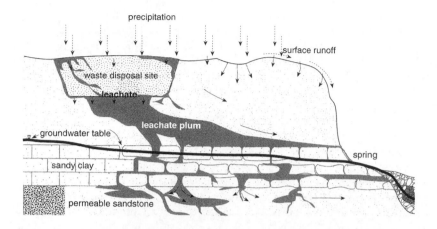

Figure 2.12 Leachate routes through waste disposal site and into the groundwater.

Predicting Leachate Production Rates

Good landfill design requires predicting the amount of leachate that will be produced. The amount of leachate can potentially affect liner leakage. It also affects the cost of postclosure care after the landfill is closed. To predict leachate formation requires water–balance calculations. The water–balance equation estimates the amount of water from rain or melting snow that can percolate through the landfill cover. Over a period of time, the volume of percolating water nearly equals the volume of leachate produced, although a lag time may occur between the percolation and the leachate production. The water balance equation is as follows:

$$PERC = P - AET \ R/O - \Delta ST \qquad (2.1)$$

where:
PERC = percolation
P = precipitation
AET = actual evapotranspiration
R/O = runoff
ΔST = change in soil moisture

The EPA, in cooperation with the Army Corps of Engineers Waterways Experiment Laboratory, has prepared a computer program that calculates the water balance, titled, Hydrologic Evaluation of Landfill Performance model version 3.0. This model has weather records and data files and offers options for predicting leachate generation under many combinations of cover conditions.

Air Pollution

Particulate emission from the incinerator increases with increased undergrate airflow. It is difficult to determine the proper amount of air needed below and above the grate, because the amount of air varies with the type of refuse to be burned and the type of refuse varies continually. Particulate matter includes organics, such as organic acids; and inorganics, such as aluminum oxide, silicon dioxide, potassium oxide, calcium oxide, iron oxide, titanium oxide, zinc oxide, sodium oxide, and magnesium oxide. The flue gases leaving the incinerator include the following organic gases: organic acids, esters, aldehydes, ketones, alcohols, hydrocarbons, polynuclear hydrocarbons, halogenated hydrocarbons, carbon monoxide, and carbon dioxide. The flue gases also contain the following inorganic gases: ammonia, nitrogen oxides, sulfur oxides, fluorine, hydrogen fluoride, chlorine, hydrogen chloride, and hydrogen cyanide.

Excess air has an effect on gaseous pollutants. The oxides of nitrogen are produced when large quantities of excess air exist at high temperatures. Gaseous pollutants appear to decrease with an increase in underfire air rates, whereas particulate matter appears to increase with an increase in underfire air rates. Other air pollutants are caused by open burning, blowing dust from landfills, and so forth.

Odors are associated with the incineration of organic material and are typically due to the incomplete combustion of the material. Because waste streams are not consistent in chemistry or quality, it is important to have a mechanism to trap the odors before they go into the ambient air. Odors are defined as a sensation perceived by the human nose. An odorant is a substance that stimulates an olfactory response, which is the sensation of smell. The odorant may be intense, pervasive, or acceptable depending on the nature of the odor. Typically, those odors coming from incinerators are not acceptable.

Water Pollution

Groundwater pollution by solid waste disposal is and has been a serious concern, especially in those areas where water is scarce and groundwater is essential. Most research has been conducted concerning the effects of leachate on surface waters. We do know that landfills leach into surface water and abandoned mines leach into surface or underground waters, eventually causing contamination of these water areas. However, research on the underground water supply is very definitely lacking. It is not known how much and what types of chemicals or gaseous products of decomposition are carried down to the water tables and through the water systems. A further concern involves the use of the ocean for dumping. Millions of tons of dredging materials have been hauled out to sea and dumped. We have no real understanding of what effect this has on the ocean environment and what long-range effect this will have on the fish life, which is so important to the economy and is a protein source for many people.

Solid Waste and Disease

Public health officials have always assumed that solid waste was an excellent source of disease. As a result of this, solid waste disposal laws have been acted on and resulting potential hazards have been reduced in many situations. Unfortunately, there is still a considerable lack of good research data on the spread of disease, although recently ecological and epidemiological data gathering has been started. This section of the solid waste chapter is concerned with the potential for disease and attempts to establish how disease may be spread through the storage, handling, treatment, and disposal of solid waste.

Solid waste can produce undesirable effects on humans by biological, chemical, physical, mechanical, or psychological means. Pathogens found in human feces are a biological threat. Chemical hazards are contributed by industrial waste or municipal waste. Flammable materials, which may be found in a variety of waste, cause a physical hazard of fires or explosions. Broken glass, pieces of metal, sharp-edged bricks, and other sharp-edged objects are mechanical hazards. Psychological problems are caused by the threat to the destruction of property, unsightliness of waste disposal, and depression that goes along with a cluttered dirty environment. In attempting to trace disease from solid waste to humans, it is necessary to determine what potentially harmful agents are present in the waste; whether harmful by-products are created in the waste; disease or chemical hazards attributable to solid waste; routes by which the disease agents or chemicals may affect humans; and whether an interruption of the pathway can result in a disruption of the disease or injury process. Some of these factors may be difficult to ascertain; however, using the intuitive powers of a scientist and the understanding that potential hazards may exist, the environmental health practitioner does not need to seek ultimate scientific justification for elimination of potential hazards.

General Routes of Transmission

Biological vectors, such as insects and rodents, may directly or indirectly transmit disease agents from solid waste to humans. Vectors multiply within or around the solid waste, which is often close to humans, and the vector then becomes a potential source of transmission. For example, flies transmit enteric organisms from fecal materials to humans. They easily transmit bacteria, viruses, protozoa, or helminths. As a second example, rats and mice are found in abundant quantities wherever solid waste is deposited. These vectors not only have a source of habitat and food but also readily transmit a variety of diseases to humans.

Studies of samples of incinerated residue indicate that large numbers of microorganisms, including coliform, fecal coliform, heat-resistant spore formers, and certain enteric pathogens, continue to exist after incineration. This suggests that many incinerators are not operating effectively. Some of the organisms found include salmonella, *Staphylococcus aureus*, *Diplococcus pneumoniae*, *Klebsiella pneumoniae*, and alpha hemolytic streptococcus. It is obvious that not only the inadequately burned residue but also the dust that gathers within the incinerators and the quenching water must be analyzed to determine whether microorganisms are present.

Pathogenic bacteria survive in composting material, particularly if sewage sludge has been mixed with the municipal refuse. A significant number of coliforms appear when the temperature of the composting process drops. *Mycobacterium tuberculosis* is normally destroyed by day 14 of composting if the temperature averages 149°F. If the compost has a temperature of 130°F or higher for at least 30 days, polio virus becomes deactivated. However, because research in this area is limited, considerable concern still remains about the potential of pathogenic organisms surviving in various types of compost.

Cysts, ova, and larvae parasites exist in compost. Helminth ova and larvae are found in various stages of stockpiled compost. In one study, two out of five samples of marketed compost revealed viable ova of *Trichuris trichuria* and *Strongyloides*. In addition, viable hookworm larvae were found. Because people walk barefoot and the larvae penetrate the foot, a potential source of infestation exists. Considerable additional studies should be conducted in this area to determine the potential micro-biological and parasitological hazards of composts.

Solid waste often constitutes a physical or mechanical hazard, because it contains flammables, explosives, or asphyxiating gases. These gases readily move through the soils, creating hazards to humans or their dwellings. Solid waste can cause injury by mechanical means.

Solid waste evolves during the combustion process. The particulates of burning coal or incinerators not only are a direct hazard in themselves but also carry with them pathogenic organisms, fungi, viruses, and spores. Direct contact with the solid waste is another source of disease. This occurs during the collection and disposal of solid waste by the workers, and may also occur during the household disposal and storage of solid waste. The potential of exposure is enormous and the exposure itself creates distinct hazards due to the proximity of individuals to the hazards.

Solid waste or the products of solid waste may leach into groundwater or aquifers that are used as a source of potable drinking water. In addition, solvents and other chemicals may permeate the groundwater supply. The disease-causing agents in solid waste, which may include anything from inorganic and organic chemicals to viruses and bacteria, can leach into the water, can be consumed by fish, and can travel through the food chain. The waste material may also be ingested by animals that are used as human food, and in this way can travel through the food chain to humans. Fertilizers and pesticides are particular hazards in any food consumed by humans.

Insect-Borne Diseases

Flies breed in large numbers in human and animal excreta, and in sewage sludge. Flies are highly adaptable and can readily change breeding habits. Flies breed in domestic animal manure, dog and cat droppings, backyard compost piles, spilled garbage, and so forth. The fly posing the greatest problem to humans is *Musca domestica*, or the housefly. This insect is frequently found in privies and is probably the fly that spreads the greatest number of diseases to humans. Although numerous sources of fly breeding exist, including dead animals and all types of excreta, the greatest production of flies occurs in garbage. The difficulty with flies is that they are in very close proximity to humans. They travel to a house and within the house, and land on food or other substances consumed by humans. Flies are shown to transmit, under controlled conditions, typhoid fever, swine fever, staphylococcus, and anthrax. In addition, they transmit cholera, various types of diarrhea, polio, dysentery, paratyphoid fever, and various foodborne diseases. The housefly can readily travel up to 3 miles from its source. It may spread disease to large urban communities.

Mosquitoes are known transmitters of numerous diseases, including dengue fever, encephalitis, filariasis, malaria, and yellow fever. Because many mosquitoes breed in artificial containers, they are frequently associated with areas of poor solid waste disposal. By properly disposing of the waste, water cannot accumulate and the breeding place is eliminated.

Animal Fecal Waste

Waste fungal diseases are spread by disposing of avian manures and feathers in landfills, because this material contains pathogenic soil fungi. Once the soil becomes contaminated, it continues as a reservoir of mycotic infections for humans for many years to come.

Direct Contact

It is possible for waste disposal workers to come in contact with a variety of infectious materials. Although anthrax is not frequently found in the United States, it can be transmitted from an infected carcass of the animal to a worker.

Disasters

During periods of disaster, it is estimated that enteric diseases are only second to respiratory diseases in incidence. The incidence is related to crowding and lack of facilities for good personal hygiene, especially a lack of hand washing facilities and procedures, which helps contribute to the spread of these diseases. However, during the course of a disaster, enormous quantities of solid waste are produced. It is not clear whether the solid waste becomes a source of the spread of disease or whether the solid waste is simply a function of the disaster or both. However, once again, using intuitive public health principles, it would seem that where large quantities of solid waste are present, where flies and mosquitoes can breed, and where rats can grow and breed in huge quantities, the disease spread by these vectors could easily affect humans. Because the solid waste contains chemicals, an additional potential hazard exists.

Chemical Hazards in the Human Environment

Endocrine Disruptors

The Food Quality Protection Act and the amendments to the Safe Drinking Water Act, both passed in 1996 by Congress, required that the EPA develop a screening procedure for endocrine disruptors by August 1998, and implement the screening and testing by August 1999.

Chemicals may interfere with the functioning of the very complex endocrine system. These endocrine disruptors can mimic a natural hormone thereby confusing the body into overresponding to the hormone. They may also block the effects of a hormone in various parts of the body normally sensitive to it. They may directly stimulate or inhibit the endocrine system leading to overproduction or under production of hormones. Some of the more environmentally persistent chemicals that have raised concerns about possible hormonal effects include PCBs, dichlorodiphenyltrichloroethane (DDT), chlordane, aldrin and dieldrin, endrin, kepone, toxaphene, and 1,4,5-T, etc. The foods that contain these potential chemical problems include freshwater fish, animal fat, etc. Food containers also need to be evaluated.

On-Site Incineration of Dioxin-Like Compounds

Waste chemicals such as dibenzodioxins, dibenzofurans, chlorophenols, solvents, and herbicides, including Agent Orange stored on-site in landfills and incinerated there, may lead to excess exposure in community residents within 25 miles of the storage and combustion area. Studies of these residents have indicated this.

Combustion of Hazardous Waste

Cardiac and respiratory symptoms have been associated with the combustion of hazardous wastes in semirural communities. Acute respiratory symptoms including morning cough or wheezing in the chest were higher in the incinerator communities than in control communities. Symptoms, which could be either cardiac or pulmonary in nature, such as tightness in the chest or shortness of breath were also significantly more prevalent among the residents of the incinerator communities. The same results have been found in urban communities living near hazardous waste sites.

Apparently, there is a greater risk to poor individuals of certain race and ethnic populations than the general community, because the Superfund sites, as well as the hazardous waste disposal sites tend to be located more frequently where these people live. Read section on environmental justice.

Currently, data are used on the physical growth and development of children, which accesses the impact of environmental influences, including toxic exposures. Growth impairment is a common response to chronic, low-level exposures to lead, PCBs, mercury, radiation, and possibly noise stress. Weight loss is one measure of toxicity.

The incidence of adverse health effects in the population to hazardous waste is significant. Some of these adverse health effects are as follows:

1. Cancer risk in the general population may be increased by the consumption of agricultural commodities, such as meat and milk, which are produced in the vicinity of hazardous waste combustors. Dioxins may be stored in the fat of these products, with the greatest risk potentially in the consumption of milk or other dairy products.
2. Cancer risk in the local population near hazardous waste combustors may be increased by exposure to carcinogenic compounds, such as dioxin, arsenic, beryllium, cadmium, chromium, and nickel. These compounds can enter the body by either the oral or the respiratory route or both.
3. Children who live near hazardous waste combustors are exposed to lead emissions through diet, inhalation, and incidental soil ingestion. Children that already have

elevated blood lead levels may have a further body burden because of this environmental exposure.

4. Emissions of particulate matter can increase the morbidity and mortality of community residents. The smaller sizes, below 2.5 and 10 μm in diameter are especially significant.
5. Endocrine disruptors have already been discussed.
6. Exposure to VOCs in contaminated water or just living close to the waste disposal site increases the risk of birth defects, stroke, and general illness.

Hazardous chemicals may cause liver, kidney, lung, heart, and skin damage. They may also affect the reproductive, nervous, cardiovascular, hematologic, and immune systems. Each hazardous waste site must be examined individually to determine whether a complex of chemicals may have affected people. Approximately 41 million people live within a 4-mile radius of the 1193 National Priority List sites. The risk to the public is categorized as follows:

- Urgent hazard, 1%
- Health hazard, 33%
- Indeterminate hazard, 20%
- No apparent hazard, 39%
- No hazard, 7%

Injuries and Occupational Hazards

Solid waste disposal consists of a number of operations and occupations, including the use of trucks, incinerators, bulldozers, and other kinds of equipment, as well as hand sorting and other hand operations. The exposure to hazards must be high. Studies show that high accident frequency rates occur in individuals involved in solid waste collection and disposal. An excess in arthritis and muscle and tendon disease also appears to occur. The hazardous materials present in ordinary solid waste can certainly cause problems of dust, fires, contamination, explosions, and mechanical injuries. Numerous cuts and abrasions, and injuries due to the lifting of heavy loads, are part of the overall picture. Sewage sludge disposal workers are exposed to hepatitis and other enteric diseases, as well as leptospirosis.

The labor force used in waste management includes roughly 60,000 men and women, who lift the cans, remove the trash, operate the landfills and incinerators, as well as the recycling centers. About 11% may experience a serious accident each year. Most of the accidents result in back and muscle injuries. The majority of the workforce who are employed by government are not covered by the Occupational Safety and Health Act. The training of these individuals is haphazard, although they are required to operate expensive, complex, and dangerous machinery. At the landfill or incinerator gate, they are responsible for keeping toxic, infectious, and explosive materials out of the waste stream. Further, at the hazardous waste site, personnel may be exposed to chemicals, fires and explosions, oxygen deficiency, ionizing radiation, biological hazards, safety hazards, electrical hazards, heat stress, cold exposure, and noise.

One of the major distinguishing factors between the hazardous waste site and other occupational settings is the uncontrolled condition of the site. Even when the site is under control, the danger may exist if the materials are not properly handled. A large variety and number of substances may be present at any given site. Hundreds or thousands of chemicals may be present in a multitude of different situations. Frequently, the identity of these substances on a site are unknown, particularly during initial stages of an investigation of a hazardous waste site. Workers are also subjected to dangers posed by the disorderly physical environment of hazardous waste sites and the stress of working in protective clothing. The working environment may pose an immediate danger to life or health, may not be immediately obvious or identifiable, may vary according to the location on the site and the task performed, and may change as site activities progress.

Chemical Exposure and Occupational Hazards

Preventing exposure to toxic chemicals is a primary concern at hazardous waste sites. Most sites contain a variety of chemical substances in gaseous, liquid, or solid form. These substances can enter the unprotected body by inhalation, skin absorption, and ingestion, or through a puncture wound (injection). The contaminant can cause damage at the point of contact, may be distributed throughout the body and cause somatic affects, or may be toxic in a given area of the body. Acute or chronic exposure may occur from working with chemicals. Acute exposures would occur during or shortly after working in an environment with a sufficiently high concentration of the contaminant to cause a major reaction by the individual. The concentration required to cause this type of reaction varies from chemical to chemical and from person to person.

Chronic exposure refers to working in low concentrations of the contaminant over a long period of time. Chronic symptoms or diseases may occur months or years after the individual worked in the environment. The toxic effect may be temporary and reversible, or may be permanent, causing disability or death. Some chemicals cause burning, coughing, nausea, tearing of the eyes, or rashes. Other chemicals may affect the respiratory system or may cause cancer. Still other chemicals affect the central nervous system and the peripheral nervous system. The effect of the exposure depends on the chemical concentration, its route of entry, duration of the exposure and individual smoking habits, alcohol consumption, use of medication, nutrition, age, and sex.

Inhalation is an important route of exposure at the hazardous waste site. The lungs are extremely vulnerable to chemical agents. Even if the lungs are not directly affected, the chemical may pass through the alveoli into the bloodstream, where it is transported to other parts of the body. Chemicals may not be easily detected because they are colorless, odorless, and their toxic effects do not occur immediately. Chemicals may also enter the respiratory tract through punctured eardrums caused by accidents.

Direct contact of the skin and eyes by hazardous substances is an important route of exposure. Some chemicals directly injure the skin, whereas others pass through the skin into the bloodstream where they are transported to organs that are affected by the chemical. Skin absorption is enhanced by abrasions, cuts, heat, and moisture. The eye is especially vulnerable, because airborne chemicals can dissolve in the moist surface and can be carried to the rest of the body through the bloodstream, because the capillaries are very close to the surface of the eye.

Ingestion is a less significant route of exposure at the hazardous waste site, unless the individual deliberately chews gum or tobacco, drinks, eats, smokes cigarettes, or applies cosmetics at the site. Injection of the chemicals occurs when the body is punctured by sharp objects.

Explosions and Fire

Explosions and fires at hazardous waste sites are caused by:

1. Chemical reactions that produce explosions, fires, or heat
2. Ignition of explosives or flammable chemicals
3. Ignition of materials due to oxygen enrichment
4. Agitation of shock- or friction-sensitive compounds
5. Sudden release of materials under pressure

Explosions and fires may occur spontaneously. However, they more frequently occur from activities at the hazardous waste site, such as moving drums, accidentally mixing incompatible chemicals, or introducing a spark from equipment into an explosive or flammable environment. At the hazardous waste site, the explosions and fires not only pose the hazards of intense heat, open flame, smoke inhalation, and flying objects, but also may cause the release of toxic chemicals into the environment.

Oxygen Deficiency

The oxygen content of normal air at sea level is about 21%. Physiological effects of oxygen deficiency in humans are apparent when the oxygen concentration in the air decreases to 16%. These effects include impaired attention, judgment, and coordination, and increased breathing and heart rate. Oxygen concentrations lower than 16% can result in nausea, vomiting, brain damage, heart damage, unconsciousness, and death. To avoid error, the oxygen concentration should not be below 19.5%. Oxygen deficiency may occur when oxygen is displaced by another gas during a chemical reaction, including combustion reactions. This is especially true in low-lying areas and in containers or confined spaces.

Ionizing Radiation

Radioactive materials emit potentially harmful radiation, including alpha, beta, and gamma radiation. Alpha radiation is stopped by clothing and the outer limits of the skin. However, alpha radiation, if taken internally, may be extremely hazardous. Therefore, the individual must be protected against inhalation or ingestion of the material. Radiation can cause harmful beta burns of the skin and damage to the subsurface blood system. Beta radiation is also hazardous if inhaled or ingested. It is essential to use appropriate protective clothing and scrupulous personal hygiene and decontamination. Gamma radiation readily passes through clothing and human tissue, and causes serious permanent damage to the body. Chemically protective clothing is of no value. However, respiratory and other protective equipment can help keep the radiation-emitting materials from entering the body by inhalation, injection, or skin absorption.

Biological Hazards

Waste from hospitals and research facilities may contain organisms that cause disease. Etiological agents may be dispersed into the environment through water and air. Other biological hazards may be present at the hazardous waste site, including materials from biological hazards producing sources such as insects, animals, and disposable materials. Protective clothing and respiratory equipment are essential to reduce the chance of exposure, and a thorough washing of potentially exposed body parts is important.

Safety Hazards

Hazardous waste sites may contain numerous safety hazards. Of course, these safety hazards may also be present in municipal solid waste sites. The hazards include holes or ditches; precariously positioned objects, such as drums or boards; sharp objects such as nails, metal pieces, and broken glass; slippery surfaces; steep grades; uneven terrain; and unstable surfaces, such as walls that may cave in, or flooring that may give way. The presence of the heavy equipment used in these areas may also create additional hazards for the workers.

Electrical Hazards

Overhead power lines, downed electrical wires, and buried cables pose the danger of shock or electrocution to workers if they come into contact with them or sever them during their work at the hazardous work site. Electrical equipment used at the site may also pose a hazard to the workers. All work should be suspended during electrical storms, and all wires should be protected and away from the workers.

Heat Stress

Heat stress is a major hazard, especially for workers wearing protective clothing. The materials that shield the body from chemical exposure also limit the removal of body heat and moisture. The personal protective clothing can create a hazardous condition. Depending on the ambient conditions and the kind of work done, heat stress can occur within as little as 15 min and may pose great danger to the workers' health. In the early stages, heat stress causes rashes, cramps, discomfort, and drowsiness, which result in impaired ability to function. Continued heat stress can lead to heatstroke and death. Appropriate training and frequent monitoring of personnel who wear protective clothing, proper scheduling of work and rest periods, and frequent replacement of fluids can protect against this hazard.

Cold Exposure

Cold injury due to frostbite and hypothermia occurs at low temperatures and impairs the individual's ability to work. Cold injury can be extremely hazardous when the windchill factor is very low. To guard against cold exposure, it is necessary to wear appropriate clothing, have warm shelter available, schedule work and rest periods, and monitor workers' physical conditions.

Noise

Work around large equipment often creates excessive noise. This may cause workers to become startled, annoyed, or distracted. It may cause physical damage to the ear, pain, and temporary or permanent hearing loss. It may interfere with speech communication and increase the potential hazards, because individuals are unable to be warned of dangers. Employees should not be subjected to more than 90 dBA (decibels on the A-weighted scale) for an 8-hour period of time. If individuals' exposure exceeds 85 dBA, they must be monitored by means of a hearing conservation program.

Hazardous Waste

All chemicals must be considered hazardous. The chemical products are present in dust and liquid form. The leaching action of rain causes the chemicals to move through the soil and into water supplies. The chemicals can be hazardous to the individual by contact with the skin, and also through accidental ingestion or inhalation of the material. Where the chemicals become too hazardous to deal with, they may be dumped into deep wells. It is possible that, in the drilling process for deep well dumping, the aquifers may also become contaminated by the hazardous chemicals.

Types of Hazardous Waste

Hazardous waste includes those that are ignitable, such as solvents, and can cause fires; corrosive, such as acids and can corrode metals drums; reactive, which are unstable and can cause explosions; and toxic, which are harmful or fatal when ingested or absorbed (Figure 2.13). The rate of hazardous waste generation is growing annually from increased production and consumption, bans, and cancellations on toxic substances and energy requirements. The generation of nonradioactive hazardous waste in 1995 was 279 million tons from 20,000 hazardous waste generators, regulated under RCRA. About 40% of these wastes by weight are inorganic material and about 60% are organic. Over 90% occur in liquid or semiliquid form. Hazardous waste disposal in landfill sites is increasing. Vapor-suppressing foam can be used in place of the 6-in. temporary dirt cover at the hazardous waste site. Because

Figure 2.13 Types of hazardous waste.

waste contaminates air and water, stringent air and water pollution controls have been put on other types of disposal.

Inadequate management of hazardous wastes has caused adverse public health and environmental problems. These problems may be short range, immediate, or chronic. By the year 2000, the EPA had taken over 2000 actions, called records of decision, on hazardous waste sites. The National Priorities List cites 1193 sites. These actions address immediate threats to air, land, water, and people.

Hazardous waste is any waste or combination of wastes that poses a substantial present or potential hazard to human health or living organisms. The sources of hazardous waste are scattered throughout the country. The waste comes from industry and the federal government, including the Department of Energy, Department of Defense, and Department of Agriculture. The waste may also come from hospitals and research laboratories. About 70% of the industrial hazardous waste comes from the mid-Atlantic, Great Lakes, and Gulf Coast areas of the United States. Sources of radioactive wastes are nuclear power generation, fuel-reprocessing facilities, medical research and developmental laboratories, industrial laboratories, and industrial usage. A special problem is related to uranium mill tailings. Nearly all the nonradioactive hazardous waste produced in the United States is toxic. Toxicity is

the ability of waste material to produce injury on contact, with or without accumulation, in a susceptible site in or on the human body. Most toxic wastes belong to
the inorganic toxic metals, such as salts, acids, or bases. The synthetic organic
materials are flammable or explosive. These categories overlap. A single substance
can be flammable, explosive, and also organic.

About 25% of the metals used today are toxic. The concentration in chemical
form of the toxic metals determines what their potential hazard is to health and to
environment. Some metals that are important to life in very low concentrations, such
as fluorides, are toxic in higher concentrations. The largest quantity of toxic metals
are produced in mining, metallurgy, electroplating, and metal-finishing industries.
Arsenic is collected from the smelting of copper, lead, zinc, or other ores that
naturally contain arsenic. Approximately 40,000 tons of arsenic are produced in
these ways each year.

Synthetic organic compounds include the halogenated hydrocarbon pesticides,
polychlorinated biphenyls and phenols. An estimated 6000 tons of synthetic organic
pesticide wastes are produced yearly. Many governmental agencies, as well as private
firms, have banned pesticides that should be discarded.

The flammable waste consists mainly of contaminated organic solvents, oils,
pesticides, plasticizers, complex organic sludges, and chemicals that have not met
normal production specifications. Highly flammable waste creates very serious disposal hazards. These hazards are increased by the mode of transportation and
handling.

Explosive waste includes obsolete explosives from the Department of Defense,
manufacturing wastes from the explosive-producing industries, and contaminated
industrial gases. The former practice of sinking ships containing obsolete munitions
has been discontinued. Better techniques have to be worked out for these hazardous
explosives. In the process of producing explosives, organic chemicals, such as
ammonia, nitric acid, and sulfuric acid, are utilized or are inadvertent by-products.
These wastes are difficult to dispose of in an adequate manner.

Most of the radioactive wastes consist of nonradioactive materials contaminated
with radionuclides. The concentration varies from a few parts per billion to as much
as 50% of the total waste. Radioactive wastes are categorized as low- or high-level
wastes depending on the concentration of the radionuclides. A long-term hazard is
not proportional to the level of radioactivity, but instead to the specific toxicity of
the material and its decay rate. The most significant radionuclides, from the viewpoint of solid waste management, are those that have half-lives of from months to
hundreds or thousands of years. This means that the waste may be radioactive for
extremely long periods of time.

High-level radioactive wastes refer to those wastes requiring special techniques
for eliminating the heat that is produced during radioactive decay. The heat can be
a substantial factor in causing changes in the concrete storage areas. A biological
hazard of the radioactive waste is due to the effects of penetrating and ionizing
radiation, instead of the toxicity of the chemicals themselves. The hazard from certain
radionuclides is more acute than the most toxic chemicals by sometimes as much
as 600%. This hazard cannot be neutralized by a chemical reaction or by any current
techniques. The only effective means is storage of the radionuclides under carefully

controlled conditions to ensure that they are isolated during the period of radioactive decay. Radionuclides are present in gaseous, liquid, or solid forms. Most of the problems that we might encounter are related to air and water pollution, because the solid forms, unless they become airborne, are not of particular concern as long as they are properly protected.

Biological waste includes pathological hospital waste and biological warfare agents. The pathological waste in hospitals is usually less infectious than the agents of biological warfare. However, both types of waste are infectious. Approximately 170,000 tons of pathological hospital waste are generated annually. This is approximately 4% of the total annual hospital waste. This waste includes malignant or benign tissue from autopsies, biopsies, or surgical procedures; animal carcasses and their wastes; hypodermic needles; materials and drugs that have not met specifications or have become outdated; microbiological wastes; infected bandages, sputum cups, and other materials of this type.

Biological warfare agents include any type of antipersonnel agent such as *Bacillus anthrax*. The quantities of these agents produced are limited in the future, because a halt has been called to the production and stockpiling of biological warfare agents. Chemical warfare agents include mustard gas, as well as other gases. These have been discontinued. However, significant stockpiles of these agents have to be treated and eventually discarded.

Pesticides are one specific type of community solid waste problem. Pesticides include, but are not limited to, dimethyldichlorovinyl phosphate (DDVP), dieldrin, chlordane, diazinon, parathion, and heptachlor. The residue of these pesticides is found in a huge variety of containers utilized by consumers in their homes and gardens. In addition, pesticides are introduced into the solid waste stream in processing and from pest control operators. It is obvious that the major current technique of burying pesticides in landfills is a highly inadequate method of disposing of them, because they stay in the ground for many years, and finally seep into and poison water supplies. There have been cases of individuals who have walked across plots of ground in agricultural areas and have become poisoned by a pesticide buried years before. The best technique for disposing of pesticides is by proper use of incinerators with adequate air pollution devices attached to the smokestacks. All other techniques contain certain hazards.

A number of factors are going to increase the quantity of hazardous wastes generated in the future and may have a very definite effect on disposal requirements. These factors include production and consumption rates, legislative and regulatory actions, energy requirements, and recycling. Production and consumption rates are increasing at a rate of 4 to 6% each year, whereas resource recovery from solid waste is declining. As a result of this, hazardous waste may increase. These wastes include dyes, pigments, heavy metals, and pesticides. As product material content changes, the problem of hazardous waste increases.

Hazardous Waste Transportation

Hazardous waste poses a serious transportation problem. The Hazardous Material Transportation Act gives the Department of Transportation the authority to regulate the movement of substances that may pose a threat to health, safety, property, or environment when transported by air, highway, rail, or water. Bulk containers shipped by water are controlled by the U.S. Coast Guard. All transporters must have EPA identification numbers. A uniform hazardous waste manifest must accompany all shipments. The shipper must provide the following information:

- Proper shipping name
- Hazard class
- Identification number
- Labels
- Packaging
- Markings
- Placards
- Shipping documentation

Waste Discharge Hazards

Improper disposal of hazardous materials may result in the materials discharged into surface waters. This can destroy aquatic and animal life. The land and ground-waters have been contaminated from improper storage and handling techniques, accidents during transportation, and indiscriminant disposal. Many chemical wastes have been disposed of by burial, since this was the easiest approach to use. Unplanned releases of hazardous substances into the environment are wide-ranging and can seriously harm the areas and cause significant health hazards. The Office of Technology Assessment estimates more than 400,000 potential hazardous waste sites in the United States. The cost of clean up could range between 500 billion and one trillion dollars.

Organic lead waste from manufacturing processes in the San Francisco Bay area amounts to about 50 tons/year. The waste is disposed of in ponds at an industrial waste disposal site. In an attempt to process this waste to recover alkyl lead, plant employees became intoxicated by the alkyl lead. In another situation, the plant employees were affected, as well as the employees in nearby industries, and even the toll collectors on a bridge used by the trucks carrying the material to the plant. The material was placed in holding basins.

An industry in Houston, TX became aware of the fact that its hazardous wastes were seeping into the Houston ship channel; the chemicals involved were cyanide, which was produced at a rate of 25.4 lb/day; phenols, which were produced at 2.1 lb/day; and sulfides and ammonia. The toxic waste came from blast furnace gas from the coke plants. Cyanide levels of 0.05 mg/l are lethal to shrimp and small fish found in the area.

In Mosco Mills, MO, an insecticide company rinsed and cleaned its truck by dumping unused endrin into the river. This killed 100,000 fish and closed off this area for fishing for an entire year. A manufacturing company in Waterloo, IA burned technical grade phosdrin and contaminated the plant area. The list of poisonings and problems goes on and on and on. They include phenol discharges in Kansas City, MO; hydrocarbon gases into the rivers; cyanide discharges from an air force base; poisoning of local water supplies by arsenic and other materials; phosphate spillage;

hazardous hydrocarbons in the Ohio River; and accidental release of the gas methyl isocyanate in the 1980s in Bhopal, India, which killed over 2800 people.

The water supplies of Toone and Teague, TN were contaminated with organic compounds that leached from a nearby landfill. About 6 years earlier, the landfill contained some 350,000 drums of pesticide, many of which were leaking. The landfill was closed because of the hazardous waste in the water supply. The water supply was no longer usable.

At least 1500 drums containing waste, mostly from metal-finishing operations, were buried near Byron, IL for an unknown period of time. Surface water, as well as the soil and groundwater were contaminated with cyanides, heavy metals, phenols, and other materials.

About 17,000 drums littered a 7-acre site in Kentucky, known as the "Valley of the Drums." This location is about 25 miles south of Louisville. Some 6000 drums were full and many of them were oozing and leaking their contents into the ground. Analysis of the soil and surface water showed some 2000 organic chemicals and 30 metals present. Waste containing HCB was disposed of in a landfill near Darrow and Geismar, LA. The HCB vaporized and accumulated in cattle over a 100-mi^2 area.

A fire broke out at a disposal site in Chester, PA where 30,000 to 50,000 drums of industrial waste had been received over a 3-year period. The smoke forced the closing of the Commodore Barry Bridge and 45 firemen required medical treatment as a result of the chemical fumes.

Love Canal is the most infamous of the many hazardous waste problems that have occurred. Love Canal located near Niagara Falls was the site of chemical waste that had been buried for a quarter of a century. The drums holding the waste corroded, their contents percolated through the soil into yards and basements, which forced the evacuation of over 200 families. About 89 chemicals were identified, several of them suspected as carcinogens. In the 1980s, many of these families had individuals who came down with leukemia or other forms of cancer. In the 1990s, many of these still suffered from a variety of chemically induced diseases.

In 1984, the EPA began a series of expensive emergency removal actions at the Baird and McGuire Superfund site, which was a 20-acre area south of Boston in Holbrook, MA. For over 70 years, the company formulated and produced soaps, disinfectants, floor waxes, pesticides, and herbicides. Over 100,000 gal of hazardous materials were stored in a number of tanks at the site. This action has been repeated regularly into the 21st century.

A truck carrying a 40,000-lb load of bottled concentrated acids crashed on Interstate 70 near Wheeling, WV. Toxic fumes were generated by the interaction of the acids, endangering nearby schools, residences, and a trout stream. The damage was so extensive that, although the acids were neutralized to prevent further damage, the highway had to be torn up and replaced.

Uncontrolled disposal sites containing hazardous waste and other contaminants have presented numerous serious environmental problems in the United States. The sites have contaminated groundwater, led to explosions, and presented other types of dangers to the individuals and to the environment. Most of the abandoned or inactive waste sites and many of the active hazardous waste facilities have had hazardous wastes escape at some point. Approximately 40,000 potentially contaminated sites that may pose a threat to human health or the environment have been reported to the EPA. At about 1200 of these sites 18 of the substances found are known carcinogens, and 30 of the chemicals possess systemic toxicities.

Underground Storage Tanks

Effective December 22, 1998, all owners of USTs had to upgrade, replace, or close those tanks that did not meet the EPA technical standards for protection against spills, overfills, and corrosion. The original orders had been issued more than 8 years previously and an extension was not granted by the U.S. EPA. Under Subtitle 1 of the Resource Conservation and Recovery Act, Congress had directed the U.S. EPA to establish regulatory programs that would prevent, detect, and clean up releases from UST systems containing petroleum or hazardous substances.

By the year 2000, the various states and EPA had made significant progress in implementing a comprehensive regulatory UST program. Over 1.3 million federally regulated substandard tanks had been permanently closed. During the past year, approximately 400,000 releases from federally regulated tank sites were reported, with work already begun on approximately 346,000 of these sites. However, a very serious problem still exists. The Congress in Subtitle 1 excluded from the program several types of tanks. These included farm and residential tanks of 1100 gal or less capacity holding motor fuel used for noncommercial purposes and tanks storing heating oil on the premises. In the late 1980s, approximately 2.7 million of these heating oil tanks and an estimated 300,000 farm and 100,000 residential tanks were in existence. No current estimate is available about the number of tanks that have been excluded, but they still constitute a large number and are likely to easily exceed the number of federally regulated USTs.

Currently, approximately 85% of the regulated tanks are in compliance with the spill, overflow, and corrosion protection portion of the regulations. However, the various states estimate that the operational compliance rate with the leak detection requirements are approximately 60%, and therefore the statistics are open to dispute.

The EPA compliance plan has four major categories as follows:

1. Active registered contains approximately 722,000 tanks.
2. Abandoned registered contains approximately 38,000 tanks. These tanks were abandoned largely because owners and operators were financially unable to meet spill, overflow, and corrosion protection requirements. They still constitute a potential hazard.
3. Active unregistered contains approximately 38,000 tanks. The problem is to identify these tanks and where they are located.
4. Abandoned unregistered contains approximately 152,000 tanks. This is the most diverse and complicated group of tanks. They have never been registered and have not been in use for an indefinite period of time. USTs taken out of operation before the effective date of the regulations, December 22, 1988, were not required to meet the prevention and detection requirements. Tanks taken out of operation before 1974 were not even required to be registered. However, all these tanks provide a future potential serious threat to the environment and the health of people.

Originally, the tanks including oil, were placed underground as a fire prevention measure. Products released from these leaking tanks can threaten groundwater supplies, damage sewer lines and buried cables, poison crops, and lead to fires and explosions.

Many of the petroleum tank systems were installed during the oil boom of the 1950s and 1960s. By 1985, studies of tank age distribution indicated that approximately one third of the existing motor fuel storage tanks were over 20 years old or of unknown age. Most of these aging tank systems were constructed of bare steel, not protected against corrosion, and were near the end of their useful lives. Many of these old tank systems have already started to leak. Adding to the problem of the old tanks still in use are the thousands of gas stations with buried tanks that closed during the oil crisis of the 1970s. These abandoned tanks frequently were not closed properly and may be sources of pollution to the groundwater supply. Tank integrity testing can be accomplished by doing a soil–vapor analysis, soil sampling, and groundwater sampling.

Although federal regulations do not include small home heating oil tanks, many states have adopted special regulations for these systems. It is even more urgent now to test all oil tanks, because their life expectancy is only 10 to 15 years. Various states have found an increasing problem with leakage from varied heating oil tanks. The storage tanks usually fail from rust perforation due to the effects of water inside the tank, water in combination with sulfur in the fuel, bacterial action, and other factors. Tanks that are abandoned without removing the remaining fuel, without confirming leakage, without cleaning the tank, and without using an approved filler cause potential hazards. Additional hazards may occur due to cave-ins.

Toxics Release Inventory

The Toxics Release Inventory (TRI) includes data for on-site and off-site releases of chemicals from a facility, on-site waste management of the chemicals, and transfers of the chemicals in waste to off-site locations for further management (Figure 2.14). The TRI data reflect releases of chemicals, and other waste management chemicals, but not exposure of the public to these chemicals. However, the data can be used as a starting point in evaluating exposures that may result from releases and other waste management activities that involve toxic chemicals. The TRI list consists of chemicals that vary widely in their ability to produce toxic effects. Some high-volume releases of less toxic chemicals may be a more serious problem than lower volume releases of more toxic chemicals.

The level of exposure to a chemical is based on a variety of conditions, including:

1. The potential degradation or persistence of a chemical in the environment is crucial. The longer a chemical stays in the environment, the greater the opportunity for exposure to humans. Microorganisms may degrade some chemicals such as methanol into less toxic ones. Metals are persistent and not degradable on release.
2. Bioconcentration of a chemical in the food chain, such as mercury, can potentially cause significant problems, although only small amounts have been released.
3. The environmental medium (air, water, land, or underground injection) helps determine the level of exposure to people. Releases of a chemical into the air can affect those in close proximity or those downwind from the facility. The chemicals

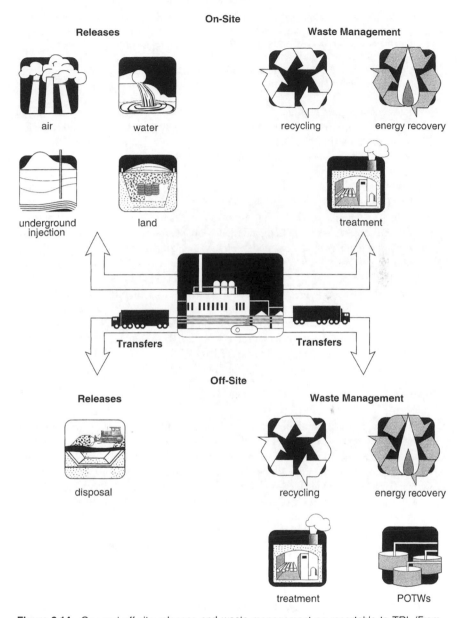

Figure 2.14 On- and off-site releases and waste management as reportable to TRI. (From U.S. EPA. *National Overview of 1996 Toxics Release Inventory.*)

may also contaminate the land or water bodies. Chemicals released in water may affect people downstream, who use the water for drinking, bathing, cooking, or recreational purposes. Injection of toxic chemicals into wells, if not properly protected, can affect the ground water supply.

4. The quantity of toxic chemical that ultimately enters the environment depends on transportation, disposal, treatment, energy recovery, or recycling techniques. The remainder after various types of treatment may still be hazardous and therefore a concern for the population.
5. On-site waste management depends on appropriate handling of the chemical or chemicals during disposal, treatment, energy recovery, or recycling.

TRI reporting sites include over 21,500 facilities, with 2.43 billion lb of toxic chemicals. Air emissions alone accounted for 1.45 billion lb. Underground injection accounted for 204 million lb. Discharges into water bodies accounted for 173 million lb. The remainder went to landfills after appropriate treatment. The top states for total releases were Texas, Louisiana, Ohio, Pennsylvania, Indiana, Illinois, Tennessee, and Alabama.

The TRI data lists specific chemicals including use, toxicity, and environmental fate; groups of chemicals of special concern, such as metals; chemicals identified or suspected as carcinogens; chemicals that may adversely affect children's health; and pesticides. Zinc, manganese, lead, mercury, and copper are of particular concern.

Children are especially responsive to toxic chemicals, because of their body weight relative to food and air intake, their play patterns resulting in higher outdoor exposures, their developing systems, and their relative inability to identify and protect themselves from exposure to toxins. In 1994, the EPA added 286 toxic chemicals to the TRI list, because of their potential adverse affects on children. These chemicals may cause structural abnormalities, reduced birth weight, and nonviable births, and create permanent problems in the growing child. Roughly one third of these chemicals can cause developmental effects.

Oil Pollution

The *Exxon Valdez* incident, where a large quantity of oil was spilled into the waters of Alaska, resulted in the Oil Pollution Act of 1990, National Oil and Hazardous Substances Pollution Contingency Plan, and Oil Spill Liability Trust Fund. The first National Contingency Plan was developed and published in 1968 in response to a massive oil spill from the oil tanker Torrey Canyon. The latest revisions to this plan were finalized in 1994, to reflect the oil spill provisions of the Oil Pollution Act of 1990. The potential for oil contamination is great, because about 120 billion gal of crude oil and other petroleum products are imported each year, and spillage can occur at every point during production, distribution, transportation, and usage.

Materials-Handling Technologies

Hazardous waste sites contain different types of materials that require physical separation, classification, and decontamination. The materials or debris are frequently contaminated with hazardous chemicals. Debris is any unused, unwanted, or discarded solid or liquid that requires staging, loading, transportation, pretreating, treatment, or disposal. Other materials handled include soil, sludge, asbestos, etc. Equipment may be needed for these materials and debris as well as for site preparation, pretreatment, and ultimate disposal. The diversity of materials, debris, and contaminants found on hazardous waste sites requires an evaluation be done of each waste site to determine what the content is and what may be needed to resolve the problems. The determination of the equipment to be used is based on:

1. Type and quantity of contaminants present
2. Amount and type of contaminants on the disposal site
3. Type of removal or remedial action needed
4. Treatment processes used on-site
5. General site characteristics such as climate, soil type, amount of moisture, and topography

The most frequent contaminants are:

- Trichloroethylene
- Lead
- Chromium
- Polychlorinated biphenyls
- Heavy metals
- Tetrachloroethylene
- Benzene
- Toluene
- VOCs
- Arsenic
- Cadmium
- 1,1,1-Trichloroethane
- Copper
- Zinc
- Vinyl chloride
- Xylene
- Chloroform
- Phenols
- 1,1-Dichlorethane
- Waste solvents
- Cyanides
- Nickel
- 1,1-Dichloroethylene
- Ethylbenzene
- Methylene chloride

Special handling is needed for textiles, glass, paper, metal, plastic, rubber, wood, construction debris, soil, sludge, liquids, and asbestos. Some of the equipment used includes backhoes, front-end loaders, bulldozers, forklifts, cranes, various pumps, vacuum units, vibrating screens, tire shredders, hand tools, pressure washers, air hammers, chain saws, cutting torches, etc.

Material at hazardous waste sites should be handled as follows:

- Identify debris and material to be handled
- Analyze for contamination
- If contaminated, segregate and pretreat and treat by contaminant class

- If not contaminated, segregate by type and modify size for transportation
- Recycle if possible
- Final disposal

At hazardous waste sites the following materials handling procedures may occur:

- Excavation and removal
- Dredging
- Pumping
- Size and volume reduction
- Separation and dewatering
- Use of conveying systems
- Use of storage containers and bulking tanks
- Use of containment
- Drum handling and removal
- Compaction
- Asbestos remediation
- Handling of low-level radiation
- Emission control
- Decontamination of equipment

Hazardous Waste Site and Waste Characteristics

When hazardous waste is removed to a disposal site, the following information is needed to make a proper assessment for control purposes. The site characteristics to be determined are site volume, area, and configuration; disposal methods; climate as it relates to precipitation, temperature, and evaporation; soil texture, permeability, moisture, and slope; drainage; vegetation; depth to bedrock and aquicludes; degree of contamination; direction and rate of groundwater; distance to drinking water wells and surface water; and existing land use. The waste characteristics include quantity, chemical composition, carcinogenicity, mutagenicity, toxicity, persistence, biodegradability, radioactivity, ignitability, reactivity, corrosiveness, treatability, infectiousness, solubility, volatility, safe levels in the environment, and compatibility with other chemicals.

Small Quantity Generators

Hazardous waste is also generated in small quantities from a series of businesses and industries. The cleaning and maintenance industry produces acids, bases, and solvents. Chemical manufacturers produce acids, bases, cyanide wastes, heavy metals, other inorganics, ignitable wastes, pesticides, and solvents. The construction industry produces acids, bases, ignitable wastes, and solvents. Educational and vocational shops produce acids, bases, ignitable wastes, pesticides, reactives, and solvents. Equipment repair shops produce acids, bases, ignitable wastes, and solvents. Funeral parlors produce solvents and formaldehyde. Laboratories produce acids, bases, heavy metals, inorganics, ignitable wastes, reactives, and solids. Laundries and dry cleaners produce dry-cleaning filtration residues and solvents.

Household Hazardous Waste

Household hazardous waste is a serious problem today, because the chemicals are either dumped into the trash or down the toilet. Americans generate 1.6 million tons of household hazardous waste each year. The average home can accumulate 100 lb of hazardous waste in cellars, garages, and closets. If the home is on an on-site sewage system, groundwater contamination may occur from the leaching of the chemicals down to the groundwater. Further leachate from landfills may also cause groundwater or surface water contamination. The discharge from sewage treatment plants for the homes that are on public sewage may cause problems in the stream in which the effluent is discharged. In the home, hazardous waste may be corrosive. Examples of this would be drain openers, oven cleaners, toilet bowl cleaners, ammonia and ammonia based cleaners, lye, pool acids, and photographic chemicals.

The chemicals may be flammable. Examples of this would be floor and furniture polish, spot removers and dry-cleaning fluids, disinfectants, air fresheners, aerosols, automotive waste oil, enamel- or oil-based paint, paint solvents, or thinners. The chemicals may be irritants, such as glass cleaners or air fresheners. The chemicals may be poisonous, such as antifreeze, pesticides, and herbicides; or the chemicals may be an oxidizer, such as chlorine. Car batteries are a special problem.

Asbestos Wastes

Asbestos is manufactured by mining the ore deposit and separating the fibers from the nonasbestos rock. The process of separating asbestos fiber, grading, and packaging the fibers is called milling. The mines generate waste that is piled adjacent to the mine. The mills generate asbestos containing waste on the air-cleaning controlled devices. Asbestos products are manufactured by combining the milled asbestos fibers with binders, fillers, and other materials. The resultant mixture, which may be dry or wet, is molded, formed, or sprayed, and then cured and dried. Manufactured products may then be fabricated by another manufacturer, the installer, or final consumer. Manufacturing and fabricating operations generate empty asbestos shipping containers; process waste, such as cuttings, trimmings, and off-specification reject material; housekeeping waste from sweeping or vacuuming; and pollution-control device waste from capture systems. Further, a significant quantity of asbestos-containing waste has been generated through the removal of friable (easily crumbled or pulverized asbestos material from buildings). All buildings prior to demolition or renovation must meet the federal regulations for the removal of asbestos. Asbestos must also be removed from schools.

Waste Streams

The following hazardous materials were found in the nation's waste streams: ammonium chromate, ammonium dichromate, antimony pentafluoride, antimony trifluoride, arsenic trichloride, arsenic trioxide, cadmium alloys, cadmium chloride, cadmium cyanide, cadmium nitrate, cadmium oxide, cadmium phosphate, copper arsenate, lead arsenate, mercury cyanide, mercury, potassium cyanide, silver cyanide, sodium arsenate, thallium compounds, zinc cyanide, fluorine, dieldrin, parathion, aldrin, chlordane, ethylene bromide, organic fluorides, and methyl bromide. At least 140 separate hazardous wastes have been identified in water, and it is certain that many more exist there and many more will be added over the years unless effective measures are taken to limit the types of waste discharge hazards.

Recycling

The health hazards associated with recycling include the possibility of the release of a variety of deadly organic or inorganic poisons and the release of a variety of hazardous metals. For example, it is quite possible that recyclable material could be contaminated with pesticides, acids, bases, or metals, such as beryllium. PCBs have been found in recycled paper. HCB has been found in the sublimation of land-disposed material. Organic mercury has been found in process waste.

Because the recycling of animal waste has been a useful means of providing fertilizer for the land, as well as crude protein, this has been frequently used as a technique for animal waste disposal. Certain environmental health hazards, however, are associated with this recycling. They include drug and hormone residues, metals, microbial agents, and transmission or induction of drug resistance. Residues of veterinary drugs have been found in manure. These drugs have contained metals, antibiotics, arsenicals, and hormones. Copper toxicity has been seen in sheep fed diets made up of the litter of broilers. Another problem is the continuing growth of salmonella infections. Because poultry litter and cattle manure contain a large number of pathogens that can be transmitted to humans, when these are utilized in the feed of other animals, eventually humans become the recipient of the disease.

Aesthetics

Even if solid waste were not a hazard, physically, chemically, or microbiologically, there still would be an aesthetic problem for the population. In areas with substantial quantities of solid waste, the environment loses its pleasant qualities. To walk along a beach and find hundreds or thousands of beer cans, along with other trash and garbage, destroys the pleasure of the sea and the rocks. A similar situation occurs within the cities. A comfortable or pleasant city street is ruined by the appearance of discarded solid waste, vacant lots piled with waste 2 and 3 ft high, abandoned automobiles, broken glass, and improperly stored solid waste. A neat and clean home gives an individual a comfortable feeling; when garbage and trash are scattered outside, the individual loses the pleasure of the area and in effect loses a good part of the environment that contributes to improved emotional and physical health.

Special Problems

Plastics

Disposal of polyethylene, as well as all plastics, is extremely difficult. In a sanitary landfill, the polyethylene can remain intact for many years. It is resistant to many chemicals, and no known bacteria can attack it rapidly enough to make its disposal effective. When the polyethylene is in rigid form, it is difficult to compact, creating landfill problems. It is not converted in composts, it is difficult to pulverize, and it creates problems in incinerators. Most conventional incinerators are designed to burn materials emitting much less heat than polyethylene. When heavier plastics are burned, they may form a slag or a molten mass on the furnace grate and reduce the amount of air available. The problems may increase as the use of plastics increase.

On-Site Incineration

The disadvantages of on-site incineration are that it produces strong burning garbage odors, has a poor residue quality, takes considerable time, causes substantial problems with wet refuse, does not achieve the necessary gas temperature level to properly incinerate all the material burned, causes fires, and produces such air pollutants as fly ash and smoke.

Other problems relating to on-site incinerators result from the use of flue-fed incinerators. Individuals do not understand the inherent limitations of these incinerators. There is little agreement among air pollution control officials as to what performance requirements should be expected of on-site incinerators. As a result, the existing codes reflect a diversity of opinions and approaches. Some of the problems with flue-fed incinerators, which are used primarily in apartment houses, originate in the design and construction of these units. This design problem leads to an excess of combustibles found in the residue, plus incompletely oxidized materials, such as highly odorous aldehydes, organic acids, and esters; small amounts of nitrogen oxides; and small amounts of sulfur compounds. Fly ash and other particulate material are also discharged into the atmosphere. Smoke and odorous gases escape not only from the chimney but also into the corridors.

Composting

A disadvantage of composting is that it may become a breeding place for flies and rats. The compost area must be cleaned out every 8 to 14 months, and the residue may be quite odorous.

Backyard Burning

Backyard burning in the past has been a volume-reduction technique used by many people. Although it does an adequate job of reducing volume, it produces air pollutants and odors; attracts insects, rodents, and other animals; and still leaves a residue that must be taken to a disposal site.

Garbage Grinders

The several disadvantages to garbage grinders include large bones, fibrous materials, and paper containing food particles cannot be fed into the garbage grinder and must be discarded with the rest of the solid waste. The grinder has a decided effect on the municipal system, because the garbage is transported by water to the sewage treatment plant. Grinders cannot be used without creating additional problems if the home is on a septic tank. The sewage treatment plant must be increased by 70 to 100% in the primary treatment phase to handle the additional sludge if the entire community uses garbage grinders. Also, the BOD increases at a level of about 50% per capita. Fats increase the BOD by 60% per capita.

Pulping

The disadvantages of pulping include the initial expense, the specialized equipment needed, the amount of water used, and the possibility of chutes clogging.

Compaction

The disadvantages of compaction include the requirement for compressed air in certain models, the necessity for having wheeled containers to move the solid waste, and the potential for chutes clogging.

Collection

When snow occurs, schedules are interrupted, frozen materials may attach to the containers, and depth of the snow plus the icy conditions may hinder the collection of the solid waste. Rain adds weight to the contents and also increases the possibility of accidents. Excessive heat slows the collection of the solid waste and both debilitates the workers and causes breakdowns in the equipment. Topography is a special problem, because the collection of solid waste must be determined by the type and nature of the land traversed by the vehicles.

Open Dumps

In 1969, in the United States, 77% of all collected solid waste went to 14,000 open dumps, 13% went to sanitary landfills, and approximately 10% went to incinerators. Land disposal, which was the most widely used technique, in many cases created extensive fly, rodent, and odor problems. Frequently, dumps spontaneously caught fire and created not only safety hazards but also air pollution problems. All open dumps are now illegal.

Several common errors were made in the closing of dumps. They included the underestimation of the increase in the volume of solid waste, because part of the solid waste had been burned in the past; an improper policing of the areas, which allows some of the individuals who had dumped in the past to continue to do so; an extremely serious problem of rats moving out of the dump area and into adjoining areas or homes; the problem of underground fires occurring when fires have not been completely extinguished; and the hazards to equipment operators, who may turn over equipment in areas where voids have been created because solid waste had been burned.

Several other concerns must be taken into account in closing dumps. It is necessary to know that the dump area was turned into a sanitary landfill and if so where and how the sanitary landfill was operated. It is necessary to know where the water table is, the type of soil that is utilized, and the ultimate use of the dump site. Further, wells for sampling should be sunk and leachate should be treated.

Redeveloping Brownfields

Brownfields are contaminated or abandoned industrial sites. Problems of cleanup, liability, and cost creates difficulties in dealing with these vacant parcels found in cities and towns. Superfund and other laws create serious concerns about redevelopment of these contaminated properties. The Brownfields Action Agenda is a comprehensive approach to reduce the barriers created by many laws and encourage redevelopment of these properties. In January 1995, to provide the EPA, states, and localities with useful information and new strategies for promoting environmental cleanup through redevelopment, 60 pilot projects at brownfields sites were started. The first project in Cleveland, OH leveraged 4.2 million in private investment, created 200 jobs, and generated more than $1 million in new payroll taxes for the city. By the end of 1995, EPA had removed more than 24,000 sites, nearly two thirds from the Superfund site database. Many of these were found not to be seriously contaminated whereas others were cleaned up under state programs.

The Brownfields National Partnership Action Agenda contains 129 commitments made by the partners, government agencies, communities, businesses, and nongovernmental organizations, to take specific actions to accelerate the assessment, cleanup, and redevelopment of brownfields. Important completed commitments are as follows:

1. More funding for brownfield assessments, cleanups, and redevelopment
2. Changed policies that remove unreasonable barriers to brownfield assessments, cleanups, and redevelopment
3. Support to more than 35 state and tribal voluntary cleanup programs
4. Expanded community outreach to increase brownfields pilot applications
5. Additional multipartner efforts to transfer information about brownfield technology, expertise, and remedies
6. More interagency agreements to reduce redundancy in brownfields activities
7. New publications that provide guidance, tools, and other resources for communities interested in working on their own brownfields

These programs have increased sharply into the 21st century. They help protect human health and the environment, while providing employment, training, environmental skills, cleanup, and redevelopment of properties that leads to an improvement in a community. Public participation is a key component in this program. The recycling of our land is an important issue in many cities and towns. It acts as a catalyst for sustainable economic development.

Federal Facilities

Contaminated federal facilities are one of the most important areas where mismanagement of toxics and hazardous materials occurred in the past. These facilities can be grouped into three major categories:

1. Nuclear weapons complexes
2. Nonnuclear industrial contamination sites resulting from federal operations
3. Land managed by federal agencies with contamination from governmental or private activities

The most difficult and costly problems are connected to the Department of Energy facilities that developed, produced, and tested nuclear weapons. These sites contain radiological and mixed waste that create unique technological and practical impediments to prompt remediation and restoration. There are 15 major Department of Energy facilities and a dozen or so smaller ones, with 6 of the major facilities producing about 80% of the problems. They are Hanford, Savannah River, Oak Ridge, Fernald, Idaho National Engineering Laboratory, and Rocky Flats. The federal government is spending over $9 billion a year to clean up federal facilities. Over the next 75 years, the cost of cleanup at Department of Energy facilities alone is estimated to be between $200 and $300 billion in 1995 dollars.

Nuclear Waste

The Department of Energy and Nuclear Regulatory Commission recognize the following six major types of waste:

1. Spent nuclear fuel is fuel rods permanently withdrawn from a nuclear reactor because they can no longer effectively sustain a nuclear chain reaction. The spent fuel contains some relatively short-lived fission products as well as long-lived radionuclides such as plutonium, which remains radioactive for tens of thousands of years. Spent fuel accounts for about 95% of all accumulated radioactivity. U.S. commercial nuclear reactors currently generate about 2000 tons of spent fuel annually. This is in addition to the approximately 40,000 tons of spent fuel stored at about 70 power plants sites around the country.
2. High-level waste includes highly radioactive residue created by spent fuel reprocessing, mostly used for defense purposes in the United States. Isolation of this material needs to be for 10,000 years or more.
3. Transuranic waste is relatively low-activity waste with some long-lived elements heavier than uranium, primarily plutonium, which is generated almost entirely by nuclear weapons production processes. Disposal needs to be long-term isolation.
4. Uranium mill tailings are sandlike residues from the processing of uranium ore. The mill tailings have low radioactivity, but a large volume of material.
5. Low-level waste, classified as none of the preceding, makes up more than 85% of the U.S. total, but less than $1/10$ of 1% of the radioactivity.
6. Mixed waste is made up of high-level, low-level, or transuranic waste, which contains hazardous nonradioactive waste.

Public Acceptance

Because solid waste disposal is a negative activity, it becomes an emotional problem or issue within a community. Although people complain about their electric, gas, water, and telephone bills, they recognize that these utilities are needed to have a properly operating home and community environment. However, solid waste disposal, which in effect is also a service, is not well accepted by the community. The worst problem is a lack of adequate communication between the community, the health department, and the other official agencies involved in solid waste management.

Economics

Recycling is affected by the economy and economic factors, which are linked to a need for the material and tax deductions for depleting resources. Constant fluctuation of prices for recycled materials causes a problem.

The cost of operation is extremely important. The community utilizes the least costly method of solid waste disposal. If good public health measures are to be instituted, the cost of solid waste collection and disposal must be kept at a minimum. Therefore, cost data must be collected on all phases of the operation, from labor to equipment, from collection to disposal, from utility cost to administrative cost, and from initial capital investment to final usage value of the land.

POTENTIAL FOR INTERVENTION

The potential for intervention varies with the type of solid waste. In general, with the exception of certain types of hazardous wastes, the potential is good. Intervention consists of isolation, substitution, shielding, treatment, and prevention. Isolation occurs when hazardous substances are removed to specialized landfills and are buried in special leakproof containers. Substitution is the use of a less hazardous for a more hazardous process. In general, this occurs when hazardous materials can be put through a proper pyrolysis process, where adequate heat is applied and all gases are reburned and then scrubbed. Mechanical equipment may also be used in place of hand operations. In the shielding process, the workers wear gloves, safety goggles if necessary, and proper clothing. Treatment is given to all injured or contaminated individuals. Treatment is also a technique used to reduce the level of toxicity, size, and composition of materials, and to neutralize such substances as acids and bases. Prevention is used to reduce the potential spread of disease and the time period during which solid waste is produced. Prevention also occurs when an industrial process is altered to reduce the waste generated.

RESOURCES

Scientific and technical resources are found in agricultural production corporations, building construction corporations, chemical and allied products manufacturing, and new product manufacturing plants. Community groups involved are basically at the local level, with some associations at the state level. Governmental resources include all state health departments, environmental protection agencies, many local health departments, National Bureau of Standards, Department of Defense, EPA, Bureau of Mines, NASA, and National Science Foundation.

Resources include the office of Emergency and Remedial Response Superfund of the EPA, Washington, D.C.; American Institute of Chemists, 7315 Wisconsin Ave., Bethesda, MD; Chemical Manufacturers' Association, 2501 M Street, N.W., Washington, D.C.; Chemical Waste Transportation Council, 1730 Rhode Island Ave., N.W., Suite 1000, Washington, D.C.; Clean Sites, Inc., 1199 North Fairfax St., Alexandria, VA; Hazardous Waste Services Association, 133 New Hampshire Ave., Suite 1100, Washington, D.C.; Hazardous Waste Treatment Council, 1919 Pennsylvania Ave., N.W., Suite 300, Washington, D.C.; Industrial Chemical Research Association, 1811 Monroe, Dearborn, MI; International Society for Chemical Ecology, 101 T.H. Morgan Building, University of Kentucky, Lexington, KY; National Agricultural Chemicals Association, 1155 15th St., N.W., Washington, D.C.; Natural Resources Defense Council, 122 E. 42nd St., New York, NY; Sierra Club, 730 Polk St., San Francisco, CA; Synthetic Organic Chemical Manufacturers' Association, 1330 Connecticut Ave., N.W., Washington, D.C.; Council for Solid Waste Solutions, 1725 K St. N.W., Suite 400, Washington, D.C., 1-800-2-HELP-90. The EPA has produced many fine documents in a variety of areas, including "Guide for Industrial Waste Management."

Other resources include the RCRA hotline, which is 1-800-424-9346; National Pesticide Telecommunications Network, 1-800-858-7378; United Air Toxics Web site; U.S. EPA Office of Solid Waste, 401 M St., S.W., Washington, D.C.; Association of State and Territorial Solid Waste Management Officials, Suite 345, Hall of the States, 444 North Capitol St., N.W., Washington, D.C.; EPA regional offices in the 10 EPA regions; and state hazardous waste agencies, which are found in all states in the Union.

STANDARDS, PRACTICES, AND TECHNIQUES

Individual Storage

When considering how to store material in residential areas, it is important that the containers be of a manageable size, usually a 20- to 30-gal capacity. The containers should also be a convenient shape. Where solid waste material is stored in any area in which rats or mice are present, it is important that metal cans, that is, galvanized steel cans, be utilized. The reason for this is that rats can readily eat through plastic, paper, and even lightweight rubber containers. Plastic containers and lightweight rubber containers cause an added problem because they can be easily upset and may roll for considerable distances if wind is present. It is well to line the galvanized metal container with a plastic sack to keep the container as clean as possible. The lids to the containers must fit snugly; otherwise, rats can knock them off, climb in, and obtain an adequate food supply. In any case, whether single service, kraft paper bags, or plastic liners are used for holding solid waste, it is important that the contents be protected against the weather, be prevented from scattering, and be stored so that insects or other vectors cannot get at the solid waste.

Where containers are utilized in an apartment complex, the containers should be in a rack at least 12 in. off the ground to provide for adequate cleaning of spillage under the slotted rack. Containers must be maintained in a clean condition or they become an insect and rodent attractant. This can be done not only by using the previously mentioned liners but also by washing containers regularly and taking the leftover solid waste and placing it in another container. Solid waste may be stored out of doors on racks in proper containers as in enclosed trash sheds, or it may be stored in the garage or basement. Generally, the site must be convenient to the homeowner, as well as to the collector of the solid waste. The site that is least odorous and least apt to attract insects and rodents is the most desirable. The solid waste should be picked up regularly. During the summer, two collections are needed per week of any solid waste containing garbage or just a simple garbage collection. During the winter, a minimum of once a week collection is necessary. The reason for these time periods is to prevent flies from breeding and also to reduce potential odor problems.

Community Storage

In most communities the solid waste is collected from the residents, the various apartment complexes, and so forth on a regular basis. The storage of the waste is an individual task. In some communities, however, metal containers such as large metal boxes are placed at strategic positions, and individuals are then able to put their solid waste into the containers.

The community's main function, however, is not in the storage of the solid waste on a large scale, but instead in supervising proper storage of solid waste so that the solid waste may be stored and collected properly. The community then further supervises proper disposal techniques.

Industrial Storage

Industrial storage varies tremendously with the type of industrial operation and the nature and quantity of solid waste generated. It also varies with the shape, volume, and construction of the type of solid waste generated by the particular industry. In some industries, such as coal mining, the solid waste is left at the site. In other industries, the solid waste is hauled off to various types of disposal systems.

Dump Closing and Conversion to Landfills

Once the decision is made to close dumps, citizen support is gained, and new sites and collection techniques are developed; then the technical phase of dump closing begins. A typical dump-closing operation includes fencing in or restricting unauthorized access to the dump site, placing necessary informational signs advising that the dump is closed, indicating where a landfill or other proper solid waste disposal sites are available, assigning a responsible manager to the dump site, stopping the burning of solid waste, stopping scavenging by individuals, controlling the immediately preventable water pollution problems, closing the dump to incoming refuse or establishing a specific spot on the dump for sanitary landfill operation during the closing period, controlling insects and rodents prior to the actual dump closing, providing necessary drainage, properly grading the land, cleaning all land of junk and debris, compacting and covering the material, planting grass seed in the area, and maintaining the area free of litter.

Various methods are utilized in closing dumps. The trench method is used when a high water table is not present. A trench is dug in the ground, and the loose solid waste is pushed together into the trench, compacted, and covered with the excavated material. A minimum of 2 ft of soil cover is needed at the top of the trench. A second trench is then dug in the ground and the process is repeated.

Another technique used for dump conversion is the area method. This is utilized where high water tables are present. A pile of soil is placed at one part of the site. The solid waste is then packed against the earth and compacted. A 2-ft final cover is placed on the compacted solid waste. In this operation, a working face is present. The working face should be compacted each night as much as possible and 6 in. of soil should be utilized as a face cover. The technique is repeated until all solid waste has been compacted and covered. The bank method is a modification of the area method, where the bank of a slope is used as the pile of soil against which the solid waste is compacted. After the solid waste has been compacted, a 6-in. soil cover is placed on the working face. The working slope in this technique should not exceed a 3:1 ratio.

The wet land method is used where the dump is in a marshland or in a river or other watercourse. The solid waste is removed from the water and stockpiled. Impervious material is then placed in the water area above the saturation zone and compacted tightly. The solid waste is dumped back on the impervious material and compacted. The final area has a 6-in. face cover, a 2-ft final soil cover, and a maximum 3:1 slope. In all four techniques, proper grading is necessary to prevent water from seeping into the solid waste material.

As has been mentioned, the scattering of rats is a substantial potential problem. It is necessary that rat control start a minimum of 1 month prior to the dump closing. Initially, gassing of rat burrows should be utilized. Large quantities of anticoagulants should be placed in areas where rats can readily get at it. As the dump is closed down, the rats increasingly seek out the rat poison. It is essential that the rat-baiting stations, both dry and liquid, be kept completely full at all times.

A modification of the dump-closing technique, in the handling of the solid waste itself, is to use grinders. Grinders have the advantage of reducing volume, eliminating voids, permitting easier handling and easier compaction of materials, and homogenizing the material. Another alternate technique is baling. This is used in an open dump where the quantity of refuse is so great that it cannot be disposed on the actual site under ordinary conditions. Baling equipment reduces the volume and helps control dust and odors.

Sanitary Landfills

To determine whether land and soil pollution can be averted, it is necessary to determine the geology, the size, the cost, and the proximity to existing dwellings and subdivisions; the availability of adequate quantities of cover dirt; and whether individuals within the area will accept the landfill site.

Problems of sanitary landfills are the siting of new ones, generation of methane gas, pollution of ground- and surface waters by leachate, and generation of odors. Another serious concern is keeping inappropriate materials out of landfills.

Site Preparation

Prior to the actual operation of the sanitary landfill, certain preliminary preparations are necessary. They are the clearing and cleaning of the site; constructing necessary access roads; providing necessary utilities and drainage; providing employee facilities and equipment facilities, including a communications system, a fire protection system, an equipment maintenance system, and all necessary safety and first aid equipment; and providing adequate movable and stationary fencing.

At the landfill site, guard rails, bumper logs, guide and directional signs, identification, public relations, and information signs are needed to facilitate the landfill operation. Restricted access is necessary to prevent illegal dumping, salvaging, and vandalism, and to promote good relationships with the community.

Safe landfill management techniques include the use of liners; application of cover materials; leachate collection, removal, and treatment; groundwater monitoring; and control over the materials placed in the landfill.

Leachate Management

RCRA subtitle D regulations establish a timetable for incorporating liners, leachate control systems, and final cover systems into the design of new municipal solid waste landfills. In addition, groundwater monitoring systems are required for new, existing, and lateral expansions of existing landfills. Groundwater protection performance standards are required for a series of chemicals including metals and organic compounds.

Landfill Liners

Clay liners require very low permeability. Flexible membrane liners are better because they overcome known leakage from clay liners. The design considerations of the flexible liners ensure compatibility with the waste, structural soundness, good seaming, firm base, freedom of debris or sharp objects underneath, construction quality control, and protection of the liner after construction. A typical flexible membrane liner thickness is 0.030 to 0.080 in. A concern about the synthetic liners is that certain chemicals may interact with the liner and affect its integrity (Figure 2.15).

Leachate Collection Systems

The effectiveness of the leachate collection system is based on the design of the liner and the collection pipes. The slope of the liner should be at least 2% and preferably 4% or more to promote lateral flow of leachate to collection pipes. The pipe should be sloped at 1% minimum to ensure leachate flow and prevent accumulation

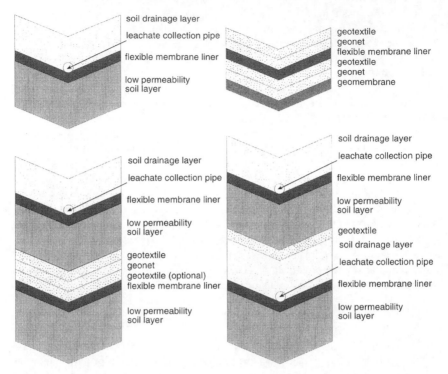

Figure 2.15   Examples of landfill liner systems. (From *Decision Maker's Guide to Solid Waste Management,* V2, EPA 600, Office of Solid Waste, Municipal and Solid Waste Division, U.S. Environmental Protection Agency, Washington, D.C., 1995, pp. 9–38.)

of low spots along the pipeline.The pipe is placed in a trench or directly on the liner at the low points. The trench should be backfilled with gravel and the pipe should be well supported to avoid crushing.

Leachate Treatment Processes

Leachate treatment includes on-site treatment, discharge to the municipal sewage treatment plant, or combination of these approaches. Some studies have indicated that leachate recirculation can increase the rate of waste stabilization, improve leachate quality, and increase the quality and quantity of methane gas production. If the leachate is sent to a municipal treatment plant, it is essential that the plant should not be overloaded. Leachate strengths are significantly greater than normal municipal wastewaters. Leachate of high BOD, such as those typically coming from a young landfill, are most effectively handled with anaerobic biological treatment. This can result in a 90% or more reduction in BOD. Leachate of medium BOD levels may be treated in aerobic biological systems, such as activated sludge. There is natural attenuation of leachate, which occurs when it flows through the soil into

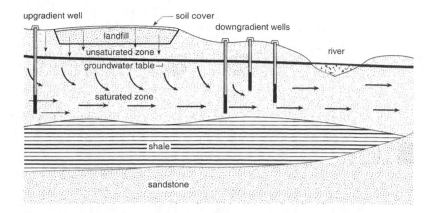

Figure 2.16 A traditional groundwater monitoring system. (From GAO, 1995.)

the underlying formations. Most of the heavy metals, such as lead, arsenic, zinc, cadmium, and mercury are retained by the soil.

Monitoring Wells

Groundwater monitoring systems determine water quality at a facility and whether contaminants have been released through the base of the landfill. Monitoring wells are cased in a manner that maintains the integrity of the borehole and meets design specifications. The number, spacing, and depths of the wells are based on site-specific characteristics. The wells are constructed to facilitate the collection of groundwater samples. The casing, seals, and grout protect the integrity of a borehole and minimize the chance of getting water from different zones. Appropriate detection equipment is utilized to check for VOCs and metals (Figure 2.16).

Area Method of Landfill Disposal

In the area method, the solid wastes are placed on top of the land, spread out, and then compacted into thin layers. This is covered with a layer of soil and then more waste is spread and compacted. Finally, the cover layer of soil is placed on the area. The daily earth cover utilized should be a minimum of 6 in. of compacted soil. The final earth cover should be a minimum of 2 ft of compacted soil.

Trench Method

In the trench method, the waste is spread and compacted in a trench. The cover material is taken from the excavated trench and spread on top of the compacted solid waste. Material that is not utilized on a daily basis can be stockpiled and then later utilized for final cover. Cohesive soils, such as glacial till or a clay silt, are desirable in a trench operation, because they compact well and make adequate walls for cells. The trench can be as deep as soil and groundwater conditions permit. The width should be twice the width of the compacting equipment.

A cell is composed of a mound of earth against which the compacted waste is placed. The soil then covers the waste and is compacted into the mound. Normally the material is spread in layers up to 2 ft thick and covered with 6 in. of soil. The weight of the overlying material, which should be at least 800 lb/yd^3, keeps the elastic materials from spreading. It is important that a narrow working face is used. Generally, the working face should be wide enough to handle the backlog of trucks waiting to dump, but never should exceed 150 ft. The best height for a cell varies with the advisor. Some believe that 8 ft or less is the best height. Some think that 16 ft or even greater should be utilized. The most important point is that the material should be compacted as well as possible to prevent an ultimate settling of the landfill site. A cell should be roughly square and sloped on the sides at a 20 to 30° angle.

A combination method of the area and trench methods may be used depending on the landfill site. This is determined by the slope of the ground and the techniques used in actual compaction and cover. (Figures 2.17 to 2.19 show cell construction and sanitary landfill methods.)

Superfund

CERCLA, also known as Superfund, was enacted by Congress on December 11, 1980. This law created a tax on the chemical and petroleum industries and provided broad federal authority to respond directly to releases or threatened releases of hazardous substances that may endanger public health or the environment. It established prohibitions and requirements concerning closed and abandoned hazardous waste sites, provided for liability of persons responsible for releases of hazardous waste at these sites, and established a trust fund for cleanup when no responsible party could be identified. The law provided for the National Contingency Plan, and the NPL. The cleanup process consists of the following:

1. Preliminary assessment through investigation of site conditions
2. Use of a screening mechanism to place sites on the NPL, which consists of the most serious places identified for possible long-term remedial response
3. Remedial investigation and feasibility study, which determines the nature and extent of contamination
4. Record of decision, which explains the cleanup alternatives
5. Construction completion, which identifies the completion of cleanup activities
6. Operation and maintenance, which is conducted after site actions are complete to ensure that all actions are effective and operating properly
7. Removal of sites from the NPL

The NPL includes high-priority Superfund sites located across the United States and its territories. The Record of Decision (ROD) contains remedial actions addressing the source of contamination, such as soil, sludge, sediments, solid-matrix wastes, and non-aqueous-phase liquids.

Superfund cleanups are very complex. The EPA has specialists at each of the ten regional offices, who are responsible for cleanup activities. They work with

CELL CONSTRUCTION
Layer Thickness

Solid waste spread in 2-foot layers
compacted to 1-foot thickness

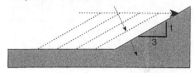

1. This technique requires the initial construction of a 3:1 slope or berm. Refuse is deposited at base of slope, spread upward in 2-foot layers and then compacted to about a 1-foot thickness.

Layer Thickness

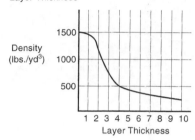

2. This recommended practice is based on field determinations which show that an optimum density is achieved by using a 2-foot thickness.

No. of Passes

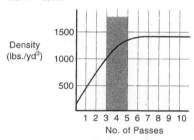

3. To achieve this optimum density requires about 5 passes over each layer of refuse.

Figure 2.17 Cell construction in a sanitary landfill. (From U.S. Environmental Protection Agency. *Elements of Solid Waste Management* (Training course manual, prepared by the Training Academy, Systems Management Division Office of Solid Waste Management Programs, Section II, Sanitary Landfill II, 1973, p. 1.)

experts in science, engineering, public health, management, law, community relations, and many other fields. They are responsible for keeping the public from exposure to contaminated air, direct contact with hazardous waste, contaminated drinking water, fires or explosions from hazardous waste, food chain contamination by hazardous waste, contaminated groundwater, contaminated soil, and contaminated surface water.

The EPA has attempted to reinvent the Superfund, because it has been subject to numerous criticisms over the years. People who live near the Superfund sites have often been frustrated by the slow cleanup process. Many people subject to Superfund

CELL CONSTRUCTION
Working Face Cover

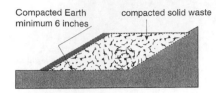

Compacted Earth   compacted solid waste
minimum 6 inches

4. Building of cell continues (as outlined in step 1 above) until the day's incoming refuse is compacted in place or desired length is reached. The working face is then covered with 6" of compacted soil.

CELL CONSTRUCTION
Final Top Cover

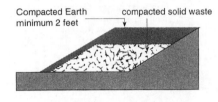

Compacted Earth   compacted solid waste
minimum 2 feet

5. Top of cell is covered by no less than 2 feet of compacted earth. Additional mounding can be provided to allow for settlement and graded to prevent ponding on surface.

CELL CONSTRUCTION
Cell Height

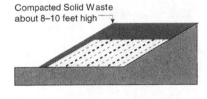

Compacted Solid Waste
about 8–10 feet high

6. Cell height is measured vertically and is normally 8–10 feet. This will vary and in some cases may be greater, depending on the skill of the operator and amount of refuse being handled.

CELL CONSTRUCTION
Intermediate Top Cover

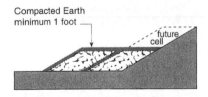

Compacted Earth
minimum 1 foot   future
                                             cell

7. If additional lifts (layers of cells) are to be placed above, an intermediate cover of 1 foot of compacted earth can be provided.

Figure 2.18   Cell construction in a sanitary landfill. (From U.S. Environmental Protection Agency. *Elements of Solid Waste Management* (Training course manual, prepared by the Training Academy, Systems Management Division Office of Solid Waste Management Programs, Section II, Sanitary Landfill II, 1973, p. 2.)

liability complain that enforcement actions are not fair and that cleanup costs are excessive. The EPA through administrative improvement have increased enforcement fairness, reduced transportation costs, improved cleanup effectiveness and consistency,

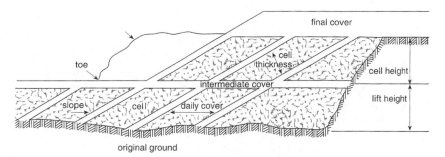

original ground

Figure 2.19   The cell is the common building block in sanitary landfilling. Solid waste is spread and compacted in layers within a confined area. At the end of each working day, or more frequently, it is covered completely with a thin continuous layer of soil, which is then also compacted. The compacted waste and soil constitute a cell. A series of adjoining cells makes up a lift. The completed fill consists of one or more lifts. (From Brennan, D.R. and Keller, D., *Sanitary Landfill Design and Operation.* U.S. Environmental Protection Agency, U.S. Government Printing Office, Washington, D.C., 1972, p. 27.)

expanded public involvement, and enhanced the states' role in the program. Small-volume contributors of hazardous waste have been given relief from many of the requirements.

National Oil and Hazardous Substances Pollution Contingency Plan

The National Oil and Hazardous Substances Pollution Contingency Plan (National Contingency Plan) is the federal government's means of responding to both oil spills and hazardous substance releases. The regulation applies to non-transportation-related facilities with a total aboveground oil storage capacity greater than 1320 gal, or total underground oil storage capacity greater than 42,000 gal. The facilities must prepare a plan certified by a registered professional engineer on how to deal with potential oils spills.

Underground Storage Tanks

According to the CFR, an UST is defined as any one or a combination of tanks that have 10% or more of volume below the surface of the ground in which they are installed. Releases from the tank systems include releases from the tanks, piping, pumps, as well as from spills and overfills. Leak detection equipment must be utilized, and all leaks must be corrected. If serious contamination occurs, the appropriate authorities must be notified. More than 300,000 releases have been reported from UST systems. These releases have been caused by leaks, spills, and overfills. Many have posed serious threats to human health and environment. Fumes and vapors can travel under the ground and collect in basements, utility vaults, and parking garages where they can cause explosions, fires, etc. Gasoline, leaking from service stations, is one of the most common sources of groundwater pollution.

Incinerators

Unloading, Storage, and Charging Operations

Unloading operations take place in an enclosed area approximately 58% of the time and in an open area about 42% of the time. The open areas are becoming more popular because they are cheaper to construct.

Their disadvantage is that they are exposed to adverse weather conditions. The floor surfaces of the unloading area should be impervious, should be easily cleaned, and should contain floor drains. The size of the area is determined by the number of discharge spaces needed to avoid truck tie-ups during peak periods. Separate entrance and exit doors should be provided for this. The storage pits are constructed of reinforced concrete, are durable, are rodent proof, and have easily cleaned surfaces. Armor plating is recommended to protect against damage by crane buckets. The recommended pit capacity is 100 to 150% of the daily rated incinerator capacity. Because refuse is usually delivered within a 7- to 8-hour period regardless of the time of the incinerator operation, additional space for storage is recommended. The pit should not be larger than 150%, because it will never be fully emptied and this can create problems at the plant.

In the event of breakdowns or overload, plans should be established to remove the excess solid waste to other incinerators or sanitary landfills. The problems encountered within the pits include excessive dust from unloading and crane operations, odors, rodents, fires, and explosions. Municipal refuse is composed of many substances that are relatively light and difficult to move. The material is put into the incinerator by push charging or drop charging. Push charging, which is not as common as drop charging, occurs when the refuse is discharged from the truck or an overhead crane directly into the furnace. Pushing is done with hydraulic rams and dozers, or by hand. Drop charging, by contrast, is charging from the top of the furnace instead of the side as in push charging, and is done by a hopper. The refuse is fed in an even flow.

Combustion Chambers

The principles of combustion include time, temperature, turbulence, and air. The functions of the incinerator furnace are to move the solid waste through the combustion area, dump the residue after the burning takes place, dry the solid waste before combustion takes place, retain the solid waste until there has been a reduction of combustibles, retain and mix the combustion gases and particulates until a satisfactory combustion has been achieved, and maintain a sufficient temperature to achieve the desired combustion of materials while using an appropriate amount of air.

The types of incinerator furnaces include the batch-fed, intermittent-fed, and continuous-fed furnaces. The batch-fed furnace has either fixed or movable grates. The advantage of this unit is that it is cheaper in initial cost and it can accept smaller quantities without problems. The disadvantages include poor temperature control, high maintenance cost, poor residue quality, and considerable air pollution potential.

Intermittent feed may be carried out in a batch-fed furnace. Continuous-feed furnaces have movable grates. The purpose of the grate is to seal the furnace when the level of the solid waste drops below the gate and down into the furnace. The column of solid waste in the chute provides an air seal and therefore better combustion. The advantages include larger capacity, better temperature and air supply control, greater flexibility in the operating rate, and much less thermal shock to the furnace refractories. Refactory brick is a nonmetallic substance used to line furnaces because it can endure high temperatures. It also resists abrasion, pressure, chemical attack, and rapid changes in temperature. The disadvantages include higher initial cost and higher degree of skilled labor.

The type of grates are important in the operation of the incinerator. They are the fixed-rate grate, which is used for batch feeding and requires hand stoking to remove the residue; conical grates, which are equipped with moving arms for batch or intermittent feeding; traveling grates, which are a continuous belt on which the burning takes place (mixing and breaking up occurs when a series of these grates are present in the furnace); rocking grates, which move and break up the material as it passes through the furnace; reciprocating grates, which are alternately moving horizontal grate sections, slide back and forth and break up the material in the combustion chamber; and rotary kilns, which are excellent for drying and igniting the solid waste and provide an excellent residue.

The combustion chamber uses a refractory brick lining to maintain the incinerator temperature. The temperatures are usually kept at a range of 1400 to 1800°F to ensure that all odor-producing materials are burned. The minimum temperature needed for adequate incineration of materials is 1400°F. Unfortunately, some of the materials create a tremendous amount of heat release that raises the temperature to abnormal levels. Temperatures rise as high as 3000°F. This damages the incinerator and impinges the flame. It is important to recognize this problem when designing the incinerator.

Two types of air are used in the furnace. Underfire air passes through the grate and is used to cool the grates and support combustion on the grate. Overfire air is injected through either the roof or the sidewalls; and is used to provide air for the combustion of the gases, mixing, turbulence, and then cooling of the furnace. In any case the air supply must be flexible so that it can be used to the best advantage.

Infectious and Medical Waste

In the fall of 1988, Congress passed the Medical Waste Tracking Act. The law called for a 2-year demonstration tracking program for medical waste. The program was to be a first step in controlling the irresponsible disposal of medical waste. Congress recognized that medical waste required special handling and disposal and that experience would be gained from a pilot program in the proper management of this waste. The Medical Waste Tracking Act amended RCRA by adding subtitle J. It defined medical waste as "any solid waste which is generated in the diagnosis, treatment, or immunization of human beings or animals, in research pertaining thereto, or in the production or testing of biologicals." The Medical Waste Tracking

Act required the EPA to establish a 2-year demonstration program in New York, New Jersey, Connecticut, and the states bordering the Great Lakes. The program was to provide generators of medical waste a uniform method for tracking these wastes to help ensure that the wastes were disposed of properly. The EPA also directed that standards be set for safely separating, packaging, and labeling these wastes before they were shipped to authorized treatment or disposal facilities.

The EPA recommended that a responsible person or committee at the institution prepare an Infectious Waste Management Plan outlining policies and procedures for the management of infectious waste. The plan included the following: designation, segregation, packaging, storage, transport, treatment, disposal, contingency planning, and staff training. Infectious waste included a variety of potential waste categories. These waste categories were isolation waste; cultures and stocks of infectious agents and associated biologicals, such as specimens from medical and pathology laboratories, as well as waste from the production of biologicals and discarded live and attenuated vaccines; human blood and blood products such as waste blood, serum, plasma, and blood products; pathological waste such as tissues, organs, body parts, blood, and body fluids removed during surgery, autopsy, and biopsy; contaminated sharps, such as contaminated hypodermic needles, syringes, scalpel blades, pipettes, and broken glass; contaminated animal carcasses, body parts, and bedding from animals that were intentionally exposed to pathogens; waste from surgery and autopsy, such as soiled dressings, sponges, drapes, lavage tubes, drainage sets, underpads, and surgical gloves; miscellaneous laboratory waste, such as specimen containers, slides, and coverslips, disposal gloves, lab coats, and aprons; dialysis unit waste, such as tubing, filters, disposal sheets, towels, gloves, aprons, and lab coats; contaminated equipment such as equipment used in patient care, medical laboratories, research, and production and testing of certain pharmaceuticals.

The EPA recommended the segregation of infectious waste at the point of origin. Also the segregation of infectious waste with multiple hazards was important. Distinctive clearly marked containers or plastic bags were needed for infectious waste. The universal biological hazard symbol was to be used on infectious waste containers.

The EPA recommended selection of packaging materials that were appropriate for the kinds of waste that were generated. Plastic bags that were impervious, tear resistant, and distinctive in color or markings appropriately folded or tied at the top, could be used for many types of solid or semisolid infectious waste. Liquid waste needed to be placed in capped or tightly stoppered bottles or flasks. Puncture-resistant containers were necessary for sharps. Compaction of infectious waste or packaging of infectious waste could not be done before treatment.

The EPA recommended that infectious waste be stored for a minimal time. The packaging materials excluded all insects and rodents. There was to be limited access to the storage area, and universal biological hazard symbols were to be posted on the area, as well as on waste containers, freezers, or refrigerators.

The EPA recommended that mechanical loading devices that may rupture packaged waste be avoided. The carts used to transfer the waste within the facility had to be frequently disinfected. All infectious waste was to be placed in rigid or semirigid containers before transporting the waste off site. The transport of infectious waste had to be done in closed leakproof trucks or dumpsters.

Treatment was defined by the EPA as any method, technique, or process designed to change the biological character or composition of the waste. The EPA recommended that standard operating procedures be established for treating infectious waste. All infectious waste was to be monitored during treatment processes and afterward for rapid, efficient, and effective treatment. Biological indicators were to be used to monitor the treatment when needed. The EPA recommended that waste from surgery and autopsy should be incinerated or steam sterilized. Miscellaneous laboratory wastes were to be either incinerated or steam sterilized. Dialysis unit wastes were to be either incinerated or steam sterilized. Contaminated equipment was to be incinerated, steam sterilized, or gas–vapor sterilized. Other techniques of treatment included chemical disinfection and sterilization by irradiation. In any case, biological indicators for monitoring treatment processes, such as steam sterilization, incineration, and thermal inactivation, were to be used where possible.

The U.S. Pharmacopoeia (USP) recommended the use of biological indicators for various types of treatment processes. The indicators provided an instantaneous indication, usually by a chemically induced color change of the material at a certain temperature. Some of these indicators were not suitable for use in monitoring the sterilization process, because the treatment includes not only an elevation in temperature but also a certain pressure of steam and a specific time period.

Steam sterilization of infectious waste utilizes saturated steam within a pressure vessel, known as a steam sterilizer, autoclave, or retort, at temperatures sufficient to kill infectious agents present in the waste (Figure 2.20). The two general types of steam sterilizers are a gravity displacement type in which the displaced air flows out the drain, and a steam-activated exhaust valve and the prevacuum type in which a vacuum is pulled to remove the air before steam is introduced into the chamber. In both types, the air is replaced with pressurized steam as the temperature of the treatment chamber increases. Treatment by steam sterilization is time and temperature dependent. It is, therefore, essential that the entire waste load be exposed to the necessary temperature for the appropriate period of time.

Factors that cause incomplete displacement of air include the use of heat-resistant plastic bags, which may exclude steam or trapped air; the use of deep containers, which may prevent the displacement of air from the bottom; and improper loading, which may prevent free circulation of steam within the chamber. In treating infectious waste with steam, it is necessary to know the type of waste, the packaging and containers used, and the volume of the waste load and its configuration in the treatment chamber. Infectious waste with low density, such as plastics, is amenable to steam sterilization. High-density waste, such as large body parts and large quantities of animal bedding and fluids, inhibit direct steam penetration and require longer sterilization time. Plastic bags, metal pans, bottles, and flasks are used as containers in steam sterilization. The volume of the waste is an important factor in steam sterilization, because it may be difficult to obtain sterilizing temperatures in large loads. It is more efficient to autoclave a large quantity of waste in smaller loads.

Many infectious wastes have multiple hazards, such as toxicity, radioactivity, or containment of hazardous chemicals. These wastes should not be steam sterilized. Certain antineoplastic drugs that are used to inhibit cell division in certain cancer cells should not be steam sterilized, because they become volatilized by the steam

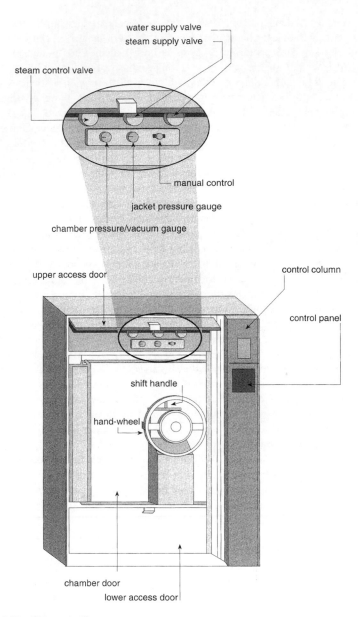

Figure 2.20   Steam sterilizer.

and present a potential hazard. A recording thermometer, as well as an indicating thermometer, should be placed on all steam sterilizers and should be operational.

Combustion is a process that converts infectious materials into noncombustible residue or ash. The product gases are vented to the atmosphere through the incinerator stack. The treatment residue may be disposed of in a sanitary landfill. Combustion provides the advantage of greatly reducing the mass and volume of the waste. This may be as much as 95% in well-run specialized incinerators. Combustion is especially advantageous with pathological waste and contaminated sharps, because it renders the body parts unrecognizable and the sharps unusable.

Incinerators that are properly designed, maintained, and operated are effective in killing organisms that are present in the infectious waste. If the incinerator is not operating properly, viable pathogenic organisms may be released to the environment from the stack emissions, the residue ash, or the wastewater. Combustion of infectious waste is based on the variation of the waste composition, the waste feed rate, and the combustion temperature. The waste composition affects combustion because of moisture content and heating value. The waste feed rate is essential to keep from overloading the system. The combustion temperature must be at an optimum level to combine the combustion air and fuel with the pathological substances to be combusted, to achieve a good burn.

Infectious waste contains multiple hazards as mentioned earlier. It should be combusted only in an incinerator that provides high temperatures for long periods of time to ensure complete destruction. The plastic content of the waste can cause problems with combustion and must be considered before the temperature is determined. Many incinerators can be damaged by temperature surges caused by the combustion of large quantities of plastic. Polyvinyl chloride and other chlorinated plastics may be present in the waste. The combustion products of these plastics include hydrochloric acid, which is corrosive to the incinerator and may damage the refractory brick lining, the chamber, and the stack. The minimum temperature for operating an infectious waste incinerator is 1600°F in the secondary combustion chamber. The minimum residence time is 1 sec.

Thermal inactivation includes treatment methods that utilize heat transfer and provide conditions that reduce the presence of infectious agents and waste. This is usually used for treating large quantities of infectious waste, such as in industrial applications. Thermal inactivation of liquid infectious waste is carried out in a vessel of sufficient size to contain the liquid waste generated during the operating period. The waste is preheated by heat exchangers or heat may be applied by a steam jacket that surrounds the vessel. Heating is continued until a predetermined temperature is achieved and maintained for a designated period of time. Mixing may be appropriate to homogenize the waste and temperature during loading and heat application steps of the treatment cycle. The temperature and holding time depends on the nature of the pathogens present in the waste. After the treatment cycle is complete, the contents of the vessel or tank are discharged into the sewer.

Thermal inactivation of solid infectious waste or dry heat treatment may be carried out in an oven that is operated by electricity. Dry heat is less efficient than steam heat, and therefore higher temperatures and longer treatment cycles are necessary. A typical cycle for dry heat sterilization is treatment at 320 to 338°F for 2 to 4 hours.

Gas–vapor sterilization may be used for treating certain infectious waste. The sterilizing agent is a gaseous or vaporized chemical. The two most commonly used chemicals are ethylene oxide and formaldehyde. Sufficient evidence is available that these chemicals are probable human carcinogens and that it is necessary to use them under strict control. Formaldehyde should only be used by persons trained in the use of formaldehyde as a gaseous sterilant.

Chemical treatment is most appropriate for liquid waste and may also be used in treatment of solid infectious waste. To have appropriate chemical treatment, it is necessary to know the type of microorganisms, degree of contamination, amount of protein material present, type of disinfectant, concentration and quantity of disinfectant, contact time, temperature, pH, and mixing requirements.

The treatment of infectious waste can be done by the use of ionizing radiation. The advantages of this method are low electricity requirements, no steam requirements, no residual heat in treated waste, and good system performance. The disadvantages are high capital cost, requirement for highly trained personnel, large space requirements, and problem of the ultimate disposal of the decayed radiation source.

The Medical Waste Tracking Act passed in 1988, was for a 2-year experimental period. The regulations drafted under this act expired on June 21, 1999. The information gathered led the EPA to conclude that the disease-causing potential is greatest at the point of generation of the waste and that it tapers off from this point. It is more of an occupational concern that a risk to the general public.

Currently, over 90% of potentially infectious medical waste is incinerated. In August 1997, EPA promulgated regulations governing the emissions from medical waste incinerators. These regulations included stringent air emission guidelines for states to use in developing plans to reduce air pollution from medical waste incinerators built on or before June 20 1996; and final air emission standards for medical waste incinerators built after June 20, 1996. Many of the 2400 existing medical waste incinerators have been discontinued. Alternatives, previously mentioned, are in use to replace medical waste incinerators.

Today, numerous federal agencies including the EPA regulate infectious medical waste. The Occupational Safety and Health Administration (OSHA) regulates items potentially contaminated with bloodborne pathogens. The Department Of Transportation regulates shipments of potentially infectious medical waste. The Nuclear Regulatory Commission regulates some types of radioactive medical waste. The Food and Drug Administration (FDA) regulates medical devices such as sharps containers that are designed to safely contain used needles. The U.S. Postal Service regulates medical waste shipped through the mail. The EPA Office of Solid Waste regulates other aspects of infectious medical waste.

The EPA on June 24, 1998, entered into a voluntary partnership with the American Hospital Association and its member hospitals to:

1. Virtually eliminate mercury waste generated by hospitals by 2005
2. Reduce overall hospital waste volume by 33% by 2005, and 50% by 2010
3. Jointly identify additional substances to target for pollution prevention and waste reduction

Hog Feeding

If hog feeding is permitted in the locality, the garbage must be segregated from the rest of the solid waste and collected separately. Draining of the garbage is highly desirable, because it enhances storage and collection. The garbage must be heated to an internal temperature of 137°F before it can be fed to the hogs. The materials must be placed on special concrete platforms, which can then be properly washed. The garbage remnants must be removed and the liquid waste must be properly treated before it enters a sewage treatment process or a stream.

MODES OF SURVEILLANCE AND EVALUATION

The solid waste management system is evaluated by developing an accurate data system concerning crew size, equipment cost per day, personnel cost per day, total cost per day, and cost per ton. It also includes an evaluation of the organization and supervision of the solid waste disposal programs, average weight and composition of material collected per day, average time spent collecting per day, and numbers of homes and businesses serviced. Influences on solid waste management include population trends and densities, and dwelling unit densities; and transportation networks, including types of streets and expressways, width of the streets, type of topography, distance between pickups, distance from pickup of solid waste to disposal sites, and time required to go from pickup to disposal and return to pickup. The preceding information is placed in a computer system and a model is established for the actual performance of the waste disposal program. A model is created for efficient pickup, put into operation, and once again evaluated. Weather conditions may become a significant contributing factor to the solid waste collection and disposal system. However, the experience of trained professionals is as important, or at times more important, than a computer model because of the many variables in solid waste collection. This evaluation is necessary to determine the cost of a given operation and the best means of altering the operation to obtain greater efficiency.

An evaluation of the solid waste disposal sites prior to their use is important for understanding the migration of chlorides, total organic carbons, chromium, copper, and cadmium in the soil. Where these chemicals are present and are migrating in a large quantity, it would be improper to utilize the soil for solid waste disposal sites. Because biological, chemical, and physical processes take place in the solid waste, and because this leads to the development of gases and leachate that may move contaminants to the water tables or surface bodies of water, studies are needed. The soil-sampling procedures consist of making test borings and analyzing the soil at various depths.

Leachate is collected from the subsurface soil by use of wells or tiezometers placed in drilled holes. A tiezometer is a screen or permeable plastic tip fastened to the end of a tube that is installed in a boring. The space above the tip is sealed so that the water level measurements or water samples are restricted in the bottom part of the boring below the seal in the annulus. A well point without a seal is similar to a tiezometer. As the leachate permeates through the soil, the collection device captures it. Leachate is also gathered from springs, trenches at the bottom of the solid waste disposal site, or areas where the leachate is starting to come forth in a body of water. The leachate is then analyzed for its chemical content. The leaching test is important, because environmental contamination cannot occur until the waste leaves the solid waste disposal site and starts to migrate through the soil and into the water. This is difficult to evaluate in the case of hazardous waste, because these wastes are not homogeneous mixtures and may range in consistency from a liquid to a sludge to a solid at the point of generation and at the point of disposal.

Environmental health practitioners also evaluate the operation of each of the land disposal sites, incinerators, and other types of disposal, as well as the collection and storage techniques. They determine whether the operations meet local and state codes, whether the operations are creating nuisances or hazards, and whether the procedures are contributing to air, land, water, insect, rodent, or injury problems.

Ground-penetrating radar is under study as a tool for determining anthropogenic objects such as 55-gal drums being placed under the ground surface, and for detecting and defining the extent of contaminated soil or groundwater.

Geographic Information System-Based Tool

All the geocoded Superfund and TRI sites from the EPA Landview 3 database in three urban counties in Washington State were developed into a geographic information system (GIS). A spatial overlay used to estimate population, various socioeconomic variables, and death rates from several causes, was used on a census block population base. A profile was developed to help facilitate comparison between areas where there were Superfund/TRI sites and random points inside the study counties. This technique assesses the characteristics of neighborhoods located around areas that contain toxic waste or facilities that emit toxic waste. The object is to determine whether different rates of disease or death exist in exposed populations and whether any differences can be attributed to the exposure. Other factors evaluated include indicators of socioeconomic status, age, race, and national origin.

Environmental Effects on the Ocean

Very little is known about the immediate or long-range effects of dumping solid and liquid waste into the ocean. One basic concern is the dumping of raw or treated sewage sludge in areas where shellfish concentrate the bacteria or viruses and cause human disease. To conduct a proper study, it is necessary to evaluate the effects of dumping each of the foreign substances into the ocean to determine whether there is a change in the characteristics of the marine life; and whether the water in the dumping area picked up the pollutants and moved them farther out to some spot where a hazard may be caused. These baseline data are necessary to make a final determination of the effects of dumping into the ocean.

Monitoring of Marine Waste Disposal

Solid waste in the ocean can be monitored through inspection or surveillance at the site by ships or planes of the regulatory agencies. It may also be monitored by requiring that adequate records be kept and that the records reveal not only the type of solid waste but also the potential hazards of the solid waste, the place where the waste is to be dumped, and the quantity of each of the components of the solid waste. Monitoring, along with special research, can give us a much clearer understanding of ocean dumping.

CONTROLS

Landfills

See section in this chapter under Standards, Practices, and Techniques.

Incinerators

Under subpart 0 of RCRA, a trial burn for alternate data must be used to determine whether the incinerator meets the following three performance standards:

1. Of each principal organic hazardous constituent (POHC) specified in the permit, 99.99% must be destroyed or removed from the incinerator.
2. Hydrogen chloride (HCl) emissions must be minimized.
3. Particulate emissions must be limited.

Incinerators are required to meet federal codes. However, currently many of them are still not in compliance. The types of particulate control equipment used include the settling chamber, which has an efficiency of 40 to 60%; inertial cyclones, with an efficiency of 75 to 90% with particles less than 10 μm; multicyclones, with an efficiency of 90 to 98% with particles above 5 μm; scrubbers, which have an efficiency of 90 to 99%, with particles down to 1 μm; baghouse filters, which have an efficiency of 98 to 99.9% down to submicron size; and electrostatic precipitators, which have an efficiency of 90 to 99.9% down to submicron size. Gases are controlled by burning within the incinerator or immediately above the incinerator grate, dispersion into the atmosphere, and use of the dilution technique.

Instruments are necessary within the incinerator operation to control the combustion process through the control of combustion air, charging rate, use of auxiliary fuel, and grate speed. Instrumentation is also needed for protection of equipment and personnel from overheating of furnaces and ducts, and from failures of quench water and electrical power; and for use with safety alarms. The equipment is needed to protect the environment and to automate and monitor the operation. All equipment must produce not only a written record but also the audible signals in the event of emergencies.

Odors from incineration processes may be controlled through fume incineration (afterburner) or by packed tower (absorber). Odors have also been controlled through catalytic oxidation, adsorption, dilution, and masking. Burning is the ultimate odor control technique, because the majority of odor-forming compounds are organic. When these compounds are oxidized, they form carbon dioxide, water vapor, and other compounds, which contain no odors. A temperature of 1400°F maintained for at least 0.5 sec satisfactorily and completely destroys materials that cause odors.

Catalytic oxidation is a relatively low temperature incineration technique. The catalytic agent is inserted in the gas stream and thus allows the destruction of the contaminant at a lower temperature. When the contaminant is destroyed, the odor is no longer available. Absorption is the process where the odor is brought into solution through a chemical bond and, therefore, the odor is no longer available. Adsorption is a process where the odor is brought into contact with a substance such as activated carbon. This is a physical reaction where the odor is trapped in the material. Dilution is a technique where small amounts of the odorous emissions are released into the air. Masking is a technique in which a substance is used to either counteract or neutralize the odor.

Control of Air Emissions

A wide variety of control technologies are used to control air emissions from municipal waste combustors. Particulate matter along with the metals that have adsorbed onto them, are most frequently removed by use of electrostatic precipitators or fabric filters. Cyclones, electrified gravel beds, and Venturi scrubbers may also be used. The control of acid gas emissions is most frequently handled through spray drying or dry sorbent injection.

The electrostatic precipitators consist of a series of high-voltage discharge electrodes and grounded metal plates through which the flue gas flows. Negatively charged ions formed by the high-voltage field (corona) attach to the particulate matter in the flue gas, causing the charged particles to migrate toward and to be collected on the grounded plates. The greater the amount of collection plate area, the greater is the efficiency of the unit in collecting particulate matter. Rapping, washing, or other techniques are used to remove the dust layer. Small particles generally have lower migration velocities than large particles and are therefore harder to collect.

Fabric filters are used for particulate matter and metals control, especially in combination with acid gas control and flue gas cooling. Fabric filters (baghouses) remove particulate matter by passing flue gas through a porous fabric that has been sewn into a cylindrical bag. Multiple individual filter bags are arranged in a special compartment. A complete fabric folder consists of 4 to 16 individual compartments that can be operated independently. As a flue gas flows through the filter bags, particulates are collected on the filter surface through inertial impaction. The collected particulates build up on the bag, forming a filter cake. As the thickness of the filter cake increases, the pressure drop across the bag increases. When the pressure drop becomes excessive, the compartment is taken off-line, mechanically cleaned, and put back on-line.

Spray dryers are most frequently used for acid gas control in the United States. In the spray drying process, lime slurry is injected into the spray dryer through a rotary atomizer or dual-fluid nozzles. The water in the slurry evaporates to cool the flue gas, and the lime reacts with acid gases to form Ca salts that can be removed by a control device.

Dry sorbent injection is primarily used to control acid gas emissions. Duct sorbent injection involves injecting dry alkali sorbents into flue gas downstream of the combustor outlet and upstream of the particulate matter control device. Furnace sorbent injection refers to the injection of sorbent directly into the combustor.

Wet scrubbers, including spray towers, centrifugal scrubbers, and Venturi scrubbers, are used to control acid gas emissions from the combustors. It is not anticipated that many new units will be built in the United States.

The control of nitrogen oxides can be accomplished through either combustion controls or add-on controls. Combustion controls include staged combustion, low excess air, and flue gas recirculation. Add-on controls include selective noncatalytic reduction, selective catalytic reduction, and gas burning. Combustion controls involve the control of temperature or oxygen to reduce nitrogen oxides formation. In selective noncatalytic reduction, ammonia or urea is injected into the furnace along with chemical additives to reduce nitrogen oxides to nitrogen without the use of catalysts. In selective catalytic reduction, ammonia is injected into the flue gas downstream of the boiler where it mixes with the nitrogen oxides and passes through a catalyst bed, where the nitrogen oxides are reduced to nitrogen by a reaction with the ammonia.

Unlike other metals, mercury exists as a vapor at typical combustor operating temperatures. The collection of mercury is highly variable. Mercury control is best when you have good particulate matter control, low temperatures, and sufficient level of carbon in the fly ash. Control technologies include the injection of activated carbon or sodium sulfide into the flue gas prior to the acid gas control system, or through the use of activated carbon filters.

Industrial and commercial combustors that generally burn between 50 and 4000 lb/hr need emission control systems including gas-fired afterburners, scrubbers, or both. They must meet the same emission limits as those required for municipal waste combustors.

Trench combustors or air curtain incinerators, forcibly project a curtain of air across a pit in which open burning occurs. The air curtain is intended to increase combustion efficiency, reduce smoke, and reduce particulate matter emissions. Underfire air is also used to increase combustion efficiency. Trench combustors are used to burn wood wastes, yard wastes, and clean lumber. Domestic combustors are used for apartment complexes, residential buildings, or other multiple use facilities. They usually have special burners to help in combustion.

Vitrification

Vitrification technologies are used for remediating hazardous waste sites and for treating high-level radiation wastes. The purpose of this technology is to immobilize metals and destroy organics by pyrolysis.

Hazardous Waste

Deep-Well Injection

Over 1 billion tons (where a metric tonne equals 2200 lb) of dilute, aqueous waste are deep-well injected each year. The majority of deep wells are located on-site, so wastes are disposed of at the plant where they are generated (Figure 2.21). A small number of commercial firms operate deep-well systems. Most deep-well systems are concentrated along the Gulf Coast of Texas and Louisiana. The EPA has prohibited

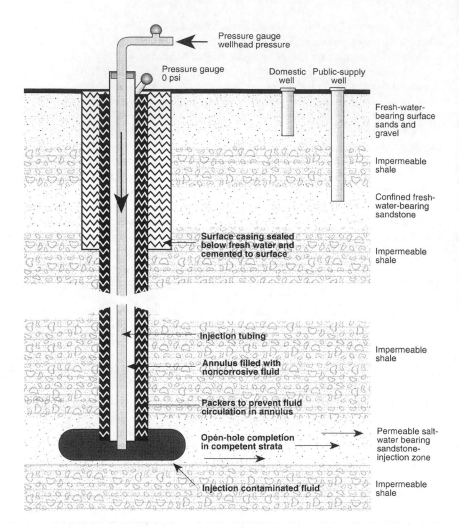

Figure 2.21   Features of a deep-well injection system.

the underground injection of certain solvents, dioxins, and California list waste (see Hazardous and Solid Waste Amendments Program for a definition of the California list). The rules also prohibit the injection of concentrated halogenated organic compounds (HOCs) waste.

GIS Technology

Prior to GIS technology, data needed for enforcement actions were stored in separate tables, as hand-drawn sketches, and U.S. Geological Survey topographic sheets. It was difficult to combine all this information in an accurate and precise manner and to analyze the material effectively. In the U.S. EPA, Region 2, Underground Injection Control Section, GIS technology was used to locate the most vulnerable groundwater areas on Long Island, NY. By using a risk rankings scheme based on a modified Human Health Risk Index formula developed by the EPA Region 6, they were able to do large-scale water risk assessment. They ranked the risks on maps, which were the product of population density rank, land-use rank, and percolation travel time to the water table. They established the risk ranking by zip codes.

The Underground Injection Control Program as mandated by Part C of the Safe Drinking Water Act, 42 U.S.C. is administered by the EPA to regulate wells to protect underground sources of drinking water. According to Title 40 of the CFR, a well is defined as either a dug hole, or a bored, drilled, or driven shaft with a depth greater than its largest surface dimension. Well injection is defined as the subsurface emplacement of fluids through a well. Because of the nature of the soil, the underground aquifer system on Long Island, the multitude of potential sources contaminating the groundwater supply, and the groundwater flow system, the GIS system allows for prioritizing potential threats and populations involved. This permits the efficient use of valuable resources to resolve the greatest number of problems.

Class 5 injection wells, typically shallow disposal systems used to place a variety of fluids below the land surface, into or above underground sources of drinking water, can be a serious route of contamination of the water. Over 1 million systems exist in the United States today. They are located in every state, especially in unsewered areas where a population is likely to use groundwater for its drinking water source. The greatest source of water pollution is due to motor vehicle waste disposal wells, industrial waste disposal wells, and large-capacity cesspools. All these systems are now regulated under the EPA Underground Injection Control Program. Existing cesspools are to be totally phased out, whereas new ones are prohibited. Industrial waste disposal wells are prohibited from exceeding drinking water standards or other health-based limits at the point of injection.

TREATMENT

The objectives of the treatment of hazardous waste or hazardous waste sites are destruction of hazardous substances, recovery for reuse, or conversion into innocuous substances that cannot cause a hazard to people, animals, or environment. Several techniques are usually needed to completely treat these wastes. In some cases, the hazardous wastes cannot be treated or destroyed and therefore must be stored in such a way that they cannot become a hazard. Current technology is not necessarily such that we have found the answer to complete destruction of hazardous wastes.

However, there are acceptable techniques, which include physical treatment processes used to concentrate waste brines and to remove solid organics and ammonia from aqueous solutions. Flocculation, sedimentation, and filtration are widely used as a means of precipitating solids from the liquid form. Further, absorption is utilized to remove soluble organic material from liquid waste streams. The chemical treatment process is used to neutralize waste, to precipitate material, and to carry out oxidation–reduction. The thermal treatment process is used for destroying or converting solid or liquid combustible hazardous waste. Pyrolysis is used to convert the hazardous wastes into useful products, such as fuel gases and coke. Biological treatment is used for biodegrading the organic waste. However, the limitations of these systems must be understood. The disposal techniques utilized may still consist of landfills and deep-well injection. Either of these may be hazardous.

Physical treatment consists of reverse osmosis, dialysis, electrodialysis, evaporation, carbon adsorption, ammonia stripping, filtration, sedimentation, and flocculation. Reverse osmosis is the physical movement of a solvent across the membrane boundary, where external pressure is applied to the side of less solvent concentration so that the solvent can flow in the opposite direction. This allows the solvent to be extracted from the solution. Almost any dissolved solid can be treated by reverse osmosis, provided that the concentrations are not too high and that the pH can be adjusted to a range of 3 to 8.

Dialysis is a process where various substances in solution, having widely different molecular weights, can be separated by solute diffusion through semipermeable membranes. This is accomplished by the difference in chemical activity of the transferred material on both sides of the membrane. Dialysis is used effectively in the textile industry.

Electrodialysis is similar to dialysis. It differs in its dependence on an electric field that is the driving force in the separation. It is possible to separate ionized forms of material from the nonionized solution. Such ionizable substances as lead nitrate and sodium phosphate can be removed in this process.

Evaporation is the technique of removing solvents as vapor from a solution or slurry. This occurs when the solvent is brought to its boiling point by the use of heat energy. The vapor may or may not be recovered, depending on how valuable it is. Evaporation is one of the most versatile processes used, but it is also very costly.

Carbon adsorption is a process in which a substance is brought into contact with a solid and held at its surface or internally by physical or chemical forces. Activated carbon is the most frequently used of the adsorbing materials. The adsorption material is then desorbed to remove the contaminant.

Ammonia stripping is the removal of ammonia from alkaline aqueous wastes using steam at atmospheric pressure. The steam is introduced at the top of a packed or bubble cap tray-type column. Ammonia, because of its high partial pressure over alkaline solutions, is readily condensed and then can be reused. The process may also be used to remove other volatile and organic contaminants from waste streams.

Filtration is the physical removal of solid materials from an aqueous waste stream by using a filter medium. The filter medium is composed of a variety of materials. Most solid materials are removed in this way.

Sedimentation, or settling, is a process used to separate aqueous waste streams from the particles suspended in them. The material is placed in a tank and the particles are allowed to settle out. This process is widely used.

Flocculation is a process used when fine particles are present in the aqueous stream. Material is added to the waste stream to cause the fine particles to settle. The process is used to further cleanse aqueous solutions from which larger particles have already been removed.

Chemical treatment consists of ion exchange, neutralization, oxidation, reduction, precipitation, and calcination. Ion exchange is the reversible interchange of ions between solid and liquid phases where no permanent change takes place in the solid structure. It is a technique of collecting and concentrating undesirable materials. It utilizes resins that react with either cations or anions. The technique has been utilized for many years to remove trace metals and cyanides, and also to remove fluorides, nitrates, and manganese from drinking water. The contaminants are either removed or recycled.

Neutralization is a process where excessively acid or alkaline wastes are neutralized prior to discharge in effluents. Some of the methods used include mixing the waste to get a near neutral pH of 7, passing the acid waste through lime slurries, passing the acid waste through beds of limestone, adding proper amounts of caustic soda or soda ash to acid wastewaters, blowing the waste boiler flue gas through alkaline waste, adding compressed carbon dioxide to alkaline waste, and adding sulfuric acid to alkaline waste. Neutralization is used most effectively in precipitating the hydroxides or hydrous oxides of heavy metals, and also in precipitating calcium sulfate.

Oxidation is a process in which the waste-containing reductants are oxidized with chlorine, hypochlorites, ozone, peroxide, or other oxidizing agents. This is used for treating cyanides and other reductants. Reduction is a process in which the waste-containing oxidants are treated with sulfur dioxide to reduce the oxidants to less noxious forms. Sulfite salts and ferrous sulfite are also utilized. This technique is used to treat chromium-6 and other oxidants. Precipitation is a process of separating the solid constituents from the aqueous waste by combining with a chemical in a chemical process producing an insoluble constituent. This is primarily used for precipitating heavy metals. Calcination is a process of heating to a high temperature without fusing. It is used primarily in the processing of high-level radioactive waste.

Thermal treatment consists basically of incineration and pyrolysis. Thermal treatment is a controlled process to change waste into less bulky, toxic, and noxious materials. The 11 basic types of incineration units include open pit, open burning, multiple chamber, multiple hearth, rotary kiln, fluidized bed, liquid combustors, catalytic combustors, afterburners, gas combustors, and stack flares. A variety of solids, liquids, and gases may be incinerated. It is important that secondary abatement equipment is used to capture any pollutants escaping from the incineration process.

Incineration is a proven method for the permanent destruction of solid hazardous waste. Hazardous waste may include soils contaminated with chemicals; by-products of petrochemical factories, residues, and coal gasification; pesticides; halocarbons; and other compounds. The solid hazardous waste may have sufficient heating value to sustain combustion, or may need to have additional heat to make the combustion process work in the case of contaminated soil. Contaminated soil may also include heavy metals, such as lead, zinc, copper, and vanadium. Incineration is a permanent solution to hazardous waste disposal, providing it is carried out in a proper manner that meets the regulatory requirements of 99.99% destruction of the contaminants. Some of the more common hazardous compounds that can be incinerated are benzene, toluene, xylene, carbon tetrachloride, chloroform, trichloroethylene, perchlorethylene, acetone, naphthalene, aldrin, dieldrin, chlordane, dichlorodiphenyldichloroethane (DDD), DDT, malathion, and mirex.

The process of incineration of these wastes from a hazardous waste site may be very complex because, along with all the chemicals that have been mentioned, PCBs, explosives, heavy metals, and oils may also be present. A careful analysis of the waste site is needed to determine how the materials should be incinerated and how much heat is needed to carry out the job properly.

Solid hazardous waste combustion consists of the volatilization of the solids through pyrolysis, melting, and boiling, as well as sublimation. Melting, boiling, and sublimation are heavily influenced by the soil or inert materials that are present. Sublimation is the passing of the material from the solid to the vapor state by heating and then again condensing to the solid form. Once volatiles are produced, they go through a gas-phase reaction by mixing with incoming air or chlorine present in the hazardous waste to form carbon dioxide, water, and hydrochloric acid. Gas and solid reactions occur through pyrolysis where hydrogen, oxygen, nitrogen, and sulfur are produced. Temperatures used for hazardous waste combustion must be high enough to meet the requirements of the various laws, but temperatures must be taken into consideration, because the combustion of the waste may create extremely high additional temperatures that could be harmful to the incinerator. The potential air-borne emissions from hazardous waste combustion include particulates, sulfur dioxide, hydrogen chloride, oxides of nitrogen, carbon monoxide, hydrocarbons, dioxins, and furans. Further, other organic compounds may be emitted to the air. Rotary kiln incinerators and fluidized-bed incineration systems are used for hazardous waste destruction.

Biological treatment consists of activated sludge, aerated lagoons, trickling filters, and waste-stabilization ponds. Activated sludge is a process in which the biologically active growths are continuously circulated and come in contact with the organic waste in the presence of oxygen. Oxygen is supplied to this system by injecting air bubbles, which keeps the entire system in a turbulent condition. The activated sludge contains microorganisms that feed on the organic waste.

Oxygen is very important for the growth of these organisms. The incoming waste water is mixed with the recycled activated sludge, and the mixture is aerated for several hours in special aeration tanks. During this period, oxidation takes place, which is responsible for the removal or destruction of much of the organic matter. The effluent from the aeration tank is then passed to a sedimentation tank where the flocculated microorganisms or sludge settle. Flocculation had been taking place in the aeration tank. Activated sludge is used extensively in treating refinery, petrochemical, and biodegradable organic waste waters.

The aerated lagoon is usually a basin that is 6 to 17 ft deep, in which organic waste stabilization takes place. Oxygen is provided mechanically through the mass. The aerated lagoon is an economical way of treating industrial waste where it is not necessary to get high-quality effluents.

The trickling filter is an artificial bed of rocks or other porous material through which the liquid containing the settled organic waste is percolated. The waste is brought into contact with air and biological growths. The liquid is applied intermittently or continuously over the top of the surface of the filter by the means of a distributor. The filtered liquid is then collected and discharged at the bottom. The primary removal of the organic material is a result of an adsorption process similar to activated sludge that occurs at the surfaces of the biological growths or slimes covering the filter media. Trickling filters are used extensively in the treatment of acetaldehyde, acetic acid, acetone, acrolein, alcohols, benzene, butadiene, chlorinated hydrocarbons, cyanides, epichlorohydrin, formaldehyde, formic acid, ketones, monoethanolamines, phenolics, turpenes, ammonia, ammonium nitrate, resins, and rocket fuels.

Waste stabilization ponds are large, shallow basins usually 2 to 4 ft in depth, where the wastewater is purified when it is stored under climatic conditions favoring the growth of algae. The organics are converted to inorganics or stabilized because of the metabolic activity of bacteria, algae, and surface aeration. These ponds have been used in treating steel mill wastes.

It is necessary to determine where to dispose of hazardous waste that cannot be altered. This is the material that either cannot be processed or is left over from processing and still has a degree of hazard associated with it. The ideal hazardous waste disposal facility is a secure landfill with a double liner, where appropriate equipment and structures are used for burial of hazardous solid waste. Surveillance must be made on a regular basis to determine that the contaminants are not getting into the land or water. Special measures should be utilized in the backfilling process.

PROGRAMS

Employee Protection Program

The Employee Protection Program consists of planning and organization, training, medical programs, site characterization, air monitoring, personal protective equipment, site control, decontamination, handling drums and other containers, and site emergencies.

The Site and Safety Plan should identify the key personnel and alternates responsible for site safety. It should also describe the risk associated with each operation, the training program for general and specific situations, the protective clothing and equipment to be worn, the medical surveillance requirements, the periodic way of monitoring and environmental sampling, the actions needed to contain contaminated materials, the site control measures, and the decontamination procedures of personnel and equipment.

Training programs should be presented to general site workers, on-site management and supervisors, health and safety staff, and visitors when needed. The complexity of the program would vary with the type of individuals. In any case, the following topics should be considered for training: biology, chemistry, and physics of hazardous materials; toxicology; industrial hygiene; rights and responsibilities of workers under OSHA; monitoring equipment; hazard evaluation; site safety plan; standard operating procedures; engineering controls; personal protective clothing and equipment; medical program; decontamination; legal and regulatory aspects; and emergencies. A record should be kept of all training and the individuals involved in the training efforts.

Medical programs are important to try to prevent problems and then, where necessary, to protect the individuals. Prevention consists of surveillance, such as preemployment screening, periodic medical exams, and follow-up exams when appropriate, and termination exams. Treatment should be provided for emergencies and nonemergencies on a case-by-case basis. Record keeping is extremely important. Program review should be carried out periodically. Make sure that the individual efforts are conducted in an appropriate manner.

Site characterization provides the information needed to identify site hazards and to select worker protection programs. During site surveys, it is possible to determine the kinds and numbers of large containers or tanks that must be entered. Also determine the enclosed phases, such as buildings and trenches that must be entered. Potentially explosive or flammable situations may be indicated by bulging drums, effervescence, gas generation, or reading of specific instruments. Extremely hazardous materials, such as cyanide, phosgene, or radiation sources, should be determined. Look for visible vapor clouds. Areas where biological indicators, such as dead animals or vegetation, should be located and the reason for this destruction should be identified.

During the site survey, interviews should be conducted; and records should be searched about the site, meteorologic conditions, terrain, geologic and hydrologic data, proximity to population centers, accessibility by air and roads, pathways of dispersion of contaminants, hazardous substances involved (especially important), and chemical and physical properties. If possible, review the site from the air as well as from the land. Note any special labels, markings, or placards on containers or vehicles; level of deterioration or damage to these containers or vehicles; and any other unusual conditions, such as clouds, discolored liquids, oil slicks, vapors, or other suspicious substances.

The ambient air and the site of the perimeter should be monitored for toxic substances, combustible and inflammable gases or vapors, oxygen deficiency, ionizing radiation, and any other special problems if known. Note any unusual odors. Collect and analyze off-site samples of soil, drinking water, groundwater, site runoff, and surface water. On-site evaluation is essential to determine the types of containers, impoundments, and storage systems; their composition; and location. Also, note the physical condition of the materials, identify natural wind barriers, and note any additional safety hazards. All the site characterization information should be documented and assessed before any type of program is put into operation.

Air monitoring consists of use of on-site or direct-reading instruments and taking of samples for laboratory analysis. Direct-reading instruments may rapidly detect flammable or explosive atmospheres, oxygen deficiency, certain gases and vapors, and ionizing radiation. Some direct-reading instruments include the combustible gas indicator, which measures the concentration of the combustible gas or vapor; the flame ionization detector with gas chromatography option, which measures many organic gases or vapors; the gamma radiation survey instrument, which measures environmental gamma radiation; the portable infrared spectrophotometer, which measures concentrations of many gases and vapors in the air; the ultraviolet photoionization detector, which measures many organic and some inorganic acids and vapors; the direct-reading colorimetric indicator tube, which measures concentrations of specific gases and vapors; and the oxygen meter, which measures the percentage of oxygen in the air. Samples may be collected for anions, such as bromide, chloride, fluoride, nitrate, phosphate, and sulfate. Samples may also be taken for aliphatic amines, asbestos, metals, organics, nitrosamines, particulates, PCBs, and pesticides. At all hazardous waste sites, it is essential to also understand and measure the variables, such as temperatures, wind speed, rainfall, moisture, vapor emissions, and work activities. (See Chapter 12, Volume I and Chapter 9, this volume for instrumentation.)

Personal protective equipment includes general protective equipment, eye and face equipment, noise exposure equipment, respiratory equipment, head equipment, foot equipment, and electrical protective equipment. The respiratory equipment may be self-contained breathing apparatus, supplied-air respirators, or air-purifying respirators. The self-contained breathing apparatus and the air-purifying respirators are also differentiated by the face plate, which may be a positive-pressure respirator, a negative-pressure respirator, or a full-face piece mask.

The self-contained breathing apparatus provides the highest available level of protection against airborne contaminants and oxygen deficiency. Its problem is that is it bulky, has a limited air supply, and may impair movement in confined places. The positive-pressure supplied-air respirator enables the person to work longer than in the other type, it is less bulky and heavy than the other type, and it still protects against most airborne contaminants. However, this respirator is not approved for use in atmospheres immediately dangerous to life or health or in oxygen-deficient atmospheres.

The air purifying respirator enhances mobility and is lighter than the other equipment. However, it cannot be used in immediate dangerous to life or health situations or an oxygen-deficient atmosphere. Protective clothing may include fully encapsulating suits, nonencapsulating suits, aprons, leggings, sleeve protectors, gloves, and blast and fragmentation suits. The head is protected by safety helmets, which protect against chemical splashes and particulates. The eyes and face are protected by facial splash hoods, safety glasses, goggles, and sweat bands. The ears are protected by earplugs and muffs or headphones. The hands and arms are protected by gloves and sleeves; the feet are protected by safety boots. Individuals involved in hazardous waste site work should also carry a knife, flashlight or lantern, personal dosimeter, personal locator beacon, two-way radio, safety belts, harnesses, and lifelines.

Site control should include the development of a very accurate site map; appropriate site preparation; establishment of site work zones, including exclusion zones, contamination reduction zones, and support zones; buddy system; site security; communication systems; and safe work practices. The exclusion zone is the area where contamination either exists or could exist. In this area, all site characterizations should be mapped, photographed, and sampled where needed. Wells should be installed for groundwater monitoring. Cleanup work, such as drum identification, drum staging, and material bulking, should take place here. The contamination reduction zone is the transition area between the contaminated area and the clean area. The clean area is the support zone where additional assistance, materials, etc. should be available.

The hotline is the boundary of the exclusion zone. The hotline is determined by visually surveying the site and determining the locations of hazardous substances, drainage, leachate and spilled material, and visual discolorations. Data should be evaluated from the initial site survey concerning the presence of combustible gases, organic and inorganic gases, particulates or vapors, and ionizing radiation. The soil and water should be sampled and evaluated. An appropriate distance should be determined to prevent an explosion or fire from affecting personnel outside the exclusion zone. Also to be considered are how far the personnel must travel to and from the exclusion zone, physical area needed for site operations, meteorologic conditions, and potential for contaminants to be blown into the area. The hotline should be secured and marked.

Decontamination is the process of removing or neutralizing contaminants that have accumulated on personnel and equipment. The decontamination plan should establish a number of decontamination stations, determine the equipment to be used, determine the procedure to be used, establish clean areas, and establish means of disposing of clothing and equipment that are not completely decontaminated. Standard operating procedures should be used to minimize contact with the waste. Proper work practices should be established and stressed. Remote sampling, handling, and container opening should be done whenever feasible. Protect monitoring and sampling instruments by placing them in appropriate bags with openings for sampling parts and sensors. Disposable outer garments and disposable equipment should be used where appropriate. Equipment and tools should be covered with a strippable coating, which can be removed during decontamination. Contaminants either are located on the surface of personal protective equipment or can permeate into the personal protective equipment. Surface contaminants may be easy to detect and to remove. Permeated contaminants may be very difficult or impossible to detect and remove.

Five factors that affect the extent of permeation are contact time, concentration, temperature, size of the contaminants molecules and pore space, and physical state of the waste. Some of the decontamination methods used include rinsing with water under pressure, chemical leaching and extraction, evaporation, use of pressurized air jets, scrubbing and scraping, and use of steam jets. For surfaces that have been permeated, it may be necessary to dispose of the materials in a special manner. Chemical detoxification can be carried out through halogen stripping, neutralization, oxidation–reduction, and thermal degradation. Disinfection or sterilization can be carried out through the use of chemicals, dry heat, use of gas and vapors, irradiation, or steam.

The handling of drums and other containers can be extremely hazardous. Accidents may occur that result in detonations, fires, explosions, vapor generation, and physical injury. The appropriate procedure for handling drums depends on the contents of the drums. Symbols, words, or other marks on the drums, for example, *radioactive*, *explosive*, *corrosive*, *toxic*, and *inflammable*, should be noted (Figure 2.22). Signs of deteriorating, such as corrosion, rust, and leaks as well as swelling or bulging of the drums should be noted. The drums should be placed into five categories, including radioactive, leaking-deteriorated, bulging, explosive-shock-sensitive, and small volume of laboratory waste or dangerous materials. Special configurations should be noted. Appropriate procedures and proper numbers of individuals should be used in all cases of handling hazardous waste.

Site emergency procedures should be planned and thoroughly understood by everyone before working on the site. The causes of emergencies at hazardous waste sites may be worker related or waste related. Worker-related emergencies may range from minor accidents and chemical exposure to heatstroke, failure of personal

Figure 2.22 Some labels for identification of hazardous contents in storage containers.

protective equipment, physical injuries, and electrical burns. Waste-related emergencies include fire, explosion, leaking of hazardous materials, release of toxic vapors, reaction of incompatible chemicals, collapse of the containers, and discovery of radioactive materials. The emergency response team must be well trained and well organized. Communications is the necessary link between the various groups and ties the program together.

Municipal Solid Waste Program

A municipal solid waste program consists of the following parts, which have been previously discussed:

1. Waste collection
2. Operation of transfer stations
3. Transport of waste from transfer stations to waste management facilities
4. Waste processing or disposal at waste management facilities
5. Sale of by-products

Solid waste management can be enhanced through recycling, composting, and turning waste into energy. These are all means of reducing the quantity of waste that has to be landfilled or incinerated. Good programs are community based and well accepted by the people.

Pollution Prevention Program

The Pollution Prevention Act of 1990 and the EPA 1991 Pollution Prevention Strategy make clear that prevention is the first priority within environmental management. Proper management includes prevention, treatment, and disposal in an appropriate manner. The pollution prevention policy states:

1. Pollution should be prevented or reduced at the source whenever feasible.
2. Pollution that cannot be prevented should be recycled in an environmentally safe manner when feasible.
3. Pollution that cannot be prevented or recycled should be treated in an environmentally safe manner.
4. Disposal should only be used as a last resort and should be conducted in an environmentally safe manner.

Pollution prevention is source reduction that leads to an increased efficiency in the use of raw materials, energy, water, or other resources; and protection of natural resources by conservation. This leads to a reduction in hazards to public health and environment associated with the release of pollutants of all types. The program includes the modification of equipment, technology, process, or procedures; reformulation or redesign of products; substitution of raw materials; and improvements in housekeeping, maintenance, training, and inventory control.

In the agricultural sector, the pollution prevention program leads to:

1. Reduction of the use of water and chemicals
2. Adoption of less environmentally harmful pesticides or cultivation of crop strains with natural resistance to pest

3. Protection of sensitive areas
4. Increased efficiency in energy use
5. Substitution of environmentally benign fuel sources
6. Design changes that reduce the demand for energy

The pollution prevention program is carried forth through the use of regulations and compliance utilizing permits, inspections, and enforcement. It also involves the use of state and local partnerships, private partnerships, and federal partnerships. The program is enhanced through the use of the Pollution Prevention Information Clearinghouse, which is a free, nonregulatory service of the EPA dedicated to reducing or eliminating industrial pollutants through technology transfer, education, and public awareness. Selected EPA documents, pamphlets, information packets, and fact sheets on pollution prevention can be obtained. In addition, the program provides for the Pollution Prevention Special Collection, which is located in the Office of Prevention, Pesticides and Toxic Substances Chemical Library. This collection includes pollution prevention training materials, conference proceedings, periodicals, and federal and state government publications.

Used Oil Program

The Used Oil Program of the EPA is involved in teaching people how to collect, recycle, and use over again an estimated 380 million gal of used oil. Used oil can be recycled by: reconditioning on-site and reusing; inserting as feedstock into a petroleum refinery; and re-refining the oil by removing impurities, so that it can be used as a base stock for new lubricating oil. It may also be processed and burned for energy recovery.

Underground Storage Tank Program

As of September 30, 1999, the EPA estimates that 85% of the USTs were in compliance with the 1998 requirements for spill, overfill, and corrosion protection; and 60% of the tanks were in compliance with the leak detection requirements. The program consists of carrying out appropriate inspections of the remaining tanks not in compliance and ordering appropriate actions. In addition, it is necessary to investigate every facility that has a registered UST that has not been permanently closed. Further, unregistered tanks that are both active and abandoned need to be found and inspected. Real estate transactions can be useful in furthering the objectives of the program by providing information on parcels of property that may be contaminated.

INNOVATIVE TECHNOLOGIES

Innovative and alternative treatment technologies are under development in hazardous waste disposal as a result of the Superfund Amendments and Reauthorization Act (SARA). Some of these technologies are discussed next.

Pyretron® Oxygen Burner — The Pyretron® technology involves an oxygen–air–fuel burner and uses advanced fuel injection and mixing concepts to burn waste. The burner operation is computer controlled to automatically adjust the amount of oxygen to sudden changes in the heating value of the waste. The Pyretron technology developed by American Combustion Technologies Inc. achieved greater than 99.99% destruction in the removal efficiencies of all principal organic hazardous constituents.

Vapor Extraction System — The vapor extraction system uses a low temperature, fluidized bed to remove organic and volatile inorganic compounds from soils, sediments, and sludges. The contaminated materials are fed into a cocurrent fluidized bed, where they are well mixed with hot gases at about 320°F in the gas-fired heater. The gas treatment system removes dust and organic vapors from the gas stream. The vapor extraction system, which was developed by American Toxic Disposal Inc., can remove volatile and semivolatile organics, including PCBs; polynuclear aromatic hydrocarbons (PAHs); and pentachlorophenol (PCP); volatile inorganics; and some pesticides from soil, sludge, and sediment.

Integrated Vapor Extraction and Steam Vacuum Stripping — This system, developed by AWD Technologies Inc., simultaneously treats groundwater and soil contaminated with VOCs. The system consists of two processes: a vacuum stripping tower that uses low-pressure steam to treat contaminated groundwater and soil gas–vapor extraction–reinjection process to treat contaminated soil. The system is of high efficiency and typically reduces up to 99.99% of VOCs from water.

Soil Washing System — This system, developed by Biotrol Inc., is a volume reduction method for treating excavated soils and may be used when the soils are predominantly sand and gravel. The system is based on the principle that the contaminants are associated primarily with soil components finer than 200 mesh, which includes fine silts, clays, and soil organic matter. The system uses attrition scrubbing to disintegrate or break up soil aggregates, resulting in the liberation of highly contaminated fine particles in course sand and gravel.

Solvent Extraction — This system, developed by CF Systems Corporation, uses liquefied gas solvent to extract organics, such as hydrocarbons, oil, and grease from wastewater or contaminated sludges and soils. Carbon dioxide is used for aqueous solutions, whereas propane or butane is used for sediment, sludges, and soils that are semisolids. The solvent separates more than 99.00% of the organics from the waste. The technology can be used on waste that contains carbon tetrachloride, chloroform, benzene, naphthalene, gasoline, vinyl acetate, furfural, butyric acid, high organic acids, dichloroethane, oils and grease, xylene, toluene, methylacetate, acetone, higher alcohols, butanol, propanol, phenol, heptane, PCBs, and other complex organics. In the laboratory extraction, efficiencies of 99.99% have been obtained for volatile and semivolatile organics in aqueous and semisolid waste.

Solidification and Stabilization — This process, developed by Chemfix Technologies Inc., is an inorganic system in which soluble silicates and silicate setting agents react with polyvalent metal ions and certain other waste components to produce a chemically and physically stable solid material. The system is suitable for contaminated soils, sludges, and other solid waste. It may be used for electroplating waste, electric arc furnace dust, and municipal sewage sludge containing heavy metals, such as aluminum, antimony, arsenic, barium, beryllium, cadmium, chromium, iron, lead, manganese, mercury, nickel, selenium, silver, thallium, and zinc.

X-Trax™ Low-Temperature Desorption — This process, developed by Chemical Waste Management, is a low-temperature 500 to 800°F thermal separation process designed to remove organic contaminants from soils, sludges, and other solid media. The process is used to treat soils contaminated with PCBs or other organic contaminants.

Carver–Greenfield Process for Extraction of Oil Waste — This process, developed by Dehydro-Tech Corporation, is a continuous evaporation used to separate materials into their constituent solid, oil (including oil-soluble substances), and water phases. The process is especially good for oil-soluble hazardous organics that are concentrated in the oil phase. The carrier oil is mixed with waste sludge or soil and placed in an evaporation system to remove any water. A carrier oil at the boiling point of 400°F is typically used. The process can be used to treat sludges, soils, and other water-bearing waste containing oil-soluble hazardous compounds including PCBs and dioxins.

Membrane Microfiltration — This system, developed by the E.I. duPont DeNemours and Obelrin Filter companies, is designed to remove solid particles from liquid waste forming filter cakes, typically ranging from 40 to 60% solids. The filter material is a thin, durable plastic fabric with tiny openings about 1/10 μm in diameter that allows water or other liquids, along with solid particles smaller than the openings, to flow through them. The technology can be used for hazardous waste suspensions, especially liquid-, heavy metal-, and cyanide-bearing waste. The system is supposed to treat any type of solids including inorganics, organics, and oily waste.

In Situ Biological Treatment — This bioremediation technology, developed by the Ecova Corporation, is designed to biodegrade chlorinated and nonchlorinated organic contaminants by the use of aerobic bacteria. The contaminants become the carbon source of the bacteria. The technology is supposed to be applied to the biotreatment of soil and water on-site as well as the treatment of contaminated groundwater. In addition, the process can be applied to sludge, sediment, and other types of materials contaminated with organic constituents.

Leaching and Microfiltration — This system was developed by the Epoc Water Company. In this process, soils and sludges are decontaminated by leaching and microfiltration. The three main steps involved include the chemical leaching of the metals in the waste, separating the solids in the waste using a specially designed automatic tubular filter press, and precipitating of the metals using a special microfiltration method. The system can separate particles as small as 0.1 μm. It can be used to decontaminate sludges or soils that contain heavy metals, including barium, cadmium, chromium, lead, molybdenum, mercury, nickel, selenium, silver, and zinc.

Chemical Oxidation and Cyanide Destruction — This system, developed by Exxon Chemicals Inc. and Rio Linda Chemical Company, uses chlorine dioxide generated on-site by a special process to oxidize organically contaminated aqueous waste streams, and simple and complex cyanide in waste or solid media. This technology can be used on aqueous waste, soils, or any leachable solid media contaminated with organic compounds. It can also be applied to groundwater contaminated with pesticides or cyanide, sludges containing cyanide, PCBs, or other organics and industrial wastewater similar to refinery wastewater.

In Situ **Vitrification** — This system, developed by Geosafe Corporation, uses an electrical network to melt soil or sludge at temperatures of 1600 to 2000°C, thereby destroying organic pollutants by pyrolysis. Inorganic pollutants are immobilized within the vitrified mass. This system can be used to destroy or remove organics or to immobilize inorganics in contaminated soils or sludges.

Solidification and Stabilization — This technology, developed by Hazcon Inc., immobilizes contaminants in soil by binding them to concrete-like, leach-resistant mass. This process can solidify contaminated material with high concentrations of up to 25% organics.

Flame Reactor — This system was developed by the Horsehead Resource Development Company Inc. The technology is a hydrocarbon-fueled, flash smelting system that treats residues and waste containing metals. The reactor processes waste with a very hot >2000°C reducing gas produced in the combustion of solid or gaseous hydrocarbon fuels in oxygen-enriched air. This system has been successfully tested on electric-arc furnace dust, lead blasted furnace slag, iron residues, zinc plant leach residues, and purification residues, and brass mill dust and fumes.

In Situ **Solidification and Stabilization Process** — This process, developed by International Waste Technologies-Geo-Con Inc., immobilizes organic and inorganic compounds in wet or dry soils, using reagents (additives) to produce a cementlike mass. The process has been laboratory tested on soils containing PCBs, PCP, refinery waste, and chlorinated and nitrated hydrocarbons.

Liquid and Solid Contact Digestion — This process, developed by Motec Inc., utilizes liquid–solid contact digestion technology to biologically degrade organic waste. Organic materials and water are placed in a high-energy environment, where the organic constituents are then biodegraded by acclimated microorganisms. The technology is used to treat halogenated and nonhalogenated organic compounds and some pesticides and herbicides.

Circulating Fluidized Bed Combustor — This process, developed by Ogden Environmental Services, uses high-velocity air to entrain circulating solids and create a highly turbulent combustion zone for the efficient destruction of toxic hydrocarbons. The technology may be used on soils, slurries, and sludges contaminated with halogenated and nonhalogenated hydrocarbons. It has also been applied to soils contaminated with PCBs and fuel oil.

Soil Washing and Catalytic Ozone Oxidation — This process, developed by Ozonics Recycling Corporation, is designed to treat soils with organic and inorganic contaminants. The technology, which is a two-stage process, extracts the contaminants from the soil and then oxidizes them. Oxidation involves ozone, ultraviolet, and ultrasound. The process may be applied to soils, solids, sludges, leachates, and groundwater containing organics — such as PCBs, PCP, pesticides, and herbicides — dioxins, and inorganics such as cyanides.

Chemtact™ Gaseous Waste Treatment — This process, developed by Quad Environmental Technologies Corporation, uses gas scrubber technology to remove gaseous organic and inorganic contaminants through efficient gas-liquid contact. The technology may be used on gaseous waste streams containing a wide variety of organic or inorganic contaminants, but is best suited for VOCs.

Solvent Extraction — This process, developed by the Resources Conservation Company, is potentially effective in treating oily sludges and soils contaminated with hydrocarbons by separating the sludges into three fractions: oil, water, and solids. As the fractions separate, contaminants are partitioned into specific phases. The system may be used for most organics or oily contaminants in sludges or soils, including PCBs.

Plasma Reactor — The system, developed by Retech Inc., is a thermal treatment technology that uses the heat from a plasma torch to create a molten bath, which is used to detoxify contaminated soils. Organic contaminants are vaporized and react at very high temperatures to form innocuous products. This system is best used for soils and sludges contaminated with metals and hard-to-destroy organic compounds.

In Situ **Solidification and Stabilization** — This system, developed by the S.M.W. Seiko Inc., involves the *in-situ* fixation stabilization and solidification of contaminated soils. The system is best used for soils contaminated with metals and semivolatile organic compounds (SVOCs), such as pesticides, PCBs, phenols, and PANs.

Solidification and Stabilization — This system, developed by Separation and Recovery Systems Inc., uses lime to neutralize sludges with high levels of hydrocarbons. The system may be used for acidic sludges containing at least 5% hydrocarbons, such as those produced in the remanufacturing of lubricating oils. It can also stabilize waste containing up to 80% organics.

Infrared Thermal Destruction — This system was developed by Shirco Infrared Systems. This electric infrared incineration technology is a mobile thermal processing system that uses electrically powered silicon carbide rods to heat organic waste to combustion temperatures. Remaining combustibles are incinerated in an afterburner. This system may be used for soils or sediments with organic contaminants, and liquid organic wastes that have been treated by mixing with sand or soil.

Solidification and Stabilization with Silicate Compounds — This system, developed by Silicate Technology Corporation, uses silicate compounds and may consist of two separate technologies: (1) one technology is used to fix and solidify organics and inorganics contained in contaminated soils and sludges; (2) another technology removes organics from contaminated water. This system may be used on soils and sludges to remove metals, cyanides, fluorides, arsenates, ammonia, chromates, and selenium.

Solidification and Stabilization — This system, developed by Soliditech Inc., immobilizes contaminants in soils and sludges by binding them in a concrete like, leach-resistant matrix. This system is located for treating soils and sludges contaminated with organic compounds, metals, inorganic compounds, oil, and grease.

Steam Injection and Vacuum Extraction — This system, developed by Solvent Services Inc., is used *in situ* to remove VOCs and SVOCs from contaminated soil. Steam is forced through the soil to thermally enhance the vacuum extraction process. The system is used to treat soil contaminated with VOCs and SVOCs in total concentrations ranging from 10 ppb to 100,000 ppm by weight.

In Situ **Vacuum Extraction** — This system, developed by Terra Vac Inc., is a process of removing and venting VOCs from the vadose or unsaturated zone of soils. These compounds may be removed from the vadose zone before they contaminate groundwater. The system may be used on organic compounds that are volatile or semivolatile at ambient temperatures in soils and groundwater.

In Situ Steam and Air Stripping — This system, developed by Toxic Treatments (USA) Inc., uses *in-situ* steam and air stripping of volatile organics from contaminated soil. The system may be used on organic contaminants such as hydrocarbons and solvents, with sufficient partial pressure in the soil.

Ultraviolet Radiation and Oxidation — The system, developed by Ultrox International, uses ultraviolet radiation–oxidation processes, ozone and hydrogen peroxide to destroy toxic organic compounds, especially chlorinated hydrocarbons in water. This system may be used for contaminated groundwater, industrial wastewater, and leachates containing halogenated solvents, phenol, PCP, pesticides, PCBs, and other organic compounds.

Solidification and Stabilization — This system, developed by Wastech Inc., uses proprietary bonding agents to soils, sludges, and liquid waste containing VOCs or SVOCs and inorganic contaminants to fix the pollutants within the waste. The treated waste is then mixed with cement materials and placed in a stabilizing matrix. The system may be used for a wide variety of waste streams consisting of soils, sludges, and raw organic streams, such as lubricating oil, aromatic solvents, evaporator bottoms, chelating agents, and ion exchange resins.

Pact® Wet Air Oxidation — This system, developed by Zimpro-Passavant Inc., is a treatment method that combines two technologies: the Pact Treatment System and wet air oxidation, which use powdered activated carbon combined with conventional biological treatment, such as an activated sludge system to treat liquid waste containing toxic organic compounds. The system may be used to treat a variety of industrial wastewaters, dye production wastewaters, pharmaceutical wastewaters, refinery waste waters, and synthetic fuels wastewaters as well as contaminated groundwater and mixed industrial–municipal wastewater.

Microbial Degradation of Polymers in Plastic Materials — This method is tested by a number of polymer scientists. One such process using bacteria and fungi is currently in an experimental state at the University of Illinois at Urbana-Champaign in the Department of Agronomy.

Gas-Phase Chemical Reduction Process — This process, developed by ELI ECO Logic International Inc., is a gas-phase reduction reaction of hydrogen with organic and chlorinated organic compounds at elevated temperatures to convert aqueous and oily hazardous contaminants into a hydrocarbon-rich gas product with primary components of hydrogen, nitrogen, methane, carbon monoxide, water, and light hydrocarbons. The process is used to treat aqueous and oil waste streams contaminated with PCBs, PAH, chlorinated dioxins and dibenzofurans, chlorinated solvents, chlorobenzenes, and chlorophenols.

Cross-Flow Pervaporation System — This process, developed by Wastewater Technology Centre, involves removing VOCs from contaminated water. Permeable membranes that preferentially adsorb VOCs are used to partition VOCs from the contaminated water. The VOCs diffuse from the membrane–water interface through the membrane and are drawn under vacuum. The process is used to treat aqueous waste streams (in groundwater, lagoons, leachate, and rinse water) contaminated with VOCs such as solvents, degreasers, and gasoline. The technology is applicable to the types of wastes currently treated by carbon adsorption, air stripping, and steam stripping.

Rochem Disc Tube Module System — This process, developed by Rochem Separation Systems, Inc., uses membrane separation systems to treat a range of aqueous solutions, from seawater to leachates containing organic solvents. The system uses osmosis through a semipermeable membrane to separate pure water from contaminated liquids. It is used to treat sanitary landfill leachate containing both organic and inorganic chemical species, water-soluble oil wastes used in metal fabricating and manufacturing industries, solvent–water mixtures, and oil–water mixtures generated during washing operations at metal fabricating facilities.

Membrane Filtration and Bioremediation — This process, developed by SBP Technologies, Inc., is a hazardous waste treatment system consisting of (1) a membrane filtration unit that extracts and concentrates contaminants from groundwater, surface water, wash water, or slurries; and (2) a bioremediation system that treats concentrated groundwater, wash water, and soil slurries. The process is used to concentrate contaminants and reduce the volume of contaminated materials from a number of waste streams including contaminated groundwater, surface water, storm water, landfill leachates, and industrial process wastewater.

***In Situ* Steam Enhanced Extraction Process** — This process, developed by Udell Technologies, Inc., removes VOCs and SVOCs from contaminated soils both above and below the water table. Steam is forced through the soil by injection wells to thermally enhance the vapor and liquid extraction processes. The process is used to treat hydrocarbons such as gasoline, diesel, and jet fuel; solvents such as trichloroethene, trichloroethane, and dichlorobenzene; or a mixture of these compounds that are contaminating soil.

Debris Washing System — This process, developed by Risk Engineering Laboratory and IT Corporation, is used for on-site decontamination of metallic and masonry debris at CERCLA sites by spraying, washing, and rinsing processes. The process is used on various types of debris (scrap metal, masonry, or other solid debris such as stones) contaminated with hazardous chemicals such as pesticides, dioxins, PCBs, or hazardous metals.

Microwave Disinfection System — This process, developed by ABB Sanitec, Inc., is an alternative solution to handling medical waste, combining shredding, steam injection, and microwaves, which takes place inside a compact, enclosed unit. It is used to reduce the volume of hazardous waste, moisten it at high temperatures, and then heat all particles by use of microwaves. The steam is extracted through a high-efficiency particulate air (HEPA) filter, trapping any potential harmful airborne pathogens. The treatment end product volume has been reduced by 80%.

SOLID AND HAZARDOUS WASTE LAWS AND REGULATIONS

The major acts in solid waste that are discussed include the Solid Waste Disposal Act of 1965, Resource Recovery Act of 1970, RCRA, revisions of RCRA in 1980, revisions of RCRA in 1984 (known as the Hazardous and Solid Waste Amendments

[HSWA]), CERCLA, or Superfund, of 1980, and Superfund Amendments and Reauthorization Act (SARA), of 1986. To distinguish between these acts, the RCRA promotes "cradle-to-grave" management for hazardous waste from the point of generation to the final disposal location. RCRA regulates the handling, transport, and disposal of hazardous waste generated by more than 20,000 generators (Figure 2.23). The HSWA altered the focus of waste management to require EPA to concentrate on permitting land disposal facilities and eventually phasing out land disposal of some waste. The CERLCA was passed to set up a comprehensive program to clean up the worst abandoned or inactive waste sites already existing in the country. The law requires that those responsible carry out the cleanup. SARA was passed to strengthen the EPA mandate to focus on permanent cleanups at Superfund sites. Also when CERCLA was amended by SARA, included in the new law was the Emergency Planning and Community Right-to-Know Act, known as Title III.

Other environmental laws controlling hazardous substances include:

1. Clean Air Act, which regulates the emission of hazardous air pollutants
2. Clean Water Act, which regulates the discharge of hazardous pollutants into the nation's surface waters
3. Marine Protection Research and Sanctuaries Act, which regulates waste disposal at sea (also called the Ocean Dumping Act)
4. OSHA, which regulates hazards in the work place, including working exposure to hazardous substances
5. Safe Drinking Water Act, which regulates contaminant levels in drinking water
6. Toxic Substances Control Act (TSCA), which regulates the manufacture, use, and disposal of chemical substances
7. Hazardous Materials Transportation Act, which regulates the transportation of hazardous materials
8. Atomic Energy Act (Nuclear Regulatory Commission), which regulates the nuclear energy production and nuclear waste disposal
9. Surface Mining Control and Reclamation Act, which regulates the environmental aspects of mining, especially coal and reclamation
10. Deep-Well Injection Regulation of 1988, which was required by the Hazardous and Solid Waste Amendments (regulation requires that all underground injection of certain solvents, dioxins, and other chemicals to be determined by the EPA have to go to other forms of treatment instead of deep-well injection)
11. Medical Waste Tracking Act of 1988, which requires the EPA to list medical wastes to be covered by the program (listed medical waste to be tracked from the point of generation to the point of disposal)
12. Mercury-Containing and Rechargeable Battery Management Act, which facilitates the recycling of nickel–cadmium and certain small sealed lead–acid rechargeable batteries and phases out the use of mercury
13. Land Disposal Program Flexibility Act of 1996, which directs the administrator of EPA to provide additional help to approved states for any landfill that receives 20 tons or less of municipal waste per day

The Solid Waste Disposal Act of 1965 enabled the federal government to fund research and technical assistance for state and local planners. In 1970, the program was expanded by the enactment of the Resource Recovery Act. This law promoted

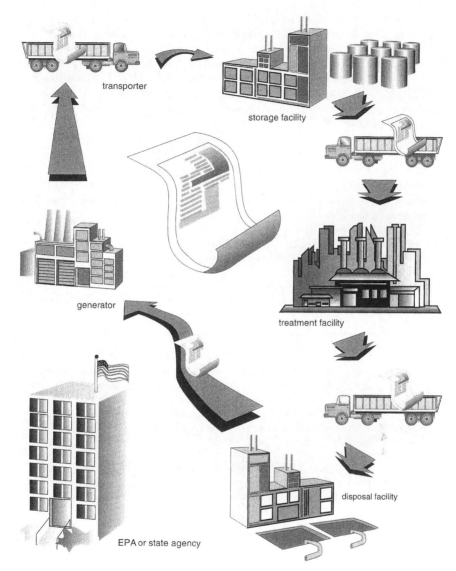

Figure 2.23 RCRA tracking system for hazardous waste. (Adapted from *Solving the Hazard-ous Waste Problem.* Office of Solid Waste, U.S. Environmental Protection Agency, U.S. Printing Office, Washington, D.C., 1986, p. 11.)

the development of sanitary landfills and encouraged communities to shift from the disposal of material toward conservation, recycling, and new control technologies.

As a greater awareness of the disposal of chemical and industrial waste occurred, Congress passed the RCRA in 1976. RCRA promoted cradle-to-grave management for hazardous waste from the point of generation to the final location of disposal.

The program set up requirements for hazardous waste generators, transporters, treatment, storage, and disposal facilities. RCRA was amended by HSWA in 1984. The amendments alter the focus of waste management. HSWA required the EPA to focus on setting up permitting for land disposal facilities and eventually phasing out the land disposal of certain wastes. It expanded the RCRA-regulated community to include businesses that generate small amounts of hazardous waste. Recycling and waste minimization provisions were also included in the act to reduce the amount of waste generated.

HSWA also added provisions for USTs containing petroleum and some hazardous substances. Congress gave the EPA the responsibility under RCRA for regulating the storage of gasoline and other commercial products, instead of only waste materials.

In September of 1988, the U.S. EPA issued a draft agenda for action that set out a series of national goals and objectives relating to solid waste management. The goals were as follows:

1. The United States must find a safe and permanent way to eliminate the gap between waste generation and the available capacity in landfills, incinerators, and secondary materials markets.
2. Source reductions and recycling are the preferred options for closing the gap in reducing the amount and toxicity of waste that must be landfilled or incinerated.
3. All waste management practices should be made safer.

The objectives were:

1. Increase the waste planning and management information available to states, local communities, waste handlers, citizens, and industry, and increase data collection for research and development.
2. Increase effective planning by waste handlers, local communities, and states.
3. Increase source reduction activities by the manufacturing industry, government, and citizens, or increase recycling by government and by individual and corporate citizens.
4. Improve the safety of municipal solid waste combustion to protect the human health in the environment.
5. Increase the safety of landfills to protect human health and the environment.

The EPA at the same time issued a set of proposed regulations establishing minimum standards for operating landfills. The EPA also issues rules concerning air pollution and ash disposal practices for refuse combustors. It is encouraging governmental agencies to buy the recycled goods.

In 1991, the President signed an executive order, Federal Recycling and Procurement Policy, which requires that federal agencies increase recycling efforts and encourage market demand for recycled products.

In 1994, the President issued Executive Order 12898, which focuses federal agency attention on environmental justice. Environmental justice is the fair treatment for all people of all races, cultures, and incomes, concerning the development of environmental laws, regulations, and policies. There is a concern that minority populations or low-income populations have a disproportionate amount of the adverse health and environmental effects due to serious environmental problems.

In 1995, the administration fine-tuned federal policies on hazardous waste by emphasizing the targeting of regulations toward higher risk chemicals, reducing the economic burden of the program, and encouraging greater community participation. A new rule exempted from RCRA subtitle C regulations certain waste that did not pose a significant public health threat, resulting in substantial savings to businesses handling this low-risk waste and allowing more time to focus on greater hazards to public health and the environment.

In 1995, the Marine Protection, Research, and Sanctuaries Act (Ocean Dumping Act) regulations were updated. The EPA developed ocean dumping criteria to be used in evaluating permit applications and designated recommended sites for ocean dumping. The Corps of Engineers is permitted to dump dredged material subject to the EPA concurrence and use of EPA dumping criteria. The factors considered are need for dumping; effect of dumping on human health and welfare; effect of dumping on fish, wildlife, and shorelines; effect of dumping on marine ecosystems; persistence and permanence of effects; and effect of dumping of certain volumes and concentrations. The dumping of sewage sludge and industrial waste is unlawful.

In 1996, Congress enacted the Land Disposal Program Flexibility Act, which reversed the requirement of the EPA to issue stringent and costly treatment requirements for certain low-risk wastes that are already regulated under the Clean Water Act or the Safe Drinking Water Act. Also in 1996, the Mercury Containing and Rechargeable Battery Management Act established uniform national labeling, storage, and transportation standards to encourage environmentally sound recycling of batteries, and to prohibit the sale of certain mercury-containing batteries.

The administration supported reforms to promote greater community participation in environmental protection, such as in East St. Louis, IL. This is a highly industrialized area suffering from a variety of environmental problems. Residents and companies have been working together to reduce the level of TRI chemicals. Local hospital staffs state that the elimination of chemical spills and various industrial sites has resulted in hundreds fewer emergency room admissions.

In 1997, the president issued Executive Order 13045, Protection of Children from Environmental Health Risks and Safety Risks. This executive order is both economically significant, and protective of the health of children.

State Government

States have a role to play in solid waste management activities. Not all states are involved in all aspects of the solid waste program. However, recommended items to be handled by the state are surveying all existing practices; determining the immediate and long-range needs for communities in solid waste disposal; establishing and enforcing minimum standards and necessary guidelines to meet these standards; coordinating air, land, and water pollution control activities along with planning activities and solid waste disposal; providing financial and technical assistance where needed to local governmental units; encouraging local governments to cooperate within the state and with neighboring states; providing a continuous public health education program for solid waste management; providing enabling legislation allowing various governmental units within the state to formulate their own rules, regulations, ordinances, and codes. Above all, the states must be involved in capacity limitations and siting controversies in solid waste management.

At present, every state has minimum technology standards for the operation and maintenance of municipal solid waste landfills. In addition to this, 44 states and territories have implemented landfill location standards; 51 states and territories had closure and 42 had postclosure air requirements; 42 states and territories require groundwater monitoring; 23 states and territories require leachate monitoring; and 10 states require surface water monitoring.

Solid waste disposal capacity is declining in many areas of the country, especially in the Northeast and in other areas with high population densities. States such as Connecticut, Massachusetts, and Florida are rapidly running out of landfill space.

New Jersey, California, Florida, Vermont, Oregon, Minnesota, and others have established comprehensive environmental regulation of landfills, developed planning and siting mechanisms, set recycling goals, and established state procurement rules favoring recycled materials, banned or restricted certain types of packaging, established rules governing disposal of medical waste, and taken numerous other steps to gain control of the solid waste issue. Some states have begun to develop regulations governing incinerator ash disposal. The federal laws have put the state governments at the forefront of planning and regulating; therefore, not only the states mentioned but others must become deeply involved in all issues related to the solid and hazardous waste situation.

Local Government

Local authorities must ultimately resolve the problem of solid waste, because solid waste is generally a local area problem, instead of a total state and federal problem. The legal basis for local governments establishing laws is based on the enabling authority from the state and the enforcement of certain federal regulations. The local government should have legal authority to establish area-wide solid waste management programs; adopt requirements for storage, transportation, processing, and disposal of solid waste including abandoned automobiles, industrial, agricultural, and community mixed waste; establish a definite agency for regulation or operation and control of solid waste materials; establish contracts with individuals, private corporations, or various governmental agencies to carry out the necessary programs; acquire land by eminent domain, if necessary, for solid waste facilities; abate the nuisances and health hazards caused by improper handling, storage, and collection of solid waste; and establish a system of financing of solid waste programs.

It should be noted that laws are only as good as the enforcement that follows them. To enforce a solid waste disposal program and ensure that the ordinances, rules, and regulations are obeyed, it is necessary to stringently apply the penalties to the individuals who violate the laws.

The most exciting and forward looking recycling programs, the best run landfills and energy recovery facilities, and some of the worst landfills and incinerators are found in local areas run by local entities. The major issues on the local level are planning and siting, regionalization, financing, regulatory compliance, and choice of technology to be used. Because a proposal to site a landfill or incinerator causes an enormous reaction from people living nearby, the whole question of siting is a political issue, beside the technical issue of the type of geology, water bodies, wetlands and floodplains, transportation, screening, and aesthetic considerations, which need to be accounted for.

Capital budgeting is another serious challenge for local decision makers in connection with landfills, incinerators and ash disposal facilities. Capital budgeting must account for cradle-to-grave handling of the solid waste and must see to it that enough finances are available for any eventualities.

Private Sector

The private sector is an important part of the solid waste industry. Private companies haul garbage, own and operate landfills, run recycling facilities, and build and operate trash combustors and compost. This is often done under contract to local governments. The private sector may provide more professional staff and technical capabilities, share the risk of solid waste disposal, make available capital finance, provide a faster turn around time for construction of new facilities, and establish more flexible labor agreements. Many private providers deal directly with businesses and residences, as well as provide services under contract to local governments. Many communities are contracting their municipal solid waste management to the private sector.

Nonprofit organizations have been successful in the solid waste business especially in the area of recycling, where they have proved the feasibility of approaches that the private sector later adopted.

RCRA Programs

The three main RCRA programs include the Solid Waste Program, which is subtitle D, the Hazardous Waste Program, which is subtitle C, and the underground storage tank program, which is subtitle I. The three goals of RCRA are:

1. To protect human health and the environment
2. To reduce waste, conserve energy, and natural resources
3. To reduce or eliminate the generation of hazardous waste as expeditiously as possible

The Solid Waste Program goals are to encourage environmentally sound waste management practices, to maximize the reuse of recoverable resources, and to foster resource conservation. The two main parts of the program are to establish technical standards for solid waste disposal facilities and solid waste management grant programs for states on a voluntary basis. The technical standards and open dump criteria cover the areas of floodplains, endangered species, surface water, groundwater, waste application limits for land used in the production of food-chain crops, disease transmission, air, and safety. The minimal technical standards require double liners for landfills, groundwater monitoring wells, and capture and treatment of all leachate. A major facet of the program is the establishment of regulations for state plans. The purpose of these regulations is to assist states in developing and implementing EPA-approved solid waste management plans. Such plans help ensure environmentally sound solid waste management and disposal, resource conservation, and maximum utilization of valuable resources. In developing a solid waste management plan, the state must go through a number of steps, as detailed in 40 CFR, part 255. This requires the government to select or establish an agency to develop the state plan. The State must then gain EPA approval by complying with the standards. The standards set forth several points:

1. Identify the responsibilities of state, local, and regional authorities in implementing the plan.
2. Describe a regulatory approach that prohibits the establishment of new open dumps, provides for the closing or upgrading of all open dumps, and establishes any state regulatory authorities.
3. Ensure that state or local governments can establish long-term contracts, force the supply of solid waste toward resource recovery facilities, secure long-term markets for material and energy recovery, and conserve materials and energy by reducing waste volume. The state must also detail the combination of practices that are necessary to use or dispose of solid waste in an environmentally sound manner.

Hazardous Waste Cradle-to-Grave Management

The cradle-to-grave management effort required by RCRA involves five basic elements:

1. Identification of generators and the types of waste that they produce is needed.
2. Tracking the waste by means of a uniform manifest describing the waste, its quantity, the generator and receiver, and how it is transported is needed.
3. Permitting is necessary for all hazardous waste treatment, storage, and disposal facilities.
4. Restrictions and controls on hazardous waste facilities are needed. They must follow the EPA rules and guidance specifying acceptable conditions for disposal, treatment, and storage of hazardous waste.
5. Enforcement and compliance on generators, transporters, and facilities are needed so that they comply with the regulations. The hazardous waste manifest trail starts with the generator and goes to the transporter storage facility, transporter treatment facility, transporter disposal facility, transporter, and back to the generator. A one-page manifest must accompany every waste shipment. The resulting paper trail documents the progress, treatment, storage, and disposal of the waste. A missing form alerts the generator to investigate, which may mean calling in the state agency or EPA.

Underground Storage Tanks

USTs are covered under subtitle I of RCRA. UST is defined as any tank with at least 10% of its volume buried below ground including any pipes attached to the tank, which is used for storing petroleum products (including gasoline and crude oil), and any substance defined as hazardous under Superfund. However, RCRA does not regulate tanks storing hazardous waste. Tanks not included in this program, besides those for hazardous waste storage, are farm and residual tanks with a holding capacity of less than 1100 gal of oil, on-site tanks storing heating oil, septic tanks, pipelines (regulated under other laws), surface impoundments, systems for collecting storm water and wastewater, flow-through process tanks, and liquid traps or associated gathering lines related to operations in the oil and natural gas industry.

The UST program has five parts, which includes a ban on unprotected new tanks, modification program, regulatory program, state authorization, and inspections and enforcements. All new tanks must be protected against corrosion, constructed of noncorrosive material, or steel clad with noncorrosive material; or must be designed to prevent the release of the storage substances. The material used in the construction or lining of the tank must be compatible with the substance to be stored. The EPA is assisting the states in managing USTs. A large number of these tanks need to be replaced. The EPA is helping the states get started on cleaning up the contaminated land and water where leaks have already occurred and helping ensure that new tanks have protection to prevent future leaks.

In May 1995, the EPA promulgated the Universal Waste Rule to reduce the amount of hazardous wastes entering the municipal solid waste stream, to encourage the recycling and proper disposal of certain common hazardous wastes, and to reduce the regulatory burden on businesses that generate these wastes. Regulations were made simpler. The rule recognizes that some common hazardous wastes, such as nickel–cadmium rechargeable batteries, do not require the full amount of hazardous wastes regulatory requirements. The rule also eases the regulatory burden on handlers and transporters by streamlining a number of RCRA hazardous waste collection and management requirements, including those related to notification, labeling and marking, accumulation time limits, employee training, and off-site shipment. The rule extends the amount of time that certain businesses can accumulate materials on-site, and also allows certain companies to transport the materials with a common carrier, instead of the hazardous waste transporter. The rule does not automatically apply to all states, and therefore each state must seek authorization from the EPA.

In 1997, the EPA issued rules concerning the Land Disposal Program Flexibility Act of 1996. The rules provided for additional flexibility for alternate frequencies of daily cover, methane monitoring, infiltration layers for final cover, and means for demonstrating financial assurance. This provided owners and operators of small municipal solid waste landfills the opportunity to reduce the cost.

Over the last several years, the EPA has issued a series of rules and documents to help carry out RCRA. In 1999, the EPA modified the Notification of Regulated Waste Activity and RCRA Hazardous Waste Part A Permit Application Modification. The agency also drafted, the 1999 Hazardous-waste Report; Office of Solid Waste Burden Reduction Project; agency information collection activities; continuing collection; comment requests; general hazardous waste facility standards, etc. Rules have been issued on disposal of nonhazardous waste; financial assurance mechanisms; generation and transportation of waste; identification of hazardous waste; land disposal restrictions on military munitions and mining waste; municipal solid waste landfills; recycling of hazardous and nonhazardous waste; treatment of hazardous waste; and waste minimization.

Hazardous and Solid Waste Amendments Program

HSWA is using a phased approach to prohibit the land disposal of all untreated hazardous waste. By 1990, any waste listed as hazardous on the date of the enactment of HSWA was prohibited from land disposal, unless the EPA had published treatment standards for the waste or a petition had been approved that demonstrated that there would be no migration of hazardous constituents from the disposal unit as long as the waste remained hazardous. HSWA has identified solvents and dioxins as restricted and the California list as restricted. The California list includes liquid hazardous waste containing metals, three cyanides, PCBs, low pH wastes, and liquid and nonliquid hazardous waste containing HOCs above specified levels. The EPA has also specified treatment standards for PCB- and HOC-containing waste. All other listed hazardous wastes, which are approximately 450 in number, were divided into thirds based on volume and toxicity. The HSWA minimum technology standards require the improvement of the ways in which existing and new landfills, as well as surface impoundments, are constructed and operated. These standards include installing two or more liners, the leachate collection system, and a groundwater monitoring system. Many of the surface impoundments are to be closed because the systems cannot meet the minimum technology requirements. These hazardous wastewaters may be shifted to treatment tanks. The sludges from treatment tanks are then to be subjected to RCRA when removed from the tanks if they are listed as a hazardous waste or they exhibit a hazardous characteristic. If the treated wastewaters are discharged into surface water or to publicly owned treatment works, the discharge is to be subject to the Clean Water Act.

Superfund (CERCLA)

The Superfund was started in 1980 because of a serious problem with existing hazardous waste disposal sites. CERCLA authorized $1.6 billion over 5 years for a comprehensive program to clean up the worst abandoned or inactive waste sites in the country. The reauthorization of CERCLA is known as SARA of 1986. These amendments provided $8.5 billion for the cleanup program and an additional $500 million for the cleanup of leaks from USTs. Under CERCLA, responsible parties clean up sites themselves with EPA or state oversight. The EPA and the states also can start actions to clean up sites after attempts and negotiations with the companies that have created the problem have failed or if an emergency occurs. The government lets the EPA later sue the companies to cover the cost of the cleanups.

The EPA designed a process to implement CERCLA. This process starts with discovering the site. Then the basic steps are to assess the nature and degree of contamination, to determine the relative threat, to analyze the potential cleanup alternatives, and to take actions to clean up the site. This process also provides for emergency cleanup of contamination. If a site is determined to present an immediate danger to public health, welfare, or environment, the EPA uses funds available to take any one of several actions necessary to alleviate the threat. The EPA may provide an alternate water supply, put up fencing, remove discrete sources of contamination on the surface (such as drums), or in severe cases temporarily relocate people living there until the danger is eliminated. Over 35,000 potentially contaminated sites have been identified.

The EPA then does a preliminary assessment to determine whether an imminent threat exists that would require immediate emergency attention. If the preliminary assessment indicates that more detailed investigation is necessary, site inspection is conducted. The site inspections are evaluated using the hazard ranking system. This is a technical evaluation that considers how contamination at a site could affect people or the environment. The number of potential people exposed to the chemical contamination is factored into the criteria. This system helps set up cleanup priorities for the Superfund.

At present, sites that have scored 28.5 or above using the hazard ranking system, are placed on the NPL. This is the EPA official list of hazardous waste sites that need attention under the Superfund. Further, each state may propose a top-priority site. Since 1980, over 18,000 sites have been found to require no further action from the EPA. Over 30,000 site evaluations have been conducted; and 1200 sites have been listed or proposed for listing on the NPL. Except for any removal actions, listing on the NPL is a prerequisite for cleanup activities that would use federal Superfund money.

When sites are included on the NPL, further in-depth study is used to define the nature and extent of contamination. After the public has had an opportunity to comment on the in-depth study, the most effective remedy for long-term cleanup is determined. Before a remedy is selected, the EPA or the state determine whether the remedy complies with federal and state cleanup standards, the cost effectiveness of the action, and the mandate to use treatment to the maximum extent practical. Throughout the Superfund process, the EPA tries to identify the companies or individuals whose waste caused or contributed to the problem and either to have them do the cleanup or to help pay for the cleanup.

Superfund Amendment and Reauthorization Act

SARA reauthorized the Superfund Program for 5 years. It increased the size of the fund to $8.5 billion, strengthened and expanded the cleanup program, focused on the need for emergency preparedness and community right to know, and changed the tax structure for financing the fund. The new Superfund under SARA stressed permanent remedies and treatments for recycling technologies in cleaning up hazardous waste sites. It set specific cleanup goals and standards, provided new enforcement for authorities and responsibilities, increased state involvement in every phase of the Superfund program, and increased the attention to community and state emergency preparedness activities. The new Superfund also increased the focus on human health problems posed by hazardous waste sites, encouraged greater citizen participation in making decisions on how sites should be cleaned up, and expanded research and training activities to promote the development of alternative and innovative treatment technologies. The subject of the emergency preparedness and community right to know is discussed in Chapter 8, Terrorism and Environmental Health Emergencies. SARA also established the Leaking Underground Storage Tank Trust Fund and Program.

The authorization was formally extended to 1994, and has been on a continuing basis since then. As of 2000, Congress has introduced a number of bills to modify the original act. The most controversial element of the law is its broad liability scheme. The generators of the hazardous substances, the transporters who selected the site, and the owners and operators past and present, are all liable. The average cost of cleanup is $30 million per site.

Reducing Toxic Risks through Voluntary Action

The 33/50 program is an EPA initiative designed to reduce toxic waste through a voluntary effort by industry. Industry reduced toxic wastes by 33% by 1991 and 50% by 1994. The 50% reduction eliminated 700 million lb of toxic waste. More than 6000 TRI chemical sites and 1300 corporations were involved. The chemicals targeted were many of those found at hazardous waste sites. Companies are continuing to voluntarily cut toxic wastes.

Small Quantity Generator

A small quantity generator of hazardous waste is considered to be one that generates no more than 100 kg (about 220 lb or 25 gal) of hazardous waste and no more than 1 kg (about 2 lb) of acutely hazardous waste in any calendar month. Federal law requires that these individuals identify all hazardous waste generated; send the waste to a hazardous waste facility, a landfill, or another facility approved by the state for industrial or municipal waste; and never accumulate more than 1000 kg of hazardous waste on their property. Individuals who generate between 100 and 1000 kg/month or between 220 and 2200 lb or about 25 to 300 gal of hazardous waste and no more than 1 kg of acutely hazardous waste per month are considered to be a 100 to 1000 kg/month generator. Generators of 1000 kg/month or more, about 2200 lb or 300 gal or more of hazardous waste, or more than 1 kg of acutely hazardous waste in any month are listed as generators of 1000 kg/month or more. One barrel equals about 200 kg of hazardous waste, which is about 55 gal. The individuals in all three categories must determine the amount of hazardous waste they are generating and then obtain a U.S. EPA identification number. The number may be used on all manifests involving the transportation, storage, and disposal of the hazardous waste.

The three most important concerns in managing small quantities of hazardous waste on-site are:

1. Comply with storage time, quantity and handling requirements for containers and tanks.
2. Obtain a storage treatment or disposal permit if storage treatment or disposal of the hazardous waste is done on-site in a manner requiring the permit.
3. Take adequate precautions to prevent accidents and be prepared to handle them properly in the event they do occur.

When hazardous waste is stored on-site, no more than 6000 kg may be stored for up to 180 days, or up to 270 days if the waste must be shipped to a treatment, storage, or disposal facility that is located over 200 miles away. If the time limits are exceeded or the quantity limits are exceeded, the operation is to be listed as a storage facility and therefore has to meet all the RCRA storage requirements. In any case, the containers must be kept in good condition, must be handled carefully, and must not leak. The containers must not be ruptured, leak, or be corroded as a result of the hazardous waste that is stored. They should be closed except during filling or during emptying. Each week, the container must be checked for leaks or corrosion. When ignitable or reactive wastes are stored, they must be done in such a way that they are completely away from all processes.

Household Hazardous Waste

A household hazardous waste program is initiated by first conducting a waste stream study in determining the quantity and type of household hazardous waste that needs to be collected. The household hazardous waste may be collected on a routine basis with house to house pickup, on an occasional basis, or by providing a central place where individuals can bring their household hazardous waste on a weekly or monthly basis. Typically, the fire station might be a good place for the pickup of hazardous waste. Public awareness and public participation is the key to the success of this program.

Asbestos

Asbestos, after it has been removed from buildings or has come from manufacturing and fabrication, needs to be disposed of in a landfill in a proper manner. The asbestos is transported in special vehicles in which the transporter requires a chain-of-custody form signed by the generator. This form should include the name and address of the generator, name and address of the pickup site, estimated quantity of asbestos waste, types of containers used, and destination of the waste. The transporter must ensure that the asbestos waste is properly contained in leakproof containers with appropriate labels, and that the outside of the containers are not contaminated with asbestos debris. The asbestos disposal involves the isolation of asbestos material to prevent fiber release to the air or water. Landfilling is recommended as an environmentally sound isolation method because asbestos fibers are virtually immobile in soil. A disposal facility must meet EPA requirements, which state that no visible emissions to the air may occur during disposal, or emissions that do occur within 24 hours of covering of the waste must be extremely minimal. A 6-in. soil cover of nonasbestos material is required on the asbestos waste. Some asbestos waste disposers require that the waste is placed, bagged into 55-gal metal drums, and then put into landfills. The landfill must be approved for disposal and must maintain all the rules and regulations of the EPA and the appropriate state bodies. Before asbestos is unloaded from the truck, a water spray must be used to contain any possible dust. Public access must be completely controlled in these areas and record keeping must be of the highest quality.

Medical Surveillance

Because of the extreme hazard to individuals working in the hazardous waste storage, transportation, and disposal fields, it is essential that all individuals receive preplacement medical examinations to assess employee fitness. The physician determines the physical demands of the work; physical demands associated with the extensive use of personal protective equipment; problem of confining garments and gloves, which may result in heat stress and skin irritation; problems related to the use of respirators; and individual concerns related to the workers' health condition.

During the cleanup of a site there may always be a possibility of a major accident or emergency. Therefore, physicians must be aware of conditions, such as seizure disorders or insulin-dependent diabetes; severe or poorly controlled heart and lung disease; disorders of the liver, kidney, and bloodstream; and psychiatric disorders among the workers. These types of individuals would be at high risk from any type of exposure to chemicals. Laboratory testing can be used to assess liver and kidney function, and the integrity of the blood-forming elements in the bone marrow. Lung function tests, and chest x-rays are extremely important. Cardiovascular evaluation, including electrocardiograms, are helpful in establishing the health of the individual. The measurement of chemical substances in the blood, urine, hair, or fat can establish what the individual has been subjected to. The medical surveillance not only includes the initial examination but also periodic examinations and epidemiological studies to determine short-term and chronic health effects. The physician must be ready to deal with a multitude of very serious emergency situations.

The Public

The RCRA program gives the public numerous opportunities to get involved in all phases of the solid waste and hazardous waste programs. The EPA provides information and solicits comments on all proposed and final agency actions relating to the development of regulations. The agency incorporates public comments into the decision-making process. The EPA has established an appeals process for certain agency decisions. RCRA guarantees participation in program implementation and enforcement, as well as access to information. The general policies related to public participation were codified as 40 CFR, part 25 in February of 1979. The objectives of these regulations are:

1. Making sure the public understands the RCRA program and proposed changes to it
2. Responding to public concerns and taking them into account
3. Developing a close link among the EPA, states, and public
4. Providing opportunities for public participation beyond what is required whenever feasible

To achieve these objectives, the EPA provides free copies of reports, alerts interested and affected parties of upcoming public hearings, and establishes EPA-funded advisory groups when an issue warrants sustained input from a group of citizens. Agency requirements include:

1. Considering public comments pertaining to permit violations
2. Notifying the public of the intent to issue a permit
3. Allowing 45 days for public comment on a permanent application
4. Notifying the public of proposed major modifications to an operating permit

To make the public participation system work, the EPA:

1. Identifies public concerns early in a permitting process
2. Encourages the exchange of information among the EPA, state, and community
3. Creates open and equal access to the permitting process
4. Anticipates conflicts and provides an early means of resolution

The EPA established the Office of Ombudsman under the HSWA. The primary function of the office was to create a central clearing house for public concerns on matters relating to the implementation and enforcement of RCRA. The office was given a 4-year life span, ending on November 8, 1988.

Planning

Plans are based on careful study and evaluation of needs, projected needs, types of waste to be disposed of, solid waste disposal sites, and various aspects of the program. Planning is carried out by the local agencies in cooperation with the state and with various private contractors, as well as with professionals operating in solid waste management. Planning includes:

1. A description should be provided of the physical factors, including size, area, topography, geology, climate, air, water, and land resources.
2. A description should be prepared of the population and land use at present and the projected population and land use for the next 20 years.
3. A survey should be conducted of the actual solid wastes that are collected within the community to determine the amount of garbage, refuse, ashes, bulky waste, street refuse, dead animals, abandoned vehicles, construction and demolition waste, industrial refuse, special wastes such as hazardous materials and security materials, animal and agricultural waste, and sewage treatment residues.
4. A study should also be made of the existing collection and disposal practices of both public and private agencies, which includes information on the owners, operators, their location, size, hours of operation, adequacy of operation, and life expectancy of operation. Once the existing collection and disposal practices have been evaluated and all the previous information has been evaluated, it is necessary to determine what regulations are available and what regulations need to be passed before a good program can go into effect.
5. Seasonal variations and other local problems of the existing solid waste management service areas should be determined.
6. The quality of storage practices of solid wastes and identification of storage practices need to be improved.
7. Identification and determination of the capacity and the service quality and any other attributes of all solid waste collection systems, including public and private, should take place.

8. Determinations of the numbers and types of acceptable on-site disposal and reduction methods and the unacceptable ones, such as open burning, should be made.

9. All disposal, reclamation, reduction, and transfer sites and facilities operated by public and private agencies should be identified and their remaining life, cost, and acceptability should be determined.

10. The weight of all solid waste generated, transported, and disposed of in the given area, as well as all local, political, economic, and social factors affecting a solid waste management program, should be determined.

11. All the previous data should be projected and the type of program that will be needed for at least the next 20 years should be determined.

This systems approach includes not only riding with the solid waste collection people but also evaluating everything that can affect a solid waste program. Only after making this type of evaluation can the professional advisor use the systems approach to determine what program or combinations of services should be suggested for a given area and what is the most expedient and economical way of removing the waste from the specified area.

Organization

The organization of a solid waste program must become the responsibility of a given public agency. This agency must have the authority to establish policy, develop public information, budget, plan and review, draft, adopt, and enforce standards and operational techniques for solid waste management. It must be able to implement, supervise, and evaluate these functions. It is well to have a strong manager, and not a political appointee, at the head of this program. This manager will be responsible for an enormous amount of budgeting, must show flexibility in moving individuals from one area to another, and must enforce the laws when necessary. Although the private sector will definitely assume a portion of the responsibility, or even possibly carry out the entire responsibility of collection and disposal, the function of government is to ensure that it works in an effective manner to protect the health of the public.

Financing

A solid waste program is financed by collecting taxes and special charges; issuing general obligation and revenue bonds; refunding bonds for lower interest rates; issuing liens against property for delinquent taxes or charges; increasing or eliminating debt limitations for incinerators, sanitary landfills, transfer stations, and compost plant bonds; exempting from the debt limit revenue bonds that are secured by service charges; accepting grants-in-aid from state and federal agencies; licensing private waste operators and using the fees to offset the operation of the disposal system; collecting service charges from the tax-exempt property; and collecting a special tax for solid waste. Additional revenue may come from utilizing the solid waste for energy, recycling materials, and utilizing the land that was once a sanitary landfill for profit-making operations.

Public Support

Throughout this entire chapter, a hidden factor remains that was not thoroughly and completely discussed. Governments can plan and agencies can work, but unless the public is willing to support a good solid waste program, nothing will occur. It is up to governmental agencies to work with community groups who are interested in doing something about solid waste problems in their communities. It is important to include people who do not want to spend money, as well as those who are interested in a solution, are apathetic, or are politically opposed to the program. It is necessary, with the help of newspapers, radio, and TV, to get everyone to support the community effort to clean up the environment. The public relations phase of solid waste management is as important as any other phase of the program.

SUMMARY

The collection and proper disposal of solid waste is essential for the operation of a society that wants to avoid disease and injury and to have a pleasant, aesthetic environment. The means of disposal include sanitary landfills, incinerators, recycling, and use as a source of energy.

A sanitary landfill should not create adverse effects on the physical environment if it is properly planned and designed and if the site selection and site preparation are handled professionally. It is obvious that the landfill should be developed and stabilized as quickly as possible, using proper cover as an aid in achieving this goal. Until more efficient ways can be found, the sanitary landfill is still one of the least expensive and most acceptable means of disposing of solid waste.

Incinerators, when functioning properly, are very effective in reducing the volume of solid waste and volatile material. If the incinerator residue and wastewater are properly treated and then disposed of, and if the air emission system is properly controlled and monitored, the incinerator becomes a valuable adjunct to the composite picture of solid waste disposal.

A vast potential exists for the use of discarded materials. Considerable research is needed in the coming years to find those techniques that are economically feasible and that may be of value to society. In this era of continuing energy crises and establishment of cartels by many foreign countries, it would be wise for the American public to use one of the best natural resources that exists, that is, our solid waste.

RESEARCH NEEDS

Research is needed to determine the long-term problems related to the contamination of the soil and water by leachate. How much zinc, copper, nickel, lead, cadmium, selenium, mercury, boron, and arsenic are leached into the water or are absorbed by plants? What are the potential health hazards related to bacterial, viral, and parasitic organisms? Is it safe to recycle animal waste into feeds for other animals? What are the problems related to drugs, hormone residues, metals in feed supplements, and resistant organisms?

Epidemiological studies are needed to determine whether a link exists between metals and such diseases as osteoporosis, Parkinson's, and Alzheimer's in the middle-aged and elderly. Better animal models of disease caused by metals, and animals that respond to metals in the same way as humans do, need to be developed. It is necessary to determine what the role of the skeleton is in measuring metals, and how it acts as a reservoir for metals and as a target for lead and cadmium. What is the difference in effects of chronic and acute exposure to solid and hazardous waste between children and adults? What is the mechanism of metal toxicity and how is it best to be treated? What is the relationship between nutritionally essential metals (such as calcium, iron, zinc, and selenium) and toxicity of metals (such as lead, mercury, and cadmium).

Agricultural chemicals increase the quality and quantity of our diets. However, these chemicals can damage the environment and people. Epidemiological studies of chronic or low-level exposure to agricultural chemicals by women, children, elderly people, and migrant workers are needed. What is the potential for delayed neuro- and immunotoxicity? Studies are needed to identify more of the biochemical and pathophysiological specific biomarkers of exposure. Studies of cancer and non-cancer end points of chemical exposure are needed. What is the mechanism of injury or disease and how can this information be used in prevention, control, and treatment?

Additional research is needed in the conversion of solid waste to energy to make the technique so economically feasible that it would become the preferred method of waste disposal and provide needed sources of energy for our society. Particularly in large metropolitan areas, research should be conducted to determine whether solid wastes can be turned into a fluid and shipped by pipeline to other areas for processing into energy.

A substantial data gap exists in the geohydrologic processes that exist at the soil–water–air–hazardous waste interface. Very little is known or fully understood about the underlying processes in the disposal of hazardous waste. Additional knowledge is needed about how chemicals move through the soil, how they are transformed, and how they may move into the air or water environments.

Little knowledge is available about the quantitative importance of fundamental soil processes. What happens to the residuals? How much of them can be assimilated? How much of them can be destroyed and how much of them can move through the environment. The transfer rates of volatilized chlorinated organics from hazardous waste sites, sewage treatment plants, or contaminated soil need to be further studied.

Additional studies are needed on the movement of high concentrations of complex mixtures of waste through soil from hazardous waste sites or accidental spills. The laboratory modules in this area are generally on individual chemicals, instead of high concentrations of complex chemical mixtures.

Additional research is needed to determine the health impact of dioxins, furans, lead, mercury, and other chemicals from incineration emissions. Studies are needed on the additive, multiple, and synergistic impact of mixed chemical exposures. What are the sensitivity patterns of chemicals related to age, gender, race, or ethnic backgrounds? How do socioeconomic stresses affect levels of disease from chemicals? How does existing disease or poor nutrition affect levels of disease from chemicals in waste materials?

What are the problems related to handling of ash from solid waste combustion. What should be done with the ash? Should it be handled as a hazardous waste, requiring special processing and high-cost landfilling, or should it be placed in normal municipal landfills? Is it appropriate to mix bottom ash and fly ash, or is this a case of impermissible pollution dilution? Can the toxins in the ash be reduced or locked in, so that they do not injure people? Is it feasible from an economic and environmental point of view to create landfills dedicated only to ash from solid waste combustion?

Recycling, source reduction, and reuse of waste materials need to be researched further to find better techniques to make these phases of waste management more economical. Land disposal, treatment, and incineration techniques need to be enhanced. However, the major key to future waste management belongs to the area of innovative and alternative technologies that need to be developed for the handling, transportation, treatment, and disposal of waste. The Superfund Innovative Technology Evaluation (SITE) Program developed by the EPA is a beginning. However, many additional innovative technologies must be developed that can become cost effective.

CHAPTER 3

Private and Public Water Supplies

BACKGROUND AND STATUS

Water is in short supply because the amount of rain and snow basically remains constant but population and usage per person are both increasing sharply. The population of the United States was 150 million in 1950. Today, 171,000 public water systems serve approximately 287 million people, and the numbers continue to increase. Water usage in the United States is ten times greater today than it was in 1900. Increased use is due to more bathrooms, garbage disposals, home laundries, lawn sprinklers, and so forth. Industry has increased its water usage 1450% since 1900. It currently uses about 218 billion gal/day (bgd), with thermoelectric power plants using 190 bgd of the total amount. Agriculture has increased its water usage by over 700% since 1900. It currently uses about 143 bgd. The total water usage in the United States is estimated at about 400 bgd of freshwater and 60.3 bgd of saline water.

Private and Public Water Supplies

Drinking water is provided to over 249 million Americans by 54,700 community water supply systems and to nonresidential locations such as campgrounds, schools, and factories by 116,000 small-scale suppliers. The rest of Americans are served by private wells. About 42% of all Americans use water supplies drawn from groundwater. Untreated water drawn from groundwater and surface waters, such as lakes and rivers, and used as a drinking water supply, can contain contaminants that pose acute and chronic threats to human health.

Water must be reused or we may be faced with an inadequate supply to satisfy all the needs of our citizens, businesses, and industries. Water reuse is complicated by water pollution. Water is polluted by many sources and exists in many forms. It may appear as oil slicks, excess aquatic weeds, decline in sport fishing, and increase in carp, sludge worms, and other forms of life that readily tolerate pollution. Water pollution is detrimental to health, recreation, aesthetics, commercial fishing, agriculture, and water supplies (industrial, municipal, and private).

To avoid duplication, the entire area of water quality control is subdivided into specialized areas in the coming chapters. This chapter covers individual and public water supplies. Chapter 4 discusses artificial and natural swimming areas. Chapter 5 covers plumbing and plumbing hazards. Chapter 6 examines on-site sewage disposal, public sewage disposal, and soil analysis. Chapter 7 is a comprehensive chapter discussing the overall problem of water pollution control and water quality. Chapters 3 to 6 are purposely limited, as much as possible, to their specific topics. Chapter 7 is very comprehensive and should be used as a resource for finding answers to questions that may arise after the earlier chapters.

The EPA evaluated data from outbreaks of waterborne disease for a 24-year period. These data related to groundwater in community systems as well as non-community systems. The community systems outbreaks were due to source contamination, 70; untreated groundwater, 31; disinfected groundwater, 36; filtered groundwater, 3; distribution system contamination, 35; inadequate control of chemical feed, 5; and unknown, 3. The noncommunity systems outbreaks were due to source contamination, 217; untreated groundwater, 127; disinfected groundwater, 90; filtered groundwater, 0; distribution system contamination, 15; inadequate control of chemical feed, 3; and unknown, 8. Many outbreaks are not reported.

Etiological Agent	Outbreaks in Community Systems	Outbreaks in Noncommunity Systems
Campylobacteria	5	5
Chemical	17	5
Cryptosporidium	3	1
Escherichia coli	1	1
E. histolytica	1	0
Gastroenteritis undefined	40	183
Giardia	15	7
Hepatitis A	8	9
Norwalk virus	4	12
Salmonella	5	1
Salmonella nontyphoid	4	0
Shigella	10	17
Salmonella typhi	0	1

One especially serious outbreak occurred in 1993, in Milwaukee, WI, when 400,000 people experienced intestinal illness from *Cryptosporidium*. More than 4000 were hospitalized and 50 people died. Typhoid fever and cholera epidemics were common 100 years ago in the United States. The advent of water treatment and chlorination helped resolve many of these epidemics. However, the nation must be alert to new microorganisms as well as a reoccurrence of old ones.

Laws and Water Quality

Congress gave the EPA, the states, and the Indian tribe governments broad authority to deal with water pollution. The Clean Water Act (CWA) of 1972 established a goal to restore and maintain chemical, physical, and biological integrity of the nation's water. The EPA has developed regulations and programs to reduce pollutants from entering all surface waters, including lakes, rivers, estuaries, oceans, and wetlands. Congress then passed the Water Quality Act of 1987, which reauthorized and strengthened the CWA. The amendments ensure continued support for municipal sewage treatment plants, initiated a new state–federal program to control non-point-source pollution, and accelerated the tightening of controls on toxic pollutants. The EPA is now involved in finding creative alternatives to resolving water quality problems.

The Safe Drinking Water Act (SDWA) of 1974 created major authority for protecting drinking water resources. This act was passed as the result of the discovery of numerous impurities in the drinking water of the country. Asbestos fibers were found in the water supply of Duluth, MN and 66 organic chemicals, some of which were carcinogenic, were found in a New Orleans drinking water supply. The SDWA was amended in 1986 by the SDWA amendments. These amendments required establishment of additional drinking water standards, as well as protection of underground sources of drinking water from underground disposal of fluids. The amendments also established two major groundwater protection programs: the Wellhead Protection Program to protect areas around public drinking water wells and the Sole-Source Aquifer Demonstration Program to protect unique underground water supplies.

Major sources of groundwater contamination were reported by the states to the EPA. These sources were as follows:

1. Of the 46 states reporting septic tanks, 9 reported them as a primary source of pollution.
2. Of the 43 states reporting underground storage tanks (USTs), 13 reported them as a primary source of pollution.
3. Of the 41 states reporting agricultural activities, 6 reported them as a primary source of pollution.
4. Of the 34 states reporting on-site landfills, 5 reported them as a primary source of pollution.
5. Of the 33 states reporting surface impoundments, 2 reported them as a primary source of pollution.
6. Of the 32 states reporting municipal landfills, 1 was reported as a primary source of pollution.
7. Of the 29 states reporting abandoned waste sites, 3 were reported as a primary source of pollution.
8. Of the 22 reported oil and gas brine pits, 2 were reported as a primary source of pollution.
9. Of the 19 states reporting saltwater intrusion, 4 reported this as a primary source of pollution.
10. Of the 18 states reporting other landfills, none were reported as a primary source of pollution.
11. Of the 16 states reporting road salting, 1 reported this as a primary source of pollution.
12. Of the 12 states reporting land application of sludge, none was reported as a primary source of pollution.
13. Of the 12 states reporting regulated waste sites, 1 was reported as a primary source of pollution.

14. Of the 11 states reporting mining activities, 1 was reported as a primary source of pollution.
15. Of the 9 states reporting underground injection wells, none were reported as a primary source of pollution.
16. Of the 2 states reporting construction activities, none were reported as a primary source of pollution.

In the 1992 National Governor's Association Survey, 965 hazardous waste sites had restricted groundwater use and 13,656 wells were closed or restricted, with 73% restricted because of organic chemicals present. A 35% increase had occurred from 1989 to 1992. As can be seen, many potential sources of groundwater contamination exist, resulting from commercial and household activities. Hundreds of chemicals have been found in groundwater.

When the SDWA of 1974 became law, the federal government had little reliable information about water suppliers, quality of the drinking water provided, treatment techniques used, or even about water system owners. This information was available for about 19,200 community water systems. Today, information is available on 171,000 systems. The EPA has established standards defining the maximum extent to which contaminants are allowed (maximum contaminant levels [MCLs]).

Safe Drinking Water Act Amendments of 1996

The SDWA amendments of 1996 established a new emphasis on preventing contamination problems through protection of water sources and enhancement of water system management. The amendments also set up a state revolving fund system to provide money to communities to improve their drinking water facilities, with $1 billion each year authorized through 2003. The funds can be used for voluntary source water quality protection partnerships with public water systems, local governments, and private companies. The funds also can be used for enhancing the capacity of the water systems, and implementing appropriate changes.

States were required to develop and implement Source Water Assessment Programs to analyze existing and potential threats to the quality of the drinking water throughout the state. A state program includes:

1. Delineating the source water protection area
2. Conducting a contaminant source inventory
3. Determining the susceptibility of the water supply to contamination from the inventoried sources
4. Releasing the results of the assessments to the public

A significant part of the 1996 amendments was to recognize the importance of public involvement in addressing and preventing threats to drinking water quality. The EPA issued regulations requiring all community water systems to prepare at least once a year a report with information about their source of water and the level of contamination in the drinking water. The requirements that the EPA promulgate standards for 25 additional contaminants every 3 years was repealed. This mandate was impossible to meet within the required time frames. All new standards now undergo a cost-benefit analysis. The 1996 amendments required the establishment of a priority list of unregulated contaminants, a streamlining of the enforcement process, increase in penalties, and requirements that the EPA promulgate rules on arsenic, enhanced surface water treatment incorporating standards for cryptosporidium, and a radon standard using the new standard-setting authorities. The EPA was directed to conduct research on sensitive subpopulations that may experience greater adverse health effects from drinking water contaminants.

SCIENTIFIC, TECHNOLOGICAL, AND GENERAL INFORMATION

Hydrologic Cycle

The hydrologic cycle consists of inflows, outflows, and storage. Precipitation from the clouds comes to the earth in the form of rain, snow, sleet, dew, fog, and hail. When the precipitation falls, a part of it may evaporate and return immediately to the atmosphere. Precipitation that is in excess of the amount that wets a surface, or evaporates, is a potential source of water. Plants, trees, or other green materials also resupply the clouds through transpiration. Additional moisture evaporates from rivers, lakes, streams, and other bodies of water. The water that arrives on the ground either may replenish the soil moisture that is needed by growing plants, and then eventually returned to the atmosphere by transpiration, or may penetrate below the root zone and fill voids within the earth to form a zone known as the zone of saturation. (This is known as groundwater.) It may run off the ground into streams, rivers, and oceans. When the water in the hydrologic cycle penetrates the subsoil and goes into the water table, it may be taken out by wells or springs, and utilized in homes. It then carries waste material. The water portion eventually comes back to the streams, rivers, oceans, and clouds. The underground water may also flow into underground rivers, which may surface at some point, or may simply stay in the ground until the water is ready to be utilized (Figure 3.1).

Groundwater Quality

Groundwater is extremely vital to our economy and our society. It is used for public and private water supply systems, for irrigation and livestock watering, and for industrial, commercial, mining, and thermoelectric power production. In many parts of the United States, groundwater is the only reliable source of drinking and irrigation water. Section 106 E of the CWA requests that each state monitor groundwater quality and report the findings to Congress in their 305 B, state water quality reports. The evaluation of the country's groundwater quality is complex and is based on a variety of different generalized overviews. The overviews are most frequently drawn from suspected contamination sites and finished water quality data from public supply systems. It is necessary to determine the quality of water from specific aquifers or hydrogeologic settings within the state. Despite the variations in reporting

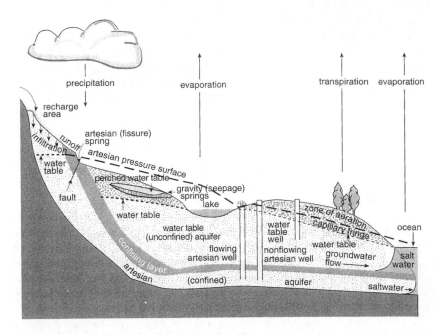

Figure 3.1   The hydrologic cycle. (From Water Supply Division, Office of Water Programs, U.S. Environmental Protection Agency, *Manual of Individual Water Supply Systems,* U.S. Government Printing Office, Washington, D.C., 1973, p. 2.)

style in the state water quality reports, it is a first step in improving the assessment of groundwater quality and eventually groundwater quality.

Human Impact on Water Cycle

Although Earth's water supply remains constant, people are capable of altering the cycle. The increase in population, rising standards of living, and growth of industry have made greater demands on the environment. The activities of people may affect the quantity, quality, and availability of natural water resources in any given area. A large population not only uses more water but also discharges more wastewater. Water supplies are overloaded with hazardous chemicals and bacteria from domestic, agricultural, and industrial waste, where intensive use of pesticides, herbicides, fertilizers, and industrial chemicals are utilized. Poor irrigation increases soil salinity and evaporation. The previous factors reduce the availability of potable water. Large cities accompanied by large suburban areas affect local climate and hydrology. Urbanization accelerates drainage of water through road drains and cities sewer systems, also causing an increase in urban flooding. The rates of infiltration, evaporation, and transpiration are altered from the natural setting, reducing the replenishment of groundwater aquifers. All these factors have very negative consequences for river watersheds, lake levels, aquifers, and environment in general.

Hydrogeology

Hydrology is the science of water occurrence, movement, and transport. Hydrogeology is the part of hydrology that deals with the occurrence, movement, and quality of water beneath Earth's surface. Because hydrogeology deals with water in a complex subsurface environment, it is a complex science.

The water-bearing properties of rocks are based on the types of rocks and the openings in the rocks that serve as reservoirs or conduits of the groundwater, such as the interstices, which are found between rock particles, including sand and gravel. Cracks, fissures, crevices, and caverns also act as reservoirs or conduits. Important water-bearing properties include the porosity of the rock, the specific yield of water, and the permeability of water.

The rocks that comprise the crust of the earth are divided into igneous, sedimentary, and metamorphic. Igneous rocks are derived from hot magma deep in the earth. They include granite, coarse crystalline rocks, dikes, sills, basalt, and other lava rocks or fragmented volcanic materials.

Sedimentary rocks are composed of chemical precipitates and rock fragments deposited by water, ice, or wind. This group includes deposits of gravel, sand, silt, clay, sandstone, siltstone, shale, limestone, gypsum, and salt. Metamorphic rocks are derived from both igneous and sedimentary rocks. These metamorphic rocks have been altered by heat and pressure, at great depths. This group includes gneiss, schist, quartzite, slate, and marble.

Although the openings in the rocks are usually small, the total amount of water that can be stored underground is large. The aquifer, which is the water-bearing formation, gives the most water where clean coarse sand and gravel, coarse porous sandstone, or cavernous limestone and broken lava rock are found. Some limestones are very dense and therefore little water is present. A specific hazard is associated with limestone, because cracks and crevices may be many miles long, and water coming from a limestone area may have been polluted 25 or more miles away. Most of the igneous and metamorphic rocks are hard and dense, have low permeability, and therefore give little water. Silts and clays are very poor sources for water retrieval.

Two physical properties of the rock types making up an aquifer control the amount and movement of groundwater. The first, which is porosity, is defined as the percentage of the total volume of the rock consisting of voids. Porosity determines how much water a rock body can hold. The second factor, which is permeability, describes the size of the pores and the degree of interconnection between them. This combination controls the ease with which water can flow through the rock. Permeable layers of rock permit the easy passage of water. In contrast, impermeable layers of rock allow water to pass through it only with great difficulty, or not at all.

The groundwater supply, or zone of saturation, at its upper surface is called the water table. If an overlying impermeable formation confines the water in the zone of saturation under pressures greater than atmospheric pressure, the groundwater is known as an artesian water supply. The water may flow by itself, once it has been tapped, to the surface of the land, because of the greater pressure within the aquifer. For this to occur, the water in the well must stand above the top of the aquifer. In other situations, a porous material may be above the water table, which contains some water that has invaded the smaller void spaces by capillary action. This is known as the capillary fringe. It is not a source of water supply, because it does not drain freely by gravity.

The two major subsurface zones are the vadose zone of suspended water and the aquifer. The vadose zone consists of the soil waterbelt and soil moisture, the intermediate belt and pore spaces partially filled with water, and the capillary fringe. Then comes the groundwater, which is the aquifer or water under hydrostatic pressure.

An aquiclude is an impermeable formation and does not furnish a water supply. The water table continuously changes, depending on the amount of water withdrawn by wells, or resulting from the water table occasionally intersecting the surface of the ground or the bed of a stream, lake, or ocean. Groundwater is constantly in motion, although the movement is slow. The amount of water in a water table is also based on the amount of rainfall that recharges the underground water supply. The greater the amount of water seeping down to the water table, the higher is the level of the water table and vice versa. Most of the fluctuation in water tables, where water is not withdrawn for water usage, is based on the season and the amount of rainfall. It is important, when planning to use an underground water supply, to obtain a solid understanding of the geology and water source from qualified groundwater engineers, geologists, and contractors who are familiar with the area. Data may also be provided by the U.S. Geological Survey (USGS), federal and state agencies, and the National Water Well Association.

Groundwater development for drinking water use or any water use obviously is based on the water-bearing formations. Wells may be nonartesian, water table wells, or artesian wells. Springs may be gravity or artesian springs. Nonartesian wells are those that penetrate the formations in which groundwater is found in water tables. The pumping of the well lowers the water table in the vicinity of the well, while the water moves toward the well and up toward the ground surface under the artificially created pressure. Artesian wells have already been discussed.

Gravity springs may percolate laterally through permeable material that overlays an impermeable stratum and therefore comes to the surface. They also occur when the land surface intersects the water table. This type of spring is particularly controlled by seasonal fluctuations of groundwater storage. They can make satisfactory individual water supply systems if they are properly stored and treated. Artesian springs come from artesian aquifers. They may occur when a rupture or fault occurs in the overlying formation of the aquifer. The spring may also occur if the topography of the land is such that the land is at a lower point than the aquifer itself.

Artesian springs are usually more dependable than gravity springs, but the amount of water produced is based on the amount of water pumped by wells from the same aquifer. Seepage springs are those that seep out of sand, gravel, or other material. These springs may be large or small; may be heavily contaminated with organic material, iron, or other contaminants; and are especially affected by surface runoff. Tubular springs come from solution channels, limestone caverns, and soluble rocks that occur in glacial drift. The water comes through large openings. The water penetrates sand or fine-grained material and may be free of contamination. If the water receives surface water from septic tanks or other sources, it may be unsafe.

Fissure springs come from the joints, cleavage, or fault planes. The water passes through the breaks within the rock. The springs may be uncontaminated if the water is of a deep source origin, or contaminated if it is of a strata close to the surface.

Surface water may be the result of direct runoff, effluent from industries, private homes, municipal treatment plants, sewage treatment plants, or groundwater sources that have come to the surface. In addition, surface water may come from controlled catchment areas, ponds, lakes, surface streams, irrigation canals, or oceans.

The selection and use of surface water sources as private or public water supplies must be carefully examined to determine the danger of contamination by disease-causing organisms and the danger of chemical or radiological hazards. At times the surface water source is used for processing, gardening, firefighting, and irrigation; and therefore does not receive the kind of treatment that potable water supplies receive. It is essential to ensure that the water sources are not mixed. To determine whether the surface water source can be used for drinking water, careful studies must be made of the water use before the water enters the area where it is to be utilized for drinking water.

Studies must also be made of the hydrologic, geologic, and meteorologic factors affecting the source of surface water. It is necessary to determine the yield and protection of the drainage area. The drainage area, or watershed, must be protected against all forms of potential hazards. The yield depends on total annual precipitation, seasonal distribution of precipitation, normal annual or monthly variations of precipitation, normal annual and monthly evaporation and transpiration rates, moisture requirements and infiltration rates, amount of runoff, and any available records that show long-term water levels from the potential watershed.

All surface water must be considered to be contaminated and therefore must go through an approved filtration and treatment process prior to use as a potable drinking water source.

Geographic Information System

A geographic information system (GIS) is a computer system capable of assembling, storing, manipulating, and displaying geographically referenced information, that is, data identified according to their locations. The GIS includes operating personnel and data that go into the system. The GIS is both a database system with specific capabilities for spatially referenced data and a set of operations for analysis of the data. The GIS technology can be used for scientific investigations, resource management, and development planning.

The GIS can use information from many different sources, in many different forms that can help with analysis. The primary requirement for the source data is that the locations for the variables are known. Location may be noted by x, y, and z coordinates of longitude, latitude, and elevation. Other systems used may be zip codes or highway mile markers. A GIS can also convert existing digital information, which may not yet be in map form, into a format it can recognize and use. Census or hydrologic tabular data can be converted into a maplike format. An example of the previous usage is the relating of information concerning the rainfall in a state to aerial photographs of a county to determine which wetlands dry out at certain times of the year.

A GIS can use the information in a map, even if not already in digital form, by various techniques that capture the information, such as digitizing the map or hand tracing with a computer mouse to collect the coordinates of features. Electronic scanning devices can also convert map lines and points to digits.

A GIS can make it possible to link and integrate information that is difficult to associate in any other way. It can use combinations of mapped variables to build a set of new variables. For example, by using GIS technology and water company billing information, it is possible to simulate the discharge of materials from the septic systems in a neighborhood into a body of water. The amount of water a person uses roughly predicts the material that can be discharged into the septic systems and then into water bodies.

Projection is a fundamental component of map making. A GIS can use the information from existing maps and the processing power of the computer to transform digital information from a variety of sources to make a common projection.

A GIS can be used to depict two- and three-dimensional characteristics of Earth's surface, subsurface, and atmosphere from information points. It can quickly generate a map with lines that indicate rainfall amounts.

Spatial or Geospatial

Spatial, or geospatial, is defined as information that identifies the geographic location and characteristics of natural or constructed features and boundaries on Earth. This information may be derived from remote sensing, mapping, surveying technologies, etc. Information may be gathered from aerial and satellite photographs. It is distributed by the Earth Resources Observations System (EROS) Data Center. This center manages the repository of multiagency National Aerial Photography Program photos at a scale of 1:40,000 in color-infrared or black and white; the National High Altitude Aerial Photography Program at a scale of 1:58,000 for color-infrared and 1:80,000 for black and white; and aerial photos at various scales from the USGS mapping projects as well as other federal agencies such as the National Aeronautics and Space Administration, Bureau of Reclamation, Environmental Protection Agency (EPA), and U.S. Army Corps of Engineers.

National mapping and remotely sensed data from the U.S. Geological Service can be provided as follows:

1. Digital elevation models are digital records of terrain elevations for ground positions regularly spaced apart, such as 7.5-, 15-, and 30-min intervals, and 1 degree.
2. Multiple aerial photographs cover the United States, in addition to satellite hand-held photos along paths in the United States and other parts of the world.
3. Digital line graphs are representations by points, lines, and areas of planimetric information derived from 7.5- and 15-min scale topographic quadrangle maps as well as 30- by 60-min intermediate scale maps, and 1:2 million scale sectional national atlas maps.
4. Digital orthophoto quadrangles are digital images of an aerial photograph in which displacements caused by the camera and the terrain have been removed, combining the image characteristics of a photograph with the geometric qualities of a map.

5. Digital raster graphics are scanned images of a USGS standard series topographic map, including all map information.
6. Digital satellite images are images of the United States, including satellite advanced very high resolution radiometer useful for tracking vegetation growth on a regional basis.
7. Geologic data consists of earthquake, volcano, and landslide hazards research, geologic framework and process studies, global change research, marine and coastal geologic surveys, and mineral and energy resource surveys.

The EPA provides a series of services utilizing GIS data to aid in environmental protection and research. They are as follows:

1. Maps on demand is a group of different applications, enabling users to generate maps displaying environmental information for the entire United States at the national, state, and county level.
2. The National Shape File Repository, containing data from the USGS, U.S. Department of Transportation, EPA Spatial Data Library System, etc., provides users with interactive GIS functionality. This allows them to view spatial data at the national, state, and county levels, as well as displaying multiple spatial layers, zooming, panning, identifying features, and querying single envirofacts points.
3. Locational reference tables are a repository for latitude and longitude information that has been documented through the EPA.
4. Spatial data library system is a repository for the new and older EPA geospatial data holdings.
5. The Geospatial Data Clearinghouse enables users to determine which EPA geospatial data exist, find the data they need, evaluate the data's usefulness for their applications, and obtain or order the data.

EnviroMapper

EnviroMapper, an Internet GIS-assisted environmental analysis tool, generates maps and reports on various types of environmental information, including drinking water, water discharge permits, toxic air releases, hazardous waste, Superfund sites, environmental profiles, and watershed indicators. EnviroMapper also links to text reports, which provides even more information, can be used to view and query EPA-regulated facility information stored in the Envirofacts Warehouse; and can also be used to view environmental statistics, profiles, trends, environmental information for U.S. metropolitan areas, watersheds, and Superfund National Priorities List (NPL) sites. This analysis tool also provides users with interactive GIS functions using EPA geospatial data for the United States, and allows users to view the data from the national level clear down to within 1 mi of a given site.

GIS Uses

With a GIS you can point it at a location or at objects on the screen, and retrieve recorded information from off-screen files. By using scanned aerial photographs as a visual guide you can ask a GIS about the geology or hydrology of an area or even how close you are to a specific environmentally sensitive area. A GIS can recognize and analyze the spatial relationships among map phenomena. It can simulate the route that pollutants may take along a linear network. A GIS, using map overlays, can produce new maps that rank an area according to its sensitivity to damage from nearby businesses, factories, or residential areas. The critical components of the GIS are the data outputs. A GIS has the ability to produce graphics on a screen or on paper that show the results of analysis to people who make decisions. Wall maps and other graphics can be generated. This allows the individual to view and therefore readily visualize the results of analyses or simulations of potential events. The USGS works in cooperation with local and state governments to develop information for decision making.

In Connecticut, the U.S. Geological Service digitized more than 40 map layers for a watershed to assist in determining where to dig a new well and how to avoid areas of potential contamination. The service used 7.5-minute topographic quadrangle maps for this purpose. To prepare the analysis, digital maps of the water service areas were stored in the GIS. By using the buffer function in the GIS, a half-mile zone was drawn around the water company service area. Potential well sites were at least 100 meters away from unsuitable streams. Point sources of pollution, recorded by the Connecticut Department of Natural Resources, were avoided by placing a 500-meter buffer zone around each point. The various map layers were combined to produce a new map of areas suitable for well sites. The map of superficial geology showed the earth materials that were above the bedrock. By knowing the type of soil and the previous conditions shown by the accumulated maps, they were able to find a proper well within 0.5 mi of the plant.

The GIS has been used for innumerable projects. Some of them include:

1. An analysis of the emergency response planning to an earthquake was conducted in the Wasatch Fault zone of the Wasatch Mountains in north central Utah, where a GIS was used to combine road network and earth science information to analyze the effect of an earthquake on the response time of fire rescue squads.
2. An analysis was completed of a land swap between the National Forest Service and a mining company seeking developmental rights to a mineral deposit in the Prescott National Forest of Arizona, where the GIS illustrated the dramatic changes to the topography that mining would cause.
3. A three-dimensional view was recorded that looked down at the San Andreas Fault, where two types of data were combined in a GIS to produce a perspective view or a portion of San Mateo County, California, using a digital elevation model consisting of surface elevations recorded on a 30-meter horizontal grid.
4. The relationship between health outcomes and the environment was projected by use of a GIS format with data drawn from a community environmental health profile combined with local, state, and national data, and patient household data along with an environmental health exposure form.
5. The National Spatial Data Infrastructure for Finding and Assessing USGS spatial data related to water resources was used in various areas of the country.

Global Positioning System

The global positioning system (GPS) is a means of facilitating the collection and analysis of spatial data through handheld units, which pinpoint locations by longitude, latitude, and altitude. GPS consists of three parts. They are the 24 satellites, a portable receiver, and the control center on Earth. Each satellite carries a computer and a very accurate atomic clock. The control center calculates each satellite orbit for each week, predicts ionospheric conditions over that time (called the ephemeris), and then uploads the information into the satellite computer. The satellite can determine where it is in the sky at any given microsecond by consulting its clock and ephemeris. The receiver has a less accurate clock. When the receiver is activated it listens for satellites that are scheduled to be above the horizon. Each satellite has its own radio frequency. The receiver subtracts the received time from the satellite from the time of its internal clock, which gives a distance appearing to be a sphere around the satellite several hundred miles in diameter. The next signal gives a somewhat different sphere, and the intersection of the two is a circle that passes through Earth. Three signals are needed simultaneously for latitude and longitude and a fourth one is required for fixing altitude. GPS-calculated positions are within 100 meters of where a surveyor would place it. For greater accuracy it is necessary to get signals from three satellites 120° apart around the horizon, and the fourth satellite directly overhead.

GIS uses a set of GPS locations to create detailed computer maps, and satellite-derived remotely sensed data. The Centers for Disease Control and Prevention (CDC) research station in western Kenya has scientists using GPS to map 7500 households, rivers, roads, and medical facilities within a 75 mi^3 area. They are linking the map to an epidemiological database, which shows how many cases of malaria occurred at each household, whether the malaria strains were drug resistant, whether mosquito breeding grounds were present, and whether children died. Epidemiologists plan to use the material to determine whether proximity to mosquito breeding grounds increases child mortality; proximity to a medical facility decreases child mortality; drug resistance is spreading in a predictable pattern; and public health officials can use corrective measures by incorporating the discovered data to target intensive vector control efforts in households that harbor large numbers of mosquitoes.

Water Treatment

Most well water is treated with chlorine for the removal of bacterial contamination, and water softeners for the removal of hardness. Occasionally, ultraviolet (UV) light, bromine, or iodine is utilized. The finished water should have a level of 0.4 ppm of available chlorine.

The following discussion is applicable to individual and to public water supply systems. The water treatment process is used for wells, surface water, infiltration galleries, springs, and cisterns. It should never be assumed that raw water, no matter what its original source, is free of contaminants. Therefore, water treatment is utilized to modify or supplement the natural process of decay or dilution of contaminants in the flowing water. Water treatment should reduce to acceptable levels or remove entirely all contaminants present in the water, chemical, microbiological, physical, and radiological. The techniques utilized are sedimentation, coagulation–flocculation, filtration, disinfection, conditioning, softening, fluoridation, removal of tastes and odors, corrosion control, algae control, and aeration.

Sedimentation is the settling out of comparatively heavy suspended material in water because of gravity. The settling takes place in a quiet pond or a specially constructed tank basin. A minimum 24-hour detention time is necessary to have a significant reduction in suspended matter. The incoming water should be distributed evenly across the width of the basin as it flows toward the outlet. Baffles are usually used to reduce high velocities. The cleaning of the tank is facilitated by two separate sections.

Coagulation is the forming of flocculent particles in a liquid by adding chemicals. Alum (hydrated aluminum sulfate) is mixed with turbid water and then allowed to remain quiet. The larger particles or floc settle to the bottom of the basin. Adjustment of pH may be necessary for sedimentation.

Filtration is the removal of suspended material from water as it passes through beds of porous material. The amount of removal depends on the character and size of the filter media, the thickness of the porous media, and the size and quantity of the suspended solids. Filtration cannot completely remove all bacteria; therefore, it is necessary to use additional treatment processes on the effluent.

Slow sand filters are beds of fine sand through which water travels at an average rate of 0.05 gal/min/ft^2 of filter area. Pressure sand filters are filters through which water passes at a rate of 2+ gal/min/ft^2 of filter area. Frequent backwashing is needed in pressure sand filters. Diatomaceous earth filters are composed of a layer of diatomaceous filter media through which water is passed at a rate of 2+ gal/min/ft^2 of filter area. This filter material must be replaced periodically. Porous stone, ceramic, or unglazed porcelain filters are utilized in households and may be attached to faucets. Slow sand filters require a minimum of maintenance and can be readily adapted to an individual water supply. The necessary cleaning time varies with the water turbidity and is accomplished by removing 1 in. of sand from the surface of the filter and either discarding, or washing it and reusing it. The sand should be hard, durable grains that are free of clay, loam, dirt, and organic matter. For best results in the slow sand filter, the water should pass through at a rate of 60 to 80 gal/day/ft^2 of filter bed surface. This can be accomplished by adjusting the incoming water through the use of valves.

Disinfecting the water is the most important water treatment process utilized to destroy all pathogenic bacteria or other harmful organisms. However, proper disinfection cannot occur unless the organic material and other materials are removed prior to the disinfecting process. Disinfecting agents may consist of chlorine, ultraviolet light, or iodine. The most frequent treatment agent is chlorine. Chlorine concentration should be 1 mg/l, which in water is equivalent to 1 ppm. Chlorine feed or dosage is the amount of chlorine added to the water system in milligrams per liter. Chlorine demand is the chlorine added to the water that combines with the impurities and is unavailable as a disinfecting agent. Combined chlorine is the chlorine that combines with ammonia nitrogen to form chlorine compounds that may have biocidal properties. This gives the combined chlorine residual. If ammonia is not present in the water, the chlorine that remains in the water after the chlorine demand has been satisfied is called free chlorine residual.

Chlorine contact time is the amount of time that elapses between the introduction of the chlorine and the use of the water. The biocidal or the killing effect on microorganisms is based on the chlorine concentration, the type of chlorine residual, the contact time between the organisms and chlorine, the temperature of the water, the pH, and the presence or absence of organic matter. The higher the concentration is, the more effective the disinfection. Free chlorine is much more effective than combined chlorine. The longer the time and at room temperatures, the more effective the disinfection. The lower the pH, the more effective the disinfection. Chlorine is obtained from sodium or calcium hypochlorite, prepared hypochlorite solutions, household bleach or tablets, and gas.

Residual chlorine exists in water as a chlorine compound of the organic material and ammonia, or as both combined and free available chlorine residual. The total available residual is composed of the combined available chlorine residual and the free available chlorine residual. The orthotolidine-arsenite test (OTA) is used to determine the amount of free available chlorine. When orthotolidine reagent is added to water containing chlorine, a greenish-yellow color results. The intensity of the color is measured against a chart, which then helps determine the amount of free available residual chlorine in the water.

Reactions

1. $Cl_2 + H_2O \rightarrow HOCl + H^+ + Cl^-$
   (hypochlorous acid; 1° disinfectant)
2. $HOCl \rightarrow H^+ + OCl^-$
   (hypochlorite ion; 2° disinfectant)
3. HOCl reacts with $NH_3 \rightarrow$ chloramines:

$$NH_3 + HOCl \rightarrow NH_2Cl + HOH$$

$$NH_2Cl + HOCl \rightarrow NHCl_2 + HOH$$

$$NHCl_2 + HOCl \rightarrow NCl_3 + HOH$$

4. Summary
   - Chlorine is added to water $\rightarrow$ hypochlorus acid + hypochlorite anion.
   - HOCl reacts with reducing agents and ammonia $\rightarrow$ chloramines.
   - Microbes are oxidized $\rightarrow$ death.
   - Free chlorine residual results.

Dosage and Concentration Terminology

1. Dose = amount added to the water.
2. Demand = amount needed to oxidize the pathogenic microbes.
3. Free residual = [HOCl] + [OCl⁻].
4. Combined residual = [chloramines].
5. Breakpoint = chloramines oxidize $\rightarrow N_2O + N_2 + NCl_3$ and residual is present as HOCl.

Environmental Concerns

Formation of potentially carcinogenic trichloromethanes and ethanes:

$$CH_4 + Cl_2 \rightarrow CHCl_3 \text{ (trichloromethane; chloroform)}$$

$$C_2H_6 + Cl_2 \rightarrow CCl_3CH_3 \text{ (1,1,1-trichloroethane)}$$

Ultraviolet light is an effective bactericide for water, providing the water passes in a thin film under the tube and the tube is kept perfectly clean. The quantity of radiation is also dependent on the turbidity, color, and dissolved iron salts present. An uninterrupted source of electric power is necessary. Water may also be disinfected by the use of bromine, iodine, and ozone. Conditioning is a technique used to remove undesirable materials from water and to add those materials that improve water quality.

Metals may be removed by automatic chlorination and filtration. The chlorine chemically oxidizes the iron or manganese and forms a precipitate. It also kills iron bacteria. The filter then removes the precipitates. Aeration followed by filtration or ion exchange, or treatment with potassium permanganate followed by filtration also removes iron and manganese.

Water softening is the removal of minerals, primarily calcium and magnesium, which cause hardness. Softening of hard water reduces the amount of soap needed, scale on cooking utensils or laundry basins, and hardening formations on the interior pipes or water tanks; and increases heat transfer efficiency through the walls of heating elements or exchange units of water tanks.

Corrosion control is important to the continuous and efficient operation of the individual water system and to the delivery of properly conditioned waters that will not present a health hazard. Corrosion is an electrochemical reaction in which metal deteriorates or is destroyed when it comes in contact with air, water, or soil. When the reaction occurs, the flow of electric current goes from the corroding portion of the metal toward the electrolyte or conductor of electricity, such as the water or soil. The water characteristics affecting corrosion are acidity, where acidic water is corrosive; and conductivity, which is a measure of the amount of dissolved mineral salts. An increase in conductivity increases corrosion and oxygen content, where the amount of dissolved oxygen in the water promotes corrosion by destroying the thin, protective hydrogen film found on the surfaces of metals immersed in water. The rate of corrosion is also increased by carbon dioxide, which forms carbonic acid, attacks metal surfaces, and increases water temperature.

Water Distribution

Water is distributed through a series of pipes from the pressure tank to the various points where water is used. The pipes may be copper, plastic, or various other materials. Plastic pipes for the cold water system are usually simple to install and have a low initial cost. However, it should be certified by the National Sanitation Foundation as nontoxic and nontaste producing. Plastic pipes can be crushed by equipment or attacked by rodents. The distribution system should be installed carefully in trenches separated from any sewage or other types of systems.

Well Yield

When a new well has been completed and its quality determined, it is necessary to stabilize the appropriate quantity of yield for the house. Although many state codes require a minimum of 300 gal/hr for a well yield, it is strongly recommended that the minimum capacity be 600 gal/hr for the average home system. If the larger amount cannot be realized, it is necessary to have larger storage capacity to provide for peak periods during household usage.

Ponds or Lakes

A pond or lake is usually an inadequate or unacceptable private drinking water supply. However, if the watershed is properly selected and water treatment is properly carried out, the water may be usable. The watershed must draw water from a clean and preferably grassed area that is free of barns, septic tank systems, and privies; away from livestock; and protected against erosion.

Springs

Springs are selected because of their capacity to provide proper amounts of good-quality water throughout the year and because they are protected from contamination. Because springs can readily become contaminated, the second feature must be checked very carefully. The proper development of a spring is carried out by encasing it in an open bottom watertight basin that intercepts the source at the bedrock or provides a system of pipes that lead into the bedrock. Provision must be made for proper cleanout and emptying of the tank when necessary and also for overflow into an alternate distribution area or storage area when needed. It should be understood that the spring can dry out, and therefore an alternate source of water must be available for the individual. It is extremely essential, in fact mandatory, that the two water systems are not hooked up to avoid a detrimental cross connection that can contaminate either of the sources.

These waters are usually listed in three classes. Class A water comes from uninhabited or sparsely inhabited areas at or near the point of rainfall or snowmelt, is collected in a storage reservoir, and is clean and clear enough to be distributed to customers after disinfection. Class B water is impounded from an area not densely inhabited and allowed to flow from storage in a natural stream to the point of withdrawal, and requires varying degrees of treatment plus disinfection. Class C water has flowed in a natural stream before storage for a considerable distance and has received polluting materials from municipalities, industries, or agricultural areas. This water may be confined in a reservoir for purposes of storage, and then later used for public drinking supply. It requires complete treatment.

Some of the difficulties of getting pure water are that the watershed is owned by various people, or individuals or animals may intrude into the watershed, thereby contaminating the water. Special precautions are needed in the watershed area. This area should be such that heavy rainfall does not excessively increase turbidity in the storage reservoir. If there is swampland, it should be small and should not affect the color of the impounded water. If any pollution exists at all in the watershed, the source of pollution should be treated before it is allowed to enter the watershed area. Population density of the watershed area should be determined yearly and should be kept as low as possible. If recreation is permitted in the upstream reservoirs, the permission should be by permit and should be revoked if problems occur. Where chlorination or other disinfecting techniques are the only means of treatment, standby units must be available in the event of breakdown of the disinfecting equipment.

Water Treatment Plants

Almost all surface waters should receive complete treatment before they are used as a potable water source. This treatment includes initial removal of large materials, chemical treatment by coagulation, sedimentation, filtration, and disinfection. The filtration systems utilized may be diatomaceous earth, slow sand filters, or occasionally rapid sand filters. The design and construction of the water treatment plants vary with local circumstances (Figure 3.2). Each plant must be capable of handling the characteristics of the treated water to meet all standards and must also be adaptable if the water source changes in initial quality.

Intake Delivery Quantities and Quality

The intake is the area where a continuous adequate quantity of the best available grade of raw water enters into the water treatment plant. The location of the intake depends on the stream or lake bottom, currents, and potential sources of pollution. To account for varying environmental influences, the intake unit should be designed and built to permit the withdrawal of raw water at various levels or locations. The intake capacity, including the pumping facilities, must provide sufficient raw water for the treatment plant to operate at all times, including peak periods. The amount of finished water in storage helps to determine the intake capacity. The intake capacity usually equals the average rate of demand on a maximum day. Additional facilities must be provided in case of mechanical equipment breakdown. If pump priming is necessary, care must be taken that a cross connection cannot exist between the raw water and the potable water supply. The intake areas should ensure continuous raw water, whether floods, icing, plugging, high winds, power failure, or boat damage occurs. An intake should be inaccessible to trespassers, and adequate toilet facilities for the personnel at the intake station should be provided so that their waste does not get mixed into the raw water supply. Automatic warning systems must go into effect if the automatic or semiautomatic pumping stations stop operating. Sufficient additional capacity must be available for the time when the plant has to expand.

Location

The plant location must ensure that conduits, basins, and other structures do not influence the treatment plant by infiltrating sewage, surface, or other types of drainage. The plant should be located on high ground, above flood levels, and definitely

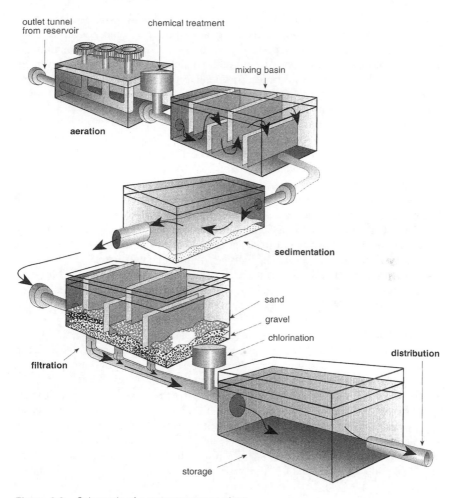

Figure 3.2   Schematic of a water treatment plant.

out of the floodplain. The plant must be structurally sound, have an adequate auxiliary power source, and additional pumping equipment to ensure continuous drainage removal under emergency conditions. Facilities must be provided to remove filter wash water plant waste and sanitary waste during floods. All drainage and sewage lines from the plant facilities must have backflow preventers to keep raw water from entering through submerged outlets.

Reservoirs

Reservoirs are utilized for presettling, and also separate reservoirs are used for finished water storage. Presettling reservoirs have a function of removing turbidity by sedimentation. In special cases coagulants or chlorine are added to help in the actual settling of the materials. The presettling reservoirs must be located away from potential floodwaters and should have adequate capacity to remove sand, silt, and clay. The shape of the reservoir and the inlet and outlet design must minimize a potential short circuiting of water from inlet to outlet without settling. A convenient rapid removal of sludge is necessary to maintain the efficiency of the presettling reservoir. A duplicate reservoir should be available as a bypass when highly polluted waters of varying quality are going to be utilized, when special treatment is necessary, or when the initial reservoir is cleaned.

The finished water reservoirs should be covered if they are located below adjacent structures or below ground elevation. Adequate protection must be given to this storage basin to prevent contaminants from draining into the water or to prevent groundwater from entering the water storage area. Numerous environmental concerns exist in stored water. Eutrophication, which is a process of providing nutrients in lake areas that aid in the production of nuisance organisms such as algae, causes serious problems within the water storage area. Nitrogen and phosphorus are essential to the growth of these algae. Fertilizers from agricultural or private home gardens or lawns provide these needed chemicals. The light penetration into the impounded waters and its effect on photosynthesis are of concern. Optimum light intensities are not necessary to support healthy algae. In fact, light intensities above the optimum are detrimental to algae. It would therefore seem that one of the controls for algae would be to reduce the turbidity in the water and allow as much light to enter it as possible.

Impounded waters must be evaluated for the amount of oxygen production by plankton or amount of dissolved oxygen present, carbonate equilibrium, nutrients and nutrient-removal techniques, nitrogen, phosphorus, sulfur, iron, manganese, pollutants, and various organic materials. Methane is produced in the bottom muds of marshes and swamps. It is also produced at the bottom of large reservoirs. Color is a problem in reservoirs. It is a result of the physical–chemical action of various coloring matter plus the relationship between color and inorganic constituents of the water.

Thermal stratification occurs in impounded bodies of water because of density differences (Figure 3.3). The cool water settles to the bottom and the warmer water rises to the top. In spring and fall, the water literally turns over and causes a mixing, which may result in higher levels of contamination and therefore a greater need for chlorine. Coagulation and sedimentation basins follow the presettling reservoir. Their function is preparation of water for filtration. Generally, coagulating agents are added to the water in these basins and the water is gently agitated to promote flocculation. The effluent, which is relatively clear water, then passes on to the remainder of the water treatment process. A weir is placed at the end of the sedimentation basin. The flow of water over the weir should be less than 20,000 gal/day/ft of weir to prevent surging and disruption within the sedimentation basin.

Chemical Feeding for Water Treatment

Water treatment plants must be provided with equipment that accurately measures and adds chemicals used for coagulation or other purposes. A reserve unit for chemical feeding is necessary in the event that the unit breaks down. The chemical feeding unit should have continuous recording devices and alarm systems in the

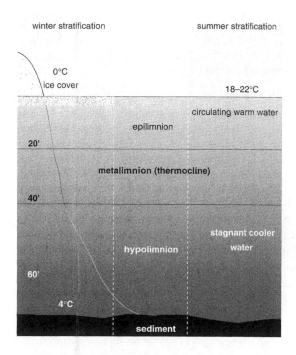

Figure 3.3   Thermal stratification in a reservoir.

event the system becomes disabled. The flow of chemicals must be adjusted to the needs of the water. Therefore, the raw water should be evaluated on a regular basis, depending on the amount of water processed. The raw water evaluation then is utilized in determining the adjustment of the amount of chemicals needed to maintain adequate treatment on a 24-hour basis. Adequate quantities of chemicals must be stored to prevent shortages or to meet increased needs. At least a 30-day supply of chemicals should be on hand. The actual amounts change from month to month based on the raw water quality and varying demands for the water. All chemicals should be rotated so that the incoming inventory always goes to the back and the older chemicals are utilized first.

Chemical water treatment consists of a variety of chemical usages. They include removal of phosphorus; removal of taste, odors, and colors; reduction of various inorganic compounds present in the drinking water supply; removal or reduction of a variety of organic chemicals that may have serious potential health significance in drinking water; reduction of hardness in the water; chlorination of water to destroy bacteria and viruses; and use of bromine or iodine, ozone, and other chemicals as disinfecting materials.

Activated carbon, or activated charcoal, is used particularly in the removal of many of the substances that cause taste, odor, and color. This carbon is also utilized in the removal, by adsorption, of organic materials prior to the chlorination of the water supply. Chlorinating the hydrocarbons in water leads to a potentially dangerous situation where potentially carcinogenic chlorinated hydrocarbons called trihalomethanes and trihaloethanes are formed within the water system.

Filtration Treatment Plants

The three kinds of filtration treatment plants are conventional filtration, direct filtration, and lime softening with conventional filtration. All these plants remove turbidity-causing suspended solids and some mineral matter from drinking water supplies.

A conventional filtration plant for drinking water treatment removes suspended solids and some dissolved mineral matter. It also destroys pathogenic microorganisms in the water supply. In this type of plant, the raw water is pumped to a rapid mix tank in which the chemical alum and a polymer in solution are added to enhance flocculation. The flocculation basin allows adequate time for the suspended solids to aggregate into larger particles or flocs, which then can be removed by the downstream treatment steps of sedimentation and filtration. Sedimentation basins are circular or rectangular. The flocs are allowed to settle there. The basins can be made of either concrete or steel, depending on the size. The waste sludge of solids and water is removed and discharged to a municipal sewer or hauled to a landfill for disposal. The clarified water then moves forward to the filter unit. The filters consist of one or more steel or concrete basins containing granular material such as graded sands, anthracite, and garnet. Solids are strained from the water as it passes through the filters. When the pressure drops through the filters to a sizable degree, then the system is backwashed and the backwash water goes to waste.

In direct filtration, the sedimentation step from before is deleted. Direct filtration is used for any drinking water supply where the suspended solids level are sufficiently low to allow for reasonable backwash cycles on the filter units. An upper limit to the influence of suspended solid concentration that can be tolerated exists, and that limit can only be determined by testing.

In the lime softening plant, the system is the same as before, except lime is substituted for other chemicals and a recarbonation step is added after sedimentation. A lime softening plant is usually used to treat raw water with higher concentrations of dissolved minerals such as calcium and magnesium. This type of plant also can be used toward a greater removal of toxic mineral substances. The lime can be added directly to the influent raw water as a solid or as a premixed water slurry. Recarbonation is the addition of gaseous carbon dioxide to a lime-treated water to neutralize the excess alkalinity resulting from the lime addition.

Hypochlorination of Water

The value of using calcium hypochlorite in water treatment is that it is a powerful oxidizing agent and dissolves in water to give hypochlorous acid and hypochlorite ions in varying ratios, depending on the pH. This destroys microorganisms, organic material, and algae, depending on the quantities utilized. Frequently, raw water is chlorinated prior to the actual treatment process. Great care should be taken to avoid chlorinating existing hydrocarbons that may be flowing downstream in surface water or that may have seeped into groundwater. The use of activated charcoal prior to chlorination is a valuable instrument in removing organic compounds. Chlorine is also used in the raw water to assist in removal of color and some of the material that is causing turbidity. Obviously, turbidity due to clay, silt, or other materials of this type is not affected by chlorination. Chlorination is widely used when cross connections or interruptions in water supplies have occurred. Chlorine in sizable quantities can eliminate organic material and kill the potential organisms present. Chlorine is also introduced after the cleaning of reservoirs, basins, tanks, wells, and pipes in the distribution system. Algae are controlled by introducing chlorine at a rate of 1.5 ppm.

Breakpoint chlorination is the amount of chlorine needed in water to reach a level where the combined available residual chlorine disappears and the free available chlorine appears. Breakpoint chlorination is used where a substantial organic contamination occurs. Chlorine demand is the difference between the amount of chlorine applied and the amount of residual chlorine remaining at the end of the contact period. The demand varies with the quality of the utilized water.

Hypochlorination is also used in sewage treatment for disinfection of effluents, odor control, slime control, control of filter flies, biological oxygen demand (BOD) reduction, and chemical precipitation of ferric chloride.

DISINFECTION

Ozone

Ozone is used to remove color, taste, and odors from drinking water. This agent also oxidizes and allows for the removal of iron and manganese, as well as helping in the removal of turbidity. Ozone is a powerful oxidizing agent that works over a wide pH and temperature range. It is excellent for destroying viruses, bacteria, and amoebae cysts. Ozone can work 3100 times faster than chlorine in the disinfection process and leaves no lasting residual in the water.

Ozone is generated on-site by passing air or oxygen through an electrical arc. The ozone generated is fed to the dissolver chamber, where it is well mixed with a sidestream of the water under treatment. The solution then flows into a contact chamber, where it mixes with the mainstream water flow. The required contact time is about 15 min in the contact chamber.

Chloramination

In chloramination, chlorine and ammonia are mixed together in the water solution to form chloramines, which act as a disinfectant. Chloramination does not form trihalomethanes as does direct chlorination. The system is designed for either aqueous ammonia or ammonia feeding.

Aeration

Aeration is a process that is used for the removal of volatile organic materials from drinking water. Flowing streams of air and water are put into contact with each other, so that the volatile organic materials are evaporated into the airstream and removed from the water. Aeration can be carried out in towers or aeration basins to provide the necessary contact time between air and water. The aeration basin is usually constructed of concrete. The aeration tower is similar to a water cooling tower, which is used in large air-conditioning systems. With basin aeration, the water enters one or more open concrete-type basins and compressed air is fed into diffuser pipes in the bottom of the basins. The air bubbles strip the organic compounds from the water as they rise to the surface. The cooling towers consist of fiberglass-covered metal framework containing a plastic packing medium. As the water enters through the top of the tower, it flows downward through the packing and comes in contact with air, which is flowing upward. Organic materials are stripped from the water and leave with the exit airstream. After the water has been completely treated, it is disinfected to provide a safe potable water supply for individuals. In most situations, chlorine again is the usual technique for disinfecting water.

Other disinfecting techniques include the use of ionic silver, ozone, chlorine dioxide, chloramination, bromine, iodine, and ultraviolet light. Usually these other techniques are used in small water supplies, individual water supplies, or swimming pools. In any case, disinfecting agents must come in contact with all particles of the water treated; be effective, despite the conditions of treatment or the characteristics of the water treated, such as color, turbidity, pH, total dissolved solids, and temperature; be nontoxic; have a residual action to protect the distribution system; be readily measured in water; destroy virtually all microorganisms, including bacteria and viruses; and be a practical means of disinfection.

Chlorine

Chlorine may be added directly to the water as chlorine gas or indirectly as a sodium hypochlorite solution. The chlorination system includes chlorine storage and feed equipment. The chlorinator is a metering device and a system in which the chlorine mixes with a small sidestream of water. After passing through the chlorinator, the sidestream joins the main flow of the water supply. The dosage of chlorine needed to disinfect the water depends on the individual characteristics of the water, such as pH and possible organic impurities.

Chlorine Dioxide

Chlorine dioxide is used for disinfection of drinking water in the same way chlorine is. The feed equipment is essentially the same. Chlorine dioxide gas is usually generated on-site by mixing a high-strength chlorine solution with a high-strength acidified sodium chloride solution. These solutions are fed through a mixing chamber that is called a generator. The generator is a plastic cylinder containing a loose porcelain fill material. The detention time in the generator is about 2 minutes or less. The gas evolving from the solution then is fed through a device that is similar to the chlorinator.

EMERGENCY REPAIRS TO WATER DISTRIBUTION SYSTEMS

Water distribution systems have been mentioned several times. Beyond the breaks that may occur in a system and the installation of new portions of the system, other problems may occur. Valves leak and may contribute to, or be the cause of, contamination of potable drinking water. The installation, character, and service of the valves should be recorded. Obviously, all valves should be readily accessible. Dead ends are parts of systems where the water literally comes to a stop. The dead end is usually put into a system because it is thought that eventually the system may expand in a given area. Unfortunately, this is a place where the water may become quiescent and contamination may occur, leading to problems in the overall water system.

Emergency supplies and materials, including a variety of pipes and pipe sizes and valves, must be available in the event of a breakdown in the water system. The emergency repair equipment varies from portable air compressors, to jacks and braces, to hammers, to cleanout tongs. When an emergency repair is needed, the area should be clamped off, the water shut off, that portion of the system drained, the broken part removed, a clean part installed in its place, and then the area should be disinfected.

Vessel Water Points

Public water is loaded aboard ships, buses, airplanes, trains, and so on. It is essential in all cases that the water comes from an approved source and is placed in sanitized holding containers. The pipes carrying the flow of the water from its source to the sanitized container should be installed in such a way that backflow cannot occur or nonpotable water cannot seep or be added accidentally into the storage container. It is absolutely essential that the water distribution system, the piping system, and the receiving system are plainly marked, and that they are totally separated when not in use. At all times the water distribution system must be at a higher pressure than the final receiving vessel that must subsequently be chlorinated at all the appropriate pipes. It must be thoroughly cleaned and sanitized before storage, and once again cleaned and sanitized before reuse.

WATER SUPPLY PROBLEMS

The problems of the individual or community water supplies include water quality that does not meet existing standards; facilities that are not operated in accordance with accepted standards; inadequate or incomplete standards because of a lack of knowledge of the relationship between certain pollutants and contaminants found in water and their effect on health; inadequate knowledge and consideration of long-term planning and development of water and related land resources as they might affect the health of the population; inadequate water treatment facilities; inadequately trained operators; and inadequate surveillance of contaminants of public water supplies.

Environmental Hazards

The environmental hazards discussed in this section are elaborated on in Chapter 7 on water quality and water pollution control.

Physical Problems

All water utilized as a drinking water source must be free of turbidity, color, taste, odor, and foamability, and should have a desirable temperature. Turbidity is the presence of suspended materials such as clay, silt, finely divided organic material, plankton, and other inorganic material. Turbidity, although not a hazard itself, may be an indication that pollution has been introduced into the water.

Color is caused by dissolved organic material from decaying vegetation and also comes from certain inorganic material. Color may be produced by some algae or aquatic microorganisms. Although color is not objectionable in itself, it is aesthetically objectionable, increases COD, and therefore should be removed.

Taste and odor are caused by foreign materials in the water, such as organic compounds, inorganic salts, or dissolved gases. Taste and odors come either from water collecting dissolved materials as it passes through a particular area or from some source of pollution. Again, taste and odor are only an aesthetic problem, but can be an indication of a potential source of contamination.

Foamability is due to the adding of detergents to water. Alkylbenzene sulfonate (ABS) degrades slowly in nature. The more rapidly degradable material known as linear alkylate sulfonate (LAS), has been substituted in most detergents. However, even this does not degrade very rapidly, especially when oxygen is not present, such as in septic tanks and tile field systems. Although the foam itself is not hazardous, it is an indication that hazardous materials or materials from a sewage origin may be present. Therefore, any frothing is considered unacceptable.

Hardness is due to the presence of calcium and magnesium compounds. Water that exceeds 120 mg/l is considered hard water and results in the formation of scum and lime deposits in water heaters and pipes and reduces the ability of soaps to lather. Iron in groundwater can adversely affect the taste of water, corrode water mains, and stain clothing and plumbing. Once iron is exposed to the air, it forms a precipitate and possibly gives the water a displeasing red color. Manganese, like iron, forms a precipitate that may impart bad taste, stain materials, and form clogging deposits in pipes and filters. At a concentration of over 500 ppm, chlorides give a salty taste to water. Chlorides also corrode pipes and fixtures. Sulfates at concentrations of about 250 ppm can have a laxative effect on people. High levels of sulfates form slimes, encrustations, and odorous waters. Nitrates impart a bitter taste to water at levels of 20 to 50 ppm. Nitrate levels of about 25 ppm often indicate contamination of groundwater from human sources such as animal waste, inorganic compounds, and chemical fertilizers.

Chemical Hazards

Chemicals, organic and inorganic, come from a variety of sources. They are used by industry, businesses, governmental agencies, and private citizens. Further, by-products from various industries are eventually disposed on land or are incinerated. In either case, the chemicals can enter both surface and groundwater supplies. It is estimated that about 142,000 tons of hazardous wastes are generated every day in the United States at about 750,000 different locations. These wastes go to about 50,000 different sites. Eventually they may leach through the soil into the aquifer. In recent years a number of organic contaminants have been found in numerous locations in the United States. It is now possible to detect these chemicals in parts per billion. Toxic effects may occur when individuals are exposed to one or more of these chemicals at very low levels over long periods of time. Although it is estimated that only 0.5 to 2% of the U.S. groundwater supplies are contaminated, much of the contamination occurs in areas where this is the only source of drinking water.

About 50% of the wells sampled contained one or more pesticides, with the highest levels in shallow groundwater beneath agricultural or urban areas, and the lowest levels in major aquifers, which are generally deeper. The herbicides atrazine and its breakdown products, prometon, and simazine have been found in 5% of all wells sampled in specific areas. Once pesticides are applied they may be taken up by plants; evaporate into the atmosphere; become drift; be ingested by insects, worms, and microorganisms; adhere to soil particles; dissolve in irrigation water or rainwater; and flow to a surface body of water or percolate down to the groundwater. This is accomplished through leaching, which is the vertical movement of water and solutes through the soil profile and dependent on the following four primary factors:

1. Properties of the chemical
2. Properties of the soil
3. Site conditions including rainfall and depth to groundwater
4. Management practices, including method and rate of application

The properties of the chemical include persistence, adsorption, solubility, and volatility. Persistence describes how long the chemical can maintain its structure. Chemicals are degraded at different rates by soil microorganisms, chemical reactions, and sunlight. Persistence is usually measured in terms of half-life. Adsorption describes how tightly a compound becomes attached to soil particles. The less the chemical is adsorbed, the greater its potential for mobility in the soil. Solubility is the tendency of the chemical to dissolve in a solvent, in this case water. Volatility is the tendency for a liquid or solid to change into a gas, and therefore not to be available for leaching.

The properties of the soil include permeability, soil texture, organic matter, and soil moisture. Permeability is a measure of how fast water can move vertically through the soil. It is affected by the texture and structure of the soil. Soil texture describes the relative percentage of sand, silt, and clay. Soil structure describes how the soil is aggregated. Soils with a loose structure allow water to move more readily. Organic matter helps in the adsorption of the chemical. Soil moisture affects how fast the water travels through the soil.

Site conditions include rainfall, depth to groundwater, sinkholes, and depth to bedrock. The amount of rainfall affects the flow of chemicals into the groundwater. The depth from the surface to the groundwater affects the flow of chemicals, which may cause problems. Sinkholes or cracked stone allows the chemicals to flow readily. Bedrock stops the flow.

Management practices related to the use of pesticides are affected by type of pesticide, application rate, equipment calibration, use of integrated pest management techniques, prevention of spills, time of application (preferably not before heavy rain), and appropriate storage and disposal of the chemical.

Inorganic chemicals of concern include arsenic, cadmium, lead, and mercury. These chemicals come from a variety of sources. Arsenic comes from the manufacture and use of insecticides, weed killers, fungicides, wood preservatives, electrical semiconductors, and alloys; and in the smelting of ores. It is hazardous to the kidneys, skin, lungs, and lymphatic system. Cadmium comes from the production of nickel–cadmium batteries, metal plating, plastics, and synthetics. It is hazardous to the respiratory system, kidneys, and bones.

Lead is used in the production of batteries, ammunition, sheet lead, solder, pipes, paint (lead-based paint is present in older houses), ceramic products, roofing materials, and caulking. Lead typically gets into drinking water through lead pipes or solder from joints of the pipes. The lead dissolves continuously into the water. It is hazardous to the digestive system, reproductive system, blood, central nervous system, kidneys, and the mental and physical development of exposed children.

Mercury is used during the manufacture of inorganic and organic compounds for use as pesticides, antiseptics, germicides, skin preparations, amalgams, mildewproof paints, batteries, lamps, felt, etc. Mercury is hazardous to the skin, respiratory system, central nervous system, kidneys, and eyes.

In 1991, the U.S. Congress appropriated funds for the U.S. Geological Service to begin the National Water-Quality Assessment Program. This program, which is currently in operation, focuses on water quality in more than 50 major river basins and aquifer systems. Studies have determined that fertilizers and pesticides have degraded the quality of streams and shallow groundwater in agricultural areas, and have resulted in high concentrations of nitrogen in nearly half of the streams sampled in agricultural areas. Concentrations of nitrate exceeded the U.S. EPA drinking water standard 10 mg/l in 15% of samples collected in shallow groundwater, in agricultural and urban land. Nitrates are converted within the body to nitrites by bacterial action. The nitrites react with hemoglobin to cause a condition known as methemoglobinemia, in which hemoglobin loses the ability to carry oxygen. This is particularly dangerous in infants, because they have a lower pH in their gastric juices, permitting better growth conditions for the bacteria. This disease has been known to kill infants. It also can affect children up to 5 years of age.

Another contaminant of drinking water supplies is deicing salt. The deicing salt, which is sodium chloride, becomes a particular problem when it seeps into an underground water supply or runs into a surface water supply. Deicing salt may also include calcium chloride and contaminants, such as ferric ferrocyanide, sodium ferrocyanide, sodium hexametaphosphate, and chromate salts. The ferric ferrocyanide is used to keep the salt from caking while in storage. The sodium hexametaphosphate is used to inhibit corrosion of automobiles.

Salt may also come from water softeners. Most homes use on-site regeneration systems where the salt solution is discharged into a well or septic system. The salt then enters the groundwater supply.

Chlorinated hydrocarbons are a serious hazard to humans. Most of them are found in pesticides. Unfortunately, there is a cumulative affect. By using dichloro-diphenyltrichloroethane (DDT) as an example, studies show that people in various parts of the world have a 9 to 15 ppm level of DDT stored in their bodies.

The polychlorinated biphenyls (PCBs) are important contaminants of our environment; they are persistent and insoluble in water but soluble in animal fat; and are used in many industrial processes and products, such as plasticizers, insulating fluids, rubber, electrical products, and brake linings. PCBs have been recognized as an occupational hazard for many years and as a public health hazard since the 1940s. The PCBs are stored in fat tissue. In primates, they cause loss of hair, skin lesions, liver changes, and reproductive losses. In rodents, they produce cancer. In employees exposed to them, PCBs cause nausea, dizziness, eye irritation, nasal irritation, asthmatic bronchitis, dermatitis, and acne.

Oil is another chemical contaminant because it contains hydrocarbons that may be toxic or carcinogenic. It is passed through the food chain to humans by marine life. Excessive fluorides cause fluorosis, which is a mottling of the teeth. In extreme amounts, it also causes bone problems; and, of course, in very large amounts, above 6 or 7 ppm, it is extremely toxic.

Iron is found in small amounts in water. It is usually washed from the soil into the water. Iron is soluble or insoluble: soluble creates reddish brown particles when a glass of cold water (containing the iron) sits, and insoluble produces red water that pours into the glass. Iron stains fixtures and glassware and is a nuisance beyond 0.3 mg/l. The water may also have a metallic taste or odor. Iron bacteria combine dissolved iron or manganese with oxygen and use it to form rust-colored deposits. These bacteria cause a brown slime that builds up on well screens, pipes, and plumbing fixtures; and cause odors, corrode plumbing, reduce well yields, and increase the chances of sulfur bacteria problems.

Manganese causes aesthetic and economic damage, and may also have some physiological effects on individuals. It impairs the taste of beverages, including coffee and tea.

Sodium is a particular problem because it frequently is added to the water by water softeners in the ion-exchange method. Sodium can be very dangerous for individuals who have heart, kidney, or circulatory diseases or who have potential hypertension. Sulfates, found in water leached from deposits of magnesium sulfate or sodium sulfate, produce laxative effects.

Zinc causes an undesirable taste in drinking water. Zinc also is a surface and groundwater pollution problem that has developed from existing and abandoned mining operations. Alkalinity of water increases through bicarbonates, carbonates, or hydroxides. Hardness creates scales on kettles, heating coils, or cooking utensils, thereby reducing fuel effectiveness by cutting down on proper heating. Carbonates cause temporary hardness that may be removed by heating the water. The bicarbonates present are broken down by the heat, and the carbonates precipitate out as insoluble material. The noncarbonate, or permanent hardness, is due to the presence of calcium sulfate and calcium chloride, or magnesium sulfate and magnesium chloride. These must be removed by other measures.

Groundwater and Surface Water Relationships

The interrelationships between groundwater and surface water is of great importance because surface water may contaminate groundwater. Heavy rainfall may carry chemicals deep into the soil where the soil is made up of sand and gravel. Further crevices in limestone and other rocks may facilitate contamination. The chemical and bacteriological quality of the groundwater may vary with the depth of the aquifer, the usage of the land, and the presence of waste disposal sites.

Groundwater Contamination

Contamination may come from land burial of municipal wastes and surface or near-surface disposal of sludge, agricultural chemicals, and accidental spills.

Broad categories of actual or potential sources of contamination are known. These categories include the following:

1. Hazardous waste sites are known as potential candidates for the Superfund NPL.
2. Millions of septic systems (about 33% of the people in the country use septic systems) are potential contaminators of the groundwater supply. Septic systems are the largest volumetric source of effluent discharged into the groundwater zone.
3. Over 180,000 surface impoundments (pits, ponds, and lagoons) could leach into the groundwater supply.
4. Hazardous waste land disposal facilities and municipal and other landfills may be potential sources of contamination.
5. Several million USTs exist, many of which are thought to be leaking and can cause potential hazards.
6. Thousands of underground injection wells exist.
7. Millions of tons of pesticides and fertilizers are spread on the ground, primarily in rural areas, and may be sources of leaching into the groundwater.
8. Accidental spills occur each year because of the large volume of toxic material transported and the number of vehicles having accidents.
9. Animal feed lots are another source of contamination.
10. Storm water infiltration contaminates groundwater with nutrients such as nitrates and phosphates, pesticides, other organics, viruses, heavy metals, and salts from winter traffic.
11. Flooding contaminates groundwater supplies.

The major concern of the EPA is with toxic chemicals, such as the synthetic organic chemicals that are used in plastics, solvents, pesticides, paints, dyes, varnishes, and ink. Some of the community public water systems and private wells that furnish 50% of Americans with drinking water are known to be contaminated with these substances. Trichloroethylene, which is a known carcinogen, is a frequently found contaminant. The EPA has also found that about 60 pesticides have been detected in 30 states at various levels of contamination. Although in most cases the pesticide levels were below health advisory levels, it still is a serious problem. At least 13 organic chemicals that are confirmed animal or human carcinogens have been detected in drinking water wells. Ethylene dibromide has been found in wells in Florida.

Contaminants in Public Water Systems Related to Class 5 Injection Wells

Class 5 injection wells may possibly cause contamination of public water systems. These contaminants are known to be associated with effluent from class 5 industrial and automotive waste disposal wells. The contaminants include inorganic compounds and volatile organic compounds (VOCs) (Tables 3.1 and 3.2).

Many other compounds may contaminate water supplies from the previously mentioned sources. All the inorganic compounds were detected in groundwater systems in various states above the minimum reporting levels. Cyanide was detected in the lowest percentage of systems and barium was detected in the highest percentage of systems. All the VOCs were detected in groundwater systems in various states above the minimum reporting level. Vinyl chloride was detected in the lowest percentage of systems, and methylene chloride was detected in the highest percentage of systems. All the preceding chemicals were also detected in surface water systems.

Table 3.1   Class 5 Inorganic Chemicals

Name of Chemical	Source and Uses	Potential Health Effects
Arsenic	Natural deposits, smelters, pesticides, electronic wastes	Skin, nervous system toxicity
Barium	Natural deposits, pigments, epoxy sealants, coal remnants	Circulatory system
Cadmium	Natural deposits, galvanized pipe corrosion, batteries, paints	Kidney
Chromium	Natural deposits, mining, electroplating, pigments	Liver, kidney, circulatory system
Cyanide	Electroplating, steel, plastics, mining, fertilizer	Thyroid, nervous system
Lead	Pipes, solder, paint, batteries	Brain damage, kidney damage, central nervous system
Mercury	Natural deposits, batteries, electrical switches	Kidney, nervous system
Selenium	Natural deposits, mining, smelting, coal or oil combustion, paint	Liver
Silver	Photographs, jewelry, solder	Skin discoloration, kidneys

Table 3.2   Class 5 Organic Chemicals

Name of Chemical	Source and Uses	Potential Health Effects
Benzene	Gas, drugs, paint, plastics, solvent	Cancer
Carbon tetrachloride	Solvents and their products	Cancer
Chlorobenzene	Waste solvent from metal degreasing	Nervous system and liver
1,2-Dichloroethane	Fumigant, paint, solvent	Cancer
1,1-Dichloroethene	Plastics, dyes, perfumes, paints	Cancer, liver, and kidneys
Ethylbenzene	Gasoline, insecticides, solvent, chemical wastes	Liver, kidneys, and nervous system
Methylene chloride	Paint stripper, metal degreaser, propellant	Cancer
Methyl ethyl ketone	Solvent, cement, adhesives	Decreased fetal birth weight
Tetrachloroethylene	Dry cleaning, solvent	Cancer
Toluene	Gasoline additive, solvent, etc.	Liver, kidneys, nervous and circulatory systems
1,1,1-Trichloroethane	Adhesives, aerosols, textiles, paints, inks, degreasers, solvent	Liver, nervous system
Trichloroethane (TCE)	Textiles, adhesives, degreasers	Cancer
Vinyl chloride	Plastics, PVC pipe, solvent breakdown	Cancer
Xylenes	Paints, inks, detergents, solvents by-product of gasoline refining	Liver, kidneys, and nervous system

Drinking Water in the Home

Today special concerns exist about copper, lead, radionuclides, microbiological contaminants, and disinfection by-products at the faucet in the house. Copper above 1.3 mg/l may be a relatively common cause of diarrhea, abdominal cramps, and nausea. The copper can leach from pipes or containers.

The primary source of lead in drinking water is the corrosion of plumbing materials, such as lead service lines and lead solders in water distribution systems, and in houses and larger buildings. Almost all public water systems serve households with lead solders of varying ages. Most faucets are made of materials that can contribute some lead to the drinking water. The health effects related to the ingestion of lead are extremely serious, especially in young children. Lead leads to impaired blood formation, brain damage, increased blood pressure, premature birth, low birth weight, and nervous system disorders.

Disinfection by-products are produced in water treatment by the chemical reactions of disinfectants with naturally occurring or synthetic organic materials present in untreated water. Many disinfection by-products may pose health risks. These risks tend to be related to long-term exposure at low levels of contaminants. Because disinfectants are essential to safe drinking water, it is necessary to find ways to minimize the risks of the by-products.

Polycyclic Aromatic Hydrocarbons

The polycyclic aromatic hydrocarbons (PAHs) found in groundwater and surface water are organic toxicants that are most frequently found in industrial waste impoundments and solid waste disposal sites. The priority pollutants listed by the

EPA are naphthalene, 2-chloronaphthalene, acenaphthylene, acenaphthene, fluorene, phenanthrene, anthracene, fluoranthene, pyrene, chrysene, benzo(a)anthracene, benzo(b)fluoranthene, benzo(k)fluoranthene, benzo(a)pyrene, indeno(1,2,3-cd)pyrene, dibenzo(a,h)anthracene, and benzo(ghi)perylene. The arenes that are EPA priority pollutants are benzene, benzidine, chlorobenzene, hexachlorobenzene, 1,2,4-trichlorobenzene, 1,2-dichlorobenzene, 1,3-dichlorobenzene, 1,4-dichlorobenzene, 3,3-dichlorobenzidine, phenol, toluene, 2,4-dimethylphenol, 2,4-dinitrotoluene, 2,6-dinitrotoluene, ethylbenzene, nitrobenzene, 4,6-dinitro-o-cresol, 2,4-dinitrophenol, 2-nitrophenol, 4-nitrophenol, pentachlorophenol, and toxaphene. The arenes that are on the EPA priority list are N-nitrosodimethylamine, N-nitrosodiphenylanine, and N-nitrosodi-n-propylanine.

These polycyclic (polynuclear) aromatic hydrocarbons are produced in greatest quantity by the manufacture, transport, combustion, and use of petroleum hydrocarbons by industry. Anthracene is used in producing dyes, and naphthalene is used as a feedstock in dyes and pharmaceuticals. The PAHs enter the aquatic environment from a variety of sources including airborne, industrial, storm water runoff, and point source pollution. Arenes, primarily benzene and its derivatives, are used extensively in industry, as well as by consumers. Benzene may be found in laboratory solvents, paint strippers, rubber cement, gasoline, engine oil flushing, etc. The arenes get into water through the improper disposal of benzene and its derivatives. N-nitrosodiphenylamine is used as an accelerator in vulcanizing rubber. Nitrosamines can be formed when primary and secondary amines react with nitrosating agents.

Volatile Organic Compounds

VOCs of health concern are trichloroethylene, carbon tetrachloride, 1,2-dichloroethane, vinyl chloride, benzene, 1,1-dichloroethylene, p-dichlorobenzene, and 1,1,1-trichloroethane. Trichloroethylene is a common metal cleaning and dry-cleaning fluid. It gets in the drinking water by the improper disposal of waste. The chemical has been shown to cause cancer in laboratory animals, such as rats and mice, when the animals are exposed to high levels over their lifetime. Carbon tetrachloride is a health concern. This chemical, which was once used in the household as a cleaning fluid, generally gets into the drinking water by improper waste disposal. It has been shown to cause cancer in laboratory animals, such as rats and mice, when the animals are exposed to high levels over their lifetimes. 1,2-Dichloroethane is used as a cleaning fluid for fats, oils, waxes, and resins. It generally gets into the drinking water from improper waste disposal. This chemical has been shown to cause cancer in laboratory animals, such as rats and mice, when the animals are exposed to high levels over their lifetimes.

Vinyl chloride is used in industry and is found in drinking water as a result of the breakdown of related solvents. The solvents are used as cleaners and degreasers of metals, and generally get into drinking water by improper waste disposal. This chemical has been associated with significantly increased risks of cancer among certain industrial workers who were exposed to relatively large amounts of the chemical during their working career. The chemical has also been shown to cause cancer in laboratory animals when the animals are exposed at high levels over their lifetimes.

Benzene is used as a solvent and degreaser of metals. It is also a major component of gasoline. Drinking water contamination generally results from leaking underground gasoline and petroleum tanks or improper waste disposal. The chemical has been associated with significantly increased risks of leukemia among certain industrial workers who were exposed to relatively large amounts of the chemical during their working careers. The chemical has also been shown to cause cancer in laboratory animals when the animals are exposed at high levels over their lifetime.

1,1-Dichloroethylene is used in industry and is found in drinking water as a result of the breakdown of related solvents. The solvents are used as cleaners and degreasers of metals and generally get into drinking water by improper waste disposal. The chemical has been shown to cause liver and kidney damage in laboratory animals, such as rats and mice, when the animals are exposed at high levels over their lifetimes. p-Dichlorobenzene is a component of deodorizers, mothballs, and pesticides. It generally gets into the drinking water by improper waste disposal. The chemical has been shown to cause liver and kidney damage in laboratory animals, such as rats and mice, when the animals are exposed to high levels over their lifetime.

1,1,1-Trichloroethane is used as a cleaner and degreaser of metals. It generally gets into the drinking water by improper waste disposal. This chemical has been shown to damage the liver, nervous system, and circulatory system of laboratory animals, such as rats and mice, when the animals are exposed to high levels over their lifetime.

VOCs are a class of organic or synthetic chemical compounds that have the ability to vaporize into the air in a gaseous state. Volatilization is the evaporative loss of a chemical from the evaporative surface based on the vapor pressure of the chemical and the ambient environment. Volatilization is an important source of airborne pollution.

In addition to the sources of the individual chemicals previously discussed, several other potential sources of VOCs are:

- Industry, through the industrial discharge of materials into water or the spreading of the materials on the land
- Landfill leachates
- Septic tank degreasers
- Household detergents, window cleaners, oven cleaners, etc.
- Sewer leaks
- Accidental spills
- Water from cleaning and rinsing of tanks and machinery
- Leaking storage tanks
- Treated waste introduced into aquifers as groundwater recharge

Several factors increase the vulnerability of a well to VOC contamination. They are as follows:

- Distance between well and source of contamination
- Amount of VOC that has been dumped or spilled
- Depth of well casing
- Local geology

Gasoline Oxygenate

Methyl-*tert*-butyl ether (MTBE) is a gasoline oxygenate. When it is added to gasoline, it increases the oxygen level in gasoline and decreases the carbon monoxide emissions and ozone levels in the atmosphere. MTBE moves rapidly into water. It is the second most frequently detectable VOC in shallow urban groundwater, and is less biodegradable than common gasoline compounds. The USGS sampled storm water in 16 cities in metropolitan areas and found concentrations of 62 VOCs including MTBE. It was the seventh most frequently detected VOC in storm water. MTBE air exposure has caused the symptoms of headaches, dizziness, irritated eyes, coughing, nausea, and disorientation. Research is continuing on the potential health effects of drinking water containing MTBE. The EPA has issued a draft lifetime advisory for MTBE in drinking water that ranges from 20 to 200 µg/l.

Natural Organic Compounds

Naturally occurring organic compounds come from decaying weeds, leaves, and trees that can lead to formation of color, taste, and odor problems, and oxygen depletion. These compounds may also interfere with the water treatment process by creating halogenated compounds with chlorine.

Risk Assessment

Because a number of organic chemicals have been detected in groundwater samples taken from various locations in the United States, it is necessary to do an appropriate risk assessment to determine what should be done, and how it should be done to protect the public from potential chemical hazards. The questions that must be addressed are as follows:

1. What is the toxicity of the chemical?
2. What are the chances that the chemical might contaminate groundwater at a specific location?
3. How much, if any, of the chemical can be consumed safely without acute or chronic health effects?
4. How many people and what kind of people are likely to be affected and for what duration of time?

Risk assessment is the use of the factual information to define the health effects of exposure of individuals or populations to hazardous substances. The four inter-related components of risk assessment are as follows:

1. Hazard identification is a qualitative determination of the nature of the toxicity of the chemical, including carcinogenicity, to animals or people.
2. Dose-response assessment is a quantitative evaluation of the relationship between the dose, magnitude of exposure, and incidence of toxic effects.

3. Exposure assessment is the identification of the various quantities and time of exposure to a given chemical by all possible routes of entry.
4. Risk characterization is an estimate of the probable incidence of any adverse health effects in any given population likely to be exposed to a chemical.

Risk management in relationship to water would include closing contaminated wells, providing alternate drinking water supplies, installing special water treatment systems, and banning or restricting the use of the chemical.

Some of the problems connected to risk assessment relate to lack of understanding of toxicity and of data, interacting toxic effects caused by multiple toxins, and difficulties in determining the duration and exposure to the toxic compound.

Biological Factors

Water for drinking or cooking purposes must be free of disease-producing organisms, including bacteria, protozoa, viruses, and helminths (worms). Because it is difficult to keep track of the human disease carriers, it is necessary to ensure that water sources do not become contaminated and that good treatment measures are utilized to protect against biological contamination. Unfortunately, the specific disease-producing organisms present in water are not readily identifiable. The techniques for bacteriological examination are complex and time consuming. The coliform test is most widely used as an indicator of fecal contamination. Coliform not only inhabit the human intestinal tract, but are also found in most domestic animals, birds, and certain wild species. Unfortunately, viruses (including the polio-producing virus, the enteroviruses causing heart disease in infants, influenza, and the virus producing infectious hepatitis) are found in water. Small quantities of viruses transmit infection. Chlorination may not necessarily eliminate viral material in water.

Waterborne Disease Outbreaks — 1991 to 1992

In 1991 and 1992, over 17,000 people, in 34 outbreaks of disease in 17 states were reported ill from drinking contaminated water. *Giardia lamblia* or *Cryptosporidium* were responsible for 71% of the cases where the etiologic agent was determined. The etiologic agents were bacteria, viruses, and chemicals.

Giardia is a protozoan causing giardiasis. This disease may be asymptomatic or associated with chronic diarrhea, abdominal cramps, fatigue, weight loss, and poor absorption of fat soluble vitamins. Incubation time is usually 7 to 10 days. The organism is found worldwide, with children more frequently infected than adults.

Cryptosporidiosis is a protozoan infection of the gastrointestinal tract associated with intestinal symptoms such as copious watery diarrhea, abdominal cramps, malabsorption, and weight loss; incubation time is 2 to 14 days with an average of 7 days and the duration of infection is 10 to 14 days of watery diarrhea followed by continuing shedding of oocyst for an additional 14 to 21 days. It is caused by *Cryptosoporidium* sp., an aflagellar intestinal protozoan with motile sporozoites that are released in the host intestinal tract when the oocyst outer wall disintegrates.

Sporozoites implant on the host epithelium and undergo sexual and asexual development. The parasite host relationship is unique in that it is intracellular, that is, enveloped by the host cell membrane, but extracytoplasmic to the cytoplasm. Sporozoites develop into trophozoites and subsequently go through asexual multiplication, formation of macrogametes and microgametes, fertilization, and oocyst formation. Newly formed oocysts that are expelled in the feces are immediately infective.

Cryptosopordium, able to develop wholly in one host, thereby creating tremendous potential for infection, is found worldwide in day care centers and among immunocompromised individuals as a nosocomial infection; and in the general population through outbreaks of waterborne disease with a reported level of 4 to 7% of the population. *Cryptosopordium* is also found in up to 65% of the population in areas of outbreaks of disease, and 10 to 15% asymptomatic individuals carrying the parasite. This protozoan also causes outbreaks of waterborne disease where the source may be surface water; water treated by rapid sand filtration and chlorination; and water contaminated by heavy rainfall and runoff, or by animals (especially cattle). The reservoir of infection includes people and domestic animals. The infection is transmitted by ingestion of oocysts in water contaminated by feces, person to person, and self-recontamination; and is communicable for the entire period of the infection and at least 2 to 3 additional weeks. Susceptibility is high, especially in target populations such as day care centers and among the immunosuppressed. The infection is controlled through proper disposal of feces, rigorous personal hygiene, closing of contaminated water supplies, thoroughly cleaning filters, and superchlorination.

Waterborne Disease Outbreaks — 1997 and 1998

In the year 2000, the final report for waterborne disease outbreaks in 1997 and 1998 was issued, with 13 states reporting 17 outbreaks associated with drinking water that resulted in over 2000 people becoming ill. No deaths were reported. The organism or chemical that caused the illness was identified in 70% of the outbreaks. The outbreaks were linked to groundwater sources in 88% of the cases. Drinking water outbreaks associated with surface water decreased from about 32% during 1995 and 1996 to roughly 12% in 1997 and 1998. Outbreaks associated with groundwater sources increased from approximately 59% to approximately 88%. Florida reported the greatest number of outbreaks.

The outbreaks of gastroenteritis due to parasites increased for drinking water. The outbreaks of known etiology were due to *Giardia*, with one in New York affecting 50 people, and one in Oregon affecting 100 people. In New York, the individuals used a surface water supply that was chlorinated but unfiltered. In Oregon, the individuals used a campground where the water supply was a combination of groundwater from an untreated well and a chlorinated spring.

Two outbreaks of copper poisoning were reported in Florida. Two people became ill after consuming fruit drink made with tap water, in the first outbreak. Improper wiring and plumbing caused leaching of copper from restaurant piping. In the second outbreak, elevated levels of copper in tap water caused gastrointestinal illness in 35 people in one community. A defective check valve and a power outage caused a malfunction in a water treatment facility, which released high levels of sulfuric acid, causing corrosion in the pipes and allowing leaching of copper into the system. Waterborne chemical poisonings are probably underreported to the CDC.

The nonidentified etiologic agents were associated with about 29% of the outbreaks of waterborne disease. Four of the outbreaks, incubation periods, durations, and symptoms were consistent with viral contamination. Other outbreaks were related to improperly functioning or nonexisting chlorinators. Previously, in 1996, an outbreak of gastroenteritis from drinking tap water at an elementary school in Florida resulted in 594 students and members of the staff becoming ill with gastrointestinal symptoms. No known cause of the outbreak was found.

From 1971 to 1998, there were 691 outbreaks of waterborne-disease associated with drinking water reported to the CDC. The largest number of outbreaks occurred in 1980, with the greatest number of problems attributed to community water systems. Unfortunately, many outbreaks of waterborne disease are not reported and therefore are not corrected. The reporting system is strictly on a voluntary basis.

Escherichia coli *O157:H7 and* Campylobacter *Outbreaks — 1999*

An outbreak of waterborne disease at the Washington County Fair, New York, August 23 to 29, 1999 resulted in at least 921 people reporting diarrhea. Stool cultures showed *E. coli* O157:H7 in the feces of 116 people, and *C. jejuni* in 13 of these people existing as a coinfection; 65 people were hospitalized, and 11 children developed hemolytic uremic syndrome. Cases of diarrheal illness among fair attendees were reported from 14 New York counties and 4 different states. Drinking water or beverages made with water from a suspect well was associated with the illness. When control for water consumption was considered, other exposures, such as eating food at the fair and contact with manure, were not significantly associated with illness. Subsequently, the New York State Public Health Laboratory demonstrated the presence of *E. coli* in the well water samples.

The outbreaks due to bacteria were caused by *E. coli* O157:H7, and *Shigella sonnei*. The *E. coli* outbreaks occurred in Wyoming, Illinois, and Washington. In Wyoming, 157 people became ill after they drank from an untreated community water system supplied by a spring and two wells. The outbreak resulted from fecal contamination by wildlife near the spring. In Illinois, three people became ill when they drank water from an untreated well located near a cow pasture. In Washington, four people became ill when they drank water at a trailer park that was associated with a chlorinated ground water supply. The means of contamination could not be determined. The *Shigella* outbreak occurred in Minnesota at a local fair supplied by a community water system, with 83 people becoming ill. The exact cause of the outbreak was not determined.

Causes of Waterborne Outbreaks of Disease

Untreated spring water, untreated surface water, inadequately or interrupted disinfection of surface water, inadequate or interrupted disinfection of spring water, untreated well water, and inadequately or interrupted disinfection of well water are responsible for outbreaks of waterborne disease. Cross connections, back siphonage, contamination of household plumbing, corrosion of household plumbing, contamination of water mains during construction, contamination of broken or leaking mains, contamination of distribution and storage facilities, and contamination of cisterns or private storage tanks also contribute to the outbreaks. Further outbreaks are attributed to use of water that was not intended for drinking, contaminated ice, contaminated bottle water, and ingestion while swimming.

The epidemiological characteristics and patterns of waterborne disease generally indicate that if a disease appears to be widespread and there does not seem to be a single point source, then water must be considered as a possible source for the disease. Although water is not a medium for pathogenic growth, it is a means of transmission of the pathogen to the place where the individual is able to consume it and therefore starts the outbreak of disease. Usually lower attack rates occur in waterborne outbreaks than in foodborne outbreaks of disease. Except for infants who may be breast fed or may receive boiled water, people of all ages and both sexes are affected in a waterborne outbreak.

Radiological Contaminants

The development and use of nuclear energy as a power source and the mining of radioactive materials have increased the amount of radioactivity present in water and potentially in drinking water supplies. Obviously radioactive material is hazardous to human beings, and therefore the concentrations must be severely limited.

Radionuclides are radioactive isotopes that emit radiation as they decay. The most significant radionuclides in drinking water are radium, uranium, and radon, all of which occur naturally. Radium most frequently occurs in the Midwest and Appalachian regions. Slow moving groundwater can remove radium from rocks, because it dissolves minerals as it passes. Radium can cause bone cancer. Natural uranium occurs most frequently in the Rocky Mountains. Radon, in gas form, occurs most frequently in the Northeast. Recent surveys, however, show that radon may also be occurring in other parts of the country. The highest concentrations of dissolved radon gas occurs in groundwater flowing through granite or granite sand and gravel formations.

Ingestion of uranium and radium in drinking water may cause cancer of the bone and kidney. Radon is usually inhaled, whereas radium and uranium are ingested. Radon is typically released into the air during showers, baths, and other activities, such as washing clothes and dishes. The main health risk of radon is lung cancer. The radionuclides in drinking water occur primarily in groundwater, because naturally occurring radionuclides are seldom found in surface waters such as rivers, lakes, and streams.

Abandoned Wells

Unsealed abandoned wells are a serious potential hazard, not only to small children but also to the public. The abandoned well becomes contaminated by flooding or by a variety of problems occurring as a result of pesticides, agricultural washings, or any other physical, chemical, or microbiological material that seeps through the abandoned well into the underground water supply. The basic concept of sealing the well is to restore the geologic conditions that existed before the well was constructed. This can be done by filling the well and surrounding areas with concrete, cement grout, or clays. Abandoned wells should never be used for disposal of sewage or other wastes.

Well Problems

A well may perform properly for a reasonable period of time and then suddenly develop a problem within the water system, and the quantity or quality of the water deteriorates. When the quality of the water deteriorates, it is due to an increase in minerals, nitrates, bacterial contamination, turbidity, surface water pollution, over-pumping, or site problems. Calcium, sulfate, and magnesium salts that border an aquifer may enter the well after water has been drawn from it for a period of time. These salts may also be present in water softener waste that is discharged into the ground and eventually seeps into the groundwater supply. Excessive nitrates usually seep from feedlots and septic tank systems. The nitrates percolate through the soil into the aquifer. Bacterial contamination can occur from artificial or natural sources. It results from septic tank effluent, animal feces, or decay and seepage of waste material from wild birds. Turbidity occurs especially in shallow limestone aquifers after heavy rains when the sediment in the area is washed into the limestone aquifer. Limestone is a poor material from which to draw water, because contamination flows for 25 or more miles through the aquifer without natural decontamination.

Surface waters infiltrate or leach into underground aquifers. Most surface waters are contaminated and are therefore bound to cause well problems. Overpumping of a well or the placement of several wells in an area where overpumping of an aquifer occurs increases the distance from which the water is drawn to provide adequate water in the wells. This sharp draw-down brings water from 1 mi to several miles away. Ordinarily the well would bring water only a few hundred feet from the source of the well. Increased travel distance increases the chances of pollution. Water quality is also affected by cracks in the casing, surface water or flood waters at the well head, contaminated pumps, and cross connections (Figure 3.4).

Well failure, which is a reduction or total halting of water from the well, is due to a variety of causes, including mechanical encrusting, in which the well is physically plugged by silt, clay, and fine sand; bacterial infection, in which the pipe becomes clogged with slimy growths; chemical encrusting, in which precipitation of calcium and manganese carbonates plugs the well; and fine sand movement. Pump and motor failures are due to a variety of problems and should be analyzed by professionals, who can best understand the problem and take the necessary action.

Impact on Other Problems

Since our country is experiencing problems in obtaining adequate quantities of pure drinking water, it has now become necessary to reduce the pollutants entering water. This has resulted in the necessary installation of pollution control equipment and in changes in a variety of operations. The water available for drinking is of

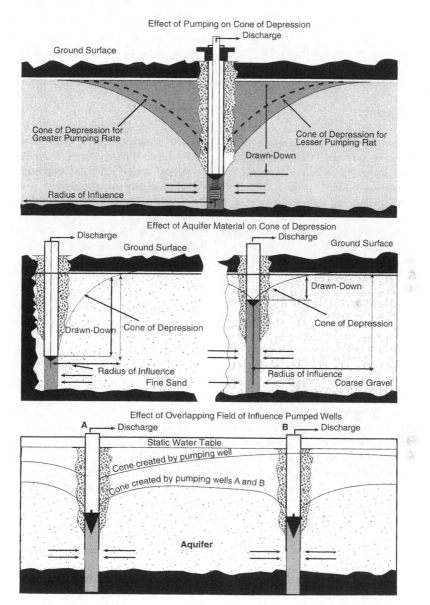

Figure 3.4 Pumping effects on aquifers. (From Water Supply Division, Office of Water Programs, U.S. Environmental Protection Agency, *Manual of Individual Water Supply Systems,* U.S. Government Printing Office, Washington, D.C., 1973, p. 29.)

primary importance. Therefore, industrial plants, sewage treatment plants, communities, solid waste disposal operations, and so forth must be altered to avoid contaminating the water. Decisions must be made about the quantity of water that can be utilized for manufacturing and processing, agriculture, and human and animal consumption. As the supply drops, the impact of the necessity of obtaining better drinking water will cause a profound change in our society. Plants may be dislocated, older operations may go out of business, water may have to be piped in the same manner as oil, and therefore there may also be a dislocation in the economy.

Economics

The economic problems are tied into cost of securing potable water, cost of treatment and distribution, and finally, cost of disposal. Where water is becoming a scarce resource, cost is increasing, because deeper wells must be sunk and the communities must go greater distances to obtain necessary water supplies. In some areas, desalination is becoming economically feasible, because the water is scarce and better desalination programs and processes are under development. Cost has increased because of inflation. In the coming years, the cost of water will continue to increase sharply.

The estimates of costs involved in meeting water quality goals for the construction of municipal treatment facilities and urban storm water problems are in the tens of billions of dollars for water treatment and hundreds of billions of dollars for storm water control. The costs of industrial pollution abatement in nonpoint sources cannot be evaluated at this time.

POTENTIAL FOR INTERVENTION

The potential for intervention varies with the area of the country. Where water comes from good sources and where adequate treatment plants are available, the potential for intervention is excellent for various bacteria, fair for viruses, and good for a variety of chemicals. For well water that is highly contaminated, the potential for intervention varies with the type of plant used for treatment. Under ideal situations, all contaminants can be removed from water. Under practical field situations, this varies with water sources, water treatment processes, quantities of water, and initial contaminants present. The potential for intervention includes isolation, substitution, shielding, treatment, and prevention. Isolation is a technique by which the watersheds are protected from contamination by humans, industries, or animals. This makes the initial well water of good quality. Isolation can also refer to water drawn from artesian aquifers.

In substitution, water from better sources is used. For example, the water from a deep well would be substituted for surface water supplies. Shielding is accomplished when sewage from on-site sewage disposal systems or other sources are kept considerable distances from water supplies that will be used for potable water. Treatment is one of the most important techniques used in the protection of human drinking water. Various treatment techniques are discussed under the area of controls. Prevention is the removal of sources of pollution from areas that would allow pollutants to enter the raw water supply used for drinking purposes. This would include many of the water pollution controls instituted by industry.

RESOURCES

Scientific and technical resources include the School of Public Health University of North Carolina, New Mexico State University, National Water Well Association, Texas A & M University, American Public Health Association, and National Environmental Health Association.

The civic resources include the National Drinking Water Advisory Council, Concerned Inc., Environmental Defense Fund, International Association for Pollution Control, National Water Resources Association, Water Pollution Control Federation, and Water Quality Association.

Governmental resources include the Department of Agriculture, National Bureau of Standards, Department of Defense, U.S. EPA, Department of Health and Human Services, National Science Foundation, and U.S. Geological Service. Additional resources include the Federal Water Quality Association, c/o W.H. 547 U.S. EPA, 41 M Street, S.W., Washington, D.C.; the Ground Water Management Districts Association, 1125 Maze Road, Colby, KS; and state and local health departments.

STANDARDS, PRACTICES, AND PROCEDURES

Standards for Safe Drinking Water

The SDWA of 1974, established a program for protecting the nation's drinking water. This program was implemented by the EPA and called for primary drinking water standards that were developed by the National Academy of Sciences. The standards upgraded the 1962 Public Health Service Drinking Water Standards. The SDWA was amended in 1986 and 1996, allowing the EPA to set drinking water standards. In the year 2000, the EPA has set new standards for radon, revised current radionuclides regulations, set a new standard for uranium, and revised the existing standard for arsenic. The EPA is identifying ways to protect groundwater from microbial contamination, and is monitoring public water systems serving at least 100,000 people, for up to 30 unregulated contaminants.

Drinking water standards are regulations that the EPA establishes to control the level of contamination in the nation's drinking water. The SDWA requires a multiple barrier approach to drinking water protection. This includes:

1. Assessing and protecting drinking water sources
2. Protecting wells and collections systems
3. Ensuring that water is treated by qualified operators
4. Ensuring the integrity of distribution systems
5. Establishing appropriate drinking water standards
6. Ensuring the availability of information to the public on the quality of their drinking water

There are two categories of drinking water standards. The National Primary Drinking Water Regulation, or primary standard, is a legally enforceable standard that applies to public water systems, which limits the levels of specific contaminants that can adversely affect public health. The National Secondary Drinking Water Regulation secondary standard is a nonenforceable guideline concerning contaminants that may cause cosmetic effects, such as tooth discoloration, or aesthetic effects such as taste, odors, or color in drinking water. Drinking water standards applies to public water systems, which provide water for human consumption through at least 15 service connections, or regularly serve at least 25 people.

In establishing standards, the EPA considers the input from many different individuals and groups. The National Drinking Water Advisory Council, which is a 15-member committee, was created by the SDWA, for the purpose of assisting the EPA. The EPA in setting a standard must identify drinking water problems, establish priorities, and set standards, while considering public input throughout the process. The National Drinking Water Contamination Candidate List helps identify the contaminants not regulated under the SDWA, which should be considered for further action. During the coming years, the EPA will select contaminants from this list and make necessary determinations as to whether they should be regulated. As of August 1999, a new National Contaminant Occurrence Database is storing data on regulated and unregulated chemical, radiological, microbiological, and physical contaminants likely to occur in finished, raw, and source waters of public water systems of the United States. These data are available to the public.

When establishing a National Primary Drinking Water Regulation, the EPA reviews health effects studies and then sets a Maximum Contaminant Level Goal (MCLG), which is the maximum level of a contaminant in drinking water at which no known or anticipated adverse effect on the health of persons would occur and which allows for a margin of safety. MCLGs are nonenforceable public health goals. When determining an MCLG, the EPA considers the risk to sensitive groups such as infants, children, elderly people, and those with compromised immune systems. For noncarcinogens, not including microbial contaminants, the MCLG on the reference dose is an estimate of the amount of a chemical that a person can be exposed to on a daily basis, but is not anticipated to cause adverse health effects over a person's lifetime. For carcinogens, there is no safe dose, and therefore the MCLG is zero. For microbial contaminants that may present a public health risk, the MCLG is set at zero, because ingesting one organism might cause adverse health effects.

The MCL is an enforceable standard that is as close to the MCLG as possible. The MCL is the maximum permissible level of the contaminant in water that may be achieved with the best available technology treatment techniques. When no reliable method is economically or technically feasible to measure a contaminant at very low concentrations, a treatment technique (TT) is set instead of an MCL. A treatment technique is an enforceable procedure or level of technological performance that public water systems must follow to ensure control of a contaminant. The Primary Drinking Water Regulations are broken up into four sections. They are inorganic chemicals, organic chemicals, radionuclides, and microorganisms. The chemicals will be listed either as MCL or TT in milligrams per liter unless otherwise noted.

The inorganic chemicals include: antimony, 0.006; arsenic, 0.05*; asbestos, 7 million fibers per liter; barium, 2; beryllium, 0.004; cadmium, 0.005; chromium

---

* The proposed standard for arsenic is 0.005.

(total), 0.1; copper, 1.3 TT; cyanide, 0.2; fluoride, 4.0; lead 0.015 TT; inorganic mercury, 0.002; nitrate (measured as nitrogen), 10; nitrite (measured as nitrogen), 1; selenium, 0.05; and thallium, 0.002.

The organic chemicals include: acrylamide, 0.05% dosed at 1 mg/l or equivalent; alachlor, 0.002; atrazine, 0.003; benzene, 0.005; benzo(a)pyrene, 0.0002; carbofuran, 0.04; carbon tetrachloride, 0.005; chlordane, 0.002; chlorobenzene, 0.1; 2,4-dichloro-phenoxyacetic acid (2,4-D), 0.07; dalapon, 0.2; 1,2-dibromo-3-chloropropane, 0.0002; o-dichlorobenzene, 0.6; 1,2-dichloroethane, 0.005; 1-1-dichloroethylene, 0.007; cis-1,2-dichloroethylene, 0.07; $trans$-1,2-dichloroethylene, 0.1; dichloromethane, 0.005; 1-2-dichloropropane, 0.005; di(2-ethylhexyl)adipate, 0.4; di(2-ethylhexyl)phthalate, 0.006; dinoseb, 0.007; dioxin (2, 3, 7, 8-TCDD), 0.00000003; diquat, 0.02; endothall, 0.1; endrin, 0.002; epichlorohydrin, 0.01% dosed at 20 mg/l or equivalent; ethyl-benzene, 0.7; ethelyne dibromide, 0.00005; glyphosate, 0.7; heptachlor, 0.0004; heptachlor epoxide, 0.0002; hexachlorobenzene, 0.001; hexachlorocyclopentadiene, 0.05; lindane, 0.00002; methoxychlor, 0.04; oxamyl(vydate), 0.2; polychlorinated biphenyls (PCBs), 0.0005; pentachlorophenol, 0.001; picloram, 0.5; simazine, 0.004; styrene, 0.1; tetrachloroethylene, 0.005; toluene, 1; total trihalomethanes (TTHMs), 0.10; toxaphene, 0.003; 2,4,5-TP (silvex), 0.05; 1,2,4-trichlorobenzene, 0.07; 1,1,1-trichloroethane, 0.2; 1,1, 2-trichloroethane, 0.005; trichloroethylene, 0.005; vinyl chloride, 0.002; and xylenes (total), 10.

The radionuclides include beta particles and photon emitters, 4 mrem/year; gross alpha particle activity, 15 pCi/l; radium 226 and 228 combined, 5 pCi/l.*

The microorganisms include *Giardia lamblia*, 99.9% killed and inactivated; heterotrophic plate count, no more than 500 bacterial colonies per milliliter; *Legionella*, no limit, but EPA believes that if *Giardia* and viruses are inactivated, *Legionella* can also be controlled; total coliforms including *E. coli* and fecal coliform, no more than 5.0% of samples of total coliform can be positive in a month (see Total Coliform Sampling Requirements, Table 3.3); turbidity, 5 nephelolometric turbidity units (NTU) maximum; viruses, 99.99% killed and inactivated.

Secondary drinking water regulations include aluminum, 0.05 to 0.2 mg/l; chlo-ride, 250 mg/l; color, 15 color units; copper, 1.0 mg/l; corrosivity, noncorrosive; fluoride, 2.0 mg/l; foaming agents, 0.5 mg/l; odor, 3 threshold odor number; pH, 6.5 to 8.5; silver, 0.10 mg/l; sulfate, 250 mg/l; total dissolved solids, 500 mg/l; zinc, 5 mg/l.

The source of the previous regulations, dated in the year 2000, is the Current Drinking Water Standards from the National Primary and Secondary Drinking Water Regulations, Office of Water, Office of Groundwater and Drinking Water, U.S. EPA. Because the standards are continuously updated, it is essential to contact the EPA periodically to ascertain if changes have been made in any of the regulations.

---

* The final effective date of the rule will be August of the year 2003, and if the multimedia mitigation approach is used, it will be August 2005. The multimedia mitigation program uses an alternate maximum contaminant level for radon in drinking water, which allows the states to reduce exposure to radon in indoor air by an amount at least equal to the reduction that would have been achieved by meeting the maximum contaminant level. The cost of the option would be lower.

Table 3.3   Total Coliform Sampling Requirements According
to Population Served

Population Served	Minimum No. of Routine Samples Per Month[a]	Population Served	Minimum No. of Routine Samples Per Month
25–1000[b]	1[c]	59,001–70,000	70
1001–2500	2	70,001–83,000	80
2501–3300	3	83,001–96,000	90
3301–4100	4	96,001–130,000	100
4101–4900	5	130,001–220,000	120
4901–5800	6	220,001–320,000	150
5801–6700	7	320,001–450,000	180
6701–7600	8	450,001–600,000	210
7601–8500	9	600,001–780,000	240
8501–12,900	10	780,001–970,000	270
12,901–17,200	15	970,001–1,230,000	300
17,201–21,500	20	1,230,001–1,520,000	330
21,501–25,000	25	1,520,001–1,850,000	360
25,001–33,000	30	1,850,001–2,270,000	390
33,001–41,000	40	2,270,001–3,020,000	420
41,001–50,000	50	3,020,001–3,060,000	450
50,001–59,000	60	3,960,001 or more	480

[a] In lieu of the frequency specified, a Non-Community Water System (NCWS) using ground water and serving 1000 persons or fewer may monitor at a lesser frequency specified by the state until a sanitary survey is conducted and reviewed by the state. Thereafter, NCWSs using ground water and serving 1000 persons or fewer must monitor in each calendar quarter during which the system provides water to the public, unless the state determines that some other frequency is more appropriate and notifies the system (in writing). Five years after promulgation, NCWSs using ground water and serving 1000 persons or fewer must monitor at least once/year.

A NCWS using surface water, or ground water under the direct influence of surface water, regardless of the number of persons served, must monitor at the same frequency as a like-sized Community Water System (CWS). A NCWS using ground water and serving more than 1000 persons during any month must monitor at the same frequency as a like-sized CWS, except that the state may reduce the monitoring frequency for any month the system serves 1000 persons or fewer.

[b] Includes public water systems which have at least 15 service connections, but serve fewer than 25 persons.

[c] For a CWS serving 25–1000 persons, the state may reduce this sampling frequency, if a sanitary survey conducted in the last five years indicates that the water system is supplied solely by a protected ground-water source and is free of sanitary defects. However, in no case may the state reduce the frequency to less than once/quarter.

Source: Criteria and Standards Division Office of Drinking Water, U.S. Environmental Protection Agency, Drinking Water Regulations under the Safe Drinking Water Act, Washington, D.C., 1990, p. 220.

Plumbing

The subject of plumbing is discussed in Chapter 5. Because plumbing is applicable to all areas of water and sewage, it is a duplication of efforts to discuss it in each of the chapters unless some special item arises.

Individual Water Supply Systems

Dug or Bored Wells

Dug and bored wells are usually shallow and hand excavated. They are difficult to maintain, because groundwater seems to be frequently contaminated with surface water and its source is so close to the surface that many impurities percolate into this type of aquifer. The dug well permits the flow of water in relatively large quantities. The walls of the dug well are usually curbed or lined with rocks, bricks, wood, or concrete in an attempt to prevent surface water from entering and to prevent the well from caving in. Whenever possible, other types of wells should be used because of the possibilities of contamination. Bored wells are usually constructed by using earth augers that are turned either by hand or by power equipment. Such wells have a practical depth of 100 ft, whereas the dug well has a practical depth of approximately 25 ft. The bored wells may have a diameter of 2 to 30 in. depending on the type of earth that the auger is going through and the necessities of larger diameters in the bored well.

Problems characteristic of dug wells are also characteristic of bored wells. Bored wells are cased with vitrified tile, concrete pipe, iron, or steel casing. In either case, there should be a minimum of exterior grouting or casing between the well and the soil. The grouting goes down a minimum of 25 ft or until it reaches impervious material. The casing should also extend 12 in. above the ground surface and should be at least 6 in. thick (Figure 3.5).

Driven Wells

Driven wells are constructed by driving a point, which has been fitted with a series of successive pipe sections, into the ground. The drive point is usually made of forged or cast steel. Drive points vary from 1.25 to 2 in. in diameter. The well is driven usually when water can be reached at depths of about 50 to 60 ft and if the water is 25 ft or more below the surface. The value of the driven well is that it goes through a variety of porous soils, such as alluvial deposits of high permeability. The presence of coarse gravel or boulders may interfere with the sinking of the well point and may damage the wire mesh jacket. Well drive points are available in a variety of designs and materials. The drive point may be driven down by the use of a sledge hammer or a special driving device. An additional value to the driven well is that it prevents more of the possible surface contamination from getting into the well, because it is one unit driven down through the ground. However, the area around the driven well must still be protected. The yield of water is usually moderate.

A variation of the driven well is the jetted well. This is a means of sinking well points by jetting or washing in. A source of water and a pressure pump to start the operation are required. The water is forced under pressure down the riser pipe and emerges from a special washing point. The well point and pipe are then lowered as material is loosened by the jetting. The pipe of the jetted well is often used as the suction pipe for the pump. The problem of the jetted well, even though its installation is more

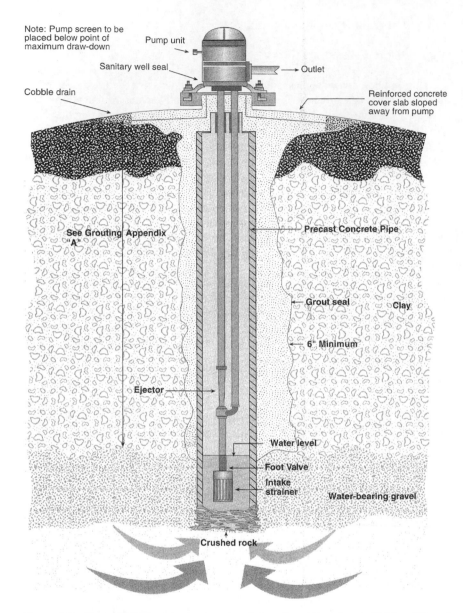

Note: Pump screen to be placed below point of maximum draw-down

Pump unit

Sanitary well seal

Outlet

Cobble drain

Reinforced concrete cover slab sloped away from pump

See Grouting Appendix "A"

Precast Concrete Pipe

Grout seal

Clay

6" Minimum

Ejector

Water level

Foot Valve

Intake strainer

Water-bearing gravel

Crushed rock

Figure 3.5   Dug well with two-pipe jet pump installation. (From Water Supply Division, Office of Water Programs, U.S. Environmental Protection Agency, *Manual of Individual Water Supply Systems,* U.S. Government Printing Office, Washington, D.C., 1973, p. 34.)

rapid, is that surface water may contaminate the underground water supply; special precautions should be taken to avoid this. The space between the well casing and the ground must be filled with cement grouting down to a distance of 25 ft (Figure 3.6).

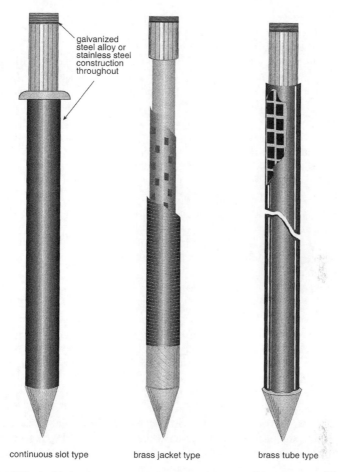

Figure 3.6   Different kinds of drive well points. (From Water Supply Division, Office of Water Programs, U.S. Environmental Protection Agency, *Manual of Individual Water Supply Systems,* U.S. Government Printing Office, Washington, D.C., 1973, p. 35.)

Drilled Wells

Drilled wells are the must efficient and best type of wells used. The drilled well is constructed by the use of rotary hydraulic drilling or percussion drilling. The method used depends on the geology of the site and the kinds of equipment available. In hydraulic drilling, the drilling equipment, which consists of a derrick and a hoist, penetrates down through the ground with a drilling bit at the end of the pipe. The bit breaks up the material as it rotates and advances. The drilling fluid, which is pumped down the drill pipe, picks up the cuttings and carries them up the space between the rotating pipe and the wall of the hole. The mixture of mud and cuttings is discharged to a settling pit, where the cuttings drop to the bottom and the fluid is recirculated through the drill pipe. When the hole is completed, the drill pipe is withdrawn and the casing is enplaced. The casing has a screen at the end of it. The space between the casing and the ground is then grouted to a level of at least 25 ft below the surface or to a solid formation. Grouting may extend much farther down, depending on the type of soil encountered initially. The air rotary drilling method, which is a variation on the hydraulic drill method, is the same as the hydraulic system, except that air is used instead of mud or water. The air system allows for rapid penetration through the ground. The penetration rate ranges from 20 to 30 ft/hr in very hard rock.

The down-hole pneumatic hammer combines the percussion effect of drilling, cable tool drilling, and rotary movement of rotary drilling. The tool bit is equipped with a tungsten–carbide insert at the cutting surfaces, which makes it very resistant to abrasion. Drilled wells are installed when greater volume, diameter, and depth are needed in a well. The casing or pipe is driven down as the well is drilled and extends either to water-bearing sand or to rock. When the well goes down to the rock, the casing or pipe is driven down until the rock is reached and the drilling continues into the rock and to the water supply. Drilled wells extend hundreds of feet down into the ground.

Location of Wells

The location of the well is extremely important. It should be located so that it is not subject to flooding or to drainage from sources of pollution. It should be fenced in to keep livestock away, situated on higher ground than sources of pollution, and located in such a way that pollutants cannot flow into the well. Wells should be located the following minimum distances from the following sources of potential pollution: cast iron sewer pipe, 10 ft; sewer and drains constructed of less leak resistant materials, 50 ft; septic tanks and privies, 50 ft; tile fields or absorption fields and seepage pits that are not over 3 ft deep, 100 ft; barns, stables, and manure piles, 50 ft; lakes, streams, ponds, and ditches, 100 ft; property lines (dug or bored wells), 25 ft; property lines (drilled or driven wells), 15 ft; houses or other dwellings, 3 ft.

Well Construction

There are numerous potential problems related to wells. In the construction of the well, major points of concern that should be taken into consideration are the sources of well water; a good location; protection near the top of the well, which consists of eliminating low spots where standing water can accumulate or drain into the well; protection below the ground, which consists of a casing of at least 25 ft below the surface or 10 ft into rock; curb and platform construction at the top of the well, which consists of a platform with a slope of 0.25 in./ft from the casing to the outer edge and a curbing so that water cannot run under the platform, at least a 5-ft diameter with 2.5 ft of concrete platform extending from the well, and 6 in. thick construction composed of reinforced concrete.

The well basically consists of a hole that has been made into the ground, a series of pipes that are inserted through the hole, a screen at the end of the pipes to keep sand and silt from coming up through the well and clogging it, a grouting around the pipe to keep surface water from entering the underground water supply, a platform above the well to protect the well, and a pump that is used for moving the water from the underground supply.

Well construction is based on the type of geologic formation that the well is going to go through. Test wells are utilized when it is necessary to determine in unconsolidated formations, such as gravel and sand formations, the proper depth of the permanent well. The results of the yield and quality of water are analyzed before a permanent well is sunk. In the unconsolidated formation, the well may have several different forms. However, in all cases where the unconsolidated formation extends the full depth of the well, a permanent casing should be installed down to the pumping level in the finished well. The pumping level is the point to which the water is lowered for water usage any time during pumping. Under no conditions should there be less than 20 ft of permanent casing utilized. (Table 3.4 lists suitability of wells by geologic conditions.)

Table 3.4 Suitability of Well Construction Methods to Different Geologic Conditions

Characteristics	Dug	Bored	Driven	Percussion	Drilled Rotary Hydraulic	Air	Jetted
Range of practical depths (general order of magnitude)	0–50 ft	0–100 ft	0–50 ft	0–1000 ft	0–1000 ft	0–750 ft	0–100 ft
Diameter	3–20 ft	2–30 in.	1¼–2 in.	4–18 in.	4–24 in.	4–10 in.	2–12 in.
Type of geologic formation							
Clay	Yes	Yes	Yes	Yes	Yes	No	Yes
Silt	Yes	Yes	Yes	Yes	Yes	No	Yes
Sand	Yes	Yes	Yes	Yes	Yes	No	Yes
Gravel	Yes	Yes	Fine	Yes	Yes	No	¼ in. pea gravel
Cemented gravel	Yes	No	No	Yes	Yes	No	No
Boulders	Yes	Yes, if less than well diameter	No	Yes, when in bedding	(Difficult)	No	No
Sandstone	Yes, if soft and/or fractured	Yes, if soft and/or fractured	Thin layers only	Yes	Yes	Yes	No
Limestone			No	Yes	Yes	Yes	No
Dense igneous rock	No	No	No	Yes	Yes	Yes	No

Note: The ranges of values in this table are based on general conditions. They may be exceeded for specific areas or conditions.

From Water Supply Division, Office of Water Programs, U.S. Environmental Protection Agency, *Manual of Individual Water Supply Systems,* U.S. Government Printing Office, Washington, D.C., 1973, p. 31.

When wells are installed in limestone or creviced formations, which are also called consolidated formations, the well should go through at least 30 ft of solid rock and the rock should extend to a radius of 0.25 mi. If these criteria cannot be met, a chlorinator or other means of disinfection should be used. The casing should extend at least 40 ft below ground level when the covering rock is less than 30 ft thick. The diameter of the drilled hole should be a minimum of 2 in. greater than the outer diameter of the casing. The space between the ground and the casing should be filled with cement grouting under pressure.

In the case of artesian wells, the well must be so constructed that the water flow can be controlled. The protective casing and outer grouting should be installed and allowed to set or harden before the drill hole is extended into the water-bearing formation. The inner casing should be watertight and connected to the protective casing.

All wells, with the exception of those that go into artesian or deep rock formations, need some type of screening or slotting to keep sand and gravel from getting into the well and the water supply. The types of screens vary considerably with the companies producing them. The screen must be structurally strong and corrosion resistant. Gravel-packed wells are built by placing gravel between the outside of the well screen and the face of the undisturbed water-bearing formation. This is particularly useful when dealing with material of uniform grain size.

The joints of pipes must be welded or threaded so that the seal is watertight. The well casing must be strong and water resistant. All well vent openings should be piped in a watertight manner to a point at least 24 in. above any known floodwater level and at least 12 in. above the surface of the ground. The openings should be a minimum of 0.25 in. in diameter, but sufficient in size to avoid clogging. In the event that the vent gases are toxic or flammable, they must be maintained in such a way that they cannot produce a hazard. Well cementing or grouting is utilized not only to keep surface water from entering the underground water supply but also to prevent corrosive water from attacking the casing.

One of the most critical zones for corrosion is the zone of aeration, or the unsaturated region above the water table. The combined action of atmospheric oxygen and organic acids in soil accelerates the rate of corrosion. If pumping the well affects the water table, then the lowest anticipated drawdown level of the water table must become the point below which the grouting extends. This should be a cemented area of at least 10 ft below anticipated drawdown. It is necessary to line up all the pipes and the well carefully for the well to work in an adequate manner. Therefore, care should be taken in the well alignment.

Well development means the process of moving silt and fine sand from the formation adjacent to the well screen. This can be accomplished either by surging water rapidly and reversing water flow at the bottom of the well, by using a plunger; or by high-velocity hydraulic jetting, where water under pressure ejected from orifices passes through the slot openings and violently agitates the aquifer material. The sand grains that are finer than the slot size move through the screen and either settle to the bottom of the well or are washed out of the top through a well overflow. If they are at the bottom, they can be removed. Conventional, centrifugal, or piston pumps can be used for this purpose. Pressures of at least 100 lb/in.[2] and preferably

150 lb/in.^{2}, are utilized. The ultimate objective is to get the fine sand and silt away from the well screen by either removing them at the surface or having them settle down to the bottom of the formation. This allows for proper and adequate water removal from the well without danger of clogging (Figure 3.7 illustrates complete well construction and pump house).

Well Pumps

Pumps are used in individual water supply systems to move the water from the aquifer to the point of need. The usual pumps include positive displacement, centrifugal, and jet. The total *head* is a combination of the lift, which is the vertical distance in feet that the water is pumped to the water surface, and the elevation, which is the vertical distance in feet from the pump to the highest point of water delivery. The friction loss within the piping system depends on rate of flow, length, size, type of pipe, number and type of fittings, various corrosion factors, and pressure desired at the outlet. Each pound of pressure, that is, each pound per square inch, will raise water 2.3 ft. When calculating the needed pressure, it is necessary to take into account that on an average a loss of 0.434 lb/in.^{2} for each foot of height above the pressure tank results from pressure losses due to pipe friction. To find the total head in a water system, it is necessary to do the following: determine the amount of lift, determine the difference in elevation, make a diagram of the water main and branch pipelines, determine the flow rates needed at each of the outlets, select the proper pipe sizes for both the main and branch lines, determine the friction loss within the pipes and fittings, determine the pressure needed, and determine the total head.

The positive displacement pump forces or displaces the water through a pumping mechanism. Various types of positive displacement pumps are available. The pump consists of a mechanical device that moves a plunger back and forth in a closely fitting cylinder. The plunger is driven by a power source, and the power motion is then converted from a rotating action to a reciprocating motion by the combined work of a speed reducer crank and a connecting rod. The cylinder, which is composed of a cylinder wall, plunger, and check valve, is located near or below the static water level to eliminate the possibility of having to prime the pump. The actual pumping begins when the water enters the cylinder through a check valve. When the piston moves, the check valve closes; this forces the water through a check valve to a plunger, where the water is eventually forced, stroke by stroke, to the surface and into the discharge pipe.

Another type of positive displacement pump is the helical or spiral rotor. In this pump, the rotor consists of a shaft with a spiral surface that rotates in a rubber sleeve. As the shaft turns, it pockets or traps the water between the shaft and the sleeve, and forces it upward. Other types of positive displacement pumps include the regenerative turbine pump that has a rotating wheel or propeller.

Centrifugal pumps contain a rotating impeller on a shaft that is turned by a power source. The rotating impeller increases the velocity of the water and discharges it into the surrounding casing, which slows down the flow of the water and converts velocity to pressure. Turbine pumps are vertical-drive pumps that have one or more

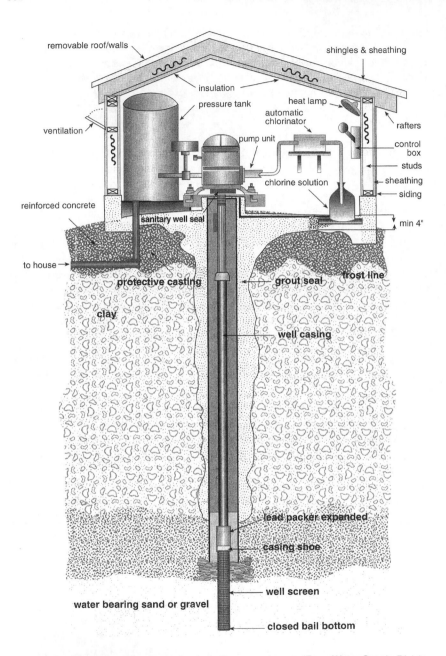

Figure 3.7   Complete well construction including pump house. (From Water Supply Division, Office of Water Programs, U.S. Environmental Protection Agency, *Manual of Individual Water Supply Systems,* U.S. Government Printing Office, Washington, D.C., 1973, p. 108.)

stages where a vertical shaft connects the pumping assembly to a drive mechanism located above the pumping assembly. The discharge casing, pump housing, and inlet screens are suspended from the pump basin at the ground surface.

Submersible pumps are centrifugal pumps driven by a close coupled electrical motor constructed for submerged operation as a single unit. Electrical wiring is waterproofed and the electrical control must be properly grounded. The pump and motor assembly are supported by the discharge pipe. Jet pumps are a combination of centrifugal and ejector pumps. A part of the water that is discharged from the centrifugal pump is diverted through a nozzle and the Venturi tube. A pressure zone lower than that of the surrounding area exists in the Venturi tube, which therefore causes water from the well to flow into the area of reduced pressure. The velocity of the water from the nozzle pushes it through the pipe toward the surface, where the centrifugal pump can complete the work by lifting the water by suction. The centrifugal pump then forces the water into the distribution system (Figure 3.8).

Submersible multistaged pumps have a practical lift of up to 1000 ft. They produce a smooth and even flow of water but, as a disadvantage, must be pulled from the well when repairs are needed on the pump or motor. This pump is easily damaged by sandy water. The system works by a series of impellers. It usually requires at least a 4-in. casing. The jet or ejector pump has a practical lift of up to 22 ft in a shallow well and 85 ft in a deep well. Their advantages are that they have few moving parts and produce high capacity at low heads. Their disadvantages are that they are readily damaged by sandy water and that the amount of water returned to the injector has to increase with the amount of increased lift. Of the total water, 50% is pumped at a 50-ft lift.

The centrifugal pump used in a shallow well has a practical lift of up to 15 ft. The advantages of this pump are that it produces a smooth, even flow of water and also requires service infrequently. The disadvantages are that the pump loses its prime easily and its efficiency depends on operating under the designed heads and speeds. It is very efficient for capacities exceeding 50 gal/min and pressures less than 65 lb/in.^{2}.

The reciprocating or piston pump in shallow wells has a practical lift of up to 22 ft. Its advantages are that it can pump water that has small amounts of sand in it and it can be installed over small-diameter wells. Its disadvantages are that it has a pulsating discharge and can cause vibration and noise. The pump contains a piston that is driven within a chamber and develops a vacuum. The water fills the vacuum and is then forced into the water system. In a deep well, the reciprocating or piston pump can pump up to a practical lift of 600 ft. It has the same advantage as the shallow well type of reciprocating or piston pump, and the same disadvantages.

The deep-well turbine multistaged pump has a practical lifting power of up to 1500 ft. It operates in the same manner as a centrifugal pump, except more impellers are mounted close together on a vertical shaft. The advantages are that the pump produces a smooth and even flow, is easy to frostproof, and is in a straight and vertical well casing. The disadvantage is that the pump must be pulled from the well for repair.

The submersible or helical rotor pump has a practical lift of up to 1000 ft. The advantages of this pump are that it produces a smooth and even flow of water, is easy

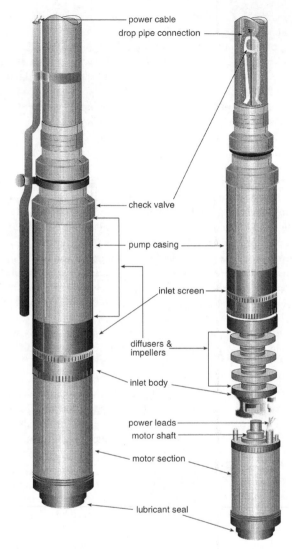

Figure 3.8 Exploded view of a submersible pump. (From Water Supply Division, Office of Water Programs, U.S. Environmental Protection Agency, *Manual of Individual Water Supply Systems,* U.S. Government Printing Office, Washington, D.C., 1973, p. 95.)

to frostproof, and pumps sand with less pump damage than any other type of pump. The disadvantage is that the pump or motor must be pulled from the pump for repair.

The pitless well adapter is an assembly of parts that permits the water to pass through the wall of the well casing or extension, and provides access to the well and the parts of the water system within the well. It also provides transportation of the water and the protection of the well and water from surface water or surface contamination.

The pitless well adapter has eliminated the need for the well pit, which historically has been the place where the well water became contaminated. It is a specially designed connection between the underground horizontal discharge pipe and the vertical casing pipe or the watertight casing. It extends 8 in. or more above ground level, which is a safe height for prevention of pollutants. The underground section of this discharge pipe is permanently installed, and it is not necessary to move it when repairing the pump or cleaning the well. Numerous varieties and models of pitless adaptors and units are available. To determine the most acceptable model, the environmental health practitioner should consult the codes of the state and local health department and also the National Sanitation Foundation criteria (Figure 3.9).

The design of the pump head is very important, because there is an extreme necessity to prevent pollution of the water supply by lubricants, maintenance materials, dusts, grain, birds, flies, or animals. In addition, the pump base must be set up in such a way as to facilitate the installation of a sanitary well seal within the well cover. The installation of the pumping portion of the assembly must be below the static water level to avoid priming. The system must be designed for frost protection, and the overall design of the pump must be such that necessary maintenance and repair can be carried out readily. The only check valve should be between the pump and storage. It should be located within the well. This ensures that the discharge line at any portion in contact with the soil can remain under positive system pressure, whether the pump is or is not operating. No check valve should be at the pressure tank.

When selecting a pump for a given well, it is necessary to determine the yield of the well or of the water source, daily needs and high frequency demand, size of the pressure or storage tank, size and alignment of the well casing, total operating head pressure of the pump during normal delivery, difference in elevation between ground level and water level in the well during the pumping process at the lowest well level (dry season), availability and type of power to be used, availability of replacement parts for the pump and the ease of maintenance, initial cost and economy of the operation, reliability of the equipment, and availability of maintenance workers.

Storage

Storage of water should occur in pressure tanks. The pressure develops by forcing the water into the tank until the air in the tank is compressed to a preset pressure. When the outlet is opened, the pressure forces the water out. Obviously, as the water level in the tank drops, the pressure drops. As the pressure reaches a low preset value, the pump starts to work and forces more water into the pressure tank. Pressure tanks have pressure control switches that stop or start the pump at predetermined tank pressures. The switch can be adjusted to determine the cut-in or cut-out pressure. The volume of the tank depends on the well yield, pump capacity, maximum water requirements, and operating pressures. Most tanks usually only use 10 to 30% of their volume for storage; the rest is used to develop pressure for forcing the water. Usually a pressure tank should be ten times the pumping rate in gallons per minute. Pressure tanks need air controls to maintain adequate amounts of air within the water system. Water may be pumped into a water tower that is above the water usage area. If so, the water then flows by gravity to the area of usage.

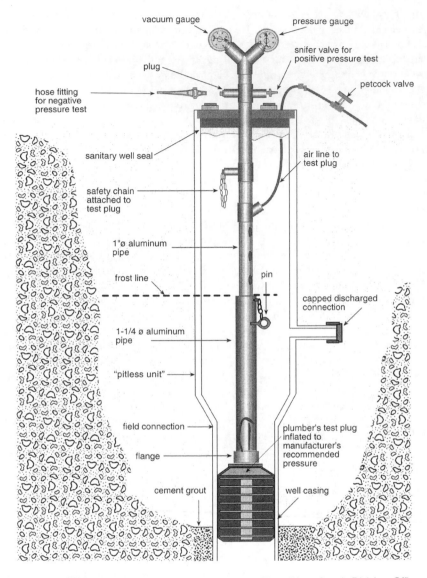

Figure 3.9   Pitless adapter and unit testing equipment. (From Water Supply Division, Office of Water Programs, U.S. Environmental Protection Agency, *Manual of Individual Water Supply Systems,* U.S. Government Printing Office, Washington, D.C., 1973, p. 117.)

Well Testing

Wells are tested for yield and drawdown and also for the installation of suitable pumping equipment. A pumping test includes determination of the volume of water pumped per minute or per hour, depth to the pumping level at the completion of several pumpings, recovery of the water level after all pumping ceases, and length of time the well is pumped at each rate during the test. Because water levels differ according to the season of the year, especially in wells that draw from water tables instead of artesian sources, it is preferable that tests are conducted at the end of the dry season. A test at this time produces more accurate volume and quality ratings for the water source. The individual well should be test pumped at a constant pumping rate for at least 24 hours. If the water level does not return within a 24-hour period, the water bearing formation should not be listed as dependable.

The static level in a well is the point that the water reaches below the ground surface. The pumping level is the point to which the water drops in the well during the pumping process. The drawdown is the difference between the static level and the pumping level. A serious problem occurs when several wells utilize the same water table and the draw-down is considerable. As a result, water is pulled from areas farther and farther away from the well site, and water that ordinarily is not contaminated may become contaminated. The well yield is the volume of water per unit of time, usually rated in gallons per minute or gallons per hour. This is sometimes referred to as the pumping rate. The specific capacity is the well yield per unit of drawdown. This is usually expressed in gallons per minute per foot of drawdown.

Well Disinfection

All newly constructed wells must be disinfected because equipment, material, and surface drainage that is introduced during the construction phase of the well can readily cause contamination. An effective technique is the addition of 70% available chlorine, which is found in calcium hypochlorite. The chemical can be purchased in either tablet or granular form from chemical supply houses or swimming pool supply houses. The calcium hypochlorite should be reduced to a dosage of 100 mg/l of available chlorine in the well water (100 ppm). Although 50 ppm may be required as a minimum by some codes, 100 ppm is recommended for safety. This is roughly equivalent to mixing 2 oz of the 70% hypochlorite with 2 qt of water 10 to 15 min prior to use and then pouring the clear liquid containing the chlorine into 100 gal of water. Other sources of chlorine include sodium hypochlorite, which varies from 12 to 15% by volume, and household bleach that is 5.25% available chlorine. Also 2 qt of household bleach can be added to 100 gal of water. It is important that the chlorine solution remain in the well for at least 24 hours. After 24 hours, the well should be pumped until all traces of chlorine are flushed out.

Pond or Lakes

The pond or lake should be at least 8 ft deep and have a minimum depth of water storage of 3 ft. It must be fenced against livestock and kept free of weeds, algae, and floating debris. The pond or lake intake must draw water of the highest possible quality, and should be located between 12 and 18 in. below the water surface. The intake should take the water to an area where it undergoes proper water treatment. Water treatment consists of the use of settling basin, filtration unit, clear water storage, and disinfection. In the settling basin, large particles of turbidity settle.

The effluent may then be treated with alum to further settle suspended material. pH adjustment may be necessary, because the pH is lowered by the alum. The filtration system consists of slow sand filters, diatomaceous earth filters, or rapid sand filters. These forms of filtration are discussed in this chapter. Clean water storage occurs in a special storage area after the water has been filtered. Disinfection is most important prior to the time that the water is used for livestock or human consumption. It is essential that adequate time is permitted for proper contact between the chlorine and the water prior to use. This subject is discussed in detail in this chapter. The remainder of the water treatment process for removal of taste and odors, iron and manganese, water weeds, and so forth has also been discussed previously.

Springs

The springhouse or tank must be constructed of impermeable materials, such as concrete or impermeable clay. The tank or house should be aerated but screened to prevent the entrance of birds, rodents, and so forth. It should be locked to keep children from entering the springwater storage area. The excess drain pipe must be above normal ground or flood levels by at least 6 in. and the discharge end must be screened. All springs should be protected against barnyard waste, sewers, septic tanks, or cesspools. If the spring comes from a limestone formation, it is easily contaminated through sinkholes and large openings in the ground. If the water passes through the tubular channels in glacial drift, it also becomes contaminated by groundwater that travels for long distances. All springwater should have the following precautions installed as part of the water system: removal of all surface drainage from the site by the use of ditches, construction of fences to keep livestock away from the uphill and immediate areas of the water source, provision of access to the tank or springhouse for maintenance and cleaning, and monitoring of the quality of the spring water on a periodic basis. If the water becomes turbid after a rainstorm, this is an indication that surface runoff is leaching into the spring. Springwater should definitely be disinfected by chlorine (Figure 3.10).

Cistern

The location of the cistern is determined by convenience and quality protection. However, in no case should it be closer than 50 ft to any sewage disposal system or should the sewage disposal system be on higher ground than the cistern. If the water is collected from roof surfaces, the cistern should be adjacent to the building, but not in a basement that can flood. Cisterns may be placed beneath the ground for protection against freezing. The size of the cistern depends on the size of the family and the length of time between heavy rainfalls. Usually the size of a cistern is based on two thirds of the amount of annual rainfall, which then is utilized as a potable water source. Cisterns should be composed of concrete with watertight construction. Manholes should have a watertight curb with edges that project at least 4 in. above the level of the surrounding surface. The cover should overlap the curb by a minimum of 2 in. A pipe leads from the water source to the cistern. All pipes, including the inlet pipe and the outlet and waste pipes, must be screened against the entrance of

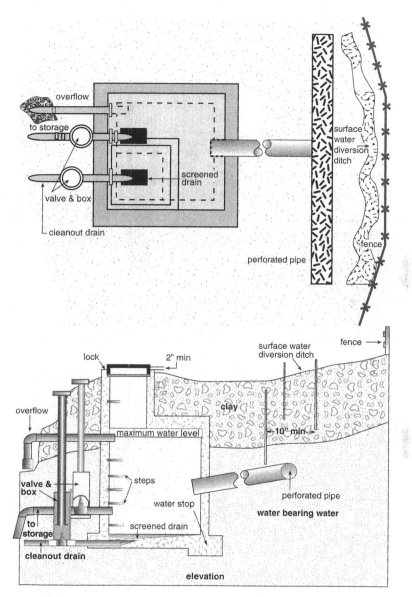

Figure 3.10 Spring protection. (From Water Supply Division, Office of Water Programs, U.S. Environmental Protection Agency, *Manual of Individual Water Supply Systems,* U.S. Government Printing Office, Washington, D.C., 1973, p. 57.)

insects, rodents, and birds. The water travels through a filtration system that is a slow sand filter, a rapid sand filter, or a diatomaceous earth filter and is chlorinated before use (Figure 3.11).

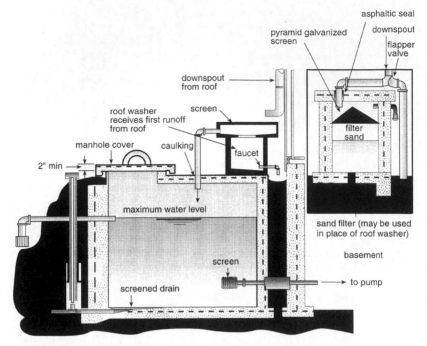

Figure 3.11 Cistern. (From Water Supply Division, Office of Water Programs, U.S. Environmental Protection Agency, *Manual of Individual Water Supply Systems,* U.S. Government Printing Office, Washington, D.C., 1973, p. 65.)

Infiltration Gallery

The gallery is constructed by building an underdrained sand filter trench parallel to the stream bed about 10 ft from the high water mark. The sand filter is located in a trench with a minimum width of 30 in., a depth of about 10 ft, and perforated or open joint tile laid in a bed of gravel about 12 in. in thickness. About 4 in. of graded gravel is across the top of the tile to support the filtering material at the bottom of the trench. The embedded tile is covered with clean, coarse sand to a minimum depth of 24 in., and the rest of the trench is backfilled with impervious material. The collection tile carries the water into a watertight concrete basin, where it is diverted or pumped to the distribution system following chlorination. Occasionally, where debris and turbidity are a problem, a modified infiltration gallery, which is a combination of a slow sand filter and an infiltration gallery, is used in the stream bed. A dam is placed across a stream to slow the water. The filter is installed in the pool behind the dam by laying perforated pipe in a bed of graded gravel that is covered by at least 24 in. of clean coarse sand. Despite the fact that infiltration galleries are utilized only in uninhabited areas, pathogenic bacteria, in addition to soil bacteria, are found in the water. It is important that this water is also chlorinated before use.

Water Vending Machines

The water vending industry sales are over 84 million a year and growing at a rate of 7% each year. The machines are designed to reduce odors, tastes, and dissolved solids. Activated carbon filters are utilized. The water source must be potable and the final product must be safe.

Emergency Disinfection

In the event of emergency water contamination, purification is achieved by heat or certain chemicals. The technique involves straining the water through a clean cloth into a container and then boiling it for at least 10 min prior to drinking. The boiled water can then be poured back and forth to reaerate and improve its taste. If boiling is not possible, strain the water carefully and then use either chlorine or iodine. Liquid chlorine is obtained from ordinary household laundry bleach, which is 5.25% available chlorine. Two drops should be added to 1 qt of clear water after straining, and 4 drops to 1 qt of cloudy water after straining. Mix thoroughly and allow to stand for 30 min. If the chlorine odor is still detectable, allow it to stand for 15 more minutes. Tincture of iodine, which is 2% iodine, may be added at a concentration of 5 drops to 1 qt of clear water or 10 drops to 1 qt of cloudy water. Allow to stand for 30 min prior to use. Iodine or chlorine tablets are also available from stores.

Slow Sand Filters

The slow sand filter is effectively used to treat certain types of relatively clean water. The filter is preferably covered. It operates normally at a level of 4 million gal/acre/day. Sudden increases in the filtration rate affect the operation of the slow sand filter and cause problems in the final effluent. Several units should be available because the unit must be serviced every 20 to 60 days. In the use of this filter, the water is frequently pretreated by sedimentation and coagulation before it reaches the filter. The filter operates on the basis of a combination of mechanical removal of materials, including bacteria, and also by the biological principle. The amount of chlorine used is based on the needs in the clear well storage system. The filter is composed of an upper portion of 2 to 4 ft of graded sand, a lower portion of 12 in. of graded gravel from a size of 2 in. in diameter at the bottom to $1/16$ in. at the top of the 12 in., and a system of tile underlying the gravel. The effective size of the sand is 0.2 to 0.4 mm.

Rapid Sand Filters

The usual water treatment process used in rapid sand filtration (Figure 3.12) consists of the following steps: the raw water enters a racked screen or bar screen, which removes all heavy material, grit, and sand. A flash mix of a coagulant is utilized for a period of 1 min. (If appropriate coagulants are not used, *Giardia* cysts can pass through the filter.) Flocculation occurs in the next tank for a period of 20 to 40 min. The tanks have slow moving paddles to keep the water gently in motion.

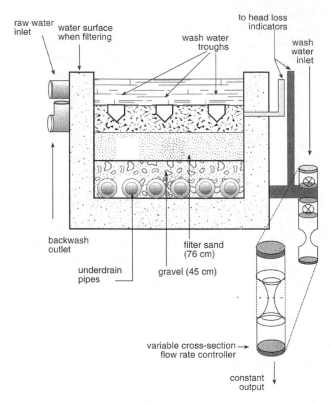

Figure 3.12 Rapid sand filter.

The effluent flows into a sedimentation basin, where it is detained for a period of 2 to 4 hours. The sludge from the basin is removed to a sludge drying and disposal area. The sludge is composed of settling floc containing the coagulant and impurities. The effluent flows from the sedimentation basin over a series of rapid sand filters. The rapid sand filter, also known as a rapid granular filter, is preferably of the gravity type and should be open to permit ready and continuous inspection. The depth, effective size, and uniformity of the materials are determined by the desired yield and the filter efficiency. In general, the rapid sand filter is operated at a rate of 1.5 to 3.0 gal/ft^2/min. They are usually backwashed and cleaned every 12 to 40 hours. The backwashing is determined by the head loss that occurs. If the water is of a higher quality, the filtration rate may also be higher. The rate of application is also expressed at a level of 125 to 190 million gal/acre/day. The rapid sand filter operates approximately 60 times as fast and gives about 60 times as much water as the slow sand filter. It operates basically on a mechanical process, instead of the mechanical and biological process of the slow sand filter. The effluent from the sand filters flows into a clear well where the water is disinfected. From the clear well the water moves into the distribution system. Prior to distribution it may be necessary to adjust the pH, if the water is too acidic.

Alternate Systems

A diatomaceous earth filter is used in place of the sand filters in the system just discussed. Another system is the aerated sand filter, in which the effluent from the sedimentation basins is sprinkled through mechanical devices over the sand filters to allow air to penetrate the water oxidizing any organic materials present.

Chlorination Equipment

Chlorination equipment must be selected, installed, and operated to achieve a continuous and effective disinfection under all conceivable conditions. Standby units must be ready to go into operation immediately if the original units break down. The determination of chlorine dosages is based on the quality of the water. The operator should consult with the various health or environmental protection agencies to get proper instruction in adequate chlorination. Adequate supplies including extra supplies in case of an emergency to satisfy at least a 30-day demand should be available at the water treatment plant. Free residual chlorine must be provided in the water system. Regular and frequent tests must be made, depending on the size of the unit, to determine whether proper chlorination is occurring. A minimum of 0.4 ppm or mg/l, should be available in the distribution system. If chloramines are used, the residual should be 1.0 to 2.0 ppm or mg/l. The routine samples should not only be collected at the water treatment plant but also at representative points in the water distribution system. The contact time of chlorine with water should be at least 15 min. Tests at the minimum contact time should be made on an hourly basis. If the contact time is several hours, then the tests should be run at least once every 8 hours. All data concerning the disinfecting process must be recorded as follows: rate of flow and volume of water treated per unit time (this should be a continuous basis); gross weight of chlorine cylinders or containers in use and the weight at the end of a preselected time, usually 24 hours or less (this should be a continuous record); pounds of chlorine used in a selected time period of 24 hours or less; gallons of water treated in a preselected time period of 24 hours or less; applied dosage for the selected time period; chlorinator control settings; time and location of sampling; and results of the residual chlorine tests. It is highly recommended that the water have a contact period of at least 30 min at a level of 0.4 ppm of free chlorine. Free residual is based on the temperature, contact time, pH, amount of chlorine that reacts with ammonia, organic nitrogen compounds present, and other substances that form combined chlorine. An increase in the amount of chlorine past the breakpoint results in an increase in the free chlorine residual.

Water Mains

Water mains should be disinfected not only at time of installation, but also if they become contaminated, because they carry the finished potable water to the ultimate consumer. The water pipes laying on the ground before installation may readily become contaminated or the water mains may crack or be broken for a variety of reasons. Water main disinfection consists of preliminary flushing of the water main at a velocity of 3 ft/sec to scour the inside of the pipes; disinfecting, preferably by use of chlorine; applying chlorine at the point where the problem occurs; utilizing at least 50 ppm of chlorine for the disinfecting process; maintaining a contact period of a minimum of 24 hours but preferably 48 to 72 hours; and flushing of the main prior to its use, as a means of distribution of potable water.

Fluoridation

Fluorine is added to the water at a level of 1 ppm to prevent dental caries and to strengthen the brittle bones of older people. The fluoridation process is evaluated at the same time as the plant evaluation occurs. Treatment records for fluorine must include the amounts of water treated each 24 hours; of fluoride compound used, and of fluoride ion used; theoretical dosage in milligrams of fluoride ion per liter of water treated; setting on the dosing machine; amount of fluoride compound on hand; average maximum daily air temperature; and fluoride ion concentration measured on a regular basis.

MODES OF SURVEILLANCE AND EVALUATION

The environmental survey is an extremely important part of the entire planning and approval process for water supplies. Obviously water samples must be taken to determine the impurities or contaminants found in the water. However, water samples may not truly reflect the potential hazards. Therefore, it is incumbent on the individuals conducting the environmental survey to determine whether there are any possible health hazards, and to assess their present and future importance. The environmental health practitioner must have a good understanding of engineering concepts and epidemiology of waterborne diseases. Although not an engineer, the environmental health practitioner should have access to proper consulting services in the evaluation, use, and construction of the water source, usage technique, and equipment. An environmental survey of groundwater supplies should include the following items: the character of the local geology including the slope of the ground surface; nature of the soil and underlying porous strata, including whether there is clay, sand, gravel, rock, or limestone; coarseness of sand or gravel, thickness of the water-bearing stratum, and depth of the water table; location, log, and construction details of local wells in use and those abandoned must also be considered; slope of the water table as determined from other wells if possible, if not, by the slope of the ground surface; size of the drainage area that contributes to the water supply; nature, distance, and direction of local sources of various pollutants; potential contamination of the well because of flooding or surface drainage, and the methods of protecting against these problems; techniques used to protect the water supply against waste from sewage treatment, waste disposal sites, and other forms of water pollution; and well construction specifications including depth of the well, casing diameter, thickness, type of material, and distance from the surface. Consideration also should be given to the type of screen or perforation material, its diameter, construction, location, length, seal, kind of material utilized, and means of placement; type of protection to be used at the top of the well including the sanitary well seal, height of the casing above the ground floor or flood level, protection of the well vent, and protection of the well from erosion and animals; construction of the pump housing including the type of floors, drains, pumps, capacity of pumps, and drawdown of pumps when operating; availability of unsafe water supplies that might be utilized if properly treated; and type of disinfection of the water supply including equipment, means of supervision, types of test kits, and laboratory controls.

Environmental surveys should also be conducted where surface water supplies are in use on either an individual or a community basis. These surveys should include the type of surface geology including the soils and rocks; type of vegetation, forest, and cultivated and irrigated land; amount of salt present and its effect on water use for irrigation; present and projected user population for the next 20 years; kind of sewage disposal and whether the effluent enters the watershed; character and efficiency of sewage treatment; proximity of fecal pollution sources to the intake system of the water supply; location, sources, and type of industrial or mining waste that may contaminate the watershed; adequacy of the supply; wind direction, velocity, pollution drift, and the amount of sunshine for lakes or reservoir water supplies; and nature and quality of the raw water, including the amount of coliform, algae, turbidity, color, odors, and chemicals. These surveys also should include the minimal period of detention of water in reservoirs or storage basins; minimum time taken by water to flow from sources of pollution to the reservoir and through the reservoir intake system; shape of the reservoir, and currents that may be induced by wind or reservoir discharge; protective measures used on reservoirs to control fishing, boating, swimming, wading, and animal use; efficiency and schedule of policing of watersheds; treatment of water including the adequacy, effectiveness, contact period, supervision, amount of free chlorine residual, and availability of parts when needed; and types of pumping facilities including the pumphouses, pump capacity, standby units, and storage facilities.

The two recommended surveys presented in this chapter illustrate that some very serious decisions must be made concerning the source of the water, how it is contaminated, the kind of treatment that will be used, and the ultimate product, which is high quality and adequate quantities of potable drinking water.

Water Treatment Plant Surveys

The environmental survey is used in conjunction with laboratory analysis to determine the following basic data needed for submission of a report and plan for the establishment of a water treatment system:

1. Field and office survey of the water and its environment, from its source to the consumer
2. Physical description of the water system, including all physical features that account for supply, treatment, storage, and delivery
3. Analysis of at least a 1 year set of records of bacterial, chemical, physical, and radiological concerns
4. Analysis of operating records showing present capacity, water demands, amount of water needed to meet peak demands, and anticipated future growth for the next 20 years

5. Review of management and operation techniques, training, experience, and capabilities of the personnel
6. Review of the treatment plant, laboratory equipment, procedures, and qualifications of all personnel
7. Review of state and local regulations concerning water, waste, water pollution, and plumbing
8. Complete summary and analysis of pertinent facts relating to health hazards

All field surveys should be documented, carefully showing the location of potential hazards, present hazards, and system flow. The survey engineer should not only be a professional engineer but also should have considerable knowledge concerning water supply, water quality control, water hazards, environmental health, and potable water supplies and their sources. The survey report should include the name and owner of the supply; and descriptions of the sources and catchment areas, of the engineering plans of the storage available before and after treatment, and of the system including data about installation of existing waterworks and proposed extensions. Necessary engineering drawings should accompany this report. These environmental surveys should be submitted to the state health department or other state agency for evaluation, changes, and ultimate approval.

Mapping Program for Groundwater Supplies

The Michigan Department of Public Health has developed a project on mapping for potable water wells. The project is a joint venture of the Land Subdivision and Planning Section and the Section of Groundwater Quality Control of the Michigan Department of Public Health. The project has been shared with local environmental programs. The agency recommends that a county map be set up by townships and sections that are based primarily on well records on file. The map is modified as needed by information supplied by well drillers, experience of the environmental health staff, and outside sources, such as the USGS Division of the Department of Natural Resources. It is recommended that the basic map be updated at least semiannually by adding and altering the basic information present. The map is an indicator that can be used by the environmental health specialist to determine groundwater conditions that exist and should be investigated or at least considered prior to starting subdivisions, mobile home parks, apartment houses, motels, restaurants, campgrounds, or even individual homes.

The intent of the map is to accumulate groundwater information and to build on it in the future as additional data become available. The scale of the map is recommended to be 1 in./mi. The map should have at least townships and sections showing on it. It is also helpful to have subdivisions and lakeside development included. The primary source of data is well records in the files of the health department. Although each record should be read individually, the composite results should be put on the map. It is advisable to list the logs that were used to evaluate a particular section, and to select and mark the place on each log as it is used, to indicate it has been considered in the preparation of the basic map. It is necessary to be able to reproduce the data. The mapping is more successful if individual well drillers are contacted to gain information from their records on soil formations, and understanding of their soil interpretations and how they evaluate the soils. Prior to the installation of large-capacity wells, such as public, industrial, or irrigation wells, test wells should be installed and pumped. This information is also valuable. The USGS may also assist in the development of the map.

The map should include the colors red, yellow, and green, and the open areas should not have color. Secondary indicators would use the numbers 1 to 4. Red would indicate an evidence of problems in the section for which there are not present solutions. Yellow would indicate groundwater problems for which solutions or adjustments are known that may be made through special requirements in the construction or location of the wells. Green would indicate no apparent groundwater problems evident in the section. The open or blank area would indicate insufficient information in the area to make any meaningful evaluation.

The number 1 would indicate evidence of problems in the sections relevant to the amounts that ordinary domestic wells can yield. Wells that yield less than 8 gal/min including dry holes are low-yield indicators. Number 2 would indicate that quality problems have been found, such as coliform, nitrates, chlorides, and hydrogen sulfites (not including iron or hardness). The number 3 would relate to aquifer protection. This number indicates the existence of an aquifer that does not have the protection of at least 15 ft of impervious soils in a known continuous quantity over the aquifer. The number 4 signifies the existence, in the section, of bedrock at less than 25 ft.

In areas designated as red only a multiple water supply condition may exist. For example, conditions may exist that require the red indicator, but other problems may need to be indicated that do have a solution. Therefore, any number circled in the red section means that the problem indicated by that number is subject to solution. All mapping is usually done by section. Bedrock depths can be estimated for a township as follows:

1. An estimate of 37S would indicate 37 ft to sandstone.
2. An estimate of 50Sh indicates 50 ft to shale.
3. An estimate of 66L indicates 66 ft to limestone.

If a variation exists in one section from the rest of what occurs in the section, this can be indicated by putting an X prior to the number. For example, a section colored green X-1 shows the location of a single well with a yield (1) problem. To show the number of well records used in a section rating, use the proper number in a bracket in the extreme southeast section corner. For example, put a 22 in a bracket. This inventory can be of great value to the health department in trying to determine the type of water supply that may be available for use from groundwater sources.

In Ohio, the DRASTIC mapping system is used to evaluate the pollution potential of an area assuming a contaminant is introduced at the surface with the mobility of water and is flushed into the groundwater by precipitation. The system consists of two major elements: the designation of mappable units called hydrogeologic setting and a relative rating system for pollution potential. Hydrogeologic settings incorporate the major hydrogeologic effectors that effect and control groundwater movement such as depth to water, net recharge, aquifer media, soil media, topography, impact of the vadose zone media, and hydraulic conductivity of the aquifer.

Water Sampling

Public Supplies

All water quality tests should be made in accordance with the current edition of *Standard Methods for the Examination of Water and Waste Water*, prepared and published jointly by the American Public Health Association, American Water Works Association, and Water Pollution Control Federation (WPCF). The schedule of laboratory testing varies with the volume of water and the character of the water being treated. Tests must satisfy the bacteriological, physical, chemical, and radiological requirements of the current National Primary Drinking Water Regulations. Under ordinary circumstances, where the water quality of a small plant is good at the outset, a test should be run a minimum of once every 24 hours, 7 days a week. In a large plant or in cases where the water is of changing quality, tests should be run on an hourly basis, 24 hours a day, 7 days a week.

Private Supplies

Water samples should be taken prior to the completion of the well and should, therefore, make up part of the record of the well plan and approval. At this point it may be possible to determine if the water meets the existing maximum concentrations allowable under drinking water standards of chemicals, radiological materials, and physical and biological contaminants. Where these standards are not met, the area must be abandoned, the water table from which the sample is drawn must be abandoned for drinking water purposes, or a treatment process must be instituted to make the necessary corrections of the problems noted in the water sample. Water samples are also taken immediately after the well has been initially disinfected and after all chlorine has been removed from the water. When this water sample is taken, it is necessary once again to make chemical, biological, physical, and radiological analyses to determine whether the water qualifies as a drinking water source. If it does not, proper steps must be taken before the water can be approved.

Depending on the changing situations or emergencies, such as floods, dry wells, unusual potential pollutants, and so forth, water samples should be taken at the time of the emergency and for a reasonable period of time thereafter. It is difficult to specify how many samples should be taken because contamination tends to flow in slugs and not in a straight line. Therefore, when a positive sample is taken, there is little doubt that the water is contaminated, but if the sample is negative, there is no guarantee that the next sample will not show contamination. If everything appears to be operating properly, the water should be sampled at least once a year. It is also sampled when an alteration is made to the pumps or the water system.

Waterborne Disease Investigation

Waterborne diseases are investigated by environmental health practitioners, who make a detailed survey of the potential sources of disease. (Figure 3.13 gives an evaluation form.) These practitioners then evaluate the survey and the water samples, make necessary recommendations, and issue appropriate orders for correction of health violations.

CONTROLS

Groundwater Protections Programs

Various localities are attempting to protect their groundwater supplies by developing or utilizing existing laws or tools and techniques. Zoning ordinances are used to control open spaces, conservation and recreation areas, and aquifer recharge or well protection zones. They are prohibiting placement of hazardous waste and controlling landfill locations. Subdivisions are regulated for lot sizes, slopes, drainage, and protection of vulnerable areas. Site plan reviews include environmental surveys, necessary permits, and water and wastewater concerns. Design standards require proper building codes, setbacks, structural requirements, drainage, septic standards, and protection against sinkholes. Operating standards require alternative waste treatment systems where needed, proper storage and transportation of hazardous materials, and use of underground storage tanks (USTs). Pesticide management plans are used and necessary permits are required for groundwater discharge. Public education is provided at all levels and for all concerns of environmental health and protection. Groundwater monitoring includes wells at landfills and critical locations; regular testing for contaminants; regular testing for microorganisms; evaluation of wastewater treatment plants; and self-inspection in all water use and reuse activities. Water conservation consists of flow control devices, retrofitting devices, and recycling wastewater. Development of emergency response plans and implementation are essential. Policies are developed for storm water management and classification of aquifers and wetlands.

Governmental

U.S. EPA Programs

The EPA has issued and continuously modified groundwater protection strategies to provide an integrated framework for the many EPA laws affecting the protection of groundwater. Federal and state institutions have worked together to try to protect the quality of groundwater supplies. The EPA has provided financial assistance to the states to help them develop and carry out the strategies. All 50 states have developed overall state strategies in groundwater protection. This strategy was made into a law as part of the SDWA amendments of 1986, as amended in 1996. Wellhead protection from contaminants was deemed necessary to protect the health of the public. Wellhead protection areas are defined as the surface and subsurface areas surrounding a water well or well field that supply a public water system. These areas must not be contaminated by normal or exceptional human activity. The wellhead area is a radius of 2 mi from the actual well. A public water system is a system that provides to the public piped water for human consumption. This system serves at least 15 connections or regularly serves an average of at least 25 individuals daily

DEPARTMENT OF
HEALTH, EDUCATION AND WELFARE
PUBLIC HEALTH SERVICE
CENTERS FOR DISEASE CONTROL
BUREAU OF EPIDEMIOLOGY
ATLANTA, GEORGIA 30333

Pretest

1. Where did the outbreak occur?	2. Date of outbreak: (Date of onset of 1st case)
State_____(1-2) City or Town_____County_____	_____(3-8)

3. Indicate actual (a) or estimated	4. History of exposed persons:	5. Incubation period (hours):
(e) numbers	No. histories obtained (18-20)	Shortest____(40-42) Longest____(43-45)
Persons exposed (9-11)	No. persons with symptoms (21-23)	Median____(46-48)
Persons ill (12-14)	Nausea (24-26) Diarrhea (33-35)	
Hospitalized (15-16)	Vomiting (27-29) Fever (36-38)	6. Duration of illness (hours):
Fatal cases (17)	Cramps (30-32)	Shortest____(49-51) Longest____(52-54)
	Other, specify (39) _____	Median ____(55-57)

7. Epidemiologic data (e.g., attack rates [number ill/number exposed] for persons who did or did not eat or drink specific food items or water, attack rate by quantity of water consumed, anecdotal information)*   (58)

ITEMS SERVED	NUMBER OF PERSONS WHO ATE OR DRANK SPECIFIED FOOD OR WATER				NUMBER OF PERSONS WHO DID NOT EAT OR DRINK SPECIFIED FOOD OR WATER			
	ILL	NOT ILL	TOTAL	PERCENT ILL	ILL	NOT ILL	TOTAL	PERCENT ILL

8. Vehicle responsible (item incriminated by epidemiologic evidence): (59-60)

9. Water supply characteristics   (A) Type of water supply**   (61)
         ☐ Municipal or community supply (Name)
         ☐ Individual household supply
         ☐ Semi-public water supply
          ☐ Institution, school, church
          ☐ Camp, recreational area
         ☐ Other
         ☐ Bottled water

 (B) Water source (check all applicable) (62-65)   (C) Treatment provided (circle treatment of each source checked in B)
  ☐ Well   a   b   c   d   a. no treatment
  ☐ Spring   a   b   c   d   b. disinfection only
  ☐ Lake, pond   a   b   c   d   c. purification plant–coagulation, settling, filtration,
  ☐ River, stream   a   b   c   d disinfection, (circle those applicable)
                   d. other

10. Point where contamination occurred: (66)

 ☐ Raw water source   ☐ Treatment plant   ☐ Distribution system

* See HSM 4.245 INCDC) Investigation of a Foodborne Outbreak, item 7.
** Municipal or community water supplies are public or investor-owned utilities. Individual water supplies are wells or springs used by single residences. Semipublic water systems are individual-type water supplies serving a group of residences or locations where the general public is likely to have access to drinking water. These locations include schools, camps, parks, resorts, hotels, industries, institutions, subdivisions, trailer parks, etc., that do not obtain water from a municipal water system but have developed and maintain their own water supply.

CDC 4.461
2-75

Figure 3.13 Investigation of a waterborne outbreak. (From Centers for Disease Control, U.S. Department of Health, Education and Welfare, *Foodborne and Waterborne Disease Outbreaks,* Annual Summary 1975, HEW Pub. No. (CDC) 76-8185, September, 1976, pp. 59–60.)

at least 60 days out of the year. The term includes any collection, treatment, storage, and distribution facilities under the control of the operator of this system and used primarily in connection with this system. A community water system is a public water system that serves at least 15 service connections used by year-round residents or regularly serves at least 25 year-round residents.

11. Water specimens examined: (67)
Specify by "X" whether water examined was original (drunk at time of outbreak) or check up (collected before or after outbreak occurred)

ITEM	ORIGINAL	CHECK UP	DATE	FINDINGS		BACTERIOLOGIC TECHNIQUE
				Quantitative	Qualitative	(e.g., fermentation tube, membrane filter
Example: Tap water	X		6/12/74	10 fecal coliforms per 100 ml.		
Raw water		X	6/12/74	23 total coliforms per 100 ml.		

12. Treatment records: (Indicate method used to determine chlorine residual)
Example: Chlorine residual – One sample from treatment plant
effluent on 6/11/74 – trace of free
chlorine

Three samples from distribution system
on 6/12/74 – no residual found

13. Specimens from patients examined (stool, vomitus, etc.) (68)

SPECIMEN	NO. PERSONS	FINDINGS
Example: Stool	11	8 Salmonella typhi
		3 Negative

14. Unusual occurrence of events:
Example: Repair of water main 6/11/74; pit contaminated
with sewage, no main disinfection. Turbid water
reported by consumers 6/12/74.

15. Factors contributing to outbreak (check all applicable)
- ☐ Overflow of sewage
- ☐ Seepage of sewage
- ☐ Flooding, heavy rains
- ☐ Use of untreated water
- ☐ Use of supplementary source
- ☐ Water inadequately treated
- ☐ Interruption of disinfection
- ☐ Inadequate disinfection
- ☐ Deficiencies in other treatment processes
- ☐ Cross connection
- ☐ Back siphonage
- ☐ Contamination of mains during construction or repair
- ☐ Improper construction, location of well/spring
- ☐ Use of water not intended for drinking
- ☐ Contamination of storage facility
- ☐ Contamination through creviced limestone or fissured rock
- ☐ Other (specify)_____

16. Etiology: (69-70)   (71)
Pathogen _____ Suspected 1
Chemical _____ Confirmed .... 2 (Circle one)
Other _____ Unknown 3

17. Remarks: Briefly describe aspects of the investigation not covered above, such as unusual age or sex distribution; unusual circumstances
leading to contamination of water; epidemic curve; control measures implemented; etc. (Attach additional page if necessary)

Name of reporting agency: (72)

Investigating Official: Date of investigation:

Note: Epidemic and Laboratory assistance for the investigation of a waterborne outbreak is available upon request by the State
Health Department to the Centers for Disease Control, Atlanta, GA 30333.

To improve national surveillance, please send a copy of this report to: Centers for Disease Control
Atn: Enteric Diseases Branch, Bacterial Diseases Division
Bureau of Epidemiology
Atlanta, GA 30333
Submitted copies should include as much information as possible, but the completion of every item is not required.

CDC 4.461 (Back)
2-75

Figure 3.13   (Continued.)

A second goal of the groundwater strategy is to develop a better understanding of the groundwater contamination problems that are of national concern, including pesticides, USTs, and other sources of contamination. Appropriate action then needs to be taken. The EPA has conducted national surveys of agricultural chemicals, pesticides, and fertilizers in drinking water and developed an agricultural chemicals in groundwater strategy in cooperation with the U.S. Department of Agriculture.

A third goal of the groundwater protection strategy is to create a consistent and rational policy for the protection and cleanup of groundwater. A fourth goal is to coordinate the groundwater strategy internally within the EPA, because numerous EPA acts, including Resource Conservation Recovery Act (RCRA) and Comprehensive Environmental Response, Compensation, and Liability Act (CERCLA), have portions of the laws that relate to groundwater contamination.

The EPA is building a structure in the various states and Indian tribal governments to protect their own groundwater, because they have a primary responsibility for groundwater protection and should maintain their traditional role in managing water and land use.

Small water supply systems serving between 2500 and 3300 people account for the vast majority of the country's public water suppliers. They represent the majority of water systems in the nation that are not in compliance with the national drinking water standards. Many small communities cannot afford to purchase the equipment necessary for treating the drinking water adequately or hire experienced operators to maintain the drinking water system.

Underground injection wells that are used to dispose of solid and hazardous waste and other fluids can contaminate underground sources of drinking water. The EPA is particularly concerned about three types of wells: class 1 wells, which are used for the disposal of hazardous waste; and class 2 and class 5 wells, which are used for injecting solid waste into or above underground sources of drinking water.

Regulations have been developed to implement the SDWA for coliform bacteria, turbidity, and a number of inorganic, organic, and radioactive chemicals. The regulations require periodic monitoring of public water supplies for the specified contaminants and notification of water users when any of the standards are exceeded.

The problem related to surface waters as they become a source for drinking water purposes is discussed in Chapter 7 on water pollution.

The primary supervision of private and small public water supplies is usually carried out by the local health agency, although the state health agency may be the responsible party. The primary health authority controlling public water supplies is the state health or environmental protection agency. Congress is the primary authority in establishing the interstate water quality standards, and the EPA is the administrative agency enforcing these laws.

Source Protection

Drinking water, which may come from groundwater or surface water is a vulnerable natural resource, which needs to be protected before contamination can cause significant public health problems and high cost. Contamination needs to be prevented from traveling through the groundwater or surface water into the raw water used as a drinking water source. Compared to the cost of cleanup after contamination and treating sick individuals, it is less expensive to install appropriate safeguards for raw water sources, which means taking positive steps to manage potential sources of contaminants and contingency planning for the future by determining alternate sources of drinking water. The 1996 SDWA amendments establishes a new focus on source water protection by requiring the states to implement source water assessment and protection programs. This act also provides for a cost-effective protection of drinking water quality, state flexibility, and citizen involvement. The source water assessment programs state requirements include:

1. Identifying the areas that supply public drinking water
2. Inventorying contaminants and assess water system susceptibility to contamination
3. Informing the public of the results

EPA Assistance

The states are provided with expertise and guidance from the EPA, other states, and other federal agencies. The EPA under the 1996 SDWA provides to the states the following:

1. Drinking water state revolving fund
2. Guidance in operator certification
3. Assistance for small systems and capacity development
4. Microbial and disinfection by-products rules
5. Consumer confidence reporting rules
6. Public notification of violation of the drinking water standard
7. Laboratories and monitoring capability
8. Variances and exemptions to allow public water systems to use less costly technology
9. Guidelines for water conservation
10. Enforcement through the EPA Office of Enforcement and Compliance Assurance

Supervision

Water treatment plants must be under the supervision of full-time, technically trained supervisors. For small plants, part-time trained supervisors are permissible, providing that they are available for emergency situations and that they inspect the plant at least twice a week and in the event of emergencies. The different types of certification are based on the kind of water supplies to be supervised. Certification differs for class A water supplies, which require a complete treatment process; for class B water supplies, which are treated only for disinfecting purposes; and class C water supplies, which are specially designated by the health department or the EPA. In any case, the supervisor must have an excellent understanding of the problems that may occur and the techniques to be utilized to correct these problems.

Specific Controls

The most effective methods of removing arsenic^{3+} and arsenic^{5+} are by using ferric sulfate coagulation at a pH of 6 to 8, alum coagulation at a pH of 6 to 7, or excess lime softening. Oxidation prior to water treatment also removes arsenic^{3+}. Barium is removed through lime softening at a pH of 10 to 11 or through ion exchange. Cadmium^{3+} is removed through ferric sulfate coagulation above a pH of 8, lime softening or excess lime softening. Chromium^{3+} is removed through ferric sulfate coagulation at a pH of 6 to 9, alum coagulation at a pH of 7 to 9, or excess lime softening. Chromium^{6+} is removed by ferric sulfate coagulation at a pH of 7 to 9.5. Fluorides are removed through ion exchange with activated alumina. Lead is removed by ferric sulfate coagulation at a pH of 6 to 9, alum coagulation at a pH of 6 to 9, or lime softening or excess lime softening. Inorganic mercury is removed through ferric sulfate coagulation at a pH of 7 to 8.

Organic mercury is removed through granular activated carbon. Nitrates are removed through ion exchange. Selenium^{4+} is removed by ferric sulfate coagulation at a pH of 6 to 7, ion exchange, or reverse osmosis. Selenium^{6+} is removed by ion exchange or reverse osmosis. Silver is removed by ferric sulfate coagulation at a pH of 7 to 9, alum coagulation at a pH of 6 to 8, or lime softening and excess lime softening. Coliform organisms are killed through chlorination, ozone, or chlorine dioxide. Endrin and lindane are removed using chlorine oxidation, powdered activated carbon, or granular activated carbon. Toxaphene is only partially removed through the use of activated carbon. Toxaphene is a particularly serious problem, because good removal processes are unavailable. 2,4-D is removed by powdered activated carbon. Silvex is assumed to be controlled by granular activated carbon. However, inadequate research is available. Methoxychlor treatment information is not available. Radioactive contaminants are reduced by lime or lime–soda softening, ion exchange softening, or reverse osmosis.

New Technologies

New technologies have been developed for the control of volatile organic contaminants in groundwater by in-well aeration techniques. Because groundwater contamination by VOCs has become common throughout the United States, studies are conducted to determine not only the kinds of contamination but also the potential techniques for eliminating the contamination. Wells have been found to contain TCE, vinyl chloride, carbon tetrachloride, tetrachloroethylene, and other organic compounds. In-well aeration has been tested as a means of removing the VOCs. This is done by using a variety of aeration equipment. Aeration equipment may consist of air-lift pumps, where compressed air is introduced by an airline into an open-ended pipe in the well, called an eductor. The aerated water in the eductor is less dense than surrounding water in the well, and therefore, is forced up the eductor and out of the well as a result of the density gradient. Mass transfer of VOCs occurs in the eductor. Air and stripped VOCs are removed in an open tank at the surface called a separator.

Based on well characterization, the air-lift pump has been studied at 130-, 200-, and 280-ft depths. The 130-ft depth coincides with 65% submergence, which is reported as optimum for air-lift pump efficiency. The maximum efficiency has been found to be 30 to 35% with the efficiency decreasing as the submergence increases down to the 280-ft level. The VOC control ranges from 90% of vinylchloride to 47% for *cis*-1,2-dichloroethylene. A secondary effect occurs in in-well aeration. The pH increases by an average of 0.4 units as carbon dioxide is stripped from the well. In-well aeration may cause an air pollution problem, because the VOCs gases are vented into the air. Water saturated with dissolved oxygen may be corrosive to some distribution system materials and could cause a problem in the system.

Studies have been conducted by the Water Engineering Research Laboratory of the EPA on a variety of types of aeration equipment. The aeration equipment when used in clean water indicates that systems that provide fine-bubble diffusion transfers oxygen most efficiently in clean water. Jet aerators transfer oxygen more efficiently than static aerators and other coarse bubble systems but not as efficiently as fine bubble diffusers. This is not necessarily true for wastewater. Aeration systems have also been utilized for the removal of radon gas from small-community water supply systems. The aeration system reduces the radon gas by about 85% in the raw water.

Radium, which has been estimated to be present in some 500 municipal water supplies, has been removed from the water by the use of sodium ion exchange and lime–soda softening. These systems generally remove 85 to 95% of the radium present in the water. Radium is found many times in water supplies that have unacceptably high concentrations of iron and sometimes manganese. Processes using oxidation and sand filtration remove the iron and manganese and remove part of the radium that is present.

Some of the innovative and alternative technologies follows.

PACT® Wastewater Treatment System — This process, developed by Zimpno Passavant Environmental Systems, Inc., is a system that combines biological treatment and powdered activated carbon (PAC) adsorption to achieve treatment standards that are not readily attainable using conventional technologies. The technology removes organic contaminants from the wastewater through biodegradation and adsorption on the PAC. It is used to treat industrial wastewaters from chemical plants, dye production facilities, pharmaceuticals, refineries, and synthetic fuel facilities, in addition to contaminated groundwater and mixed industrial and municipal wastewater.

Ultraviolet Radiation and Oxidation — This process, developed by ULTROX International, uses UV radiation, ozone (O_3), and hydrogen peroxide (H_2O_2) to destroy toxic organic compounds in water, particularly chlorinated hydrocarbons. The process oxidizes compounds that are toxic or refractory (resistant to biological oxidation) in concentrations of parts per million or parts per billion. This procedure is used to treat contaminated groundwater, industrial wastewaters, and leachates containing halogenated solvents, phenol, pentachlorophenol, and pesticides; polychlorinated biphenyls; explosives, benzene, toluene, ethylbenzene, and xylene; and methyl-*tert*-butyl ether, and other organic compounds.

Photocatalytic Water Treatment — This process, developed by Matrix Photocatalytic, Inc., efficiently removes and destroys dissolved organic contaminants from water in a solid-state, continuous flow process at ambient temperatures. The titanium dioxide (TiO_2) semiconductor catalyst, when excited by light, generates hydroxyl radicals that break the carbon bonds of hazardous organic compounds. It is used to destroy organic pollutants or to remove total organic carbon in drinking water, groundwater, and plant process water. Organic pollutants such as PCBs, polychlorinated dibenzodioxins, polychlorinated dibenzofurans, chlorinated alkenes, chlorinated phenols, chlorinated benzenes, alcohols, ketones, aldehydes, and amines can be destroyed by this technology. Inorganic pollutants such as cyanide, sulphite, and nitrite ions can be oxidized.

Biological Sorption — The AlgaSORB® sorption process, developed by Bio-Recovery Systems, Inc., is designed to remove heavy metal ions from aqueous solutions. The process is based on the natural affinity of algae cell walls for heavy metal ions. In many applications, AlgaSORB® is highly selective in capturing the metal of interest without becoming saturated by large concentrations of salts. AlgaSORB® seems particularly effective in removing mercury and uranium from groundwater. This technology is used to remove metal ions from groundwater or surface leachates that are hard or contain high levels of dissolved solids. Rinse waters from electroplating, metal finishing, and printed circuit board manufacturing industries can also be treated. Variations of the technology, some using other absorbents, may be used to recover spent acid from metal pickling lines and maintain the purity of chemical baths that use heavy metals. The system can remove heavy metals such as aluminum, cadmium, chromium, cobalt, copper, gold, iron, lead, manganese, mercury, molybdenum, nickel, platinum, silver, uranium, vanadium, and zinc.

ICB Biotreatment System — This immobilized cell bioreactor (ICB) biotreatment system, developed by Allied Signal, Inc., is an aerobic, anaerobic, or combined aerobic and anaerobic fixed-film bioreactor system designed to remove organic contaminants (including nitrogen-containing compounds and chlorinated solvents) from process wastewater, contaminated groundwater, and other aqueous streams. It is used to treat industrial wastewater and groundwater containing a wide range of organic contaminants, including PAHs, phenols, gasoline, chlorinated solvents, diesel fuel, and chlorobenzene.

Laser-Induced Photochemical Oxidative Destruction — This process, developed by Energy and Environmental Engineering, Inc., photochemically oxidizes organic compounds in wastewater using a chemical oxidant and UV radiation from an Excimer laser. The photochemical reaction can reduce saturated concentrations of organics in water to nondetectable levels. The beam energy is predominantly absorbed by the organic compound and the oxidant, making both species reactive. The process can be used as a final treatment step to reduce organic contamination in groundwater and industrial wastewaters to acceptable discharge limits. The technology is used to treat groundwater and industrial wastewater containing organics. Destruction removal efficiencies greater than 95% have been obtained for chlorobenzene, chlorophenol, and phenol. Compounds treated with UV and oxidation include:

Ethers	Pesticides
Aromatic amines	Benzene
Toluene	Ethylbenzene
Xylene	Citric acid
Complexed cyanides	Phenol
Trichloroethane	Trinitrotoluene
Polycyclic aromatics	Trichloroethene
Dichloroethane	Dioxins
Tetrachloroethene	Methylene chloride
Hydrazine	Dichloroethene
Cresols	Cyclonite
Polynitrophenols	Polychlorinated biphenyls

1,4-Dioxane	Ketones
Pentachlorophenol	Vinyl chloride
Ethylenediaminetetraacetic acid	

Deep Cleaning of Water Systems

Outbreaks of Legionnaire's disease in London created a serious concern about the increased potential for the disease. Since 1979, several hundred cases have occurred in England and Wales, with about 20 deaths each year. As a result of this, the environmental health specialists first determined how people became ill and second developed a deep cleaning water system process. The illness, in order to occur, had to have the following factors present:

1. *Legionella pneumophila* had to be present in the water system.
2. Conditions had to be right for *Legionella* to multiply.
3. The aerosols in the water source had to be transmitted through the air.
4. The person exposed to the aerosol had to be susceptible to the disease.

Although the organism is frequently present in water, conditions must be suitable to support its growth. The water has to be warm, 20 to 40°C, and stagnant, together with a source of iron such as rust, and a source of nitrogen. Also some evidence indicates that algae growth helps promote the growth of the organisms. These conditions are frequently found in cooling towers, hot water systems, and various types of tanks.

To deep clean a water system, the system must be dismantled, thoroughly cleaned and disinfected, then reassembled, and finally put into working condition. For existing systems, it is recommended that cooling towers be deep cleaned every 3 months; however, if they are well maintained, cold water storage tanks may be cleaned once a year or less. Water treatment alone cannot kill the organisms. The deep cleaning of new water systems is necessary to remove any debris that builds up during construction. This debris may be found in pipes, tanks, sump pumps, and other equipment. The debris consists of rust, deposits of oil and grease, dirt, soil, and other materials. Sludge, slime, and corrosion may also be present. An occupational hazard is potential for the individuals doing the cleaning, and therefore they must be protected from the inhalation of any aerosols present in the air. After all parts have been physically cleaned, the entire pipe work must be disinfected. Chlorine is the disinfectant of choice. After cleaning and disinfection, the system must be refurbished, especially where corrosion and disintegration of any surface coating has taken place. The maintenance of the system on a regular basis is recommended.

Lead Removal

Lead present in water in the home is a most serious concern. Water that has been in contact with the plumbing for more than 6 hours, such as water that has been in the system overnight or during the working day, should be flushed from the cold water faucet. This should be done for every faucet until the water runs cold. If the building was built prior to about 1930, service connectors may be made of lead. The water should be allowed to run for an extra 15 sec after it cools to flush out the service connectors. Flushing is important, because the longer water is exposed to lead pipes or lead solder, the greater is the possibility of lead contamination. Flushing, however, may be ineffective for high-rise buildings with large-diameter supply pipes joined with lead solder. A second step in preventing lead ingestion is to only use cold water for cooking, especially if baby formula is prepared. Reverse osmosis and distillation units are commercially available to reduce the amount of lead in the tap water in the home. Finally, if serious doubt about lead content in water remains, it is recommended that bottled water be used for any potable water consumption.

Other Controls

Usually tastes and odors are reduced or removed by aeration, activated carbon, or use of chlorine as an oxidizing agent.

Corrosion caused by acidity in the water can be controlled by adding a neutralizer. Film-forming materials, such as polyphosphates or silicates, can coat metals. Other methods used to control corrosion include the installation of dielectric or insulating unions, reducing velocities and pressures, removal of oxygen, and use of nonmetallic piping and equipment.

Algae is best controlled by the use of chlorine. However, copper sulfate is still used at a rate of 0.3 mg/l of water.

SUMMARY

The use of potable water for a variety of purposes has increased enormously over the years. It is difficult to separate actual water usage from the initial water quality. Therefore, it is necessary to carefully evaluate the raw water and the finished product.

Because waterborne diseases such as typhoid fever and cholera, as well as bacterial dysentery, amoebic dysentery, salmonellosis, and so forth, are of vital concern to us, we have developed and used disinfection techniques for over 90 years. Unfortunately, disinfection that has been so useful in eliminating bacterial diseases and certain viral diseases, has become a hazard, because the primary disinfectant used, chlorine, combines with organic chemicals that are being dumped into our raw water supplies. Surveys have shown widespread occurrences of organic materials of potential health significance in the raw water utilized for drinking in the United States. Although few of the chemicals have been tested for their potential toxicological effects, many of these have been identified as carcinogenic, mutagenic, or potentially teratogenic.

Studies have shown that lead and cadmium may be found in drinking water supplies. It is not clearly understood what function corrosive water has on releasing lead into the water supply and its subsequent relationship with blood lead levels in individuals. It is known that other inorganic contaminants that are implicated in human cancer through inhalation are present in water. They include arsenic, chromium, and nickel. It is not known whether they are cancer producing in the water medium. Nitrites combine with secondary and tertiary amines to form nitrosamines, a class of potent chemical carcinogens. Not only have the viruses that cause infectious hepatitis been found in water, but also the polio, coxsackie, echo, and adenoviruses have been detected. Little is known about the interference of particulate matter and its removal, or the removal of viruses from drinking water. Unfortunately, because water is reused in such large amounts, effluents from sewage are bound to be introduced into the raw water and obviously contain numerous viruses and bacteria.

Drinking water contaminants not only are a definite problem in causing bacterial and viral diseases, but also contain the etiologic agents of cardiovascular disease and cancer. We have reached a point in our society where it is essential not only to construct a complete profile of contaminants in the raw drinking water but also to profile the finished product. In-depth research is needed to determine immediate and long-term effects of potential contaminants and how best to eliminate these contaminants if they constitute a hazard to our population. Many sources of potential contamination still exist, including:

1. Hazardous waste sites are known as potential candidates for the Superfund NPL.
2. Millions of septic systems are potential contaminators of the groundwater supply.
3. Tens of thousands of surface impoundments (pits, ponds, and lagoons) could leach into the groundwater supply.
4. Hazardous waste land disposal facilities and municipal and other landfills may be potential sources of contamination.
5. Millions of USTs are in existence, of which hundreds of thousands are estimated to be leaking and can cause potential hazards.
6. Thousands of underground injection wells exist.
7. Millions of tons of pesticides and fertilizers are spread on the ground primarily in the rural areas, and may be sources of leaching into the groundwater.

The major concern of the EPA is with toxic chemicals, such as the synthetic organic chemicals that are used in plastics, solvents, pesticides, paints, dyes, varnishes, and ink. Some of the tens of thousands of community public water systems and millions of private wells that furnish drinking water are known to be contaminated with these substances.

RESEARCH NEEDS

Qualitative and quantitative techniques of analysis must continue to be enhanced to detect important classes of waterborne toxicants, such as the polynuclear aromatic hydrocarbons, amines, and nitrosamines in small quantities. Studies are needed to determine effective alternate purification techniques to eliminate waterborne pathogens from raw water. Research is necessary to determine the extent of formation of chlorinated hydrocarbons in drinking water. Reexamination of disinfectant techniques is necessary. Techniques are needed to obtain a profile of the sources and relative contributions of various organic toxicants in surface and groundwater.

Because cardiovascular disease and cancer may be caused by the chemicals in drinking water, it is necessary to study these chemicals to determine whether a direct cause–effect relationship exists. The roles of cadmium, sodium, and other elements in influencing cardiovascular disease require further study. Epidemiological studies should be conducted of the contribution of lead from drinking water to the total body burden of lead. Research is needed to determine the extent to which viruses are associated with colloidal and particulate matter in water.

Research programs are needed to increase the knowledge of monitoring, fate, and transport of pollutants, aquifer restoration, source control, and health effects of exposure to contaminated groundwater. It is also necessary to be able to integrate the research findings of state and local governments and the private sector through technical assistance and training programs. Additional study is needed in the area of agriculture chemicals in groundwater. It is necessary to determine the long-term effects of these agricultural chemicals and to develop well-reasoned plans of action to manage the problem. Aquifer restoration technology is essential to develop new techniques for restoring aquifers that have been previously contaminated. Additional research is needed concerning on-site treatment of waste and groundwater, use of microorganisms to remove the waste, and new kinds of treatments to destroy any microorganisms that may exist.

One of the greatest challenges of the Safe Drinking Water Program is communicating to the public the long-term health risk associated with drinking water with low levels of contaminants. To be able to achieve this, it becomes necessary to get a better understanding of a multitude of low-level contaminants and their long-term health effects. Although the public is geared to the short-term health problems of giardiasis, hepatitis, cholera, typhoid, and Legionnaires' disease, it is not similarly concerned with the potential hazards of organic chemicals, disinfection by-products, or radionuclides. Two other major areas of research include (1) underground injection and the best ways of preventing contamination of the groundwater from this source and (2) pesticides in drinking water supplies.

Long-term research is needed for the fundamental processes involved in the subsurface soil–groundwater environment. Specific research is needed to better describe the transport and transformation of organic compounds, considering hydraulic, biotic, and abiotic aspects in both homogeneous and heterogeneous soil matrices. Water plays a dominant role in the soil environment, but we need a better quantitative understanding of the variation of the sorption and vapor pressure of organic compounds with soil moisture, particularly when the moisture is below 10%. We need to know more about the fundamentals of hydraulics of geochemistry for low-permeable materials. More information is needed about the hydrologic properties of crystalline rocks and the importance of fracture related to deep injection wells.

CHAPTER 4

Swimming Areas

BACKGROUND AND STATUS

Swimming water is potable water used for recreational purposes. Humans do not live within the water environment. Therefore, each time they enter it they are at risk from a bacteriological, chemical, and safety point of view. Obviously, while in the water humans inadvertently consume a portion of the water and therefore are exposed to disease or injury. Swimming in effect is communal bathing, with fecal and urinary accidents, and various orifice washings. Swimming pool water can be compared to dilute sewage, because it is contaminated by the swimmer's skin, mucus, feces, urine, dirt, and a vast variety of contaminants that may come into the water from a variety of sources. It is therefore necessary for the swimming pool or natural bathing area to be frequently tested, properly constructed, and adequately protected by means of good filtration and good disinfection techniques.

There are a vast variety of swimming areas. They include fill-and-draw pools, where the water is put into the pool and then emptied at regular intervals; flow-through pools, where the water comes into the pool and then passes out at the other end on a continuous basis; recirculating pools, where the water is filtered and chlorinated, and then sent back to the pool again; therapeutic pools and whirlpools, which are used for treatment purposes in institutions; saunas, which are specially constructed small pools utilized in clubs, motels, or hotels; individual home pools, which may vary from a child's 5-ft-diameter pool to a concrete pool installed in the ground; outdoor swimming areas, such as ponds, lakes, rivers, and oceans, which exist naturally; and outdoor areas that have been artificially created to contain water for recreational use.

Millions of people swim each year in municipally operated swimming pools, hotels, country clubs, athletic clubs, YMCAs, military reservations, summer camps, beaches, and anywhere where some small body of enclosed water can be found. The number of public and private pools has grown tremendously and continuously. The situation is such that many thousands of skilled and trained pool operators are needed. Pools are increasingly becoming contaminated because of the large numbers of individuals who are utilizing these facilities. The recirculation systems and filtering systems in many areas are incapable of handling the numbers of swimmers who utilize the facilities. Chlorine cannot be kept in adequate concentrations with many of the existing mechanical-feed chlorinators. The lack of trained personnel has resulted in innumerable pool problems. In most areas, many of the pools should be shut down because of the violations of good pool practice.

Many beaches have been closed and others will be closed because of water pollution. Some of the beach areas receive discharges from septic tanks, sewage plants, and industrial plants that put forth acids, alkalies, solvents, greases, and oils. Many trace minerals, such as lead, copper, mercury, zinc, nickel, chromium, and cobalt, can be found in these waters. Animal wastes, as well as chemical wastes and agricultural runoff, are further polluting these open bodies of water where swimming beaches exist. One third of the U.S. shoreline is either continuously or intermittently unacceptable for swimming because of severe bacterial contamination. The situation is deteriorating instead of improving. Added to the existing problems are oil spills, which contaminate large areas of beaches, and disposal of hazardous medical wastes.

SCIENTIFIC, TECHNOLOGICAL, AND GENERAL INFORMATION

Water Treatment

The water treatment process is a series of steps in which the water is removed from the pool, treated, and returned in a safe manner (Figure 4.1). These steps include water leaving the main drains and going to the water treatment plant; use of a hair catcher to remove coarse material such as hair, lint, and leaves; addition of freshwater in a makeup tank to replace the water that has been splashed out, evaporated, or may be lost during the filtering or backwashing process; and use of pumps to move water through the rest of the process. Also included in the water treatment process are the addition of alum or filter aid material to assist in filtration; addition of soda ash to maintain proper pH; filtration through sand filters or diatomaceous earth filters; backwashing by reversing the flow of the water through the filters and then removing the filtered material to the sewer; addition of chlorine or bromine as a disinfectant; heating of water, where necessary, for proper temperature; and recirculating the water back to the pool.

Sources of Water Supply

Water may come from approved wells that are properly protected or from lakes and rivers. Lake and river water must be treated before it can be used in a swimming pool. Obviously it cannot be treated as such when individuals are swimming in the open lake area. The difficulties encountered in water supply are similar to those already discussed in the areas of private and public water supply. All waste materials from the pool and all contaminated pool water should be discharged into a sewage treatment system for proper treatment before it is allowed to go into a body of water.

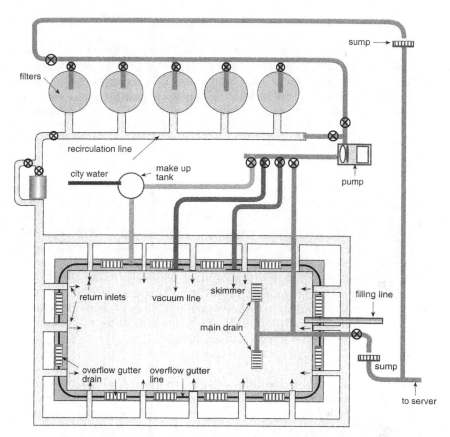

Figure 4.1 Swimming pool piping system. (From Michael, J.M., *Swimming Pools; Disease Control through Proper Design and Operation,* Training Manual Communicable Disease Center Training Branch, DHEW, U.S. Government Printing Office, Washington, D.C., 1959, p. 19.)

In any case, the diatomaceous earth filter material and any sand that may come off the sand filters should be allowed to settle and concentrate before the effluent is permitted to go into the sewer. The settled material should then be disposed of in the same way that solid waste is handled.

Pool Hydraulic System

The fundamentals of hydraulics are discussed in Chapter 5 on plumbing. For the purposes of the swimming pool section, the concern is with the peculiarities involved in moving water within the distribution system of the swimming pool environment. The pool is designed for recirculation and filtration. As has already been mentioned, many steps along the way are necessary but can lower the pressure of the water going through the pipes. The rate of water flow is determined by the water volume in the pool basin and the number of times it has to flow through the pipes and the filters.

The effectiveness of the hydraulic distribution depends on the technique used to introduce the filtered water uniformly into the pool basin, the extent to which it is introduced, and the methods of water distribution. It is necessary to dispense the water uniformly and to avoid all short circuits by having the water withdrawn uniformly at a point as far away as possible from the entry of the water into the pool. There is concern not only about the initial system and proper movement of water but also in the maintenance of the system. Dirty filters, dirty hair catchers, clogged openings, and discharge points, as well as clogged lines reduce water pressure.

In pools it may be necessary to provide hydrostatic relief valves to protect the pool shell. When the pool is empty, water that may have accumulated beneath the shell could possibly cause the shell to rise and thereby cause damage. In the design of the pool, it is important to have a thorough understanding of the geology of the ground beneath it, as well as the potential water conditions, to determine the number of hydrostatic relief valves needed. The pump must satisfy the maximum peak operational conditions. However, contaminated water may enter through these valves. The pump should have an impellor that is one size smaller than the largest diameter that the pump casing can hold. The pump should be 1 ft below the operating water level, although typically it is usually several feet above the main drain.

Disinfection

Swimming pool water may be disinfected by the use of chlorine, bromine, iodine, silver salts, ultraviolet radiation (UV), or ozone. Chlorine present in gaseous form is the cheapest to use. It is also good for any pool over 100,000 gal and is reasonably safe, providing the operator understands the hazards associated with the gas. The disadvantages of chlorine gas are that separate rooms are needed for storage and operation, the initial and operating costs are fairly high, and experienced individuals are needed to operate and maintain the units.

Free chlorine residual is most important in determining the effective killing power of chlorine in swimming pool water. The combined chlorine residual is that chlorine that combines with anhydrous ammonia or with organic matter in the pool water to form disinfecting compounds known as chloramines. Chloramines plus residual chloramines take about 100 times the contact period to kill bacteria as free chlorine.

Bromination is a technique often utilized in pool disinfection. Bromine is a heavy brown liquid weighing about three times the weight of water. It is volatile at room temperatures and the gas form of it is very toxic.

Iodine at room temperature is crystalline and brownish in color. Apparently it causes less eye irritation as compared with chlorine. Potassium iodide put into the water releases iodine gradually. Because iodine has the greatest atomic weight of the four halogens, it is least soluble in water, is least hydrolyzed, has the slowest oxidation potential, and also reacts more slowly with organic compounds. Indications are that low iodine residuals should be more stable and therefore persist longer in the presence of organic or other oxidizable matter than the other halogens. Iodine appears to have an effective germicidal range, which crosses a wide variety of pH values; does not combine with ammonia to form iodoamines; and appears to be somewhat less dependent on bathing loads and the organic matter present when used in the form of the hypoiodite ion. Iodine removes odors and tastes and reduces eye irritation. However, considerably more investigation should be carried out on its usage.

Silver salts, such as argeral and silver nitrate have been experimentally used in pool water. However, this technique is very expensive and very slow acting. UV radiation can kill bacteria if the water is passed in a thin film immediately under the banks of UV lamps. There is a serious question as to the effectiveness of this procedure, because the effective bacterial kill varies enormously with the age and strength of the lamps, the amount of time of exposure of the water to the UV light, and the fact that turbidity can block the penetration of the rays. The process is also quite expensive. Ozone is an excellent oxidation agent, but the excess gases must be removed before the water is used.

Swimming Pool Chemistry

Chlorine is the most commonly used bactericide in pools. For the chlorine to function properly in the destruction of organic materials and as a bactericide, the pH of the water must be within a range of 7.2 to 7.8. When chlorine is added to water, it forms two acids. They are hypochlorous ($HOCl$) and hydrochloric acid (HCl). Hydrochloric acid is a useless by-product of chlorination and has to be neutralized by adding 1.25 to 1.5 lb of soda ash for each pound of chlorine added. Hypochlorous acid exists in either the molecular or the ionized state. The pH of the water determines the amount in each state. At a pH of 7.0, the acid is 72% molecular. At a pH of 7.5, the acid is 50% molecular. At a pH of 8.0, the acid is 21% molecular. Hypochlorous acid in its molecular form is an effective bactericide. In its ionized form it kills bacteria much more slowly. Chlorine is a much less effective bactericide at high pH. Hypochlorous acid in either the molecular or the ionized state is referred to as *free chlorine*, in cold water. The amount of hypochlorous acid that stays in the water uncombined with ammonia is called the *free residual chlorine*. This free residual chlorine must be maintained to have adequate disinfection. Free residual chlorine dissipates rapidly in bright sunlight, at high temperatures, and if weather conditions are windy.

The combination of chlorine and ammonia is called *combined chlorine*. This occurs when the chlorine in the hypochlorous acid has oxidized organic matter and some inorganic substances. It then reacts with ammonia to form compounds called chloramines. If the concentration of hypochlorous acid is too low to bring the reaction to completion, the ammonia cannot be completely oxidized and dissipated. Chloramines are bactericides, but the rate of disinfection is 60 to 100 times slower than that of free chlorine.

The total residual chlorine is the total concentration of residual free chlorine plus the residual of combined chlorine. It is possible to maintain the bacterial standards of the pool with total residual chlorine; however, the concentration range is 2.0 to 2.5 ppm. This is a poor practice, because the chloramines cause eye irritations and produce objectionable chlorine odors.

Stabilized chlorine is free chlorine combined with cyanuric acid. This is a combined chlorine that has the stability of chloramines in sunlight and greater bacterial effectiveness than the chloramines, but less than free chlorine. When testing for free residual chlorine, the stabilized chlorine responds to the test as if it were free chlorine. The concentration of cyanurate should be kept about 50 to 60 ppm. The upper limit is 100 ppm.

Breakpoint chlorination occurs when chloride is added to water containing ammonia and forms chloramines. If more chlorine is added to the water, the total residual chlorine continues to rise until the concentration reaches a point that forces the reaction with ammonia to go rapidly to completion. Nitrogen and chlorine are then released from the water and the apparent residual chlorine decreases. The point at which the residual suddenly drops is called the *breakpoint*. If enough chlorine is added to pass the breakpoint, all combined chlorine compounds disappear, eye irritation potential and chlorine odors disappear, and the chlorine that remains in the water is in the free state. The breakpoint occurs at different concentrations in different waters. Superchlorination usually exceeds the breakpoint.

In the case of chlorine or any of the disinfectants, the bacteria are oxidized and organic materials present are oxidized. As mentioned earlier, some environmental concerns in the use of chlorine are the potential for the production of trichloromethane, chloroform, and 1,1-trichloroethane.

Chemistry of Ozone in Water

In clean water, such as the influent to pools or spas, the major factors that affect the chemistry of ozone are solubility, decomposition of ozone, pH, temperature, and transfer efficiency of gaseous ozone into water. These factors are also a concern in the chemistry of ozone and contaminated pool and spa water. The oxidation or disinfection benefits from ozone depend on the amount of ozone transferred to the pool water during contact. The solubility of the ozone is a function of the partial pressure of ozone in the air above the water, which is determined by the concentration of the ozone in the air. Efficient gas or liquid contact is essential for the transfer of appropriate amounts of ozone to the water to bring about oxidation or disinfection. There should be a minimum ozone concentration in the air of 18 g/m^3 of air or 1.5% by weight.

Ozone is generated by passing an oxygen-containing gas through a high energy electrical apparatus (corona discharge) or through a high energy radiation source (UV radiation). The ozone concentrations of 1 or 2% by weight that are obtained by corona discharge are far greater than the amount produced by UV radiation. When ozone is dissolved in water, it can react with water contaminants in two distinct ways: either by direct action as the ozone molecule or by indirect, free radical reactions. In strongly acid solutions, the direct reaction is most common. When the pH is above 7, the radical reactions are most common. When ozone is present in water above pH 7, the following may be found: hydroxyl, free radicals (OH)·, hydroxide ions $(OH)^-$, perhydroxide ions (HO_2^-), the anion of (H_2O_2), perhydroxyl free radicals $(HO_2)\cdot$, superoxide ions $(O_2)^-$, and ozonide anions $(O_3)^-$. The most reactive of these radicals and ions is the hydroxyl free radical, which has an oxidation potential higher than that of the ozone molecule itself. Its presence is highly desirable in swimming pool water for the oxidation of contaminants. Many of the anions participate in chain reactions that produce hydroxyl free radicals in the presence of pool water contaminants.

In clean raw drinking water at a pH of about 8, approximately half the ozone is decomposed within about 10 min, producing hydroxyl free radicals. The ozone molecule and the hydroxyl free radicals perform oxidation and disinfection work in the water; however, the half-life of hydroxyl free radicals is a microsecond and therefore only the ozone molecule disinfects in swimming pool water with a pH range of 7.2 to 8.0.

At the temperature of pool and spa waters, the solubility of ozone is adequate to provide the desired amount of oxidation or disinfection. The ionized water is then degassed to remove as much air–ozone gases as possible. This is accomplished by passing the ozone remaining in the gas and liquid phases over elemental carbon, which is a strong reducing agent.

Ozone partially oxidizes organic compounds. Many of the organic compounds are very difficult to oxidize. Ozone can destroy viruses, fungi, amoebae, and cysts, depending on the concentration of the disinfectant and the contact time in minutes. Ozone reacts with inorganic contaminants. The reaction of ozone with ammonia is extremely slow. The reaction of ozone with iron that may be found in the water results in the precipitation of the gelatinous ferric hydroxide. The ferric hydroxide adsorbs some of the polar organic materials that may be present. Ozone reacts with soluble manganese in the water to produce insoluble manganese oxydihydroxide. Ozone oxidizes sulfide ions to the sulfate form, which is innocuous.

Ozone, when combined with smaller amounts of chlorine, such as free chlorine residual at 0.2 to 0.5 mg/l in a pool, acts effectively to help produce high-quality water by minimizing the use of chlorine. The chloramines that are produced when chlorine combines with organic nitrogen compounds are destroyed by the ozone, and free residual chlorine is released again. However, where the chlorine residual is up about 1 ppm it tends to destroy the ozone that is present.

When ozone is used with a bromide ion process in chemical flocculation to improve water quality, the ozone acts as a flocculating agent and oxidant.

Swimming Pool Calculations

1. Pool turnover rate

   $$\text{Hours per turnover} = \frac{\text{pool capacity (gallons)}}{60 \times \text{pump capacity (gpm)}}$$

   Ex. pool capacity = 100,000 gal

   then $\dfrac{100,000 \text{ g}}{60 \times 40 \text{ gpm}} = 4.17$ h per turnover

   Pump capacity = 400 gpm

2. Maximum swimmer load

$$\text{Maximum swimmer load} = \frac{\text{pump flow rate in gpm} \times 60 \times 24\,\text{h}}{\text{number of gal of water per person}}$$

Ex. pump flow rate = 300 gpm

Number of gallons per person = 600 gal

then Maximum swimmer load = $\dfrac{300 \times 60 \times 24}{600} = 720$

3. Pool leak and makeup water

Amount of makeup water = inches of water lost from pool × pool surface area in square feet × 0.625

Ex. 6 in. lost

Pool 40 × 75 ft = 3000 ft^2

Amount of lost water or makeup water needed = 6 × 3000 × 0.625 = 11,250 gal

4. Flow rate for filter

Flow rate for filter = square feet of filter area × rate of flow in gpm/ft^2

Ex. filter area = 200 ft^2

Rate of flow = 2 gpm

200 × 2 = 400 gpm

Therefore, a pump with this capacity to be used

Therapeutic Pools, Saunas, Spas, and Hot Tubs

The use of water in physical therapy is an important part of the treatment and curative procedures used for patients. It is possible to carry out exercises, to stimulate muscles, and also to slough off dead skin or debride decubitus ulcers. Water tanks are utilized frequently in curative procedures for burn victims. The types of therapeutic tanks used include swimming pools, whirlpools, Hubbard tanks, and moist-air cabinets.

It is necessary to operate the equipment in such a way that it is not only clean but also sanitized. In the case of therapeutic pools, all the normal pool operations apply. In addition, the pool should be vacuumed and the hair catcher emptied, as well as the filters backwashed, at least twice a week. If an infected patient has used the pool, the process should be carried out immediately after this water use. The pool decks should be scrubbed down daily with a good detergent disinfectant, rinsed, and then sanitized using a compatible disinfectant. On a biweekly basis, therapeutic pools, saunas, spas, and hot tubs should be emptied; thoroughly scrubbed with a good detergent disinfectant; and rinsed thoroughly and then refilled prior to the next use. It should be remembered that not only the infected patient is at risk in picking up additional infections, but also the noninfected person is at risk in becoming infected. Therapeutic pools, as well as all other pools, need to be properly tested. This is discussed in a subsequent section.

Bathing Beaches

Bathing beaches, or natural and semiartificial swimming areas with dirt or sand bottoms, are subject to varying degrees of contamination, because such bodies of water receive a varied amount of surface water drainage, in addition to the contaminants that may come from the actual water source. A bathing beach is defined as a beach and bathing area offered to the public for bathing or swimming. It does not include a swimming pool.

Areas that are not natural ponds and lakes have been excavated to provide swimming beaches. These areas generally fall into the category of an artificial swimming pool. The artificial swimming pools, if utilized, have to conform to the requirements of swimming pools. Ponds and lakes may be created by damming up a stream or waterway or by having a large excavation, such as a sand and gravel pit, filled with water.

SWIMMING AREA PROBLEMS

Pool Problems

Swimming pool problems are basically broken down into six areas: operator, design and construction defects, water quality, inadequate service and maintenance, disease, and safety hazards. The operator is the key individual involved in ensuring that the pool can be effectively utilized and properly protected. Unfortunately, because pool operators may also be involved in many other functions, they either may do a poor job or may delegate the responsibility to others who have inadequate knowledge and training, and who also may do a poor job of taking care of the pool and its equipment. Prior to each swimming season, pool operators or their designated subordinates should be required to take a swimming pool course from the local or state health department.

The design and construction of many existing pools are such that they have poor flow of water, poor turnover, poor filtration, and poor chlorination. It is essential to have the pool thoroughly evaluated by the environmental health practitioner and to have necessary corrections made. Water quality should be a subject of immense concern to the operator and to the public. Algae, which are plant growths, may grow uncontrollably if permitted to get a good start. The algae may grow along the sides and bottom of the pool and along side ladders. Algae create a high chlorine demand and therefore makes chlorine residual maintenance difficult. They increase the turbidity of the water, create slipping hazards, act as an agent to allow bacteria to grow, and also produce odors. As a result of the algae taking carbon dioxide out of the water for its growth process, there is an increase in the pH. Algae may be controlled through routine chlorination, if it is kept at a proper level.

However, once algae have started to grow, three kinds of treatments may be utilized. One is the addition of copper sulfate at a rate of 10 lb per million gal of water. The problem with the use of copper sulfate is that in hard water the copper is precipitated. It may discolor swimmers' suits and hair, and may also produce an inky precipitate when hydrogen sulfide is present. Copper sulfate is the least recommended technique. A much more acceptable technique, and one that works very well is to utilize a strong chlorine residual by superchlorinating the water. The superchlorination kills off the initial algae and the high residual of 1 to 2 ppm reduces the opportunity for algae to start growing again. The problem is that maintaining water above 1 ppm can cause discomfort to swimmers. During superchlorination, at least 10 ppm of free chlorine residual should be present. Obviously, this should not be done during normal swimming hours.

As a last resort, if the previous techniques do not work, it is recommended that the pool be scrubbed with a 5% hypochlorite solution and then thoroughly rinsed off. Good, durable rubber-based pool paints help to prevent the problem of algae growth.

Cloudy water may occur as a result of either poor filtration or the filter medium backing up into the swimming pool. It may be also caused by hard water. All these factors should be checked to determine the cause. Colored water is usually caused by dissolved metals in the water. Iron may make the water appear red or brown, copper may make it appear blue-green, and manganese may make it appear brown or black. These problems can be corrected by adjusting the pH to 7.2 to 7.6, superchlorinating with sodium hypochlorite or calcium hypochlorite, running the filter continuously, adding alum to the skimmer before the sand filter at a rate of 2 oz/ft^2 of filter surface area, backwashing the filter, and then readjusting the pH. Green water is apparently due to either algae, which have already been discussed, or copper in the pool. The copper may come from corroding pool fixtures due to high acidic water. This may be corrected by stabilizing the incoming water to a pH of 7.2 to 7.6.

The burning of eyes is due to too little or too much chlorine. Chloramines, which are a combination of chlorine and various nitrogen compounds, may cause eye burning when there is not enough chlorine and the pH is too low. The chloramines are created from the action of chlorine on urine, perspiration, and products of rainfall. The chloramines can be destroyed through superchlorination.

Unpleasant odors come from the chlorine or various combinations of chloramines or from water that contains sulfur. Other problems encountered include too high or too low a pH, the results of which have already been discussed.

Scale and stains may occur in pools because of any of the previously mentioned metals or from carbonates. A major cause is inadequate servicing and maintenance. The job is usually done in the fastest and cheapest way possible, because the pool is only open for a short period of time. Unfortunately, this inadequate service and maintenance not only leads to poor water quality, but also to accidents. Pool accidents are due to loose diving boards, poorly secured gates, cracks in concrete, slippery walks and floors, improper chemical storage, electrical hazards, and inexperience or overexuberance of the swimmers.

Problems encountered in the use of chlorine solutions include the deterioration of the solution in sunlight and under warm conditions, and hardness of the pool water, which may react with the calcium from the calcium hypochlorite and form calcium deposits at the point where the chlorinator tube enters the water line. Another problem is that the liquid requires more space than the gas. Also, individuals coming into contact with chlorine can have skin irritations or their clothing may be bleached. Apparently bromine does not control *Pseudomonas aeruginosa* in the hot water found in spas and whirlpools.

Pool water fails to meet bacteriological standards when there is inadequate disinfection or the presence of algae, leaves, and organic matter; rough pool surfaces permit harborage of organic matter; improper rate of filtration or filters are not cleaned; swimmers are not using shower facilities as they should; surface skimmers are either improperly working or not working at all; and sampling techniques are incorrect or are improperly conducted.

Bathing Beaches

Microbial Characteristics of Beaches

Nearly 4000 beach closings and swimming advisories are issued annually by local and state governments. Most problems with beach pollution have been reported for ocean, bay, and Great Lakes beaches. These problems are not limited to coastal areas. Most of the beach closings in the United States are due to the presence of high levels of harmful microorganisms found in untreated or partially treated sewage that gets into the beach area. Most of the sewage comes from combined sewer overflows, sanitary sewer overflows, and malfunctioning sewage treatment plants. Untreated storm water runoff from cities and rural areas also is a significant source of beach water pollution. In addition, important local sources include waste from boats and from malfunctioning septic tanks. People who are swimming in water near storm drains are 50% more likely to develop a variety of symptoms than those who swim farther away. Symptoms of disease including fever, nausea, gastroenteritis, flulike symptoms, sore throat, fever, coughing, etc. were caused by a variety of microorganisms. The route of entry was probably through ingestion of the contaminated water.

The postchlorination levels of coliform were reduced by 99.999% by treatment plants. However, viruses that could cause gastroenteritis can survive the chlorinating process. Other potential sources of contamination at the beach area included improperly refrigerated foods that were kept at very poor temperatures for periods of time, restrooms lacking soap, and diaper changing in restrooms where parents did not have the opportunity to clean and wash their hands, and then handled food. Other health problems include respiratory and skin conditions.

The contaminants found in these waters may be microbiological, chemical, physical, or even possibly radiological in nature. If part of the water source is from storm sewers, it is possible that individual sewage systems may be hooked into the storm sewer and may cause additional potential hazards. Further, where bathing beaches are found, generally, the houses closest to the beach area have individual sewage systems. In older areas especially, many of the sewage systems overflow and the waste ends up in the water. Animal waste, fertilizers, pesticides, and dirt are washed down from farmlands into the water.

Organisms that can cause infections of the skin, eyes, ears, nose, throat, and lungs have been found in natural-type swimming areas. Other organisms that cause a variety of enteric diseases, including typhoid fever, bacillary and amoebic dysentery, infectious hepatitis, and salmonellosis have also been found in waters that have been highly contaminated. However, because it is difficult to test for these organisms, generally the coliform test is utilized as an indicator of pollution from the intestinal tract of warm-blooded animals, including humans. A new source of problems are protozoan parasites. Outbreaks of cyptosporidiosis and giardiasis have occurred.

Rivers and streams are not considered to be satisfactory sites for swimming areas, because they are subject to a large amount of runoff from soils and to discharge from tile systems and sewage treatment plants. Many of these areas also have muddy bottoms, which make the water turbid and therefore create a safety hazard. The dammed-up type of pond or lake may be a problem because of runoff. The sand and gravel pit may have better water quality and clarity. However, the pit may be extremely deep and have sharp dropoffs, which can make the area unsafe.

Therapeutic Baths

Equipment utilized in these areas are needed for lifting and carrying, and for holding. The equipment becomes normally contaminated through the general movement of people. Further, the equipment becomes contaminated because of infected patients. The type of contamination varies with the patient.

Spas and Hot Tubs

Generally, spas and hot tubs are not drained, cleaned, or filled after each use. They may include hydrojet circulation, hot water, cold water mineral baths, air-induction systems, or some combinations of these. These spas and tubs are shallow in depth and are not intended for swimming or diving. They do, however, have closed cycle water systems that should be designed for water circulation, filtration, heating, disinfection, and overflow systems.

Drugs and alcohol should never be used by anyone going into high water temperatures. Individuals may drown if they fall asleep in the spa or hot tub. The high temperatures of the spas or hot tubs, combined with a small amount of alcohol in the bloodstream, tends to accelerate drowsiness. Safe disinfection residuals and pH levels are quickly depleted during periods of use because of the high temperature, high water turnover, and large number of individuals using the facility. Injuries can occur when drain grates are broken or missing. It is easy to slip, trip, or fall on a wet interior and deck surface in the spa or hot tub.

Pregnant women, elderly people, and individuals suffering from heart disease, diabetes, or high or low blood pressure should not enter the spa or hot tub without prior medical consultation and permission from their doctor. Individuals who are on tranquilizers or other drugs that cause drowsiness or drugs that may raise or lower blood pressure should not use the spa or hot tub. Long exposure to the hot water may result in nausea, dizziness, or fainting. It is recommended that individuals spend a maximum of 10 to 15 min and then leave the water and cool down before returning for another brief period of time.

Potential Health Problems in Recreational Water

Diseases of concern that may be spread from swimmer to swimmer include typhoid fever, paratyphoid fever, amoebic dysentery, bacillary dysentery, cryptosporidiosis, and giardiasis. In addition, it is suspected that a variety of respiratory diseases including colds, sinusitis, and septic sore throats might be more readily spread in swimming areas due to close contact and availability of the organisms. Eye, ear, nose, throat, and skin infections may occur because of the presence of the organisms that create the infections and also because of the use of excessive amounts of water-treatment chemicals. In locker rooms, such diseases as athletes foot, impetigo, and various types of dermatitis may be spread from bodies, towels, or surfaces.

A vast variety of inorganic and organic chemicals enter the pool in the potable water supply. Because in many cases chlorine is the disinfectant of choice, the chlorination of water carrying hydrocarbons could cause the production of chlorinated hydrocarbons, which are hazardous to humans.

Recent Outbreaks of Disease

In August 1999, individuals who were attendees at an interactive water fountain at a beachside park in Florida became ill from *Shigella sonnei* and *Cryptosporidium parvum*. All 38 people who became ill had entered the water fountain. It was determined in an environmental assessment that the fountain used recirculated water that drained from the wet deck and play area floor into an underground reservoir. The recirculated water volume was 3380 gal and the minimum flow rate through the recirculation system was 115 gal/min. The turnover rate was 30 min, which was required by state code for interactive water features. The recirculated water passed through a hydrochlorite tablet chlorination system before it was pumped back to the reservoir and then to several high-pressure fountain nozzles at ground level throughout the play area. No filtration system had been installed. The fountain was popular with diaper and toddler-aged children who usually stood directly over the nozzles. Chlorine levels were not monitored, and the hydrochlorite tablets were depleted within 7 to 10 days of use. They had not been replaced in a timely fashion. Since 1989, about 170 outbreaks have been associated with recreational water, including swimming pools, water parks, fountains, hot tubs, spas, lakes, rivers, and oceans.

During 1997 and 1998, a total of 18 states reported 32 outbreaks causing illness in an estimated 2128 people, associated with recreational water. Seven of these outbreaks occurred in Wisconsin and four in Minnesota. Of the outbreaks, 90% were of known etiology, with 50% of the outbreaks caused by parasites, 22% caused by bacteria, and 11% caused by viruses; 47% were associated with freshwater and 53% with treated water. All the outbreaks caused by parasites were due to *Cryptosporidium*. The bacterial outbreaks were caused by *Escherichia coli* O157:H7. Typically gastroenteritis caused by bacteria occurred in nonchlorinated or improperly chlorinated water.

Current methods of water treatment may not always ensure protection against protozoan infections in swimming pools. The inactivation of these cysts may require greater than 15 min, depending on such factors as temperature, pH, and chlorine concentration. Under normal operating conditions for pools, at least several days would be needed to inactivate *Cryptosporidium* oocysts.

Most of the reported outbreaks of dermatitis associated with hot tubs, whirlpools, and swimming pools are associated with inadequate operation and maintenance procedures. They are preventable if the pH is 7.2 to 7.8, the free residual chlorine levels in the range of 2.0 to 5.0 mg/l. In swimming pools they are preventable if the pH is 7.2 to 8.2, and chlorine levels are 0.4 to 1.0 mg/l. *Pseudomonas* is the typical etiologic agent.

Specific incidents of disease have been traced to various types of water used for swimming. In Glenwood Springs, CO, 262 cases of granulomatous lesions of the skin were found. The pool used was a fill-and-draw pool. The disease was caused by a bacterium of anal origin, which creates a tumor or nodule. Control is carried out through the use of proper disinfecting residuals. Granuloma has also occurred in northern Florida, Sweden, Britain, and California. The organism most frequently identified has been *Mycobacterium valeni*.

In Saint Paul, MN, skin infections were reported from a heated whirlpool located at a local motel. Of the 60 people involved, 56 were 13-year-old children. Apparently, the skin infections came from both the swimming pool and the adjacent heated whirlpool. The infection was caused by *Pseudomonas aeruginosa*. Defective equipment and inadequate disinfection had taken place at the pools. Debris was found at the bottom of the pool and the floor was cracked. Although the free available chlorine residual was 0.2 ppm and the pH was 7.8, this still did not kill off the causative organisms. The skimmer was not functioning in the smaller pool. In the larger pool, the clarity of the water was fair; however, the free available chlorine residual was 3.0 ppm and the pH 7.8 to 8.0. The water temperature was 82°F. Both skimmers were nonfunctional and one of the pools contained diatomaceous earth sludge accumulation. It is evident from these studies that, even with chlorinated water, an opportunity still exists for contamination by organisms. Carpeting that was immediately adjacent to the pool may have been a contributing factor to the bacterial contamination of the pool, because some of the carpets had positive *Pseudomonas*. The symptoms ranged from mild to severe skin rashes in 36 people, to sore throats, sore eyes, rhinitis, malaise, nausea and vomiting, and infected skin lesions in others.

Studies conducted in Tennessee indicated that *Mycobacterium marinum* could be found in swimming pool water. It was also found that the organism was quite resistant to free available chlorine in a range of 1.5 ppm. In Oakland County, Michigan, an outbreak of gastroenteritis at a campground lake was probably caused by a Norwalk-type virus.

The previous studies indicate that considerable potential exists for health problems in pools and other swimming areas. Many of these problems probably go unrecognized because individuals are affected, instead of groups. It certainly is the concern of the environmental health practitioner not only to conduct further research into the potential microbiological and chemical hazards due to pool water but also to insist that pools be properly constructed, properly designed, properly maintained and operated, and properly disinfected at all times. There is no reason for people to have any of the previously mentioned diseases or to suffer from boils, conjunctivitis, impetigo, or any other waterborne disease that may be transmitted at the time when they are seeking pleasure through the recreational use of water.

Acute respiratory distress has occurred in bathers at indoor pools. The irritating effects may be due to chloramines, chlorine dioxide, or ozone. Inadequate ventilation adds to the problem.

Today the environmentalist is challenged by the sharp increase in the number of pools. More than 1 million permanent pools and 1 million home pools plus 4 million portable pools and 10 million wading pools are found in the United States. This compares to 1947, when there were approximately 10,000 permanent pools and 2500 home pools. In many cases the water environment does not appear to be an attractive one for the environmentalist to work in, because it involves a routine procedure that must be followed week in and week out during the swimming season. However, the sheer routine nature of inspections, studies, and sampling is the greatest safeguard against the outbreak of serious diseases.

The basic bacteriological analysis may not be sufficient in new types of water recreation areas. Although pools may be operating with apparently satisfactory free chlorine concentrations, pseudomonas and staphylococci may be found. *Pseudomonas aeruginosa* has been linked to contaminated swimming pools. It can be a special problem in hydrotherapy pools and spa baths, where the water is warm, as well as in leisure pools. Pseudomonas may be found on filters, mats, and pool furniture. *Staphylococcus aureus*, which commonly inhabits the skin and nasopharynx in 10% of the human population, easily gets into swimming pools.

Potential Safety Problems

The greatest potential hazard within the swimming area is the swimming pool accident. Unfortunately, an appalling number of accidents and drownings occur, and also supervision of bathers is unfortunately lacking. Some specific environmental problems that cause accidents include loose diving boards, badly located water slides, projecting pipes, improperly installed or maintained electrical equipment, use of glass bottles in the pool area, slippery and slimy pool decks and pool surfaces, and slippery walls and floors. Chlorine, especially chlorine gas, may be a hazard if leakage occurs and if maintenance personnel or pool supervisors do not understand how to handle the potential hazard.

Over 4000 people die from swimming accidents. The U.S. Coast Guard reports over 6000 crashes of recreational vehicles each year. Of the fatal accidents, 50% are alcohol related. Countless millions are injured during the course of recreational water use.

The peak period of fatalities occurs between May and September, with July the peak month. Most deaths occur on weekends; the fewest deaths occur on Wednesdays. Most deaths occur between noon and six in the evening. Over one half of the home drownings were infants and children under 10 years of age. Fatalities are caused by inadequate knowledge of swimming skills, inadequate supervision of swimmers, drinking by adults, poor pool design, and lack of barriers against trespassing individuals.

Impact on Other Problems

The huge numbers of individuals seeking water sports are creating severe problems in disposal of solid waste, abuse and misuse of sewage systems and water systems, and destruction of usable park and beach areas.

Economics

At present it is not feasible to give a cost analysis of the problems related to swimming areas. The data are incomplete.

POTENTIAL FOR INTERVENTION

Intervention consists of isolation, substitution, shielding, treatment, and prevention. Isolation occurs when sources of pollution are physically kept away from swimming pool areas. Substitution is a technique where larger or more efficient filtration systems and mechanical chlorinators are used in place of existing equipment. Shielding does not apply. Treatment, one of the primary techniques used for intervention, consists of proper filtration systems, as well as proper chlorination systems. Prevention is achieved through using good water supplies and by having trained operators constantly monitor the swimming area, backwash when needed, and keep all equipment in proper operating conditions. Prevention is also achieved by determining the sources of pollution that affect an area and then eliminating the pollution. The potential for intervention in pools is excellent. The potential for intervention on beaches varies with the type of beach and the potential contaminants coming from a variety of sources.

RESOURCES

Scientific and technical resources consist of the American Public Health Association, National Environmental Health Association, National Sanitation Foundation, and National Swimming Pool Institute. Civic resources include the many groups interested in water pollution control. Governmental resources include local and state health departments: the National Science Foundation, Environmental Protection Agency (EPA), and Centers for Disease Control and Prevention (CDC).

STANDARDS, PRACTICES, AND TECHNIQUES

Pool Design and Construction

Proper pool design and construction are necessary to provide an aesthetically pleasing recreational area that is free of disease potential and injury potential. (Figure 4.2 gives a typical plan of pool service facilities.)

Size and Capacity

The pool is composed of a shallow area and a deep area, and is fitted either with an overflow gutter that runs around the periphery of the pool or, in newer pools, surface skimmers that are recessed into the wall. The overflow gutter or surface

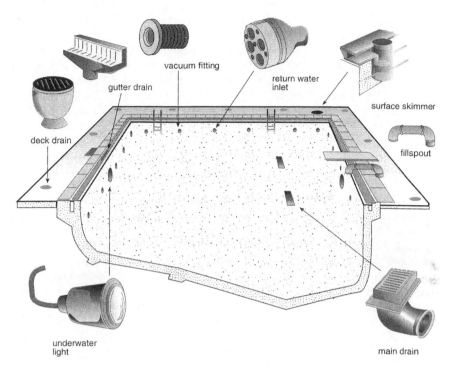

Figure 4.2   Typical plan of pool service facilities. Longitudinal section through pool showing fittings. (From Michael, J.M., *Swimming Pools; Disease Control through Proper Design and Operation,* Training Manual Communicable Disease Center Training Branch, DHEW, U.S. Government Printing Office, Washington, D.C., 1959, p. 17.)

skimmer removes dirt and oil-laden film from the surface of the swimming pool on a continuous basis. The surface skimmer is preferred, because the skimmer, if properly applied, can do an excellent job of maintaining a clean water surface and is readily cleanable. The sides and walls of a pool are straight, but the bottom of a pool gradually slopes toward the deepest part close to the diving board and then slopes up toward the wall. The pool facility includes the bathhouse or locker room, equipment room, chlorine room, area for spectators, and pool itself.

In determining the capacity of a pool, various items must be taken into consideration. Primarily, pool usage determines not only the size and capacity but also the shape and depth of the pool. For each person, 27 ft^2 of water surface area should be provided; that is, a pool that is 75 × 45 ft would accommodate a capacity of 135 people. In the shallow water area, under 5 ft in depth for recreational swimming, there should be a minimum of 14 ft^2 per person. For swimming instruction, it should vary between 20 ft^2 per person indoors to 25 ft^2 per person outdoors for the advanced swimmer and 40 ft^2 per person indoors to 45 ft^2 per person outdoors for the beginning swimmer. In the deep water area, which is over 5 ft for recreational swimming, the requirement is a minimum of 20 ft^2 per person indoors and a minimum of 25 ft^2 per person outdoors.

Capacity for advanced swimmers in deep water should be a minimum of 25 ft^2 per person indoors and 30 ft^2 per person outdoors. In the diving area, the requirement is a minimum of 300 ft^2 per diving board. In the diving area, a maximum of 12 persons are permitted within 10 ft of the diving board. When designing the pool, it is necessary to determine whether competitive swimming or recreational swimming will be the major function of the pool. If it is recreational swimming, then 80% of the water should be less than 5 ft in depth. Steps are needed to allow people to easily get in and out of the pool. The deck area should be two to three times as great as the water area for indoor pools and a ratio of 3:1 for outdoor pools.

Shape and Depth of Pool

Most pools used for athletic programs or pools that were built in the past are rectangular in shape. The slope of the pool from the 5-ft marker was not permitted to exceed 7%. However, modern pools are being built in T, L, I, H, and oval shapes. The water depths and sizes vary considerably. However, in pools where recreational swimming takes place, the minimum depth has to be 2½ ft. In competitive pools, the minimum depth is usually 3½ ft. In junior wading pools, the depth ranges from 6 to 18 in., with a slope of 1 in. in 15 in. In lake or main pools, the same 6.6% slope after 5 ft is maintained. In no case should a sharp dropoff occur from the shallow toward the deep end; it should be gradual. When ropes are used in pools to separate shallow from deep sections, the rope should be placed prior to the change in depth of the pool.

Overflow Systems

The surface skimmer, which has already been mentioned, is the best means of removing surface dirt and other debris, including insects, leaves, oils, and other suspended matter, from the pool water. The container devices should be of corrosion-resistant materials and should be made up of the housing, a weir, strainer basket, an equalizer, a cover and mounting ring, the trimmer valves, and the vacuum cleaner connections. The housing should be resistant to deformation during installation and must withstand the water pressure developed by the weight of water within it. It should withstand the crushing pressure developed by a vacuum of 25 inHg with a safety factor of 1½. A smooth flow of water over the entire effective weir length, which is at least 7½ in. at the entrance throat, must be built into the housing. If a circular weir is used, it has to accommodate at least a 4-in. diameter weir with a clearance of at least 2 in. between the weir lip and the side of the housing. The weir should be automatically adjustable and operate freely with continuous action at a flow rate of 30 gal/min. The strainer basket must be easily removable and made of material that is easily cleanable. It should have maximum openings of ¼ in. in diameter, and a retentive capacity of at least 160 in.^{3} within a minimum of 25% of the open area. Equalizers are provided to prevent a loss of prime in the skimmers. The cover and mounting ring are removable so that the service person can readily get at the basket. If trimmer valves are utilized as part of the surface skimmer, they must not interfere with the performance of the unit. Where vacuum cleaner connections are an integral part of the skimmer, they should be convenient for use and not interfere with normal operation when not in use. Use of a surface skimmer is undoubtedly a far better technique than any of the open gutter systems used in the past.

Underwater Lights and Electrical Systems

All wiring must conform to the various codes of the state and local fire marshals and other state and local agencies. The wiring also should conform to appropriate national recommended codes. Underwater lighting should not be less than 0.5 W/ft^2 of swimming pool surface area. The bottom must be visible without a glare problem. In indoor pools, a minimum of 100 lumens/ft^2 of surface area of the pool is required; and in outdoor pools, a minimum of 60 lumens/ft^2 of surface area of the pool. Lighting fixtures at the shallow end within the pool should be 3 ft deep. The lighting fixtures in the deep end should be 6 to 8 ft deep. The lights should be inclined slightly downward at a 3 to 5° level to avoid annoying surface reflections. White is the recommended color of lights, but amber is used in outdoor pools at night to reduce insect attraction. Colored lights are not recommended, because they cause a visual distortion in the water.

Construction

When constructing a pool, the area is first laid out by an engineer, the boundaries are staked, a finish grade level is determined by the use of a transit, and the finished grade level is utilized for measurement of the various pool depths prior to the actual excavation. After the pool dimensions have been determined, the thickness of the wall is added in, as well as the working room needed outside of the wall. The area is then staked out and excavated. Rough excavation is usually carried out by means of a front-end loader. The equipment needs about 8- to 10-ft-wide excess areas. In finishing, the loader can cut within 6 in. of the pool perimeter. Afterward, the shaping and finishing must be done by hand. The final finishing must be done carefully, because the wall material follows the exact lines of the excavated area. This occurs when gunite and fiberglass are used in the pool as pool materials. Where concrete block, poured concrete, or steel and vinyl liner pools are utilized, the area is over-excavated and then backfilled. Care must be taken in backfilling, because settling of the backfilled material may cause a weakness in the pool structure. Various problems must be overcome during construction. If water is encountered, difficulty may occur in moving the front-end loader; the walls may sag; finishing may be poor; and the water, if it is under sufficient pressure, may actually cause the pool to rise, as a bathtub would if it were placed in a large body of water under considerable pressure. Obviously, this type of soil creates a poor site for a pool. Where rock is encountered, air hammers are needed and operation becomes expensive, but the rock can be used as a solid foundation for the pool. Where sand is encountered, the walls may cave in. It is possible to prop up the walls carefully and work in this type of area. However, care must be taken that sand cannot wash out from under the pool. Therefore, a good solid foundation must be placed.

Materials

A variety of materials are used in pool construction. They include concrete, steel, vinyl liners, and fiberglass. Concrete pools are built by use of the gunite technique, poured concrete technique, hand-packed concrete, or use of concrete blocks. In the gunite technique, the only framing needed is a level beam. After excavation, the bottom is filled with a layer of crushed rock. Wire, 6 × 6 in. screen of 10 gauge, is placed on the crushed rock, and steel is placed over the wire. A mixture of sand and cement is made into a dense mix and then applied to the material that has already been utilized to frame the pool. It is important that the shell be of even thickness, that all loose sand be removed, and that the gunite be pushed in behind the steel. When the concrete is dry enough to walk on, the finishers then smooth the walls and floors.

When poured concrete pools are made, forms must be put in place and supported. Steel is placed in to help strengthen the walls. The steel is usually used to lock the inside forms in place. A mixture of cement, sand, and rock is used. The mixture contains no more than 6 gal of water to 1 gal of cement. The floor is poured 4 to 6 in. thick with 8 in. in the corners. After the forms have been removed, the walls are backfilled. The last step is to bend the projecting steel rods to reinforce the beams. In hand-packed concrete, the same materials are used as in poured concrete, but most of the forms are eliminated.

When using concrete blocks, the operation is similar to poured concrete, except the blocks serve as the forms and are actually incorporated into the pool structure. The footings are usually 12 × 12 in. or 8 × 24 in., and are reinforced with steel rods. The footings must be level and solid. The first row of blocks is placed on the poured footing. Again all blocks must be level. Grout is used to provide a solid joint with the footings. The blocks are then set one on top of another until the top of the wall is reached. All areas of the pool must be grouted.

Steel is designed to stand alone, be buried, or be partially exposed. The steel used for pools is ¼ in. thick, and it is constructed in either straight or curved panels. When using steel, an access for trucks is needed for carrying the steel wall and floor sections. The trucks must be able to come fairly close to the pool, and a crane must be fairly close to the pool.

Vinyl liners are used for pools. The vinyl liners are put in the structured shell. Poured concrete is usually used first where the walls are 8 in. thick and reinforcing steel is placed between the walls. Concrete blocks, wood panels, or steel or aluminum shells may also be utilized. The vinyl liner is then installed within the shell.

Fiberglass is now used in some pool installations. The installation creates some problems, because the seam must be watertight. Usually the sections are overlapped and then bolted in place. Polyester resins are placed between the fiberglass sheets to form a watertight bond.

Bathhouses

The bathhouse or locker room should be located as close to the pool as possible. If the pool is outdoors, it should help protect the pool against the prevailing winds. The only entrance to the pool should be through the bathhouse or locker rooms. The facilities should include a lobby, basket storage area, dressing rooms, and toilet facilities. Usually the area of this facility is about one-third the area of the pool. The dressing rooms need to provide 3½ ft² for men and 7 ft² for women. Because normal attendance at pools is roughly two men to each woman, the dressing room areas should be equal. The floors need to be smooth, impervious, and nonslip. They should be sloped to drains at a rate of ¼ in./ft. The rooms must be ventilated, have heating that keeps the temperature between 70 and 75°F, and have lighting that provides a minimum level of 10 ft located 3 ft from the floor. The shower areas should contain one shower head for every 40 swimmers. The shower water should be controlled at 90°F and flow at a rate of 3 gal/min per showerhead. Toilet facilities for men include one commode and one urinal for every 60 men, and one commode for every 40 women. Sinks must contain hot and cold running water, with soap and individual towels available. One sink should be provided for every 60 persons.

Pool Plans Review

All pools must be evaluated and their plans approved prior to construction by the state or local health departments. Prior to operation, the pool is again evaluated to determine whether all the specifications have been met and whether the pool is ready to be put into operation. The following information must be provided, along with a set of plans and specifications for the pool: (1) statement on the nature, scope, and special requirements for space, equipment, and facilities; (2) a listing of the proper depths to be used for the various age groups and for various activities; (3) provision of at least 9 ft of water for 1-meter boards, 12 ft of water for 3-meter boards, and 15 ft of water for 10-meter boards; (4) adequate numbers of swimming instructors and lifesavers; (5) an explanation of the overflow system and how it will be serviced; (6) a listing of adequate storage for maintenance materials separate from instructional materials and equipment; (7) an indication of how diving boards and lifeguard chairs will be anchored; (8) the means of access to the filter room; (9) the pitch to the drains within the pool, on the pool deck and on the floors of the showers and dressing rooms; (10) the distance between the diving board and the side walls; (11) the type and amount of lifesaving equipment; (12) the type and amount of pool-cleaning equipment; (13) the types and level of illumination for underwater lights; (14) the type of pool heater and its capacity; (15) the expected demand on the recirculation-filtration system and the anticipated future demands; (16) provision of separate space for gas chlorinators; and (17) provision for waste-water to be moved to waste instead of recirculated.

In indoor pools the following concerns must also be listed: (1) the means of mechanical ventilation; (2) provision of and type of adequate acoustical treatment of walls and ceilings; (3) overhead clearance of 16 ft for 1-meter boards and 16 ft for 3-meter boards; (4) provision of 100 foot candles of light; (5) types of wall-base covers that will be used to facilitate cleaning; (6) the type of temperature control in the pool room for both water and air; and (7) the type of service trenches and the deck of the pool.

For outdoor pools, the following information is needed: (1) location of the pool, which should be away from railroad tracks, heavy industry, trees, and open dusty

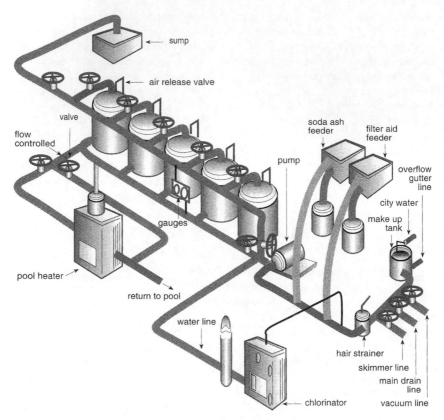

Figure 4.3   Swimming pool equipment room. (From Michael, J.M., *Swimming Pools; Disease Control through Proper Design and Operation,* Training Manual Communicable Disease Center Training Branch, DHEW, U.S. Government Printing Office, Washington, D.C., 1959, p. 21.)

fields; (2) distance from sand and grass to the pool; (3) type of fence, its height, and the means of securing it; (4) type of subsurface drainage being provided; (5) amount of deck space for sunbathing; (6) types and locations of outdoor lights; (7) provision of nonslip material for the deck of the pool; and (8) provision of an eating area separated from the pool deck.

Pool Equipment

See Figures 4.1, 4.2, and 4.3.

Hair Catcher

The hair catcher or hair strainer is an important part of the water treatment process. The unit should always precede the motor in the operation. It should be opened at least once a week for inspection and should be thoroughly cleaned with a stiff wire brush or washed out. A single-screw clamp is easier to unfasten than a bolt-type fastening device, and therefore the single-screw clamp is preferred. The hair strainer can also be utilized in emergencies for adding disinfectants, coagulants, algicides, or soda ash for pH control if the normal dispensing equipment breaks down.

Makeup Tank

The makeup tank, or balance tank, has water added to it from a pipe above, which is at least two diameters of the pipe above the tank itself. This avoids a possible cross connection by providing a necessary air gap. The water in the makeup tank is then utilized to maintain the pool water level. In the event of emergency, this tank may also be used for adding a variety of chemicals to the pool water.

Pumps

The pumps and motors must be protected against corrosion. Excessive water or chemicals can cause the pump to become partially oxidized and therefore begin to erode and become ineffective. Galvanic corrosion can occur if different metals are used in the pump. This can be minimized by using anodic metal, such as cast iron, for the pump housing, and a cathodic metal, such as bronze, for the impellor. The pump is lubricated frequently by having controlled leakage of water. Priming of the pump excludes air, which may interfere with proper operation. The pumps and motors should be lubricated according to the manufacturer's instructions. Pumps can fail because of lack of priming, wrong direction of the impellor rotation, or too low a motor speed. The pumping capacity can be reduced because of air pockets or leaks in the suction line, clogged impellors, or excessively worn impellors. Mechanical troubles and noise may result from poor alignment of the motor pump shaft, a bent shaft, or damaged bearings.

Heaters

Pool heaters are used to raise the temperature of the water to 75°F. In an indoor pool, the air temperature should not be more than 8°F warmer or 2°F colder than the water in the pool when in use. Actually the best temperature of the air is 5°F higher than the water temperature. The pool heater can become full of scales as a result of chemical action. When this occurs, a 50:50 mixture of hydrochloric acid and water can descale the unit. When the chemical foaming action stops, the material should be completely flushed out of the heater unit and then the unit should be again flushed very carefully before it is put back into operation.

Bacterial, Chemical, and Physical Standards

Whenever chlorine, calcium hypochlorite, or other chlorine compounds are utilized without the use of ammonia, the chlorine should be a minimum of 1.0 mg/l or 1.0 ppm of free residual chlorine. The recommended pH is 7.2 to 7.8. Unfortunately, the addition of alum or aluminum sulfate, which is used as an aid in filtration, tends to raise the pH into a more alkaline zone. A free chlorine residual of 2.0 mg/l or bromine residual of at least 4.0 mg/l should be maintained when the spa or hot tub is in use.

Water must have a clarity at a level such that a black disc 6 in. in diameter placed on a white field can be clearly visible from the sidewalks of the pool.

The bacterial quality of swimming pool water must conform to the following standards: (1) the total count must not have more than 15% of the samples containing 200 bacteria or more per milliliter, in any 2-month period of time; (2) no more than 15% of the standard samples must be used for determining coliforms, showing gas in any of the 5 to 10 ml portions over a period of time; and (3) the arithmetic mean density must not exceed one coliform organism per 50 ml of water with use of the membrane filter technique.

Water Treatment

A minimum turnover of water in the pool should be at least three times every 24 hours or once every 8 hours. In junior pools, the turnover rate should be a minimum of once every 4 hours. In shallow wading pools, the minimum should be a turnover of once every hour. The best system utilized has a double pump arrangement with each pump capable of an 8-hour rate.

Within the water treatment system itself, it is necessary to replace a minimum of 20 gal of water per bather. This is the water that is splashed out, evaporated, or lost somewhere in the water filtration process. The surge tank, which is used to supply water to the pool, must be set up in such a manner that too large a surge cannot occur and therefore the excess water cannot be lost to the overflow drains or to the surfaces.

Sand Filters

The basic type of sand filter used today is a pressure sand filter (Figure 4.4). In the past, in older pools, gravity sand filters were also utilized. The pressure sand filter is a closed tank that contains the filtering media. The filter media is made up of filter sand, free of carbonates or other foreign material, with an effective size of 0.4 to 0.55 mm, and with a uniformity coefficient not exceeding 1.75 mm. The filter sand must be at least a minimum of 20 in. in depth. Where gravel is used to support the sand filter, it must be rounded material free of limestone, clay, and other materials. At least four layers that are properly graded are needed to prevent intermixing. The total supporting material must be a minimum of 10 in. above the lower distributor openings. These filters are supplied with a filtering mat of coagulant material. They should be able to operate 24 hours a day at a rate of 3 gal/min/ft^2. When the difference in gauge pressure between the influent pressure and the effluent pressure reaches 5 lb/in.^{2}, the filter should be backwashed. The backwash rate is 15 gal/ft^2/min for

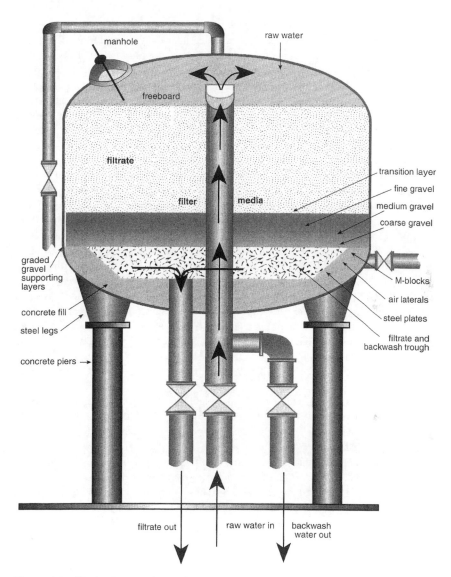

manhole

raw water

freeboard

filtrate

filter   media

transition layer

fine gravel

medium gravel

coarse gravel

graded
gravel
supporting
layers

M-blocks

air laterals

concrete fill

steel plates

steel legs

filtrate and
backwash trough

concrete piers →

filtrate out   raw water in   backwash
water out

Figure 4.4   The basic type of sand filter: pressure sand filter.

about 5 min or until the backwash water appears clear in the glass viewing chamber for at least 2 min. At this rate, filter sand should not be washed out of the filter itself.

The advantages of the sand filters are that they are fairly easy to operate, they can be placed outdoors, the operating cost is moderate, they are fairly easy to backwash, they need minimum maintenance, the turnover rate of the water is good, and the initial cost is comparatively low. The disadvantages of the sand filters are that they need a fairly large area, they require a qualified swimming pool operator to operate the unit, and they require more space than the diatomaceous earth filter.

Certain operational problems occur in the use of the sand filters. They include air binding, which prevents the water from covering the entire top of the sand, because air is trapped at the top of the shell above the filter media. This can be corrected by use of automatic or manual air release valves. A mechanical loss of filtering mat occurs when the system has been shut down for extended periods of time. Personnel may have difficulty reading the pressure differential, because the gauges may be operating improperly or improperly calibrated gauges are utilized. There is a rate of flow change through filters. Although the pump is running at the same speed, different amounts of water can pass through the filter throughout the filter run. A rate of flow controller should be installed prior to the filters to correct this. Filter media difficulties may occur when the media either is mixing throughout its various depths or too much of the media has been lost during backwashing. This may also be due to a dirty sand surface.

Backwashing is an integral part of the entire water treatment process, because so much depends on the filtering system removing the various impurities. It is necessary to check the backwash pump and to make sure that it is operating properly. It may also be necessary to use a caustic soda solution on top of the sand filter and then to slowly backwash the filter to remove the various contaminants that may be clogging the filter. If this technique is used, the filter must be carefully washed before it is put back into operation. Occasionally, filter beds that are underdrained become obstructed by accumulations of dirt and sand held together by organic growths. This can be corrected by applying 2 oz of calcium hypochlorite per square foot of filter surface, allowing the hypochlorite solution to soak into the material, backwashing slowly, and then giving the filter a complete and thorough rinsing before putting it back into use. Eventually it may be necessary to replace the filter media itself. Anthraf, which is anthracite, is utilized at times as a filter media. The backwash rate here is 9 to 12 gal/min/ft^2.

It is recommended that the standards titled, Sand-Type Filters for Swimming Pools and Diatomite-Type Filters for Swimming Pools, as well as other swimming pool standards put out by the National Sanitation Foundation in Ann Arbor, MI, be consulted for the latest changes in the types of filtering systems and media utilized.

Diatomaceous Earth Filters

Diatomaceous earth filters are filtering systems that use a hydrous form of silica or opal composed of the shells of diatoms, which are microscopic unicellular algae whose shells have been impregnated with silica. Several million of these shells may be found in 1 in. of diatomite. The material becomes an excellent filtering material and is utilized not only in swimming pools, but also for filtering and clarifying sugar and syrups. The largest deposits in the United States are found in Santa Barbara County, California where the beds are 1000 ft thick and spread over several square miles.

The diatomaceous earth system either may be under pressure or run on a vacuum. A slurry of diatomaceous earth is added to the filtering system and is slowly permitted to coat an inner core or a series of inner plates. The filter spectrum is tested at a filter rate of 2 gal/min/ft^2 when using 0.15 lb of the test filter aid per square foot of filter area. The filter aid that passes during the first minute of flow must not exceed a turbidity level of 10 ppm. The precoating operation and the backwashing operation are extremely important in the use of diatomaceous earth filters. During precoating, a uniform distribution of the material on the core that is in use is needed. Unless backwashing proceeds properly, the material may be flushed back into the pool and therefore cause a tremendous mess. This may occur when inexperienced individuals are applying the filter aid or are backwashing the system. The filters are designed to handle 2 to 3 gal/ft^2/min while in operation 24 hours a day. The pressure filters are backwashed when the differential between influent and effluent pressures are in a range of 25 to 50 lb/in.^{2}, depending on the type used. The vacuum filters are backwashed when the vacuum gauge reads 10 to 15 inHg.

The advantages of the diatomaceous earth filter are that a considerable amount of space is saved, the water is clearer and cleaner, the initial cost is reasonable, and the operating cost is fairly low.

The disadvantages of the diatomaceous earth filter include high maintenance cost, high operating cost, difficulty in operating, larger pumps required, need to backwash more frequently, and difficulty in determining whether the entire filtering media have been removed during backwashing.

Some of the operational problems that occur include shorter filter runs than with sand filters; clogged filter elements; diatomite and organic matter clogging, which can be relieved by adding calgon, allowing it to stand for 2 hours, and then scrubbing with a wire brush; iron clogging, which can be removed by soaking in hydrochloric acid, with the system flushed very carefully afterward; manganese, calcium, and magnesium clogging, which can also be relieved by soaking in hydrochloric acid and then rinsing thoroughly; loss of the filter material as a result of shutting down the system even for a short period of time; and failure of the filter elements through either breakage or other difficulties.

It should be recognized that in many smaller pools especially, diatomaceous earth is preferred to sand filters because of cost saving, and space saving. To ensure that the system functions as well as possible, it is necessary to consider the following major points: utilizing the correct grade of filter aid and properly applying this material; operating the system at a rate of 2½ gal/min/ft^2 of filter area, because too high a flow can reduce the filter runs and too low a flow can reduce the filter media on the filter element; routine cleaning of strainers, surfaces, and filter elements; routine maintenance of pumps, motors, and pump shafts; routine repair of leaks; and inspection and cleaning the air relief system on a routine basis (Figures 4.5 and 4.6).

Cartridge Filter System

A cartridge filter system consists of a container that has a series of cartridges in it. Each cartridge filter module is used for filtering the swimming pool water. When the cartridges become dirty, they are removed from the filter and are rinsed thoroughly with a garden hose. The cartridges are then soaked for at least 8 hours in a

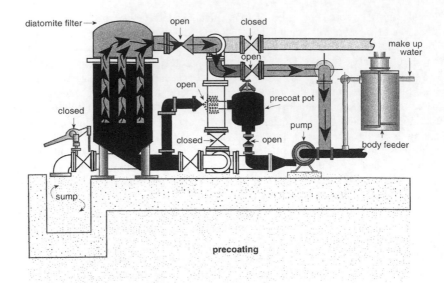

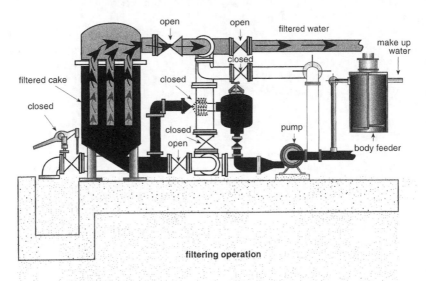

Figure 4.5   The diatomite filter. (From Michael, J.M., *Swimming Pools; Disease Control through Proper Design and Operation,* Training Manual Communicable Disease Center Training Branch, DHEW, U.S. Government Printing Office, Washington, D.C., 1959, p. 37.)

solution of 1 lb of trisodium phosphate per 3 gal of water. The cartridges are then washed off. This removes oils and dirt. Next the cartridges are soaked for at least 2 hours in a solution of 1 gal of muriatic acid per 3 gal of water. This is done in a plastic container. Then the cartridges are washed off with water. This removes the

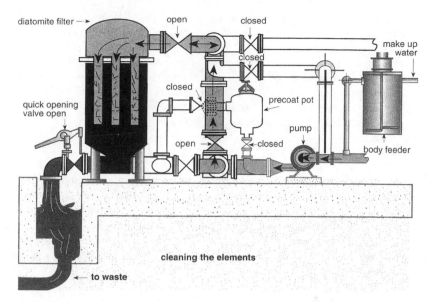

Figure 4.6 Cleaning the pool elements. (From Michael, J.M., *Swimming Pools; Disease Control through Proper Design and Operation,* Training Manual Communicable Disease Center Training Branch, DHEW, U.S. Government Printing Office, Washington, D.C., 1959, p. 38.)

calcium deposits. The cartridges can then be placed back into the filter tank ready for reuse.

Disinfecting Chemicals

Chlorine

In handling chlorine gas containers, it is important to store them under cover, because they have fusible plugs that melt at 160°F. Storage should also be in ventilated surface areas, because chlorine is heavier than air and creates more hazardous conditions underground. When the tank is in use, it is secured tightly to the wall or to the floor so that it is not knocked over. Protective caps are utilized on the cylinder when it is transported. When the cylinder is hooked up to the chlorinator, it must be done in such a way that all leaks are prevented. New gaskets should be used on all connections, and the shutoff valve should be completely secured prior to the time of the hookup. The temperature of the chlorine cylinder must not be higher than the chlorine feed lines, because chlorine hydrate may be formed and the lines may clog.

Liquid chlorination may be produced by using calcium hypochlorite, which is a white, dry, granular compound, or may also be produced from the tablet form. The chlorine yield equals 70% of the weight of the calcium hypochlorite. It is easier and safer to use than chlorine gas. Sodium hypochlorite may be found in liquid solutions, such as household bleaches. The cost is slightly higher. However, the liquid, which is in a 5.25% available chlorine solution, is ready to be utilized.

In preparing calcium hypochlorite solutions that contain 1% chlorine or 10,000 ppm, 19 oz of 70% available chlorine–calcium hypochlorite is mixed with 10 gal of water. The material is stirred in the water and allowed to settle. When the precipitate goes to the bottom, the supernatant fluid is siphoned off and is then ready for use in chlorination of the water supply. It is important that the hypochlorinators be cleaned thoroughly on a daily basis, and if necessary, treated with dilute hydrochloric acid and then thoroughly flushed and rinsed. Replacement parts should always be available to prevent downtime on the chlorinator.

Bromine

The residual for bromine required in a pool is 4.0 ppm to be as effective as chlorine in killing bacteria. Bromine costs two to four times as much as chlorine. Although it is supposed to cause less eye irritation, its value is still widely disputed by experts in the field. Bromine can be tested by using a chlorine test kit. If the reading on the chlorine test kit is 2.0 ppm, the correct reading for bromine is twice as much, or 4.0 ppm. The test indicates some difficulty, because chlorine is present with the bromine.

Iodine

The recommended level for iodine is 1.0 ppm.

Ozone

The ozone treatment process consists of the treated water going from the pool to a balance tank where makeup water is added and some water is disposed of; then the water goes to treatment, heater, chlorine, and pH adjustment and back to the pool. The treatment process consists of flocculation, filtration, ozonation at a level of 1 mg/l dosage; and a contact time of 2 minutes to degassing to remove the excess ozone to granular activated carbon (GAC) filtration and ozone gas destruction to the heater, chlorination, pH adjustment, and back to the pool.

Chemical Feeding and Proper pH Control

In addition to the disinfecting agents that are used in water, various other chemicals are used to either raise or lower the pH, and to cause coagulation in the water treatment process. Further, special problems, such as algae, are controlled by the use of chemicals. The pH in swimming pool water should range from 7.2 to 7.8 to have the most effective operation of the pool with the least irritation to the swimmers. The pH is the hydrogen ion concentration. A pH value of 7 is neutral, below 7 is acid, and above 7 is alkaline. The pH level is important because in coagulation, when using sand filters and coagulating chemicals to produce a proper floc, the optimum pH should be 7.2 to 7.6. In the use of chlorine in disinfection, if the pH becomes too high, the effectiveness of the chlorine is greatly reduced. In any event, if the pH is below 7 or above 8.4, swimmers may feel skin and eye irritations.

The factors that affect pH levels are additions of disinfectants, chlorine gas, or bromine, all of which reduce the pH, or hypochlorites, which raise the pH. Further, coagulants such as alum lower the pH. The makeup water must be taken into account when adjusting pH, because if the water is either highly alkaline or highly acid, additional chemicals must be utilized.

To raise the pH in a pool when it is below 7, soda ash (sodium carbonate) is fed into the pool, either through the recirculation system or with briquets, which are large tablets fed manually into the pool. Caustic soda, which is sodium hydroxide or lye, is more hazardous to use, but more effective in raising pH. However, lye is only recommended when serious acid water problems exist, and then the final water must be carefully analyzed after it has been allowed to become thoroughly mixed. Lime, which is calcium hydroxide, raises pH, but is not recommended for pool water, because it makes the water turbid.

To lower the pH, certain acids may be used. Again, great care must be taken in the use of the acids because they may make the water too acid or they may become too concentrated in one area. Dilute hydrochloric or sulfuric acid, which is potentially hazardous to the operator, swimmers, or equipment, has been used for lowering the pH. These acids can interfere with alum coagulation and increase turbidity. The dilute hydrochloric acid is also known as muriatic acid.

Sodium bisulfate (sodium acid sulfate) is safer to use than acids for lowering pH and also has the additional benefit of removing lime deposits from the chemical feed lines, filter sand, and piping. Sodium bicarbonate is used for alkalinity adjustment; is long lasting, inexpensive, and easy to use; and is used for making extensive adjustments in alkalinity, but is slower acting than soda ash. Cyanuric acid is utilized in some pools along with chlorinated isocyanurates, not only as a disinfectant, but also as a pool stabilizer. The chlorine residual is kept at 1.0 ppm at a pH range of 7.2 to 7.8 and a concentration of cyanuric acid of 10 to 100 ppm.

To obtain proper coagulation, which precipitates the material present in water, alum (either in the form of aluminum sulfate, which is most commonly used), ammonium aluminum sulfate, potassium aluminum sulfate, or special filter alums (which are more expensive and specially marketed by certain swimming pool companies), are mixed with pool water on the suction side of the recirculation pump to form the necessary precipitate. The feeding technique used with alum is important, because the place where the alum is inserted, amount of alum used, and time of contact all affect the actual filtering of the pool water. If too much alum is used, then the filter runs are shortened because the filter surfaces become more readily clogged. If too little alum is used, the fine material may pass back into the pool and cause turbidity. Care must be taken to check the feeding devices to make sure that they do not become clogged.

Other Standards

Other swimming pool standards include UV light plus hydrogen peroxide, 40.0 ppm; total dissolved solids, 2500 ppm; spa maximum temperature, 105°F; calcium hardness, 200 ppm; and total alkalinity, 100 to 125 ppm.

Saturation Index

The saturation index is a formula using arbitrary numeric values assigned to each of the five balanced water factors to produce an index figure to determine and adjust the balance of swimming pool waters. The factors are pH, temperature, total alkalinity, calcium hardness, and total dissolved solids. An index of –0.5 to +0.5 is considered adequately balanced.

Maintenance

Beside the maintenance of equipment utilized in the water treatment process, it is also necessary to have adequate pool and pool area maintenance performed on a regular basis. Cracks in pools must be repaired with a watertight material to prevent water loss, serious damage to the pool, and accident hazards. If the pool loses 1 in. of water over a 1000 ft^2 pool surface area, 625 gal of water will have been lost. When repairs are made, it is necessary to use special hydraulic cement compounds, avoiding the use of tar or asphalt fillers. The tar and asphalt are not only unsightly, but also ooze into the pool water.

Pool cleaning not only is a function of the water treatment process, but also must be carried out by use of suction cleaners and cleaning of the skimmers. The suction cleaning system should be separate from the other water systems. It should be utilized to remove the materials that have fallen on the surface of the pool, as well as those that have settled to the bottom.

Pool painting is necessary in certain types of pools to reduce algae penetration, to protect the pool walls and floors against chemical attack, and to make a more pleasant environment. The most suitable paints are rubber-based paints, because they are the paints that deteriorate the least. However, previous layers of paint must be removed before new rubber-based paints are utilized effectively.

In the off-season, all the recirculation equipment should be carefully drained and inspected, oiled, lubricated, serviced, and covered, where possible, to protect them. When the pool proper has water in it, wooden poles should be placed within the water against the sides of the pool to protect them against ice thrust pressure. All other areas of the pool should be thoroughly cleaned and serviced before locking up the area for the next swimming season.

Bathing Beaches

Where an area is approved for wading and swimming, it should have a sand bottom from at least the edge of the water to about 5 ft in depth. A sand beach is also desirable. A diversion ditch should be constructed to divert any drainage that may come into the swimming area and also to divert the surface runoff from flowing into the swimming area. The wading and swimming areas should be plainly desig-nated by using floats or other barriers. The swimmers must be kept within the designated boundaries by lifeguards, who must be on duty whenever the area is in use. Eating areas or food-serving areas should be separated from the beach area by a suitable barrier or fence. Trash containers should be placed conveniently within the eating area and also at the entrance gate to the beach. Food, especially glassware, should not be permitted on the beach itself.

The water used in the swimming area should be clear. In natural areas, low turbidities are frequently found in ponds or lakes. High levels of turbidity are caused by surface runoff or by bathers stirring up sediment on the bottom of the natural bathing area. The standard of clarity should be such that a black disk, 12 in. in diameter on a white field can be visible at a minimum depth of 4 ft from the sides of the swimming area.

The bacterial standards allowed for bathing areas vary from state to state. In any case, they exceed those that are permitted in swimming pools. In the state of Illinois, a bathing beach may be classified as being bacteriologically unsafe if the fecal coliform count exceeds an average of 200 per 100 ml sample collected during a 30-day period, or 400 per 100 ml sample in 10% of the total number of samples collected during a 30-day period. In the state of Michigan, the maximum total coliform geometric average must not exceed 1000 per 100 ml sample in any series of 10 consecutive samples. It must not contain more than 2 samples that exceed 5000 per 100 ml, and the maximum fecal coliform must not exceed a geometric average of 100 per 100 ml sample. The chemical determinations taken must follow standard methods and show that the water is free of chemical substances that can create toxic reactions or irritations to the skin or membranes of the bather or swimmer. The physical determinations must show the water to be free from turbidity, color, deposits, growths, oils, or greases, which may be a safety hazard, a hazard to health, or a nuisance to the bather. In the state of Michigan, the health officer has the right to close the bathing area if an evaluation including an environmental health survey indicates that the beach is bacteriologically, biologically, chemically, or physically hazardous to the bathers or swimmers.

Although it is difficult to provide good water quality in many natural bathing areas, a number of techniques can be utilized to improve water quality. They include the installation and proper use of adequate bathhouse facilities to minimize the adding of various pollutants to the water by the swimmers and the installation and proper use of high-capacity disinfecting equipment to maintain a chlorine residual in the water where the individuals are swimming or bathing. The water circulation and treatment can be carried out by having a pump installed to suction clean water out of the lake at an area away from the swimming area and then pumping it into the swimming area at inlets no farther than 8 ft apart. The inlet openings should be located under the water surface along the shoreline, and in shallow water at about a 12- to 18-in. depth. The pump capacity should permit displacement of the water in the swimming and wading zone every 6 to 8 hours. Attached to the equipment should be chlorine gas-feeding equipment that can pump chlorine into the water at a rate that maintains a free residual value. The actual amount of the chlorine needed depends on water turbidity. It may be necessary to feed from 10 to 40 lb of chlorine per day for each 100,000 gal of water contained in the 0- to 6-ft deep swimming and wading areas. This would correspond to a dosage rate of from 3 to 12 ppm.

MODES OF SURVEILLANCE AND EVALUATION

Plans Review and Approval

When evaluating pool plans, the environmental health practitioner has to determine the capacity of the pool in the nonswimming area, the swimming area, and the diving area. The plan must then be rated according to maximum bather load. It is also necessary to indicate the kinds of filters and whether they can do the job adequately, pumps, types of alum feeders that will work on a 24-hour basis, types of soda ash feeders that work on a 24-hour basis, type of chlorinator to work on a 24-hour basis, and location of the main drains and return inlets. It is quite important that the plans be approved only if the filtration can definitely work properly and if the pump sizes and various chemical feeders are able to adequately handle the chemicals that are to be utilized. The pool filtering equipment should be able to recirculate the water at least once every 6 to 8 hours to obtain proper filtering of the water. The plans should indicate the kinds of pressure and rate of flow gauges that are available and where they will be placed. There should also be a diagram of the entire system from where the water enters the hair strainer to where it finally goes back into the pool. The environmentalist then determines whether the original plan of the pool is adequate and, if it is, grants permission for construction. It is important that the geology of the land and the water source be considered in the initial plan.

National Beach Program

In May 1997, the EPA administrator formally announced the BEACH Program to strengthen U.S. beach water protection programs and water quality standards, better inform the public, and promote scientific research to further protect the health of the public. This program is part of the Clean Water Action Plan issued by the President and his administration. In the spring of 1999, the EPA conducted the second annual National Health Protection Survey of Beaches. This study does not include all beaches in the United States, because it was of a voluntary nature. In 26 states, 1403 beaches were identified and information was sent in concerning them. Of these beaches, more than 350 had an advisory or closing in 1998. Another 935 beaches had water quality monitoring programs. A wide variety of standards were used in the monitoring of water quality at the beaches. The average period standard test showed: total coliform, 49; fecal coliform, 99; *E. coli*, 34; enterococci, 49; other, 0.

The major components of the beach program include strengthening beach standards and testing; providing faster laboratory test methods; predicting pollution by setting up models based on facts; investing in health and methods research; and informing the public. The EPA is also involved in other programs to help reduce substantially possible pollutants. From 1972 to 1996, the EPA awarded approximately $70 billion to municipalities to assist in the construction and improvement of wastewater treatment plants. The EPA is attempting to end sewage overflows in communities with outdated sewer systems. The EPA is implementing a national storm water program to reduce urban runoff. The EPA also works with the U.S. Coast Guard to improve sewage and other waste disposal from recreational boats and other vessels. The Beaches Environmental Assessment Closure and Health Act is under consideration by the U.S. Congress.

Local Beach Program

A sampling program for recreational waters exists in Oakland County, Michigan. It is extremely well run and provides adequate data on a yearly basis for determining the kinds of environmental problems as well as the bacterial load present in the lakes within that county.

Random samples taken of recreational waters are meaningless and cannot be properly interpreted. They only reflect the bacteriological quality of the water at the time and place of sampling. The bacteriological quality of the surface water is affected by such factors as use, nearby residential developments, wind direction, weather conditions, temperature, season of the year, storm water runoffs, and migrating birds. All these conditions must be considered when developing a sampling program for evaluating recreational waters.

The sampling program should consist of the preparation of area plans, setting up of sampling stations, actual sampling and collection, computing of results, and necessary comments concerning water quality and what may be done to improve it. The area plan consists of an area survey that includes the name of the body of water and the township and county, a drawn outline of the shoreline, and names and locations on a map of roads and houses. This should include those houses that have individual water supplies and individual sewage disposal. Determination of whether the beach is public or private should be made. The approximate extent of the area to be placed under surveillance in actual feet and the approximate distance between sampling stations and the location of sampling stations should be included, as well as the location of storm sewer outlets. The names and addresses of persons interested in the recreational area project and those interested in maintaining a good beach should be added.

Sampling stations are set up at a maximum distance of 500 ft apart, with a minimum of two sampling stations per beach area. Additional stations are established 500 ft from each side of the beach area.

For sample collection to be effective, at least ten consecutive samples must be taken at each station. A sample should be collected at least twice a week, and the sample must meet the laboratory technicians' operating schedule. In most health departments, sampling is carried out Sunday through Wednesday, to allow adequate time for results to be read. Water samples should be taken immediately to a laboratory. If some delay occurs, the water should be refrigerated. However, the sample should never be more than 24 hours old.

The sample should be collected at a depth of 3 ft or more to avoid picking up the material from the top or bottom of the area. The sample that is collected in lakes or pools is placed in a sterile bottle that contains sodium thiosulfate. The bottle is opened at the time of filling. The water is collected by using a sweeping arc under the surface of the water. The sample results are determined by using the most probable number (MPN) index. This indicates the most probable number of coliform organisms per 100 ml of sample.

The survey combined with the samples should give a reasonable idea of the water quality from a bacteriological standpoint. From a chemical standpoint, the water would have to be analyzed for each of the chemicals that are shown in the requirements for potable water supply. It should be recognized that beaches may be unsatisfactory, not only from a bacteriological, chemical, or physical standpoint but also from heavy growths of weeds and mucky bottoms.

Water Testing

Pool water test kits give a determination of the swimming pool water and indicate what changes in chemical quantities are needed. A pH and available chlorine test kit is necessary to measure these two vital components of good water quality. Kits are also available that measure total alkalinity, hardness, calcium hardness, and cyanuric acid in the water. The chemicals in the test kit are usually in liquid form but are also found in tablet form. The chemicals deteriorate with age and therefore should be changed frequently. They are inexpensive and readily replaceable. The test chemicals should not be kept any longer than one swimming pool season. Before utilizing the test kit, the instructions should be read carefully. The samples should be taken from both the deep and shallow ends of the pools. They should be taken as far below the surface as possible and should not be taken from areas close to the inlet pipes. Always rinse the collection container and companion tube several times prior to testing and also after testing.

The two most frequently used tests for total chlorine are the orthotolidine (OT) test and the diethyl-p-phenylene-diamine (DPD). At normal pool water temperatures, OT reacts with the total chlorine content to produce a series of colors that range from light yellow to deep orange. If the very low level of total chlorine is present at around 0.1 to 0.2 ppm, a light yellowish color appears. At higher concentrations of chlorine, 0.8 to 1.2 ppm, stronger yellow shades of color appear. At very high levels, 5.0 to 20.0 ppm amber, orange, and red colors are produced. The orthotolidine reacts with the water to measure only the free available chlorine. The OT test is carried out in the following manner: rinse the test tubes in the water to be tested, fill the test tubes to the marked line, add the OT reagent in the amount specified, cap the test tube and mix the contents, allow 3 min for maximum color development, and compare the resulting color with the color standards of known value. The reading on the kit is the total residual chlorine.

When testing for free chlorine, the DPD test is used. It reacts only with free chlorine at low levels of 0.1 to 0.2 ppm to produce a light pink color. At 1.0 to 1.5 ppm, a varying of shades from pink to red form. At levels above 5.0 ppm, deep red colors form.

The colors occur in the OT test because chlorine with other oxidizing agents in acid solutions produces yellow-colored compounds. The proper development of the color depends on the sample having a low pH value. Extremely alkaline pool waters may produce green or blue instead of the typical yellow to orange. Such alkaline waters must be neutralized with a dilute hydrochloric acid solution before making the test. A chlorine residual flash test may be carried out with a pocket comparator. The two test tubes have 10 ml of water in them and are then placed in the comparator. About 20 drops, 1 ml, of OT is then added to the test tube in the right-hand opening and shaken. Within 10 sec the color should appear and should read anywhere from 0.1 to 1.0 ppm. The OT-arsenite chlorine residual test is also a flash test, similar to the previous one, but 5 g of sodium arsenite in 1 liter of distilled water is utilized. Of this mixture, 0.5 ml is added to the 10 ml of test water in the test tube. It is mixed, and then 0.5 ml of OT is added. Compare color standards against a comparison tube, as previously stated. You can then record your false residual. Next put 0.5 ml of OT in a test tube and add 10 ml of the test water. Immediately add 0.5 ml of sodium arsenite solution. Compare the test solution with the color standards and record the results as free chlorine plus false residual. The difference between the first reading and the second reading is the free chlorine residual.

The pH test is made by adding a specific amount of an extremely sensitive organic dye, known as phenol red, to the water sample. The phenol red produces a characteristic color according to the degree of acidity or alkalinity of the sample. Phenol red is most widely used in a pH range from 6.8 to 8.2. The presence of chlorine in the test sample may cause an indirect color to develop, which will provide a false reading. Therefore, in testing pool water, an indicator reagent that includes an antichlorine bleaching material, such as several drops of sodium thiosulfate, should be used to eliminate the chlorine problem. The pH test is carried out the same way as the OT test. The exceptions are the difference of reagent, and, of course, the obvious difference in color, which show on the standard and the tube. The colormetric determination of pH is based on the ability of certain organic materials to change color in varying hydrogen ion concentrations.

In an emergency pH test, brom thymol blue indicator solution is added to 10 ml of water. The amount added is approximately the contents of a small medicine dropper. The water is then shaken and allowed to stand. Dark blue indicates 7.6 pH, pure light blue about 7.0, green about 6.8, greenish yellow about 6.4, and yellow about 6.0. Water may also be tested for alkalinity by titrating a sample of the pool water with a standardized acid reagent: the acid reagent may be in tablet form. At the endpoint where there is a sharp change in color, the alkalinity may be determined.

Hardness of water can be determined by titrating the water with a standard reagent. To the titration tube containing the water sample, add five drops of buffer reagent and shake. Then add five drops of indicator reagent or one indicator tablet and shake. Fill a microburette with the hardness titration reagent. Add hardness titration reagent until the red color changes to blue. The hardness is read in parts per million from the scale on the barrel of the microburette.

A cyanuric acid test may be utilized, because cyanuric acid is used as a chlorine-extending agent. The desired level of cyanuric acid is between 25 and 50 ppm. Below 25 ppm the chlorine-extending property is rapidly reduced. The upper limit is 100 ppm. The test is carried out by mixing a standard amount of pool water and an equal amount of standardized melamine solution and measuring the turbidity that results. Low concentrations of cyanuric acid cause finely divided particles to appear in the mixture, giving it a hazy appearance. Higher concentrations cause a formation of larger particles, giving it a milky appearance. During the test, the temperature of the test sample and the test reagent must be kept at 70°F or below.

A variety of test kits produced by many firms are on the market. The test kits come in all shapes, sizes, and prices. The health department should determine which is most effective and most usable for the particular county or state health department.

Inspection Techniques

A swimming pool inspection or survey is extremely important to determine the actual physical problems that may exist and to determine the techniques of operation and the maintenance procedures utilized at the pool. Structural and equipment defects must be noted, discussed with the pool operator, and corrected immediately. Inspections on all pools should be carried out at a minimum of once a week during the peak summer season. On indoor pools during the rest of the year, it is adequate to make an inspection once a month. However, these minimum inspection periods are utilized only on a routine basis. Where a pool is having problems, it may be necessary to return each day; or if the problems cannot be resolved, the pool should be closed down until they are resolved. The inspections should be made at all times, including periods of minimum as well as maximum use. During the course of the inspection, the pool, pool structure, water treatment structure, various buildings, water quality, records, and number of bathers should be examined and recorded. Also, the environmental health practitioner should take chlorine and pH tests, along with the lifeguard on duty, at both the shallow and deep end. If the lifeguards on duty have test kits that are not accurate, they should be given new materials. If the kits are still not accurate, they should be required to obtain an accurate test kit within a 24-hour period. A key point to evaluate is record keeping, because a record should be kept of the pool pH and chlorine levels for each hour of each day of operation. In addition, the filters should be checked to ensure that backwashing is not needed. The chlorinator should be checked to ensure that it is operating properly. It should be determined whether there are adequate supplies of disinfecting and other chemicals kept on hand. All safety equipment and safety procedures should be evaluated. It is recommended that where pay phones are used, coins should be taped to the phone so that they are readily available in the event of an emergency. Where a 911 system is in use, this is not necessary. The collection of a water sample at the time of the swimming pool inspection is vital. The sample results should be correlated with the actual inspection results.

CONTROLS

Methods of preventing disease not only include the previously mentioned techniques, especially the proper treatment and disinfection of water, but also include the exclusion of sick individuals who have nasal drainage, discharging ears, or skin infections from the pool; thorough cleaning and disinfecting of the locker rooms or bath houses; prohibition of the common use of suits, towels, and drinking cups; provision of individual towels for the swimmers; enforcement of showers prior to entering the swimming area; and elimination of foot baths (they cause more problems than they resolve). In addition, methods should include keeping all surface runoff from entering the pool; cleaning skimmers and vacuuming the pool regularly; backwashing the filters whenever necessary and on a frequent basis; and providing small children breaks so that they can use toilet facilities, instead of using the pool as one large toilet facility.

Pool accidents or other types of swimming accidents could be reasonably controlled if fences were utilized in areas where children could get into the pools. A 4½ ft fence is adequate for toddlers and small children, and a 5- to 6-ft fence is adequate for older children. In any case, pools should have monitoring systems where alarms can be activated if someone enters the swimming pool area when authorized personnel are not present. Life preservers, life jackets, and ropes are of very limited value in pools. In lake situations they can be of value, providing there are trained personnel around to use them. Usually long poles, 12 ft in length, are best used in a pool situation, because the individual can grab hold of it and be pulled to the pool side. In any case, there should be a minimum of one trained lifeguard for every 75 bathers or for every 2000 ft^2 of pool surface, whichever may be the lesser of the two. A trained lifeguard means one who has been certified by the American Red Cross as a senior lifesaver. Other types of emergency equipment available at the pool should include first aid kits, telephones, and blankets; and in the case of a beach area, boats and ring buoys. All senior lifesavers should be highly trained in cardiopulmonary resuscitation techniques. Lifeguard assistants should be available as additional help when needed and also as relief for the lifeguard during the application of proper resuscitation techniques.

Chlorine

Because chlorine can be hazardous, a leak-detection system should be set up. One of the most effective techniques is to have a bottle of ammonia at the entrance of the room. The ammonia bottle is then opened near the potential chlorine leak. If chlorine is present, a white cloud of ammonium chloride forms. If a strong chlorine leak is occurring, the individuals should not enter unless they have a properly fitted and proper type of gas mask; a second individual should be available in the event that the service person is overcome by the chlorine fumes. Good safe practices prevent disasters. If individuals are exposed to chlorine, they should be removed to fresh air as quickly as possible, kept warm, and taken to a hospital as soon as possible for treatment.

SUMMARY

Swimming is a much sought after means of recreation in our society today. Vast numbers of people are making use of our water environment, which is continuously used for other purposes. As a result of the variety of water uses and the constant pressure of people on swimming areas, these areas have increasingly become contaminated. In the ensuing years, the problems will increase sharply. Along with this increase, there will be an increase in disease due to microorganisms and a potential for disease due to a vast number of chemicals that are entering the swimming water areas. Continued surveillance by public health authorities will be needed to prevent large outbreaks of waterborne disease due to swimming.

RESEARCH NEEDS

There is a need to determine the level of coliform that may be considered hazardous in swimming areas other than swimming pools; and a further need to determine the kinds of filtration systems, filtration mechanisms, and disinfecting mechanisms that are most useful in controlling bacterial and viral contaminants. Studies should be made of the relationship of a variety of chemicals found in water and the health of individuals who are using the water for swimming.

CHAPTER 5

Plumbing

BACKGROUND AND STATUS

One of the most important systems developed to protect the health of people and to provide people with a better way of life has been the system of plumbing, which is the piping of potable water to its ultimate use and the draining away of waste materials to a variety of treatment processes. It is essential in our modern society to recognize that plumbing is to society what the circulatory system is to people. It is a system that must function efficiently to avoid epidemics and to avoid chemical pollution. Good health practices require that plumbing in a community be free of cross connections, backflow connections, submerged inlets, and poor venting. It also must transport a good quality of potable water in adequate quantities to service our modern society. One of the great difficulties that we face as a society is that older existing plumbing may deteriorate and may create health hazards; also, repairs of plumbing may be carried out in such a way that they can create health hazards.

Plumbing is the practice, materials, and fixtures used in installing, maintaining, and altering of pipes, fixtures, appliances, and appurtenances utilized for potable water supply, sanitary or storm drainage, and venting systems. Plumbing does not include the drilling of water wells, installing water softening equipment, or manufacture or sale of plumbing fixtures, appliances, equipment, or hardware. Plumbing systems consist of an adequate potable water supply system, a safe adequate drainage system, and ample fixtures and equipment.

Public health personnel have long been concerned with cross connections, backflow connections, and submerged inlets in plumbing systems and public drinking water supply distribution systems. These cross connections make possible the contamination of potable water with nonpotable water or contaminated water. To prevent contamination and potential disease, the only proper precaution is to eliminate all the possible links and channels where potable water may become contaminated. Cross connections may occur when individuals installing the plumbing are not aware of the danger and may not realize that water can reverse its direction. In fact, it may even go uphill. In addition, the valves may fail or may be carelessly left open. To combat this problem, installers must understand the hydraulic and potential contamination factors that can cause environmental health hazards. They must also know what types of standard backflow prevention devices and methods are utilized, and how to obtain the materials and install them properly.

This chapter on plumbing is not meant to be an overall plumbing guide. It is not even meant to list all the potential hazards. However, it should provide sufficient material and diagrams to help the environmental health practitioner have a better understanding of plumbing and its effects on health.

The current status of the plumbing problem is very difficult to ascertain because data are lacking in this area. However, it can be assumed that plumbing systems in many areas are rapidly deteriorating because of the age of the structures. Many individuals fail to pay adequate attention to the enormous potential hazard of disease and injury due to microbiological, chemical, or physical agents.

SCIENTIFIC, TECHNOLOGICAL, AND GENERAL INFORMATION

Basic Principles of Plumbing Related to Environmental Health

Certain basic principles of environmental health and safety are satisfied through the proper design, installation, and maintenance of a good plumbing system. Although the details of plumbing construction may vary, the health of the public must be protected. The following 22 items cover these principles:

1. All occupied premises used for human habitation must have a potable source of water, preferably a treated public water supply, which is safe and protected against the hazards of backflow or back siphonage.
2. Water must be provided at all use outlets in the plumbing fixtures at pressures and at quantities that make the system functional for human use. Undue noise should be eliminated.
3. Hot and cold running water is required at all plumbing fixtures that are normally used for washing or cleaning.
4. Water should be utilized in quantities consistent with proper performance and cleaning, and yet in compliance with necessary water conservation measures.
5. Devices used for heating and storing of water must be designed and installed to avoid the danger of overheating or explosion.
6. Public sewers should be utilized whenever possible.
7. Each family dwelling unit must have at least one sink and stool in a bathroom and a separate sink for the kitchen, as well as a bathtub or shower.
8. The drainage system has to be designed, constructed, and maintained in such a way that it cannot clog readily; but if it does clog, adequate cleanouts must be provided.
9. The piping within the plumbing system must be of durable material and free of defects to give a long expected life of service.
10. Each fixture directly connected to the drainage system has to be equipped with a liquid seal cap to trap gases created in the sewer system.
11. The trap seals must be protected in the design of the drainage system by providing adequate circulation of air in all pipes and by eliminating the possibility of siphonage, aspiration, or breaking of the trap seals under ordinary use conditions.

12. All gases and unwanted odors must be vented to the outer air.
13. The plumbing system must be put under pressure to test for leaks and defects in the workmanship or material.
14. Materials that clog or help clog pipes produce explosive mixtures, destroy the pipes or their joints, or interfere with the sewage disposal process must not be put into the building drainage system.
15. All food, water, and sterile goods must be protected against backflow of sewage or by leaking sewage from overhead pipes.
16. All bathrooms must be properly lighted and ventilated.
17. Where on-site sewage disposal systems are necessary, they must conform to local and state standards and be able to perform under unusual conditions.
18. If a plumbing drainage system is subject to backflow of sewage from the public sewer, adequate protective measures should be used to keep the sewage from entering the building.
19. Plumbing systems must be maintained in a safe and serviceable state.
20. Plumbing fixtures have to be accessible for cleaning and for maintenance.
21. Plumbing must be installed in such a way that the structural members, such as walls and floors, are not damaged.
22. Sewage or other waste must not be discharged onto the surface, into storm water drains, or into the subsurface water supply. (See Figure 5.1 for typical house drain installation.)

Principles of Hydraulics

Hydraulics describes the principles of liquid flow in quantitative terms. An understanding of these principles is necessary for the environmental health practitioner to know how liquids flow into pipes. Water at rest exerts a pressure that is in proportion to its depth. The greater the depth is, the greater the pressure. The pressure at the bottom of a column of water is proportional to the height of the water in the column. Pressure is a force per unit area expressed in either pounds per square inch, feet of water, or inches of mercury. The pressure equals the depth of water times the specific weight of water. The specific weight of water is 62.4 lb/ft^3. The formula for pressure is $P = y \times H$. At atmospheric pressure, 14.7 lb/in.^{2} equals 33.9 ft of water and 29.92 inHg. Atmospheric pressure is not the same as gauge pressure. Gauge pressure is that pressure that is read on some mechanical device. Atmospheric pressure is the pressure that is found at 1 atm at sea level, usually recorded at 60°F. Absolute pressure is a combination of gauge pressure and atmospheric pressure. Therefore, if the atmospheric pressure is 14.7 lb/in.^{2}, and the gauge pressure is 10 lb/in.^{2} the absolute pressure would be 24.7 lb/in.^{2}.

When you are working with a vacuum, the following example might apply. If you drew a vacuum of 10 lb/in.^{2} and were trying to find the absolute pressure, you would subtract 10 lb/in.^{2} from 14.7 lb/in.^{2}. The absolute pressure within the vacuum area would be 4.7 lb/in.^{2}. The same types of calculations can be applied to the other means of stating pressures, such as in inches of mercury or feet of water (Figure 5.2). Pressure may be measured by using pressure gauges or manometers.

All liquid flow is classified either as laminar flow or turbulent flow. In the laminar state, all the fluid particles flow along the same path with no lateral mixing. In the turbulent state, there is a mixing of the particles. The Reynolds number makes it

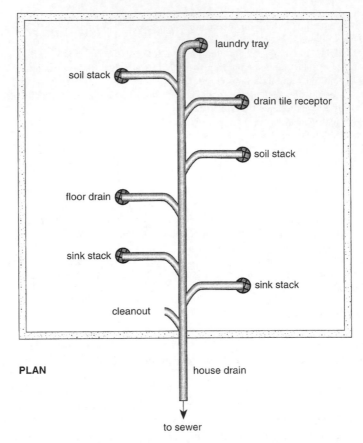

PLAN

house drain

to sewer

Figure 5.1   Typical house drain installation. (From *Basic Housing Inspection,* Training Course
Manual, U.S. Department of Health, Education, and Welfare, Public Health Service, Environmental Health Service, Environmental Control Administration, Cincinnati, OH, March 1976, pp. 6–10.)

possible to predict the state of flow of a liquid. (The Reynolds number is a dimen-
sionless number that is significant in the design of a model of any system in which
the effect of viscosity is important in controlling the velocity or flow pattern of a
fluid. The Reynolds number is equal to the density of a fluid times its velocity times
a characteristic length divided by the fluid viscosity.) Reynolds found that laminar
flow always exists below a Reynolds number of 2100. Above this, either laminar
flow or turbulent flow could occur up to 4000, depending on the initial conditions
of the nonturbulence of the fluid. Under normal temperature ranges, however, water
flows in a turbulent manner.

To determine the rate of flow in a pipe, the formula $Q = AV$ is utilized, where
Q is the rate of flow in cubic feet per second or gallons per minute; A is the cross-
sectional area of the pipe, usually in square feet; V is the average velocity in the
pipe, usually in feet per second. When the diameter of the pipe changes, the velocity

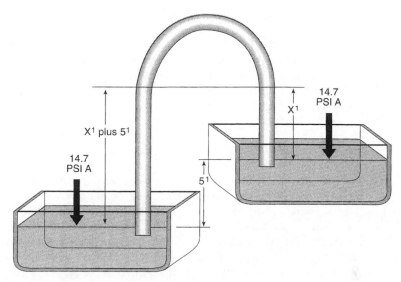

Figure 5.2   Pressure relationships at different elevations in a continuous fluid system. (From *Water Supply and Plumbing Cross-Connections,* DHEW Publication No. 957, Rockville, MD, 1969, pp. 14–15.)

of the flow is changed. This principle is utilized in wells and in other water systems. If you want to increase the velocity of the water flowing, then simply decrease the cross-sectional area. This increase can be achieved if the quantity stays the same.

The total energy at any point in this system equals the sum of the pressure head plus the velocity head plus the potential head. Energy may be lost due to pipe friction and fittings or lost during heat transfer through the pipes.

Pipe friction interferes with the movement of the liquid particles. The greater the cause of friction, the greater is the loss of energy. This is why when a pipe becomes corroded inside as a result of either bacterial slime forming or chemical deposits, the friction caused by the material buildup reduces the flow of water, even if the initial pressure stays the same. Obviously, as the inside of the pipe narrows, velocity should increase. Part of this energy is lost to friction. Energy is also lost at the valves, elbows, bends, and other fittings. This occurs because eddy turbulence is created and superimposed on the normal turbulence. Obviously pipes should have as few fittings and bends as possible.

Liquid flow may be measured by use of a Venturi meter. The Venturi meter is a pipe that has a constricted throat section that opens up to the same diameter pipe as the original diameter. For example, the original diameter might be 5 in.; the constriction, 3 in.; and the portion past the constriction, 5 in. again. For a given quantity of water to pass through the constriction, the velocity must increase; that is, Q stays the same, A is decreased, and therefore V (velocity), must increase. As the rate of flow through the Venturi meter increases, the difference in pressure between the inlet section and the constricted section increases in proportion to the increase of the flow. Therefore, their relationship can be established between the difference in pressure and the rate of flow. This instrument is quite valuable in measuring rate of flow, because it has no moving parts and little energy is lost through the meter itself.

For more information on the principles of hydraulics, it is recommended that the reader obtain one of the numerous books written in the area of hydraulics or guides on heating, ventilation, and air conditioning that are put out by the American Society of Heating, Refrigerating, and Air-Conditioning Engineers. Water companies are also an excellent source for this material.

Cross Connections, Backflow, and Back Siphonage

A cross connection is any direct or indirect physical channel or arrangement between a potable water supply system and a substance of unknown or unsafe quality, including gases and liquids or dry material that is easily hydrated. A flow may occur from either system into the other one. A direct cross connection exists whenever a potable water supply system is physically joined to a source of unknown or unsafe substance. An indirect cross connection occurs when an unknown substance can be forced, drawn by vacuum, or otherwise introduced into a potable water supply system. Indirect connections include hose attachments and temporary bypasses of fixed proper air gaps through hoses or swing connections. Simply, a cross connection is the link that connects a source of pollution with a source of potable water.

The pressures within the public water supply distribution system constantly fluctuate, based on demands in various parts of the system. A specific water pressure cannot be maintained in various parts of the system at all times. Exceptional flow demands such as for fires, create a large concentration of water usage in a small part of the system and thereby can create low flow elsewhere and lead to contamination of the water system by direct or indirect cross connections. Water main breaks and leaks due to stress on the system, age of the system, and extreme temperatures of the ground help establish hydraulic conditions under which contaminating sources can cause potential disease or injury problems by merging with potable water supplies.

Unique and unusual events are not needed to cause contamination of the water system. Contaminants can be aspirated into the water system when even a small pinhole is present in a pipe, a loose joint or fitting is present in the water line, and the water line is submerged in a contaminant.

Cross connections are easily made by people who use the water system each day. A hose can cause a cross connection when it is used without a proper hose bib vacuum breaker, such as attaching a garden spray container full of herbicide, pesticide, or fertilizer to the hose; leaving it submerged in a swimming pool; or leaving it submerged in soapy water in a kitchen sink or bathtub. Original piping designs are often altered by maintenance personnel unfamiliar with cross-connections. Sometimes, cross connection control devices are bypassed by maintenance personnel trying to solve a clogged line problem. Tanks, vats, water cooling equipment, and water cooled processes are often directly connected to the potable water supply system without backflow protection. When a drop in pressure occurs in a potable water supply system, backflow or back siphonage can occur from all cross connections within the system.

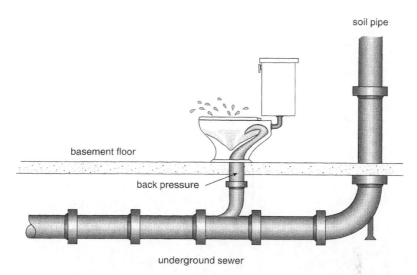

soil pipe

basement floor

back pressure

underground sewer

Figure 5.3 Loss of trap seal as a result of back pressure. (From *Basic Housing Inspection, Training Course Manual,* U.S. Department of Health, Education, and Welfare, Public Health Service, Environmental Health Service, Environmental Control Administration, Cincinnati, OH, March 1970, p. 6.20.)

Water usage has changed over the years. In the past, water was used for drinking, cooking, cleaning, and sanitary purposes. Today, water is also used for a large number of gadgets and devices that the homeowner can apply to a plumbing system to make life easier. An example would be spraying insecticides or fertilizers, washing dishes, high-pressure washing of cars, etc. Many of these devices used at home or commercially boost the pressure within the device or private water supply system to a level higher than the public water supply system, thereby creating the potential for contamination through cross connections.

The backflow connection is that point in a plumbing system through which the potable water can be contaminated by the polluted water because of a drop in pressure in the supply line, an improperly closing or leaking valve, or an increase in pressure in the polluted line. Basically, the backflow occurs at the point of cross connection (Figures 5.3 to 5.7).

Determining where cross-connections exist and eliminating them can be very difficult. First it is necessary to determine all the sources of pollution, and second to determine which factors allow the polluted water to flow in the direction of the potable water.

At sea level, air exerts a pressure of 14.7 lb/in.^{2}. If a tube were inverted into a body of water and all the air was exhausted from the tube, theoretically the water would rise in the tube 34 ft as a result of the pressure of the air on the water, which in effect would create a force that would push the water up the tube. Force, unless it is resisted in some way, produces this type of motion. The motion is always in the same direction as the force. The force that we are most concerned with in backflow is pressure. The pressure may be due to the atmosphere, as already mentioned; it may

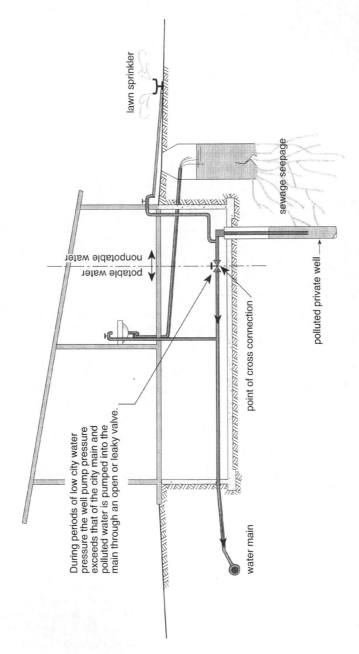

Figure 5.4 Direct cross connection. (From *Basic Housing Inspection*, Training Course Manual, U.S. Department of Health, Education, and Welfare, Public Health Service, Environmental Health Service, Environmental Control Administration, Cincinnati, OH, March 1970, p. 6.26.)

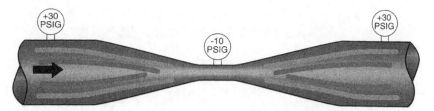

Figure 5.5 Negative pressures created by constricted flow. (From *Water Supply and Plumbing Cross-Connections,* DHEW Publication No. 957, Rockville, MD, 1969, p. 16.)

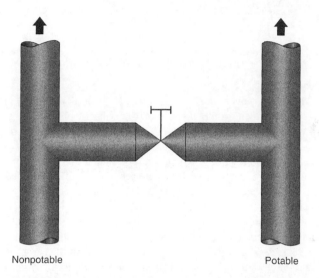

Nonpotable   Potable

Figure 5.6 Valved connection between potable water and nonpotable fluid. (From *Water Supply and Plumbing Cross-Connections,* DHEW Publication No. 957, Rockville, MD, 1969, p. 18.)

be due to the atmosphere plus additional pressure that the particular contaminant is exerting; or it may be due to the atmosphere plus additional pressure or a vacuum that may have been created on the potable side of the line. In effect, these three ways are the means by which nonpotable or contaminated waters may flow into potable or noncontaminated waters.

Back siphonage is a siphon action that occurs in a reverse direction. It is caused by the force of the atmospheric pressure exerted on the polluting liquid, forcing it toward a potable water supply system, which happens to be under negative pressure or a vacuum. To understand how siphons work, it is only necessary to consider the pressure exerted by a cubic foot of water at sea level. The average weight of this cubic foot of water at sea level is 62.4 lb/ft^3. The pressure, therefore, that is exerted is 62.4 lb/ft^2 plus the atmospheric pressure of 14.7 lb/in.^{2}. To put these in the same

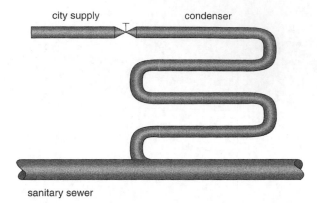

Figure 5.7   Valved connection between potable water and sanitary sewer. (From *Water Supply and Plumbing Cross-Connections,* DHEW Publication No. 957, Rockville, MD, 1969, p. 19.)

dimension, divide 144 in.^{2} into 62.4 lb and you arrive at 0.433 lb/in.^{2}. In effect, the pressure per square inch would then equal 14.7 lb/in.^{2} plus 0.433 lb/in.^{2}, which would give you a total of 15.133 psi for the water that is now bearing down on the surface at sea level. To determine the pressure of 2 ft of water, add an additional 0.433 lb/in.^{2}. The importance of this pressure is that if at some point the pressure of the potable flowing water is less than the pressure of the combined atmosphere and water, then the water would flow backward and in effect you would now have back siphonage.

One of the reasons why a submerged inlet should be banned is because if the pressure in the submerged inlet should drop, then the fluid that is present in the tank would flow back into the line instead of vice versa. This, therefore, contaminates the initial line. Submerged inlets, unfortunately, are found in many common plumbing fixtures, such as siphon jet urinals, toilets, flushing rim slop sinks, and dental cuspidors. Unfortunately, in numerous laboratories, universities, and hospitals, individuals take hoses and run the hose from the potable water supply coming from the water tap above the sink into a tank. They never once take into consideration that the water pressure in the water line may drop for a variety of reasons including breakage in the line, reduction in the pressure of the city water supply due to various problems (including firefighting), partial servicing of certain lines where others have to be shut off, etc.

Some common types of backflow problems occur around the house when the sill cock is used to fasten a hose to the outdoor or indoor water faucet. Unfortunately, the hose may be in an elevated position above the water line, it may be left submerged in swimming pools, or it may have a chemical sprayer attached to it. If a break in water pressure occurs within the water line, or the water pressure is greater in the hose than in the water line, whatever is present can flow back into the potable water supply. It is easy enough in a house to lower pressure by simply utilizing several different appliances or taking several showers at the same time.

PLUMBING PROBLEMS

Public Health Significance of Cross Connections

In the past 100 years, outbreaks of waterborne disease in the United States have been substantially reduced. However, as long as our population can emit organisms within their body waste that may cause disease, and as long as innumerable toxic chemicals are present in our environment, the potential hazards of the spread of disease and injury will always be a great concern to the public and to the environmentalist. Generally, one of the first major community projects is the development of a safe water supply that can be piped to the consumer within the home, within the institution or hotel, and within industry. Not only potential hazardous substances may enter the water supply during the piping of the water to these sites of use but also the water may become contaminated within the structure itself.

Historically, one of the worst outbreaks of amoebic dysentery in modern times occurred in Chicago in 1933, in the hotels in which people were staying who were attending the Chicago Worlds Fair. The hotels were generally old and had defective plumbing and cross connections. Back siphonage was able to occur from bathtubs and toilets, which probably contaminated the drinking water supplies. As a result of these plumbing defects, 98 individuals died and many additional thousands became extremely ill.

In the past there were numerous outbreaks of disease as well as chemical poisonings caused by cross connections. Some of these incidents included:

1. The spread of brucella from a submerged inlet caused 80 people to become ill.
2. A person trying to obtain water from a main that was out of service, leaving the valve open, thereby allowing sewage to flow from a clogged sewer line into the public water supply, resulted in 2500 people becoming ill with bacterial dysentery.
3. A drop in water pressure in drinking water lines during installation of a new city water supply allowing river water to pass through the valve connection into the drinking water, resulted in almost 500 people becoming ill with mild intestinal disorders.
4. A defective valve connecting drinking water to a fire water supply aboard a ship, resulted in almost 1200 people becoming ill with gastroenteritis.
5. The creation of a vacuum at peak periods of demand on a 3-in. water main allowing wastewater to be back siphoned through cross connections into the drinking water system, resulted in several hundred people becoming ill with gastroenteritis.
6. The use of a temporary cross connection between potable water lines and pipes containing river water for firefighting purposes, resulted in 700 people becoming ill with gastroenteritis.
7. A paint factory that used propylene glycol to keep paint from breaking down after exposure to weather caused contamination of the potable water when a valve malfunctioned.

In 1990, in Kansas, several people in a community experienced air in their water. An air compressor used to supply air under pressure to a dental office was at a higher pressure than the pressure of the water supply. This allowed a backflow in the dental office to the public water supply, with a potential for contamination of the public water.

In 1990, in Colorado, school officials closed a middle school, after an antifreeze-like chemical was found in the school's potable water supply. Ethylene glycol had backflowed into the water system from the schools hot water heating system. Nine students complaining of flulike symptoms were treated at the local hospital for ethylene glycol poisoning.

In 1991, in Arkansas, residents near a poultry farm complained that the water was discolored. The public water system had been contaminated by backflow from a chicken house. There was a water service connection between the public water system and a well. Apparently, check valves failed.

In 1991, in Michigan, residents found parasitic worms or nematodes in their water. Also rust and other debris were found in the water. The water had backflowed through a residential irrigation system into the public water supply. An atmospheric vacuum breaker on the residential irrigation system had malfunctioned.

In 1993, in Oregon, a person installed an auxiliary water system that consisted of irrigation piping supplied by water pumped from a drainage pond contaminated by sewage from a local fill area. A hose was connected to the house potable water system and then to the irrigation piping system, to help water the lawn, while the pump was being repaired. When the original pump came back, it was connected inadvertently to the potable water system and contaminated the area with drainage pond water.

In 1993, in North Carolina, a clinic complained about a strange, bitter taste and strong chemical odor in its water. The chemicals from a mixer used in x-ray development had backflowed into the clinic's potable water supply. Someone was adding water to the mixer with a garden hose, and left the hose end submerged in the tank, thus creating an indirect cross connection.

In 1994, in Los Angeles, a film crew was spraying artificial snow from a pressurized 55-gal tank. The system failed to work properly. A person connected a garden hose between the tank and a potable water system, thereby allowing 30 gal of chemical solution to backflow into the public water system.

In 1996, in Florida, a meter reader noticed that the water meter at a home was registering backward. A cross-connection had been created between the potable and reclaimed water systems at the premises, and reclaimed water was backflowing into the potable water supply. About 50,000 gal of reclaimed water backflowed into the public water system.

In 1997, two outbreaks of *Cryptosporidium* occurred. The first outbreak was in New Mexico at a group home where staff, residents, and visitors became ill after drinking from spigots supplied by chlorinated well water from an unmarked irrigation well on the property. Several people swam in the swimming pool also supplied by this well. The second outbreak occurred in Texas, where more than 160,000 gal of raw sewage spilled, and flowed through underground fissures in a creek bed and into an aquifer located near five municipal utility district wells. A total of 1004 people became ill.

In 1997, a cross connection caused an outbreak of *Shigella sonnei* in Minnesota at a local fair supplied by a community water system, with 83 people becoming ill. In 1998, two outbreaks of copper poisoning occurred in Florida. Improper wiring and plumbing procedures caused leaching of copper from the restaurants pipes and the check valve was malfunctioning. A water treatment facility released high levels of sulfuric acid, which corroded the pipes and allowed leaching of copper into the system.

Unfortunately, thousands of people become ill each year by drinking contaminated water. The level of contamination is difficult to determine because extremely poor recording of diseases exists. People who have gastrointestinal disorders, unless they become violently ill, usually treat themselves. One other problem is that an enormous number of cross connections may exist in all types of structures just waiting for the time when they can serve as a source for the spread of waterborne disease or chemical illness.

The Centers for Disease Control and Prevention (CDC) have warned that explosives more dangerous than nitroglycerine have possibly been formed by chemical reactions in plumbing in more than 15,000 hospitals and clinical laboratories in the United States. The National Institute for Occupational Safety and Health issued an alert that the most dangerous explosive chemical compound identified was sodium azide, which is formed by automatic blood-cell counters used in hospitals and clinics. Sodium azide reacts with copper, lead, brass, or solder in the plumbing system to form an accumulation of lead or copper azides, which are explosive. In recent years, gasoline and oil have infiltrated into drinking water supplies in the Midwest.

Private Water Supplies

Private systems usually serve single- or multiple-family dwellings. The waters are taken from wells, springs, cisterns, etc. The wells can be shallow or deep and provide varying amounts of water under different conditions. They may be drilled, driven, dug, or bored. The potential for contamination varies with surface conditions, drawdown in the well, protection of the well, closeness of sources of pollution such as on-site sewage, and sprinkler systems, and whether the water comes from nonpotable sources.

The water from private systems as well as public systems is used for domestic purposes, industrial purposes, and irrigation. Mixed use of potable and nonpotable water may occur in a building, thereby increasing the potential for cross contamination, if the water supplies are not properly separated. Reclaimed water systems may also be a source of great value or contamination.

Public Water Supplies

There are a variety of reasons why public water supplies may become contaminated. A direct connection may exist between the potable water and other piping systems. This occurs at times when river water is used for firefighting. A connection may exist between a well that may be nonpotable and the potable water supply. Fire hydrants containing potable water may have holes or may be damaged, and therefore may allow surface contaminants or underground contaminants to enter the potable water. Connections may exist between the public water supply and other water used for a variety of other purposes. Connections may be made to industrial establishments, shipyards, and docks where potable water can be readily contaminated with polluted water. Open reservoirs of finished water may become contaminated. Inadequately sized piping in public water supply systems may create reduced pressures and negative head conditions, which permit backflow and back siphonage. Improper maintenance contributes to the bursting of water lines, which causes contamination. Inadequate sanitization following the installation of new lines results in contamination of the public water supply. Boiler water in a hot water or a steam space-heating system that may contain chemicals may flow because of cross connections back into the potable water.

In addition, greenhouse irrigation systems and lawn sprinkling systems that may introduce chemicals, such as fertilizers, may be operating during times of low pressure in the water system. This could create back siphonage of the chemicals into the potable water. Cooling towers, which may be at considerable heights, are frequent sources of contaminated water because of antifreeze, bird droppings, insects, rodents, dust, algae, and bacteria. Because the pressure in the water tower may be greater than the pressure in the street and inadequate vacuum breakers exist, the contaminated water may flow readily into the potable water supply. In fact, this did occur in a small town in Illinois, and numerous cases of salmonellosis were reported. The hospital laundry may contaminate potable water with detergents and chemicals that are automatically injected in the wash cycle. Further, bacteria present in the laundry may go into the potable water supply if a negative pressure develops during peak laundry demands. The waste disposal system in institutions that grind the waste, including contaminated waste, may flow back into potable water unless backflow preventors are provided.

Poor Plumbing Practices and Plumbing Hazards

Common errors include incorrect connections or incorrect sealing procedures, which may occur in certain cases when two different types of plastic pipe are joined together with a cement that can fuse with only one type of pipe and not the other. Plastic pipes may be a problem, because rats can gnaw through them. If an S-trap is installed incorrectly, it is usually difficult to vent. The seal may be broken by unvented gases expanding, which may lead to possible explosions and odors. The T-type trap is a much better trap if vented properly. Where the public water supply is used in cooling the heat exchange unit, problems occur when the water pumped from the cooling tower to the heat exchange unit does not function properly because of a system pumping breakdown. Problems occur when too small a pipe is used over a long distance. Friction between the water and a pipe reduces the amount of energy in the water traveling through the pipe. Gate valves have been used incorrectly for steam when they should be used for water only. Globe valves have been used incorrectly for water, and should be used for steam only. One of the worst hazards that we deal with is the common garden hose, which is constantly abused.

Innumerable fixtures may constitute plumbing hazards. They vary from laboratory aspirators to chemical feeders to coffee urns, dishwashers, swimming pools, bathtubs, drinking fountains, garbage-can washers, icemakers, steam tables, and vegetable peelers.

Most plumbing problems leading to cross connections are caused by poor workmanship, inadequate understanding of prevention techniques by trained people, or lack of understanding by the public as they utilize the water distribution system.

This, coupled with a spotty program of cross-connection enforcement, creates many potential opportunities for disease and injury. Material failure is not a frequent plumbing problem.

Economics

The cost of plumbing problems are difficult to ascertain because of lack of data. However, because of the high cost of labor and because of the enormous potential for disease, it can be assumed that the cost is quite significant.

POTENTIAL FOR INTERVENTION

The potential for intervention includes isolation, substitution, shielding, treatment, and prevention. Isolation is the separation of potable water supplies from contaminated water supplies. Substitution is the replacement of old valves and pipes with new valves and pipes to stop contaminants from infiltrating potable water. Shielding is not an appropriate technique to be used. Treatment is the proper cleaning and chlorination of all potable water lines prior to use. Prevention is the proper installation of plumbing and proper repairs of plumbing made by qualified plumbers. The potential for intervention in new structures where plumbing has been installed properly is excellent. The potential for intervention in old structures is based on the types of individuals making repairs and the type of preventive maintenance that is carried out by the appropriate individuals. Intervention is also the training of professionals in cross-connection prevention, the education of the public in the health hazards associated with improper use of water distribution systems, and the development of properly funded, enforceable cross-connection programs.

RESOURCES

The organizations are the American Standards Association; American Society of Sanitary Engineers; American Society for Testing and Materials; American Water Works Association; Commercial Standards, Commodity Standards Division, Office of Industry and Commerce; Federal Supply Service, Standards Division, General Services Administration; National Sanitation Foundation; Plumbing and Drainage Institute; National Association of Plumbing, Heating and Cooling Contractors; International Association of Plumbing and Mechanical Officials; and Underwriters Laboratory.

STANDARDS, PRACTICES, AND TECHNIQUES

Materials

All materials, fixtures, or equipment used in the installation, repair, and alteration of any plumbing system must be approved and therefore must meet applicable approval standards. These standards were developed by the organizations listed in the previous section on resources.

Materials that have been used include cast iron, galvanized pipe, malleable iron fittings, steel pipe, iron pipe, brass fittings, brass pipe, copper pipe, lead pipe and lead traps, asbestos–cement nonpressure sewer pipe, asbestos–cement building sewer pipe, fiber sewer pipe, perforated fiber pipe, clay drain tile, reinforced and nonreinforced concrete sewer pipe, plastic water pipe, plastic soil pipe, and vitrified clay sewer pipes.

Standards are developed for materials by the International Association of Plumbing and Mechanical Officials, Research and Testing, who currently have issued the Uniform Plumbing Code, copyright 1999; and National Sanitation Foundation.

Interceptors, Separators, and Backwater Valves

Interceptors are required for oil, grease, sand, and other substances that are harmful or hazardous to the building drainage system, public sewer, or sewage treatment plant. A grease interceptor is not required for individual dwelling units or any private living quarters. The size and type of interceptor, including the grease interceptor, is determined by the local plumbing officials who have a better idea of how interceptors are utilized and where they are needed. Where grease interceptors are used, however, they should have a grease retention capacity of at least 2 lb for each gallon per minute flow of water. The rate of flow through the interceptor should be controlled. If sand or heavy solid interceptors are used, they should have a water seal of at least 6 in. and should be readily accessible for cleaning. Interceptors may be utilized in commercial laundries, bottling plants, or slaughterhouses. These interceptors must be properly vented to get rid of troublesome gases.

Separators are utilized for separation of liquids, when a mixture of light and heavy liquids having varying specific gravities are treated. They should be separated in an approved receptacle. Oil separators must have a depth of not less than 2 ft below the invert of a discharge line. The outlet opening of the separator should not have less than an 18-in. water seal. In garages where a maximum of three cars are serviced, separators should have a minimum capacity of 6 ft³. One additional cubic foot capacity should be added for each car up to ten cars in larger garages. In service stations and repair shops, the capacity of the separator should be 1 ft³ for each 100 ft² of surface to be drained into the separator, with a minimum of 6 ft³. Interceptors and separators should be maintained in proper operating condition by periodically removing accumulated grease, scum, oil, or other materials. The separator should be accessible and vented.

Backwater valves are utilized in fixtures that are subject to backflow and fixture branches below grade. All backwater valves have to be corrosion resistant and have a mechanical seal against backflow. The valve fully opened must have a capacity that is not less than that of the pipe. It should be installed in such a way that it is readily accessible for service and repair.

Plumbing Fixtures

All plumbing fixtures and drains that are used to receive or discharge liquid waste or sewage must be connected to the drainage system of the building in accordance with the requirements of local or state plumbing codes. The minimum number of plumbing fixtures and types of fixtures vary with usage of the facility.

Separate facilities must be provided for each of the sexes. All plumbing fixtures have to be readily accessible for cleaning and repair. The stools cannot be set closer than 15 in. from the center of the stool to the side of the partition or 30 in. from center of stool to center of stool.

All fixtures must be secured tightly to the floor and have watertight joints. The supply lines or fittings for the plumbing fixture have to be installed to prevent backflow. Standing water in a fixture must not overflow the fixture.

Indirect and Special Waste

A variety of indirect wastes come from food-handling establishments, bars, sterilizers, drainage outlets, and swimming pools. The food establishment may discharge its waste from dishwashers, coffeemakers, potato peelers, steam tables, walk-in freezers, and steam kettles through an air gap into a trapped and vented receptor. It is essential that an air gap be maintained to prevent backflow into these food preparation or storage facilities. If sinks, bars, or soda fountains are so located that they cannot be vented, the sink drains have to discharge through an air gap or air break into a floor drain or sink that is properly trapped and vented.

Indirect waste connections must be provided for drains, overflows, or relief vents from the water distribution system by means of an air gap. Stills, sterilizers, and similar equipment that require a waste connection must be indirectly connected by means of an air gap. Other types of devices that drain fluid away, such as ice-making machines, must also have an air gap. Cooling jackets and sprinkling systems, if emptying into the building drainage system, must only do this through an air gap.

Pipes carrying wastewater from swimming or wading pools, including pool drainage, backwash of filters, and water from floor drains that are on the walks around the pool, must be installed as an indirect water waste system. If the recirculation pump is used to discharge the waste pool water to the drainage system, the pump discharge must be installed as an indirect waste discharge to the sewer. Air gaps are needed in each of these areas to avoid contamination of the pool water.

The drains from pressure tanks, boilers, relief valves, and other equipment must discharge into the drainage system through an indirect waste discharge by means of an air gap. Any other indirect waste must be discharged in the same manner. Waste receptors or sump pumps that serve indirect waste pipes must not be installed in toilet rooms or any inaccessible or unventilated spaces, such as closets or storerooms. If the indirect waste receptor is below floor level, it has to be equipped with a running trap adjacent to it with a trap clean-out level with the floor. All indirect waste receptors need to be equipped with readily removable metal baskets into which all indirect waste pipes discharge. The plumbing receptors should be set up in such a way that the indirect waste pipes do not splash or flood the area adjacent to the receptor. Ordinary kitchen sinks or laundry trays are not to be used as receptors for indirect waste. Steam pipes must not be connected to any part of the drainage or plumbing system. Water above 140°F must not be discharged into the drainage system. These pipes have to be connected to an indirect waste receptor. When the indirect waste receptors are installed, they have to be readily accessible for flushing and cleaning. They have to meet the material, pipe sizing, and construction requirements of local and state plumbing codes.

In food- and beverage-handling establishments, the following rules apply:

1. The minimum size of the indirect waste pipe is not smaller than the drain on the unit or 1 in., and the maximum length is 15 ft.
2. Indirect drains for refrigeration coils and ice-making machines are not smaller than the drain on the unit or three quarters of an inch.
3. In walk-in coolers, the drainage line from the floor drain should discharge outside of the cooler into an approved receptor, where the overflow of the receptor is at least 6 in. below the lowest floor drain in the walk-in cooler.
4. Drains for food preparation sinks, steam kettles, potato peelers, ice cream dipper wells, and similar equipment should be no smaller than 1 in. and should be properly air gapped.
5. Bar and fountain sink drain lines should not exceed 5 ft from the fixture to the receptor.
6. Direct waste connection between a water distribution system and a drain should be through an approved air gap.

Sterilizers, stills, and similar equipment drains must not be more than 15 ft from the receptor. A proper air gap must be installed.

In the handling of special wastes, special techniques must be utilized. In no case can corrosive liquids, used acids, or other harmful chemicals that might destroy or injure a drain, sewer, soil, or waste pipe, create noxious or toxic fumes, or interfere with sewage treatment processes, be allowed to discharge into a plumbing system without becoming diluted, neutralized, or specifically treated. In most cases, these special corrosive wastes should be collected separately and disposed by contractors who are used to dealing with hazardous chemicals.

Water Supply and Distribution

All buildings that contain plumbing fixtures and that are used for human occupancy or habitation must be connected to a potable supply of cold water at a specified quantity and pressure. If the building is either a permanent residence or a place in which people are employed or go for entertainment, hot water must be provided under pressure in adequate quantities. Only potable water can be used for drinking, bathing, cooking, processing of food, medical or pharmaceutical product use, or any other use where individuals may ingest the water in one form or another.

The size of the water service pipe coming into the building depends on the number of people utilizing it, the kind of processes involved, the amount of water needed, and the location and terrain the pipe must cross from the main water line. Friction loss and flow velocity must be accounted for. The elevation of the structure is an important consideration.

Water services must be separated from the building sewer to prevent a possible contamination of the water source. The underground water service pipe and the building drain or building sewer must be a sufficient distance horizontally to prevent contamination, and must be separated by undisturbed or compacted earth. Potable water service pipes must not be located in, under, or above cesspools, septic tanks, septic tank drainage, disposal fields, or seepage pits. If the water line must cross the top of the sewer line, the bottom of the water service has to be a sufficient distance above the sewer line to prevent contamination.

However, this is not a recommended procedure. The sewer line in this case must be cast iron with mechanical joints, at least 10 ft on either side of the crossing. To protect the water service pipe as it goes through the wall of the building, adequate clearance has to be allowed. Chemical action can occur from direct contact with concrete. Further, distortion or rupture may occur from the shearing action due to settlement or due to expansion or contraction.

In the event a street water main pressure is in excess of 80 $lb/in.^2$, an approved pressure-reducing valve must be installed in the water service pipe near its entrance into the building to reduce the water pressure to 80 $lb/in.^2$ or lower, unless the water goes directly into a water pressure booster system by an elevated water gravity tank. A maximum of 80 $lb/in.^2$ should be the pressure of the outlet of any fixture. Where inadequate water pressure occurs in the street main or other source of supply, an approved booster pump and pressure tank should be utilized. Where the street water main pressure fluctuates, the building water system should be built for the minimum pressure available. If the minimum pressure is too low, then a booster pump and pressure tank should be used.

To get some idea of the required amounts of pressure and also the delivery at the point of discharge of water in certain fixtures, the following fixtures are listed: a basic faucet operates at 8 $lb/in.^2$ and delivers 2 gal/min at the point of discharge; a sink faucet operates at 8 $lb/in.^2$ and delivers 3 gal/min at the point of discharge; a bathtub faucet operates at 8 $lb/in.^2$ and delivers 4 gal/min at the point of discharge; a flush valve or a toilet operates at 15 $lb/in.^2$ and delivers from 15 to 35 gal/min of water at the point of discharge; an automatic dishwasher operates at 15 $lb/in.^2$ and delivers 3 gal/min at the point of discharge; an automatic clothes washer operates at 15 $lb/in.^2$ and delivers 3 gal/min at the point of discharge.

Hot water is an important requirement for adequate cleaning and adequate sanitization. The minimum requirements for a hot water storage tank are obviously based on the type of usage of hot water within the building and the amount of usage. In all cases, the hot water storage tank has to provide an adequate rise in temperature from proper amounts of British thermal unit input of the water-heating equipment. The hot water heater and storage tank must be of adequate size to provide sufficient hot water for all daily requirements and hourly peak loads. The storage tank must meet the construction requirements of the American Society for Mechanical Engineers or the Underwriters Laboratory. All storage tanks must be protected against excessive temperatures and pressure conditions. They should be currently marked in an accessible place with the maximum allowable working pressure. Drain cocks or valves for emptying the tanks are to be installed at the lowest point of each hot water storage tank.

In the event that a potable water system pipe or set of pipes are under repair or installed, the following techniques should be utilized for cleaning and disinfection. The pipe system is to be flushed with clean and potable water until all dirty water is gone. The system should be filled with a water solution containing chlorine at a minimum of 50 ppm and should be allowed to stand unused for 24 hours. The system may also be filled with a chlorine solution at 200 ppm and allowed to stand unused for 3 hours. The system should be flushed after the chlorine standing time until no further chlorine comes out in the water. Bacteriological tests of the water system should be made.

To determine the amount of water needed for a building, consideration must be given to its usage. The pipe sizing and the actual source of water supply, especially the quantity, depend on the minimum quantity of water needed per person per day in gallons. The following criteria should be utilized.

1. Summer cottages, boarding homes, motels without private baths, tourist camps, mobile home parks, resort camps, semipermanent work camps, motels with baths, toilets, kitchens, and self-service laundries require 50 gal of water per person per day.
2. Single family dwellings require 75 gal per person per day.
3. Multiple family dwellings require 60 gal per person per day.
4. Rooming houses require 40 gal per person per day.
5. Restaurants require 7 to 10 gal per person per day.
6. Luxury camps require 100 to 150 gal per person per day.
7. Boarding schools require 75 to 100 gal per person per day.
8. Hospitals per bed require 150 to 250 gal per person per day; other institutions per bed require 75 to 125 gal per person per day.
9. Factories require 15 to 35 gal per person per shift, exclusive of industrial use.
10. Swimming pools require 10 gal per person per day.

Peak demands can readily interfere with the usage of water in a given structure. It is very important when developing the plumbing system to take into account those special conditions that create a peak demand in any given day, or the special types of conditions, such as lawn sprinkling, that may create a peak demand during special seasons. In all cases, piping, water quantity, and storage needs should be based on peak demands.

Drainage Systems

The drainage system (Figure 5.8) must be constructed so as to handle the amount of waste that comes from each of the various types of usages. The greatest amount of waste comes from bathrooms, bathtubs, flushing sinks, urinals, and valve-operated toilets. The soil and waste stacks must not be smaller than the largest horizontal branch. The stack vent should be a minimum of 3 in. in diameter, and the underground drainage piping below a cellar or basement must not be less than 2 in. in diameter.

The horizontal drainage piping is installed in uniform alignment at a uniform slope of not less than ¼ in./ft for 3 in. diameter and less pipe, and not less than ⅛ in./ft for 4 in. or more diameter pipes. The velocity of the sewage flow should be at least 3 ft/sec. Where pipes have to change direction, there should be 45° Y joint. If the flow is directly from the horizontal to the vertical, then a sanitary T joint should be installed.

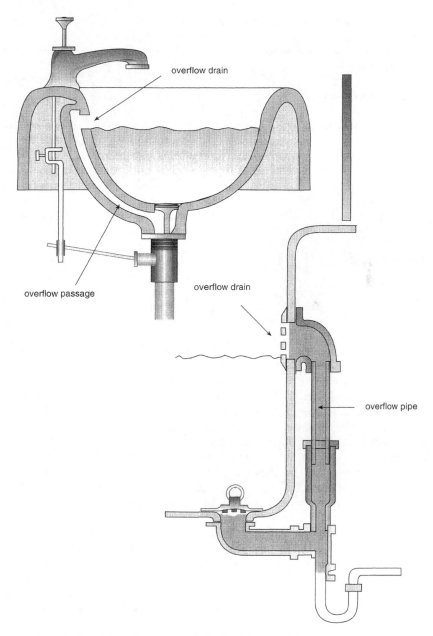

Figure 5.8 Drainage systems: overflow drains in sink and bathtub.

When drainage systems are below sewer level and therefore cannot discharge into the sewer by gravity, they should be discharged by tightly covered and properly vented sump pumps. The sump pump lifts the sewage and discharges it automatically into the sewage system.

Vents and Venting

Vents are determined by the type of sewage system within a structure. The vents are used to release or remove gases into the atmosphere. They are also used for the protection of trap seals from siphonage, aspiration, or back pressure. Every building in which plumbing is installed must have at least one main stack that maintains the same size throughout the building. The stack goes from the building drain to the open air above the roof. The vent pipe should be at least 6 in. above the roof. If the roof is used for purposes other than weather protection, the vent pipe must be at least 7 ft above the roof.

Storm Drains

Storm drains are required where roads, paved areas, yards, courts, and courtyards have to be drained (Figure 5.9). This water should go either (preferably) into a separate storm water system or into a combined sewer system. In one-family or two-family dwellings, the storm water can discharge onto the streets or lawns, as long as the storm water flows away from the building. The size of building storm drain or building storm sewer or any part of the horizontal branches is based on the actual area to be drained. As an example, a 3-in. drain can remove 34 gal/min of water from an 822 ft^2 area at a ⅛ in. slope per foot. The 3-in. drain can remove 48 gal/min of water from a 1160 ft^2 area at a ¼ in. slope, and can remove 68 gal/min of water from 1644 ft^2 in an area at a ½ in. slope. The size of the leader and conductor is based on the maximum rate of rainfall of 4 in./hr for a 5-min duration and a 10-year return period. If the rates are greater than this or less than this, the sizes may be adjusted accordingly.

The size of the roof gutters is based on the maximum projected roof area and also on the maximum rate of rainfall of 4 in./hr/5-min duration and a 10-year return period. Again, this is adjusted based on the amount of actual rainfall within an area. If a combined drain and sewer are utilized, the total pipe or combined drain must take into account the rate of sewage flow, as well as the storm water that may be entering the system. Building subdrains below the sewer level should discharge into a special sump or receiving tank, or be discharged away from the building. If the subsoil drain is subject to backwater, it should be protected by an accessible protected backwater valve. Where the storm drain is tapped into a combined sewer, there should be a trap. If it is only connected to a storm sewer, then traps are not necessary.

Medical Care Facilities

The medical care facility needs special attention, because many types of fixtures can create potential hazards to the water supply, and many types of fixtures may be contaminated by sewage. Flush rim clinic sinks, bedpan steamers, and other equipment

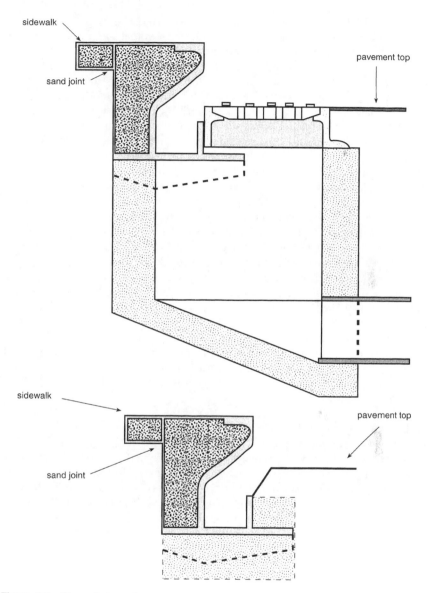

Figure 5.9 Storm drain system.

used for disposing of bedpan contents are a specific type of problem. These sinks, in effect, become toilets, and the soil from the bedpans or urinals must discharge into a facility where the waste can go into the sewer. This is accomplished by having the bedpan steamer piped to an approved receptor located in the same room through an air gap, with the waste pipe not exceeding 15 ft. Other types of sinks, such as those used in soiled utility rooms and janitor sinks, must not be used for the disposal of urine, fecal matter, or other human waste. Ice-making machines may only be installed in food pantries, diet kitchens, or other areas that are separate from the soiled utility areas. They must never be installed within the soiled utility room or in any bedpan-handling facility. Ice machines should have a drainage line that has an air gap, as a means of protection, before the water enters the drain.

All sterilizer equipment should have an American Society of Mechanical Engineering Standards symbol and date plate. Sterilizing piping and devices must be accessible for inspection and maintenance. The steam supply lines must be able to drain the condensate freely by gravity. Pressure sterilizers must be equipped with an acceptable means of condensing and cooling the exhaust steam vapors. Nonpressure sterilizers either must be automatically controlled so that the vapors are confined to the vessel, or must be equipped with an acceptable means of condensing and cooling the vapors.

Where control valves, vacuum valves, and other devices protrude from the wall in operating, emergency, recovery, examining, or delivery rooms or corridors where patients are transported on wheeled stretchers, the valves must be at an elevation that cannot allow for bumping of the patient, stretcher, or employee. If the unit must be installed at a low elevation, appropriate safety guards must be used.

In mental institutions, all piping, controls, and fittings of plumbing fixtures must be concealed or bolted to the walls. Wherever possible the fittings and pipings should be vandal-proof and out of the reach of the patients.

Ice storage chests should drain through an air gap to a receptor. As has been mentioned, bedpan washers and clinical sinks should go directly to the soil pipe and obviously should be vented. All sterilizer waste should be separated from the receptor by an appropriate air gap. Floor drains should be installed so that the entire floor area can be drained adequately.

All water sterilizer drains, including tank, valve leakage, condenser, filter, and cooling, must be installed as indirect waste with an air gap. All pressure instrument washer sterilizers must be installed in the same manner.

Water-operated aspirators must be installed only with the specific approval of the local or state health department. In operating rooms, emergency rooms, recovery rooms, delivery rooms, examining rooms, autopsy rooms, and any other locations, except laboratories where aspirators are used for removing blood, pus, or other fluids, the discharge from the aspirator must be an indirect connection to the drainage system. The suction line has to be provided with a bottle or other trap to protect the water supply. Vacuum fluid suction systems used for collecting, removing, or disposing of blood, pus, and other fluids should be provided with receptacles equipped with an overflow preventive device at each of the vacuum outlet stations. Each vacuum outlet station should be equipped with a secondary safety receptacle as an additional safeguard against fluids other than air entering into the vacuum piping system. Where central vacuum fluid suction systems are used, the system has to be equipped with collecting control tanks that can be drained and cleaned while the system is in operation. A system has to have adequate power to remain in service during a power emergency. The exhaust from the vacuum pump must discharge separately to the outside atmosphere above the roof level and must not create a nuisance or hazard.

The lowest point of the condensate riser must be trapped and discharged over an indirect waste sink. A branch has to be installed upstream from the condensate drain trap for flushing and resealing purposes. The drain and trap must be located above the lowest floor level of the building.

Radioactive material has to be disposed of in a manner that cannot create a hazard to personnel at the institution. Only on specific authority from the state health department can the institution dispose of any radioactive material into the sewage system.

All hospitals and other medical care facilities must have dual services to provide for an uninterrupted supply of water in case of a water main break. The hot water heating equipment has to furnish 6½ gal of water at 125°F/hr per bed per fixture; 4 gal of water at 180°F/hr per bed per kitchen; and 4½ gal of water at 180°F/hr per bed for the laundry. The hot water storage tanks have to have a capacity equal to 80% of the heater capacity.

Water supplies for fire protection include a standpipe system, which must be installed according to fire codes meeting the national requirements of the National Board of Fire Underwriters. Automatic sprinkler systems must be installed in hazardous areas, such as basements, paint shops, trash rooms, laundries, kitchens, and trash chutes. These sprinkler systems should comply with the state fire marshall's orders or the requirements of the National Board of Fire Underwriters.

Where oxygen and other systems are installed, the medical gas piping outlets, manifolds, and storage rooms must be installed in accordance with the requirements of the National Fire Protection Association.

Where water sterilizers, stills, or similar equipment need to be descaled or chemically treated, they must be disconnected from the water and drainage system. They then must be washed very thoroughly and rinsed thoroughly before they are permitted to be put back into the water or drainage system.

Geothermal Heat Pump Systems

A geothermal heat pump system uses a condenser circuit that is in either the earth, in the case of a closed-loop system, or in a water supply, in the case of an open-loop system, to provide heat to the dwelling or remove heat from the dwelling. In the closed-loop system, the ground acts as the condenser and the pipes from the house go out to the ground. The fluid within the closed loop consists of 20% glycol, which is antifreeze, and 80% water. The ground may vary in temperature from freezing to 60°F or more depending on the time of year and the weather conditions.

The vertical closed-loop system must be installed in such a way that groundwater is not contaminated and the intermingling of desirable and undesirable waters does not occur. Improperly installed vertical geothermal loops may, as in any other boring in the earth, constitute a hazard to health and groundwater. The basic consideration is that the installation of the vertical geothermal loops do not destroy or alter the geohydrologic conditions that existed before the boring was drilled. If the system is not properly installed, contaminated water could seep quickly down to the groundwater supply. The geothermal boring must be sealed with impermeable materials to prevent this percolation of water.

The heat-exchange loop must be constructed of pipe material especially designed for closed-loop systems, and all joints must be thermally fused. A minimum 50-ft separation is needed between vertical closed-loop excavation and a water supply well. All borings should be at least 50 to 100 ft from storm and sanitary sewer or other sources of potential contamination. The borings should be at least 100 ft from septic tank systems. All borings should be made by a licensed well driller. All borings should be filled with cement grout containing no less than 6% bentonite mixed into the grout or a thick bentonite grout placed in the base of the boring upward by pump or other acceptable procedures. Where horizontal closed-loop systems are installed, they should not be placed under sewage systems or absorption field systems. The ground in the vicinity of the boreholes should be graded to cause the run off water to drain away from the boreholes. The antifreeze solution used in the closed-loop system should be a nontoxic chemical that cannot contaminate the potable water supply.

In an open-loop system, well water that is at 55°F is used as the condenser. The water is pumped into a house and through a refrigeration process, where the heat of the water can be removed and used to heat the property. When cooling the property, the hot air can be passed across the 55°F water to raise the temperature of the water and to decrease the temperature of the air within the property. Wells used to supply open-loop groundwater heat pump systems must comply with the regulations relating to private water wells and water systems. Where open-loop systems may withdraw more than 100,000 gal/day, they must be registered with the state departments of environmental protection, environmental management, or natural resources. The wastewater from an open-loop heat pump should be disposed of in a storm water drain. Where diffusion wells are used to dispose of the heated water, they should be at least 50 ft away from the domestic water supply. This system can only be used when 25,000 gal/day or less are utilized. The airflow in temperature should under no circumstances exceed the ambient groundwater temperatures by more than 20°F with 8 to 10°F the preferred maximum temperature increase. No toxic materials may be used in the piping systems.

MODES OF SURVEILLANCE AND EVALUATION

Tests and Maintenance

New, altered, extended, or repaired systems have to be tested for leaks and other defects. Rough plumbing may be tested by use of either water or air to prove that it is watertight. The rough plumbing does not include the outside leaders and perforated or open-jointed drain tile. The plumbing should be filled to the highest point, and all openings, with the exception of the highest point, should be sealed. As each section is tested, the same procedure should be followed. Air tests can be conducted by attaching an air compressor testing apparatus to the opening and by closing all inlets and outlets to this system. There should be a uniform gauge pressure of 5 lb/in.^{2} held for a minimum of 15 min. Finished plumbing should be tested for gas and water tightness by the use of a smoke test. All the traps should be filled with water, and then a pungent thick smoke produced by a smoke machine should be introduced into the system. When the smoke appears at the stack openings on the roof, the openings should be closed and a pressure equivalent to a 1-in. water column should be built and maintained for the period of the inspection.

Where this test cannot be carried out, a peppermint test can be used instead. This is done by introducing 2 oz of oil of peppermint to the roof terminal of every line or stack to be tested. The oil of peppermint is then followed at once by 10 qt of hot (160°F) water or higher, and then the roof vents are sealed. If the odor of peppermint appears at any trap or other point in the system, then a leak exists. Obviously, all persons who have utilized oil of peppermint should be excluded from the test areas. A building sewer should be inspected and tested by inserting a test plug at the point of connection with the public sewer. The building sewer should be filled with water under a head of not less than 10 ft. The water level at the top of the test head should not drop for at least 15 min. The water supply system should be tested and proved tight under water pressure not less than the working pressure that can occur within a system.

Plumbing and drainage systems must be maintained in such a manner as to always be in compliance with the applicable local or state codes. Many of the hazards that occur are due to a lack of proper installation or maintenance of the plumbing equipment.

Surveys

The EPA Office of Inspector General conducted a survey on cross-connection controls. This was done because cross connections have resulted in significant and dramatic public health adverse effects in certain states and local communities. Of the 45 states contacted, 29 had some type of a cross-connection program, with administration and enforcement at the local level. The scope of the programs varied widely. Contamination of the potable water supply by cross connections is largely undetected, not investigated, or not sufficiently reported, because of the difficulty in identifying cross connections as the source of the contamination. Cross-connection control is not federally mandated, and does not receive appropriate attention from state and local governments because of funding shortages. Also, no federal reporting requirements are mandated for potable water contamination caused by cross connections.

Maintenance of water quality in a system may come into conflict with the fact that systems are designed to ensure hydraulic reliability. It was found that water quality deteriorated when extended lines, storage tanks and reservoirs, and interaction between the system materials and the disinfectants were present. Outbreaks of *Escherichia coli* O157:H7, *Salmonella*, etc. have been traced to cross connections.

CONTROLS

Methods of Preventing Backflow and Back Siphonage

Five basic devices are used to prevent backflow or back siphonage. They are air gaps; atmospheric vacuum breakers, which may also include hose-connection vacuum

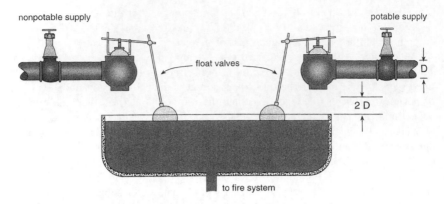

nonpotable supply

potable supply

float valves

D

2 D

to fire system

Figure 5.10 Fire system makeup tank for a dual water system. (From *Water Supply and Plumbing Cross-Connections*, DHEW Publication No. 957, Rockville, MD, 1969, p. 33.)

breakers; pressure-type vacuum breakers, which also include backflow preventors with atmospheric vents for ½-in. and ¾-in. lines; double-check valve assemblies; and reduced pressure principle backflow preventors.

An air gap is the only absolute means of eliminating the physical link between potable and nonpotable water supplies. The supply water inlet to the tank or fixture must be a minimum distance above the fixture of two times the effective opening of the pipe. In no case should less than a 1-in. distance be between the supply line and the overflow of the fixture if toxic substances are present. To provide necessary water for various operating systems, the use of surge tanks is recommended. Water fills the surge tank to the flood level and the inlet pipe is at least two times the diameter of the pipe above the flood level. As the water is needed, it can then be withdrawn from the surge pipe and a special valve in the pipe is then activated and the water is filled to its top position again. Frequently, booster pumps are needed to move water up into high buildings.

Theoretically, water will rise 34 ft because of atmospheric pressure; additional pressure is needed for the water to rise further. However, in high buildings the city water main generally does not generate adequate pressure to move the water into the upper floors as needed. To prevent the hazards of water flowing backward from the upper floors into the public water supply, once again a surge tank with an air gap feeder line is recommended (Figures 5.10 and 5.11).

An atmospheric-type vacuum breaker may only be used on connections to a nonpotable water supply. When a vacuum breaker is never subjected to back pressure, and is installed on the discharge side of the last control valve, it has to be installed above the point of usage. It cannot be used when a continuous pressure system is operating. A hose–bib vacuum breaker is an inexpensive device that is attached to a sill cock and threaded faucet. It should not be used where continuous pressure is in effect.

A pressure-type vacuum breaker can be used for connecting nonpotable systems where the vacuum breakers are not subject to back pressure. They can be used for

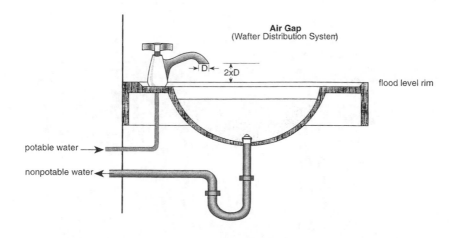

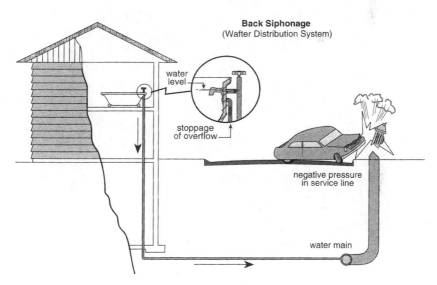

Figure 5.11   Cross connection. (From *Basic Housing Inspection,* Training Course Manual, U.S. Department of Health, Education, and Welfare, Public Health Service, Environmental Health Service, Environmental Control Administration, Cincinnati, OH, March 1970, p. 6.25.)

a continuous supply pressure. A backflow preventor with intermediate atmospheric vent is used for ½-in. and ¾-in. lines.

The fundamental theory in back-siphonage obviously is the creation either of a vacuum or of a negative pressure. If atmospheric pressure is applied to a piping system between the source of pollution and the origin of the vacuum, back siphonage can be prevented. This is what the vacuum breaker is supposed to do. Because a

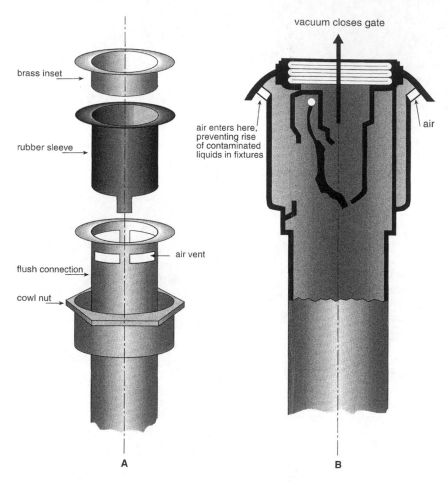

Figure 5.12   Vacuum breakers. (From *Water Supply and Plumbing Cross-Connections,* DHEW
              Publication No. 957, Rockville, MD, 1969, p. 31.)

vacuum may occur at numerous places in the piping system, it is necessary to put
the vacuum breaker as close to the source of potential contamination as possible.
However, the vacuum breaker must not be subjected to any kind of flooding
(Figures 5.12 and 5.13).

   A double-check valve assembly is used to protect all direct connections between
potable and nonpotable water, or potable water and other substances. This device is
utilized only if an air gap cannot be used, because it is difficult to provide a physical
break between the two systems. The device consists of two hydraulically or mechanically loaded pressure-reducing check valves with a pressure regulator relief valve
located between the two check valves. The flow enters a central chamber against the
pressure exerted by the loaded check valve. The supply pressure is reduced there by
a predetermined amount. The pressure in the central chamber is always maintained

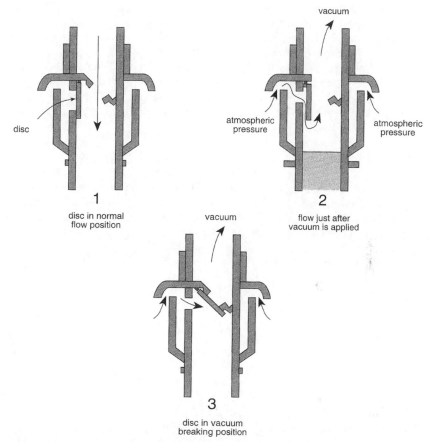

disc

1

disc in normal
flow position

vacuum

atmospheric
pressure

atmospheric
pressure

2

flow just after
vacuum is applied

vacuum

3

disc in vacuum
breaking position

Figure 5.13 Operation of a vacuum breaker. (From *Water Supply and Plumbing Cross-Connections,* DHEW Publication No. 957, Rockville, MD, 1969, p. 27.)

lower than the incoming supply pressure by operating the relief valve. Because all valves may leak as a result of wear or obstruction, the protection given by check valves is not considered adequate. Therefore, the relief valve must necessarily operate so that air from the atmosphere can flow in when needed. If one or both of the check valves or relief valves malfunction, water should start coming out of the relief port.

Another technique similar to double valves is the use of a swing connection. Two lines have a swinging ball joint, which may be connected in the event of an emergency. If the swinging ball joint is put into effect to utilize the auxiliary water supply, such as in the case of fire, the lines can become contaminated. After the water use is over, the lines should be thoroughly rinsed and chlorinated. A check valve on the potable supply, if operating properly, should keep the supply free of contamination as long as adequate pressure is exerted on the water flow. The reduced-pressure principle backflow preventors are in effect the double-check valve assembly with the additional intermediate valve that has already been discussed.

One other technique used for firefighting that incorporates the air gap method is the building of a tank that can hold the water needed for fire protection. The potable water supply line has a float valve, which triggers the inlet line located two diameters above the overflow of the tank. On the opposite side of the tank, the nonpotable water supply, which also has a float valve, is two diameters above the overflow level of the tank. As the water in the tank starts to decrease during its use for firefighting, the two float valves open the water supplies and the water passing through the air gap goes into the tank. As the tank fills, the float valves automatically shut off the water supply.

Backflow Prevention and Dentistry

In some areas regulators have required dental offices to install backflow prevention devices at the service connection or on individual dental units. Available science suggests, however, that an extremely low risk of contamination of water supplies exists from cross connections in dental units.

Federal Flush Regulations

The Energy Policy and Conservation Act of 1992 requires contractors to install ultralow flush toilets with 1.6-gal capacity in all remodeled and new homes. Nearly 50 million of these low-flow toilets had been installed, reducing the average of 19.3 gal of water per person per day for toilet use to 9.3 gal. Opponents of the act contend that the new toilets are not doing an effective job. Some members of Congress are currently pushing for the Plumbing Standards Improvement Act.

Lead Free Requirements

The 1996 amendments to the Safe Drinking Water Act required the EPA to establish a performance standard to cover the leaching of lead from any devices intended to provide water for human consumption. Products covered included kitchen, bar, and lavatory faucets; drinking fountains; water coolers, residential refrigerator ice makers, and water dispensers; and supply stops and pinpoint control valves. The EPA adopted Section 9 of American National Standards Institute (ANSI)/National Sanitation Foundation (NSF) Standard 61, Drinking Water System Components–Health Effects as the performance standard. The Safe Drinking Water Act retained the 8% lead content requirement and further mandated that faucets, drinking fountains, and other water dispensing devices must meet the performance-based leaching requirements of the standard.

Plumbing Program

Public health personnel, water works officials, plumbing inspectors, building managers, plumbers, and maintenance men all are responsible for protecting the health and safety of the public by proper use and control of plumbing. These responsibilities include proper design, installation, and maintenance of the various piping systems. These individuals are to the plumbing system what the surgeon is to the patient. They are all essential in their own way, and they all must be trained to carry out their part of the overall job. It has been stated that innumerable plumbing defects and plumbing systems are potential hazards. Luckily, few very serious problems have occurred, but not because of lack of plumbing problems. It is essential that government, industry, and labor unions establish good cross-connection control programs and good plumbing programs.

A cross-connection control program should contain substantial and proper plumbing and cross-connection control law, with good solid rules and regulations on implementation of the law; establishment of a governmental agency that has major responsibility for administering the program; provision of adequate trained staff and necessary ancillary personnel; review and inspection of all new plumbing installations; review and inspection of all corrections or improvements on existing installations; systematic surveillance of all existing installations, starting with those areas where the greatest degree of hazards occur; issuance of orders for correction of hazards and necessary follow-up; inspection and maintenance of backflow preventors; provision of adequate training in control of cross-connections for all individuals involved in this area; and good public health education programs to make the public aware of the dangers of cross connections.

Uniform Plumbing Code

The Uniform Plumbing Code, 2000 edition, was developed by the International Association of Plumbing and Mechanical Officials and supported by the American Society of Sanitary Engineers, Western Fire Chiefs Association, and United Association of Journeyman and Apprentices of the Plumbing and Pipe Fitting Industry of the United States. The National Association of Plumbing–Heating–Cooling Contractors and the Mechanical Contractors Association of America joined together with the International Association of Plumbing and Mechanical Officials to create this manual.

The value of the Uniform Plumbing Code is to bring order to the disorder in the industry as a result of widely divergent plumbing practices and the use of many different and often conflicting plumbing codes by local jurisdictions. The National Fire Protection Association is working with the previous groups on the 2003 edition to harmonize all standards.

The code discusses a variety of subjects including water supply, sewage, and cross connections. It gives a long list of unlawful connections that can contaminate the potable water supply system and how to correct them. Irrigation systems, fire protection systems, and various aspirators, ejectors, water siphons, and other such equipment are discussed with priming of nonpotable water or other fluids.

Cross connections and means of preventing backflow and back siphonage are a major consideration. The two types of indirect cross connections are under rim or submerged connections, and over rim connections. The under rim occurs when the water inlet comes into the bottom or side and is submerged in a polluted substance.

The overrim occurs when a hose is attached to a water line and is submerged in a polluted substance. A direct cross connection occurs when there is a physical connection between the two sources.

In a structure where both potable and nonpotable water are used there are color codes to distinguish between them. Potable water has a sign that has a green background with white lettering. Nonpotable water has a sign with a yellow background and black lettering that states, "Caution: nonpotable water, do not drink."

SUMMARY

Many times in the past environmental health practitioners have ignored the plumbing area, because it was not taught in colleges and seemed to be an area that was primarily the province of plumbers and plumbing inspectors. Today, however, in our current fast-moving world, where new hazards — chemical, biological, physical, and radiological — are constantly increasing at a very rapid rate, it is imperative that the environmentalist not only understand some of the basic principles of plumbing but also ensure that this vital area of the environment is not neglected.

RESEARCH NEEDS

Research needs include the development of techniques that can be rapidly taught to environmental health practitioners to give them a good understanding of plumbing problems and potential hazards. As a result of the training, they should then be able to identify these hazards prior to an outbreak of disease. Further, plumbing valves and systems should be studied to determine the best method or methods of avoiding contamination to potable drinking water supplies.

CHAPTER 6

Private and Public Sewage Disposal and Soils

BACKGROUND AND STATUS

Populations in the United States have moved from rural to metropolitan areas. This has created considerable difficulty in providing adequate sewerage systems to meet the growth. The sewerage systems built many years ago cannot handle the additional load. In many areas, communities have gone to individual septic tanks and soil absorption systems that at best are not good procedures. In 1860, of a population of 31 million, 30 million lived in unsewered areas and 1 million lived in untreated sewered areas. In 1900, of a population of 76 million, 51 million lived in unsewered areas and 25 million, in sewered areas. In 2002, of a population of 287 million, 60 million lived in unsewered areas, and 227 million, in sewered areas. As can be seen from these statistics, along with a growth in population, a considerable growth has occurred in the number of areas where sewerage systems are provided, as well as a large increase in the number of areas where sewerage systems are not provided. Although the percentage of sewer installations are much higher today compared with the total number of unsewered homes, the total population of the United States far exceeds that of 1860.

In 1957, a total of 16.4 million lb of biochemical oxygen demand (BOD) per day was entering the sanitary sewers, which was reduced to a level of discharge of 8.7 million lb. In 1973, as much as 27.1 million lb of BOD was collected per day in the sanitary sewers. After treatment, 8.6 million lb was released to the streams. By 1982, there had been a 46% reduction in BOD from municipal systems and a 71% reduction from industrial systems. Although larger amounts of sewage are collected with increasing amounts of BOD in millions of pounds, and although new treatment processes are utilized, the discharge rate after treatment has decreased because of more secondary and advanced wastewater treatment plants. In 2002, raw or insufficiently treated wastewater from municipal and industrial treatment plants still threaten water resources. These plants release excessive BOD, bacteria, nutrients, ammonia, and toxics.

Since 1972, municipal sludge has doubled to 7 million dry tons and doubled again in the year 2002. Serious economic, legal, logistic, and health problems are associated with the reuse of wastewater for agricultural, industrial, or potable purposes. Further, the sludges from waste treatment plants act as contaminants to the soil, and, if they run off to the water, as contaminants to the water.

A major concern in our society today is the proliferation of on-site sewage disposal units. Currently, there are some 25 million such units in the United States, and they are increasing each year as individuals move out into suburban–rural types of situations. A large number of these systems overflow onto the ground or into surface bodies of water, or back up into homes. These systems constitute a health hazard to the individual and to the community. The water that percolates through the soil may readily carry contaminants into the underground aquifers, contributing to additional types of health hazards through pollution of water supplies. The individual on-site sewage disposal system usually consists of a septic tank of some sort, including aerated tanks, distribution box, possible dosing tanks, and various soil absorption systems. These systems consist of subsoil absorption trenches, deep absorption trenches and seepage beds, seepage pits, subsurface trenches and filters, elevated sand mounds, sand-lined trenches and beds, and oversized seepage areas. In any case, it is important to understand that the controlling factor in individual sewage disposal is not only the soil that the effluent must enter but also the amount of water or effluent that must be absorbed.

Septage from septic tank systems are in the following range of concentrations:

1. Total solids — 1130 to 130,475
2. Total volatile solids — 350 to 71,400
3. Total suspended solids — 310 to 93,375
4. Volatile suspended solids — 95 to 51,500
5. BOD — 440 to 78,600
6. Chemical oxygen demand — 1500 to 703,000
7. Total Kjeldahl nitrogen — 65 to 1060
8. Ammonia nitrogen — 3 to 115
9. Total phosphorus — 20 to 760
10. Alkalinity — 520 to 4190
11. Grease — 210 to 23,365
12. pH — 1.5 to 12.6

The major problem today, simply stated, is one of getting rid of the effluent produced by the individuals utilizing the on-site sewage disposal system. Many variations of these systems exist. All of them are supposedly efficient. However, in actuality many of them fail, because the simple concept of getting rid of a certain amount of water in a certain amount of soil is not recognized. If the soil cannot absorb the water, it must go up, out horizontally, or back into the house.

SCIENTIFIC, TECHNOLOGICAL, AND GENERAL INFORMATION

In the coming years it will be mandatory for our country to provide huge quantities of water to efficiently operate as a society. In 1965, agriculture utilized approximately 113 billion gal/day of water. It is projected that by the year 2020, agriculture will utilize approximately 166 billion gal/day of water. In 1965, municipalities utilized an estimated 24 billion gal/day of water, and by the year 2020 will use an estimated 74 billion gal per day. Power generation in 1965 utilized an estimated 63 billion gal/day of water, and by the year 2020, power generation will utilize an estimated 414 billion gal/day. Industry utilized an estimated 46 billion gal/day of water in 1965, and by the year 2020, will use an estimated 211 billion gal/day. If the estimates hold true, then approximately 619 billion gal of additional water will be utilized each day by the year 2020. The actual estimated water usage will be over 300% above today's water usage.

Reclamation of water from tertiary treatment of wastewater can be one of the important solutions to our problem. If water can be reused many times from its point of origin until its point of disposal, eventually in the oceans, we can escape the hazards of an inadequate water supply. This is an extremely costly process. However, the end result may be worth far beyond the initial cost. The techniques utilized to improve water are discussed further in the chapter as the discussion moves into actual sewage treatment.

The techniques utilized in public sewage treatment may be the techniques used in industrial waste treatment that are discussed in Chapter 2. The major objective of wastewater reclamation is to reduce the oxygen demand of the water; to reduce the nutrient content; to avoid undesirable growths of algae; to remove final traces of suspended solids; to remove color, taste, and odor; to remove refractory materials; to remove any of the biological, chemical, or radiological contaminants causing harm to the ecological system or to humans; and to remove toxics.

Sewage

Sources of Sewage

Sewage is mostly the used water supply of a community; may be a combination of the waterborne wastes from homes, businesses, institutions, and industries; and may also contain groundwater, surface water, and storm water. The volume of sewage varies considerably with the character and water use of the community. Where a community has combined sewers, there is obviously far more sewage than if they had only sanitary sewers. Residential use may vary but still stays within a range of about 40 gal per person per day. However, when industrial waste is added, the level may jump as much as 200 gal per person per day. It is difficult to give a true estimate of how much sewage is produced by any given community. This, however, can be measured at the sewage treatment plant by evaluating the actual inflow of waste.

Human and animal wastes are composed of body discharges from toilet facilities or are washed from animal areas into the sewers. These contribute mostly to human microbiological diseases. Household wastes come from home laundry operations, bathing, kitchen waste, washing, garbage grinding, dishwashing, as well as synthetic detergents and grease. Street washings and storm flows occur when rains carry accumulated grit, sand, leaves, and other dirt and debris into the sewers, which then may end up at the sewage treatment plant. Groundwater infiltration is due to eroded sewer pipes, loose joints, collecting sewers not under pressure, and total volume of groundwater available. Industrial wastes come from a vast variety of manufacturing processes. Depending on the type and volume of industrial waste, the sewage treatment plant may operate efficiently, may be inundated by sheer quantity of liquid, or may be disabled by frothing or foaming agents, detergents, or other chemicals.

Domestic sewage is defined as human and animal waste, household waste, and groundwater infiltration. This is typical of residential areas. The sanitary sewage is defined as domestic sewage plus industrial waste. Storm sewers carry the runoff from storms and rains flowing from roofs, pavements, and other natural surfaces. Sanitary sewers only carry sanitary sewage. Combined sewage is a mixture of domestic or sanitary sewage and storm water collected in the same sewer. Industrial wastes are wastewaters from manufacturing processes that may either be disposed of separately or become part of the sanitary or combined sewage.

Appearance and Composition of Sewage

Sewage is a turbid liquid containing solid materials in suspension and solution. When fresh, it is gray in color and has a musty odor. It carries a variety of floating materials, including fecal solids, pieces of food, garbage, paper, and sticks. When old, sewage changes from gray to black, develops a foul odor, and produces black solids that float on the surface or throughout the liquid. At this point, the sewage is called septic. Sewage consists of over 99.9% water by weight. The other, roughly 0.1%, contains such solids as grit, suspended solids, or dissolved solids. The average domestic sewage contains 600 ppm of total solids. Of this group, 200 ppm are suspended solids that contain settleable solids of 120 ppm and colloidal solids of 80 ppm. The settleable solids are 90 ppm organic and 30 ppm inorganic. The colloidal solids are 55 ppm organic and 25 ppm inorganic. The 400 ppm of dissolved solids consists of 40 ppm of colloidal solids and 360 ppm of dissolved solids. These colloidal solids are 30 ppm organic and 10 ppm inorganic. These dissolved solids are 125 ppm organic and 235 ppm inorganic.

The organic solids are of animal or vegetable origin, including waste products from animal and vegetable life, dead animal matter, plant tissues or organisms, and synthetic organic compounds. These substances contain carbon, hydrogen, and oxygen, and may combine with nitrogen, sulfur, or phosphorus. The major groups are proteins, carbohydrates, and fats, plus their decomposition by-products. They may decay or be decomposed through bacterial activity and they may be combustible (Figures 6.1 to 6.3).

Inorganic solids are inert and are not subject to decomposition. They include sand, gravel, silt, and mineral salts. The strength of the sewage or the problem of breaking down the sewage is based on the organic load present. The greater the organic load is, the greater the problem of sewage decomposition.

Suspended solids are visible and in suspension in water. They can be removed from sewage by physical or mechanical means, such as sedimentation or filtration. These solids include the larger floating particles, which consist of sand, grit, clay, fecal solids, paper, pieces of wood, particles of food and garbage, and similar materials. They are roughly 70% organic and 30% inorganic, with the inorganic solids mostly sand and grit. The suspended solids are settleable solids and colloidal

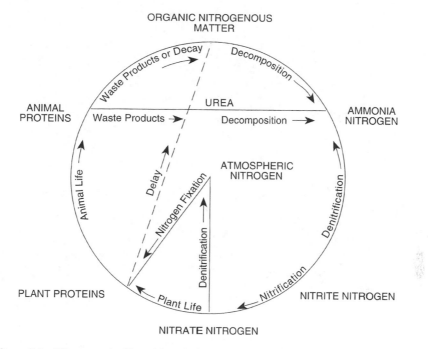

Figure 6.1 Nitrogen cycle. (From *Manual of Instruction for Sewage Treatment Plant Operators*, New York State Department of Health, Office of Professional Education, Health Education Service, Albany, NY, p. 11.)

suspended solids. The settleable solids are those of adequate size and weight to settle within a given time period, usually 1 hour. These are about 75% organic and 25% inorganic in composition. The colloidal suspended solids may settle out if a quiet period lasts longer than 1 hour, but may usually stay in suspension for several days or more. Colloidal suspended solids are not readily removed by physical or mechanical treatment. These are about 67% organic and 33% inorganic in composition.

Dissolved solids is a sewage term instead of a technical definition. Dissolved solids include some of the solids in a colloidal state. About 90% are in true solution and about 10% are colloidal. They are about 40% organic and 60% inorganic in composition.

The total solids include all solid constituents of sewage. They are the total of organic and inorganic solids, suspended and dissolved. In domestic sewage, total solids are roughly half organic and half inorganic.

Dissolved Gases

Sewage contains a small amount, but varying concentrations, of dissolved gases. These gases include oxygen, which is present in the original water supply as dissolved oxygen. Sewage also contains carbon dioxide from the decomposition of organic matter, dissolved nitrogen from the atmosphere, hydrogen sulfide from the

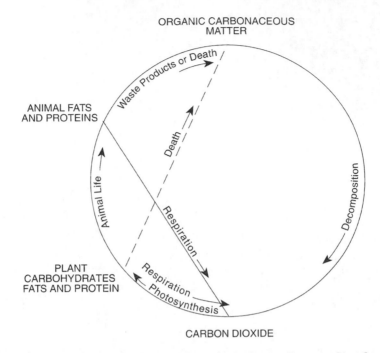

ORGANIC CARBONACEOUS
MATTER

CARBON DIOXIDE

Figure 6.2   Carbon cycle. (From *Manual of Instruction for Sewage Treatment Plant Operators,*
New York State Department of Health, Office of Professional Education, Health
Education Service, Albany, NY, p. 12.)

decomposition of organic and certain inorganic sulfur compounds, and methane
produced during sludge digestion.

Biological Composition of Sewage

Sewage contains a variety, in large numbers, of living organisms, including aerobic
and anaerobic bacteria. The aerobic bacteria help in the breaking down of organic
material when oxygen is present. The anaerobic bacteria aid in the breaking down of
organic material in digestion tanks in the absence of oxygen. The effect of bacterial
decomposition depends on not only the types of bacteria, the quantities of bacteria
and materials present, or the presence or absence of oxygen, but also the temperature,
light, moisture, and pH. In addition to the pure aerobic and anaerobic bacteria, sewage
contains facultative aerobic bacteria and facultative anaerobic bacteria. Other microorganisms present include a variety of viruses, worms, and amoeba. The biological
composition of any sewage is less important than the actual sewage treatment process.

Oxygen Demand in Sewage

BOD specifies the strength of sewage. To say that the BOD is reduced from
500 to 50 implies that there has been a 90% reduction. This means that the final

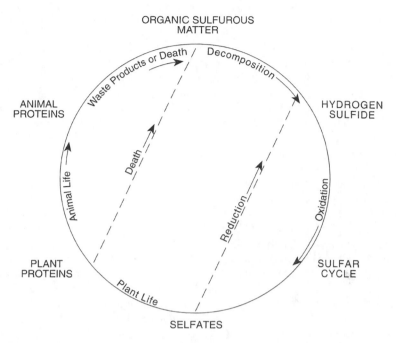

Figure 6.3 Sulfur cycle. (From *Manual of Instruction for Sewage Treatment Plant Operators,* New York State Department of Health, Office of Professional Education, Health Education Service, Albany, NY, p. 13.)

receiving body of water needs a certain degree of oxygen present as dissolved oxygen (DO) to dispose of the rest of the effluent present.

Certain substances can be degraded, but are not biodegradable. These need oxygen to chemically degrade the substance. The chemical oxygen demand (COD) then is a measure of the amount of dissolved oxygen needed in the receiving water to dispose of those substances that can be oxidized chemically but cannot be removed biologically.

DO is the actual amount of oxygen available in dissolved form in the receiving water. When the DO drops sharply, the water life is unable to continue on at a normal rate, leading to fish kills, growth of certain types of water weeds, and finally conversion of the stream into an open sewer.

Chemical Changes in Sewage Composition

Biological life produces many chemical changes in the solids present in sewage. These are biochemical changes brought about by biological growth. They not only measure the activity of the microorganisms but also measure the degree of decomposition of the solids and the effectiveness of the treatment process.

Gravity reduces the amount of suspended solids in sewage. The remaining colloidal or nonsettleable solids undergo biochemical changes, resulting in the removal of molecules of water and causing them to stick together or flocculate and form heavier or settleable solids. This material is called sludge.

In anaerobic decomposition, when the oxygen is removed from complex compounds, simpler compounds are formed. In some situations, the compounds are broken down until the final end product is a stable inorganic or organic substance. During aerobic decomposition the final end products are carbon dioxide, water, nitrates, sulfates, and phosphates. The intermediate products of biochemical decomposition include organic and inorganic acids, and gases such as hydrogen sulfide, methane, and carbon dioxide. The compounds produced depend on the type and the stages of sewage treatment. To understand the complex organic solid decomposition through carbon combined with nitrogen, sulfur, phosphorus, hydrogen, oxygen, or other forms of decomposition, it is necessary to have an understanding of the nitrogen cycle, carbon cycle, and sulfur cycle (see Figures 6.1 to 6.3). The organic compounds containing nitrogen, carbon, or sulfur pass from the dead organic matter through decomposition to products used either by plants or by animals. The three cycles illustrate nature's way of conserving matter.

Products of death become the support systems for plant and animal life. Air and water become the reservoir in which oxygen, hydrogen, nitrogen, carbon dioxide, and other gases are stored for reuse. The sewage treatment process does not alter or modify the natural process. It is only a place where the process is localized so that it can be controlled and accelerated. In the biological changes occurring in sewage, the dissolved gases are important. The DO ensures an aerobic process. When it is lacking, it ensures an anaerobic process. The hydrogen sulfide produced, which is the result of the anaerobic decomposition of sulfur-bearing compounds, exerts a corrosive action on the sewage structures. If carbon dioxide is present in excessive amounts, then acid decomposition of solids is taking place and the rate of decay can be reduced. In sludge digestion, when carbon dioxide is produced, combustible gases are not. Therefore, the methane, which acts as a fuel, is lost. In total, the chemical or biochemical changes occurring in sewage solids are measured to determine the efficiency of the process, the kind of process occurring, and the chemical compounds formed.

Decomposition and Humification of Organic Matter in Sewage and Biological Sludges

The decomposition of organic matter leads to a process of humification and, eventually, fossilization. The labile components of organic matter decompose initially, leading to an accumulation of more stable humic substances. Ratios of C to O and C to H increase as decomposition progresses from the former to the latter groups. In the strictest sense, much of the organic matter is undergoing a process that could lead to fossilization.The labile substances have a mean residence period of days to a year; the humic substances, 25 to 1400 years; and fossil forms, eons.

Humic substances are only considered metastable because the transition pathway gradually continues through a second prefossilization phase called carbonification. The humic substances are converted to fossilized forms including peat, brown coal, lignite, subbituminous, bituminous, anthracite coal, kerogens, and petroleum oils.

Under special conditions these substances could be transformed to graphite or diamond, derivatives whose coordinate points would lie at the origin of the graph — a point of maximum stability for organic matter.

During humification, labile organic matter is converted chemically into heterogenous polycarboxylated (COOH), -hydroxylated (OH), and carboxylated (C=O) aromatic polymers. Humic substances account for the characteristic dark brown and black color of decomposed organic matter. These substances, collectively called humus, consist of three moieties that can be isolated based on their solubility properties in water — humins are acid and alkaline insoluble; humic acids are alkaline but not acid soluble; fulvic acids are both acid and alkaline soluble.

Labile organic compounds, such as carbohydrates and amino acids, and more stable phenolic substances serve as precursors to humic polymers. The organic molecules undergo free-radical reactions to yield metastable heterogeneous phenolic and carboxylic polyaromatic structures. The major stable constituents, aromatic molecules, originate mainly from products of various microbial biochemical pathways.

Synthesis of humic substances, like decomposition reactions, can be catalyzed biotically or abiotically. Biotic catalysis involves intracellular enzymatic reactions. Certain bacteria and fungi synthesize humic substances that are eventually secreted into their surroundings. Some phenolics also are discharged in plant root exudates. Abiotic synthesis of humus includes catalysis by extracellular enzymes, clay colloids, and metals. Humus synthesis also occurs via autooxidation under conditions of alkaline pH. Some phenolics also are discharged in plant root exudates.

The enzymatically catalyzed reactions utilize one or a combination of three known phenoloxidases — catecholase (*o*-diphenoloxidase); laccase (*p*-diphenoloxidase), and peroxidase. Catecholase and laccase both contain copper as a prosthetic group and utilize molecular oxygen (O_2) as an electron acceptor. Peroxidase contains a heme (Fe^{3+}) prosthetic group instead and utilizes hydrogen peroxide (H_2O_2) as its electron acceptor.

Decompositional Pathways

The labile organic macromolecules in sewage and biological sludges, including proteins, carbohydrates, lipids, and nucleic acids, are decomposed into monomers through hydrolytic cleavage; after cleavage the monomers undergo oxidative reactions. Lignin is decomposed mainly through oxidative cleavage, during which its various components, such as syringyl and guaiacyl units with aliphatic side chains, also are partially to fully oxidized to gaseous end products. Thus, the macromolecules are converted into related subcompounds, minerals, humus, and gases.

Proteins (polypeptides) are polymers of amino acids linked together by amide or peptide bonds. They are the major source of nitrogen (N) and sulfur (S) during the decomposition of sludge. Hydrolytic cleavage of a protein molecule initially yields an oligopeptide plus free amino acids. Continued hydrolysis cleaves the oligopeptide into substituent monomeric amino acids.

Amino acids can be assimilated by microbes and earthworms for synthesis of biomass. However, some amino acids and even entire proteins bind with humic substances making them unavailable for metabolic activity. Dissimilation of amino acids can occur via oxidation, reduction, hydrolytic deamination, or decarboxylation. The reaction pathway followed depends on whether aerobic or anaerobic conditions prevail. The major products formed aerobically include ammonia (NH_3), carbon dioxide (CO_2), hydrogen peroxide (H_2O_2), and aliphatic acids (RCOOH). Anaerobic dissimilation of amino acids yields malodorous products such as certain amines (RNH_2) and mercaptans (RSH).

Carbohydrates (polysaccharides) are polymers of monosaccharide (sugar) units connected via glycosidic bonds. Cellulose is the most abundant carbohydrate in sludge. The bulk of the carbohydrates in sewage and biological sludges is derived from microbes, vegetation, and toilet tissue. Hydrolytic cleavage of a polysaccharide at the glycoside bonds yields oligosaccharides plus simple sugars. Oligosaccharides undergo additional hydrolysis and yield monomeric sugars. Cellulose may serve as a co-metabolite to numerous moneric organisms possibly ill-equipped to handle various recalcitrant molecules as a sole source of carbon for their energetic needs.

The lipid component of sludge consists mainly of fats, grease, and oil. A major portion of sludge lipids, including fossilized forms derived from industries and automobiles, originates from microbes and plant residues. Lipids also decompose hydrolytically and yield aliphatic alcohols (ROH) and organic acids (RCOOH). Next to lignin, lipids are the most recalcitrant macromolecules.

Ribose- and deoxyribose nucleic acids (RNA and DNA, respectively), are chains of nucleotides — sugar + N-base + phosphate — connected through phosphodiester bonds. Nucleic acids are the major source of P in sludge. Hydrolysis of a nucleic acid progresses stepwise to yield nucleotides and nucleosides.

Lignins are heterogeneous polymers of phenylpropyl units with a bound hydroxyl (OH) and one or two methoxyl (OCH_3) groups. Lignin is derived only from vascular plants. The rate of oxidative decomposition of lignin is rapid initially and then decreases due to the recalcitrance of the molecule. Numerous low molecular weight phenolics are formed during decomposition. The phenolic monomers act as building blocks for synthesis of humic substances via phenoloxidase catalysts.

The recalcitrance of lignin accounts for its increased residence time in soil. Relatively few microorganisms possess the necessary monooxygenases and dioxygenases to affect decomposition. Organification of soils, accordingly may be accelerated by applying metastable phenolic polymers such as lignin and humic substances, both of which are contained in natural organic vegetative matter.

The characteristics of the end-products derived from the dissimilation of sewage and biological sludge organic matter depend on whether the organic carbon (C), nitrogen (N), sulfur (S), and phosphorus (P) are oxidized under aerobic conditions or reduced due to anaerobic conditions. Accordingly, numerous distinct compounds are formed depending on the extent of aeration during transformation.

Factors Which Influence Decomposition

Placing anaerobically or aerobically derived nascent or metastable organic matter on soil or suspending in water and depriving the microbes of organic input as a source of energy, causes many of the microbes to autolyze and decay. Maximum acceleration of decay can be effected only under aerobic conditions, thus, mandating a need for aerobic conditions in soil, if sludge management is to be practiced effectively. During transformation, numerous microbes consume organic matter, including other microbes, in the sludge; whereas a second group of microbes consume the lysed products derived from the first group. Suitable conditions must prevail to enhance the cycle of organismal activity to maximize the rate of decay of sewage and biological sludge accordingly. Under aerobic conditions, molecular oxygen (O_2) serves as a terminal electron acceptor. Surface compaction of soil reduces the network of interstices within the soil matrix, which subsequently prevents the influx of O_2 and induces anaerobic conditions. Soil is considered anaerobic when the oxidation–reduction (redox) potential is less than +350 mV at pH 7.

The redox potential is a measured indicator of anaerobic (reduced) or aerobic (oxidized) conditions denoted by negative (–) or positive (+) net charges, respectively. In the absence of nitrate (NO_3^-) or sulfate (SO_4^{-2}) as electron acceptors, and also under anaerobic conditions, certain recalcitrant organic molecules probably cannot be decomposed. Such molecules may be totally metabolized, however, at a negative redox potential in the presence of these anions that serve as terminal electron acceptors. The rate of decay, therefore, is inversely proportional to the intensity of soil anaerobicity as related to a corresponding decrease in redox potential.

Microbial and other biological activity is minimized under dry conditions. Maximum microbial activity, as determined by CO_2 evolution, has been reported at 15% H_2O. Temperature vs. rate of decay, based on CO_2 evolution studies and remaining residue, indicate maximum microbial activity at 28 to 37°C. Survival of pathogenic microbes doubly decreases with an increase of 10°C: Q_{10} = [rate at $x°$ + 10°C/rate at x_o] = 2 or more.

Maximum decay rates occur under slightly acidic conditions, but decrease significantly when the concentration of hydrogen ions in soil drops below pH 5. The decreased rate of decay is most likely associated with decreased microbial activity. Other contributing factors, however, may be involved. Catalytic enzymes are inactivated irreversibly under highly acidic conditions. Interaction with humic substances may also decline because humus formation occurs maximally at alkaline pH, and humic substances are known to render greater stability to enzymes against microbial attack. A C-to-N ratio of approximately 10:1 is necessary for maximum turnover. In relation, concentrations of at least 1.5% N, 0.15% S, and 0.2% P are essential for a high rate of decay.

The rate of decay also varies depending on the constituents undergoing transformation. Proteins and carbohydrates decay faster than lipids, and lignin is the most recalcitrant relative to these macromolecules. The differences in rates of decay are related to the molecular structural arrangement, which, in turn, influences consumption and metabolism by different microbial species.

Sewage Disposal Concepts

Sewage treatment is a process in which the solids in sewage are partially removed and partially changed from complex, highly putrescent organic solids to mineral or relatively stable organic solids by decomposition brought about by microbial action. The amount of change is dependent on the treatment process. After treatment, the solids and the liquid effluent must still be discarded. The effluent can be either spread or sprayed over the ground through a process of irrigation, can be introduced below the surface of the ground in tile fields or other underground systems, or can be disposed of in bodies of surface water. The problem of irrigation is that flooding allows chemicals to infiltrate down to the water table and cause potential contamination if the effluent is not clear enough on disposal. Depressions in the ground readily become areas for mosquito breeding. When the material or the effluent is sprayed on top of the ground, the liquid tends to aerosolize creating additional public health hazards (Figure 6.4). Subsurface disposal, which is utilized basically for individual sewage systems, is hazardous because the more permeable the soil is, the greater the opportunity for the effluent to percolate down to a water strata. The less permeable the soil is, the greater the opportunity for the effluent to find its way to the surface or back into the residence. Adding effluent to existing bodies of water can be an effective means of disposing of the effluent, provided the effluent is of high quality. If the effluent is not of high quality, sludge accumulates in the receiving bodies of water and eutrophication may take place. Where the effluent is good, it is diluted by the receiving body of water and the DO is adequate to dispose of any of the remaining organic materials.

DO in the stream is of greatest importance for the proper functioning of biological and biochemical reactions. Oxygen is dissolved in the water from the air that comes in contact with the water surface until a point of saturation is reached. At 0°C, water contains 14.6 ppm of DO at the saturation point. As the temperatures rise, the amount of DO decreases. At 15°C the DO concentration at saturation is 10.00 ppm. Turbulent flow of streams over stones, rapids, and so forth increases the rate of reaeration, adding additional oxygen to the water.

Biochemical oxygen demand, or BOD values, shown in laboratory tests must include the amount of time and temperature used during the test. The most commonly used times and temperatures are 5 days at 20°C, which is 68°F. Self-purification is a process in which the stream proceeds by physical, chemical, and biological means to dispose of the materials, which may include suspended solids or other organic material. This is done through a rather complex arrangement, where initially suspended solids settle out and the microorganisms feed on the organic solids, causing them to decompose. These compounds are reduced to stable inorganic salts, such as nitrates, sulfates, and phosphates. These, in turn, become food for algae, which during their growth and development produce oxygen as a waste product. This oxygen plus the reaeration helps the stream to purify itself. The amount of self-purification is dependent on time, temperature, oxygen supply, and biological growths present. Self-purification usually takes place in four stages, with the stream divided into four zones that merge into each other. These are called the zones of degradation, decomposition, recovery, and clean water.

The zone of degradation is determined by visible evidence of pollution, including floating solids, pieces of garbage, sticks, papers, and fecal materials. The turbidity of the stream increases sharply as the sewage is discharged into it. The oxygen is reduced and the fish life decreases, but a substantial amount of biological life is

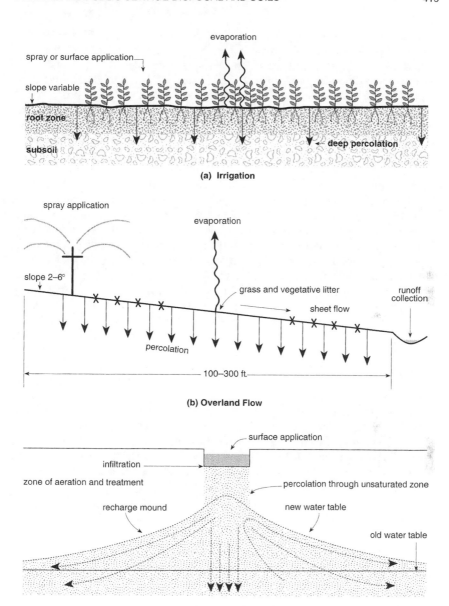

Figure 6.4 Land treatment systems use one of the three basic approaches: (a) irrigation, (b) overland flow, (c) infiltration percolation. (From Culp, G., *Environmental Pollution Control Alternatives: Municipal Wastewater,* U.S. EPS Technology Transfer, no. EPS-625/5-76-012, prepared for the U.S. Environmental Protection Agency, p. 47.)

present. The bacteria are usually those found in sewage. Fungi also are present. The microbiological life gradually exhausts the DO. If the flow of water is slow, the suspended solids tend to settle as sediment-containing sludge. The second zone is where decomposition occurs when the DO is exhausted and anaerobic decomposition or putrefaction starts. Where substantial pollution exists, this occurs very quickly. Where little pollution exists, this is a relatively mild problem. In the zone of decomposition, the water turns black as putrid odors and sedimentation occur. As the water continues to flow, the oxygen from reaeration may first equal the requirements of the stream and finally exceed it.

The zone of recovery is that area where the DO appears to increase and the organic solids decrease. The microorganisms start to die off, especially the anaerobes. Aerobic species are now present. Fish can again survive. Sedimentation of solids continues and the organic solids in the sludge banks and bottom deposits are acted on by worms and larvae to cause further decomposition. As the zone of recovery starts to improve and more oxygen enters the stream, the organic solids become the stable inorganic solids and the water once again starts to reach the point of saturation with oxygen. At this point, aerobic organisms are present in small numbers, algae are present, and fish are more abundant. The time that it takes for self-purification of a stream or the distance it must travel to pass through the four zones depends on the strength and volume of pollution, stream flow, turbulence of flow, and temperature of the water. If additional pollution is added during the process of self-purification, the process is interrupted and may have to start over again. After self-purification, pathogenic organisms and viruses may still survive, although the putrescent matter in the sewage has been destroyed. In addition, metals and other compounds may be present in the water.

Sewage treatment is a necessity to maintain or utilize the water sources available as potable water or for other domestic water uses, to prevent disease, to prevent nuisances, to provide clean water for swimming and other recreational purposes, to provide clean water for fish life, to conserve water for industrial and agricultural uses, and to keep navigable channels open for use and prevent them from becoming full of silt.

Contaminants and Groundwater

Contaminants move through soil to groundwater based on soil conditions, climate, and properties of the contaminant. The capacity of soil to degrade the contaminants is limited. Degradation occurs through diffusion and mass flow. Substances diffuse through soils based on differences in energy from one point to another. These differences may be caused by changes in concentration or temperature. Vegetation may also take out the contaminants or assimilate them, reducing the amount of material able to be transported to the groundwater.

Pesticides are a good example of contaminant removal. Pesticides may be vaporized into the atmosphere, removed with the harvested plant, run off to surface bodies of water, or leach down to the groundwater supply. Animals, insects, worms, or microorganisms in the soil can ingest and metabolize the chemicals.The chemicals may adhere to soil particles or dissolve in water.

Pathogens may be found in sewage sludge, septage, animal wastes, food-processing wastes, and septic tank effluent. Pathogens are carried in suspension with water through soil and also are removed from the soil water by filtration and adsorption. The mobility of the organisms is based on the size of the water-filled pores, and the velocity of the water in these pores.

Holding Tank Concept

Holding tanks are sealed tanks to which sewage is piped and retained. Specially equipped trucks pump the holding tank into their containers and haul the contents to sewage treatment plants. The holding tank is a temporary means of sewage disposal, pending the installation of public sewers. The tank can be utilized where existing sewage problems occur and where public sewers are to be installed shortly. The holding tank also may be utilized if builders are ready to move ahead with construction but the public sewers will not be completed for 1 to 2 years. The disposal of the sewage effluent is costly. A family of four usually generates about 1000 gal of waste in a week. The yearly cost may well range from $3500 to $5000. Holding tanks are inconvenient because the consumer has to conserve water, making it necessary to limit showers, baths, laundry usage, and so forth. The holding tank also must have regular service to avoid malfunction and overflow. In the event of severe weather, the tank must have adequate capacity to hold the waste. It is obvious that holding tanks are a last resort that should be used only under special conditions. The health department should specifically license the individuals to have holding tanks and should require that the sewage haulers, as well as holding tank owners, maintain accurate records of pickup. If any of the effluent is permitted to come to the surface of the ground or to enter any aquifer, severe penalties should be enforced.

Public Sewage Disposal

Public sewage disposal is the community's way of collecting sewage from assorted homes, businesses, and possibly industry; and then transporting it through a piping system to a sewage treatment plant, where it may be treated in a conventional way through settling and biological oxidation or through advanced wastewater treatment methods. Where the conventional method is used initially, a variety of additional treatments follow to remove the impurities before the liquid effluent is discharged into a receiving stream. The sludge is treated separately and then taken to a disposal area.

Sewage System Infrastructure

The sewage collecting system or infrastructure consists of either combined sewers or sanitary sewers. The combined sewer is a series of pipes and drains that remove sewage from homes and storm water from streets. Sanitary sewers carry the sewage from the homes or businesses directly to the treatment plant without diluting it with storm water. Each home has a sewer pipe. The sewer pipe connects to the street sewers, which connect to the main sewers; these main sewers lead to the waste treatment plant. In cities, sewage is strongest at about 10:00 A.M. and weakest at about 3:00 to 6:00 A.M. The time that the sewage remains in the sewers depends on area, contour of the sewer district, velocity of flow in the sewer, and variations of flow that occur annually, seasonally, daily, and hourly. Where a combined sewer is utilized, the storm flow may be three to six times as great as the dry weather flow.

The sewage traveling from the homes or industries to the sewage treatment plant must be moved along to prevent settling and clogging of the pipe, and production of unwanted side effects due to anaerobic action. Obviously, if the sewage always flows by gravity at a reasonable rate, it would not be necessary to use lift stations. However, because this does not occur, the lift station is utilized. Lift stations or pumping stations are specifically used where the topography varies, and as a result the hydraulic gradient is such that inadequate head exists for gravity flow through the pipes and through the treatment plants. Pumping equipment follows the same types of concepts as pumps, used to lift water. However, grit and sand may cause excessive wear of the pumps, and large objects may clog the pumps or break the impellers. Also corrosive waste may cause problems with the pumps. The lift station must protect the continuity of flow by having available duplicate units, warning devices in the event that a power shutdown occurs; devices such as screens to protect the pump, air circulation, heat, and auxiliary gas engines in the event of power failure; and sump pumps in dry wells.

Alternative Conveyance Systems

The major alternative conveyance systems or sewer systems are the pressure sewer system, vacuum sewer system, and small diameter gravity sewers. These systems are needed especially in rural areas because a conventional gravity system would be too expensive.

A pressure sewer is a small diameter pipeline that follows the profile of the ground and is buried in a shallow trench. The diameters are typically 2 to 6 in. and the pipes are buried below the frost line. Each home has a small pump to discharge the waste into the main line. It may be a grinder pump, which grinds the solids into a slurry, or a septic tank and effluent pump. The septic tank captures the solids, grit, grease, and stringy material.

The vacuum sewer was invented in the United States in 1888. It consists of three major components: the vacuum system, the collection piping, and the services. The vacuum system is similar to a conventional wastewater pumping system in design, but contains vacuum equipment in addition to the collection tank. The collection piping usually consists of 4- to 6-in. mains and piping. The services consist of a system where wastewater flows by gravity to a holding tank where air is compressed in a sensor tube connected to a vacuum valve that then opens and causes the wastewater to flow by vacuum.

The small diameter gravity sewer consists of:

- House connections to the interceptor tank
- Interceptor tanks that remove floating and settleable solids
- Service laterals that connect the interceptor tank with the collector main

- Collector mains that are small diameter pipes
- Clean-out manholes and vents for inspection and maintenance
- Lift stations when needed to handle elevation differences

The name is misleading because the main does not have to be small in diameter. The most significant feature is that primary pretreatment is provided in interceptor tanks upstream of each connection.

Pretreatment Technologies

Pretreatment technologies are used to remove or destroy substances in the influent to reduce the level of chemicals that may affect the final effluent from the sewage treatment process. Emulsion-breaking techniques enhance the efficiency of treatment units. They include chemical oxidation, carbon adsorption, hydrolysis, and chemically assisted clarification and settling.

Chemical oxidation modifies the structure of pollutants by transferring one or more electrons from the oxidant to the targeted pollutant. Alkaline chlorination uses chlorine under alkaline conditions to destroy pollutants such as cyanide and some of the active ingredients of pesticides. Toxic by-products such as chloroform may be produced. Chemical oxidation may also occur with the use of hydrogen peroxide, ozone, potassium permanganate, or ultraviolet (UV) light.

Carbon adsorption effectively removes organic chemicals from wastewater through the use of activated carbon, either in the powdered carbon form or in the larger granular carbon. After processing, the carbon is regenerated by removing the adsorbed organic compounds through steam, thermal, or physical and chemical methods.

Hydrolysis is a chemical reaction in which organic substances react with water and break into smaller and less toxic compounds. The original molecule forms two or more new molecules. Generally, increasing the temperature increases the rate of hydrolysis.

Chemical precipitation is a treatment technology in which chemicals such as sulfides, hydroxides, and carbonates react with organic and inorganic pollutants in wastewater to form insoluble precipitates. The separation process has four steps as follows:

1. Adding a chemical to the wastewater
2. Rapid mixing to distribute the chemical homogeneously throughout the wastewater
3. Slow mixing to encourage flocculation, which is the formation of the insoluble solid precipitate
4. Filtration, settling, or decanting to remove the flocculated solid particles

Conventional Sewage Treatment

Currently, municipal wastewater is usually given varying degrees of treatment before it is discharged into a receiving stream. The primary objective of treatment is reduction of oxygen-demanding materials in the water and suspended solids and

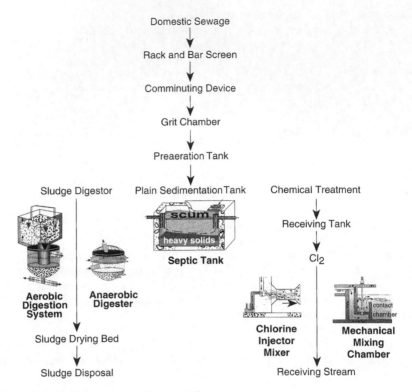

Figure 6.5   Preliminary sewage treatment flow.

bacteria. There are inherent disadvantages to parts of the conventional system. The effluent still has a high level of BOD, and removal of pollutants, as measured by the 5-day biochemical oxygen demand (BOD_5), is only about 35%.

Primary Treatment

Primary or preliminary treatment consists of the use of screens and grit chambers, comminuting devices, preaeration tanks, sedimentation tanks, chemical treatment, and discharge of the effluent to the receiving stream. The sludge is removed to a sludge digester (Figure 6.5).

The function of preliminary treatment is removal or reduction of large suspended or floating organic solids that consist of wood, cloth, paper, garbage, and fecal material. A second function is removal of heavy inorganic solids, such as sand, gravel, and possibly metallic objects, which are all called grit. The third function is removal of excessive amounts of oils or greases.

Primary sedimentation is needed to separate the solids from the liquid by settling. This prevents sludge deposits from accumulating in the streams. It also decreases the load on secondary treatment processes and makes possible the use of such devices as a trickling filter.

In still water, particles fall vertically to the bottom. In flowing water, the particles have horizontal movements, as well as downward movements. The velocity of the sewage flow is extremely important to allow for adequate settling of the materials. If the velocity is too high, the effluent can proceed rapidly to the opposite end of the tank carrying the type of material that should have been settled out in the settling tank.

One of the first and still basically utilized types of settling tank is the septic tank. The septic tank was designed to hold sewage at a very low velocity under anaerobic conditions for periods of 12 to 24 hours during which a high removal of settleable solids would occur. These solids decompose in the bottom of the tank and form gases that cause them to rise through the sewage to the surface and become a scum layer until after the gas has escaped. Then the solids settle again. During the continual flotation and resettling of solids, the sewage is carried toward the outlet. Eventually some of the solids pass through the outlet and into the tile systems, thereby helping clog the ground. Septic tanks, although they are larger in size today, still primarily serve as a settling basin and as some means of primary decomposition of waste material.

A second type of settling basin that was used in the past was the Imhoff tank. The Imhoff tank was a two-story tank in which the liquid flowed to the upper story, settling occurred, and the solids fell through a slot into the sludge compartment. Because the sewage did not come in contact with the anaerobic digesting sludge, the holding period of the tank was reduced. A gas vent and scum chamber, which in effect makes this into a three-compartment tank, was built into the Imhoff tank so the gases could be vented to the outside and the scum could be removed on a daily basis.

The plain sedimentation tank within the sewage treatment plant has a major function of removing the settleable solids from the sewage. The solids are taken continuously or frequently at intervals away from the settling tank, so that decomposition or gas formation cannot occur. The solids are removed to a sludge digester, where sludge digestion is carried on anaerobically. The fluid flowing into the sedimentation tank strikes baffles in the case of the square or rectangular tanks. This allows for an even flow across the width of the tank and maintains a constant forward velocity.

Secondary Sewage Treatment

Most of the conventional secondary sewage treatments consist of some type of biological oxidation. The trickling filter, activated sludge, and stabilization ponds are utilized to reduce the solids present and the BOD of the effluent going into the receiving stream. Separate discussions of each of these techniques are held in the ensuing sections (Figure 6.6).

Trickling Filter System

A trickling filter may be compared with the self-purification action of streams (Figure 6.7). The green growths that are present on rocks are duplicated by the

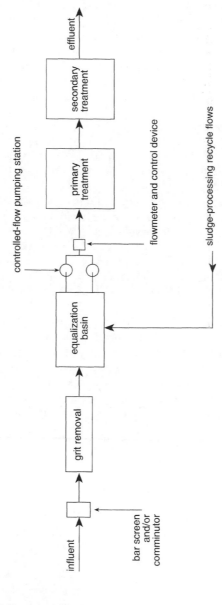

Figure 6.6 Schematic drawing for secondary sewage treatment. ((From Culp, G., *Environmental Pollution Control Alternatives: Municipal Wastewater*, U.S. EPS Technology Transfer, no. EPS-625/5-76-012, prepared for the U.S. Environmental Protection Agency, p. 53.)

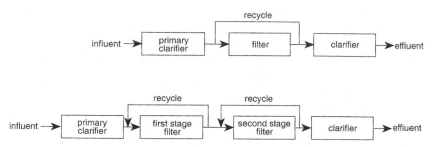

Figure 6.7 Typical one- and two-stage, trickling-filter systems. (From Culp, G., *Environmental Pollution Control Alternatives: Municipal Wastewater,* U.S. EPS Technology Transfer, no. EPS-625/5-76-012, prepared for the U.S. Environmental Protection Agency, p. 10.)

growths on the trickling filter stones. If all the stones from a stream are piled one on top of one another, a trickling filter would be developed. The trickling filter does the work in a small area of a wandering stream that may cross many miles. The word *filter* is not properly used in this term, because no straining or filter action occurs. The term *biological oxidizing bed* is more appropriate. Trickling filters are sturdy units that are not easily upset by shock loads of material. They are affected, however, by temperature; therefore, they slow their biological activity during cold weather. The trickling filter should be placed after the secondary settling process. The system converts nonsettleable colloidal dissolved solids into readily settleable solids. These solids, which are largely organic, are not removed from the sewage but are utilized by the microorganisms in their life cycle. The permanent stable material that is left eventually falls off the filter material and passes out in the filter effluent. Therefore, the effluent should be allowed to pass through an additional sedimentation stage before it is chlorinated and allowed to enter the stream. The trickling filter is composed of these three parts: bed of filter medium, underdrainage system, and distributors.

The large surface area of the filter medium provides an area where slimes and gelatinous films produced by bacteria can develop. However, it still leaves adequate voids to permit free circulation of air throughout the filter. The slimes that cover the filter contain millions of organisms. The slime acts as a place where the microorganisms can remain on the rocks instead of washed off with the sewage. It also becomes a storage place for food obtained from the sewage. The dissolved solids are absorbed into the jellylike mass. The bacteria and other organisms living in the slime eat the absorbed solids. When the storage space is full, the trickling filter is no longer capable of consuming impurities and sewage application must be stopped until the food is utilized. The amount of air present is also important in determining the amount of sewage that can be placed on the trickling filter. These organisms need air for survival, and the more food they consume the more air they need. Sloughing is the process where the old, thick growths of organisms, waste products, and by-products crack off the stones and are flushed down through the rocks. This usually occurs during the spring or fall, or during an abrupt or severe change in temperature. The air supply, which is so essential to operation during the summer months, because of downdraft can readily pass downward through the filter stones and out through the drainage system. During cold weather, especially during the winter, the air travels upward. The air is therefore drawn through the drainage system and passes up through the filter bed. The underdrainage system must provide a clear opening for air to pass through.

Ponding occurs when the spaces between stones become clogged with solids and air cannot pass through the filter. Ponding is due to excessive growths on the filter, plugging up the filter with primary tank solids, or gradual breaking down and breaking up of filter stones. Ponding may be corrected by encouraging the sloughing off of growths on the filter stone, by chlorinating, or by allowing the filter to rest. Sometimes the situation is created by overloading the filter and the filter load must be reduced. Ponding due to plugging from the primary settling tanks can be corrected by hosing under high pressure or using rods to break up the mass. If the filter stones are actually broken up, the filter stones must be replaced.

The amount of food or sewage effluent that the ordinary standard rate trickling filter can handle is 500 lb of BOD per acre foot. The distribution of this effluent of sewage is extremely important. Poor distribution causes overfeeding or overloading of part of the filter while starving another part of the filter. The effluent may be distributed by the use of fixed nozzles, rotary distributors, tipping troughs, and splash plates. The fixed nozzle and rotary distributor are most frequently used. Poor distribution may be the result of broken or clogged nozzles or clogging of distribution lines from dosing tanks to filter nozzles. It may also be due to clogged distributor orifices or improperly operating dosing tanks. Poor distribution can be corrected by regularly inspecting and maintaining the system. The filter nozzles should be flushed and cleaned daily, and the flush-out nozzles, orifices, distributors, and distribution systems should be cleaned weekly.

The underdrain system carries the sewage passing through the filter away from it for further treatment and disposal, and also provides ventilation of the filter and maintenance of the aerobic conditions. The underdrain system usually consists of precast filter blocks made from vitrified clay or concrete. This system covers the entire bottom underneath the filter. The blocks are usually rectangular in shape and have openings in the upper face equal to 20% of the surface area of the block.

The distributor system has been already partially discussed. Many types of nozzles are utilized. The siphon dosing tank is used to allow the proper amount of effluent to be put on the filters as needed.

Many variations of the trickling filter system exist. High-rate trickling filters are units that are operated with hydraulic loadings of 8.7 to 44.0 million gal/acre/day and organic loadings of 25.0 to 50.0 lb/1000 ft^3 of filter medium per day. In this system, the effluent of the trickling filter is added in the ratio of nine parts of new effluent going to the trickling filter to one part of effluent that has left the trickling filter, mixed together, and then reintroduced into the trickling filters. This increases the quantity of water that can be processed.

The biofilter (Figure 6.8) is a process that uses recirculation and high rate of application to a shallow trickling filter. The effluent of the filter or the secondary sedimentation tank passes back through the primary settling tank. Because the

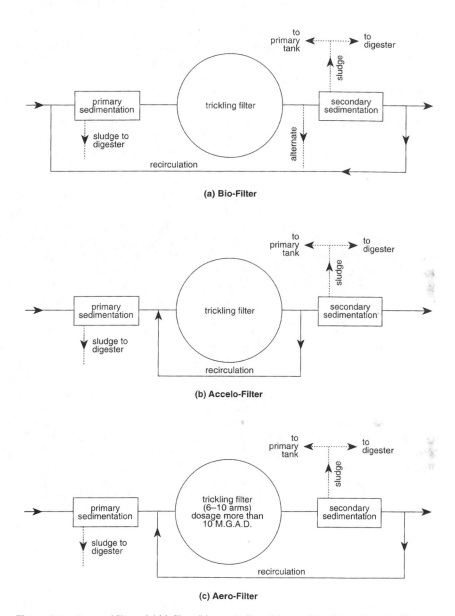

Figure 6.8 Types of filters: (a) biofilter, (b) accelo-filter, (c) aero-filter. (From *Manual of Instruction for Sewage Treatment Plant Operators,* New York State Department of Health, Office of Professional Education, Health Education Service, Albany, NY, p. 54.)

secondary settling tank sludge is usually very light, it can be fed back to the primary settling tank, where the two types of sludges are collected together and pumped to the digestor.

The acello filter (Figure 6.8) involves the recirculation of filter effluent directly back to the filter. The aero filter is a process that distributes the sewage by maintaining a continuous rainlike application of sewage over the filter bed (Figure 6.8).

Activated Sludge Process

The activated sludge treatment process is another biological treatment method that utilizes bacteria and other organisms to consume impurities in the sewage (Figure 6.9). Activated sludge consists of a flocculant material that contains the bacteria and other organisms that help purify the sewage. The activated sludge is mixed with sewage in aerators. The impurities or food are absorbed by the flocculant particles. These impurities are then consumed by the bacteria living in the floc. Prior to the activated sludge process, the sewage must pass through the primary treatment of aeration, where activated sludge and sewage are mixed together in aeration tanks (either by mechanical agitation or by the use of compressed air), and final settling. The sludge in the sewage remains in the aeration tanks for a period of time ranging from 4 to 10 hours. The sludge particles are then separated from their liquid by settling. For the process to work, activated sludge must always be fed back into the incoming sewage and mixed thoroughly. The portion of the sludge solids that is not returned to the aerators is sent to the sludge digesters.

Within the activated sludge process, the absorbed or adsorbed colloidal and dissolved organic matter, including ammonia in the sewage, is reduced to insoluble nonputrescent solids. This conversion is done through a step-by-step process. Some bacteria can attack the original complex substances and reduce them to simpler compounds as their waste products. Other bacteria use the waste products to produce even simpler by-products. Degeneration of the activated sludge or flock in sewage is a slow process. The amount formed from any volume of sewage during its period of treatment is small and inadequate for the rapid and effective treatment of the sewage, which requires a large concentration of activated sludge. These concentrations are built up by collecting the sludge produced in each volume of sewage treated and reusing it in the treatment of subsequent sewage flows. This reused sludge is known as return sludge. When a surplus of activated sludge occurs, it is permanently removed and put through sludge treatment.

The activated sludge must be kept in suspension during the period of contact with the sewage undergoing treatment. This is done through agitation and aeration, using the following steps: mix activated sludge with sewage to be treated, aerate and agitate the mixed liquor for the required length of time, separate the activated sludge from the mixed liquor, return the proper amount of activated sludge for mixture with sewage, and dispose of the excess activated sludge. A number of variations have been developed on this technique, depending on the specific kind of process that is utilized. The amount of air needed varies with BOD loading, quality of the activated sludge, concentration of solids, and desired efficiency in the removal of BOD.

In any case there should be at least 2 ppm of dissolved oxygen in the activated sludge process. When using diffused air systems, the amount of air utilized is usually

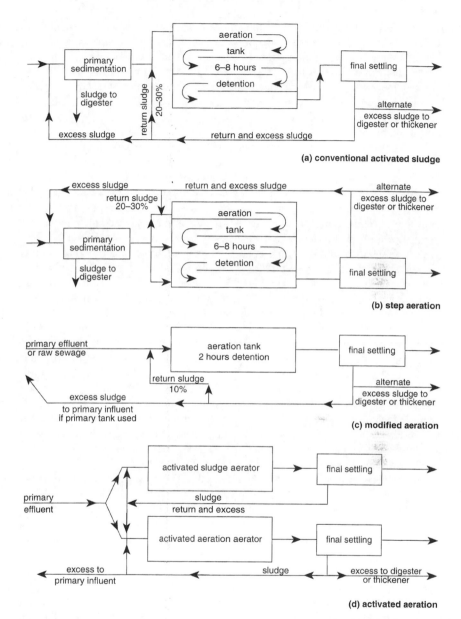

Figure 6.9 Sludge process: (a) conventional activated sludge, (b) step aeration, (c) modified aeration, (d) activated aeration. (From Culp, G., *Environmental Pollution Control Alternatives: Municipal Wastewater,* U.S. EPS Technology Transfer, no. EPS-625/5-76-012, prepared for the U.S. Environmental Protection Agency, p. 63.)

½ to 1½ ft³ of air per gallon of sewage or 1000 ft³ of air per pound of BOD to be removed. The aeration time varies again with the process. In the most conventional activated sludge process, a minimum of 6 to 8 hours is adequate for diffused aeration. For mechanical aeration 9 to 12 hours are needed. In some systems less time is required. Depending on the type of activated sludge treatment plant and the character and concentration of the sewage, the amount of return sludge may vary from 10 to 50%. In most conventional plants, the percentage is usually 20 to 30%. The mixed liquor usually contains 1000 to 2500 ppm of solids in the diffused air plants and 500 to 1500 ppm in the mechanically aerated plants. The excess sludge is usually pumped into the primary tank and from there to the sludge digestion tank.

Various modifications of the activated sludge process exist, which include step aeration, tapered aeration, modified aeration, activated aeration, contact stabilization, and aerobic digestion. In step aeration, the sewage enters the aeration tank at a number of different points, but all the return sludge is introduced at the first point of entry with or without a portion of the sewage. This means that the sludge solid concentration and mixed liquor are greatest at the first step or point of entry. In this process, about half the normal time of operation or half the aeration tank capacity is needed. In tapered aeration, the theory is that the greatest amount of air needed is during the early part of aeration. Air is therefore introduced at its highest rate at the inlet section. The advantages are supposedly better control of the process and reduction of cost. In modified aeration, which is also known as high-rate activated sludge treatment, either raw or settled sewage is mixed with about 10% of the return sludge and aerated for only 1 or 2 hours. The suspended solids of the mixed liquor are reduced to below 1000 ppm, resulting in a reduction of air requirements. By controlling the aeration period and the percentage of returned sludge, almost any degree of treatment between the primary sedimentation and the conventional activated sludge process can occur. The activated aeration process has a reduced aeration period. By taking the leftover activated sludge and transferring it to the activated aeration section, along with part of the settled sewage flow, the operation has greater flexibility. Reduction in BOD is supposed to be 80 to 85%.

In contact stabilization the biologically active sludge is brought into intimate contact with the sewage for 15 to 30 min. During this time, the activated sludges absorb and adsorb a high percentage of suspended colloidal and dissolved materials. The mixture then goes to a settling tank from which the sludge is removed to a regenerating tank or stabilized and regenerated by aeration. This is supposedly valuable in treating industrial waste and avoiding shock loads to all the activated sludge. Aerobic digestion or total oxidation is used as a continuous process where a flow of raw sewage or waste is aerated vigorously in a tank that contains an entire day's flow. The aerated sewage then passes through a conventional settling tank, where the clarified effluent overflows and the settled sludge is recirculated back to the aeration compartment at high rates. This process is extremely sensitive to sudden changes in volume and character of the waste, but it may be applied to small installations.

The advantages of the activated sludge process is that it provides, when operating properly, effluents with extremely low BODs. The operation of the activated sludge plant is quite elastic and therefore can be manipulated properly by the operators of the plant. When starting an activated sludge plant, it is necessary to get "seed activated sludge," which is kept warm and is less than 4 days old, and bring it to the new plant to get the plant in operation.

Rotating Biological Contactors

Rotating biological contactors, reactors, and surfaces have closely spaced plastic disk drums, which are rotated partially submerged in wastewater that has already gone through primary treatment. It is best when trash and grit are first removed. The biomass that forms on the area of the disk surfaces is wetted by the effluent and kept in aerobic condition by rotations and air contact. The process is similar to a trickling filter. Some of the growth is sloughed off or stripped off of the disk as it passes through the moving wastewater. This growth goes to a secondary settling tank, where it is removed. An effluent that has a better BOD quality and reduced nitrification can be achieved when the pH is at 8.4.

Contact Aeration Process

The contact aeration process, like other secondary treatment processes, depends on aerobic, biological organisms to break down the complex putrescent organic materials in sewage to simpler and more stable forms. The organisms, as in the trickling filter process, are stationary and are attached to a fixed medium. However, they are continually submerged and supplied with air by means similar to the diffused air system in the activated sludge process. The process, like the trickling filter, can withstand shock loads and removes 90% or more of the BOD and suspended solids.

Intermittent Sand Filters

Intermittent sand filters are specially prepared beds of sand from which effluents from primary treatment or from trickling filters or secondary settling tanks are applied intermittently. The effluent passes through the filter to an underdrainage system. The filter bed is composed of clean sand overlying clean graded gravel. The sand is at least 24 in. deep, and the gravel is placed in three layers around the underdrains and 6 in. over the top of the underdrains. The sand should have an effective size of 0.3 to 0.6 mm and a uniformity coefficient not exceeding 3.5. If the intermittent sand filter is used following primary treatment, the amount of effluent put on the intermittent filter should not exceed 125,000/gal/acre/day. If the effluent is coming from either trickling filters or secondary settling tanks, the effluent should not exceed 500,000 gal/acre/day. The intermittent sand filter is a true filter that strains out and retains the fine suspended solids, but it also acts as an oxidizing unit. Most of the filtering and oxidation occurs at or near the surface of the sand. The straining is due to the fine nature of the sand. The oxidation is affected by the aerobic microorganisms present. In operation, the filter should be allowed to empty itself and to obtain a fresh air supply periodically. This is done by intermittent dosing of the sewage onto the filter at a rate of two to six times a day. The quantities covering the surface should not exceed 2 to 3 in. Two or more units are usually utilized in rotation. The sand filters must have the top layer of sand replaced periodically or they become clogged. The pooling of waters on a surface should not be permitted, because it does cause problems. The well-operated intermittent sand filter should produce an effluent almost completely oxidized and nitrified. There should be a removal of 95% or more of the BOD and suspended solids from the raw sewage (Figure 6.10).

Stabilization Ponds

Most stabilization or oxidation ponds are 2 to 4 ft deep with a continuous flow through the pond. They are designed for a loading of 1 acre per 400 persons, 50 lb BOD per acre per day, or 15 lb BOD per acre foot per day with a detention period greater than 30 days. The natural soil must be impervious so that seepage cannot occur. The stabilization pond process consists of two steps. Carbonaceous matter in the sewage is first broken down by aerobic organisms with the formation of carbon dioxide. The carbon dioxide is then used by algae in photosynthesis, which produces oxygen. The organic matter in the sewage, with oxygen present, goes through aerobic decomposition. This system has been used for treatment of raw sewage or secondary treatment for settled sewage. However, today some very serious concerns exist about the growth and development of mosquitoes in the oxidation pond areas. These particular units may be potentially hazardous (Figure 6.11).

Chlorination of Sewage

Sewage is chlorinated to assist in the sewage treatment process. Because chlorine is so extremely active and reacts with so many different compounds, it is necessary to take into account, when adding chlorine, the amount of chlorine that can be utilized by various impurities. Chlorine is utilized as a disinfectant. It is also utilized to prevent sewage decomposition, to aid in plant operation, and to aid in the reduction of biochemical oxygen demand. Ozone is also used in this process before the discharge of the effluent to the stream.

Sludge Digestion, Treatment, and Disposal

Sewage sludge, which is the mixture of sewage and settled solids, may be a combination of primary, secondary, excess activated, or chemical sludge. It may be either raw or fresh, digested, elutriated, dewatered, or dried. It may contain a variety of microorganisms. *Clostridium perfringens* has been found in sludge at disposal sites. Sludge is treated to facilitate its final disposal. The objectives of sludge treatment are reduction of volume of material by removing as much liquid as possible and decomposition of highly putrescent organic matter. The quantity and composition of sludge vary with the nature of the sewage and the sewage treatment process used. The sludge from plain sedimentation tanks is basically settleable solids in the raw sewage. It is usually gray and odorous, and contains feces, garbage, and other

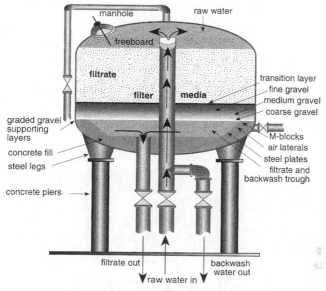

pressure filter-filter cycle schematic

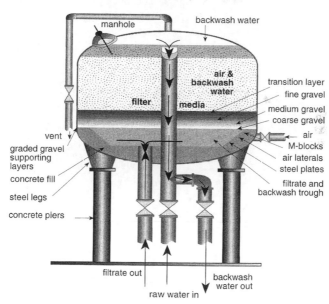

pressure filter-backwash cycle schematic

Figure 6.10 Intermittent sand filters. (From Culp, G., *Environmental Pollution Control Alternatives: Municipal Wastewater,* U.S. EPS Technology Transfer, no. EPS-625/5-76-012, prepared for the U.S. Environmental Protection Agency, p. 31.)

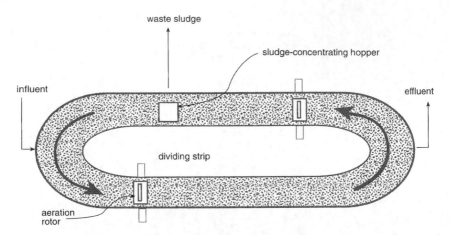

Figure 6.11 Oxidation ditch. (From Culp, G., *Environmental Pollution Control Alternatives: Municipal Wastewater,* U.S. EPS Technology Transfer, no. EPS-625/5-76-012, prepared for the U.S. Environmental Protection Agency, p. 18.)

debris. Sludge from secondary settling tanks is usually dark brown and flocculant. It can become septic and odorous. Sludge and chemical precipitation is usually black and may be objectionable in odor. It can also be decomposed further. Sludge treatment can be accomplished by thickening, digesting with or without heat, drying on sand beds that are open or covered, chemical conditioning, elutriation, vacuum filtration, heat drying, incineration, or wet oxidation (Figures 6.12 and 6.13). Thickening is a process in which thin sludges become concentrated into more dense sludges. Much of the water is eliminated by pumping the sludge from a settling tank

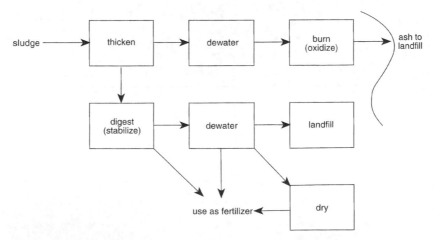

Figure 6.12 Basic sludge-handling alternatives. (From Culp, G., *Environmental Pollution Control Alternatives: Municipal Wastewater,* U.S. EPS Technology Transfer, no. EPS-625/5-76-012, prepared for the U.S. Environmental Protection Agency, p. 54.)

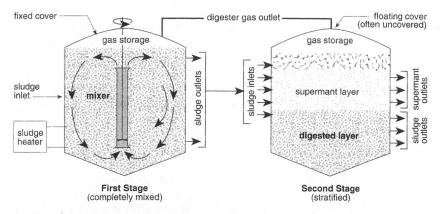

Figure 6.13 Schematic of two-stage digestion process. (From Culp, G., *Environmental Pollution Control Alternatives: Municipal Wastewater,* U.S. EPS Technology Transfer, no. EPS-625/5-76-012, prepared for the U.S. Environmental Protection Agency, p. 60.)

into a thickener that has a low overflow rate, so that a lot of excess water overflows and sludge solids concentrate on the bottom. A sludge with solid content of 10% can be produced in this way.

Digestion is the reduction in volume and the decomposition of highly putrescent organic matter to relatively stable or inert organic and inorganic compounds. Sludge digestion is usually done by anaerobic organisms in the absence of free oxygen. Because the solid matter in raw sludge is about 70% organic and 30% inorganic, the anaerobic organisms break down the complex molecular structure of the solids, set free the water that has been bound in the molecules, and thereby obtain the necessary oxygen and food for growth. The microorganisms first attack the soluble or dissolved solids, such as the sugars, and form organic acids and gases, such as carbon dioxide and hydrogen sulfide. The pH drops to a range of about 5.1 to 6.8. During the second stage, the organisms live best in an acid environment, acid digestion occurs on the organic acids, and nitrogenous compounds are attacked and liquefied at a much slower rate. In the third stage, known as the period of intense digestion, stabilization, and gasification, the more resistant nitrogenous materials, such as the proteins, amino acids, and others, are broken down. The pH rises to a range of 6.8 to 7.4. Large volumes of gas containing about 65% or higher methane are produced. Methane can then be used as a fuel. The remaining solids are either relatively stable or slowly putrescent. From time to time a supernatant digester liquor must be removed and fresh solids are added.

A sharp drop occurs in organic material at the end of the digestion process. The well-digested sludge is black in color, has a tarry odor, and appears to be granular. The digestion process usually continues for about a 30-day period. The final sludge product was disposed of in the ocean until 1991, after which it was illegal, as stipulated by the Marine Protection, Research and Sanctuaries Act amendments of 1988. The sludge product also may be disposed of on land by either burial or fill, utilized as a fertilizer or soil conditioner, or burned.

In a properly operated digestion procedure, 12 ft^3 of gas are produced per pound of volatile matter destroyed. This gas is 70% methane and 30% noncombustibles, including carbon dioxide. It has a heat value of 650 Btu/ft^3. A production of 7800 Btu/lb of volatile matter is destroyed. Because manufactured city gas is about 500 Btu/ft^3 and natural gas is 1000 Btu/ft^3, this indicates that the gas production is certainly within the heat values utilized by individuals. The sludge gas is usually used for operation and heating of the sewage treatment plant and the digestion tanks. Because the gas is hazardous, the same precautions must be taken as in the use of other types of gases.

Biosolids, the new name for treated sewage sludge, are primarily treated wastewater materials from municipal wastewater treatment plants and are suitable for recycling as a soil amendment. Sewage sludge now refers to untreated primary and secondary organic solids from the wastewater treatment process. Biosolids are a slow-release nitrogen fertilizer with low concentrations of plant nutrients such as nitrogen, phosphorus, potassium, zinc, and iron. The Environmental Protection Agency (EPA) in 1993 set the 503 rule for the management of all biosolids generated during municipal wastewater treatment. The biosolids are tested for dioxins. The biosolids are used for fertilizing agricultural crops, gardens, and parks. They are also used at mine sites to establish sustainable vegetation in highly disturbed soil. In forestry, biosolids promote rapid timber growth.

Advanced Water Treatment

In our era of increasing shortages of water supply, the luxury of having water flow away from a given area can no longer be allowed. Today reuse of this precious natural resource is essential. Deliberate reuse of treated municipal waste waters is a very definite way of satisfying this need for water. Tertiary treatment or advanced waste treatment is the answer to the reuse problem.

A complete advanced wastewater treatment plant exists in South Lake Tahoe, CA. It continuously produces an effluent that exceeds all the requirements of all regulatory agencies. It meets all drinking water standards. The system consists of ten major components, including conventional primary treatment, completely mixed activated sludge secondary treatment, chemical coagulation and absorption by use of lime, nitrogen removal by air stripping of ammonia, filtration through mixed media separation beds, granular activated carbon adsorption, disinfection by chlorination, coagulant recovery by decalcination, thermal activated carbon regeneration, and sludge incineration. The system has a tremendous efficiency. It removes 99.4% of BOD; 96.4% of COD; 99.1% of phosphorus; 100% of suspended solids, color, odor, coliform bacteria, and viruses; and 99.9% of the turbidity (Figure 6.14).

In our modern society it is necessary to eliminate the complex organic compounds, metallic salts, and other exotic chemicals found in natural waterways and even in drinking water supplies. Conventional wastewater treatment processes are incapable of removing these types of contaminants from our water. These resistant, or "refractory," materials range from simple inorganic salts to highly complex synthetic organic chemicals. The sources of these chemicals are the by-products of

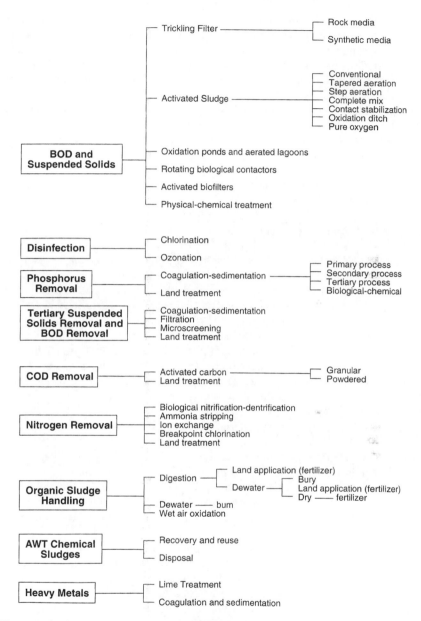

Figure 6.14 Summary of available alternatives relative to requirements for pollutant removal. (From Culp, G., *Environmental Pollution Control Alternatives: Municipal Wastewater,* U.S. EPS Technology Transfer, no. EPS-625/5-76-012, prepared for the U.S. Environmental Protection Agency, p. 75 [modified].)

modern technology and the industrial and commercial waste effluents that come from this technology, as well as the effluents that come from American households. Although many substances are biodegradeable, many others are not. As a result of the new kind of problem, it is necessary to come up with a series of techniques that can remove all undesirable biological and chemical substances from the water so that it can be utilized. Advanced waste treatment is the concept that can be utilized. Adsorption, foam separation, electrodialysis, reverse osmosis, distillation, freezing (including eutectic freezing), ion exchange, solvent extraction, oxidation, and highly improved suspended solids removal are some of the processes discussed in advanced waste treatment. In addition, a tremendous need exists to remove the various nutrients, including phosphates, nitrates, and ammonia. The degree of purification required can be determined by the type of waste, reuse intended, and potential health hazards that exist.

Suspended Solids Removal

Because the removal of suspended solids is a function of primary treatment, secondary treatment, and advanced waste treatment, the material discussed crosses all these treatment techniques. Although some of the material may duplicate earlier chapters, the current presentation is meant to give a comprehensive view of the latest methods of removal of all solids from the effluent before it goes into the receiving stream (Figure 6.15 for solids interrelationships in water).

The removal of suspended solids from wastewater is accomplished through chemical process, gravity systems, physical straining, and in-depth filtration. Wastewater solids refer to the total settleable, suspended, colloidal, and dissolved solids. Suspended and dissolved solids are differentiated by using standard analytic filtration through a standardized filter. Colloidal solids are particles with the size range of 1 to 500 μm. These represent the intermediate fraction between the suspended and dissolved solids.

Standard methods of determination are not readily available in describing these solids. Bacteria, viruses, phages, and other cellular debris fall into the colloidal group. Solids can also be differentiated into organic or inorganic, hydrophilic, and hydrophobic categories. The average suspended solids in municipal wastewaters equals 200 mg/l. However, this may vary tremendously with the type of municipal wastewater available.

The removal of suspended solids is controlled by equalization of the flow of the sewage, coagulation of wastewater, chemical feeding, chemical process, gravity systems, physical straining processes, and deep-feed filtration.

The wastewater entering most sewage treatment plants varies significantly throughout the day in the rate of flow and in the concentration of contaminants. This is due to the habits of the population and the hydraulic characteristics of the sewage system. As a result, the quality of the effluent entering the receiving stream may also vary considerably. The design of most of the major portions of the treatment plant is based on the average rate of flow. The conduits, siphons, and distributing mechanisms in the plant are designed on the basis of the extreme rate of flow. Where a plant receives sewage mixed with storm water through combined sewers, the sharp

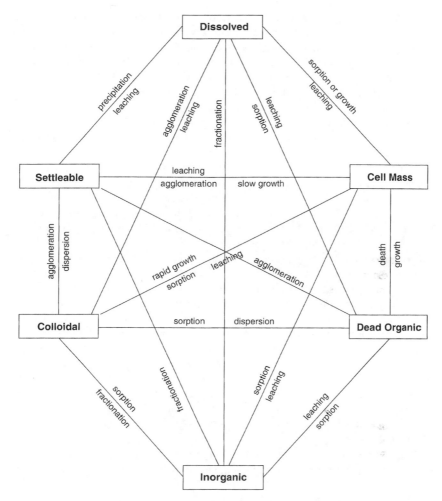

Figure 6.15 Interrelationships of solids in water. (From *Industrial Waste Surveys, Training Manual,* U.S. Environmental Protection Agency, Office of Water Programs, Washington, D.C., January 1973, pp. 15–20.)

increase in liquid must be considered further. Flow equalization has been used in industrial waste treatment for considerable periods of time. However, this is a process that has not been well utilized in municipal wastewater treatment. Objectives of the equalization are obtainment of a constant flow rate that can be treated in a specific manner. Another objective is to distribute the shock loads of either toxic or process-inhibiting substances that may affect the sewage treatment system. Numerous economic and environmental reasons exist for creating a proper flow equalization system.

Where clarifiers are used, if a large surge occurs, the influent overflows the clarifier and either lessens its ability to operate or inactivates it. Where microscreen units are utilized within limits, the microscreen is unaffected by a change in the flow rate. Tube settlers are only slightly affected by variations in flow. However, at times of extremely high flow, the detention time can be substantially decreased. Flow variations can have varying effects on filters, depending on the type of solids in the influent stream. Filters are usually operated at a constant rate of flow and should be protected against variations. In moving bed filters, the units are installed with level controls and the head tank automatically adjusts to the inflow rate. Flotation units are very much affected by the flow rate. Variations interfere with performance of these units. Solids clarifiers are also very sensitive to flow changes. In all cases, prior to settling, when advanced waste treatment or in effect any type of waste treatment is utilized, a flow equalization basin should be constructed and used. The proper constant flow of effluent helps provide better settling.

To provide a high-quality effluent from a waste treatment plant, it is necessary to remove a variety of organic and inorganic suspended solids that are present in the raw wastewater or have been precipitated from the solution during previous stages of the wastewater treatment. Various factors influence the rate at which wastewater solids may be removed by sedimentation. Particle size is very important. Small particles within the colloidal range generally cannot settle out during usual retention time and therefore must be brought together to form larger particles that can settle. The small particles have a very large surface area per unit weight of the solid. The forces between the particles tend to separate; bringing them together into an aggregate or larger particle is a matter of overcoming these forces. This is usually done by using chemical coagulants. The forces holding the particles together are basically electrostatic repulsion from electrical double layers surrounding particles suspended in water and physical separation by films of water adsorbed on the surfaces of the particles. The electrical charge is due to an imbalance of ions near the particle surface, which generally, in the case of domestic wastewater, causes a negative charge. The reduction of the charge is the principal function of the coagulant that is in use.

The constant random motion of small, typically submicron-sized particles due to collision of the particles of thermally agitated water molecules is called Brownian movement. Other forces involved are called van der Waals forces, which are atomic dipole interactions between atoms. The Brownian movement causes two or more particles to come within a short distance of each other and then, if the electrical forces that repel them have been reduced, the van der Waals forces will take over and pull the particles together. Van der Waals forces represent the relatively weak intermolecular, electrostatic attraction forces between two molecules as a result of complimentary regions of opposite $(-, +)$ charges. This is accomplished by the chemical coagulant that destabilizes the situation and allows for the particles to tie together, to become aggregates, and, of course, to precipitate out.

The coagulants used are alum, sodium aluminate, ferric chloride, ferric sulfate, ferrous chloride, ferrous sulfate, lime, and organic polyelectrolytes. Soda ash or clays are also used as sources of alkalinity or agents to weight material to help in the coagulation. When salts of aluminum or iron are added to water, they react with the water or alkalinity present in the water to form insoluble hydrolysis products. These hydrolysis products bring about effective coagulation. These materials that are positively charged in the neutral pH range adsorb onto the negatively charged wastewater particles, thereby reducing the repulsive forces between the particles. The coagulant may also react with phosphates and sulfates, forming hydrolysis products that contain a variety of ions. These hydrolysis products differ in their effectiveness as coagulants. It is important in adding coagulant to know the pH range, because the two of them are intimately related.

The optimum coagulation dosage and the variables present in the wastewater are determined in the laboratory by the use of a jar test. The zeta potential or particle charge may also be measured to control the coagulation process. The zeta potential is the residual charge at the interface between the layer of bound water and the mobile water phase. The coagulant may be used to remove phosphates, and therefore a third means of determining the dosage needed of a coagulant is to see how much can precipitate the phosphate within the wastewater.

The chemical processes used for suspended solids and phosphorus removal usually consist of rapid mixing of flocculation and sedimentation, or solids-contact clarifiers. In the traditional water treatment systems, whether swimming pool, drinking water, or waste treatment systems, a rapid mix usually occurs in a flocculation basin ahead of the sedimentation tanks to bring about a chemical clarification. The rapid mix provides a thorough and complete dispersal of chemicals throughout the water under treatment to allow for a uniform exposure to pollutants. Although the detention times may vary, 10 to 30 sec are usually designed into the system. Blenders in the line leading to the next step in the operation may also be utilized. The rapid mix not only causes a thorough mixing but also a rapid coagulation, because of increased turbulence. Flocculation aids in the coalescence and increase in size of particles that coagulate. Various types of flocculation devices are utilized. The solids-contact clarifiers are useful because they are smaller. The advantage of the reduced size is its ability to maintain a high concentration of previously formed chemical solids for flocculation or precipitation.

Gravity separation of solids is based on the difference between the specific gravity of the liquid and the particulate matter. The particles heavier than water settle to the bottom of the liquid, and those lighter float to the top. Sedimentation is the simplest and most widely used gravity separation process. In gravity separation, baffles are used to reduce the kinetic energy allowing the liquid to travel across the tank to the effluent weirs at a low velocity. The sludge zone at the bottom of the tank retains the solids so that a minimum of resuspension occurs, and also allows adequate time for compaction.

Primary sedimentation is most widely used in waste treatment plants. Secondary sedimentation is also widely used. In primary sedimentation, the overflow rate in the tank is measured in gallons per day per square foot. This is arrived at by measuring the number of gallons per day of flow and dividing that by the surface area of the clarifier in square feet. Whereas the primary settling is based on the Stokes law and is also determined by the width of the tank, depth of the tank, reduction of speed of influent, and amount of water flowing over the weir, secondary sedimentation must be determined differently because of a greater concentration and a lighter group of suspended solids. The Stokes law takes into consideration the density of the particles, viscosity of water, and gravitation settling rate. The settling rates are slower if allowed to settle by gravity. However, if chemicals are used, the settling rate is increased.

In the flotation process, the particulate matter is separated by allowing it to flow to the surface of the liquid. The gases produced attach to the solid to form a gas solid aggregate, which then can be skimmed off the top. Of the influent suspended solids and the BOD associated with it, 40 to 65% can be removed in this manner.

It is important that baffles, walls, or other obstructive devices are not used, and that turbulence is reduced to allow the solids to float to the surface. The effluent that exits must be submerged to prevent any interference with the froth on the surface. Skimming devices are utilized to remove froth that contains the solids. The detention time in this unit is usually 10 to 40 min. The tanks are 4 to 9 ft deep, and the overflow rate is 1 to 6 gal/min/ft^2.

Solids removal by settling in shallow basins brings about high rate settling. The problem is obtaining adequate flow distribution and sludge removal. These high-rate settlers are small, less costly to build than new clarifiers, can help upgrade overloaded plants, need less land, and are quite useful in the treatment of industrial wastewater. Generally, a series of inclined tubes are utilized for this purpose. The suspended particles settle to the bottom of the tubes and then slide outward into a collection basin, whereas the effluent continues to flow upward through the tube into a clear well. Laminar flow is necessary for the efficient and effective operation of the tube settler, because random flow produces turbulence that redistributes the solids.

The Lamella separator consists of a large group of parallel inclined plates through which the effluent, including the suspended solids, is passed. Each plate has an effective settling area equal to its projection onto a horizontal plane. By placing the plates close together, it is possible to obtain a high settling rate in a very small volume. The basic difference between the Lamella separator and the tube-settling units is that the former is fed from the top whereas the latter are fed from the bottom. In both cases, gravity causes the solids to deposit out as the liquid develops a frictional drag across the tubes or plates.

A variety of physical straining processes are utilized to remove solids from the liquid. The restrictions of the straining process are due to the straining device or to the previously removed solids that have already been deposited. Rotary screens, vibrating screens, ultrafiltration, and diatomaceous earth filters are used for this purpose. Microscreening has been used for solids removal processes for more than 20 years. The wastewater comes down or through the microscreen, depending on which process is used, and deposits the solids on it. From time to time screens are backwashed by a stream of water. Although the microscreens have small openings, the fine filtration is carried out by the previously trapped solids. Microscreening devices are usually used as a tertiary system for filtering the secondary effluents from trickling filters and activated sludge treatment. Rotary screens and vibrating screens follow the same process as the microscreens, except that they remove coarser solids. These are usually used in industrial waste treatment.

Diatomaceous earth filters have been used for clarifying secondary effluents. Although they have not been used on full-scale installations, they are utilized in other areas, such as swimming pools. The diatomaceous earth is placed in a thin layer of precoat on a porous cylinder or other porous plates. The wastewater is fed through the filter cake and the underlying material. This is accomplished by using vacuum or pressure. As filtration proceeds, the head loss through the cake increases due to the removal of solids. When a maximum head loss is achieved, the cake and solids are removed by allowing the liquid to flow in a reverse process.

All membrane processes share certain operating design features. To separate dissolved materials from water, a relatively "fine" membrane with relatively high pressures and low flow rates is required. If colloidal or suspended particles must be separated from water, then the membrane can be coarser and lower pressures used. A problem in the removal of either salt or solids is maintenance of flux on the membrane surface even if the precipitate is causing problems. Salt removal is carried out by reverse osmosis or hyperfiltration; a flux of 5 to 50 gal/day/ft^2 of membrane area at a pressure of 500 to 1500 lb/in.^{2}/gal is used. Usually 90 to 99% of total dissolved solids and 100% of suspended solids can be removed. Ultrafiltration operates at a 5 to 50 gal/day/ft^2 of flux area capacity at a pressure of 10 to 100 lb/in.^{2}/gal. It removes 100% of the suspended material.

After gravity sedimentation, the technique most frequently used is deep-bed filtration to separate solids from liquids. It has been used frequently for treating municipal or industrial water supplies, and it is now in use in wastewater treatment plants to improve the effluent from the conventional treatment plant.

The effluent from the activated sludge or trickling filter without the addition of chemical agents is applied to the deep-bed filter. The filter is also used for removal of phosphorus from secondary effluents and is used in conjunction with chemical coagulation, flocculation, and sedimentation. The water containing the suspended solids is passed through a bed of granular material until the solids are deposited on the materials. When the pressure drops excessively, the bed must be cleaned and placed back in service. At present, almost all the deep-bed filters used in waste treatment plants are of the rapid sand type.

Adsorption

Adsorption is a process in which the molecular or ionic species are accumulated on a surface and bound to the surface by forces of molecular attraction. For any adsorbent, the adsorptive capacity is roughly proportional to the surface area available. Activated carbon is a primary adsorbing agent, because it has a highly porous structure. Activated carbon can be used to reduce a variety of pollutants that are found in waste- and raw water, as well as in partially treated water. It can remove most dissolved organic materials, even if the organics have been partially degraded through a biological process. Wastewater is treated with activated carbon through the contact and regeneration systems. The water is brought into contact with the carbon by passing it through a container filled with either carbon granules or carbon slurry. When the contact time has been adequate, the maximum amount of organics and other materials have been removed. After a period of time, the adsorptive capacity of the carbon is exhausted. The carbon is then regenerated by heat through combustion of the organic adsorbate. Fresh carbon is routinely added to the system to replace that which has been lost.

Although the primary function of carbon is removal of dissolved organics, it cannot trap the small or highly polarized molecules, such as methanol, formic acid, or sugars. Although most inorganic materials are not removed by carbon, some are trapped through precipitation or biological assimilation when the biological treatment occurs. The chemicals trapped do not destroy all the biological material. Carbon can improve the quality of most effluents from the secondary biological treatment of municipal wastewater. These secondary biological effluents are liquids with less than 100 mg/l of COD and not more than 50 mg/l of suspended solids.

Activated carbon is usually used along with other techniques in the water treatment process. The effluent that leaves the process is usually of fine quality, because the majority of the contaminants have been removed and the activated carbon has been able to carry out its specific task of removal of the organics. However, unless some possible adverse side effects are controlled by the use of carbon, sulfides can be generated. As can be seen, the activated carbon is good, but as in all other systems, proper operation of the system by the sewage treatment operators is essential. The major advantages for the use of activated carbon is reduction in land area requirements to about one fourth of the needs of the conventional biological plants, a removal of a wide variety of pollutants that are nonbiodegradable, and the ability to handle changes in organic loading or surges in flow.

Oxidation

Oxidation processes are the usual biological processes taking place in wastewater. However, these processes do not completely oxidize the organic material present. Chemical oxidation has the potential for removing the rest of the organics by converting them into carbon dioxide, water, oxides of nitrogen, and oxides of other elements. Oxidation can also be used as a final means of treating wastewater after the use of activated carbon on the effluent. The techniques of oxidation include the use of ozone, hydrogen peroxide, hydroxyl free radical, molecular oxygen, and catalytic oxidation with molecular oxygen; the use of chlorine and its derivatives; and the use of oxyacids and their salts, such as potassium permanganate, and electrochemical treatment.

Ozone and hydrogen peroxide can be introduced directly into the wastewater by a corona discharge in moist air in proximity to the water that is to be treated. The corona discharge is a high-voltage, high-frequency (10,000 cps), low-temperature, electrical discharge. Several free radicals and other active oxidizing species are formed in the gas phase and transferred to the water, where oxidation occurs. This form of oxidation is also produced by radiation from radioactive sources. Radiation causes the formation of free radicals in water by breaking up the water molecules.

Foam Separation

Foam separation takes advantage of the tendency of certain surface active solutes, such as alkylbenzene sulfonate (ABS), to collect at a gas–liquid interface. Foaming provides an efficient means of generating and collecting the gas–liquid interface. When a solution contains surface-active solutes, the foam not only picks up these solutes but also may pick up some of the non-surface-active solutes attached to the surface-active ones. Adsorption occurs at the gas–liquid interface. The foam separation depends on the adsorption into the foam by the materials from the bulk of the rest of the liquid.

The success of the foaming removal is based on foam stability, requiring that the concentration of the surface layer be different than that of the bulk liquid and that the surface layer have unusually high-surface viscosity. Other important factors affecting foam stability are the concentrations in the bulk solute, pH, and temperature. The average removal of COD in experiments on effluent from secondary treatment plants processed through a foam removal system was 31%. Foam separation does not reduce the solids, but is a good additional technique for organic removal.

Distillation

Distillation is the most commonly practiced means of obtaining freshwater from seawater. Plants operating around the world develop over 100 million gal/day of freshwater. Distillation of wastewater, however, is substantially different than that of seawater. In wastewater distillation, it would be necessary to remove 95% of the water to get a concentration of 2% of the solids, with cost a significant factor.

The factors affecting distillation design are the supply of necessary heat energy, conservation or reuse of the heat energy, and design of heat-exchange equipment. The reuse of the heat energy is essential, because fuel is needed in the first place to produce the energy, and therefore the energy would go to waste unless adequately reused. When the sun is the source of heat energy, the problem is eliminated. Currently, at least in the United States, this is not a usually practiced technique. Considerably more research is needed in the distillation of wastewaters to determine whether it can be done at a practical cost.

Electrodialysis

Electrodialysis is used to produce potable water by partially removing the minerals from brackish groundwater in water-short areas of the world. In Arizona, a plant operates at a capacity of 650,000 gal/day; South Dakota has a plant operating at 250,000 gal/day. Technically, electrodialysis should be a technique that can be used to remove the minerals in wastewater. In electrodialysis, electricity is the force that separates water from minerals. A direct electrical charge is applied across a cell containing mineralized water. This causes the cations to migrate to the negative electrode and the anions to migrate to the positive electrode. If cations and anion-permeable membranes are inserted between the electrodes, mineral ions can be separated from the water. In a single pass through an electrodialysis unit, 40 to 50% of dissolved salts can be removed.

Freezing

The freezing and gas hydrate process has been used for demineralizing sea water to reclaim freshwater. This process is also a technique utilized in removing dissolved solids from municipal wastewater. Either the freezing or the hydrate process can concentrate the dissolved solids to about 1% and dispose of the residue. If greater concentration is necessary, a secondary process can be used to remove the dissolved solids in solid form. The process is accomplished by precipitating water as a solid and thereby washing the solids free of the mother liquor, which contains the impurities. The solid is then melted to release the pure water. Energy is needed for this process.

Ion Exchange

Ion exchangers are materials containing ions that can be replaced by other ions from the solution. The replaceable ion carried by the exchanger is the counterion. Carriers are called cation exchangers and anion exchangers. Once all the counterions are replaced, the exchanger is exhausted and must be restored by regeneration with a solution containing the original counterion. If the wastewater contains minerals that do not exceed 1000 to 1500 mg/l, then it is possible to utilize an ion exchange. The ion exchange is highly selective for the chloride ion. The anion resins are effective ion exchangers that do not become fouled by organics. By using this technique 50 to 60% of the COD is removed from secondary effluent.

Reverse Osmosis

Reverse osmosis is a membrane process in which water is forced to flow from a solution of high salt concentration to one of lower concentration. Natural osmosis is the opposite. Pressures of 600 to 800 lb/in.^{2} are required to cause this reversal of flow. This technique was first used to dispose of salt from brackish waters. The cellulose acetate membrane has been extremely valuable in the application of reverse osmosis for desalination of water. In this system, 99% of the minerals in brackish water can be removed. In wastewater treatment, the best membranes probably reject 50 to 75% of the inorganics and 90% of the organics with water flowing at a rate of 50 to 100 gal/ft/day. The major problem at present is the fouling of the system due to solids present. Research studies indicate that the process is excellent for the removal of many types of materials.

The preceding processes of distillation, electrodialysis, freezing, ion exchange, and reverse osmosis hold considerable potential for the future for the primary removal of inorganic materials from the effluent of secondary treatment plants. The major problems now are the cost involved and the use of energy.

Phosphate Removal

Most conventional treatment systems are biological processes and therefore are usually inefficient in the removal of phosphorus. In the United States about 400 million lb of phosphorus are discharged into the receiving streams from these treatment plants. Another 200 million lb of phosphorus are discharged from individual home units in unsewered areas. Of all the processes used, the activated sludge process is apparently most efficient in phosphorus removal, because it is most efficient in removal of solids.

Phosphorus removal is considered by many scientists to be the key nutrient in breaking the eutrophication cycle. However, the conventional secondary plants are not efficient in phosphorus removal. Phosphorus enters the plant in the highest oxidized form. Common biological systems do not reduce phosphorus, and it cannot be liberated in a gaseous form in the same manner as nitrogen, carbon, and sulfur. Biological removal is limited to the cell metabolic needs and whatever excess phosphorus is stored by the cells. To remove phosphorus from wastewaters on a sustained basis, it is necessary to choose the most efficient chemical or chemical biological methods. The strict chemical methods precipitate phosphorus either in the primary settling or in the tertiary treatment. In the chemical biological method, direct chemical dosing is carried out in the activated sludge plant. The chemically bound precipitated phosphorus is then removed with the sludge and does not become resoluble during sludge disposal.

Beside the phosphorus found in the wastewater of the municipal treatment plant, phosphorus compounds may also be entering the streams and waterways from materials used for scale control in water supplies, such as sodium hexametaphosphate. This source can contribute from 2 to 20% of the total phosphorus present in wastewater. Industrial wastewater, depending on the type of activity, can contribute large quantities of phosphorus to the receiving streams. Industrial sources may include phosphorus found in potato-processing waste, fertilizer-manufacturing waste, animal feedlot waste, certain metal-finishing waste, flour-processing waste, dairy waste, commercial laundry waste, and slaughterhouse waste. Phosphorus is also found in the fertilizers that are used on lawns and gardens in homes. This material can become part of the runoff into the receiving streams.

Phosphorus is found in wastewater in three primary forms: the orthophosphate ion, polyphosphates or condensed phosphates, and organic phosphorus compounds. The predominant form changes with the pH. The removal of phosphorus with lime results primarily from the increase in pH. Polyphosphates are considered polymers of phosphoric acid from which water has been removed. Complete hydrolysis brings about a formation of orthophosphates.

Raw sewage contains substantial amounts of all three principal forms of phosphorus. During the biological treatment, significant changes take place. As the organic materials are decomposed, the phosphorus is converted to the orthophosphates. However, inorganic phosphates are used in forming biological floc. The result is that in well-treated secondary effluent, a large portion of the phosphorus is present as orthophosphates, which is fortunate because this form of phosphorus is most easily precipitated.

Phosphorus in both the organic and inorganic forms must be removed from the raw wastewater before it is permitted to go into the receiving stream. The materials that are most practical for phosphorus removal include the ions of aluminum, iron, and calcium. The calcium is added as lime and the hydroxyl ions also aid in the phosphorus removal. The chemistry of phosphorus removal is complex. The optimum removal of phosphorus is in the range of 5.5 to 6.5, although some removal occurs above the pH of 6.5. The addition of alum lowers the pH of the wastewater because of the neutralization of alkalinity and the release of carbon dioxide. The pH reduction depends on the alkalinity of the wastewater originally. Most wastewaters are so alkaline that even large amounts of alum may not drop the pH below 6.0 to 6.5. In some cases, it is necessary to use sodium hydroxide, soda ash, or lime to get the proper pH. Sodium aluminate may also be used in phosphorus removal. However, sodium aluminate may raise the pH. Both iron in its ferrous and ferric state can be used in precipitating phosphorus. The iron salts used include ferrous sulfate, ferric sulfate, and ferric chloride. Pickle liquor, which contains a substantial amount of free sulfuric acid or hydrochloric acid, may be utilized. Iron salts are most effective at a pH of 4.5 to 5.0. However, because it is very difficult to achieve this, the optimum pH is between 7 and 8. Lime may be used to precipitate phosphorus. The calcium ion reacts with the phosphate ion in the presence of the hydroxyl ion to form hydroxyapatite. The solubility if the hydroxyapatite is so low that even with a pH of 9, a large amount of phosphorus can be removed.

The removal of phosphorus in wastewater is accomplished by modifying the conventional primary treatment process to include chemical precipitation with aluminum or iron salts. The advantages of phosphorus removal during the primary treatment include flexibility in the use of chemicals, adequate time for reaction and mixing, flocculation and removal of more suspended solids and BOD during primary settlement, and reduced amount of suspended solids and BOD going into secondary treatment. The process can be automated. The disadvantage of adding these metal salts during primary treatment alone is that a significant amount of the phosphorus may not be in the ortho form and therefore is not easily precipitated.

Nitrate Removal

The nutrients phosphorus and nitrogen that are of greatest importance to excess growth of algae and other aquatic plants are an aesthetic problem, kill fish, and finally kill the stream. Nitrogen occurs in inorganic and organic forms. In secondary effluent, the inorganic forms are the greatest nutrient problem. They are predominantly ammonia and nitrate, with a certain amount of nitrite present. Under certain conditions of excess oxygen, it is possible to convert the ammonia to nitrate biologically. This is called nitrification, and the water is then called a nitrified effluent. Nitrates not only are a nutrient, but also in high concentrations in drinking water can cause a serious condition to infants and young children, known as metahemoglobinemia. Ammonia creates additional oxygen demand, which further increases the BOD. Municipal wastewaters have a nitrogen content of 15 to 25 mg/l in the untreated and primary settled wastes. The nitrogen is divided among the organic compounds that are mostly insoluble in ammonia.

Generally, conventional biological processes can transform almost all nitrogenous compounds into ammonia and a biological sludge. Removing the ammonia is an important part of reducing the nitrate problem. Ammonia nitrogen present in effluents has the following undesirable characteristics: ammonia consumes dissolved oxygen in the receiving stream; reacts with chlorine to form chloramines, which are much less effective disinfectants than free chlorine; is toxic to fish life; is corrosive to plumbing fittings; and increases the chlorine demand at waterworks downstream of the sewage treatment plant.

A nitrified effluent that is free of ammonia has the following advantages: the nitrates provide oxygen to the sludge beds and prevent the formation of bad odors; and nitrified effluents are more effectively disinfected by chlorine and contain less soluble organic matter than before nitrification. However, the ammonia and nitrate are interchangeable nutrients for green plants and algae, as well as bacteria. If the nitrate level is too high, it stimulates undesirable aquatic growth. When the effluent is treated biologically and the nitrates are converted to nitrogen gas, the process is called denitrification. The best process at present is the change of nitrogen compounds through biological oxidation of nitrates and then denitrification. Good process control is necessary to carry out the changes that are recommended.

Nitrogen can be removed from wastewaters by column compactors. In a packed column the cell resistance time of the surface-bound slime is much greater than the hydraulic detention time. This fact, combined with a large contact surface and a short diffusion distance because of the small media used, such as sand, creates an efficient system of rapid denitrification. Methyl alcohol is used as a supplemental organic carbon source in the columnar denitrification process.

Ammonia may be removed from wastewater effluent by raising the pH to convert the ammonium ion to dissolved ammonia and by utilizing enough free air to cause desorption or stripping of the ammonia. Nitrogen is also removed by chemical methods, including the use of selected ion exchanges to remove the ammonium ions. However, this tends to be a costly process. Ammonia can be oxidized to nitrogen gas by chlorinating at the break point. Break-point chlorination also disinfects the wastewater.

Oxidation Ponds

On an experimental basis, oxidation ponds have brought about the following important changes in the final product water: pH decreased from 8.3 to 6.7, turbidity decreased from 90 to 4; total alkalinity decreased from 260 to 95, COD decreased from 250 to 50, BOD decreased from 38 to 10, and phosphates decreased from 40 to 0.25 ppm.

Package Treatment Plants

Package treatment plants are small facilities serving from tens to hundreds of homes. The plant is usually designed to handle the effluent of small subdivisions that cannot put their sewage into a large municipal sewage system. It may also be utilized for industries and large institutions. The package treatment plant can handle, depending on its size and construction, 1500 to 500,000 gal of influent per day. The package treatment plant is a sewage treatment plant and therefore must perform all the services discussed in the previous portions of this chapter.

A typical package treatment plant utilizes extended aeration or aerobic digestion. The influent is pretreated by bar screens, comminutors, and trash traps. The bar screens block the incoming sewage. The comminutors grind the sewage before passing it to the aeration chamber. Trash traps are actually holding tanks used to settle out the untreatable material and biochemically break down the organic solids before passing them to the aeration chamber. The bar screens and comminutors need frequent servicing. In the aeration chamber, the pretreated liquids enter the chamber and are thoroughly mixed and aerated with large volumes of air. The aerobic bacteria convert the sewage into clear, odorless liquids and gases. The effluent leaving the aeration chamber flows into a settling chamber, where the final phase of treatment takes place. At this point the sludge settles to the bottom of the hopper and then is pumped back to the aeration chamber for further treatment. If operated properly, the system should produce a good effluent.

Unfortunately, many of the systems are neglected and mistreated. As a result of this numerous failures occur. The system must be designed to handle unusual loads of influent. The service people should check the unit regularly to determine whether the system is operating at maximum efficiency and should then carry out a series of routine maintenance steps, including the greasing and oiling of equipment and checking of all lines. There is no question that in the event a sewage system is unavailable, a package treatment plant is the preferred technique for taking care of the sewage needs in communities and other types of groups. It is also just as certain that an integral part of the package treatment process is highly trained operators who can check and maintain the plant in an effective manner (Figure 6.16).

Innovative Technology for Biological Sludge

Vermial (earthworms) and microbial (microorganisms) management of biological sludges under dynamic conditions of temperature and seasonal changes has been studied as an innovative technique for disposal of sludges. Research has been conducted by many. In one study, sludge residues would accumulate on control plots where wet sludge had been spread, but not on the treatment plots that were inoculated with earthworms. The earthworms, in conjunction with indigenous microbial activity, appear to be a potentially efficient method for managing biological sludges and other labile organic matter that is placed on land. The turnover of sludge in treatment plots was rapid throughout the spring, summer, and fall seasons, but decreased during the winter. Earthworms are commonly considered primary consumers of detritus. However, the earthworms consume and require a combination of microbes, cellulose, and grit as sources of ingested material for maximum growth. Innovative technologies related to other sludges are discussed in Chapter 7.

SOILS

Overview

Soil is defined as any mixture of particles derived from disintegrated rock that is capable of supporting growth of vascular plants. The individual particles range in size from microscopic to 2 mm. Implicit in this definition is the condition that the rock particles provide all the approximately 27 essential elements of life, except

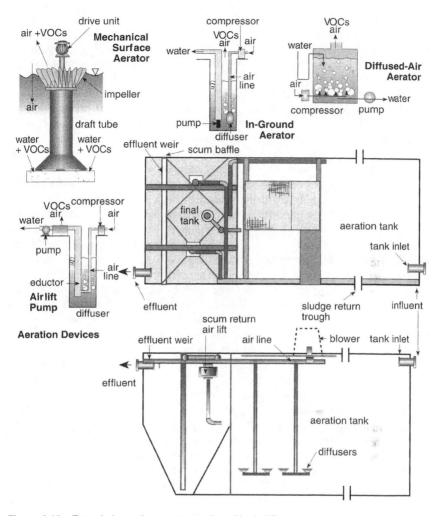

Figure 6.16 Extended aeration treatment plan with air diffusers.

carbon and oxygen, which are obtained photoautotrophically as carbon dioxide (CO_2), and hydrogen, which can be obtained from water (H_2O).

Soil is divided into inorganic (mineral) and organic (humic) fractions. Mineral soils are composed of 40 to 45% inorganic matter, 5 to 10% organic matter, 25% soil water, and 25% soil air. Soil that contains more than 20% organic matter is referred to as a histosol.

The inherent properties of soil influence the movement, residence time, and fate of the intrinsic components of relatively labile organic matter that has been applied to land. Concurrently, the interaction of organic matter via decomposition and humification reactions can alter soil properties. Accordingly, application of natural and anthropogenic organic matter, increased biotic (living) and abiotic (nonliving) enzymatic and nonenzymatic catalytic activities and formation of secondary mineral and humic colloids all contribute to perennial cycles of various biological, physical, and chemical interactions and reactions. The ultimate results are enrichment of soil, and treatment of organic matter.

Soil is all the unconsolidated material covering Earth's surface. It acts as a natural medium for plant growth. Soil has particular characteristics that result from the combined influence of climate and living matter on material that has been conditioned over a period of time. Soil is also considered any earth material that does not include embedded rock and shale. The soil varies from clay to glacial gravel and talus. The properties of soil vary with its gradation, its moisture content, its vertical position in relationship to the surface of the ground, and its geologic location and formation. Most soils were originally solid rock. Time and climate broke the rock into smaller particles, and these have been pulverized into sand-sized particles, silt-sized particles, and clay-sized particles (Figure 6.17). Chemical changes have occurred over a number of years that vary the types of soil. Soil science deals with the nature, properties, formation, classification, and mapping of soils and their behavior or response to varying uses and management.

Some of the major sources of parent material are glacial till, loess, alluvium, colluvium, lacustrine, marine, and residual. Glacial till is the debris left from slow moving glaciers that scraped away at rocks and then left rocks and material where the glaciers melted. Loess is a wind-deposited material composed mostly of silt-sized particles. They vary from a few inches to over a 100 ft in depth over bedrock. Alluvium is water-deposited material formed on bottom lands, deltas, or floodplains. These soils are usually less uniform than loess. Colluvium is parent material deposited by gravity at the base of slopes. Lacustrine is parent soil deposited in lakes. Marine is parent soil deposited in oceans. Residual is parent soil deposited in place.

Soil is not only the basis for agriculture and forestry, but also of special importance to those in environmental health, because under proper conditions it is a mechanism for allowing subsurface absorption sewage systems to operate in an effective manner. Under improper conditions, vast amounts of problems occur, including serious potential health hazards. The purpose of soil in subsurface sewage disposal is to renovate the sewage physically, chemically, and biologically before it reaches groundwater. Soil is the most important part of a septic system.The siting of the system is key to proper system performance and longevity. The study of soils in this volume is primarily concerned with the function of soil as a medium for the removal and treatment of sewage. However, removing contaminants from soil is discussed.

Soil Profile

Soils and their properties vary greatly from place to place depending on the soil-forming factors and on whether the original soil is present or fill has been added. Different layers or zones are formed that are parallel to the soil surface. This occurs as a result of the raw inherent material of the soil altered by various climatic elements

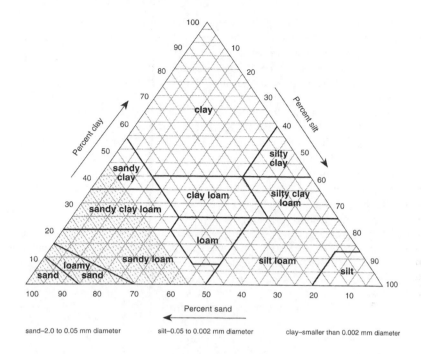

sand–2.0 to 0.05 mm diameter   silt–0.05 to 0.002 mm diameter   clay–smaller than 0.002 mm diameter

Comparison of Particle Size Scales

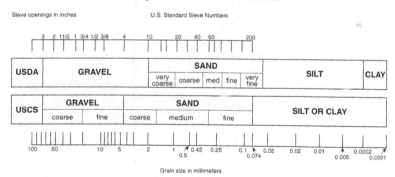

Figure 6.17 Textural classification chart (U.S. Department of Agriculture) and comparison of particle size scales. (From Brunner and Keller, *Sanitary Landfill Design and Operation,* p. 16.)

and living matter. These zones are called soil horizons. Soil profiles vary considerably by the degree to which the soil horizons vary. Young soils usually have only a few or indistinct horizons; older soils have many distinctive horizons. A soil horizon is usually differentiated from its adjacent horizon on the basis of the properties of the soil that are seen in the field. A vertical cross section of the soil horizon varying in depth from 2 to 5 ft comprises the soil profile. Generally, the horizons in the soil profile are grouped into three major divisions: horizon A, B, and C. Horizons A and B are those that have been formed by weathering in the soil-forming processes and are referred to as the solum of the soil. Horizon C is unaltered by soil-forming processes. Where an R horizon is present, it indicates that bedrock can be found there (Figure 6.18).

The A horizon includes the surface and subsurface layers of the soil. It is the original top layer, which may vary from 2 to 18 in. in depth. This horizon is also referred to as topsoil or the surface soil. If considerable farming or erosion has occurred, much of the original A horizon may be missing. The B horizon is the subsoil just below the A horizon. It is a zone of accumulation of material that has moved down from the A horizon and also has been weathered in place, to some extent. The C horizon lies below the subsoil and is usually the parent material from which the A and B horizons have been formed. The R horizon, which is the consolidated bedrock, may be limestone, sandstone, or granite. The A horizon is also known as the zone of maximum biological activity where organic straining and eluviation occurs. The B horizon is also known as the zone of eluviation (washing in), where accumulations of clay, iron, and aluminum oxide are found.

Factors of Soil Formation and Composition

The soils at any given point are determined by physical and mineralogical composition of the parent material, climate during the period of soil material accumulation and existing since accumulation, plant and animal life on and in the soil, relief or topography of the land, and length of time the forces of soil development have acted on the soil material. The parent material is composed of mineral or organic material from which the soil develops. Great elapses of time are necessary for the action of water, ice, wind, and biological activity to form soil from the parent materials. Climate is very important because temperature and humidity determine the rate at which soils are formed. Soil characteristics and type of biological processes affect the soil process. The three classes of living organisms that are important in soil formation include vegetation, animals, and microorganisms. Vegetation affects the soil through its root action and by adding organic material to the soil when the vegetation dies. Animals and insects dig into the soil and affect it mechanically. Earthworms, ants, and other such organisms dig into the soil and help to aerate it. When animals die, they furnish organic material to the soil. Microorganisms decompose organic material and cause the release of mineral elements from the parent material. The shape of the land surface affects the soil formation. If the land has steep slopes, the action of the wind and water moves the soil downhill. In floodplains, receding floodwaters deposit silt and other materials.

The three main components of the soil are solid, liquid, and gaseous forms. Solid material is mineral, such as sand, silt, clay, and coarse rock fragments. Organic matter, such as plants and animals, are also solids that decay. The liquid in the soil is water that acts as a solvent and moves through the soil transporting the dissolved solids. The gases present in the soil include oxygen, carbon dioxide, and nitrogen.

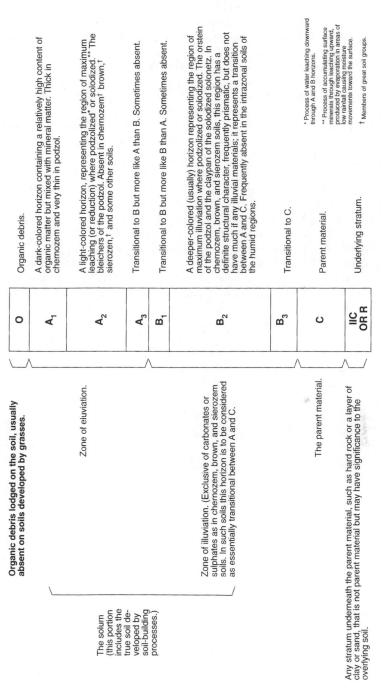

Figure 6.18 A hypothetical soil profile having all the soil horizons. (From Soil Manual for Sanitarians, 2nd Rev., U.S. Department of Agriculture Soil Conservation Service, East Lansing, MI, April 1972, p. 2.)

The factors of forming soil are so closely interrelated that it is difficult to extract one from the group and say that it is the primary motivating factor in soil formation.

Climate influences soil formation mainly through the moisture and energy that it provides to the soil environment. The weathering of rocks and the leaching of carbonates and other minerals, which are an integral part of soil formation, are controlled mainly by moisture and temperature. Rainfall appears to be the most active of the climatic agents in soil formation. As the water percolates, it distributes the carbonates, clay, weathered products, and sesquioxides of the soil profile. Rainfall is also responsible for much of the erosion that occurs. However, the total amount of rainfall is not as important as that which is absorbed by the soil. The rest runs off into surface bodies or is lost by evapotranspiration. The water that percolates down through the soil may eventually find its way down into a water table. Topography and soil texture regulate this water percolation. Runoff is based primarily on the slope and texture of the surface layer of the soil.

The cover of vegetation in a given area modifies the climatic influences. Because the vegetation and organisms present form the organic fraction of the soil, they are responsible for the dark color in many of the surface layers and subsoils of some soils. Vegetation also contributes the organic acids that cause leaching of other substances in the soil material. The redistribution of calcium is somewhat controlled by the vegetation. The type of vegetation helps determine the amount and distribution of organic matter in the soil. Grass roots contribute large amounts of organic materials to a greater depth than the roots of trees.

The parent material was formed by volcanic action, and disintegration and decomposition of large masses of solid rock that left a layer of unconsolidated mineral material over most of the surface of the earth. This layer has been moved from place to place by wind, flowing waters, or advancing glaciers, or may have remained where originally formed.

The effect of topography is seen in hilly areas, where the steep slopes tend to be drier and appear to come from a more arid climate than the depressions or level plains, which are much more moist and appear to come from a very humid climate. The time involved in the formation of the soils may be many tens of thousands of years.

Properties and Qualities of Soils

Natural Drainage, Color of Soils, and Water Tables

Soil drainage refers to the ability of the soil pores to avoid saturation. It is the amount of and speed with which water will flow from the soil. Soil drainage relates directly to the amount of air and water in the internal pores of the soil. Drainage conditions in certain cases are observed by the amount of water present on the surface of the soil, in shallow wells, or in the presence of springs or musty moist areas.

Soil drainage is a combination of runoff, internal soil drainage, and soil permeability. Runoff is based generally on the soil slope. However, the permeability is based on the type of soil and the degree of internal drainage. The internal soil drainage is the quality of the soil that allows the downward flow of excess water. It is reflected by the frequency and duration of the periods of soil saturation with water. It is determined by the texture, structure, other characteristics of the soil profile, the underlying layers, and the height of the water table. Soil drainage conditions are also inferred by the kind of vegetation and plant root penetration and by the soil colors. Soil color gives an indication of the depth of the topsoil and the depth of the saturated soil at a site. A bright uniform brown, red, and yellow set of colors are associated with good drainage. Gray, pale yellow, blue, and green colors are associated with poor drainage and lack of aeration. Soil mottling, which is a series of different color patterns in a soil, is associated with a fluctuating water table. Black or very dark gray surface soil is associated with soil saturation for long time periods. Metallic black coatings are usually associated with concentrations of manganese, which is formed as a result of restricted drainage.

Where the soil is well drained, it is generally free of mottling to a level of 40 in. beneath the surface of the soil. Where it is moderately well drained, it is generally mottled in the lower B or upper C horizons, beginning anywhere from 20 to 40 in. below the soil surface. Where the soil is somewhat poorly drained, it is mottled in the upper B horizon or in the lower A horizon beginning 10 to 20 in. below the soil surface. Where it is poorly drained, the surface soil is gray or light gray, and the mottling or gray soil begins less than 10 in. from the surface. Where the soil is very poorly drained, it is black or very dark gray on the surface and mottling or graying begins at, or very near, the surface.

The water table is the zone of continuous saturation of groundwater or it may be an elevated area caused by a slowly permeable layer within or below the soil. The water table levels are influenced by the rate of evaporation and transpiration. These rates are usually high during the warm summer and low during the cold winter. As the soil loses moisture, the water table lowers. When the soil cannot lose moisture, the water table may rise. Vegetation makes a significant difference in the level or amount of evaporation transpiration. Water tables have been generally found to be lower in woodlands than in pasture or cropland areas that are of the same soil-drainage class.

Soil Texture

Texture is the coarseness or fineness of a combination of mineral particles that comprise the soil. It is the physical makeup of the soil, the mix of sand, silt, and clay held together in larger grains. Most soils fall into three classes: sand, silt, and clay. Sand is the coarsest soil, with a range in size from 2 to 0.05 mm. Silt ranges from 0.05 to 0.002 mm or 2 μm. Clay is less than 0.002 mm or 2 μm size. Sand and silt are mainly particles of quartz (SiO_2) whereas clay is composed of a combination of elements such as Si, O, Al, Fe, Mg, H, and K. Soils are classified as loams based on the percentage of each of the main fractions — sand, silt, and clay.

Incorporation of the organic fraction, humus, into soil involves the influx of soil organisms, such as microbes and earthworms, and the heterotrophic decomposition of organic matter. Humus, the ultimate end product of heterotrophic decomposition, is the major constituent of soil organic matter. The incorporation of organic C, as well as the nutrients N, S, and P, into soil appears to be a slow process. The soil may also contain coarse rock fragments that range in size and shape. Clay particles, when moist, are highly plastic. When clay is wet, it expands and becomes sticky; when dry, it shrinks and absorbs considerable energy. This absorptive quality of clay for water, gases, and soluble salts is very high.

Silt is somewhat plastic and sticky, and absorbs materials due to an adhering film of clay. However, silt does not shrink and swell on wetting as much as clay. Silt and clay in soil gives it a fine texture that slows water and air movements. When a sizable quantity of clay is present, the soil becomes sticky if too wet and hard if too dry. The expansion and contraction is larger than in sandy soils.

Sand has practically no plasticity or stickiness, and therefore is not influenced much by the moisture content. Its water-holding capacity is usually low and therefore allows water to percolate rapidly. Sand facilitates drainage and allows for good air and water movement. Soils that have a lot of sand and gravel have good drainage and aeration, and are usually loose.

It is rare to find soil samples or soil horizons that consist of one type of soil. The classes of soil texture are based on the different combinations of sand, silt, and clay found in the soil. The basic texture classes are loamy sand, sandy loam, loam, silt loam, silt, sandy clay, clay loam, silty clay loam, silty clay, and clay. Sand may be modified into fine, very fine, coarse, or very coarse grades.

Sand, which includes loamy sand, is 70 to 100% sand and less than 30% silt and clay. When moist, if squeezed, it cannot hold together readily. Sandy loam is 50 to 70% sand with less than 50% silt and 20% or less clay. When moist, it can be squeezed into a form that feels gritty. Loam is soil material that contains 7 to 27% clay, 28 to 50% silt, and less than 52% sand. When moist, it compacts readily. It has a slightly gritty feel. Silt loam contains 50% or more silt, with less than 27% clay and the rest sand. It is smooth when dry and its clods can be easily broken. When moist it takes on a form and may flow when wet. Clay loam, which contains 20 to 40% clay with less than 72% silt and the rest sand, is plastic and sticky when moist and makes a compact form that breaks readily. Its clods are hard to break when dry. Clay contains 35% or more clay, less than 65% sand, and less than 60% silt. It is sticky and plastic when moist. It makes a strong form and is very hard and resistant to breaking when dry (Table 6.1).

Permeability (Percolation)

The texture of a soil has a significant influence on the percolation rate. Sandy soils tend to have faster rates; finer textured soils tend to have slower rates. Very silty soils tend to flow and may therefore clog a soil absorption system.

The following is an estimated rate of percolation based on the general soil texture: sand, water percolates under 10 min/in.; sandy loam, water percolates in 3 to 30 min/in.; loam, water percolates in 10 to 45 min/in.; silt loam, water percolates in 30 to 90 min/in.; clay loam, water percolates over 45 min/in.; clay, water percolates over 60 min/in.

The soil permeability is one of the most important single factors in determining the suitability of a site for septic tanks and below-surface seepage fields. Permeability

Table 6.1   Soil Textural Class Names and Approximate Percentage of Sand, Silt, and Clay

General Terms	Basic Soil Textural Classes	Composition		
		% Sand	% Silt	% Clay
Coarse textured soils	Sands Coarse sand Sand Fine sand Very fine sand	+85	−15	−10
	Loamy sands Loamy coarse sand Loamy sand Loamy fine sand Loamy very fine sand	70–90	−30	−15
Moderately coarse textured soils	Sandy loams Coarse sandy loam Sandy loam Fine sandy loam Very fine sandy loam	43–85	−50	−20
Medium textured soils	Loam	23–52	28–50	7–27
	Silt loam	20–50	50–80	12–27
	Silt	—	50–80	−12
		—	+80	−12
Moderate fine-textured soils	Clay loam	20–45	15–53	27–40
	Sandy clay loam	45–80	−28	20–35
	Silty clay loam	−20	40–73	27–40
Fine-textured soils	Sandy clay	45–65	−20	35–55
	Silty clay	−20	40–60	40–60
	Clay	−45	−40	+40

Note: + = more than; − = less than.
The proportion of various sized sand particles determines the name of these textural classes.
From *Manual of Septic-Tank Practice,* U.S. DHEW, U.S. Government Printing Office, Washington, D.C., 1969, p. 19.

is a measure of the rate at which water or sewage effluent can be taken into and transmitted through the soil. The soil permeability or percolation rate is the quality of the soil that permits it to transmit water or air. It can be measured in terms of rate of flow of water through a given cross section of saturated soil per unit time. The major factor affecting permeability is the soil texture and structure. Bedrock or high water tables may also have a very definite influence on soil permeability.

Soil Structure

Soil structure is the arrangement of the individual grains or primary particles and how they fit together into secondary aggregates that comprise the mass of the soil. Texture refers to the size of the soil materials. Structure refers to the arrangement of these particles or the shape formed when the particles adhere together into an aggregate or ped. The structure is identified by its shape or type, size, and grade. The soil structure is the total sum of the particles of silt, sand, and clay that form aggregates and that are separated from adjoining aggregates or clusters of material by varying surfaces of weakness. Stability of aggregates is also maintained by interaction with soil organisms. Bacteria produce polysaccharides that cement soil particles together. Growth of soil fungi produces mycelia, which attach to soil particles and increase aggregation. Earthworms excrete feces (castings), consisting of digested soil and organic matter, that are more stable structurally than the surrounding soil. Earthworms also compress particles together during burrowing and secrete mucous exudates that possibly bind soil particles together.

Highly structured soil has low bulk density. Soil bulk density, the dry weight for a known volume of soil, is directly related to porosity. The mean density of soil minerals equals about 2.65g/cc. Typical bulk densities of soil, however, range only from 1.0 to 1.6 g/cc. Pore space accounts for the difference between the theoretical bulk density of 2.65 g/cc and any lower actual value.

The texture and structure of a loam directly affects the porosity of soil. Aggregation of soil particles forms intra- and interaggregate pores. The sizes of the particles that comprise each aggregate are inversely related to surface area and directly related to the cross-sectional area of each pore. The dimensions and surface area of pores, in turn, directly influence the rate of initial infiltration and subsequent permeation of water, air, and matter into and through soil. The permeate is attracted strongly to the surfaces of soil particles due to physicochemical bonding. The more surface area exposed, therefore, the greater is the resistance to flow.

Loams consisting of very coarse sand provide the least amount of surface area, whereas fine textured clay loams present the greatest area. In relation to this principle, sand loams have fewer but larger interstices or pores for a given volume of soil than clay loams. The larger pores have a wider cross-sectional area for flow-through with less resistance. The rate of permeation of rain, irrigation waters, and used water applied to soil, therefore, is faster through sand loams, less for silty loams, and still less for clayey loams.

Approximately 50% of the available pore space in soil is filled with air instead of water. Soil air moves through the pore space via diffusion and is continually exchanged with atmospheric air. The composition of soil air is variable depending on metabolic activity of soil organisms and movement of vapors generated from volatile synthetic organics.

Application of waste matter to soil surfaces can cause clogging. Clogging would be alleviated by adjusting the hydraulic and organic load of applied organic matter and selectively choosing a suitable loam. Indeed, one could suggest use of sand as the optimal medium due to its capacity to form a matrix of large pores for enhancing permeability. To do so, however, one would overlook the critically important characteristic of surface area. The physicochemical attraction between surface of soil particles and permeate causes resistance to flow, which permits transformation reactions catalyzed by clay and humic colloids plus associated extracellular enzymes.

Note that 1 m^3 of sand with a mean particle size of 1 mm has a solid surface area of about 480 m^2/m^3. In contrast, 1 m^3 of clay particles averaging less than 0.002 mm may have a solid surface area of approximately 2100 m^2/m^3.

Some aggregates have thin, often dark-colored surface films that may help to separate them. The four categories of aggregate shapes found in soils are:

1. *Granular and crumb* granules are common in surface soils and associated with organic matter. These granules are rounded, do not pack tightly, and therefore are quite permeable.
2. *Platy* granules are common in the lower part of the A horizon beneath the organic, enriched surface layer. The particles are arranged around a plane, generally in a horizontal manner. The amount of overlap of the individual plates in platy soil strongly influences the rate of water movement downward.
3. *Blocky shapes* are blocklike forms that are angular and subangular and are usually associated with subsoil or the B horizons. Usually a moderate to high proportion of clay found is in the blocky structure. The structure results largely from the alternate swelling, shrinking, and cracking associated with alternate wetting and drying of the clay over long periods of time. When the blocks have angular edges and corners, and they become wet and swell, their permeability is reduced substantially. In soils with somewhat less clay blocks present, the blocks are not so perfectly formed, the edges and corners are more rounded, and therefore the subangular blocks do not fit together as tightly when wet. Permeability in this group is greater than the angular blocky group.
4. *Columnar and prismatic shapes* are longer in the vertical area than the horizontal. They are usually large in size and associated with subsoils that have a high clay content. They often break down into blocky structures. The permeability of prismatic soil is generally about the same as that of the blocky soil (Figure 6.19).

The stability of the soil structure is usually classified as structureless, weak, moderate, or strong. Structureless soil does not have an observable aggregation. An example of this would be fine and medium sand. The weak soil has a poorly formed structure that does not maintain its shape when handled. Moderate soil is fairly durable when handled. It is not distinct in an undisturbed soil. Strong soil is durable when handled and definitely distinct visually in undisturbed soils. Obviously, for the purposes of subsoil sewage disposal, the tendency would be to try to get weak or structureless soil vs. strong soils. The difficulty with the structureless soil is that chemicals may percolate down to the subsoil water tables. The clay soils are usually associated with the strong classification. Silt loam textures are associated with the moderate and medium structures. In soils of similar texture, the structure determines mainly the rate of water movement.

Shrink–Swell Potential

The shrink–swell potential indicates the volume change that can be expected when a soil is wet and when it is dry. This potential is determined by the amount and type of clay present. Soils high in clay with an expanding lattice have a high or very high shrink–swell potential. Soils composed mainly of sand and silt have a low shrink–swell potential. Some clays have a much higher shrink–swell potential than others.

Most of the important physical–chemical reactions in soil take place on the surfaces of the clay particles. Clay particles, which are the smallest particles present, provide more surface area per unit of mass and therefore greatly increase chemical activity. The difference between swelling and saturation is that in swelling the soil

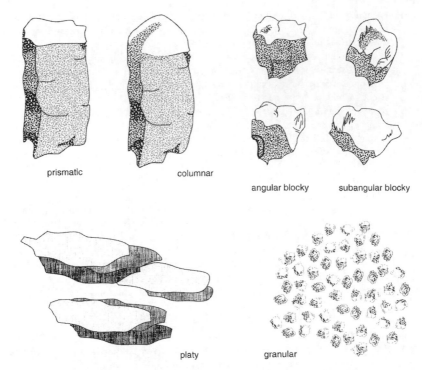

prismatic   columnar

angular blocky   subangular blocky

platy   granular

Figure 6.19   Types of soil structure. (From *Design Manual, On-Site Wastewater Treatment and Disposal Systems,* U.S. Environmental Protection Agency, Washington, D.C., October 1980, p. 36.)

can hold water but cannot be totally filled with water. In saturation, all the pores between the soil particles are filled with water.

Slope and Erosion

Soils with a gradient exceeding 10% have severe limitations for the operation of subsurface disposal fields. Far more erosion takes place in this type of situation and the system may be difficult to construct. Frequently, if the system has not been installed properly or if an impermeable layer is not far below the surface, effluent can break to the ground on the slope downward from the system.

Classification and Naming of Soils

The U.S. Department of Agriculture groups soils into seven natural drainage classes. They include excessively drained, somewhat excessively drained, well drained, moderately well drained, somewhat poorly drained, poorly drained, and very poorly drained. The determination of the natural soil drainage is made by studying the soil colors, intensity, and position in the profile. Iron is the main coloring substance of the subsoil, and the color of the iron in the soil is closely related to the amount of air present. If air is absent or only present in small quantities and when the soil becomes saturated or nearly saturated, iron exists in the ferrous or reduced state, which is gray in color. When the air supply is normal, as found in well-drained soils, the iron is in the ferric or oxidized state and is yellowish or red in color. If the soil has been frequently wet and dried, mottling occurs, which is a mixture of soil colors including gray, yellow, red, and brown. Three other classification systems are utilized including the National Cooperative Soil Survey, United Classification of Soils, and American Association of State Highway Officials Classification of Soils.

Soil correlation and naming have to do with the definition, mapping, naming, and classification of the kinds of soils in a soil survey area. The purpose of the correlation is to ensure that the soils are accurately defined when mapped and uniformly named in all soil surveys in the United States. Each of the different soils is studied and defined in terms of its properties, both physical and chemical, and is then assigned a name.

Characteristics Used to Differentiate Soils

The characteristics used to differentiate soils include not only those previously mentioned but also the types of soil horizons, depth, drainage, color, texture, structure, coarse fragments, porosity, permeability, slope, minerals present, and flooding. In classifying the soils, the smallest unit is called the pedon, which is that area of the soil that has all of the described characteristics of that soil, including the largest structure. The pedon usually ranges from 4 to 40 in. in diameter and extends from the surface to the full depth of the soil. The phase is a much larger classification that is composed of many similar pedons. Phases give information about physical characteristics of the soil, some of the chemical characteristics, soil setting in the environment, and some of the things that have happened to the soil. The defined mapping units that detail soil survey maps are mostly phases of a given soil series. The soil series is the level of classification where all soil profile characteristics apply, including the depth, texture, structure, parent material, drainage, coarse fragments, permeability, clay mineralogy, chemical reaction, base saturation, and origin. The name of a soil series is usually identified by the name of the place where it was first found.

Effluents from Septic Tanks in Soils

There are a variety of effluents from septic tanks. They include organic compounds, nitrogen, phosphorus, detergents, toxic organic compounds, bacteria, viruses, and protozoa. The organic compounds are natural and synthetic. They are expressed in terms of BOD, chemical oxygen demand, and total suspended solids. If the septic tank is designed and maintained properly, it should remove much of the organic substances from the wastewater before they enter the soil. The organics found in wastewater are biologically active and form a crust that can restrict movement of effluent into the soil. This can lead to early hydraulic failure of the septic tank system.

Nitrogen found in the septic tank effluent may be in the form of ammonia, ammonium, organic nitrogen, nitrate, and nitrite. The types of nitrogen compounds and their total nitrogen concentration in the effluent depends on the treatment in the septic tank. Under anaerobic conditions the effluent contains predominantly soluble ammonium and nitrogen in the organic form. However, under aerobic conditions the nitrogen is primarily in the form of nitrate. Denitrification, adsorption, plant uptake, and volatilization may occur in the soil. Nitrification or the conversion of ammonium-nitrogen into the nitrate form occurs in roughly the first foot below the drainfield, if the water table is not present and unsaturated conditions exist.

Phosphorus comes from two main sources, detergents and human feces. Phosphate ions are removed from the effluent through adsorption, precipitation, plant uptake, and biological immobilization. Phosphorus transport through the soil can occur in course-textured soils low in organic matter close to the water table or bedrock. Detergents are removed from the effluent in the soil through biodegradation and adsorption to soil particles. Aerobic conditions and unsaturated flow promote this. Adsorption is influenced by soil texture, mineralogy, amount of organic matter, soil solution chemistry, soil acidity, and formation of a clogging mat.

Toxic organic compounds such as chlorinated hydrocarbons may be found in septic tank cleaners or additives. They have a higher density than water and therefore sink to the bottom of the groundwater supply. They are not readily biodegradable. Pesticides, solvents, and compounds with heavy metals also may contaminate soil or groundwater when allowed to enter septic systems. Bacteria may be trapped in the soil pore spaces between the soil particles. This means of filtration is an important way of removing enteric bacteria from the effluent. A clogging mat occurs at the interface between the trench and natural soil. The mat is partially formed because of bacterial activity and helps trap enteric bacteria.

Other factors involved include the number of bacteria in the effluent, soil texture, soil wetness, loading rate, temperature, type of bacteria, and soil aeration. Viruses act differently than bacteria in the soil environment. Virus removal or inactivation in the soil may be due to filtration, precipitation, adsorption, biological enzyme attack, and natural die-off. Virus removal is controlled more by adsorption to soil particles than through filtration. Cation exchange properties of soils, mineralogy, texture, pH, temperature, etc. influence virus attenuation. Protozoa such as *Cryptosporidium parvum* and *Giardia lamblia* can be found in septic tank and raw wastewater effluent. These parasites are resistant to environmental stresses and can survive treatment techniques.

Properties of Sorption and Catalysis

The chemical properties of soil are most closely associated with soil colloids and enzymes capable of catalyzing the decomposition of organic matter and influencing the fate of its components. The property of sorption, particularly cation exchange capacity, is associated with the colloidal fraction of soil and has been a topic of much research. Soils possess three other special chemical properties, however, which have received far less attention from soil scientists. The first, free-radical interaction, is in its alpha stage of research and is associated with soil humic colloids and lignin. The second, soil enzymatic capacity, has not been studied in-depth to date, in terms of measuring units of activity. Numerous soil biochemists, however, have contributed to the alpha stage of this endeavor by showing that enzymes of all classes can be found in soil. Catalysis by clay minerals, the third special property, also has been investigated moderately, and only recently.

Soil colloids are particles that have a diameter less than 0.002 mm, have an electrical charge, and are nonsettleable in water. The mineral or clay colloids are aluminosilicates or hydrous oxides of aluminum and iron, whereas organic colloids are humified organic residues, humus. The aluminosilicate clay colloids consist of layers of silicate (Si_xO_y) tetrahedrons bound typically to layers of aluminum oxide (Al_xO_y) or aluminum hydroxide ($Al_x(OH)_y$) octahedrons. The layers form unit arrangements of clay colloids, which, based on the ratio of silicate layers to aluminum oxide or hydroxide layers, are referred to as 1:1, 2:1, and 2:2 clays.

Units of 1:1 layers combine to form stacks that are stabilized by hydrogen (H) bonds between the hydroxyl (OH) groups of the octahedral layers and the oxygen (O) atoms of the tetrahedral layers. This bonding prevents the penetration of water and adsorbates. Clays with a 1:1 structure, therefore, have a low tendency to swell when liquid sludge is applied to soil. Contrarily, H bonds do not form between stacks of 2:1 units due to adjacent layers of exposed O atoms. As a result, 2:1 clays provide interlayer spaces for an influx of water and adsorbates, and are able to swell accordingly when wetted. The 2:2 clay colloids do not swell because the interlayer space is occupied by a layer of silicate tetrahedrons.

The general formulas for the crystalline hydrous oxides of iron and aluminum are $Fe_2O_3 \cdot H_2O$ and $Al_2O_3 \cdot H_2O$, respectively. These clay colloids have less impact than the aluminosilicate clays, but are a contributing factor, nonetheless, to soil chemistry, especially in terms of anion exchange.

A third group of soil colloids, humic substances, are organic polymers that are formed during the decomposition of organic matter. The main impact of humic colloids is their buffering capacity, attributable to bound carboxyl (COOH), hydroxyl (OH), and amino (NH_2) groups.

The architecture of soil mineral and humic colloids combined with their tendency to carry a net charge, generally negative, due to bond cleavage, dissociation, or substitution, contributes to the fate of organic and inorganic compounds in soil. The fate is influenced predominantly through sorptive properties, but also through catalysis by clay minerals and free-radical interaction by humic colloids. Additionally, colloids provide a structural attachment for extracellular enzymes, the major group of catalysts in soil.

Sorption, more specifically adsorption, is a surface attraction between a compound or ion (adsorbate) and a soil colloid (adsorbent). Ionic bonding of charged atoms to the surface of soil colloids is the major mode of sorption. Ion exchange refers to the displacement of an ion bound to a soil colloid by an incoming free ion from the soil solution. Cation exchange capacity (CEC) is the sum total of exchangeable cations that a soil can adsorb and is measured in units of milliequivalents of ion bonded per 100 g dry soil (meq/100 g). Mean values as high as 400 mg/100 g have been reported for humic colloids and as high as 200 meq/100 g for clay. In fact, on average, despite the presence of only about 2% organic C in most soils worldwide, nearly 50% of buffering attributable to the upper 20 cm of soil is due to its organic content. Additionally, for each increase of 1% of the initial value of humic substances in soil, the CEC increases by 2 meq/100 g.

Anions can also be exchanged, but this property is not as prevalent as cation exchange. Nonetheless, anions can replace hydroxyl (OH) groups on soil minerals such as kaolinite, a polyhydroxylated 1:1 clay colloid. Neutral OH groups can become protonated by accepting a hydrogen ion, gaining a positive net charge, and, in turn, bonding with anions, such as orthophosphate and nitrate.

Sorption of components from decomposing organic matter in the soil also occurs via H-bonding, van der Waal forces, and complexation. Hydrogen bonds form intermolecularly between dipolar molecules. Dipolar refers to the slightly negatively and positively charged region (dipole) within a molecule, which results from an unequal sharing of an electron pair between two covalently bonded intramolecular atoms. Hydrogen bonds can connect a positive region of one dipolar molecule to a negative region of an adjacent dipolar molecule. These bonds form between either the exposed O or H atoms on soil colloids and an oppositely charged H and O region exposed on water or polar organic compounds. Water vapor competes for sorption sites on clay and humic colloids and may suppress H-bonding of organic molecules. Water may also serve as a link, however, for H-bonding between soil colloids and polar organic substances.

Van der Waal forces are weak electrostatic intermolecular forces between oppositely charged regions of induced dipoles. These forces result from the distortion of covalent bonds, causing a fluctuation of charges, and inducing corresponding opposite charges (induced dipoles) intermolecularly. The forces participate to some extent in all sorption interactions and are especially evident for sorption of high molecular weight organic substances, such as pesticides, by humic colloids.

Complexation is the incorporation of an adsorbate within the molecular structure of an adsorbent to form a complex. The bonding of heavy metals from sludge to humic colloids exemplifies a complexation reaction.

The practice of land application of animal manures and biological sludges has been gaining increasingly wider acceptance during the past 15 years. Concern of potential contamination of soil and water from heavy metals, pathogenic microbes, and recalcitrant synthetic organics, still remains, however, largely due to the threat of translocation to underlying aquifers.

The sorption mechanisms described earlier cause retention of heavy metals; pathogenic microbes; synthetic organic substances; and ions (SO_4^{-2}, N_3^-, and PO_4^{-3}) which may originate from organic matter, such as sludges, when applied to soil. Data show that the beneficial properties of sorption and degradation actually increase in sludge-amended soils. The retention of these potential contaminants in soil minimizes their translocation into ground- and surface water and maximizes their contact with soil catalysts for transformation.

The decomposition of retained microbes, labile organic matter, and synthetic organic molecules is catalyzed by clay minerals, free-radical interactions, and enzymes. To date, the property of enzymatic catalysis, intracellularly and extracellularly, appears to be the major initiator of organic decomposition. Non-enzymatic hydrolysis of synthetic organic compounds such as pesticides, however, can occur through clay-catalyzed reactions. Clay colloids can also catalyze the synthesis of humic substances.

Free-radical interactions are associated with humic substances and with lignin, a precursor or major component of many humic substances. Free radicals are atoms that contain unpaired electrons; a characteristic that accounts for their highly reactive tendency to form bonds with other atoms.

Soil enzymes originate from organisms in soil and sludge. Intracellular enzymes are incorporated in organisms and extracellular enzymes may be released from living organisms or derived from dead organisms. The activities of soil enzymes are associated with various soil constituents, such as living microbes, cellular remains, and clay and humic colloids.

Soil organisms are the biotic component of the soil triad of physical, chemical, and biological properties that collectively interact and impart to soil the tremendous capacity for decomposition and turnover of sludge. Indeed, just as a modern biologist referred to soil as "earth's placenta of life," it also may be said that soil is "earth's depot of decay."

The biotic decomposition of organic matter in soil is mainly a process in which the heterogeneous community of macro-, meso-, and microorganisms — including those endemic to both soil and the organic matter — extract nutrients, either directly or indirectly, from the decaying matter for energy and growth. The process results in several immediate and ultimate impacts, including volume reduction and stabilization of organic matter sludge, with a concomitant release of soil nutrients — C, N, S, and P.

Soil macro- and mesoorganisms enhance the process of biotic decomposition of sludge through predation, comminution, and perturbation. Microorganisms, however, are the nucleus of biotransformation reactions. Microbes are the dominant decomposers of organic and mineral matter into metabolites during cellular respiration and catabolism, while simultaneously synthesizing new cellular biomass during anabolism.

Numerous species of a common segmented worm in soil belong to the family Enchytraeidae. As slender white animals, they range in length from 1 to 10 mm and are generally 1 to 2 mm in breadth. Commonly referred to as pot worms, they consume monera, fungi, plant fragments, and other particulate matter.

Earthworms, like pot worms, are members of the phylum Annelida. These animals play a greater role in turnover of soil organic matter than any other taxon, beside bacteria through their role as the dominant tillers of soil worldwide. Earthworms also create a perturbation effect that accelerates the decompositional activity of monera, protista, fungi, and even animalia (nematodes), all of which, in turn, serve as a source of nutrients in their diet.

A variety of animals from the phylum Arthropoda also contribute to the decomposition of sludge in soil. Included are members of the classes Insecta (springtails, ants, beetles, and termites), Myriapoda (millipedes and centipedes), Arachnida (spiders and mites), and Crustacea (isopods or pill bugs). These organisms and various species of the phylum *Mollusca* (slugs and snails) are usually abundant in the upper organic horizon or litter layer of soil. These animals contribute to the heterotrophic decomposition process by moving through organic matter, thus also inducing a perturbation effect. The major contribution arises from those species with jaws (ants, beetles, mites) or other triturating organs (such as the radula in snails). Comminution of organic matter through the use of these mandibular or pharyngeal organs reduces the size of organic matter. A large group of particles, with a greater surface area collectively, is produced as a result. The increased surface facilitates heterotrophic decomposition by increasing the availability of microhabitat to microscopic members participating in the decomposition process.

Vegetation and detritus also are considered a major component of the soil eco-system. High levels of microbial and enzymatic activity are present at the rhizosphere of plants and a source of energy can be derived from detritus. As a result, the rate of decomposition may be accelerated throughout the soil ecosystem.

Reduction of Sewage Effluents by Soil

Varying soils reduce sewage effluents in a variety of ways, depending on the variations present. Generally, in well-aerated, moderately coarse to medium textured soils, the soils beneath the crust have the capacity to remove a high percentage of total phosphates, BOD, COD, ammonia–nitrogen, and various microorganisms. Aerobic conditions aid in the degradation and purification of the oxygen-demanding organic loads and help promote biological oxidation of ammonia to form nitrates. The nitrates, unless denitrified, along with chlorides and sulfates, percolate to the groundwater.

Apparently, in a well-designed properly functioning subsurface disposal system, the major environmental problem is percolation of salt ions into the groundwater. In high-density areas, this could be a problem. The minimum amount of soil needed to adequately renovate the effluent is between 2½ and 4 ft of unsaturated soil with sufficient permeability. This means that the soil below the effluent loading surface should be able to absorb at a rate greater than 60 min/in. However, this type of permeability creates other problems within the system that may cause effluent to surface or back up into the house. Therefore, the practical system has to allow for a faster rate of effluent absorption.

Sand, which is excellent for removal of biological systems when the biological systems are introduced at a reasonable rate, becomes clogged and is a poor purifier when the system is overloaded. Further, the overload may cause a channeling effect, which allows the effluent to move rapidly through the sand and thereby cause contamination of subsurface water tables.

Effects of Evapotranspiration and Climate on Effluent Disposal

Evapotranspiration depends on the amount of water at the evaporation surface; the weather, because heat is needed to change liquid into vapor; and the amount of rainfall or high humidity that may occur. Evapotranspiration is also influenced by the vegetation on the disposal field, which includes trees, bushes, and grasses.

Because grass, flowers, and deciduous shrubs become dormant during the winter, they do not transpire. In the absence of snow, they shield the surface from radiant heat and decrease transpiration. Trees and shrubs increase the amount of evapotranspiration. However, wintertime is not a good period for this effect to take place. Systems in extremely dry hot areas work quite effectively when evapotranspiration takes place; however, the same systems and types of soils in cold, rainy, or very moist conditions cannot work effectively.

Soil-Clogging Effects of Septic Tank Effluents

Soils in subsurface trenches and beds may fail because of clogging due to physical, biological, and chemical agents. Physical clogging occurs when the soil is compacted or smeared during construction, or when medium- and fine-textured soils cut off and smear when moist. Further, silt can be moved forward by liquid to clog the pores in the soil. Sandy soils are compacted by vibration. If a system is installed and not closed in time, rainfall may cause small particles to seal the bottom of the excavation. It has been shown that when a 2-mm crust is formed at a soil surface, the percolation rate may drop from 12 min/in. to as much as 2000 min/in. This type of clogging can be prevented by simply taking care during the construction of the soil absorption system and by scarring the walls and bottom of the trenches after digging them out.

Soils may also be clogged physically by a surge of sludge that comes from the septic tank into the sewage system. These suspended solids usually move along, because of the excessive amounts of water used and because of the inadequate holding time in the septic tank. Clogging may also occur when the septic tank is not serviced periodically. It is generally recommended for maximum performance that a septic tank is serviced once every 1 to 2 years. Garbage grinders should never be attached to septic tanks. However, because they are utilized when present, the tanks should be cleaned a minimum of once a year.

Biological clogging is the filling of the pores in the soil by bacterial growth and products of anaerobic decomposition. The clogging can be reversed if aerobic conditions can be brought about. It is recommended that a 20- to 30-day rest period is given to the field to allow standing effluent to be absorbed and to allow reaeration to occur. This is accomplished by having two beds with alternate dosing and resting periods.

Chemical clogging occurs from the use of synthetic detergents and a vast variety of chemicals that are generated by normal home use. Laboratory studies have shown that there is a reduction in the percolation rates in holes when a variety of detergents are utilized. It has also been noted that the soil structure may deteriorate as a result of chemical changes in the soil.

Soil Cleaning Technologies

Soil cleaning technologies are needed because of a variety of contaminants introduced into the soil by people. Some of these new technologies are as follows:

In Situ **Vacuum Extraction** — This process, developed by TerraVac, Inc., uses vacuuming for removing and treating volatile organic compounds (VOC) from the vadose or unsaturated zone of soils. These compounds can often be removed from the vadose zone before they contaminate groundwater. The process is used to treat soils containing VOCs and has successfully removed over 40 types of chemicals from soils.

Soil Recycling — This process, developed by the Toronto Harbour Commission, uses three technologies. The process removes inorganic and organic contaminants in soil to produce a reusable fill material. The first technology involves a soil washing process that reduces the volume of material to be treated by concentrating contaminants in a fine slurry mixture. The second technology removes heavy metals from the slurry through a process of metal dissolution. By using acidification and selective chelation, the metal dissolution process recovers all metals in their pure form. The third technology, chemical hydrolysis accompanied by bioslurry reactors, destroys organic contaminants concentrated in the slurry. The three integrated technologies are capable of cleaning contaminated soil for reuse on industrial sites. This process is used to remove heavy metals from soils.

Geolock and Bio-Drain Treatment Platform — This process, developed by International Environmental Technology, is a bioremediation system that treats soils *in situ.* The Geolock tank consists of high-density polyethylene (HDPE), sometimes in conjunction with a slurry wall. An underlying permeable water-bearing zone helps create inward gradient water flow conditions. The tank defines the treatment area, minimizes intrusion of off-site clean water, minimizes release of bacterial cultures to the aquifer, and maintains contaminant concentration levels that facilitate treatment. The inward gradient conditions also facilitate reverse leaching or soil washing. The application system, called Bio-Drain, is installed within the treatment area. Bio-Drain aerates the soil column and any standing water. This creates an aerobic environment in the air pores of the soil. Other gas mixtures can also be introduced to the soil column, such as air and methane mixtures used to biodegrade chlorinated organics. It is used to treat all types and concentrations of biodegradable contaminants.

Mobile Environmental Treatment System — This process, developed by Ensoltech, Inc., is a multipurpose transportable treatment unit that can continuously treat soils contaminated with organics, heavy metals, and mixed wastes. It is fitted with an UV radiation source and a vacuum suction system that uses granular activated carbon (GAC). It uses a patented polysilicate powder, LANDTREAT, as well as PETROXY, a stabilized hydrogen peroxide product of Ensotech, Inc. (Ensotech). The large surface area provided by LANDTREAT, which is added to soil before it enters the reaction chamber, adsorbs PETROXY and assists in destroying hydrocarbon contaminants. Carbon dioxide and water are treatment by-products. This system is used to treat soil contaminated with hydrocarbons, chlorinated organics, heavy metals, and mixed waste.

Fluid Extraction–Biological Degradation Process — This process, developed by the Institute of Gas Technology, is a three-step method that remediates organic contaminants in soil. It combines three distinct technologies: (1) fluid extraction, which removes the organics from contaminated solids; (2) separation, which transfers the pollutants from the extract to a biologically compatible solvent or activated carbon carrier; and (3) biological degradation, which degrades the pollutants to innocuous end products. This process is used to remove organic compounds from contaminated solids.

Septic Tanks

The septic tank is a watertight, concrete or metal (metal is not preferred) container that receives sewage from bathrooms, kitchens, and laundries (Figure 6.20). The major function of the septic tank is to protect the absorption quality of the subsoil. This is done through the removal of solids, biological action, and sludge and scum storage. The septic tank should have a minimum 1000 gal capacity, which will prevent solids that would ordinarily clog the soil from going out with the tank effluent unless an unusually large surge of water passes through the tank. However, the minimum requirements vary by state. In Indiana the following applies:

1. Two or less bedrooms — 750 gal
2. Three bedrooms — 1000 gal
3. Four bedrooms — 1250 gal
4. Five bedrooms — 1500 gal
5. Over five bedrooms — an extra 150 gal per bedroom

Within the septic tank, decomposition occurs by bacterial processes that operate in an anaerobic condition. The effluent is septic because of the anaerobic action. Sludge and scum are stored in the tank, because the sludge precipitates to the bottom of the tank and the light solids and greases rise to the top. The effluent leaving the tank is partially clarified by these processes. A large amount of the sludge and scum is liquefied through decomposition or digestion. The gas liberated from the sludge carries part of the solids to the surface, where they again accumulate with the scum. The action continues for a period of time. However, when the effluent leaves the septic tank, it has not been purified. At best, it has only been cleared of particles that may affect the leaching system in the soil. The bacterial and chemical action in the soil filter the effluent and also aid in further microbiological digestion of the sewage effluent.

One of the major reasons for subsoil sewage system failure is the accumulation of scum and particulate matter in the pores in the soil that block the effluent from entering the soil. This may be due to septic tanks that are too small, too great a usage of water in the house, or improper maintenance of the septic tank.

Aerobic Tank Systems

The aerobic tank system operates on the basis of aerobic bacteria causing degradation of solids in the presence of air (Figure 6.21). The effluent contains a lower solids content and a lower BOD. However, it is recommended that the absorption field system be the same size as the usual septic tank absorption field system, because the effluent is still hazardous. If the aerobic system is not maintained properly, if a loss of power occurs, or if an unusual surge of water passes through the tank, the system becomes disrupted and in effect becomes a septic tank.

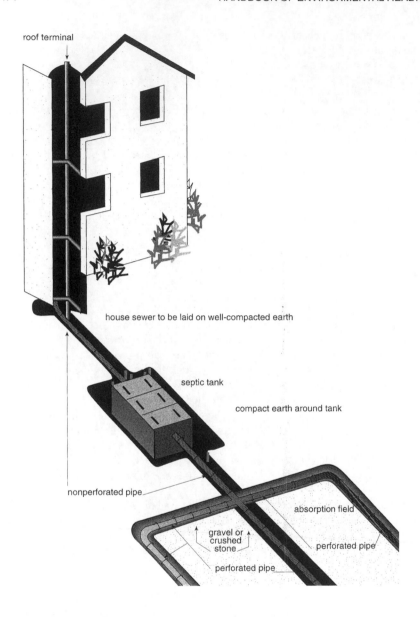

Figure 6.20 Septic tank sewage disposal system. (From *Manual of Septic-Tank Practice,* U.S. DHEW, U.S. Government Printing Office, Washington, D.C., 1969, p. 25.)

The National Sanitation Foundation (NSF) in Ann Arbor, MI has developed a set of standards that are accepted by most health departments for the type of construction and materials to be utilized in the aerobic system. Basically, the aerobic system is a three-compartment tank with a motor that drives a propeller in the middle

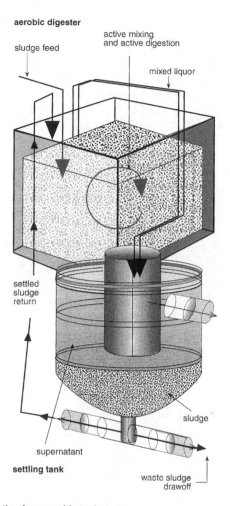

aerobic digester

active mixing
and active digestion

sludge feed

mixed liquor

settled
sludge
return

sludge

supernatant

settling tank

waste sludge
drawoff

Figure 6.21 Schematic of an aerobic tank system.

tank. In the first tank, primary treatment and settling of sludge occurs. The effluent overflows into the second tank, which is aerated and kept in motion by the motor-run unit. The aerated material then goes through an aerobic decomposition, which reduces the odors and converts much of the material to liquids and gases. In the third tank, settling occurs and finally the effluent flows out of the effluent line to the field absorption system.

The value of the aerated system is that, if properly maintained, it is able to function well in underground soil absorption systems for long periods of time. The soil, if it is already capable of removing effluent, should continue to do an effective job of effluent disposal, because the soil pores should not become clogged. However, if the aerated system is put in an area where the soil is so tight that the effluent either breaks out of the soil or comes to the surface of the soil, then it is of as little value as the standard septic tank. The aerated system should not be used as an excuse for putting a house on land where the soil cannot absorb the effluent. The problem, which has been discussed in the past, has not changed. Water is taken into a system and water must be disposed of, and if the soil cannot handle the load of water, it should not be utilized for subsurface waste disposal.

Innovative Individual Wastewater Management Systems

The aerobic treatment unit biologically treats wastewater, oxidizes organic matter, and reduces total nitrogen. The system includes flow equalization, pretreatment, aeration, clarification, filtration, and disinfection in a precast concrete tank. The three-chamber tank is installed below ground with access to grade for routine maintenance. The wastewater management system keeps BOD and solids within the treatment process, and is designed to increase the efficiencies of septic tanks, tile fields, and other treatment systems by permanently reducing loading to downstream treatment processes (Figures 6.22 and 6.23).

Distribution Box

The distribution box is a watertight tank-type box into which the effluent flows from the septic tank. The effluent is then distributed through a series of pipes that exit from the distribution box into the disposal field. It is absolutely essential that the distribution box is placed on a solid foundation and that the box is level; otherwise the effluent flows into some of the disposal lines and not into others. All the outlets should be at the same level and at least 4 in. from the bottom of the box.

Headers

A "bridal header," or double header, is used when the subsurface disposal system consists of more than six tile lines. This is a line between the septic tank and main header running parallel to the main header. The bridal header accepts the effluent from the septic tank and discharges it into the main header at two locations equally spaced on the main header. Its purpose is to more evenly distribute the effluent into the subsurface disposal field. It can be used in place of a distribution box if the tile field system is not too large.

A header receives the effluent from the septic tank and should be perpendicular to the effluent line from the septic tank. The header then distributes the effluent into the subsurface disposal tile lines in a uniform manner.

Dosing Tank

When the amount of sewage exceeds the amount that can be disposed of reasonably in 500 linear feet of tile, a dosing tank should be used with the septic tank. The dosing tank is set up in such a manner that it has an automatic siphon that discharges every 3 to 4 hours to a different disposal field. In other words, two or

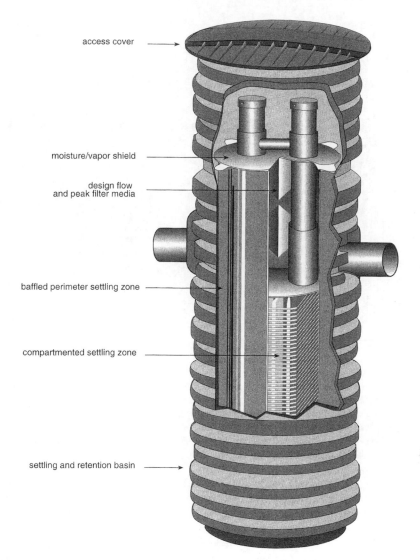

access cover

moisture/vapor shield

design flow
and peak filter media

baffled perimeter settling zone

compartmented settling zone

settling and retention basin

Figure 6.22 Bio-Kinetic® Wastewater Management System. (With permission of Norweco Equipment Company, 2001.)

more disposal fields are needed with the dosing tank. The tank discharges the effluent into field A; and then while field A is resting and absorbing the material, the tank disposes of the effluent into field B and so forth. It is necessary to ensure that the tank works properly so that all the effluent cannot flow into only one of the subsurface disposal fields.

Dosing tanks with suitable pumps may be used in an on-site wastewater disposal system as follows:

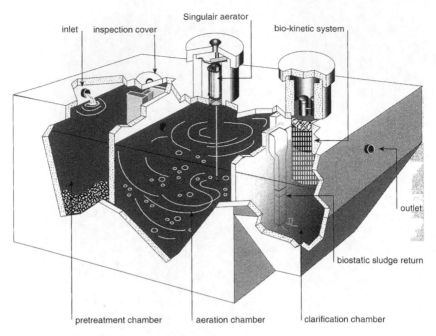

Figure 6.23 Singulair® Bio-Kinetic® Wastewater Treatment System. (With permission of Norweco Equipment Company, 2001.)

1. To overcome pipe friction losses and static head between a septic tank and
   absorption field
2. To provide more even distribution of septic tank effluent in large absorption fields
   where trenches have a total length greater than 1000 ft (lineal)
3. To provide the designed pressure and discharge rate in gallons per minute through
   a small diameter distribution system

Dosing can be carried out by means of submersible effluent pumps or siphons.
If the system is on a sloping site, siphons may be used. Large absorption systems
may be divided into smaller dose systems, which allow a more even distribution of
wastewater in the trenches or the bed. Also small dosing tanks and lower capacity
pumps can be utilized. If a system requires 1500 ft² of trench bottom, a dosing
system might be considered. The usable capacity of the dosing tank is related to the
number of doses required per day and the average daily flow of wastewater into the
system. The tank and the piping connecting into the tank must be sealed to prevent
groundwater intrusion and resultant overloading of the absorption field.

Soil Absorption System

The soil absorption system utilized depends on the soil conditions that have been
previously discussed and the amount of room available on the lot where the structure
is placed. Absorption systems include absorption trenches, which are subsoil and

may either be regular or serial; seepage beds, which are subsoil; deep absorption trench and seepage beds; seepage pits; subsurface sand filters; elevated sand mounds; sand-lined trenches and beds; and oversized seepage areas.

Privies

The sanitary privy is utilized in those areas where it is impractical to have on-site sewage disposal. Generally, the sanitary privy should not be utilized. However, if it is properly constructed and maintained and kept flyproof, then it is an acceptable means of sewage disposal.

Type of Septage

Septage is mostly organic in nature and highly variable in content with significant levels of grease, grit, hair, and debris. The liquids and solids in the septic tank and cesspool have an offensive odor, and contain viruses, bacteria, parasites, and various other potentially toxic substances. In the septic tank, concentrated BOD, solids, nutrients, variable toxics, inorganics, pathogens, oil, and grease are found. Aerobic tanks contain variable BOD, inorganics, odor, pathogens, and concentrated solids.

The septage is removed periodically from the tanks and discarded, may be spray irrigated after pretreatment, and can be used on steep or rough land. Particular concern must be taken to avoid the spread of offensive odors by varying winds. Septage may also be incorporated in the subsurface soil, thereby reducing odors and potential health risks. Septage has been buried as well as placed in sanitary landfills. Septage may be treated further through use of stabilization lagoons, chlorine oxidation, aerobic digestion, anaerobic digestion, biological and chemical treatment, conditioning and stabilization, composting, dewatering, and incineration.

SEWAGE DISPOSAL PROBLEMS

Public Health Implications

Fresh raw sewage carries large numbers of microorganisms found in the human intestinal tract. These microorganisms include many of the pathogens that can be transmitted from humans through sewage to water or food and back to humans again (Figure 6.24). Further, the effluent contains the vast variety of chemicals previously mentioned in this chapter. Some of the microorganisms found include bacteria causing cholera, typhoid fever, and paratyphoid fever. Viruses, including the enteroviruses that have been known to cause outbreaks of gastroenteritis and those that cause infectious hepatitis, have been found in water from a source that was originally contaminated by sewage. Polio viruses have also been identified in sewage. Protozoa that are responsible for amoebic dysentery have been transmitted through sewage to humans. Many of the worm infections have also been transmitted through sewage to humans. The coliform test has been utilized as a means of determining whether a body of water is contaminated with sewage. The coliform test is effective,

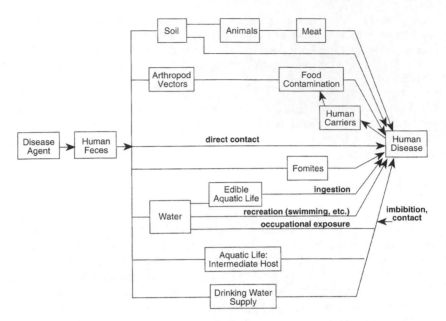

Figure 6.24   Human fecal waste and human disease pathways (postulated). (From Hanks, T.G., *Solid Waste/Disease Relationships: A Literature Survey,* Bureau of Disease Prevention and Environmental Control, Cincinnati, OH, 1967, p. 78.)

because these organisms are found in the human intestinal tract. However, the organisms may also be found in the intestinal tract of animals or in the soil. Unfortunately, the coliform test is only an indicator-type test; it does not reveal whether a specific organism that causes disease is present.

Although we know a wide range of resistant viruses exists (including polio viruses, five of the coxsackie viruses, six of the echo viruses, and three of the adenoviruses), and although numerous outbreaks of waterborne disease still occur, inadequate studies have been made to determine whether current techniques can destroy pathogenic microorganisms, including viruses.

It has been shown that pathogens can survive in soil systems. When *Salmonella typhimurium* organisms were deposited on soil, they were recovered from 33% of the soil samples 12 weeks later. In one study *Salmonella typhi* ranged in its survival in soil from a minimum of 24 hours to a maximum of 2 years in frozen moist soil. Very little information is available on the survival time of viruses on soils. However, one study indicated that enteroviruses survived for 170 days in a sandy soil. Pathogens can move through soil systems by percolation. The size of the pathogen as well as the porosity of the soil, is a determining factor in the actual movement. *Ascaris* have traveled between 30 and 60 cm in sand. Coliforms may travel up to 65 meters in coarse gravel, which would indicate that any of the bacteria of that size may do the same thing. Viruses have not been studied in adequate detail to determine rates of movement. Survival times of pathogenic organisms in the soil are as follows: *Ascaris,* which are part of the parasitic worm group, may survive 2.5 to 7 years in soil; *Entamoeba* cysts, which are protozoa, may survive 40 to 170 days in soil; *Salmonella* may survive 1 to 85 days; coliform may survive 100 to 150 days; and mycobacterium can survive 5 to 15 months. For enteroviruses, no information is available on total survival time; it may be many years.

Chemical contaminants are found in the soil for very long periods of time, ranging anywhere from 82 to 425 days for chemical contaminants coming from septic tank effluents. The organic compounds may be chemically oxidized to more stable compounds during the time period mentioned. The inorganic compounds may continue to filter down to water-bearing strata over a continuous period of time without having an adequate quantity removed. Two of the most serious substances moving downward into the water table are nitrogen and its compounds, and phosphorus and its compounds. Nitrate–N levels in some aquifers where on-site sewage systems are used have been found to equal or exceed the standard of 10 mg/l.

Groundwater

Pathogenic bacteria and viruses can move into groundwater used for drinking-water supplies. These organisms may come from sewage in on-site systems, although the downward movement of the bacteria and viruses may be slowed by absorption, the chemical binding between surfaces, and the fact that the organism may be weak, especially in sandy soils. The organisms can then be resuspended in water moving through the soil pores and continue downward to the groundwater supply. Adsorbed viruses may be released during heavy rainfall when the water percolates rapidly through the soil. Bacterial and viral survival is prolonged when saturated and anaerobic conditions occur beneath the absorption trenches. Higher water tables are the main cause of this problem. In some soils, the effluent moves very rapidly because of the permeability of the soil. Cold temperatures enhance the persistence of bacteria and viruses.

Planning

Many of today's sewage problems are due to yesterday's improper planning or lack of planning. Many of the older communities had developed lots that were too small for individual sewage and were in areas where public sewage was unavailable. Communities had also been developed in floodplains, on poor soils, or in other undesirable areas. The unplanned growth of the last 35 to 40 years has resulted in sizable problems in disposing of sewage and in acquiring necessary water. In many cases, there was no anticipation of the direction in which future building would expand and the rate in which houses, factories, shopping centers, summer homes, and industries would be built. Further, the amount of wastewater generated has increased tremendously over this period of time. Today, homes are routinely equipped with automatic washers, garbage grinders, dishwashers, and lawn sprinkling systems. Many communities are therefore faced with the perplexing problem of what to do with their wastewater. In some cases, the situation is so acute that environmental health problems and environmental pollution problems have occurred. Unfortunately, the solution to these problems is not a simple one.

Where sewers were available, they were not adequate in size to handle the rate of flow of liquid. Within the sewer, as the oxygen was depleted, the anaerobic bacteria took over and hydrogen sulfide was produced. Hydrogen sulfide reacted with oxygen from the air to form sulfuric acid, which corroded the sewer pipes.

Concerns include the infiltration and the inflow of extraneous waste into water systems from excess outside water. This additional load on the sewage system has resulted in flooded basements, excessive power costs for pumping, overloading lift station facilities, overloading treatment plant facilities, excessive cost of sewage treatment, obnoxious odors and health hazards, and street settling due to excessive pumping up of water in manholes to relieve pressure within the pipes. These problems of infiltration and inflow have been relieved at times by running sewage directly to the receiving stream or to ditches. However, this unfortunately adds raw sewage to the streams, and is unlawful.

Secondary Treatment Problems

Some of the problems associated with activated sludge are that DO must be present at all times in the treated sewage and in the final settling tanks; optimum rate of returning activated sludge varies from installation to installation and with different load factors; optimum suspended solids in the aeration tanks vary considerably; sludge held too long in the final settling tank becomes septic, loses its activity, and becomes a problem; periodic or sudden organic overloads may cause great difficulty to the system; industrial waste, including toxic chemicals, can ruin the system; and sludge bulking may occur. This happens when the character of the sludge changes so that its ability to settle decreases and a sharp increase in the sludge volume index occurs. The sludge is then carried away in the effluent and therefore adds organic load to the receiving waters. The sludge bulking may be due to excessive storm flow as rainwater, which results in shortened aeration periods, short-circuiting of the aeration tanks, either too little or too much solids in the aeration tanks, interruption of the recirculating sludge, and large amounts of fungi in the sludge. Frothing is a formation of a thick layer of froth over the surface of the aeration tank. The causes of this are not definitely known, but seem to be somewhat tied into the use of synthetic detergents.

The disadvantages of the activated sludge process are that activated sludge is sensitive and the process is more easily disrupted than the trickling filter process; operation of the plant requires careful supervision; toxic waste and slugs of strong sewage can upset the process; and contact periods in the aerators are much greater than the less than 1 min required on trickling filters.

Sludge

Depending on the sources (i.e., municipal vs. industrial), sludge and treated wastewater effluents may contain toxic metallic and nonmetallic elements, toxic organic substances, and pathogenic microorganisms. Some of the elements that have been found in wastewater and sewage sludge include boron, cadmium, cobalt, copper, chromium, mercury, nickel, lead, selenium, and zinc. Zinc, copper, nickel, and boron may be toxic to plants. Sheep, which are very sensitive to copper, may be affected by excess copper. Mercury, copper, and zinc may enter the food cycle. Lead and other metals may also enter the food cycle and be hazardous to animals and people, depending on the concentration.

Problems have occurred since chlorinated hydrocarbons, including pesticides, polychlorinated biphenyls (PCBs), and so forth, have been found in our water distribution system. These may be coming from our sewage treatment plants as well as from industry.

Wastewater sludge and treated effluents have shown a large amount of microbial organisms. It is not certain how long pathogenic bacteria, viruses, and parasites may survive, especially when wastewater is sprayed over fields, thereby creating aerosols that may be inhaled.

As the population has moved into urban and suburban areas that are part of the metropolitan urban area, the use of individual sewage systems has increased to an extraordinary degree. Approximately one fourth of the new homes under construction in suburban or rural areas are utilizing individual sewage disposal systems. These systems malfunction frequently and discharge directly or indirectly to aquifers and surface bodies of water.

The pollutants causing oxygen demand are generally measured in BOD or COD. These pollutants deplete the DO in the water. Unless the stream can assimilate additional oxygen, the loss of DO creates difficulties in the stream. Nitrogen and phosphorus are the primary nutrients for algae. When these chemicals are present, algae growth increases substantially.

Eutrophication is a process by which streams and lakes may be converted into swamps and eventually meadows because of enrichment from nutrients, such as nitrogen and phosphorus, that have been introduced into streams. This is a naturally occurring process. However, humans have accelerated the process by allowing agricultural fertilizers and wastewater discharges, which contain these chemicals, to eventually find their way into the water system. Suspended solids from wastewater effluents eventually settle in quiescent reservoirs, streams, or pools. The solids form sludge deposits on the bottoms of these water areas. If the decomposition of this sludge takes place where oxygen is present, it creates a tremendous oxygen demand. Where oxygen is absent, it causes bad odors. In any case, turbidity is increased and aquatic life may be affected.

Refractory substances are those substances that contribute to the total organic carbon content of water but do not generally represent BOD or COD. These include the synthetic detergents and many of the organic pesticides. If the substances are not carcinogenic, they still may cause undesirable side effects, such as foaming or frothing, in the waterways.

Package Treatment Plant

The great difficulties attached to the package treatment plant are that generally financial support is inadequate and, most of all, untrained operators inadequately supervise the plant. In various studies it has been shown that package treatment plants were out of service for times varying from 1 hour to an entire year. Sedimentation units were out of service in 11% of the plants in one study. Trickling filters and digesters were reported out of service from 1 hour to 9 months. There have been reports of bypass of untreated sewage for periods ranging from 6 hours to 300 days. Unfortunately, unless the system is maintained by a trained operator who understands the operation of the package treatment plant, there may be considerable difficulty in attempting to maintain it in the proper manner to prevent the mistreatment and misuse of this means of sewage disposal.

On-Site Sewage Disposal Problems

The major shortcoming of the elevated sand mound is that it was developed in a specific type of soil in Wisconsin and it was not tested under normal operating conditions that would be found in a home in other areas of the country. As a result, the sand mound may become ineffective and sewage may seep from the sides of the system if the soil beneath the sand mound is unable to accept the amount of effluent that is supposed to percolate through it. If the people in the home utilize large quantities of water at any given time, the water may cause a short-circuiting effect in the sand and quickly move down to the ground surface level. Although the sand mound system may be acceptable in some areas, under certain conditions considerable caution should be utilized in permitting the installation of this system. Once again, the major problem of on-site sewage disposal is the removal of water into the soil. The soil can only accept a certain amount of effluent. Once the land is at full capacity, the water has only one other channel.

Chlorination

In many areas where subsurface sand filters, filter trenches, elevated sand mounds, or sand-lined trenches and beds are utilized, it is required that any effluent escaping from the system is chlorinated. The usual requirement is maintenance of a residual of 0.5 to 1.0 ppm after the effluent has been thoroughly mixed with the chlorine and has been allowed to stay in contact with the chlorine for at least 15 min during peak flow. Unfortunately, in many of these situations, the effluent is not properly chlorinated, because the owners of the properties have very little understanding of the use of chlorine on effluent and further do not take the time or utilize the money necessary to purchase chlorine and to put it into the system. Maintenance of the system is the problem.

Existing System Malfunctions

One of the worst problems faced by the environmental health practitioner is handling overflowing sewage effluent complaints. Unfortunately, the cause of the complaint is rooted in the original design of the system and land used, or in the maintenance of the system. Sewage frequently backs up in the system and flows into the house, breaks through the surface, passes directly into the water table, is connected to storm water drains, or finds its way into surface bodies of water. The first sign of malfunction may occur when the toilets or other water-using appliances do not function. This may be caused by high water table, slowly permeable soil, organic or fine material in mound fill, improper disposal of soluents or grease in septic tank, failure of dosing pump, line cloggage, full septic tank, broken pipes, inoperative distribution boxes, boxes that are not level, or improperly operating drainage system in the field.

If the distribution box is full, the problem is usually in the field and a major replacement job is necessary. Wet or damp spots may be noted in the drain field area. When the subsurface absorption field does not function properly, the tank may not have been pumped out regularly and the material may have been allowed to move outward into the distribution field itself. A common additional problem is caused by allowing installation of garbage disposals where septic tank systems are in use. Other problems are caused by allowing downspouts to propel water across the surface of absorption fields or foundation drains to discharge above the surface across the distribution field or be tied directly into the distribution field. Occasionally, the water table rises unexpectedly and floods the field. A special problem is related to overflowing sewage in seasonally used septic systems in beach communities. The houses typically are on sites where high water tables are located and in close proximity to the water. The soil tends to be saturated.

Research indicates that one of the reasons for system failure is a gradual formation of impermeable clogged or crusted soil that occurs in the seepage bed area. The flow of water through this area becomes severely restricted or even eliminated. It has been suggested that physical, chemical, and biological phenomena occur in progression or jointly to produce the zones of gradually decreasing permeability. The physical structure of the soil may be clogged as water moves small particles of soil into pores, thereby causing a physical closing of the pores. Further, most normal septic tank effluents contain about 140 to 150 ppm of suspended solids. These solids again become soil cloggers. The effluent from a septic tank may contain as many as 12,000 organisms per milliliter. These bacterial cells act in the same manner as small particles and again may cause soil blockage. Because the septic tank effluent is not equally distributed throughout the system, only certain portions of the soil may be clogged.

However, as the system becomes more and more clogged, the amount of actual effluent that can seep away decreases sharply. With the reduction of seepage, the aerobic conditions in the bed turn into anaerobic conditions. Anaerobic conditions result in production of sulfides, which are toxic to most microorganisms that ordinarily would be degrading the organic material present in the bed. Some of the free sulfur is then converted to insoluble sulfur and causes further blockage of the soil pores. The bulk of the sulfur combines with iron, manganese, nickel, copper, magnesium, and zinc, and produces an insoluble inorganic sulfide that coats the gravel with a black slime. The insoluble sulfides contribute further to the blockage of the soil capillaries. Anaerobic bacteria produce polysaccharide slimes or gums that may form an impermeable layer in the bacterial zone, thereby causing clogging.

Impact on Other Problems

Sewage disposal on an on-site basis can readily contaminate the water supplies from underground sources. It can also be a source of pollution to surface bodies of water. Sewage that has surfaced from the on-site systems becomes a source of mosquito breeding, an unsightly and odorous mess, a reservoir of infection for children, and causes a lowering of property values.

In the area of public sewage, land that might be used for other purposes must be preempted for the sewage treatment plant. The discharges from the plant may contaminate the receiving bodies. Other problems include possible diversion of existing wastewater flow in the event of overload, effect of flooding and plant malfunctions in emergency situations, sludge treatment and disposal, and impact on the land and the air. If the sludge is dried properly it may be disposed of in landfills. However, all the chemicals, especially the heavy metals present, can eventually seep through the landfill into underground water supplies. The sewage treatment plants may create noise, aerosols, and odors. In the area of sewage, considerable impact occurs on all other areas.

Economics

The economic problem associated with sewage disposal is one of the predominant concerns in trying to obtain proper removal of sewage. The costs of building and operating sewage treatment plants are extremely high. In areas where considerable numbers of homes are located, the individual homeowner may have to pay substantial dollars, depending on the front footage of the house, for a hookup to the sewer lines. The building of the sewage treatment plant and the sewer lines is financed through various bond issues. These bonds must be repaid with a substantial amount of interest. Although there are areas of the country where sewage flows into streams or onto the ground, municipal sewage treatment may not be available for years. The ability of the community to build a sewage treatment plant, the topography of the land, the availability of a final receiving stream, and the distance between properties controls the cost of installation and either makes a municipal treatment project feasible or infeasible.

POTENTIAL FOR INTERVENTION

The potential for intervention varies with the individual on-site sewage system and with the public sewage system. In the case of the on-site sewage system, the type of soil and the size of the property, as well as the amount of liquid put into the ground are the controlling factors. Where the soil is poor, potential is poor. Where the soil is excellent and adequate room exists for additional on-site sewage disposal systems, the potential may be good. In any case, on-site sewage disposal is typically a stopgap technique to be utilized until such time as public sewage treatment is available. The potential for municipal treatment varies with the type of plant and the type of process within the plant. In a tertiary treatment plant where special techniques are used to remove all contaminants from the effluent, the potential for intervention is excellent. The potential for intervention decreases with the lesser types of treatment plants and the quantity of water that must be handled.

Intervention consists of isolation, substitution, shielding, treatment, and prevention. In isolation on the individual system, the sewage is kept away from all potential sources of potable water and from properties and houses. In substitution, the municipal treatment plant should be utilized whenever possible in place of the on-site sewage disposal system. Shielding does not apply. In treatment, many chemicals are utilized within the public sewage system treatment plants to remove contaminants from the water. Chlorine is used as a means of final disinfection. In prevention, adequate planning is needed prior to the time that houses are built to determine whether the given area should have on-site sewage systems and also to determine whether small package treatment plants are feasible. In the case of public sewage systems, prevention consists of proper maintenance of sewer lines in streets, proper operation of the package treatment plants, and proper planning for expansion of the systems when additional burdens are placed on them.

RESOURCES

Scientific and technical resources consist of the NSF, American Public Health Association, National Environmental Health Association, Research Triangle in North Carolina, Soil Science Society of America, and National Onsite Wastewater Recycling Association. Civic resources consist of the Environmental Defense Fund, International Association for Pollution Control, National Water Resources Association, Water Pollution Control Federation, and Water Quality Association. Governmental resources consist of all state health departments, most county health departments, and local health departments; Department of Agriculture; Maritime Administration; National Bureau of Standards; Department of Defense; Energy Research and Development Administration; EPA; Department of Health and Human Services; National Science Foundation; and U.S. Coast Guard. Additional resources include the National Environmental Training Association, 8687 Via De Ventura, Suite 214, Scottsdale, AZ.

STANDARDS, PRACTICES, AND TECHNIQUES

Municipal Treatment

A variety of pipes are utilized in sewers, including concrete, cement, asbestos, clay, and plastic-lined asbestos pipes. In the past, iron pipes have been utilized. However, they tend to rust, develop holes, and allow water to infiltrate or sewage to flow out of the sewer piping into the surrounding ground. The concrete pipe is made of a mixture of portland cement, which is limestone, shale or both. The material is ground into a powder, heated to high temperatures, and then dehydrated. When moistened, the material becomes extremely hard and forms a strong bond with the sand and aggregate that is added to the cement.

Cement asbestos pipe came into production after World War II. This is a strong, lightweight pipe, which has enormous usage. It is made of about 80% portland cement and 20% asbestos fiber. The fiber gives the pipe additional strength. Plastic-lined asbestos cement piping couplings were approved in 1964 by the NSF in Ann Arbor, MI. The specifications for the materials are available from them. One of the chief values of the plastic-lined asbestos cement pipe is its resistance to a solution of 5% sulfuric acid by weight and other acids. This has protected the pipes from acid reaction.

The clay pipe was first utilized 5000 years ago by the Babylonians. The clay pipe is chemically inert, which makes it excellent for use when chemicals are present. The clay is heated to 2000°F, and then is vitrified, which means fused, into a uniform glassy state. The disadvantage of the vitrified clay pipe is that it is much more easily broken than concrete pipes.

Many cast-iron pipes still exist in sewer systems in older cities. These pipes are subject to attack by chemicals and oxygen, and therefore must be replaced periodically or they constitute a hazard. When planning a sewage system, the engineers must attempt to plan for at least a 20-year period. This gives a fairly adequate guarantee that the sewage system can work during this period of time.

The general design of the lift station includes screening and pump protection, wet wells, pumps with automatic float control, on and off switches, recording gauges and meters, and proper ventilation. Building structure, pumps, and meters must be checked periodically. Screens must be cleaned regularly and grit accumulations removed from the wet wells. Sewage pumps must be so constructed as to avoid small passageways and to provide easy access to interior parts. The lift station should not be in a floodplain. It should be in a low area where the pumps can be effectively used to continue to move the sewage along.

Primary Treatment

The racks and bar screens consist of bars that are usually spaced ¾ to 6 in. apart. These provide clear openings of 1 to 2 in. The screens are usually set at a 45 to 60° angle. The screens are an inexpensive means of removing coarse material that can cause damage within the plant itself. Beyond the racks and bar screens are fine screens with openings of ⅛ in. or less, which are utilized to remove additional solid materials. The comminuting devices, which are grinders, cutters, or shredders, may be electrically driven revolving drums with cutting blades. The sewage passes through the cutting screen and all solids are reduced to sizes from ³/₁₀ to ³/₈ in. These are very effective in removing the kinds of materials that cause problems in the sewage treatment process. It is very important to remove the materials from the bar racks on a daily basis to prevent clogging and to clean the sludge, grease, and other deposits from the screen chamber very frequently. The screened materials should be removed to tightly covered cans for disposal. The screenings should be promptly buried in at least 6 in. of dirt or incinerated, because they rapidly decompose, produce obnoxious odors, and enhance the breeding of rats and insects.

The grit chamber is a type of settling chamber. The function of the grit chamber is to settle out the inert heavy material and to allow the sewage solids to flow through to the primary settling tank. Settling is controlled by the velocity of the material based on specific gravity, size of particle, temperature of the water, and length of time within the grit chamber. The detention time in the grit chamber is controlled by the velocity of horizontal flow, which is regulated by the discharge weir. The discharge weir is simply a horizontal object over which the liquid flows at a given speed. Grit chambers should be designed in such a way that the heavy particles can settle out before the liquid reaches the end of the chamber. The organic sewage solids should continue on into the primary settling basin. The velocity of flow in the grit chamber is usually 1 ft/sec for a 45- to 60-ft tank.

Proper cleaning and operation of the grit chamber is essential. If the chamber is hand cleaned, the materials should be removed before the efficiency of the operation is hampered. If the unit is properly operated, the grit could then be cleaned and used for fill. As mentioned, because most grit is not cleaned, it should be placed in a landfill and covered. Hand-cleaned sewage operations should have at least two hand-cleaning units. Mechanically cleaned areas should have at least one mechanically cleaned unit with a bypass to allow for adequate time for cleaning the grit chambers. The grit chamber can be mechanically cleaned by scrapers or buckets. Generally, where combined sewage is treated, 1.0 to 4.0 ft^3 of grit may be expected from each million gallons of sewage.

The preaeration of sewage is utilized before primary treatment to obtain a greater removal of suspended solids in the sedimentation tanks, to help remove grease and oil carried in the sewage, to freshen up the septic sewage prior to further treatment, and to reduce the BOD. Preaeration can be accomplished by introducing air into the sewage for 20 to 30 min during the flow. This can be done by forcing compressed air into the sewage at a rate of 0.10 ft^3/gal of sewage over a 30-min aeration period. Mechanical agitation is necessary to ensure that the compressed air comes in contact with all surfaces. The detention period, preferably should be at least 45 min, even though 30 min of compressed air is adequate for aeration. This assists in the collection, or flocculation, of the lighter suspended solids into heavier masses that settle more readily in the sedimentation tanks. This also helps to separate the grease and oil from the sewage and sewage solids and to carry them to the surface, where they can be removed by skimming.

In the sedimentation tank, baffles are normally placed 2 to 3 ft from the wall, run the width of the tank, and are submerged at least 18 to 24 in. The outlet devices are overflow weirs that permit a given amount of effluent to flow forward. In circular tanks, the flow is introduced at the center and decreases in forward velocity as the particles approach the weir, which is placed around the circumference. The sedimentation tank provides a 2-hour detention period for the settleable solids to settle. Removal of suspended solids is influenced by the nature of the solids and by the quantity present. It is reasonable to expect an average removal of 50 to 60% of suspended solids. The sedimentation tanks may be cleaned mechanically by the use of scraper arms. Scraper arms serve to move the floating solids of grease and oil to a scum collector at the outlet end of the tank. An intermediate process may be added to the primary treatment process by allowing the effluent that leaves the settling tank to flow into a second settling tank, where chemicals are added to produce floc.

Aluminum sulfate, ferrous sulfate, ferric sulfate, and ferric chloride, with or without lime, are used as flocculating agents. The purpose of the chemicals is to further reduce the amount of solids present in the effluent. If the chemicals are mixed gently for 15 to 30 min with the effluent, flocculation occurs. Chemical treatment can reduce suspended solids by 90% and BOD by 70%. In some plants, chlorination is utilized on the effluent to further purify it prior to release to the receiving stream.

Sludge Digestion

Although sludge digestion may proceed at almost any temperature, rapid changes in temperature are detrimental. At 75°F, the time of digestion is 35 days; at 85°F, it is 26 days. Digestion between the temperatures of 85 and 95°F is considered the best range. Heating of the raw sludge to a uniform temperature aids in sludge–digestion process. Sludge digestion must be watched by the operator to ensure that the process is moving along properly. When the digested material is removed, it is placed on sand-drying beds. The drying bed consists of a layer of graded gravel, 12 in. deep, over a layer of clean sand, 6 to 9 in. deep. Open-joint tile underdrains are placed beneath the gravel layer with at least 6 in. of gravel cover and spaced not more than 20 ft apart. Open beds are most commonly used. The drying of the sludge occurs as a result of drainage and evaporation. A major portion of the drainage occurs in the first 12 to 18 hours. Further drying is due to evaporation. Alum may be used with 1 lb of alum per 100 to 300 gal of sludge. When the sludge has dried adequately, which takes about 3 weeks to produce an 8-in. layer or 4 weeks to produce a 10-in. layer, the material is scraped from the bed and discarded.

The best time to remove the sludge to drying beds depends on the subsequent treatment and the moisture content of the sludge on the beds. If the sludge contains 60 to 70% moisture, it can be removed by shovels or forks. However, when it has been reduced to 40% moisture, it weighs only about half as much. If it gets down to 10% moisture, it becomes very dusty and difficult to handle. In most situations, the sludge is removed by hand, although in some large plants mechanical units are available. Material may be placed in wheelbarrows and then dumped into pickup or dump trucks. The final disposition of the sludge is to a landfill or stockpiled over the winter and then used for certain agricultural purposes. Sludge lagoons are utilized to avoid the problems involved in sludge drying. These lagoons are excavated areas where the digested sludge is allowed to drain and dry over a period of months, a year, or more. The depths of the lagoon vary from 2 to 6 ft. The material is removed from a lagoon after a period of time and is put in land disposal.

Secondary Treatment

Trickling Filters

The type of media used in trickling filters is controlled by local availability. Field stone, gravel, broken stone, glass furnace slag, and anthracite coal are used for this purpose. Whatever material is used, it should be hard, clean, and free of dust and insoluble materials. The material should be roughly cubical in shape to prevent compacting and should have a size that passes a 4½-in.^{2} screen but be retained by a 2-in.^{2} screen. The depth of the filter should be a minimum of 5 ft and a maximum of 7 ft. The bed may be either rectangular or circular. Rectangular beds are most common where the sewage is distributed by fixed spray nozzles. Circular beds are used most often with rotating distributors.

Sludge is conditioned by the use of chemicals. Sulfuric acid, alum, ferrous sulfate, and ferric chloride with or without lime have been utilized. The cost of the chemical is the determining factor. The chemical lowers the pH of the sludge to a point where small particles coagulate into larger ones and the water is driven upward readily.

Elutriation, or purification of sludge, is accomplished by washing with either water or effluent from the sewage treatment plant. This process is used to eliminate excessive amounts of amino and ammonia compounds, thereby reducing the coagulant demand. Elutriation is used as a pretreatment before chemical coagulation. The sludge is mixed with the plant effluent for often less than 20 sec, using mechanical or diffused air agitation. The mixture settles and the supernatant fluid returns to the sewage treatment process. This treatment is carried out in tanks similar to sedimentation tanks. The advantages of elutriation includes a reduction by 65 to 80% of the amount of conditioning chemicals required, a lower ash content in the filter cake, and the use of little or no lime in the conditioning chemical. Heat drying is used on sludge that is turned into fertilizer. The moisture content must be reduced to below 10%. If the sludge can be incinerated, it must be dried to a point where it ignites and burns. The units utilized for heat drying include the rotary kiln dryer, the flash dryer, the spray dryer, and the multihearth furnace.

Incineration is used to dispose of sludge. In all incinerators, the gases of combustion must be kept at 1400°F until they are completely burned. This prevents odor nuisance. Raw primary sludge contains about 70% volatile solids and produces 7800 Btu/lb of dry solids. Once combustion has started, sludge burns without supplementary fuel. Usually excess heat is produced. Digested sludge may or may not require supplementary fuel. Incineration has the advantages of economy, freedom from odor, independence from weather conditions, and considerable reduction in the volume and weight of the end product.

In the wet oxidation process, the sewage sludge is reduced in size by comminutors to pass through ¼-in. openings into a preheated (180°F) storage mixing tank. The sludge is then fed by high-pressure pumps into a pipeline in which air is introduced at 1200 to 1800 lb/in.². The air sludge mixture is then raised to 400°F until it enters a vertical reactor for upward flow. The oxygen from the air combines with organic matter in the sludge, oxidizing it into an ash. The temperature in a reactor raises to about 500°F. The liquid portion of the effluent carries the ash removed by sedimentation into tanks or lagoons. The hot gases and steam may be utilized to drive a turbine.

Activated Carbon

In some typical performance operations in the use of activated carbon, raw influent with a COD of 305 mg/l was reduced to 46 mg/l in the clarifier effluent and 13 mg/l in the carbon effluent. After the primary treatment, the influent, which has a COD of 235 mg/l, was reduced to 177 mg/l of COD in the clarifier effluent, and the final carbon effluent dropped to 44 mg/l. In a secondary treatment process with an influent of 40 mg/l of COD, the clarifier effluent showed 25 mg/l of COD and the carbon effluent showed 10 mg/l. Carbon treatment of wastewater can be carried out by the use of downflow or upflow procedures. In the downflow carbon beds, adsorption and filtration of wastewater occurs. The hydraulic loading varies from 2 to 19 gal/min/ft^2. At the lower levels suspended solids are filtered out, and at the higher flow levels the suspended solids may penetrate into the bed. Obviously, the bed must be backwashed periodically. Biological growth takes place on the surface of the carbon granules and helps to destroy any biologically degradable materials, but also clogs the bed. To prevent the development of sulfides, the beds should be kept aerobic by maintaining dissolved oxygen in the feed and backwash water.

In the upflow process, if the effluent is applied at less than 2 gal/min/ft^2, the bed of carbon remains substantially packed at the bottom of the column. At a level of 4 to 7 gal/min/ft^2, the bed becomes partially expanded. At higher rates, the carbon is lifted and packed against the top of the column, which means that it can operate in the same way as a packed bed. If this occurs, suspended solids are collected on the bottom of the bed. Unless adequate pretreatment for removal of solids has occurred, this causes considerable problems in rejuvenating the bed.

The main advantage of a gravity flow system is that it may be operated either downflow or upflow without the use of large pumps and therefore reduces cost. The pressurized system operates at a higher rate of flow and therefore passes more effluent in shorter periods of time. Carbon beds may be in a single or multiple arrangement. In the multiple arrangement, the material passes through one bed and then goes through subsequent beds, removing as many of the impurities as possible.

A substantial amount of materials may be converted into activated adsorbent, usually made up of raw porous materials of some carbon origin. They include wood, charcoal, coal, peat, lignite, sawdust, coconut shells, and petroleum residues. Most activated carbon is carried out by a steam activation process at temperatures of 750 to 900°C, in an oxygen-depleted atmosphere. Once activated, the product is crushed or aggregated, graded, washed with acid, and then washed with water before packaging. The porous structure of the raw material is converted into a highly developed porous structure with submicroscopic networks of irregular pores within a graphite crystalline matrix. The great degree of adsorptive character of the activated carbon is a result of the porous structure providing an enormous surface for adsorption. Under ideal conditions, the adsorption capacity of the activated carbon is exhausted when dissolved organics cease to be removed from the liquid phase. However, the ideal situation does not occur in municipal wastewater.

A wide variety of molecular weights of dissolved organics exists, so that the gross organic removal cannot be predicted on theoretical grounds. Wastewater is also a biological fluid, and some of the organics may be degraded through the biological activity that occurs on the carbon granule. This secondary phase of biological activity effectively increases the adsorption capacity over what it would be expected to be if biological activity did not exist. Carbon usually loses some of its adsorptive capacity during regeneration, because the combustion of the adsorbed organics is never really complete. Also some ash accumulates at the carbon pores.

The most important factor affecting adsorption in columns is the contact time. The amount of time depends on the affinity of the carbon for a particular solute, degree of ionization, competition among solutes for adsorption sites, molecular sizes present in the solution, surface area of the carbon, and distribution of the pore sizes in the carbon. The actual time utilized should be more than what is considered proper for adequate removal. Contact time usually varies from 15 to 40 min on an empty bed of activated carbon. Hydraulic loading has no effect on the adsorption in a range of 2 to 10 gal/min/ft^2. However, a buildup of head loss occurs where the hydraulic loading is greatest and a system must be recharged more frequently. A suggested 4 gal/min/ft^2 is the usual flow of liquid.

In the design of a carbon adsorption system in a secondary treatment plant, 500 to 1800 lb of carbon should be allowed for each million gal of effluent treated. In a tertiary treatment plant, 250 to 350 lb of carbon should be allowed for each million gallon of effluent treated. In both cases, the hydraulic loading is 2 to 10 gal/min/ft^2, the contact time on the empty bed is 15 to 40 min, and the backwash rate is 15 to 20 min/ft^2.

Phosphorus Removal

The phosphorus removal technique used during primary treatment is as follows. Add the iron or aluminum salts to the raw sewage and thoroughly mix. If a base is needed, add it at least 10 sec later. Allow for a 5-min reaction time. Add anionic organic polymer and flash mix for 20 to 60 sec. Use mechanical or air flocculation for 1 to 5 min. Allow the flocculated wastewater to gently enter the sedimentation units.

Utilizing iron salts in the phosphorus removal process in primary treatment results in the removal of 60 to 91% of the phosphorus. Adding iron in secondary plants results in 75 to 93% removal of the phosphorus. Only half the iron dose is required when adding a base and a polymer. An increased removal of suspended solids and BOD occurs by adding iron salts.

When a trickling filter or activated sludge unit is used in the secondary treatment process following chemical precipitation in the primary unit, an additional 10 to 15% of the phosphorus can be removed. The use of various chemicals for removing phosphorus increases the amount of primary sludge and decreases the amount of secondary sludge. The sludge conditioning must be altered with the addition of iron and additional salts.

When phosphorus precipitation is accomplished by adding lime to raw wastewater in conventional plants that use an activated sludge process, the phosphorus removal is usually about 80%. The pH ranges from 9.5 to 10. The removal of the phosphorus in primary treatment results in increased BOD and suspended solids removal, and therefore reduces the organic load in the secondary treatment facility. The use of iron and aluminum salts in the effluent during the trickling filter does not seem to cause any functional problems, although the filter may slough off more than usual.

Package Treatment Plant

In the package treatment plant, laboratory analysis should be made to determine 5-day BOD, suspended and volatile suspended solids, alkalinity, ammonia–nitrogen, total soluble phosphate, COD, nitrate–nitrogen, and DO, all in milligrams per liter; and temperature of the influent and effluent. These are discussed in some detail in Chapter 9.

Soil Surveys

Agencies Involved

Soil surveys are conducted on a nationwide basis in conjunction with each individual state. It is a joint effort of the Soil Conservation Service and Forest Service agencies of the U.S. Department of Agriculture and usually of the various state agricultural experimental stations. The responsibilities for defining and naming the different soils, conducting the surveys, and maintaining the mapping standards, as well as publication of the soil surveys, are a function of the Soil Conservation Service. The other agencies work with this group on a cooperative basis. Certain counties have planning commissions that furnish money to speed up the mapping process in a given area. This is why some counties have up-to-date soil surveys completed and others do not.

Nature of a Soil Survey

A soil survey is an inventory of the soil. It contains the complete set of soil maps for a county. The maps show the distribution of each kind of soil in the county and provide an interpretation of the use of each soil for different purposes. Detailed descriptions of each of the soils are also provided. Aerial photographs are often used in soil surveys, generally on a scale of 4 in. to a mile. On this map, the soil scientist draws the boundaries between different soils as well as roads, buildings, lakes, streams, bridges, quarries, townships, and section lines. The soil scientist examines the entire soil profile, measures the slope, and records the vegetation and anything on the map that might affect soil use. Usually 200 to 400 acres are covered per day.

On-site survey predictions are tested and confirmed by using an agar and spade to dig to a depth of 3 to 5 ft or more. If excavations are available, soil scientists utilize these to study the soil profile in detail. The mapping units are usually 2 to 3 acres of highly contrasting soils. Where there are small contrasting areas, such as wet spots and sandy areas, standard symbols are used to indicate them on the map. The accuracy of the soil map is determined largely by complexity and detail of the soil survey, scale of the map, and skill and experience of the mapper.

Soil Survey Map

Soil survey maps are identified by the photo index and the legal land description. After location of the area of interest on the map, that is, the parcel of land that is most important, lightly pencil the parameters of the area; check all the symbols on the map, and read a description of the symbols, using the soil map unit named; and determine the desirable or undesirable characteristics of the soil. The symbols on the map should include the slope group, soil type, and erosion class. A list of the standard map symbols are available from the Soil Conservation Service. The map symbols are important when determining the suitability of a small parcel of land for a specific use.

The advantages of using soil maps to determine whether subsoil absorption sewage systems should be installed are as follows. An interpretation can be made of whether a particular soil site is subject to intermittent high water tables. This avoids the problem of getting varying percolation rates during different seasons of the year. The suitability of the site can be based on the performance of the soil type at a variety of locations, instead of simply by utilizing one or more test holes. The soil maps help to determine site suitability for different kinds of soils regardless of the time of year. The cost of utilizing a small map is much less than a percolation test. Soil maps also provide additional information including bearing strength, erosion, flooding potential, landscaping, and plant adaptability.

The limitations of the soil map are based in good part on the size of the areas covered on the map. A soil map is usually in its smallest area a minimum of 2- to 3-acre lot sizes. Where less than this size lot is utilized, or if the lot crosses different soil mapping areas, the soils may be completely different from those that are stipulated on a soil map. Further limitations include the fact that the soil scientist only observes to a maximum depth of 5 ft. Problems occurring beneath this level could affect the soil percolation. Although soils are supposedly continuous, the soil scientist permits as much as a 15% inclusion of different soils within the small unit. These differences may occur in slope, depth, drainage, or other characteristics.

Although the soil map is an up-to-date useful tool for the environmental health practitioner, it should be utilized more for planning than for actual installation of the soil absorption system. In the planning phase for a subdivision or even for several homes or one home, the soil map can certainly rule out those areas with a serious problem for subsurface drainage. However, once a general approval is made of the area, then specific types of percolation tests or soil analysis, where the subsurface system is to be located, are necessary to anticipate how well the actual sewage disposal field can operate.

Soil Interpretations

Before a soil survey can be meaningful, it must be interpreted by a trained soil scientist in terms that the user can understand. In the preparation and use of soil interpretations, it is necessary to understand that the soil interpretation about soil behavior in use under certain conditions does not include recommendations or decisions on how to use the soil. Soil interpretations only indicate alternate uses and limitations of the soil, and pertain only to the whole soil and not to its individual properties and qualities such as texture and permeability. The interaction of all of the soil properties determines its behavior and limitations. Interpretations are based on observations of soil behavior under different uses in controlled experiments. Intensive research cannot be conducted on each individual soil.

Therefore, comparisons are made between the soils found in the fields and experimentation that has gone on in the laboratory. When the actual field permeability tests are not available for all the soils, the permeability is predicted based on the structure, texture, and other features of the soil. Soil interpretations can be based on engineering interpretations or limitations determined for specific uses. By using the interpretation sheets attached to the map, the environmental health practitioner can determine basic permeability, shrink–swell potential, amount of impermeable material, and whether the land is suitable for on-site sewage systems, sanitary landfills, and so on. Where large amounts of soil are excavated and replaced with permeable soil, it is important to determine whether the underlying strata can properly support the fill material. If it cannot, the fill material may cause the underlying strata to be compressed and in effect create a huge bathtub or storage basin.

Rating Soils for Sewage Systems

The subsoil absorption system must be set up in such a way that the effluent from a septic tank is distributed uniformly into the natural soil. The ability of the soil to absorb effluent must be determined and grouped as slight, moderate, or severe. The suitability of the soil for this filter field is determined by local experience and records of performance of existing filter fields; depth to the consolidated rock or other impervious layers; flooding; seasonal and annual groundwater level; types of horizons including arrangement and thickness; depths of the soil; drainage characteristics; color and texture; amount of coarse fragments present; structure, porosity, consistency, and permeability; parent material; mode of accumulation; minerals, free carbonates, and organic matter present; stoniness and rockiness; slope, erosion, and topographic positions; potential for flooding; salt concentrations present; usual temperature and temperature variations by seasons; and rainfall characteristics in the area.

As can be seen, the determination of soils for subsurface tile systems is a very complicated matter. When determining the performance of existing systems, it is important to evaluate the kinds, numbers of failures, and also length of time between installation and failure of existing systems. A system is considered a failure if considerable plant growth and seepage or odor occurs in the vicinity of the absorption system.

Soils that have moderate to very rapid permeability are considered to have a slight soil limitation. Soils allowing percolation at a rate of 0.60 to 1.0 in./hr are considered to have moderate soil limitations. Soils that percolate at a rate of less than 0.60 in./hr are considered to have severe soil limitations. Generally, in soil percolation tests, if the water drops at a rate faster than 1 in. in 45 min, the system, considering all other problems, should be satisfactory. If the water drops at a rate of between 45 and 60 min/in., the system may have problems and care must be taken in loading the system. If the water drops at a rate slower than 60 min/in., severe limitations exist and on-site sewage disposal should not be utilized. Loading rates in subsurface systems are shown in Table 6.2 and in aboveground systems in Table 6.3.

Where the water table, at its highest level, is 4 ft below the bottom of the trench, the soils have a slight limitation. Soils with water tables less than 4 ft below the trench have severe limitations. Impervious layers of material should be a minimum of 4 ft below the bottom of the trench. Creviced or fractured rock that has inadequate

Table 6.2 Loading Rates for Subsurface Systems (in gpd/ft²)

Soil Texture Class	Single Grain	Granular Platy[a]	Strong: Angular, Subangular, Blocky, Prismatic	Moderate: Angular, Subangular, Blocky, Prismatic	Weak: Angular, Subangular, Blocky, Prismatic	Fragipan: Very Coarse Prismatic	Structureless, Massive, Very Friable	Structureless, Massive, Compact, Firm, Very Firm
Gravel, coarse sand	>1.2	N/A	N/A	N/A	N/A	N/A	N/A	N/A
Loamy coarse sand, medium sand	1.20	1.20	N/A	N/A	1.20	N/A	N/A	N/A
Fine sand, loamy sand, loamy fine sand	0.75	0.60	N/A	0.75	0.75	N/A	0.75	N/A
Very fine sand, loamy very fine sand	0.50	0.50	N/A	0.75	0.60	N/A	0.60	N/A
Sandy loam, coarse sandy loam	N/A	0.75	N/A	0.60	0.60	0.00	0.60	N/A
Fine sandy loam, very fine sandy loam	N/A	0.75	N/A	0.60	0.60	0.00	0.60	N/A
Sandy clay loam	N/A	0.75	0.75	0.50	0.50	0.00	0.50	0.00
Loam	N/A	0.75	0.75	0.50	0.30	0.00	0.30	0.00
Silt loam	N/A	0.60	0.60	0.50	0.30	0.00	0.30	0.00
Silty clay loam, clay loam, sandy clay	N/A	0.60	0.60	0.30	0.25	0.00	0.25	0.00
Silty clay, clay	N/A	0.60	0.50	0.30	0.25	N/A	0.25	0.00
Muck	N/A	N/A	N/A	N/A	N/A	N/A	0.00	N/A
Marl bedrock	N/A	N/A	N/A	N/A	N/A	N/A	N/A	0.00

Note: N/A = not applicable.

[a] Except where platy structure has been caused by soil compaction. Platy structure caused by compaction has a loading rate of 0.00 gpd/ft².

From *Residential Sewage Disposal Systems*, Indiana State Department of Health, Indianapolis, IN, November 1990, p. 21.

Table 6.3 Loading Rates for Aboveground Systems (in gpd/ft²)

Soil Texture Class	Single Grain	Granular Platy[a]	Soil Structure Classes					
			Strong: Angular, Subangular, Blocky, Prismatic	Moderate: Angular, Subangular, Blocky, Prismatic	Weak: Angular, Subangular, Blocky, Prismatic	Fragipan: Very Coarse Prismatic	Structureless, Massive, Friable, Very Friable	Structureless, Massive, Compact, Firm, Very Firm
Gravel, coarse sand	>1.2	N/A	N/A	N/A	N/A	N/A	N/A	N/A
Loamy coarse sand, medium sand	1.20	1.20	N/A	N/A	1.20	N/A	N/A	N/A
Fine sand, loamy sand, loamy fine sand	0.60	0.60	N/A	0.60	0.60	N/A	0.60	N/A
Very fine sand, loamy very fine sand	0.50	0.50	N/A	0.50	0.50	N/A	0.50	N/A
Sandy loam, coarse sandy loam	N/A	0.60	N/A	0.60	0.60	0.00	0.60	N/A
Fine sandy loam, very fine sandy loam	N/A	0.60	N/A	0.60	0.60	0.00	0.60	N/A
Sandy clay loam	N/A	0.50	0.50	0.50	0.50	0.00	0.50	0.00
Loam	N/A	0.50	0.50	0.50	0.50	0.00	0.50	0.00
Silt loam	N/A	0.50	0.50	0.50	0.50	0.00	0.50	0.00
Silty clay loam, clay loam, sandy clay	N/A	0.25	0.25	0.25	0.25	0.00	0.25	0.00
Silty clay, clay	N/A	0.25	0.25	0.25	0.25	N/A	0.25	0.00
Muck	N/A	N/A	N/A	N/A	N/A	N/A	0.00	N/A
Marl bedrock	N/A	N/A	N/A	N/A	N/A	N/A	N/A	0.00

Note: N/A = not applicable.

[a] Except where platy structure has been caused by soil compaction. Platy structure caused by compaction has a loading rate of 0.00 gpd/ft².

From *Residential Sewage Disposal Systems*, Indiana State Department of Health, Indianapolis, IN, November 1990, p. 24.

soil cover allows sewage to travel for long distances. This material should be a minimum of 4 ft below the trench. Soils that are subject to flooding have severe limitations and should not be utilized. Where slopes of 8% are present, the field can operate properly. If the soil slopes are much higher, severe limitations occur. In certain cases where unsatisfactory soil exists in the upper 4 ft of the soil horizon, but at least 4 ft of satisfactory soil exists beneath it, the upper soil may be removed and fill material containing sand may be utilized. This can result in a properly operating system. However, simply removing 4 ft of soil from tight clay and replacing it with sand over tight clay creates a basin effect and does not permit the system to operate effectively.

On-Site Sewage Systems

The minimum distances permitted between the various parts of the sewage disposal system and other areas are as follows. The septic tank should be a minimum of 10 ft from the house and 50 ft from a well or geothermal well, with the well always uphill from the septic tank. The septic tank should be 200 ft from a public well. The distribution system or sewage disposal system should be a minimum of 100 ft from a well, 10 ft from the property line, and 50 ft from any body of water. The distribution system should be 400 ft from a public well.

Septic tanks must be watertight and constructed of materials not subject to excessive corrosion or decay. Concrete, coated metal, vitrified clay, heavyweight concrete blocks, or hard-burned bricks may be used. The best type of tank is either the cured precast or the cast-in-place reinforced concrete tank. The tank must be placed on a solid foundation and any joints that exist should be well sealed. The interior of the tank should have either a solid surface or a surface with two ¼-in. thick coats of portland cement sand plaster, in the case of block. The precast tanks should have a minimum wall thickness of 3 in. and should be reinforced. Where precast slabs are used, they should be watertight, have at least a 3-in. thickness, and be adequately reinforced. All concrete surfaces should be coated to minimize corrosion. It is recommended that each tank has two compartments, as a minimum, with the inlet tank two thirds of the total volume.

The stabilizing and leveling of the septic tank on firm ground is very important. The area around the tank should be backfilled with thin layers and then tamped in place. Settlement of the backfill may be accomplished using water, provided the wetting action occurs from the bottom upward and the tank is first filled with water to prevent floating. Access ports must be provided in each compartment of the tank for inspection and cleaning purposes. A minimum of two manhole covers of 20 in. in diameter are recommended. Where the top of the tank is more than 18 in. beneath the surface of the finished ground, inspection holes should be provided that are within 8 in. of the finished grade.

All septic tanks must have the inlet and outlet either baffled or provided with a T. The inlet T diverts the incoming sewage downward and slows it down and prevents it from shooting across the tank. The inlet and outlet T or baffle should penetrate at least 4 in. above and 12 in. below the liquid level. The outlet unit should retain the scum in the tank and also allow the maximum amount of sludge to accumulate. The bottom of the outlet device should remove an effluent as clear as possible. For any given capacity, shallow tanks function as well as deep ones. The shape of the tank is also unimportant. However, the smallest dimension recommended is 3 ft in width, and 5 ft in length. Within the tank adequate room is needed for scum accumulation. On the average, about 30% of the total scum accumulates above the liquid line. In addition, 1 in. is needed for free passage of gas back to the inlet and out the house vent pipe. Single-compartment tanks are as effective as two-compartment tanks (Figures 6.25 and 6.26).

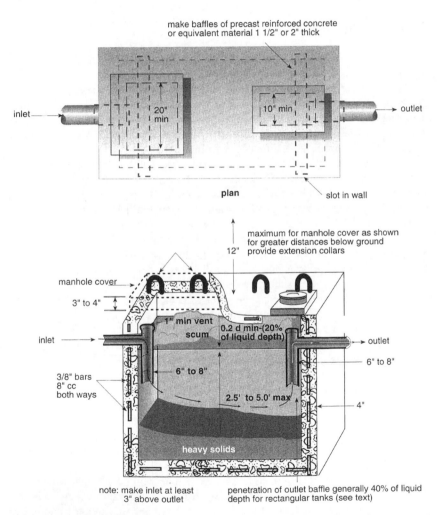

Figure 6.25   Household septic tank. (From *Manual of Septic-Tank Practice*, U.S. DHEW, U.S. Government Printing Office, Washington, D.C., 1969, p. 33.)

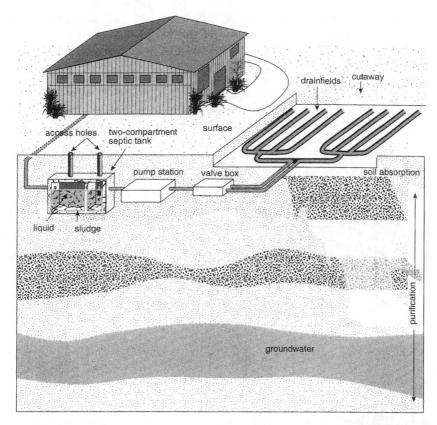

Figure 6.26   A typical large-capacity septic system installation consisting of a septic tank, an effluent distribution system (pump station and valve box), and a soil absorption system (not to scale). (From USEPA, 1997b.)

Maintenance

Septic tanks should be cleaned periodically to remove the sludge and scum before it can foul up the soil absorption system. For safest operation, it is recommended that the septic tank should be cleaned once a year. However, depending on conditions, the tank may be inspected at least once a year and then cleaned every 2 to 3 years as necessary. The scum and sludge are measured by inserting a stick into the septic tank and determining the depth of the sludge and scum. Only specialized haulers of sewage should be utilized to remove the sludge. These individuals should be licensed and inspected by the health department.

Absorption Trenches

Soil absorption trenches are a minimum of 24 in. deep and are usually 36 in. below the surface of the ground. The width of the trench varies from 18 to 36 in.

The length should be a maximum of 100 ft. Between each absorption trench there should be a minimum of 6 ft of undisturbed ground. Depth of the earth cover of lines is 12 in., but 18 in. is preferred. Within the trench, after the ground has been dug out and scratched, a minimum of 6 in. of crushed stone or gravel should be placed beneath the pipes to carry the effluent and a minimum of 2 in. of crushed stone or gravel should be placed on top of the pipes. Above the stone, untreated building paper or a 2-in. layer of hay, straw, or similar pervious material should be placed to prevent the stone from becoming clogged by the backfilling of the soil (Figures 6.27 and 6.28).

Impervious coverings should not be used because they interfere with evapotranspiration from the top of the drain field. The pipe may be 12-in. lengths of 4-in.

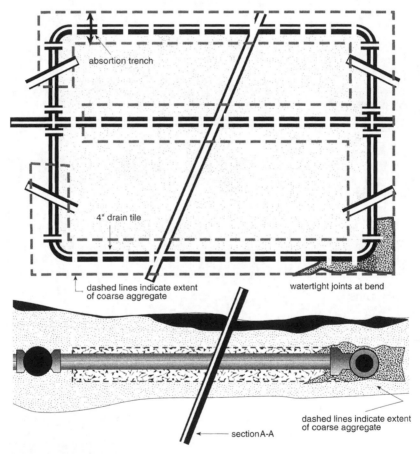

Figure 6.27   Typical layout of absorption trench. (From *Manual of Septic-Tank Practice,* U.S. DHEW, U.S. Government Printing Office, Washington, D.C., 1969, p. 11.)

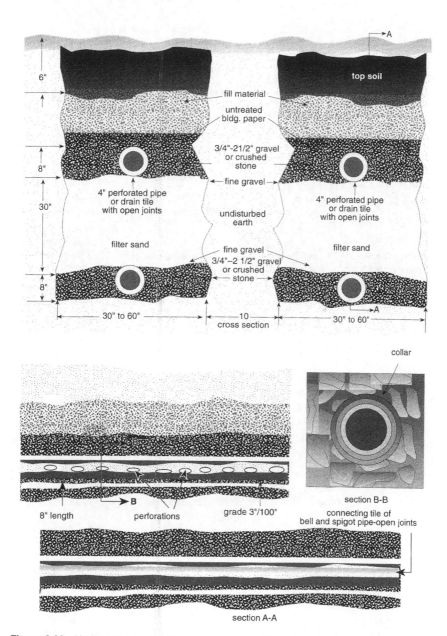

Figure 6.28 Underdrained sand-filter trench. (From *Manual of Septic-Tank Practice,* U.S. DHEW, U.S. Government Printing Office, Washington, D.C., 1969, p. 61.)

Table 6.4 Absorption Area Requirements for Private Residences

Permeability Rating	Square Feet Needed in Trench Bottom per Bedroom
2–6 in./hr	250 ft² per bedroom
1–2 in./hr	330 ft² per bedroom

From *Residential Sewage Disposal Systems,* Indiana State Department of Health, Indianapolis, IN, November 1990, p. 21.

Table 6.5 Absorption Area Requirements for Individual Residences (Provides for Garbage Grinder and Automatic Clothes Washing Machines)

Percolation Rate (Time Required for Water to Fall 1 in., in Minutes)	Required Absorption Area, in ft² per Bedroom,[a] Standard Trench,[b] Seepage Beds,[b] and Seepage Pits[c]	Percolation Rate (Time Required for Water to Fall 1 in., in Minutes)	Required Absorption Area, in ft² per Bedroom,[a] Standard Trench,[b] Seepage Beds,[b] and Seepage Pits[c]
1 or less	70	10	165
2	85	15	190
3	100	30[d]	250
4	115	45[d]	300
5	125	60[d,e]	330

Note: It is desirable to provide sufficient land area for entire new absorption system if needed in future.

[a] In every case sufficient land area should be provided for the number of bedrooms (minimum of 2) that can be reasonably anticipated, including the unfinished space available for conversion as additional bedrooms.

[b] Absorption area is figured as trench-bottom area and includes a statistical allowance for vertical side wall area.

[c] Absorption area for seepage pits is figured as effective sidewall area beneath the inlet.

[d] Unsuitable for seepage pits if over 30.

[e] Unsuitable for absorption system if over 60.

From *Manual of Septic-Tank Practice,* U.S. DHEW, U.S. Government Printing Office, Washington, D.C., 1969, p. 8.

agricultural drain pipe with ½- to 1-in. spaces between them, 2- to 3-ft lengths of vitrified clay sewer pipe with space between them, or perforated nonmetallic pipe, such as plastic. The bottom of the trenches should be a minimum of 5 ft above the highest level of the water table. (Table 6.4 and Table 6.5 can be used to determine the required number of square feet for an absorption trench.)

To avoid problems due to roots, it is best to increase the gravel or crushed stone beneath the tile to 12 in. In any case, the tile system should be as far away from big trees and their root systems as possible. If root problems still do occur, add 2 to 3 lb of copper sulfate crystals to the toilet bowl once a year and flush it down through the toilet. The solution comes into contact with the roots and destroys them. However, copper sulfate corrodes chrome, iron, and brass, and therefore, should not be allowed to come into contact with these metals. Cast iron is not appreciably affected by copper sulfate. Once the system has been completed, the ground should be hand tamped. Machine tamping may crush the trench or knock it out of alignment. To prevent rain or other surface waters from crossing the tile system, it may be necessary to run a diversion ditch around the property if the land slopes downward toward the subsurface disposal system.

Serial distribution is utilized in situations where lines are laid on sloping ground. In serial distribution, the effluent flows into the first set of trenches and fills to the top of the gravel before it overflows into another distribution box and into the second system. The value of this technique is that the first system absorbs as much effluent as it can before it overflows. It also has value in that one particular trench may not have good soil, whereas the ones beneath it may. Therefore, the first system may fill up and then flow down into the second and third and so forth. In normal distribution, the liquid is spread out uniformly to all lines at the same time and the lines that do fill with the effluent may cause the effluent to come to the top of the surface or out through the sides. It is important in serial distribution that the lines follow the contour of the earth. All other construction requirements are the same as in the regular subsurface absorption trench.

Seepage Beds

A seepage bed is an absorption system with trenches wider than 3 ft. This system is usually utilized where soil conditions are acceptable but space is inadequate for a standard tile field. The three main elements of the system include absorption surface, rock fill or packing material, and distribution system. The total absorption area must be preserved. Adequate quantities of stone must be utilized to help store the excess liquid, to distribute the effluent properly, and to prevent mechanical damage. The bed should have a minimum depth of 24 in. below the natural ground level and a minimum earth backfill cover of 12 in. The bed must have a minimum depth of 12 in. of rock with at least 6 in. below the pipe and at least 2 in. above the pipe. The bottom of the bed and distribution tile must be level. The lines used to distribute the effluent are placed no further apart than 6 ft and not further than 3 ft from the bed sidewall. When more than one bed is used, a minimum of 6 ft of undisturbed earth must be between the adjacent beds.

The advantage of the seepage bed is that it is more efficient than long narrow trenches when land is a problem. Also, modern earth-moving equipment can be used for excavating the area.

Seepage Pit

Seepage pits are used in some states where adequate seepage beds or trenches cannot be utilized because of the topsoil. The pit excavation must be at least 5 ft above the groundwater table and must not contaminate the underground water supply (Figure 6.29). The size of the seepage pit is based on percolation tests in each of the vertical strata penetrated. Where percolation rates are greater than 30 min/in. in a given stratum, that area should not be allowed to be calculated in the total area needed. The actual depth of pits may be calculated using Table 6.6. When more than

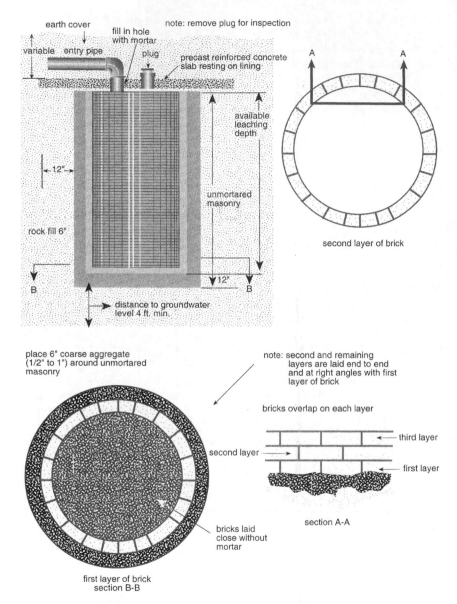

Figure 6.29 Seepage pit. (From *Manual of Septic-Tank Practice,* U.S. DHEW, U.S. Government Printing Office, Washington, D.C., 1969, p. 24.)

one seepage pit is used, they should be separated by a distance of at least three times the diameter of the largest pit. If a seepage pit is 20 ft deep, the minimum space between the pits should be 20 ft.

Table 6.6   Vertical Wall Areas of Circular Seepage Pits (in Square Feet)

Diameter of Seepage Pit (ft)	Effective Strata Depth below Flow Line (below Inlet)									
	1 ft	2 ft	3 ft	4 ft	5 ft	6 ft	7 ft	8 ft	9 ft	10 ft
3	9.4	19	28	38	47	57	66	75	85	94
4	12.6	25	38	50	63	75	88	101	113	126
5	15.7	31	47	63	79	94	110	126	141	157
6	18.8	38	57	75	94	113	132	151	170	188
7	22.0	44	66	88	110	132	154	176	198	220
8	25.1	50	75	101	126	151	176	201	226	251
9	28.3	57	85	113	141	170	198	226	254	283
10	31.4	63	94	126	157	188	220	251	283	314
11	34.6	69	104	138	173	207	212	276	311	346
12	37.7	75	113	151	188	226	264	302	339	377

Note: As an example, a pit of 5-ft diameter and 6-ft depth below the inlet has an effective area of 94 ft². A pit of 5-ft diameter and 16-ft depth has an area of 94 +157, or 251 ft².
From *Manual of Septic-Tank Practice,* U.S. DHEW, U.S. Government Printing Office, Washington, D.C., 1969, p. 23.

Construction work on pits should only be carried out on dry soils. The cutting teeth should be kept sharp and the pit should be well scratched to ensure adequate circulation. All the material should be removed from the excavation. The pit should be backfilled with clean gravel to a depth of 1 ft above the reamed ledge to provide a sound foundation for the lining. Preferred lining materials are concrete brick, block, or rings with loopholes or notches. If used, the brick and block should be laid dry and have staggered joints. Space between the blocks permits seepage from the pit into the surrounding ground. Either a brick or a flat concrete cover should be utilized and placed on the undisturbed earth extending at least 12 in. beyond the point of excavation. A 9-in. capped opening in the pit cover is needed for pit inspection.

All connecting lines from the septic tank should be of sound durable material. The pit inlet pipe should extend horizontally at least 1 ft into the pit with a T or L to divert the flow downward and to prevent a washing or eroding away of the side walls. When more than one pit is used, it should be connected in a series.

Sand Filters

Pretreatment is recommended for all sand filters. It lowers effluent strength, decreasing the burden on soil absorption fields. The filter reduces the amount of suspended solids and dissolved organic material. Microorganisms attached to the sand particles are able to aerobically digest the organic material in the wastewater. The effluent may be rerouted to the septic tank, or chlorinated and disposed. The subsurface sand filter is similar to the seepage bed, except that the filter bed contains a layer of sand added as a filter. The filter material, which is placed in the entire bed to a minimum depth of 24 in., should have a weak structure that contains between 5 and 15% clay by weight. In the subsurface sand filter, the natural earth is removed and sand is used to replace it. Beneath the sand filter is a series of underdrain collector pipes that collect the effluent and then feed it past a chlorinator before it is released. Where the total filter area is greater than 1800 ft², and where the distributor lines exceed 300 lineal feet, a dosing tank should be utilized (Figure 6.30).

Studies conducted on buried sand filters indicated that numerous systems were not operating properly because maintenance was poor, excess water was utilized, system was too small, or inadequate chlorine was used. In many cases, owners have a very poor concept of the operation and function of the buried sand filter, do not know how to utilize it properly, and certainly do not take the necessary precautions for complete and proper chlorination of any effluent that may surface.

Elevated Sand Mounds

The original mound system was developed in North Dakota in the late 1940s where it became known as the NODAK disposal system. It was researched further by the University of Wisconsin at Madison, College of Agriculture and Life Sciences, and College of Engineering under contract to the EPA and Upper Great Lakes Regional Commission (Figure 6.31). It is a soil absorption system that is elevated above the natural soil surface in a proper fill material. The sand mounds have been recommended by some authorities for use where subsoil disposal systems do not work effectively. The system consists of an absorption area composed of:

1. Sandy material comprising 330 ft² per bedroom for residence, or 1.65 ft²/gal/day for commercial, industrial, or institutional facilities using septic tanks
2. Distribution network
3. Cap
4. Topsoil

In preparing the mound, the site can have a maximum slope of 8%. If trench distribution is utilized, a 12% slope is acceptable. The area surrounding the absorption area must be graded properly to divert any surface runoff water from entering the system. The design of the sand mound is based not only on expected daily wastewater flow but also on soil characteristics. The soil beneath a sand mound must accept all liquid that is dispersed on it. The design of the mound includes an estimation of effluent flow, design of the absorption trenches, dimensions of the mounds, evaluation of limiting conditions, design of the distribution system, and size of the dosing chamber.

The fill material must be sandy material with single grains or very weak structures, which has a clay content between 5 and 15% by weight. If the soil is weak, blocky, or subangular blocky, the clay content cannot exceed 12%. Coarse fragments in the sand, larger than ¼ in., cannot exceed 15% by volume. The mound must be enclosed by a more stable, less permeable soil than the fill material and it must be compacted to prevent lateral flow. The top should be seeded. The top cover of soil over the mound should be at least 1 ft of suitable soil that is free of organic and rock masses. The elevated sand mound cannot be used in floodplain soils or in somewhat poorly, poorly, or very poorly drained soils. The sand mound also cannot be used in areas with less than 5 ft of natural soil to bedrock. The seasonal high level of groundwater must not be closer than 2 ft to the top of the ground.

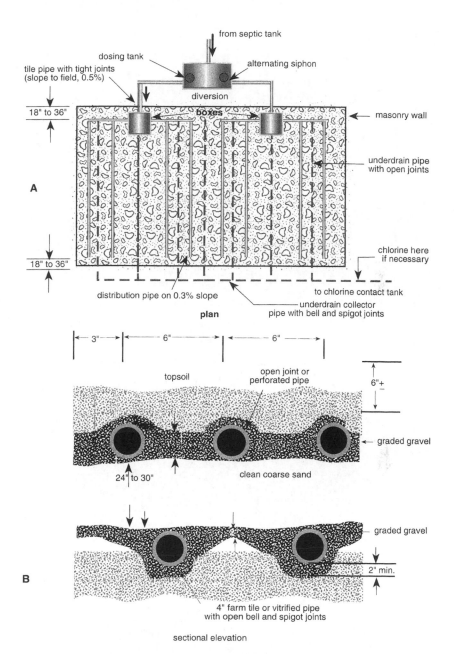

Figure 6.30 Typical plan and section of a subsurface sand filter. (From *Manual of Septic-Tank Practice*, U.S. DHEW, U.S. Government Printing Office, Washington, D.C., 1969, p. 63.)

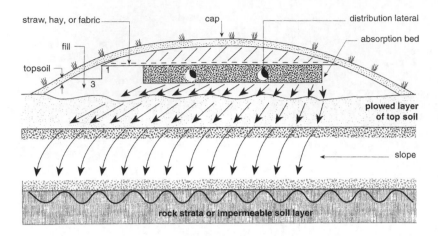

Cross Section of a Mound System for Slowly Permeable Soil on a Sloping Site

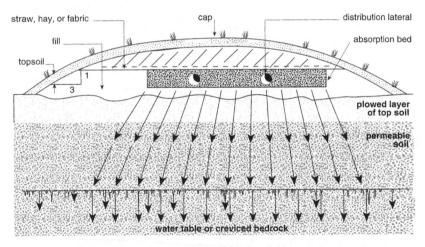

Cross Section of a Mound System for a Permeable Soil, with High Groundwater or Shallow Creviced Bedrock

Figure 6.31 Typical mound systems. (From *Design Manual, On-Site Wastewater Treatment Systems,* U.S. Environmental Protection Agency, Washington, D.C., October 1980, p. 240.)

The philosophy behind the elevated sand mound is that soils that percolate slower than 60 min/in. are unacceptable for normal subsurface sewage disposal. However, if the system is raised above the ground, a dry permeable sand layer permits percolation of the liquid onto a large surface area. This allows biological slimes to be present in the sand, instead of clogging the subsurface sewage systems, and it avoids smearing and compaction of the soil in the subsurface system. If the total bottom of the mound is designed adequately, there should be proper distribution of liquid and slow permeation of the soil. As part of the system, adequate septic tank pumping chambers are needed.

Sand-Lined Trenches and Beds

In some areas where the percolation rate is less than 6 min/in. and the system is not installed in the floodplain, sand-lined trenches or beds are utilized. The site characteristics and the design and construction are the same as in the routine subsurface disposal system. The difference is that sandy fill material is placed under the gravel, which is under the pipe, and the sandy fill material comes up the sides to the top of the gravel. The rationale is that the effluent can readily move downward instead of laterally and the ground can accept the amount of water placed into it. This creates adequate pressure of effluent above the bottom of the trench and forces the liquid through the bottom of the trench without creating a surface effluent problem. When sand-lined trenches are used, the minimum bottom width of the trench is 36 in., the minimum depth of the trench at the bottom is 36 in., and the maximum depth of the trench is 72 in. When a bed design is used, the minimum bottom width of the bed is 10 ft.

Oversized Seepage Areas

The oversized seepage area is utilized in areas where percolation rates are greater than 60 min but less than 90 min/in. and at least 4 ft of absorption field occurs below the bottom of the oversized sewage disposal area. The design and construction are the same as in standard trenches and seepage beds with the exception that 425 ft^2 per bedroom is utilized. In some, an aerobic treatment plant is required for use prior to the oversized seepage bed. The same potential problem of effluent overflow exists in this system.

Sloping Field

Where a sloping field is encountered, the septic tank effluent is pumped to the top of the slope into the first distribution box and the lines follow the contour of the land. The distribution box overflows with a solid pipe into another distribution box, which is at a lower level with the lines also following the contour of the land and down to another drop box, etc.

Shallow Placement Absorption Area

The shallow placement absorption area is an alternative method of effluent disposal for new homes constructed on sites when a limiting zone begins between 5 and 6 ft below the surface. The system cannot be used to correct a malfunctioning unit problem at an existing residence. The percolation rate on the site must range between 6 and 60 min/in. Either a standard trench or a serial distribution system may be utilized. The system is somewhat different than the typical tile field system in that from the original grade the absorption zone consists of 2 in. of crushed stone, the tile that is in 4 in. of crushed stone, and 6 in. of crushed stone beneath it for a total of 1 ft of crushed stone with the tiles coming through it. Above the original grade a fill grade of a minimum of 12 in. of good soil is brought in and put above the original grade. The fill grade must be sloped so that any water coming down on the new soil can run off to either side of the system. Beneath the 12 in. of crushed stone and the tile, it is necessary to have 5 to 6 ft of natural surface area where percolation can take place. This area must be above the highest level of groundwater or water tables and the highest level of bedrock.

Anaerobic Filter

Anaerobic filter research is under way at the University of Regina in Canada. The system consists of a septic tank and a trickling filter, with a clarifier that returns sludge to the septic tank, and an anaerobic filter containing a solid medium that supports bacterial growth. The filter follows the septic tank and allows bacteria to anaerobically reduce BOD, COD, and nitrogen. The effluent is not of high enough quality to release to a receiving stream.

Leaching Field Chambers

Leaching field chambers are an alternative to the conventional network of distribution pipes used in a drainfield. One or more chambers can be used, each connected to form a large underground cavity. The chambers are usually made of plastic and do not require gravel fill. The sides and bottom of the chamber have a network of openings to allow seepage of wastewater into the soil. These systems allow more of the soil profile to be used because the effluent is distributed more widely.

Constructed Wetlands

Constructed wetlands are techniques utilized to artificially recreate the filtering techniques of natural wetlands. The artificial wetlands consist of one or more trenches called cells, with a lining that can be either impermeable or permeable depending on the system, and contains vegetation anchored in gravel etc. The two basic designs for constructed wetlands are surface flow systems, where the wastewater passes over the medium; and subsurface flow systems, where the wastewater passes directly through the medium. Sand or gravel and sand mixtures can be used as well as gravel.

Bulrushes, cattails, reeds, and sedges are types of vegetation used in constructed wetlands. Irises also perform very well. The plant roots function to transpire oxygen and aerate the effluent. The aerobic conditions permit a large variety of microorganisms to attach themselves to the surface of the roots and to the medium. The microorganisms are the primary means of treatment, because they feed on the organic material in the effluent. The plant roots and medium help trap tiny particles. The plant roots also use part of the water of the effluent. Before the constructed wetlands is a septic tank that helps separate solids and liquids. Many unknowns exist concerning constructed wetlands design. Although large quantities of data come from thousands of articles and reports, it is difficult to sort and distill the information. The EPA has developed a North American Wetland Treatment System Database for keeping a record of engineering data. Information can be secured from the EPA Wetlands Information Hotline at 1-800-832-7828.

Drip and Spray Irrigation Systems

Irrigation systems are similar to conventional drain fields because they use the natural capacity of the land to dispose of the wastewater. The principal difference is that irrigation systems are designed to allow the water and nutrients to be used by plants. The vegetative ground cover removes the nutrients such as nitrogen from the effluent and also reduces soil erosion and maintains soil permeability. The plant roots reduce the amount of water percolating through the soil. The roots then produce oxygen and allow aerobic microorganisms to digest some of the organic matter.

Spray irrigation systems use sprinklers to distribute the effluent evenly over the surface of the ground, which allows for greater loss of water through evapotranspiration. However, the effluent may possibly distribute pathogens through the air and may also cause an odor problem. The wastewater needs to be pretreated through use of appropriate techniques including secondary pretreatment and disinfection. The systems are used occasionally for sites that are unsuitable for conventional absorption systems, due to slowly permeable soil or closeness to the water table or rock formations. A sizable amount of land is needed for this endeavor and a well engineered plan must be utilized.

Drip irrigation systems use pressure compensated drip tubing to slowly and evenly dispense the wastewater just below the soil surface but still within the root zone of the vegetation. The effluent has gone through a septic tank and into a dosing chamber, which periodically moves the effluent through a series of disk filters before entering the tubing. The system is periodically back flushed to keep the filters from clogging.

Peat Filter

The peat bed follows a septic tank system. The effluent from the septic tank is distributed over the bed and then after filtering goes through a subsurface or surface disposal system on-site. Aerobic conditions in the upper portion of the peat bed and anaerobic conditions in the gravel beneath the bed allow for nitrification and denitrification to occur. The effluent from peat systems has a relatively low pH, between 4 and 6.

Trickling Filter System

The trickling filter system consists of a septic tank followed by a trickling filter unit with recirculation of the effluent to the septic tank prior to the effluent discharge to the subsurface disposal system. The trickling filters have attached microorganisms that consume organic material during their life cycle process. The filter material includes plastic, stone, or wood. The filters do a good job in nitrification, as long as there has been sufficient removal of BOD and TSS. The system helps reduce odors.

Special Toilets

Incinerating toilets evaporate and burn the toilet wastes, producing an ash that may be disposed of with the household wastes. In this type of system the gray water is separated out. Composting toilets process feces, urine, toilet paper, carbon additive, and sometimes food wastes. The system relies on aerobic bacteria to break down the wastes. The composting toilet is a nonwater system.

Sewage Haulers

Specialized companies should be licensed for the collection, storage, transportation, and disposal of human excreta from on-site sewage disposal systems to public sewage disposal systems (Figure 6.32). This work should be carried out in a manner that cannot create a hazard to public health or a nuisance. All trucks utilized should be properly identified with the name of the company and address on the truck. The equipment should be subject to frequent inspection by the appropriate health department. The vehicle utilized should be leakproof and fly-tight. The interior and exterior sections of the pumps, hoses, tools, and so on should be carefully washed. Trucks and tanks should be equipped with either a vacuum pump or another type of pump that can prevent seepage from the diaphragm or other packing glands and that is self-priming. After pumping the septic tank or drain field system, the sewage hose should be thoroughly rinsed, drained, and stored in such a way that it cannot create a health hazard or a nuisance. The discharge nozzle should not be permitted to drip on any part of the truck or onto the roads. It should be capped when not in use.

Each year the licensed contractor must file a statement concerning the type of material carried and the methods of sewage disposal with the appropriate health department. The method of sewage disposal that is permitted depends on state health department regulations. The best type is disposal into the municipal sewer system. Other techniques utilized are discharge into sludge lagoons, sludge drying beds, or special incinerators. At times the material is permitted to be discharged into properly operating sanitary landfills. However, the latter is not a recommended practice.

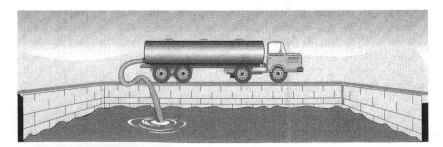

Figure 6.32   Typical sewage hauler.

Privy

It is most important that the privy be located a minimum of 100 ft from and downhill of a private water supply well. The privy should be located a minimum of 200 ft from and downhill of a public water supply well; should be at least 5 ft from any building, fence, embankment, and property line; and should be at least 50 ft from a home or business establishment. The privy should not be overhanging lakes, ponds, streams, or ditches. It also should be kept at least 100 ft away from a surface water source (Figure 6.33).

MODES OF SURVEILLANCE AND EVALUATION

Sewer System Analysis

An analysis of the sewer system infrastructure is needed periodically to determine the existence of problems that may affect the flow of the sewage or the amount treated at the wastewater treatment plant. A major concern is the amount of infiltration of inflow into the system. The following would indicate that a problem may exist and a survey is needed:

1. Greater than anticipated flow to the wastewater treatment plant
2. Flooded basements during heavy rainfall
3. Lift station overflows
4. Sewer system overflows
5. Excessive power costs at pumping stations
6. Overtaxing of lift stations
7. Hydraulic overloading of treatment facilities
8. Excessive cost of treatment
9. Aesthetic and water quality problems
10. Surcharging of manholes
11. Odor complaints
12. Structure failure
13. Corrosion due to H_2S, excessive overburden, aging, soil settlement, etc.

Infiltration is the water that enters sewers and building sewer connections from the soil through foundation drains, defective joints, broken or cracked pipes, faulty connections, etc. Inflow is the water that is discharged into existing sewer lines from roof leaders, cellar and yard area drains, commercial and industrial discharges, and drains from springs and swamps.

The sewer system study should include:

- Survey planning and cost estimating
- Physical survey
- Rainfall simulation
- Preparatory cleaning
- Internal inspection
- Preparation of report and cost

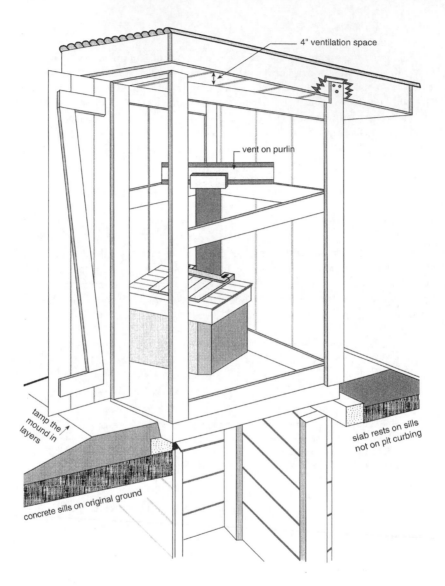

Figure 6.33   Sanitary pit privy. (From *The Sanitary Privy,* Revised Type No. IV, Supplement
No. 108, Public Health Report, U.S. DHEW, Cincinnati, OH, 1966, p. 21.)

The physical inspection includes:

- Smoke testing
- Rainfall simulation through use of dye testing, flooding, and tracing
- Building plumbing inspection

- Flow isolation
- TV inspection
- Lateral testing

Sampling and Testing

To properly operate a sewage treatment plant, laboratory tests must be performed to determine the day-to-day problems and the trend of problems that may occur. The efficiency of a treatment is also indicated by the varying tests utilized. Some of the tests include efficiency of settling by determining settleable solids, total solids, total dissolved solids, and suspended solids; efficiency of biological treatment by determining relative stability of the sewage, BOD, and total volatile solids; nature of the sewage, by determining the DO present, pH, and various other components of the sewage; chlorine demand test for chlorine requirement on the effluent entering the receiving stream; putrescence of the sewage by the use of the methylene blue test; residual chlorine test on the effluent; pH test; and coliform test (refer to Chapter 9).

The sample for testing should be taken where the sewage is well mixed. Large particles should be excluded. Deposits or other growths and floating material should also be excluded. The sample should be examined as soon as possible. Two types of samples may be taken: a catch sample, which is not representative of the average sewage, because it only reflects the condition at the time of sampling; and the integrated or composite sample. Hourly portions may be taken over a 12-hour period to prepare the composite sample. In the case of sludge, the analysis depends on the accuracy of sampling. A sampling apparatus is needed that can take samples at different depths in the sludge digestion tank.

Soil Percolation Tests

Determination of soil permeability may be carried out by means of a soil percolation test. Six or more separate test holes should be spaced uniformly over the proposed absorption field site. The hole should be 4 to 12 in. in horizontal dimension and to the depth of the proposed absorption test (Figure 6.34). A 4-in. agar can be utilized for this purpose. The test hole should be scratched on the bottom and sides with a sharp instrument to remove any smeared soil surface and to provide a natural soil interface with the water that may percolate. All loose material is removed from the hole. To protect the bottom from scouring and sediment, 2 in. of coarse sand or fine gravel are added. It is then necessary to saturate the ground by prolonged soaking by filling a hole with clear water until a minimum depth of 12 in. is reached over the gravel for a period of at least 4 hours and preferably overnight. Then determine the percolation rate 24 hours after the water has first been added to the hole. This gives the soil adequate opportunity to swell and approach the condition it faces in the wettest season of the year.

With the exception of sandy soils, the percolation rate measurement is made the day after the hole has been dug and saturated. If water still remains in the hole overnight, adjust the depth to 6 in. over the gravel and measure the drop over a

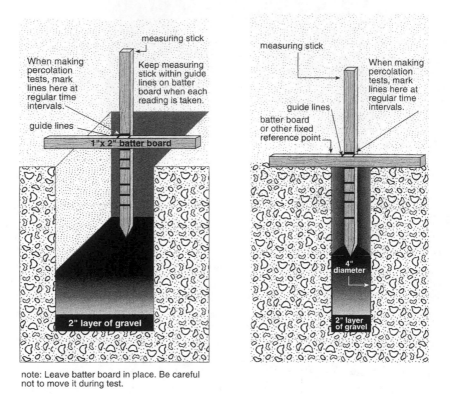

note: Leave batter board in place. Be careful
not to move it during test.

Figure 6.34   Methods of making percolation tests. (From *Manual of Septic-Tank Practice*, U.S.
DHEW, U.S. Government Printing Office, Washington, D.C., 1969, p. 7.)

30-min period. The drop is used to calculate the percolation rate. If no water remains
in the hole after the overnight swelling, add clear water to bring the depth of the
water to 6 in. above the gravel and measure over 30-min intervals for a 4-hour period,
refilling as needed. The drop that occurs during the final 30 min is used to calculate
the percolation rate. In sandy soils or where the first 6 in. seeps away in 30 min
after the overnight swelling period, the time interval measurements are taken at
10-min periods and the test is run for 1 hour. The drop that occurs in the final 10
min is used to calculate the percolation rate. Unfortunately, percolation tests may
not be accurate because of improper soaking of the ground, improper testing, or
tests not made in the actual soil absorption area but in other areas of the land.

Soil Samplers and Other Techniques

Soil samplers are on the market that resemble a hollow tube inserted into the
soil to determine the type of soil present in the area where the disposal field is to
be placed (Figure 6.35). The soil sampler accurately indicates the profile of the soils
at the point of entry into the soil and downward to the depth that it can reach. This

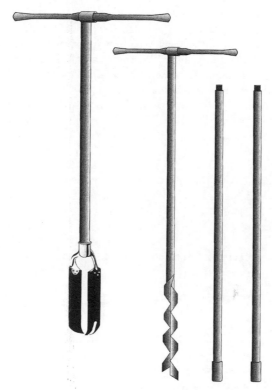

Figure 6.35 Auger and extension handle for making test borings. (From *Manual of Septic-Tank Practice,* U.S. DHEW, U.S. Government Printing Office, Washington, D.C., 1969, p. 5.)

sample is analyzed by a trained soil scientist or another individual who may be able to determine the type of drainage found in the area. Obviously, this test has some shortcomings in that the soil sample may miss parts of the field that are limited for actual percolation of water or sewage. Some health departments use backhoes to dig out the ground to the depth of the actual subsurface soil bed to determine the type of soil present. These agencies utilize the information furnished before concerning the soil texture, structure, and colors. Some health departments have experimented with digging holes with backhoes or agars down to depths greater than the subsurface disposal field and then trying to make a determination of the ability of the soil to remove liquid.

Unfortunately, many individuals seem to have their preferred system and fail to look at the possibility of combining systems. It is suggested that the soil map be utilized in the initial phase of determining whether land is usable for a subsurface sewage disposal system, provided public sewage or package treatment plants are unavailable. Soil samples would give an additional picture of the type of soils encountered. Where the soils appear to be well drained, it may not be necessary to utilize percolation tests. However, where conditions are variable, it would be best to utilize the soil map along with the soil sample and percolation tests to determine the actual capacity or the ability of the land to allow effluent to percolate. Use of all these techniques does not guarantee that the system will be failproof.

Another technique in determining percolation of soils is called soil hydraulic conductivity. It is a system that measures the hydraulic conductivity of a soil using a crust test in which sand and gypsum are mixed thoroughly with water, made into a paste, and allowed to harden on a ring. The liquid is applied through a burette until the system is purged of all air. The system is then closed. Flow of liquid is allowed to occur until an equilibrium is established. The rate of flow into the soil is governed by the crust resistance to the flow. This experimental system was developed by Fred G. Baker and Johannes Boumat. A description of it may be found in the report of the Second National Conference on Individual On-Site Waste Water Systems held by the National Sanitation Foundation in Ann Arbor, MI.

Site Evaluation

To conduct a proper site evaluation for an on-site sewage system, the evaluation must include topographic information and soil characteristics.

Topographic information includes:

- Slope
- Surface drainage characteristics
- Proposed or existing location of house and wells
- Location of other major structures
- Location of soil evaluation sites
- Topographic position of site

Soil characteristics include:

- Approximate depths of soil horizons
- Soil color, structure, and texture of each horizon
- Depth to any layer or layers which have a loading rate greater than 0.75 gal/day/ft^2
- Depth to seasonal high groundwater
- Depth to bedrock
- Soil consistence at each horizon
- Soil effervescence at each horizon
- Presence or absence of roots

Lasers for Leveling

Lasers are used for leveling when setting up on-site sewage systems. A laser allows one person to lay out an on-site sewage system and to level the system in an appropriate manner. One type of laser is a helium–neon gas laser that generates a characteristic red color. The unique characteristic of the light makes it useful for alignment purposes because it is a pure single wavelength with parallel light rays that can maintain a small diameter spot for long distances.

CONTROLS

Planning

Every municipality should be required to develop and maintain an up-to-date plan that meets the needs of the municipality for present and future sewage disposal. This plan should not be limited to the extension of existing public sewers but should also include building of new public sewerage units, smaller package treatment plants, and, as a last necessity, areas where individual sewerage systems are permitted because public sewerage would be too costly. Where individual sewerage systems are allowed in the plan, it should be stipulated at what date public sewerage systems become available within each of the areas of the community. The purpose of the official plan is to provide a means for resolving existing sewage disposal problems and for avoiding potential sewage problems when new residential or commercial development occurs.

All plans developed by a community and reviewed by the local health department should also be reviewed by the county planning authority and eventually by a state agency that should have a good firm view of anticipated growth in various areas of the state and the kinds of problems to be encountered. Obviously, in some areas the growth will be so scattered or sparse that subsurface sewage systems will be necessary. If so, information must state clearly and carefully in the plans the soil limitations of the area and the possibility of hazardous conditions occurring as a result of soil limitations. The plan should incorporate the following items:

1. An analysis of the sewerage disposal needs of the municipality is needed.
2. An evaluation of the various alternatives available to resolve existing sewage problems and to avoid potential problems is necessary.
3. A survey of existing sewerage service indicates the areas to be served by public sewers, main intercepting sewer lines, pumping stations, treatment plants, and point of discharge for sewage effluent.
4. An outline of the drainage basin and the water systems, both public and private, to be utilized is needed. The private systems should also be identified by location on the map.
5. An identification and the location of all private sewerage systems ranging from package treatment plants to on-site sewage disposal are needed. These systems should also be shown on the map.
6. A listing of all developed areas and an identification of their exact locations should be on the map.
7. Areas of anticipated growth, including residential, commercial, and industrial areas should be identified.
8. Various zoning areas, including those for subdivisions, should be identified.
9. A listing of all regulations on zoning including subdivision regulations and local, county, and regional comprehensive plans is needed.
10. An analysis of soils and proposed sewage needs in areas not served by public sewers should include evaluation of the soils, land classification system, and limitations of a system as they range from no limitations to hazardous limitations. None to slight limitations means that the soils are suitable for on-lot sewage

disposal. Moderate limitations means the soils may be suitable provided the subsoils are permeable. Severe means that the soil is not satisfactory for use because of impervious restricting layers, high water tables, periodic flooding, or other limiting factors. Hazardous means the soils generally are not suited for use because of the probability of groundwater pollution or contamination.

11. A determination is needed about those areas where growth may occur and where public health or pollution factors may increase sharply. These areas need to be sewered immediately. As part of this determination, there should be a listing of the present and potential alternatives for sewerage service that may be financed through the community. This proposed public sewerage system must contain a cost estimate. The most practical plan should be developed and a time schedule for implementation should be set forth.

12. The areas where sewerage service will be needed within a 5-year period and also within a 10-year period should be determined. As part of this determination, the location of major elements of the sewage system should be listed that will satisfy the proposed growth and development of the area.

13. A description and evaluation of existing industrial waste discharge must be determined, evaluated, and shown. It must also be determined whether the industrial waste discharge will be compatible with the existing sewerage system or how the sewage from industry will be handled to prevent problems from occurring.

The plans should be updated every 3 to 5 years and revisions and supplements should be sent through the same routing as the original plan. Very rapid growth may occur in certain areas as a result of recreational needs, expansion of malls and industrial parks, or building around the interchanges of new sections of the interstate highway system. Because this type of building may not have been anticipated in the initial plan, it should go through the same comprehensive evaluation as the original material and necessary adjustments should be made.

In the event a parcel of land is subdivided into lots that are to be used for homes, the builder must obtain necessary permits from the local health department and planning agency. The subdivision must be consistent with the initial planning. Changes in zoning of certain areas should only be permitted if it is not in conflict with the established knowledge concerning these areas. In all cases where new homes will be built, public sewers should be installed if at all possible. If for some reason public sewage cannot be utilized, an on-lot sewerage disposal system can be approved, provided that the land has been carefully and thoroughly evaluated and that it meets all the regulations and controls set forth.

Because planning is so essential to our entire way of life, the state should make it mandatory for the local agencies to fulfill their obligations. If they do not, the state should use appropriate punitive action when other actions fail. Certain areas may not have the capacity to develop a proper plan; in these cases, the municipality should request consulting services from the state.

Municipal Sewage Treatment

Techniques utilized to correct the inflow infiltration into sewer lines consist of infiltration inflow analysis, field investigations, surveys, and rehabilitation. The infiltration inflow analysis is accomplished by using a pattern interview to determine the kinds of problems that individuals are facing. This determination is made as follows: sanitary and storm sewer map study, which indicates the location and junction points, along with key manholes and pumping stations; system flow diagram, which indicates the flow; dry vs. wet weather flow determinations, which indicate the amount of sewage flow in relationship to rainfall and runoff; preliminary field survey and selected flow tests, which indicate the problems that the experienced individual may see; and determination of excessive or nonexcessive infiltration inflow, which is made by evaluating all the previous data. A plan of action, budget, and timetable for execution or change in the infiltration inflow into the sewers should then be developed.

The field investigation and survey is accomplished by conducting a physical survey and groundwater analysis, developing a rainfall simulation, preparing engineering reports and analyses, preparing a sewer cleaning program in which apparent deficiencies and amount of deposits within the sewers are identified, inspecting preselected sewers by television, preparing an evaluation survey report and analysis, and proposing a rehabilitation program.

The rehabilitation program utilized to reduce infiltration inflow of excess waters into the sewer system consists of sewer repair, pipe relining, sewer replacement, and final development of proper treatment plant design. The use of proper engineering concepts, television inspection cameras, sewer relining, and external and internal grouting equipment should substantially reduce the amount of liquid that is infiltrating into the sewer, thereby reducing this additional burden on the sewage treatment system.

Treatment Process — Chlorine

Neither the primary nor the secondary sewage treatment processes can produce an effluent free of pathogenic bacteria. For this reason, chlorine is needed to destroy the pathogenic organisms present. To achieve a final 0.1 ppm residual of chlorine after 15 min of contact, it is necessary to experiment with the amount of chlorine that is needed at varying times during the day and at night at the point where the actual effluent leaves the treatment plant. Under normal domestic sewage conditions, usually 20 to 25 ppm of chlorine are introduced into primary plant effluent; 15 ppm, into trickling filter plant effluent; 8 ppm, into activated sludge plant effluent; and 6 ppm, into sand filter effluent.

Chlorine is used in prevention of sewage decomposition to control odors, protect plant structures, and prevent sludge thickening. Odor control is achieved by adding chlorine at manholes on trunk sewers if the odors are produced in the sewer. Usually 4 to 6 ppm is adequate to achieve this control. Another technique is to add 10 ppm of chlorine at either the floor mains, pump suction wells, screen chambers, grit chambers, or settling tanks. Decomposition of sewage can produce hydrogen sulfide that may be an odor problem. In substantial concentrations, hydrogen sulfide may become a corrosion problem for various pieces of equipment. Chlorine is added to control this situation. By maintaining a chlorine residual of 1.0 mg/l or 1 ppm in the supernatant liquid where excess activated sludge or raw sludge is concentrated, the sludge does not thicken excessively and create problems.

In the plant operation, prechlorination is used in the sedimentation process to improve the sludge. Offensive odors coming from trickling filters can be controlled by chlorine. Activated sludge bulking and Imhoff tank foaming may be controlled by chlorine. Final chlorination is important and so is chlorination of the raw sewage. By adding sufficient chlorine to the raw sewage to create a residual of 0.2 to 0.5 ppm, after 15 min, a 15 to 35% reduction of BOD within the sewage may occur.

Filter Flies

Some of the techniques used to eliminate the flies are flooding the filters at 24-hour intervals, chlorination in conjunction with the flooding, removing weeds and grass from the filter area, and washing walls frequently. The use of chemical insecticides must be weighed against the residue of chemicals that may enter the receiving stream.

Plant Supervision and Certification of Operators

As in the case of the package treatment plant, the regular sewage treatment plant must have certified sewage treatment works operators who are capable of performing an effective accurate job. Operators must be trained and tested before certification to assume the job of operating a sewage treatment plant. In Illinois, certification rules stipulate that the individual must be specifically certified as a group A, B, C, or D operator. Group A, which is class one, serves a population greater than 50,000; between 20,000 and 50,000 using an activated sludge process; or between 30,000 and 50,000 using biological oxidation other than activated sludge process, chemical precipitation, separate sedimentation of sludge digestion, or highly mechanical or intricate processes. Group B, which is either class two or class one, serves a population between 20,000 and 50,000 that is not designated in group A; between 2500 and 20,000 using activated sludge; or between 5000 and 30,000 using biological oxidation other than activated sludge, separate sedimentation of sludge digestion, chemical precipitation, or highly mechanical or intricate processes. Group C, or a class three or higher operator, serves a population between 10,000 and 20,000 not designated in group B; between 300 to 2500 using activated sludge; or between 1000 and 5000 using biological oxidation other than activated sludge, separate sedimentation sludge digestion, chemical precipitation, or highly mechanical or intricate processes. Group D serves a population up to 10,000 not designated as group C and all municipal waste stabilization ponds (lagoons) treating essentially domestic waste. They are classified as class four or higher. Special groupings other than those mentioned also exist and these are considered separately by the Sanitary Water Board.

The requirements for all operators with unlimited licenses for all areas include the ability to read and write the English language; a 4-year college or university education in sanitary, chemical, civil, electrical, or mechanical engineering or its equivalent; 6 years of experience in an acceptable operation of a sewage treatment plant; 4 years of which the individual was responsible or in charge; and a score of 95 or higher in-service rating. In-service rating is a determination of the individual's knowledge and ability to establish practical competency. This is based on knowledge of sewage treatment, operation, maintenance, techniques, interpretation of laboratory tests, and knowledge of proper record keeping. Each of the individual classes has a set of specific education, experience, and service ratings. Class one is almost the same as the unlimited operators license, with the exception that the individual must have a minimum average of 75% in-service rating. Class two, three, and four continue to drop in their requirements.

Plans Review and Approval

Prior to installation of a new subsurface sewage disposal system, the installers must prepare a detailed plan for the health department. A permit can only be granted if the plan is approved. The plan must include the following:

1. Scope of the on-site sewage disposal system, including the septic tanks, holding tanks, subsurface drains, electrical devices, mechanical septic tanks, or sand beds
2. Location of the on-site sewage disposal system and the distance of the system from building foundations (minimum 10 ft), property lines (minimum 10 ft), surface bodies of water (minimum 50 ft), all private water supplies (minimum 50 ft), public water supplies (minimum 75 ft), and municipal water supplies (minimum 200 ft)
3. Protection of the water supply system
4. Sewer construction and material, including the types of pipes and joints to be used
5. Septic tank construction, including the Ls or Ts, manholes, location of the tank, and the depth below the ground surface
6. Size of the septic tank
7. Type of subsurface disposal system including the actual size of the system, the layout of the system, and the materials to be utilized in the system
8. Evaluation of the soil, including, if necessary, the geology of the soil and the groundwater sources
9. Field tile construction standards including the type of tile and the slope or grade of the tile lines that should not exceed 3 in./10 ft
10. Varying factors that affect the size of the drain field including the water table, the width of the trench, and the number of bedrooms
11. Final earth cover that can be utilized and the depth of the earth cover

Along with this previously mentioned information a detailed engineering drawing of the actual disposal system should be included. All existing grades and proposed grades of the soil should be included in the drawing. After the plan has been submitted, reviewed, and approved, the permit holder is allowed to begin installation. Prior to the time that the installation is covered, an inspection and final approval must be made by the health department. Under no condition should the contractor be permitted to cover any part of the septic tank, distribution box, lines, or actual disposal field before this inspection is made. In many cases, this practice is not followed and it becomes a serious potential hazard, because questions always remain relating to the proper installation of the system. Although the contractors are reputable, they still may make mistakes. It is the responsibility of the health department to determine whether errors exist and to seek correction before the system is covered.

Corrections in Malfunctions

It is necessary in all cases to determine the actual causes of malfunction in on-site sewerage disposal systems. Any extraneous sources of water should be dye tested to determine whether liquid is seeping to the surface or the extraneous sources are getting into the system. Where lines are clogged, they should be cleaned by conventional sewer-cleaning techniques. Broken pipes must be reported. The septic tank should be emptied by properly licensed sewage haulers. The owner should disconnect garbage disposals and if necessary reroute the plumbing from the laundry to a separate disposal area on another part of the site. The owner should severely limit the use of water in the house. If water from the roof or downspouts is crossing the system, this should be diverted. Where water is lying across the system from natural drainage of rainwater, a curtain drain should be considered for the perimeter of the property. It is essential that the sewerage system is not hooked into the curtain drain. If these techniques do not work, an alternate system should be considered on another part of the land after proper soil testing and evaluation have been conducted. If nothing else works, a holding tank may be installed and pumped out periodically. A holding tank is an extremely expensive procedure and is only really useful where municipal sewerage is expected to be installed within a fairly reasonable time period.

Soil Clogging

Laboratory studies have shown that hydrogen peroxide may restore permeability of clogged or crusted soils. Actual failed septic systems have been treated in six cases with peroxide. Three of the systems were in sandy soil, two were in glacial till, and one was in heavy clay. The septic tank was first emptied; then the beds were drained, pumped into proper sewage hauling systems, and removed. Utilized were 15 to 40 gal of commercial 50% hydrogen peroxide. Fresh water was added to the tank at a level of 300 to 600 gal. A breakthrough in the crust occurred and the soils appeared to be renovated. One system at the time of the presentation of the research study had been operating satisfactorily for 15 months. It has also been suggested that peroxide be utilized in the septic tank to prevent soil clogging. Further information concerning this technique may be obtained from John M. Harkin, Michael D. Jawson, and Fred G. Baker. These authors are professors and research scientists at the Department of Soil Science at the University of Wisconsin, Madison, WI. The Arm and Hammer Company is also conducting research in the use of baking soda in septic tanks to improve the effluent within the tank. It is certainly recommended that considerably more study be given to the entire area of septic tank effluents and clogged subsurface soil disposal systems.

Laws and Enforcement

All states have laws, rules, and regulations concerning the installation of on-site sewage disposal systems and the hazards related to overflowing sewage or contamination of streams. Where individuals are in violation of these laws, it is important that the environmental health practitioner initially works with the individual to determine the cause of the problem and to resolves it amicably. When education, motivation, and consultation works, this is the very best process to utilize. At times the amount of overflowing sewage and the land are such that no effective means of eliminating the problem exists. It is then necessary that the community work toward developing greater public sewage capacity with existing plants or establishing a package treatment plant to handle the sewage for the problem areas. Only in the event that the individuals involved will not make necessary changes should the health departments take necessary legal action, either under the general health code or under the specific rules and regulations set forth by the various states.

SUMMARY

Over the years a considerable number of subsurface sewage disposal systems have been utilized. Many companies have devised a variety of different shapes and sizes of tanks. They have also developed aerobic and anaerobic tank systems. Soil absorption systems have been developed by innumerable companies and health departments. In many areas, the given environmental health practitioner or given individual from a specific company will swear by one technique vs. another technique. However, in essence, the environmental health practitioner must never forget that on-site sewage disposal is only a stop-gap measure until a properly operated municipal or package treatment plant is used. The environmental health practitioner must understand the soils in great detail, the kind of effluent that will be allowed to percolate into these soils, and the habits of the individuals utilizing the sewage disposal system. The best technique of underground sewage disposal is the elimination of such systems. The second best or most realistic technique, because these systems continue to exist, is to properly design a system on land that can accept effluent and then to teach the homeowners how to utilize their water properly. Proper installation and maintenance of the system along with limited water usage can do more to resolve potential sewage hazards from on-site sewage disposal systems than any new fancy techniques that may be developed.

The entire approach to sewage treatment has changed drastically in the last 50 years. No longer is it enough to simply take the sewage from homes, settle it in a basin in a sewage treatment area, and then allow the effluent to enter the receiving stream. The need for water and for removal of dangerous pollutants has increased so rapidly that treatment of sewage has become an extremely important facet of environmental health controls. Chemicals that may be found in sewage continue to have a potential effect on our society. It is necessary to continuously explore sewage problems to arrive at adequate solutions to changing the public or municipal sewage treatment process.

RESEARCH NEEDS

Research is needed to determine the effects of various metals, such as zinc, copper, nickel, lead, cadmium, selenium, mercury, boron, and arsenic, on food plants, feed plants, fiber-producing plants, and forest crops. Basic studies need to be made of soil chemistry to determine the long-term consequences of metal-loading buildup and interaction. Studies are needed to determine better ways of removing toxic metals from sewage during the municipal sewage treatment process. The removal of halogenated hydrocarbons must also be studied. Studies are needed to determine the health hazards associated with given levels of bacteria, viruses, and parasites that may be present in sludge or effluent from the sewage plant. Research is needed in soils to determine the distance that microorganisms and chemicals travel through the soil after sewage or sludge has been placed into it. Intensive research is needed to determine a better way of designing individual or small-group on-site sewage disposal systems. Further intensive research is needed to determine the effects of the disposal of industrial wastes on the land.

CHAPTER 7

Water Pollution and Water Quality Controls

BACKGROUND AND STATUS

Water means different things to different people. In the preceding chapters, there have been discussions about a variety of water uses and water problems. The current chapter expands on and repeats some of the previous material, and enlarges on the entire area of water pollution and water quality control.

Water is necessary for life. It is used for navigation; as a coolant, cleanser, and dilutant; for recreational purposes; as a food resource; as a means of power; as a source of tranquil aesthetic enjoyment; as a transporter of disease; as a container for nuisances; and finally, as the once unlimited area for disposal of society's waste products. Water quality affects humans through their direct use of water. It also affects the aquatic life contained in the water.

Aquatic environments are vast in quantity and varied in quality. They consist of small streams from snow-capped mountains, brooks, lakes, small and great rivers, coastal estuaries, and oceans. All aquatic environments are affected by their altitude, latitude, area, mean depth, maximum depth, area of different depth zones, volume of different depth strata, length of shoreline, slope, drainage area, runoff rate, average inflow and outflow, detention time, water level fluctuation, number of islands or other masses present within the water, physical composition of the water bed, physical nature of the surrounding terrain, area geology, and untold number of uses and reuses by humans, animals, insects, worms, fish, plants, and trees. Water quality is affected by temperature, amount of dissolved oxygen (DO) present, pH, light, flow of water, amount of silt, oil, major nutrients, and contaminants.

Over 70% of the surface of Earth is water. Of this, 3% is freshwater found in glaciers, lakes, groundwater, and rivers, and in the atmosphere. The five Great Lakes contain about 90% of all freshwater above ground in the United States. There are 3.5 million miles of rivers and streams; 41 million acres of lakes; 34,400 square miles of estuaries not including Alaska; 101 million acres of wetlands not including Alaska; and between 170 and 200 million acres of wetlands in Alaska.

A $45 billion commercial fishing and shellfishing industry relies on clean water. Farmers irrigate about 15% of American farmland and create crops valued at approximately $70 billion a year. Each year, Americans take over 1.8 billion trips to go fishing, swimming or boating.

Federal Laws

In 1899, the Rivers and Harbors Act was the first federal legislation passed to protect the nation's waters to permit commerce. In 1948, the Water Pollution Control Act passed by Congress had the federal government offer state and local governments technical assistance and budgetary assistance to promote efforts to protect water quality. The money was used for construction of wastewater treatment facilities. In 1965, the Water Quality Act made states responsible for setting water quality standards for interstate navigable waters.

Clean Water Act — 1972

In 1972, Congress passed the Federal Water Pollution Control Act, better known as the Clean Water Act. Under the Clean Water Act, point source discharges from municipal and industrial facilities to the waters of the United States must obtain a National Pollutant Discharge Elimination System permit. This requires compliance with treatment standards based on technology and water quality. Further, because of the complexity and ecological significance of the marine ecosystems, discharges to the sea and beyond must also comply with Section 403 of the Clean Water Act, which describes the impacts from point sources on marine resources, requiring ambient monitoring programs, alternative assessments to evaluate the consequences of various disposal options, and pollution prevention techniques to reduce the amount of pollutants requiring disposal and thereby reducing the potential for harm to the marine environment. The EPA, in 1980, developed the Ocean Discharge Guidelines which include:

1. Quantities, composition, and potential bioaccumulation or persistence of the pollutants to be discharged
2. Potential transport of the pollutants by biological, physical, or chemical processes
3. Composition and vulnerability of potentially exposed biological communities
4. Importance of the receiving water area to the surrounding biological community
5. Existence of special aquatic sites
6. Potential direct or indirect impacts on human health
7. Existing or potential recreational and commercial fishing
8. Any applicable requirements of an approved Coastal Zone Management Plan
9. Other factors relating to the effects of the discharges as may be appropriate
10. Marine water quality criteria

The types of facilities with discharge permits include publicly owned treatment works (POTWs), 130; offshore oil and gas facilities, 2500; seafood processors, 300; offshore placer mining, 2; log transfer facilities, 35; seawater treatment plants, 3; sugarcane mills, 8; petroleum refineries, 3; and pulp and paper mills, 2.

Under the National Pollutant Discharge Elimination System (NPDES) program, dischargers had to achieve compliance with national minimum technology-based treatment requirements and any additional requirements needed to meet state water quality standards. Direct discharges or point source discharges are from sources such as pipes and sewers. Animal feeding operations of 1000 units or more and aquaculture projects that produce more than 20,000 lb also require permits. The discharge cannot cause long-term water quality degradation or contain contaminants at levels that are harmful to marine organisms or entire marine communities.

By the late 1960s and early 1970s, the nation's waterways had become severely polluted. An example of this was the Cuyahoga River in Ohio catching fire because of high pollutant levels.

Under the 1972 amendments to the Federal Water Pollution Control Act, standards of best practicable technology were to be utilized for water pollution control by 1977 and by 1983 standards of best available technology economically achievable were to be issued for water pollution control. The reason for all the standards and regulations was to reduce and then eliminate industrial point sources of pollution. Permits allowed a short but specific period of time to comply with the necessary regulations. If compliance was not met, the dischargers were to be in violation of the National Water Quality standards. The total identified dischargers from industrial sources were 41,454. The total number of municipal dischargers of pollutants were 20,664. Regulations for the control of discharge of 65 toxic chemicals into the surface waters of the United States were issued by the Environmental Protection Agency (EPA).

The Water Pollution Control Act made $18 billion available to municipalities to abate their point sources of pollution.

Current water issues include:

1. Increasing competition for water supplies
2. Vulnerability of surface water and shallow groundwater to drought
3. Declining groundwater levels
4. Surface water pollution from nonpoint water sources
5. Groundwater contamination from underground storage tanks, septic tanks, municipal landfills, agricultural activities, and abandoned waste sites

Other Laws

Marine Protection, Research and Sanctuaries Act — 1972

The Marine Protection Research and Sanctuaries Act helped prevent unacceptable dumping in oceans. This has resulted in the cleanup of many beaches.

Clean Water Act Amendments of 1977

The Clean Water Act amendments strengthened controls of toxic pollutants and allowed states to assume responsibility for federal programs. This has helped to reduce many of the pollutants previously found.

Clean Water Act Amendments of 1987

The Clean Water Act of 1987 authorized the spending of $18 billion over 9 years for the construction of municipal sewage treatment plants, including $9.6 billion in direct federal grants and $8.4 billion in capitalization grants to help states establish permanent means of financing for municipalities that needed help in sewage funding. The act also authorized $400 million for a new state–federal program to control nonpoint source pollution. Examples of nonpoint source pollution include oil and grease runoff from city streets, pesticide runoff from farmland, and runoff from construction sites in mining areas. The act also includes further restrictions on toxic water pollution. A new section of the act related to pollution problems in estuaries, which are the coastal water bodies where freshwater and salty water join. The EPA has been given $60 million over a 5-year period to help improve water quality and to restore fish and wildlife as well as recreational opportunities in the estuaries. A new section of the act helps the United States in its efforts to comply with the 1978 Great Lakes Water Quality Agreement with Canada to reduce or eliminate pollutants from the Great Lakes. Other areas of the act help improve water quality in lakes and authorized $85 million for the lakes including $15 million for acid rain damaged lakes.

Section 303 of the Water Quality Act of 1987 added new requirements and prohibitions to the program. All POTWs had to have primary or equivalent treatment to remove at least 30% of conventional pollutants and to meet water quality criteria. If the plant served a population of 50,000 or more, industrial sources of toxic pollutants had to implement urban treatment programs. Modified discharges were prohibited alone or in combination with pollutants from other sources. The POTWs had to provide detailed descriptions and diagrams of the treatment system and outfall configuration. They had to identify the final effluent limitations for 5-day biochemical oxygen demand (BOD_5), suspended solids, pH range, flow in cubic meters per second, toxic pollutants and pesticides in micrograms per liter, and dissolved oxygen (DO) prior to chlorination and immediate dissolved oxygen in milligrams per liter in the receiving stream. The POTWs also had to identify the diameter and length of the outfall and diffusers.

The zone of initial dilution (ZID) is the region of initial mixing surrounding or adjacent to the end of the outfall pipe or diffuser ports and includes the underlying seabed. This area may be chronically exposed to concentrations of pollutants in violation of water quality standards and criteria. The treatment plant needs to provide profiles with depths for the current discharge location and for any modified locations. The information required includes: BOD_5, dissolved oxygen, suspended solids, pH, temperature, salinity, transparency, nutrients, toxic pollutants and pesticides, fecal coliform, and representative biological communities. In estuaries, a special problem occurs at low tide when the pollutants can accumulate and become more concentrated. Health-related factors are also considered. Discharges must not have an effect on recreational areas. Monitoring programs are used to determine the effectiveness of urban pretreatment and toxics control. A program consists of the study of biological, water quality, influent, and effluent factors.

A program was established to deal with storm water discharges, including storm water permit application regulations. Storm water discharge associated with industrial activity, discharge from a large or medium municipal storm sewer system, or other potential polluters are included. A biosolids program is part of this act. The rule proposed by the EPA also covers the land application of residuals from wastewater treatment.

The 1987 law became very much involved in nonpoint source pollution. Nonpoint source pollution comes from a variety of sources. It is caused by rainfall or snow melt moving over or through the ground. It takes up and carries away natural and anthropogenic pollutants and deposits them into lakes, rivers, wetlands, coastal waters, or ground sources of drinking water. The pollutants include excess fertilizers, herbicides, insecticides, oil, grease, toxic chemicals, sediment, salt, bacteria, viruses, nutrients, and pollutants from atmospheric deposition.

The National Estuary Program was established in 1987 by amendments to the Clean Water Act to identify, restore, and protect nationally significant estuaries of the United States. This program covers a broad range of issues and involves local communities in the process. It not only involves communities in improving water quality in an estuary, but on maintaining the integrity of the whole system. That is, it involves the chemical, physical, and biological properties, as well as the economic, recreational, and aesthetic values of the estuary.

The National Coastal Water Program includes the Chesapeake Bay Program, Great Lakes Program, and Gulf of Mexico Program. The mission of these programs is to lead and empower others to protect the ecosystem for future generations.

The Clean Water Act addresses the waters that are not fishable or swimmable. States are required to identify these waters and to develop total maximum daily loads for them.

The EPA Wetlands Program encourages and enables others to act effectively in protecting and restoring the country's wetlands. It establishes national standards and a permit program to regulate the discharge of dredged or fill material into the waters of the United States.

New Laws in the Year 2000

The National Marine Sanctuaries Amendment Act of 2000, and the Coastal Barrier Resource Reauthorization Act of 2000, reauthorized the National Marine Sanctuaries Act for 5 years to continue programs in 13 national marine sanctuaries and to protect coral reefs off the northwestern Hawaiian Islands. The Oceans Act of 2000, established a Commission on Ocean Policy to make recommendations to the President and Congress to prepare a coordinated and comprehensive national ocean policy. This act is needed because half of all new developments in the United States are occurring along the coast; algae blooms are forcing beach closures and threatening marine life and human health; demand for seafood is rising; and potential for the oceans to serve as a pharmaceutical resource is endangered. Numerous conferences were held in the year 2000 on marine and atmospheric science. These conferences were called to determine how best to solve numerous problems in the oceans. Nutrient overenrichment problems are particularly severe along the mid-Atlantic Coast and the Gulf of Mexico, resulting in a depletion of DO producing excessive algal biomass, and altering marine biodiversity. U.S. coastal eutrophication is largely caused by nitrogen pollution.

By 1987, discharges of pollutants from industrial plants had largely been brought under control. More work is needed with municipalities to bring them into compliance with the Clean Water Act. All water quality had remained roughly stable between 1972 and 1982, even though a substantial growth in population, industry, and development had occurred. As much as 13% of stream miles monitored had improved; 84% of stream miles monitored had stayed the same; and 3% of stream miles had deteriorated.

In a 1990 state assessment, 70% of the rivers and 60% of the lakes assessed met water quality standards and supported fish and swimming. However, agricultural sources contributed to 60% of the impaired stream miles and 57% of the impaired lakes acres. Sewage treatment plants contributed 16% of impaired miles, whereas storm runoff contributed 11% of impaired miles. Improvement has been made in DO deficit, fecal bacteria, and phosphorus because of improved wastewater treatment. An increase in nitrogen, chloride, and dissolved solids has occurred apparently because of an increase in use of fertilizers, road salt, and other nonpoint source pollutants.

Over the last 25 years, the quality of rivers, lakes, and bays has improved dramatically. This has been a result of coordinated action between federal, state, local, tribal governments, and interested citizens groups. Today, two thirds of the nation's surveyed waters are safe for fishing and swimming. The amount of soil loss due to agricultural runoff has been reduced by 1 billion tons a year. Phosphorus and nitrogen levels in water have dropped. Examples of improvement include Greenwich Bay, where shellfish fishing is returning with a sizable net gain for the economy; Chesapeake Bay, where fishing has been restored and toxic substances has been sharply reduced; Tampa Bay, where outdated sewage treatment facilities once poured sewage directly into the bay and now have been updated, thereby cleaning the water; and Cuyahoga River, which once caught fire, but is now a place for recreation and the return of a variety of birds. Public health protection has increased as levels of toxic substances and hazardous microorganisms have decreased.

As of 2002, the Clean Water Act has not been reauthorized, and therefore the 1987 law along with the various rules and regulations are still in effect. Various members of Congress have tried to amend portions of the law, but have not been able to move these amendments into law. In the 106th Congress neither House nor Senate committees scheduled major legislative activity on the Clean Water Act, and no comprehensive reauthorization bills were introduced.

SCIENTIFIC, TECHNOLOGICAL, AND GENERAL INFORMATION

Wetlands

Wetlands are areas that are inundated or saturated by surface or groundwater at a frequency and duration adequate to support vegetation typically adapted for saturated

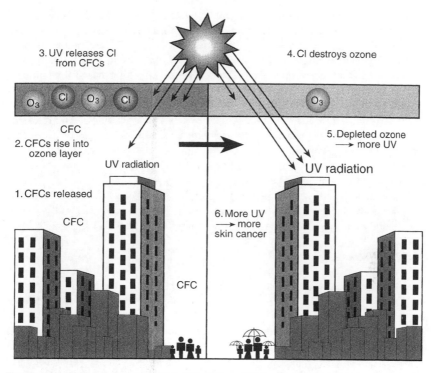

Figure 7.1   Wetlands. (Adapted from EPA Office of Water Resources.)

soil conditions. Wetlands generally include swamps, marshes, bogs, and similar areas. The wide variety of wetlands across the country are based on regional and local differences in hydrology, vegetation, water chemistry, type of soil, topography, climate, and other factors. The type of wetland is determined primarily by local hydrology and pattern of water flow. There are coastal wetlands closely linked to estuaries and inland wetlands. Inland wetlands are most common on floodplains along rivers and streams, in isolated depressions surrounded by dry land, and along the margins of lakes and ponds. Inland wetlands include marshes and wet meadows covered by grasses, sedges, rushes, and herbs. Shrub swamps as well as wooded swamps are dominated by trees, such as hardwood forests along floodplains. They may be found in various parts of the country. In wetlands water is the dominant factor determining the nature of soil development and the types of plant and animal communities living in the soil and on its surface (Figure 7.1).

Wetlands are now recognized as an important resource to people and the environment. Wetlands are among the most productive of all ecosystems. The wetland plants convert sunlight into plant material or biomass, which serve as food for many types of aquatic and terrestrial animals. The wetland plants form the base of an aquatic food chain. Wetlands are habitats for many forms of fish and wildlife. About two thirds of the U.S. major commercial fisheries use estuaries and coastal marshes as nurseries for spawning grounds. Migratory waterfowl and other birds depend on wetlands for nesting, feeding, or resting grounds. Wetlands are essential for improving and maintaining water quality in adjacent water bodies. These wetlands remove nutrients, such as nitrogen and phosphorus, and help prevent overenrichment of waters, which lead to eutrophication. The wetlands also filter harmful chemicals, such as pesticides and heavy metals, and trap suspended sediments that would produce turbidity in water. This is a natural buffer for nonpoint source pollution sources. Wetlands also help in flood control by absorbing peak flows and releasing water slowly. Along the coast they act as a buffer against the surges of hurricanes and tropical storms. Wetlands vegetation helps prevent shoreline erosion by absorbing and dissipating wave energy as well as encouraging the deposition of suspended sediments.

Unfortunately, because of neglect and wastefulness, wetlands from over 200 million acres in the continental United States have been reduced to some 101 million acres now. In addition, there are some 200 million acres of wetlands in Alaska.

Coastal Waters

The near coastal waters of the United States encompass inland waters from the coast to the head of tide, which is the farthest point inland at which the influence of the tides occur on the water level and can be detected. These waters include the bays, estuaries, coastal wetlands, and coastal ocean out to where it is no longer affected by land and water uses in the coastal drainage basin. These ecosystems support a wide range of ecological, economic, recreational, and aesthetic uses that demand good water quality. Much of the ecologically and commercially valuable species of finfish, shellfish, birds, and other wildlife live at some point in these coastal waters and wetlands. Of our country's commercial fish industry 85% is dependent on the near coastal waters. Billions of dollars are generated each year for recreational fishing, tourism, and travel; urban waterfront and private real estate development; and recreational boating, marinas, and harbors. The beaches, wetlands, and coastal oceans are used for swimming, boating, fishing, hiking, bird watching, and other recreational activities. These environments are especially susceptible to contamination, because they act as collection areas for the large quantities of pollution discharged from municipal sewage treatment plants, industrial facilities, and hazardous waste disposal sites.

Further nonpoint source runoff from agricultural lands, suburban developments, city streets, and combined sewers cause significant problems. When these areas are modified either physically or hydrologically by dredging channels, draining and filling wetlands, constructing dams, diverting freshwater, and diverting shorefront for houses, they may degrade rapidly. In some parts of our country, the growing population causes extreme pressure on these very sensitive coastal ecosystems. Major environmental problems in these areas also include toxic contamination, eutrophication, contamination by pathogens, and loss of habitat for the living organisms typically present there. Contamination has resulted in the closing of shellfish beds, bans on fishing, and bans on fish consumption.

Estuaries

An estuary is a partially enclosed body of water formed where freshwater from rivers and streams flows into the ocean, thereby mixing with the salty seawater. Although influenced by the tides, an estuary is protected from the full force of ocean waves, winds, and storms by the reefs, barrier islands, and pieces of land, mud, or sand at its boundary. As transition zones between land and water, estuaries are very useful to scientists and students, providing considerable information on biology, geology, chemistry, physics, history, and social issues.

Great Lakes

The Great Lakes are an invaluable resource to some 45 million people living in the surrounding area. The five lakes are under the jurisdiction of eight states and one province in Canada. These lakes contain 95% of the entire U.S. fresh surface water supply and 21% of the world's freshwater supply. Nutrients, toxics, pathogens from septage, sewage, and animal waste have caused substantial problems in the Great Lakes. Lake Ontario and Lake Erie, in particular, have suffered from eutrophication problems caused by an excessive input of nutrients. Toxics from pesticide runoffs, landfill leachates, and polluted sediments have been a major problem. Toxic chemicals have been found in different stages of the food web. The critical pollutants are polychlorinated biphenyls (PCBs), mirex, hexachlorobenzene, dieldrin, dichlorodiphenyltrichloroethane (DDT), dioxin, toxaphene, benzo(*a*)pyrene, mercury, and alkylated lead. Recent analyses of lake bottom sediment indicate that the major pollution time period was in the 1960s and 1970s and less pollution is now occurring.

Mixing Zones

A mixing zone is the area near a point source outfall, such as a pipe, in which concentrations of a particular pollutant mix with receiving waters. In these mixing zones, the water is allowed to exceed the water quality criteria for that pollutant. The discharge eventually becomes diluted beyond the borders of this zone. Mixing zones are part of the National Pollution Discharge Elimination System permits issued under the Clean Water Act. If the water is impaired, it cannot have mixing zones. Bioaccumulation of persistent pollutants mentioned earlier is a serious environmental threat to the Great Lakes Basin ecosystem. Mixing zones are not approved for a variety of these chemicals. About 300 major dischargers are emitting these hazardous chemicals into the water, leading to a continuation of fish advisories. The major source of pollution in the water comes from air deposition of mercury and other hazardous chemicals.

Ocean

Ocean dumping of dredged material, sewer sludge, and industrial waste has become the major source of ocean pollution. The sediments dredged from the industrialized urban harbors often are highly contaminated with heavy metals and toxic synthetic organic chemicals, such as PCBs and hydrocarbons from petroleum. The sediment, when dumped into the ocean, contains contaminants that can be then taken up through the food chain by the organisms present in the ocean. Persistent disposal of plastics from land and ships at sea have become a serious problem. The nonbiodegradable debris floats in the ocean and causes injury or death to fish, marine mammals, and birds. The debris also washes up on beaches and causes problems with the proper use of our resources.

Heat

Temperature affects the density and stratification of the water. It affects density and viscosity of sediment transportation, vapor pressure on evaporation rates, and partial pressure of gases on gas solubility, particularly oxygen and its impact on reaeration. Temperature also affects microbial reaction rate on deoxygenation by organic matter and impact on the oxygen sag. Temperature affects many physical properties of water. Water quality is most significantly affected by density, viscosity, vapor pressure, and solubility of dissolved gases. Very slight differences in density can cause a thermal stratification in quiet water bodies. The stratification process is well known in reservoirs and lakes, but the resulting change in water quality is not. Additional study is needed to understand this phenomenon. It is known, however, that in a stratified reservoir, cool incoming water can travel almost directly to the dam outlet, instead of having an opportunity to allow settling within the reservoir. The settling velocity is inversely proportional to the water viscosity and density. As the temperature rises, the viscosity and density decrease and settling rates increase.

A difference in settling velocities can contribute significantly to the amount of sediment and sludge that deposits in sluggish rivers, reservoirs, or estuaries. The evaporation rate of water increases as water temperature rises and water vapor pressure increases. Evaporation is caused by the wind and the difference in water vapor pressure between the air and the water. Water temperatures affect gas solubility. Because most living organisms depend on oxygen in one form or another, it is important that an adequate oxygen supply is present in the water. As the temperature rises in the water, the amount of dissolved oxygen in parts per million or milligrams per liter decreases. The temperature changes cause a complicated adjustment in the dynamic oxygen balance in the water and therefore compound the difficulty of relating the dissolved oxygen and other factors to oxygen demand, atmospheric reaeration, photosynthesis, diffusion, and mixing. Atmospheric nitrogen has a solubility of about one half that of oxygen.

Acidity and Alkalinity

An acid is a substance that produces a hydrogen ion in an aqueous solution. A base is a substance that produces a hydroxyl ion in an aqueous solution. In acidic water more acid materials than basic materials are present. Acidic water has to be less than 7.0, with water neutral at a pH of 7.0. Alkalinity is the condition in which more alkaline or basic materials are present than acidic materials in the water.

Alkaline water has a pH greater than 7.0. Acidity and alkalinity are determined with various colorimetric papers, pH meters, or titration devices. The pH values of some liquids are as follows: household lye, 13.7; ammonia, 11.3; baking soda, 8.3; distilled water, 7.0; orange juice, 4.2; lemon juice, 2.2; and battery acid, 0.2. The indicators utilized to determine pH are methyl yellow, which works in the range of 2.8 to 4.0; methyl orange, which works in the range of 3.1 to 4.4; methyl red, which works in the range of 4.4 to 6.2; cresol purple, which works in the range of 7.4 to 9.0; phenolphthalein, which works in the range of 8.0 to 9.6; and alizarine, which works in the range of 10.0 to 12.0. A buffer is a combination of substances that, when dissolved in water, resists a change in the pH of the water. Some chemicals used as buffering agents to maintain a pH in a given range include acetic acid or sodium acetate, boric acid plus borox, and borox plus sodium hydroxide. The pH of water affects water quality.

Conductivity

The electrical conductivity measurement of a solution determines the ability of the solution to conduct an electrical current. In concentrated solutions, large numbers of ions easily transmit currents. The commercial instruments used to measure conductivity are similar to a Wheatstone bridge. Natural water consists of numerous chemical constituents that may differ in ionic size, ability, and solubility. However, in most natural waters with less than 2000 mg/l of dissolved solids, the values of the specific conductants are closely related to the values of the dissolved solids. They range in a ratio of 0.62 to 0.70. Where there are substantially large quantities of nonionized solid material, such as organic compounds, this does not hold true. The specific conductivity is used to determine the purity of distilled and deionized water; is an estimate of the diluting factors needed for samples for quality control and analytic accuracy; and is an electrical indicator used in laboratory operations.

In agriculture, these tests determine salinity and also estimate the sodium adsorption ratio. In industry, the tests are used to estimate the corrosiveness of water and steam boilers, and to check the efficiency of boiler operations. In geology, the tests are used in geographical mapping and oil exploration. In oceanography, the tests are used to map the estuaries of ocean currents. In hydrology, the tests are used to locate new water supplies. In water quality studies, the specific conductivity aids in producing dissolved solids and concentrations of materials; and determines saltwater concentrations, the mixing efficiency of streams, the flow patterns of polluted currents, significant fluctuations in industrial wastewater effluents, and changes in the composition of the influent received by waste treatment plants.

Chemical Oxygen Demand–Biological Oxygen Demand–Dissolved Oxygen Relationships

Chemical oxygen demand (COD) is an estimate of the proportion of the sample matter that is susceptible to oxidation by a strong chemical oxidant. Other terms have been used for COD, although this is the preferred term. Oxygen absorbed (OA) is usually used in British literature; oxygen consumed (OC) and dichromate oxygen demand (DOD) are additional terms used. The COD test has the advantage of using less time, less manipulation, and has a lower equipment cost than the BOD test. COD measurement indicates the downstream damage and damage potential that sludge can cause. It is not applicable for estimating BOD, except as an experimental device.

The BOD (also known as biochemical oxygen demand) is used to determine that part of the material that can make a biological demand on the oxygen present in the receiving stream. The common 5-day BOD test is utilized as a means of stating what the level of contamination from the pollutant will be in the receiving stream. Time and temperature are essential factors in determining the BOD. BOD may also be affected by the amount of algae present and by the nitrification process.

DO is the amount of oxygen present in the stream. As has been explained previously, a certain amount of DO is necessary for fish, animal, and aquatic life. Because oxygen is essential to each of these living organisms, a drop in the oxygen level alters the existing aerobic conditions to anaerobic conditions, and putrefactive decomposition occurs that creates odorous sulfides, mercaptanes, and amines. COD, BOD, and DO tests must be run according to standard methods.

Solids Related to Water Pollution

The amount of solids present in water has been one of the major water pollution control criteria for many years because of its interrelationship to oxygen demand. Little attention has been given to this topic, because the solids have a more local effect, are difficult to measure, tend to keep moving, and are not completely understood. The solids present vary in proportion to time and conditions, from settleable solids to colloidal solids to dissolved solids.

The settleable solids are usually a mixture of organic, inorganic, trapped, dissolved, or colloidal solids plus living and dead organisms. These solids may be hydrolyzed, forming smaller molecules that become colloidal or dissolved in nature, or they may be converted into larger particles.

Colloidal solids are also a mixture of organic and inorganic material, and living or dead organisms. They come together and form settleable masses with time or under certain changing conditions. Chemical reactions can cause several variations from the colloidal solid portion.

The dissolved solids are most readily changed by biological, chemical, or physical processes. As each change occurs, a greater level of stabilization occurs as a result of the use of energy. The treatment processes help to remove many of the solids before they enter the receiving stream. The more effective the treatment is, the less the number of final solids that may affect the stream.

In the laboratory, the quality and quantity of dissolved solids, colloidal solids, and suspended solids are determined. The sum of all three become total solids. Suspended solids not only may be the result of the sewage treatment process, but also may be caused by rainfall washing particulate matter from the air or the surfaces into the water. Groundwater contributes dissolved solids that become suspended solids when oxidized. Wastewaters from sources other than municipal wastes also contribute suspended or dissolved solids.

Where the sewage systems are not properly maintained, breaks occur in the pipes leading to the treatment plant and solids may escape into the ground to a water table. Combined sewer systems remove drainage from many areas, including the sanitary sewers. In many communities, when the influent is too great for the sewage treatment plant to handle, a bypass valve is thrown and the combined sanitary sewage and storm water is permitted to go directly into the receiving stream. Standard methods should be utilized for the determination of the solids present.

Nutrients

Nutrients include nitrogen, phosphorus, carbon, and so forth. The major nutrients needed in large quantities are called macronutrients. The nutrients required in small quantities are called micronutrients. Nutrients are important because they promote a biological response that may interfere with some of the desired ways in which humans wish to utilize water. Other factors, such as temperature and light, affect the use of these nutrients and should be considered when evaluating them. Algae, bacteria, fungi, and aquatic plants are affected most directly by the nutrients. Algae consist of the phytoplankton, which are small algae suspended in water and are the basis of productivity in the aquatic environment; benthic algae are forms that are anchored to substrates of rock and bottom materials. Several forms of aquatic plants exist, including those that float and those that are rooted. Various bacteria are also found in the waters.

Certain biological laws apply to the water and have a direct bearing on the effectiveness of the nutrients that are introduced into water. Liebig's law states that essential materials available in amounts closely approaching the critical minimum tend to be a limiting factor in growth. Shelford's law states that the survival of an organism can be controlled by the quantity or quality, efficiency or excess, with respect to any of several factors that approach the limits of tolerance for that organism. The Q_{10} law states that a temperature increase of 10°C approximately doubles the metabolic processes. Photosynthesis is the fixation of the sun's energy with the production of organic matter by plants when chlorophyll is present. The general reaction is carbon dioxide plus water equals CH_2O plus oxygen.

Oxygen production is a measure of photosynthesis (Figure 7.2). For each mole of carbon dioxide reduced to organic carbon, 1 mol of free oxygen is liberated; the carbon dioxide taken up by algae has not all come from the dissolved gas. Some algae can use bicarbonate directly as a source of carbon. The removal of carbon from the water lowers the pH and causes deposits of calcium carbonate.

Nitrogen can be taken up by most algae, either as ammonium salts or as nitrates. Nitrites also may be used, if they are not present in high concentrations. At high concentrations, the nitrates inhibit growth. Some of the blue-green algae fix atmospheric nitrogen. One species of algae requires supplementary amines. The amount of nitrogen in water has a definite effect on the algae population.

Phosphates seem to be the only inorganic source of phosphorus. If the phosphorus ranges from 0.18 to 15 ppm, it is in optimum concentration for growth of algae. If it is at 20 ppm, it may start to inhibit algae growth. Algae store inorganic phosphates

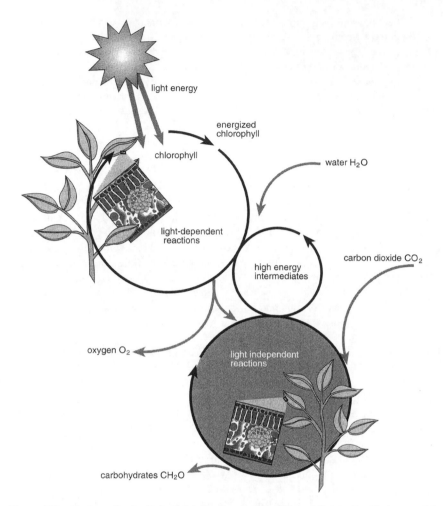

Figure 7.2   Photosynthesis. (Adapted from Cunningham, W. and Saigo, B., *Environmental Science A Global Concern,* McGraw-Hill, New York, 1999. With permission.)

in their cells. Silicon is important in the growth of algae. It is especially important in the growth of diatoms and may be the limiting growth factor in these populations.

Inorganic micronutrients include small quantities of magnesium, cobalt, manganese, molybdenum, zinc, and copper. Small quantities of organic substances include vitamin B_{12}, thiamine, and biotin.

Algae increase the total organic content of the water, thereby creating problems within the water. High algae populations produce waters that are high in turbidity, taste, and odor. The high respiratory needs cause a nocturnal oxygen deficiency. Certain algae cause taste and odors, clog filters, and in other ways interfere with the potable water process. The death of large algae populations leads to taste, odors, or both through bacterial decomposition. An oxygen deficit can result at any time. The organic sediment and sludge produced causes extreme problems to the stream.

Algae also serve as a supply of food to certain consumer organisms such as animals. With a reasonable quantity of algae, oxygen is released during the photosynthesis process and assists fungi or bacteria in breaking down pollutants. During the period of adequate light, favorable conditions may be available and visual breakdown can occur. In the darkness, cells continue to respire but consume more oxygen than they produce and therefore increase the organic load. Photosynthesis tends to occur at the surface where the light intensity is the greatest. Where poor vertical mixing occurs, stratification of water is found; and some of the strata have adequate or supersaturated oxygen, whereas others are oxygen deficient. Once nutrients enter a receiving stream, they become part of the cycle traveling through the food chain.

Organic nitrogen is nitrogen in combination with an organic radical. This nitrogen is contained in proteins, amines, amino acids, and amides, and is directly related to animals or vegetables. Most nitrogen compounds switch rapidly from one form to another by biological and chemical action. Organic nitrogen tends to hydrolyze free ammonia during storage. Organic nitrogen may come from dead animals and plant residues. Animal wastes and by-products may include urea and feces; autotrophic organisms, such as algae and bacteria; heterotrophic organisms; and food-processing wastewater from meat and milk. Other wastes include vegetable processing, pharmaceutical wastes, plastics, chemicals, and materials from the dye industry.

The nitrogen cycle best shows the relationship of hydrogen to ammonia and provides us with some understanding of the decomposition of organic materials. Ammonia is the product of the microbiological decay of animal and plant protein. It can then be used directly to produce new plant protein. Many fertilizers contain ammonia. If ammonia nitrogen is found in rural surface water, it might indicate the presence of domestic pollutants. If it is found in potable drinking water, large quantities of chlorine may be necessary to form a free chlorine residual. The chlorine initially reacts with the ammonia to form chloramines, which are somewhat bactericidal; and then finally, when all of the ammonia is consumed, a free chlorine residual is created. Nitrite nitrogen occurs in water as the intermediate stage in the biological decomposition of organic nitrogen.

Nitrate forms of nitrogen convert ammonia in aerobic situations to nitrites. Bacteria reduce nitrates to nitrites under anaerobic conditions. Nitrites are also used as a corrosion inhibitor in industrial process water. The large quantity of nitrites found in surface water indicates a source of wastewater pollution. Nitrates are converted by nitrate formers from nitrites under aerobic conditions. These nitrate formers are called nitrobacters. During electrical storms, large quantities of nitrogen gas are oxidized to form nitrates. Nitrates are also found in fertilizers. When nitrates are found in water, they usually indicate the final stages of biological stabilization. If effluents contain large quantities of nitrates as they enter the receiving streams, they can cause stress to the stream by aiding in the production of algae. Excessive amounts of nitrates in drinking water can cause a disease of infants and young children known as methemoglobinemia.

Total carbon analysis has become important, because it represents the kind and type of organic pollution that may be occurring in the stream.

Phosphorus is closely associated with water quality because of its use in the production of algae blooms. Its sequestering action interferes with coagulation. The difficulty of removing phosphorus from water is that it is unstable and changes from one form into another. Phosphorus exists commonly in the oxidized form. Most waters generally contain low levels of phosphorus (approximately 0.01 to 0.5 mg/l). The primary source of phosphorus in water is of geologic origin.

However, phosphorus is also a product of agriculture, households, and industries. Specifically, phosphorus entering streams comes from fertilizers; animal and plant residues; surfactants, which are part of cleaning agents; microbial and other cell masses; and wastewater from industrial sources where phosphorus is used for corrosion control or as an additive in scale control. Phosphorus is used as a surfactant or dispersant, and also in chemical processing of materials. Phosphorus may also be found in liquors that come from industrial cleanup operations of dusts, fumes, and stack gases. Phosphorus is available as a biological nutrient and therefore is synthesized into living masses or stored in living or dead cells. Phosphorus exists in several inorganic and organic chemical forms that may be converted from one to another under favorable conditions. From an engineering standpoint, phosphorus is related to treatment or control of water systems. Phosphorus removal is usually associated with solids removal. Solubility and temperature are major physical factors in phosphorus behavior. Soluble phosphorus is much more available than insoluble phosphorus for chemical or biological transformations. The rate of conversion from one to the other is strongly influenced by temperature.

Hydrology and land use are major factors controlling nutrient and pesticide concentrations in major rivers and in major aquifers. An increasing potential for nitrogen and phosphorus to enter streams is caused by high rainfall, snowmelt or excessive irrigation; and steeply sloping areas with insufficient vegetation; clayey and compacted soils underlain by poorly drained sediment or nonporous bedrock. Increasing potential for nitrate to enter groundwater is caused by high rainfall, snowmelt or excessive irrigation; well-drained and permeable soils; slow runoff; and low organic matter content.

Since 1972, the private and public sector has spent more than $500 billion on water pollution control directed at municipal and industrial point sources. However, the same level of concern has not been expended on nonpoint sources, which contribute to the widespread contamination of water by fertilizers and pesticides from agriculture and urban areas, and nutrients from human and animal wastes. About 15% of all shallow groundwater sampled underneath agricultural and urban areas exceeds the drinking water standard for nitrate. Many of these aquifers are used for domestic water supply. Nitrates did not appear to be a problem when the water supply came from streams or major aquifers buried deep beneath the land surface. However, in more than one half of the sampled streams, the concentrations of total nitrogen and total phosphorus are above national background concentrations. Elevated phosphorus levels can lead to excessive plant growth in freshwater environments. In three fourths of agricultural and urban streams, average annual concentrations of total phosphorus exceed the EPA desired goal for prevention of nuisance plant growth. Regional and seasonal differences in nutrient concentrations may be due to the amounts and timing of fertilizer and manure applications; and to the variety of soils, geology, climate, and land and water management practices.

In a recent study, it was shown that 90% of the nitrogen and 75% of the phosphorus found in water originated from nonpoint sources. Nonpoint source contamination comes from air pollution, agricultural lands, residential lands, livestock and waste from pets, and septic tanks. Nonpoint source contamination is the leading and most widespread cause of water quality problems.

The atmosphere is a nonpoint source of contamination. More than 3 million tons of nitrogen are deposited in the United States each year from the atmosphere, derived from natural or chemical reactions, or from combustion of fossil fuels such as gasoline. Locally, nutrients may be found in the air coming from lagoons in which manure is processed. Nearly every pesticide has been detected in air, rain, snow, or fog. Atmospheric deposition is not evenly distributed across the country. It is found more heavily in a broad band from the upper Midwest through the Northeast.

Eroded soil and sediment can transport considerable amounts of some nutrients such as organic nitrogen and phosphorus. The soil and sediment can also transfer pesticides to rivers and streams.

About 1 billion lb of pesticides are used each year. Agriculture accounts for 70 to 80% of total pesticide use. Most of the agricultural pesticides are herbicides. These chemicals along with commercial fertilizers end up in bodies of water.

Phosphorus from laundry detergents has declined as an environmental problem. This has resulted from state bans and voluntary removal of phosphates by detergent manufacturers.

WATER PROBLEMS

Natural water pollution is created through the silt that washes down along water beds due to rain or snow carrying dust, dust particles, and other materials into the water. Artificial pollution, until recently, was mostly created by domestic and simple industrial waste. The pollution problems were usually of a local nature and needed a local solution. However, as the national growth has increased, the production of goods has increased sharply and with it the production of common industrial waste has increased sharply. Further, new processes in manufacturing have produced new complex wastes that have not been easily handled by the current control technologies. The increased use of commercial fertilizers and the wide use of a large number of pesticides have contaminated many of our waters. Radioactive materials have entered waters by means of precipitation from the radioactive dust created by nuclear explosions. At present, long stretches of interstate and intrastate streams have become polluted and are continuing to be polluted.

Pollution usually is classified as either point source or nonpoint source. Nonpoint source pollution results from a vast variety of human activities, either within or outside the urban area. This includes wastewater from homes, drainage from landfills, and so on. In the rural areas, drainage comes from agricultural practices, croplands, and also any of the vast uses of pesticides. Point source pollution has come from specific industrial operations where specific materials have been dumped into the receiving waters.

Industry uses an enormous amount of water for cooling and processing. The largest amount of water utilized, which is over 90%, is for cooling purposes. The major amount is used to condense steam in thermal electric power stations. Excluding electrical utilities, industry still uses twice as much water for cooling as for processing and other purposes. Although most of the water used for cooling does not have any significant contaminants, these may be added through treatment chemicals, use of direct contact condensers, or leakage of the process into the water.

A definite problem occurs when the blowdown from an open recirculating system containing considerable hexavalent chromium or other compounds enters the water supply. Chromates and phosphates are used to control corrosion. Bromine, chlorine, copper, mercury salts, and quaternary amines are used to inhibit biological growths. The blowdown water comprises 0.5 to 2.0% of the total water used in the recirculation system. Another problem that is created in the cooling water is thermal pollution occurring as a result of the heat that has been absorbed by the water during the cooling process.

Water is used in the following ways and contains some potentially hazardous materials, such as a reactant in a chemical reaction in the production of acetylene, calcium carbide, or phosphoric acid; or such as a reaction medium to facilitate the mixing and contact of all components in the chemical pulping or sulfite process or in metal finishing baths. As a solvent, water is considered the universal solvent. Examples of this use include extraction of sugar from sliced sugar beets, hot water extraction of glues and gelatin from skin and bones, leaching of metallic ores by acid and alkaline materials, and extraction or removal of inorganic salts from crude oils. Water is also used for cleaning and washing of products, such as in the washing of fruits and vegetables, pulp washing after digestion, pulp washing after bleaching, or washing and soaking of any material where dirt, salts, and blood are to be removed; and as an intermediate and final rinse, such as in the rinsing of parts after plating or pickling, or after alkaline cleaning and metal finishing. Water is also used for washing equipment that contains grease, flour, sugar, or vast numbers of impurities; for quenching hot coke, glass furnace slag, or hot metals; for scrubbing of gases and carrying off any impurities that would have escaped from the smokestack; for transporting, thereby receiving all the contaminants of the material transported; as a processing medium, such as the flotation of ores and wet grinding; and as control of temperatures at critical stages in industrial processes, such as in the nitration of toluene to produce trinitrotoluene (TNT). As a result of these uses and various additional uses, a tremendous number of potential or actual wastewater sources enter the water stream.

Water Resource Problems

The federal government has been involved in water resources development since 1908, when a blueprint for national efforts was written for roads and canals. The U.S. Army Corps of Engineers made surveys of the Ohio and lower Mississippi Rivers in the early 1820s. In 1824, the Supreme Court, in *Gibbons vs. Ogden*, confirmed the federal government's power to protect and promote navigation under the commerce clause. In 1829, the corps had to build breakwaters and jetties to control the erosion occurring in South Carolina. In 1838, the Rivers and Harbor Acts were passed and extended the work of the Army Corps of Engineers. By the Civil War, the federal government was given $17 million a year to improve rivers, harbors, and canals. In the 1880s and 1890s, Congress directed the Army Corps of Engineers to prevent dumping and filling in the nation's harbors. By 1912 and 1913, disastrous floods that occurred along the Mississippi and Ohio river basins brought the federal government into flood control. The first federal Flood Control Act was passed in 1917. Additional laws were passed over the next 20 years. The federal government became involved with coastal erosion. In 1924, a General Survey Act authorized the President to use army engineers to survey road and canal routes of national importance. Concerted coastal erosion control began in the 1930s as a result of the 1930 Rivers and Harbors Act. In 1936, Congress passed an act for the improvement and protection of beaches along the shores of the United States.

The Homestead Act of 1862 provided 160 acres for family farms, which was fine for areas where water was plentiful but inadequate for areas of the West where water was needed. Congress then struggled with an irrigation policy and finally in 1877 passed the Desert Land Act, which allowed settlers up to 640 acres at $1.25 per acre. Through a continuous series of acts, the Congress became involved in western states in the conservation of the nation's water resources, not only for irrigation, but also for other purposes. In the 1880s, the federal government became interested in hydropower and required by 1889 that dam sites and plans be approved by the Secretary of War and the Corps of Engineers before construction. The largest hydroelectric plant up to that time, Hoover Dam, came under construction in the 1930s and then Grand Coulee Dam was constructed on the Columbia River. The Tennessee Valley Authority (TVA) provided electricity for a huge area using hydroelectric power. The TVA also was important for flood control, navigation, and reforestation. In 1944, the Flood Control Act gave the Secretary of the Interior the authority to sell power produced from federal projects. In 1968, the Wild and Scenic Rivers Act was passed to preserve prime undammed rivers. The focus of the act has now been broadened to river and corridor protection of about 10,000 river miles. All these acts, plus the additional acts on flood control and water project recreational areas, were the basis for the new environmental actions that started in 1969 with the passage of the National Environmental Policy Act (NEPA) and the establishment of the Council on Environmental Quality.

In 1973, the Water Resources Council that had been established in 1965, published Principles and Standards for planning water and related land uses. The Principles and Standards document was updated in 1980 and new guidelines were developed in 1983. Today the concept of watershed protection is used in water management and to address water supply issues.

Pollutants and Their Sources

Municipal sewage treatment plants contribute elevated BOD, bacteria, nutrients, ammonia, and toxic substances to the water stream. Industrial facilities contribute

BOD and toxic materials. Combined sewer overflows contribute BOD, bacteria, nutrients, ammonia, turbidity, total dissolved solids (TDS), and toxic materials to the water stream.

Nonpoint sources of pollutants contribute the following pollutants to the water stream:

1. Agriculture runoff: BOD, bacteria, nutrients, turbidity, TDS, and toxic materials
2. Urban runoff: BOD, bacteria, nutrients, turbidity, TDS, and toxic materials
3. Construction runoff: nutrients, turbidity, and toxic materials
4. Mining runoff: turbidity, TDS, acids, and toxic materials
5. Septic systems: BOD, bacteria, nutrients, and toxic materials
6. Landfills and spills: BOD and toxic materials
7. Silviculture runoff: BOD, nutrients, turbidity, and toxics

Pollutants in water impair or destroy aquatic light, cause potential disease outbreaks, or make the water unusable for recreational or aesthetic purposes. The pollutants restrict shellfish beds, cause bans on fishing and swimming, and cause beach closings. Algae-coated lakes and rivers are visible symptoms of such pollutants as nitrates and phosphates. Although toxic pollutants are less visible, they are the most challenging problem currently.

Municipal Waste

Raw or inadequately treated wastewater from municipal and industrial treatment plants can cause serious problems with water resources. Nutrients in the sewage create excessive growth of algae and other aquatic plants. When the plants in the water die and decay, they deplete the DO that is needed by fish and other aquatic organisms.

Sludge, which is the residue left from wastewater treatment plants, continues to be an increasing environmental problem. Since 1972, municipal sludge has doubled in volume to about 7 million dry metric tons yearly. From 1988 to the year 2000, sludge doubled again. The hazardous properties of sludge vary considerably based on potential toxic organic and inorganic chemicals and pathogenic organisms that are present in the initial wastewater.

Since 1970, the country has spent over $75 billion to construct municipal sewage treatment facilities. This extends secondary treatment to over 150 million people.

The sewers and pumps that convey wastewater from homes, factories, offices, and stores may be affected by the contents of wastewater. The industrial wastewater discharge varies by industry; however, in any case the discharge may contain a variety of toxic or otherwise harmful substances, which may include cyanide from electric plating shops and lead from the manufacture of batteries, corrosives, explosives, and oxidizing agents. All these substances can cause the system to deteriorate or create problems in the wastewater treatment plant. (See Chapter 6 for more details.)

Ocean Pollution

Ocean pollution consists of oil pollution, toxic wastes, and ocean dumping. Ocean pollution is a problem that directly affects ocean organisms and their natural balance. This pollution also may affect human health and resources.

Oil pollution is caused by spills or leakages that originate from land or rivers, which flow into the ocean. This pollution also occurs when ships transporting oil leak, crash, or sink. Some of the oil washes up on the shore and becomes tarlike lumps that can coat animals, birds, or seashores. Some of the oil may contaminate water sources in lakes, rivers, and personal water supplies. Tanker accidents account for 37 million gal of oil spilled worldwide, unless a major accident occurs. Routine maintenance of ships accounts for 137 million gal. Runoff accounts for 363 million gallons. Offshore drilling accounts for 15 million gal. Natural oil pollution, such as eroding rocks release 62 million gal.

Toxic waste is the most harmful form of pollution to marine mammals and fish, as well as humans. The toxic chemical may become part of the food chain and become concentrated. Toxic wastes come from leakage of landfills, dumps, mines, and farms. Sewage and industrial waste introduce a variety of hazardous chemicals. Farm chemicals include insecticides and herbicides as well as heavy metals such as mercury and zinc. Radioactive wastes, reactor leaks, natural radioactivity, and radio-active particles from nuclear weapons testing are dispersed in water. Medical waste include old blood vials, hypodermic needles, urine samples, and other contaminated materials.

The Marine Protection, Research, and Sanctuaries Act, Title I, known as the Ocean Dumping Act, generally prohibits transportation of material from the United States for the purpose of ocean dumping; transportation of material from anywhere for the purpose of ocean dumping by U.S. agencies, or U.S. flag vessels; and dumping of material transported from outside the United States into the U.S. territorial sea. Permits may be issued for dumping if the act cannot reasonably degrade or endanger human health, welfare, or marine environment. The EPA is charged with developing ocean dumping criteria to be used in evaluating permit applications. The EPA is also responsible for designating recommended sites for permitted ocean dumping. The dumping of sewage sludge and industrial waste is unlawful.

The ocean dumping of sludge, which had been carried out for many years, especially by the greater New York and Philadelphia metropolitan areas, was regulated under the 1988 Marine Protection and Sanctuaries amendments. On the West Coast, sludge was discharged into the ocean by pipelines, which were also regulated by the EPA under the Water Pollution Control Act.

Starting in 1992, no municipal sludge or garbage was permitted to be dumped in the ocean. In the summer of 1976, during mid-June, more than 70 miles of Long Island beaches were closed because of the influx of brown smelly garbage-filled water. Although the source could not be pinpointed, federal and state officials felt that it was a combination of untreated sewage dumped by New York and New Jersey communities, and materials discharged from ships. For 50 years, New York sludge barges had unloaded up to 150 million ft^3 annually in the waters 12 miles off Long Island shores. The huge fish kills that had resulted from this dumping not only created a serious financial problem but also indicated that a public health hazard was developing.

Over 90% of the total volume of material the United States has dumped into the oceans consisted of sediment dredged from harbors and channels. Permits are now issued by the Corp of Engineers after an EPA review for compliance with the ocean dumping regulations. The dredged material is dumped at sites designated by the EPA, and the EPA and Corp of Engineers share responsibility for monitoring the process. This system brings the United States into compliance with the London Dumping Convention, an international agreement approved by 64 countries to protect the marine environment. Under this convention, three annexes are to be considered when making decisions to issue permits for dumping. They are the black list, which is Annex I; the gray list, which is Annex II; and the other substances, which is Annex III.

The black list establishes those substances that are prohibited from dumping unless they are in trace concentrations. These substances include mercury, cadmium, and their compounds; organohalogen compounds such as pesticides and PCBs; persistent plastics; and crude oil and petroleum by-products. Also prohibited is the dumping of high level radioactive wastes and chemical warfare agents.

The gray list is a series of substances that require special permits to regulate their dumping. They include arsenic, lead, copper, zinc, and their compounds; cyanides; fluorides; organosilicon compounds; pesticides not covered in Annex I; low-level radioactive wastes; and containers and other bulky wastes.

The other substances list requires a general permit for dumping. The criteria used for determining if dredged material is acceptable for ocean dumping is based on water quality criteria, results of bioassay tests, and bioaccumulation tests. All existing information and available chemical data are considered.

National Eutrophication Study

In 1972, the EPA initiated a national eutrophication survey to identify and study lakes and reservoirs that had been affected by nutrients from municipal sewage discharges. After the Federal Water Pollution Control Act amendments of 1972 were passed, the survey was broadened to include lakes, which had also become a problem as a result of nonpoint sources. The studies made during this survey, however, were biased toward municipal wastes and therefore did not represent the conditions in all of the nation's lakes and reservoirs.

The survey found that, of the lakes studied, phosphorus was the element that most needed to be controlled to slow the rate of eutrophication. Phosphorus was the nutrient directly limiting algae production in 67% of the lakes studied. Nitrogen was the limiting nutrient in 30% of the surveyed lakes. Apparently the effluents from the sewage treatment plants were excessively high in phosphorus.

Land use, geology, soils, climate, and other geographic factors were important in determining nutrient levels in rivers and lakes. Some 1000 drainage areas that were selected for the land use study had been included in the approximately 4200 sampled drainage areas that were tributary to the survey lakes. The relationship between land use and average stream nutrient concentrations was determined for 473 drainage areas studied in the eastern United States. Streams draining areas classified as agricultural had a total phosphorus concentration of 0.135 mg/l, compared to streams draining forested areas, which had a total concentration of 0.014 mg/l.

Different regions of the country appeared to have different levels of stream nutrients present. The northeastern, or New England, states had relatively low stream nutrients, whereas the more central states of Minnesota, Michigan, and Wisconsin had generally high nutrient concentrations. Apparently a significant correlation existed between the pH characteristics in soils and nutrient concentrations. Concentrations of nutrients were generally higher in streams draining soils that were alkaline vs. streams draining soils that were mostly acidic.

Animal density at a watershed significantly influenced the stream nutrient levels. There appeared to be an increase in both phosphorus and nitrogen, depending on the number of cattle grazing in any given area.

Current Status

Over the past 28 years, despite population growth and increase in the amount of sewage entering wastewater treatment plants, BOD from the treatment plants has declined by 36%. Direct discharges of toxic pollutants have dropped dramatically since 1988 from approximately 200 million lb to approximately 50 million lb. The National Water Quality Inventory of states, territories, and tribes evaluated water quality in 19% of the nation's river and stream miles, 40% of lake areas, and 72% of estuary square miles. Standards for designated uses are met by 64% of river and stream miles, and 61% of lake, pond, and reservoir acres.

Metals, priority toxic organic chemicals, pesticides, and oil increases are among the leading persistent toxic pollutants causing water quality problems. Silt, nutrients, and oxygen-depleting substances are among the leading causes of habitat destruction.

In 1999, the National Oceanic and Atmospheric Administration (NOAA) National Estuarine Eutrophication Survey supplemented by information on nutrient maps, population projections, and land use for 138 estuaries representing over 90% of the estuarine surface area in the continental United States concluded that:

1. Of estuarine surface, 65% area is moderate to highly eutrophic.
2. Humans have a large impact on eutrophication.
3. Estuarine resource impairment affects commercial and recreational fishing and shellfishing.
4. Management requirements are dependent on present eutrophication and future susceptibility.
5. Without prevention, eutrophication may get worse over the next two decades in approximately 86 estuaries.
6. Information is still inadequate for 48 of the 138 estuaries surveyed.
7. More research needs to be done.

Currently, 28 National Estuary Programs use a combination of regulations, innovative initiatives, balanced and inclusive planning and management, scientific research and monitoring, and public education to improve these areas. Major concerns include nutrient overloading, pathogens, toxic chemicals, sediments, habitat loss and degradation, introduction of foreign species, alteration of natural flow, and decline in fish and wildlife populations.

Nonpoint Source Pollution

Nonpoint source pollution comes from many diverse sources instead of from a single industrial or sewage treatment plant. This type of pollution is caused by rainfall or snowmelt moving across the ground or through the ground. The runoff picks up and moves natural and human made pollutants and deposits into lakes, rivers, wetlands, coastal waters, or ground sources of drinking water. These pollutants include:

1. Excess fertilizers, herbicides, and insecticides from agricultural lands and residential areas
2. Oil, grease, and toxic chemicals from urban runoff and energy production
3. Sediment from improperly managed construction sites, crop and forestlands, and eroding stream banks
4. Salt from irrigation practices and acid drainage from abandoned mines
5. Bacteria and nutrients from livestock, pet wastes, and improperly operating septic tanks
6. Atmospheric deposition.

By 1988, nonpoint source pollution has been identified as a serious national concern. Levels of sediment and nutrients that have been discussed in the National Eutrophication Study had not been reduced from earlier years. Sediment, which had been the largest contributor to nonpoint source problems, caused decreased light transmission through water resulting in decreased plant reproduction, interference with feeding and mating patterns, decreased viability of aquatic life, decreased recreational and commercial values, and increased drinking water costs. Nutrients, which are now the second most common nonpoint sources of pollution, promote the premature aging of many lakes and estuaries. Pesticides and herbicides hindered photosynthesis in aquatic plants, affected aquatic reproduction, increased organisms susceptibility to environmental stress, accumulated in fish tissues, and presented a human health hazard due to fish and water consumption.

Nonpoint source pollution also became a major source of toxics, especially pesticide runoff from agricultural areas. Metals from active or abandoned mines, gasoline, and asbestos were found to come from urban areas. Bottom sediments of bodies of water also contain toxic contributors. Another source of surface water pollutants of a nonpoint category was related to air pollution. Small particles of dust containing toxic substances were deposited in surface waters through precipitation.

Nonpoint Volatile Organic Compounds in Urban Areas

Volatile organic compounds (VOCs) had been detected in the storm water and shallow groundwater from urban areas. Common compounds include toluene, xylene, tetrachloroethylene, trichloroethylene (TCE), and methyl-*tert*-butyl ether (MTBE). Urban land surfaces and air are two common nonpoint sources, with the land the primary source. VOCs are measured in stream samples as part of the U.S. Geological Survey's National Water Quality Assessment Program. The U.S. EPA and the Canadian Council of Ministers of the Environment have established water quality criteria and guidelines for 39 of the 87 organic compounds measured.

Nonpoint Semivolatile Organic Compounds

Semivolatile organic compounds (VOCs) may be found in streambed sediments. A survey of 460 locations throughout the country was used to determine the concentration of 65 SVOCs in a variety of environmental settings, with 41 different SVOCs detected. They included polycyclic aromatic hydrocarbons, phthalates, phenols, halogenated nitro- and nitros-compounds, etc.

Nonpoint Pollution from Agriculture

The United States has over 330 million acres of agricultural land. When the land is poorly managed, agricultural activities can affect water quality. The most recent National Water Quality Inventory Report states that agricultural pollution is a leading source of water quality problems in the rivers and lakes examined in this study. This pollution was caused by animal facilities, nutrients, pathogens, pesticides, and salts. Sedimentation occurs when wind or water runoff carries soil particles from a given area to a body of water.

Nonpoint Pollution from Boating and Marinas

The United States has more than 10,000 marinas. Although any given boat or marina usually releases small amounts of pollutants, the accumulated numbers due to many thousands of boats and marinas create the potential for high toxicity in the water; increased pollutant concentrations in aquatic organisms and sediments; increased rates of erosion; increased levels of nutrients; high levels of pathogens; low DO levels in water; and increased levels of metals and metal compounds. The water pollution is caused by poorly flushed waterways; boat maintenance activities; discharge of sewage from boats; storm water runoff from parking lots; and physical alteration of the shoreline, wetlands, and aquatic habitat during the construction and operation of the marina. Solvents, paint, oil, and boat cleaners containing chlorine, ammonia, and phosphates can be part of the runoff or penetrate the groundwater supply.

Nonpoint Pollution from Forestlands

Almost 500 million acres of forestland are used for production of timber in the United States. The most recent National Water Quality Inventory states that forest activities contribute about 90% of the water quality problems in the streams and rivers that were surveyed. The pollution comes from removal of vegetation, road construction and use, timber harvesting, and mechanical preparation for the planting of the trees.

Nonpoint Pollution from Households

Individual homes contribute only minor amounts of pollution, but the combined effects of an entire neighborhood can be very serious. The pollutants can cause problems related to eutrophication and sedimentation, as well as a potential for toxic problems. Urban and suburban areas are covered by sidewalks, parking lots, roads, and driveways. They reduce the level of water percolating down into the ground, and can cause high-speed runoff into storm drains that leads to bodies of water. The runoff carries with it oil, grease, and various other chemicals.

Nonpoint Pollution from Roads, Highways, and Bridges

Roads, highways, and bridges contribute a significant amount of pollutants to the country's waterways. Contamination comes from vehicles and work associated with road and highway construction and maintenance. Rainwater or melting snow washes off these surfaces, as well as roofs and other impermeable areas. The flowing water picks up dirt, dust, rubber and metal deposits, antifreeze, engine oil, pesticides and fertilizers, and various trash and other debris.

The pollutants from highway runoff includes sediment, nutrients, heavy metals, and hydrocarbons. The particulates and sediment come from vehicles, atmosphere, and maintenance of cars. Nitrogen and phosphorus come from atmosphere and fertilizer. Lead comes from soil and tire wear. Zinc comes from tire wear, motor oil, and grease. Iron comes from auto body rust, bridges, and guard rails. Copper comes from engine parts, brake linings, fungicides, and insecticides. Cadmium comes from tire wear and insecticides. Chromium comes from metal plating and engine parts. Nickel comes from fuel, lubricating oil, brake linings, and paving. Manganese comes from engine parts. Cyanide comes from deicing salt. Sodium, calcium, and chloride come from deicing salts. Sulfates come from roadbeds, fuel, and deicing salts. Hydrocarbons come from a variety of petroleum compounds.

Nonpoint Pollution from Individual Sewage Systems

Nonpoint pollution comes from the construction and maintenance of individual sewage systems. Overflowing sewage carries other pollutants into bodies of water. It creates potential health hazards. The sewage may also contribute to an increase in sediment problems.

Nonpoint Pollution from Urban Runoff

The most recent National Water Quality Inventory states that runoff from urban areas is the leading cause of problems in estuaries that were surveyed, and the third-largest source of water quality problems in lakes. Urban areas contribute to increased runoff and increased pollutant loads. New developments, as well as existing developments, have sharply increased the level of nonpoint source pollution.

National Water Quality Assessment — Nutrients

Nutrients in water are needed for productive aquatic ecosystems, but in high concentrations, nutrients can adversely affect aquatic life and human health. The most recent studies from the National Water Quality Assessment Program indicate the following major findings concerning nutrients:

1. Nutrient concentrations in water are usually related to land use in the upstream watershed or the area above the groundwater aquifer.
2. In about 12% of the domestic supply wells in agricultural areas, nutrients exceeded the standard of 10 mg/l.
3. Nitrate concentrations in groundwater are highest in agricultural areas, especially in the Northeast and Midwest, and on the West Coast.

4. Nitrate concentrations in groundwater are highest in areas of well-drained soils and where intensive cultivation of corn, cotton, or vegetables occurs.
5. Concentration of nitrates in surface water are not as high as in groundwater and rarely exceed the drinking water standard.
6. In the northeastern states, acid rain deposition may be causing elevated concentration of nitrates.
7. Ammonia and phosphorus concentrations in surface waters are highest downstream from urban areas.
8. Ammonia concentrations downstream from many urban areas and sewage treatment plants have been reduced. This has led to an increase in nitrate concentrations.

Harmful Algal Blooms

NOAA, National Science Foundation (NSF), EPA, and Office of Naval Research are jointly sponsoring a program to research harmful algal blooms commonly called red tide. *Pfiesteria* or related species have killed thousands of fish in the Chesapeake Bay and affected human health. As many as 39 people have suffered from skin rashes and other health effects, such as memory loss, as a result of this toxic microbe. Excessive nutrients, primarily from nonpoint sources, are significant factors in the growth of harmful algal blooms. The country's estuaries showed 51% with hypoxic conditions (oxygen less than 2 mg/l), and 30% with anoxic conditions. Each summer off the Louisiana coast, an area of approximately 7000 mi^2 is affected. Coastal eutrophication and harmful algal blooms are increasing and detracting from coastal economies as well as causing potential health hazards.

Pesticides

Pesticides are found frequently in streams and groundwater in all areas where they are used. One or more pesticides had been found in almost every stream sample collected as part of the National Water Quality Assessment Program. More than 95% of the samples collected from streams and almost 50% of the samples collected from wells contained at least one pesticide. Of the 83 pesticide compounds analyzed, 74 were detected at least once in streams or groundwater. Major rivers and agricultural and urban streams had the same level of insecticide concentration as other samples taken, with herbicide concentrations highest in agricultural areas, and insecticide concentrations highest in urban streams. Although individual pesticide levels generally met current regulations and criteria, the effect of multiple pesticide exposure and long time periods is a serious concern in evaluating the risk to humans. Also, the levels of pesticides differ at differing time periods and high-level exposure may not have been determined.

Urban streams had the highest frequencies of occurrence of DDT, chlordane, and dieldrin in fish and sediment, and the highest concentrations of chlordane and dieldrin. In addition, diazinon, carbaryl, chlorpyrifos, malathion, atrazine, simazine, and prometon were commonly found. Complex mixtures of pesticides occurred. The pesticides in urban areas came largely from use around homes, in gardens, parks, and commercial areas. The problem has been building for several decades.

Point Sources of Pollution

The most common point sources of pollution are industrial facilities, municipal treatment plants, and combined sewer overflows. The sources require permits. There has been a high level of compliance from these sources; however, some municipalities and industrial facilities are in noncompliance.

The major pollutants found in the Great Lakes are toxic organic chemicals mainly PCBs and DDT. Advisories have been issued concerning consumption of fish from the Great Lakes because of elevated concentrations of mercury, PCBs, pesticides, and dioxin in the tissue of fish.

Major Polluting Industries

Major polluting industries include the following:

- Aquaculture
- Centralized waste treatment
- Coal mining
- Construction and development
- Deicing of airports and roads
- Feedlots
- Industrial laundries
- Industrial waste combustors
- Iron and steel manufacturing
- Landfills

- Leather tanning and finishing
- Metal products and machinery
- Oil and gas extraction
- Pesticide formulating, packaging, and repackaging
- Pesticide chemicals manufacturing
- Pharmaceuticals manufacturing
- Pulp, paper, and paperboard
- Transportation equipment cleaning

The construction and development industry has a large impact on waterways. It creates a substantial amount of sediment and contributes large amounts of nutrients and metals to receiving streams. It impacts streambeds, habitats, shorelines, and fish populations, and increases the frequency of downstream flooding as well as aesthetic degradation.

The iron, steel, and coke manufacturing industries contribute considerable contaminants to the waterways. The manufacturing process including coke making, sintering, briquetting, iron making, steelmaking, vacuum degassing, ladle metallurgy, casting, hot forming, acid pickling, descaling, acid regeneration, cold forming, surface cleaning with acids and alkaline solutions, and surface coating all contribute to water pollution. This industry also uses a substantial amount of intake water and produces steam for power generation. In addition, industrial laundries contribute organic pollutants including solvents to water. The best way to correct this is to remove them before the materials are laundered.

Industrial waste combustors may incinerate hazardous or nonhazardous industrial materials. The pollutants come from air pollution control wastewater, slag quench wastewaters, flue gas quench wastewaters, and other wastewaters generated as a result of the combustion operations.

Landfills produce a variety of pollutants that may contaminate water (see Chapter 2). Numerous techniques are utilized to reduce the pollutants including equalization, activated sludge biological treatment, and multimedia filtration.

The leather tanning and finishing process includes the removal of hair from the product. Alkaline solutions are used for this purpose. The process must be monitored carefully to avoid high pH solutions from upsetting the wastewater treatment process in publicly owned treatment works.

Oil and gas drilling operations may use synthetic-based drilling fluids and other nonaqueous drilling fluids in the process and may discharge them into surface waters. These effluents can affect water quality.

Pesticide chemicals formulating, packaging, and repackaging may introduce hazardous chemicals directly to surface waters or indirectly to publicly owned treatment works. These pollutants must be reduced by about 99% to avoid serious water quality and potential health problems. Pharmaceutical manufacturing produces numerous waterborne chemical pollutants. Conventional, nonconventional, and priority pollutants need to be reduced sharply.

Pulp and paper mills in the United States produce 9 million tons of pulp each year and 26 billion newspapers, books, and magazines. They account for 35% of the pulp produced in the world and make up 16% of the pulp mills in the world. They produce 245,000 metric tons of toxic air pollutants that are released each year. Of the mills, 19 are associated with dioxin-based fish advisories. Wood consists of two primary components: cellulose and lignin. Cellulose is the fibrous component of wood used to make pulp and paper, and lignin holds the wood fibers together. Pulping is the process that separates the cellulose from the lignin. Pulping may be carried forth as a chemical, mechanical, or semichemical process.

The three chemical pulping methods are known as kraft, sulfite, and soda. The kraft and sulfite processes are most common. In chemical pulping, the wood is cooked in a digester at elevated pressure with a solution of chemicals to dissolve the lignin and leave behind the cellulose. This results in hazardous air pollutants such as formaldehyde, methanol, acetaldehyde, and methyl ethyl ketone, which have to be removed and may become water pollutants. In mechanical pulping, the wood is pressed against a grinder to physically separate the fibers. This is energy intensive and produces an opaque product that is weak and discolors easily when exposed to light. The semi-chemical pulping process uses a combination of chemical and mechanical methods. The wood chips are partially cooked with chemicals, and the rest of the process is done mechanically. After the wood is pulped, the pulp is washed to remove the dissolved lignin and chemicals. Next, the pulp is bleached to remove any remaining color. Chlorine, chlorine dioxide, and hypochlorite may be used in the bleaching process. Numerous chemicals including sulfur compounds, formaldehyde, chloroform, methanol, methyl ethyl ketone, dioxins, furans, and other halogenated organics may contaminate the air and the water.

Industrial Waste

Over 300,000 water-using factories in the United States have the greatest problems related to BOD and toxics. Because the character of industrial wastewater and the pollution abatement requirements vary greatly from industry to industry, it is difficult to generalize about industrial discharge. However, toxic materials, such as heavy metals and organic compounds, generally are more characteristic of certain untreated industrial wastewaters than of municipal wastewaters. Industrial wastewater may be quite acidic, neutral, or quite alkaline, whereas municipal wastewater tends to range from a pH of 6.0 to 8.0. The suspended solids in industrial wastewaters are usually quite different from those found in municipal wastewater. In the municipal wastewater, suspended solids are largely organic and they are usually associated with BOD. The solids in industrial wastewater are usually attributed to mining operations and are similar to the kinds of problems that occur from runoff, construction areas, and eroded agricultural land.

The effluent from industrial plants or municipal sewage plants contains toxic and organic wastes that may be hazardous to the aquatic environment. There is a maximum level that the stream can tolerate before problems occur. The contaminants, once the maximum level has been passed, can kill off the vulnerable organisms and the eggs of the fish, and may cause physical injury to the fish and plant life within the stream. The pollutants have a definite effect on the amount of dissolved oxygen, pH, color, and odor. The information that applies to fish life also applies readily to aquatic plant life.

In the past, water has been contaminated by industries because of their normal waste discharges, as well as accidental releases of chemicals into the water. An example of an accidental release occurred in South Charleston, WV. A 5000-lb discharge of carbon tetrachloride went into an Ohio River tributary. The Ohio River was then contaminated for more than 600 miles from Gallipolis, OH to Paducah, KY. Carbon tetrachloride is a carcinogen and can also cause damage to liver, kidneys, lungs, and the central nervous system. Cities on the Ohio River that use the river water for their drinking water supply had to close their intake gates to avoid bringing carbon tetrachloride into their water supply.

Oxygen-Depleting Waste

Organic wastes and ammonia, which come from domestic sewage and industrial waste of plant and animal origin, contribute to the reduction of oxygen in the receiving stream. The wastes are usually degraded or decomposed by bacteria. The bacteria need or utilize oxygen during this process. As the oxygen is depleted, fish and other aquatic life that depend on the oxygen to live, start to die.

Low-dissolved oxygen levels due to the oxygen demand also may significantly delay the process of self-purification of the stream. Typical industrial wastes that have high oxygen demands are sulfite waste liquors from pulp mills, effluents from canning plants, meat-packing wastes, textile scouring and dying effluents, waste from milk production, and fermentation wastes. Many other oxygen-depleting wastes exist beside these mentioned.

Toxic and Hazardous Wastes

Toxic and hazardous wastes, in contrast to oxygen-demanding wastes, have a direct lethal effect on various biological forms, whereas the oxygen-demanding wastes cause oxygen deficiency or smothering. Examples of these toxic or hazardous wastes include spent plating solutions that contain heavy metals and cyanides; acid wastes from mine drainage; and chemical manufacturing wastes, including organic materials, petrochemicals, pesticides, and textile dyes. Some specific hazardous wastes, such as kepone, aldrin, dieldrin, heptachlor, chlordane, PCBs, mercury, and lead, are discussed further in another section.

The nitrosamines occur throughout the environment. They discharge directly from the manufacturing processes. They also can be generated by the reaction of secondary amines and nitrite ions under acidic conditions. Nitrosamines had been found in the air in Baltimore, MD at levels of up to 15 times higher than expected.

Asbestos had been found in Lake Superior. The material was discharged from a company in Silver Bay, MN. The company was discharging 67,000 tons of asbestos-laden tailings into the lake on a daily basis. Asbestos-like fibers appeared in the drinking water of Duluth, MN and adjoining cities.

The most serious water pollution problems remaining are excessive levels of toxic pollutants in some streams. The Clean Water Act requires that the states identify waters that are not expected to meet water quality standards because of nonpoint source pollution and to develop programs to reduce the polluted runoff. States must also identify the toxic pollutants that are getting into the water and the responsible parties involved.

In 2001, seven additional industries, including electrical utilities, coal mining, chemical wholesalers, petroleum bulk plants and terminals, solvent recovery, and hazardous waste management facilities, reported their toxic waste output to the EPA. The total of all toxic chemicals recorded nationally is now 7.3 billion lb a year. Since the reporting of manufacturing began 13 years ago, total toxic discharges have decreased by 45%. Metal mining accounts for 48% of all discharges, or 3.5 billion lb. Electrical utilities account for 15% of all discharges, or 1.1 billion lb.

The most recent National Water Quality Inventory Report to Congress under the Section 305(b) of the Clean Water Act indicates the following:

1. The condition of 36% of rivers and streams is fair or poor.
2. Some form of pollution or habitat degradation exists in 39% of lake acres.
3. Fish consumption advisories are in effect for 97% of the Great Lakes shoreline.
4. A total of 38% of the estuary square miles are impaired for one or more uses.
5. Between 70,000 and 90,000 acres of wetlands are lost each year.
6. Leaking underground storage tanks (USTs) are the highest priority contaminants source in groundwater, with more than 300,000 confirmed releases.
7. Over 2200 fish consumption advisories are in effect, primarily due to mercury, PCBs, chlordane, dioxins, and DDT or its derivatives.

Waste Causing Physical Damage

Certain industrial wastes cause damage to the stream because of their specific physical properties, instead of because of the chemical or biological reactions that they may create. These wastes include materials from lumbering and mining operations, which produce large amounts of silt, sawdust, and other insoluble deposits; power plant discharges, which raise the temperature of the stream in the vicinity of the effluent discharge; petroleum refinery wastes, which add oils and other liquids that cannot mix readily, and which also add color and sludge to the receiving stream;

and dying operations, which can contribute large amounts of color to receiving streams.

The streams must be evaluated on the basis of the insoluble deposits that are present in terms of settleable and total solids tests. There also must be a determination of the thermal problems, oil problems, and industrial coloring problems.

Waste Producing Tastes and Odors

Many industrial wastes contain substances that contribute a variety of tastes and odors to the water supply. They include petroleum petrochemical waste; phenolics; liquid waste from synthetic rubber; organic chemicals such as dyes, medicinals, and plastics; coke quench water; creosols, acids, tars, crude oil, road oil, and gasoline; ammonia recovery gas liquors; wash water from canned goods; starches, such as corn and potatoes; acid from the sauerkraut-making process; nitrogenous wastes and lower sugars and acids from distilled beverages; yeasts from sugar beets; animal products from slaughtering and rendering containing nitrogenous and putrescent waste; fish containing amines; dairy foods containing nitrogenous and foul acid wastes; phosphate acid; and other chemicals, including cyanides, sulfides, and sulfates.

These industrial sources of odor have varying effects on different individuals. However, the odors are annoying, and they take away from the individual's ability to utilize or enjoy a given water body or water source.

Waste Containing Inorganic Dissolved Solids

Many industrial effluents contribute large quantities of the ions of sodium, potassium, magnesium, and iron. Typically, these come from tannery waste, irrigation waters, and other processes. The inorganic dissolved solids may also come from a variety of mining, oil field operations, agricultural practices, and natural sources. A vast variety of acids are discharged as waste by industry; however, the single largest source of acid is from mining operations. Many of the inorganic chemicals cause water hardness, corrode water treatment equipment, increase the cost of commercial and recreational boat maintenance, and increase the cost of waste treatment.

Plant Nutrients

Plant nutrients have been discussed in an earlier chapter and are discussed again in the section on the effects of pollution on stream life. These nutrients consist of substances in the food chain of aquatic life. The nutrients may be utilized by algae and water weeds to support or stimulate their growth. Carbon, nitrogen, and phosphorus are the three chief nutrients present. Large quantities of these are produced by sewage and also by certain industrial wastes and drainage from fertilized lands. The biological waste treatment process does not remove phosphorus, and only removes nitrogen to a limited extent. The process usually converts the organic forms of these substances into their mineral forms, making them more usable by plant life. When an excess of these nutrients overstimulates the growth of water plants, then unsightly conditions occur, unpleasant and disagreeable tastes and odors are formed, and streams begin the process of eutrophication.

Radioactive Waste

Radioactive waste may contaminate water or destroy aquatic life. Wastes may come from nature, from nuclear bomb explosions, or from nuclear reactor cooling water that has accidently entered a receiving stream. Quantities of waste may come from the uranium ore mining and refining process. Discussion in this area will probably increase in future years. Because the United States has a basic energy problem, eventually more nuclear power plants may be utilized. The problem will be proper storage of the radioactive waste from the power plants to prevent it from entering the water supply and creating a hazard.

The problem of radioactive waste management is one that still requires a considerable amount of research. At present, radioactive waste is stored in concrete vaults underground. Because of the toxicity of the wastes, these wastes must be isolated permanently.

Corrosive Waste

Corrosive wastes are highly acidic or basic wastes that may be toxic. These wastes create a tremendous problem to piping systems, ship hulls, bridge piers, and, of course, plant, fish, and animal life. Corrosive wastes not only cause physical destruction and toxicity to whatever they may come in contact with but also may utilize considerable oxygen. Typical corrosive wastes come from spent pickling solutions, alkalide discharges from the manufacture of soaps, and vast varieties of other industrial processes.

Pathogenic Waste

Pathogenic wastes are those wastes that may cause a problem to the individual by spreading disease. The category includes infectious organisms that are carried into the surface and groundwater by sewage from cities, institutions, industrial plants, and subsurface sewage disposal systems. The organisms may also be carried by runoff water into various bodies of water. Humans or animals come in contact with the organisms by drinking water, swimming, fishing, or use of the water for other recreational purposes. Typical pathogenic wastes from industry include the fecal and urinary waste materials from livestock, such as cattle, poultry, and swine; laboratory animals; materials from tanneries; pharmaceutical manufacturing waste; food processing waste; and leachate from solid waste disposal sites.

Thermal Pollution

Temperature can affect chemical reactions within the water, because heat has an important catalytic influence on chemical reactions. In general, the speed of a chemical change approximately doubles for each 10°C rise in temperature. If the chemical reaction is irreversible, the higher temperatures decrease the time required to produce the final product. Where the reaction is reversible, it comes to a halt when the rate of forward reaction equals the rate of reverse reaction. The majority of chemical reactions are brought about by temperature-sensitive enzymes. Microbial activity increases to a point with a change upward in temperatures.

The temperature affects not only the rate at which a reaction occurs but also the extent to which a reaction occurs. Changes in temperature may cause changes in ionic strength, conductivity, dissociation, solubility, and corrosion. This may bring about differing requirements in water treatment plants. Taste and odor problems are accelerated through increased temperature, which brings about more rapid chemical or biochemical action and which also depletes the oxygen more rapidly. The substances that accumulate as a result of this temperature acceleration are hydrogen sulfide, sulfur dioxide, methane, partially oxidized organic matter, iron compounds, carbonates, sulfates, and phenols. As the temperature increases in the area where material with a given BOD is discharged into the stream, an intensified action of microorganisms occurs and the BOD is removed much more rapidly, which in effect drops the dissolved oxygen excessively. This can cause a definite problem at the sag point, or the lowest level of dissolved oxygen present.

Many organisms can reproduce in polluted waters. As the temperature rises, the ability of some organisms to reproduce increases. Some of these organisms are present and reproduce in potable water as the temperature rises. These organisms include *Pseudomonas, Achromobacter, Escherichia, Aerobacter,* and *Streptococcus.*

Warm water fish survive temporarily in waters that are heated artificially to 93°F. Some fish, however, die at this temperature. In cold areas, 93°F temperatures considerably alter the number of fish available by increasing the mortality rate. Sudden changes in temperature can be harmful to all species of fish. At elevated temperatures, fish may starve because of their inability to capture food. They generally seek a zone that is best for survival and that is several degrees below the lethal temperature. A rise in temperature above 90°F also changes the population of organisms at the bottom of a water area.

The physiological effect of temperature on fish is based on the fact that most fish differ in body temperature by 0.9 to 1.8°F from the temperature of the surrounding water. Therefore, a change of temperature causes a very direct change within the body of the fish. As temperatures increase, the rate of metabolism increases. This may affect spawning or migration. A temperature rise may also change the cell chemistry by inactivating certain enzymes and accelerating others. This increase may also cause a coagulation of cell proteins, melting of cell fats, reduction of the permeability of the cell membranes, and eventually killing of the fish by toxic products of metabolism. Temperature affects the development and the distribution of fish. Further, synergistic actions occur, because as the temperature rises in the water, certain organisms increase sharply and chemicals that were toxic now become exceedingly toxic.

The major reasons for discussing the effect of polluted waters on fish, wildlife, and aquatic plant life are due to the increasing concern for necessary inexpensive sources of protein for humans, as well as the continuing concern for human beings who consume fish or animal wildlife, and thereby ingest quantities of critical materials that may be toxic or carcinogenic. In addition, a stream that has a large amount of dead fish is displeasing and detracts from the recreational and aesthetic aspects of the environment.

Dredging

Several hundred million cubic yards of sediment are dredged from waterways and ports each year to improve and maintain the country's navigation system and to maintain coastal national defense. The three management alternatives used for dredged material are open-water disposal, confined disposal, and beneficial use. Open-water disposal is a technique where the dredged material is placed in rivers, lakes, estuaries, or oceans by use of pipelines or released from barges. Confined disposal is a technique where the dredged material is placed within a diked nearshore or upland confined disposal facility by means of a pipeline or other techniques. Beneficial use involves the placement or use of dredged material for some productive purpose, such as wetland creation, island creation, or beach nourishment.

Potential environmental impacts resulting from dredged material disposal may be physical, chemical, or biological. Many of the waterways dredged are located in industrial and urban areas, where the sediments often contain contaminants from the sources.

The dredging process consists of excavation, transportation, and placement of dredged sediments in specific areas. Hydraulic dredging is the removal of loosely compacted materials by cutterheads, dustpans, hoppers, hydraulic pipeline plain suction, and sidecasters. Mechanical dredging is the removal of loose or hard compacted materials by clamshell, dipper, or ladder dredges. Hydraulic dredges remove and transport sediment in liquid slurry form. Mechanical dredges remove bottom sediment by direct application of mechanical force. The selection of the dredging process and equipment depends on the following:

- Physical characteristics of the material to be dredged
- Quantities of material to be dredged
- Dredging depth
- Distance to disposal area
- Physical environment of dredging and disposal areas
- Contamination level of sediments
- Method of disposal
- Production process required
- Type of dredges available
- Cost

Site selection is important in preventing environmental problems. Knowledge of the site characteristics is necessary for assessment of potential for physical impacts and contamination impacts. The following information is needed to make an appropriate determination of where the site should be located:

- Currents and wave climate
- Water depth and bathymetry
- Potential changes in circulation patterns or erosion patterns related to refraction of waves around the disposal mound
- Physical characteristics and size of the bottom sediment
- Salinity and temperature distributions
- Normal levels and fluctuations of background turbidity

- Chemical and biological character of the site
- Previous disposal operations
- Ability to properly monitor the disposal site
- Ability to control placement of the material
- Capacity of the site
- Public acceptability of the site

When dredged material is placed in a confined disposal area upland physical or chemical changes or both may occur. The dredged material that is initially dark in color and reduced, contains little oxygen. If the material is hydraulically placed in the confined area, the ponded water usually becomes oxygenated. This may affect the release of contaminants in the effluent. The exposed dredged material becomes oxidized and lighter in color. It may crack as it dries out and cumulations of salts may develop on the surface and then run off during periods of rain. Metal contaminants may become dissolved in the surface runoff. Sulfide compounds may become oxidized to sulfate salts, and pH may drop sharply. These chemical transformations can release complex contaminants to surface runoff, soil pore water, and leachate. Plants and animals may bioaccumulate these chemicals. Volatilization of contaminants may increase. The transfer rate for organics such as PCBs from sediment to air is faster than from water to air.

Subsurface drainage including leachate from confined areas may reach adjacent aquifers or surface waters. The leachate depending on where the material came from can cause significant environmental problems.

Beneficial use alternatives for disposal of dredging wastes can be quite successful, providing a competent and thorough study has been made of potential hazards to the environment or people. The dredging material is used to establish and manage relatively permanent and biologically productive plant and animal habitats. The material used for wetland restoration should be water saturated, reduced, and almost neutral in pH. Some of these habitats are now 50 years old and are still working successfully. Shore erosion is a major problem along the ocean and estuary beaches, and the Great Lakes. Dredged material is very useful for beach nourishment. Aquaculture is highly desirable and can be placed in dredging storage areas. Parks and recreational areas are expanded and protected by the use of dredged materials. The material may also be used for flood control. Dredged material may be used for agriculture, horticulture, forestry, strip mine reclamation, landfill cover, and industrial and commercial development.

Dredging Wastes

When a liquid discharge is made to a receiving stream, a zone of mixing is created. The water quality characteristics in the mixing zone differ from the receiving stream as a result of the effluent discharge. The materials found in the mixing zone consist of deposits that form part of the dredging wastes, scum oil, and floating debris; substances that produce objectionable color, odor, and turbidity; and substances that produce objectionable growth of nuisance plants and animals. The deposits coming from the varying sources mix with sediment from the stream and runoff materials from the soils. The total material dredged becomes a conglomeration of biological and chemical hazards. Under the 1988 Ocean Dumping Act, or Marine Protection and Sanctuaries amendments, ocean dumping by U.S. ships leaving U.S. ports is controlled by the EPA. Ocean dumping was stopped in 1991. However, the dredging wastes that are removed by the Army Corps of Engineers are still dumped into the oceans. The effect that this waste has on the oceans is not really understood at this time. It is known that the sediment contains heavy metals, chlorinated hydrocarbons, and other toxic materials at levels several times those found in the underlying water columns within the ocean.

Sedimentation Waste

Sedimentation produced by erosion is the most extensive pollutant of surface waters. It is estimated that the amount of suspended solids exceeds 700 times the suspended solids from sewage discharges. The dirty appearance of river or reservoir water after a rainstorm is usually due to sediments that wash from croplands, soils, overgrazed pastures, urban areas, and roads. The sediments increase the cost of water purification and reduce the value of the water areas for recreation. Sediments affect commercial and game fish areas, power turbines, pumping equipment, and irrigation distribution systems. The sediments may damage crops and anything that they may contact. Because erosion of land increases from four to nine times as a result of agricultural development and is increased 100 times by construction activities, there is ample reason for the tremendous increase in sediments that are found in the receiving streams. Unfortunately, along with the sediments comes a series of potential problems. The sediment reduces fish and shellfish population by blanketing fish nests and food supplies. Sediment also reduces the amount of sunlight that penetrates the water. Because sunlight is needed by the green aquatic plants to produce oxygen for normal stream balance, the sediments reduce the dissolved oxygen, thereby leading to additional problems.

The sediments have a definite effect on water quality, because they may be composed of decaying vegetation, organic deposits, materials in raw wastewater, industrial waste from paper mills, textile plants, gravel, and lumbering and mining operations. Sediments affect the color, odor, and texture of the surrounding water, and raise the BOD and COD of the stream. Beside the enormous cost of removal of sediments, additional problems are created by a vast variety of chemicals, including PCBs, other chlorinated hydrocarbons, and mercury, which tend to settle down into the bottom of the receiving stream. The sediment-bearing waste may create a problem to the water supply that is drawn from the receiving stream or it may create a problem when the area is dredged and the silt, including all the contaminants, is taken out to the ocean for dumping. Little is really known at this point concerning the effect of dumping huge quantities of sediment containing a vast number of impurities into the ocean. What may happen as the impurities are taken up through the food cycle has not been researched adequately.

In response to the Water Resources Development Act of 1992, the EPA prepared a report for Congress of the environmental health problems of sediments in the nation's waterways. Volume 1 was titled the National Sediment Quality Survey, and Volume 2 was titled Data Summaries for Watersheds Containing Areas of Probable

Concern. The EPA assembled the largest set of sediment chemistry and related biological data ever compiled into a national database. It is called the National Sediment Inventory. The database includes approximately 2 million records for more than 21,000 monitoring stations located in nearly 1363 of the 2111 watersheds in the continental United States. The reports findings are as follows:

1. Sediment contamination exists in every region and state of the country and there are 96 watersheds of concern.
2. Water bodies affected include streams, lakes, harbors, nearshore areas, and oceans. The PCBs, mercury, organochlorine pesticides, and polyaromatic hydrocarbons are the most frequent chemical indicators of sediment contamination.
3. Approximately 10% of the sediment underlying U.S. surface waters is sufficiently contaminated with toxic pollutants to pose potential risk to fish and humans and wildlife that eat fish.
4. Most of the contaminated sediment in the United States was polluted years ago by such chemicals as DDT, PCBs, and mercury.
5. Bioaccumulation is a major concern; therefore, new techniques of measurement, modeling including food chain modeling, human health-based risk assessment, and ecological-based risk assessment are needed.

Oil

In the year 2001, the total number of oil spills has decreased substantially. Between 1973 and 1999 of all the oil spills, 87% were 1 to 100 gal in volume; 46.7% came from ships or barges; 18.6%, from facilities; 17.6%, from pipelines; 7.7%, from unknown sources; and 5.8%, from nontank vessels. No spills over 1 million gal occurred between 1991 and 1999. However, because of the nature of oil transportation, size and ages of ships, and uncertain weather, new problems could easily arise. Oil spillage occurs in any of the waterways or coastal areas or on the high seas as a result of deliberate dumping; accidental spills; leaks in pipelines, drilling rigs, and storage facilities; or breakup of ships. Gasoline service stations dispose annually of more than 350 million gal of used oil. Pipelines totaling 200,000 miles carry more than a billion tons of oil. These pipelines that cross waterways and reservoirs are subject to cracking, puncturing, corrosion, or other causes of leakage.

Serious oil spillages from the past include the largest spill in 1967 of over 30 million gal of crude oil from the ship *Torre' Canyon* near England. The English spent over $8 million cleaning up after this spillage. In 1970, the tanker *Delian Apollon* ran aground and spilled over 10,000 gal of fuel oil into the bay. Over 100 mi^2 of sea became contaminated as a result. The worst period of oil problems occurred between December 15 and December 31, 1976. On December 15, 1976, the *Argo Merchant* was wrecked off Nantucket Island, MA. It dumped 7.3 million gal of oil into the sea. In the succeeding days, an explosion occurred aboard an oil tanker in Los Angeles Harbor; a hull wreckage occurred in New London Harbor, Connecticut; the *Olympic Gains* ran aground in the Delaware River; and the *Grand Zenith* disappeared south of Nova Scotia. In 1976, throughout the world, 19 tankers sank or ran aground. This doubled the total of the 1975 tonnage of oil pollution. In the first

9 months of 1976, tankers spilled nearly 200,000 tons of oil in various mishaps. In May 1977, a massive oil spill occurred when an oil derrick blew out in the North Sea.

In 1989, the *Exxon Valdez* ran aground off the coast of Alaska and created an oil slick that covered 900 mi^2 of water. It was a serious danger to the marine and bird life found in Prince William Sound. Everything that could have gone wrong went wrong. The Alaska State government, the U.S. Coast Guard, and the captain of the ship appeared to make errors that contributed to the rupture of the huge oil tanker. The cleanup cost billions of dollars and continues to cost additional hundreds of millions of dollars in lost ecological and environmentally pure areas. The birds, fish, and marine animals, such as seals and otters, that were not killed quickly by becoming coated with crude oil were in danger as the bottom oil contaminated the microorganisms and then the small fish that they eat. The catches of salmon, herring, shrimp, and crab were ruined for a considerable period of time. In 2001, the results of the spill and cleanup were a "mixed bag." A great deal of progress was made, but restoration was very painful and costly. Residents, commercial fishermen, and others felt that they did not receive adequate compensation for losses and that they were emotionally scarred by the disastrous event. Restoration is still not complete. In 1991, Iraq purposely released millions of barrels of crude oil into the Persian Gulf. No one fully understands the ultimate damage to the environment.

Some of these potentially harmful substances, such as the polycyclic aromatic hydrocarbons (PAHs), may cause cancer in laboratory animals. Some crude oil or refined oil is not made up of a single chemical or even a single classification of the chemicals, and thus poses a wide variety of potential hazards to the marine life. The fish life is exposed not only to the hydrocarbons present in the oil but also to the hydrocarbons that may combine with oxygen, nitrogen, sulfur, or other pollutant chemicals that may be present. Fish readily take up hydrocarbons into their bodies. (The hydrocarbons may be detected at less than 1 ppm in open water.) In an oil-polluted marine environment, the marine sediments may reach several thousand parts per million. The surface slicks may cut off the oxygen supply to aquatic organisms, especially in their larval form, and cause immediate kills of these organisms. A secondary effect may be the reduction of light and heat due to the surface oil. This reduction in the lower levels of the water is caused by the lack of penetration of light and heat in the water. The oils may cause the birds or mammals to drown because buoyancy is destroyed, or simply because they become intoxicated and cannot perform in a normal manner.

Crude petroleum or its products produce the hydrocarbons that accumulate in the tissues of the liver, the brain, the muscles, and other tissues and body fluids of fish. Aromatic hydrocarbons are at higher concentrations in fish. Although some fish may be able to metabolize and excrete crude oil products, many others cannot. Hepatic tumors and other liver diseases may be found among bottom-dwelling species of fish, such as English sole.

One of the biggest problems of oil contamination is the effect on the reproductive system of fish. Oil is toxic to fish eggs and larvae. These effects include alterations in embryonic activity, heart rate, premature or delayed hatching, malformed larvae, and decreased larval survival. The oil also has an effect on the taste of the fish and shellfish when adsorbed into the fatty tissue, and therefore, creates an economic loss because this source of food may be unfit for human consumption.

Unfortunately, as the demand for energy increases and as the sharp increase in imported oil continues to meet the demand, greater numbers of oil wrecks are bound to occur. In the United States alone, there are an average 30,000 arrivals a year of oil tankers. The Coast Guard estimates that in a 2-year period the United States can expect at least 6 serious oil spills and 86 tanker groundings. Three fourths of these will probably occur in the Gulf of Mexico, because the area is shallow. Whatever the ultimate solution may be, at present, enormous amounts of oil are spilled into our oceans and waterways, and this oil is contributing substantially to the serious problem of water pollution.

Mine Drainage

Mine drainage is one of the most important causes of water pollution, especially in Appalachia and Ohio basin states, as well as in some of the other mining areas of the United States. The water pollution is primarily one of chemical pollution and sedimentation. Acids are formed when water and air react with the sulfur-bearing minerals in the mines or refuse piles to form sulfuric acids and iron compounds. The acid and iron compounds drain into ponds and streams and eventually into rivers.

About 60% of the mine drainage is caused by worked and now abandoned mines. Although coal mines may not have been used for 30 to 50 years, they still may discharge large quantities of acidic waters. The acid pollution is fairly limited to the areas surrounding the coal mines; however, the suspended solids and sedimentation may move downstream considerable distances and continue to cause damage. Mine drainage seriously affects municipal and industrial water supplies; interrupts recreational uses of waters; lowers the aesthetic quality of water; and corrodes boats, gears, and other structures. Mine drainage kills fish as well as other aquatic life. The total unneutralized acid drainage from active and unused coal mines in the United States is estimated to be over 4 million tons of equivalent sulfuric acid. This estimate is for a discharge on a yearly basis. About twice this amount is actually produced, but half of it is neutralized in the mines and streams. The sediment from strip mine areas averages about 30,000 tons/mi^2 annually. This is 10 to 60 times the amount of sedimentation from agricultural lands.

Feedlot Pollution

Feedlot pollution is increasing sharply because of the number of animals that are raised in concentrated areas. The heavy concentration of the waste material, particularly manure, is extremely high in BOD demand; and when it finds its way into streams, it causes very serious problems. In addition, the animal waste adds excessive nutrients that imbalance the natural system within the water. This causes excessive aquatic plant growth and kills fish, creating further difficulties for water treatment systems because of the addition of undesirable tastes, odors, and solids.

Agriculture is the most widespread source of pollution in the country's surveyed rivers and lakes. In the 22 states listing specific types of agriculture, animal operations impacted about 35,000 miles of river. Animal feeding operations can pose a risk to water quality and public health, because of the amount of animal manure and wastewater generated. Manure and wastewater from animal feeding operations have the potential to contribute pollutants such as oxygen-demanding substances, ammonia, nutrients (nitrogen, phosphorus), sediment, pathogens, heavy metals, hormones, and antibiotics. Manure is the primary source of pollution. U.S. animal waste production is 13 times greater on a dry-weight basis than human sanitary waste. Sources of manure include direct discharge, open feedlots, pastures, treatment and storage lagoons, manure stockpiles, and land application fields. Other animal wastes include dead animals, flush water, bedding, feed wastes, and wash water.

Animal feeding operations contribute a considerable amount of pollution to the 40% of the country's waterways that do not meet the goals for fishing, swimming, or both. This in part is due to the 1.2 million livestock and poultry operations in the United States. About 450,000 of these agricultural operations nationwide confine animals. The trend is to larger and fewer farms where the pollutants are more concentrated. This trend will continue as a result of the February 1998 Clean Water Action Plan approved by the President. The plan provides a blueprint for restoring and protecting water quality across the country. There are 111 specific actions to expand and strengthen existing efforts to protect water quality. Beside improving sewage treatment, controlling industrial waste, and protecting recreational waters, it also identifies polluted runoff, including animal wastes, as a very serious problem.

Waste from Watercraft

In the various marinas in our country, a serious problem of wastes coming from watercraft exists. Over 46,000 federally registered commercial vessels, over 65,000 unregistered commercial fishing vessels, over 1600 federally owned vessels, and over 8 million recreational watercraft use the navigable waters of the United States. The potential sewage load from these vessels is equivalent to the amount of raw sewage that would come from more than 500,000 people. In major harbors, this constitutes a water pollution hazard that may be particularly important to shellfish harvesting, recreation, and also the water utilized for drinking supplies. An increase in concentration of fecal coliform bacteria in water has occurred in areas where vessels discharge. This problem is especially bad in lakes. The nutrients found in sewage encourage algal growth, smother coral reefs, decrease animal and plant diversity, and affect fishing and swimming. Ships contribute other materials by discharging bilge and ballast water, which contain oils and a variety of other substances.

Irrigation

When new irrigation schemes are introduced into previously dry areas, disease tends to follow the path of the water. The changing patterns of disease are due to microclimatic and other environmental changes. Arthropods, reservoir animals, and other intermediate hosts are involved in the transmission of many waterborne parasitic diseases. Insect vectors are especially important.

PUBLIC HEALTH ASPECTS OF WATER POLLUTION

Whereas microbiological hazards used to be the major concern associated with water quality, today the situation has become much more complex. Water treatment plants and the chlorination process have eliminated many of the past waterborne outbreaks of disease due to microbiological factors. However, the potential for disease due to microorganisms is ever present, with actual disease outbreaks continuing to occur. From time to time, new organisms seem to create a new potential for disease.

The modern world is one of chemicals that when used properly are of great value, but when misused or when not thoroughly understood in their use become a direct hazard to humans. In addition to microbiological problems, chemical problems and potential hazards from radiation are creating additional public health concerns.

BIOLOGICAL HAZARDS

Bacteria

Organisms associated with intestinal waste from warm-blooded animals found in wastewater and sludge may be of major pathogenic concern or of minor pathogenic concern as transported through sludge. Those of major pathogenic concern include *Campylobacter jejuni*, which may also be found in a nonhuman reservoir of cattle, dogs, cats, and poultry; *Escherichia coli* (pathogenic strains); *Leptospira* spp., which may also be found in domestic and wild mammals and rats; *Salmonella paratyphi*; *S. typhi*; *Salmonella* spp., found in domestic and wild mammals, birds, and turtles; *Shigella sonnei*, *S. flexneri*, *S. boydii*, and *S. dysenteriae*; *Vibrio cholerae*; *Yersinia enterocolitica,* also found in wild and domestic birds and mammals; and *Y. pseudotuberculosis.*

Pathogenic bacteria of minor concern as transported through sludge include *Aeromonas* spp., *Bacillus* spp., *Clostridium perfringens, Coxiella burnetii, Enterobacter* spp., *Erysipelothrix* spp., *Francisella tularensis, Klebsiella* spp., *Legionella pneumophila, Listeria monocytogenes, Mycobacterium tuberculosi, Mycobacterium* spp., *Proteus* spp., *Pseudomonas aeruginosa, Serratia* spp., *Staphylococcus aureus,* and *Streptococcus* spp.

Many of these organisms in themselves are major problems, depending on the route of entry into the body and the mode of transmission. *Campylobacter jejuni* causes acute gastroenteritis with diarrhea. It may be as prevalent as salmonella and shigella, and has been isolated from the stools of 48% of patients with diarrhea. The pathogenic strains of *E. coli* are of three types. They are enterotoxigenic, enteropathogenic, and enteroinvasive. All produce acute diarrhea, but by different mechanisms. The fatality rates may range up to 40% in newborns. Outbreaks occur occasionally in nurseries and institutions, and are commonly found among travelers in developing countries. *Leptospira* spp. are bacteria excreted in the urine of domestic and wild animals. Rats are the primary source of the *Leptospira* to enter municipal wastewater. Leptospirosis is a group of diseases caused by the bacteria that gives the symptoms of fever, headache, chills, severe malaise, vomiting, muscular aches, and conjunctivitis; and occasionally meningitis, jaundice, renal insufficiency, hemolytic anemia, and skin mucous membrane hemorrhage.

Routinely, tests are taken of the bacteriological quality of water by measuring the indicator groups, such as coliform and certain subgroupings, fecal streptococci and certain subgroupings, and other miscellaneous indicators of pollution. If these organisms are present, it is assumed that the water has been contaminated by warm-blooded animals and therefore may be a hazard to humans. Sources of *E. coli* are now determined by chromosomal DNA matching.

The term *enteric bacteria* includes all those facultative bacteria whose natural habitat is in the intestinal tract of humans and animals, including members of several different families. These bacteria are all Gram-negative, non-spore-forming rods. The Enterobacteriaceae family includes the Escherichieae, made up of *Escherichia*, *Edwardsiella*, *Citrobacter*, *Salmonella*, and *Arozona*; the Klebsielleae, made up of *Klebsiella*, *Enterobacter*, *Hafnia*, and *Serratia*; the Proteae, made up of *Proteus*; Yersinieae, made up of *Yersinia*; and Erwinieae, made up of *Erwinia*. Obligate anaerobic bacteria, which constitute 95 to 99% of the intestinal flora, are usually not included in the term *anaerobic bacteria*. The term *total coliform* defines the indicator that is looked for as an indication of contamination of the water.

Pathogenic bacteria found in the feces of infected persons, including the following, are of great concern:

- *Campylobacter jejuni*
- *Escherichia coli* (enteropathogenic strains)
- *Salmonella paratyphi* (A, B, C)
- *S. typhi*
- *Salmonella* small species
- *Shigella sonnei, S. flexneri, S. boydii, S. dysenteriae*
- *Vibrio cholerae*
- *Yersinia enterocolitica, Y. pseudotuberculosis*

The previous organisms, when placed into the soil through sludge application or through the dispersal of effluent from septic tanks, survive based on a series of specific conditions. These conditions are moisture content, the higher the moisture the greater the survival time is; moisture-holding capacity, the greater the holding capacity, the greater the survival rate is; temperature, the lower the temperature, the greater the survival rate is; pH, the lower the pH (3 to 5), the shorter the survival rate is; sunlight, the more sunlight at the surface, the shorter the survival rate is; organic matter, the more organic matter, the longer the survival rate is; and soil microorganisms, the greater the number of competing organisms, the shorter the survival rate is.

Fecal streptococci are excellent indicators of human waste pollution. These organisms are also called enterococci, group D streptococci, and *Streptococcus fecalis*. Fecal streptococci are associated with human feces and therefore could be good indicator organisms. However, the difficulty with using these organisms is that a special medium is needed for the growth of all the varieties of streptococci that fall into the fecal strep group. The pH range of the fecal strep must be kept at 4 to 9, and the samples must be analyzed as quickly as possible. Storm water and combined sewers are known to carry a variety of fecal strep in substantial numbers. Generally, if surface waters contain fecal strep and few or no other warm-blooded animals are present, the problem can be attributed to humans.

The advantages of fecal strep tests are that because these organisms have a more limited survival time in the aquatic environment, if they are present, the environment has probably been recently polluted with human waste or other warm-blooded animal waste. The disadvantage is that fecal strep may come from the throat or skin, and therefore may not be an indicator of fecal material present. This is especially true in artificial pools. The normal *S. fecalis* is ubiquitous, and therefore significance is limited in finding this organism. However, the atypical *S. fecalis* that hydrolyzes starches is a better indicator of pollution from human waste.

Shigellosis or bacillary dysentery, again described in the food section in Volume 1, Chapter 3, is rarely fatal, but it can create a serious hazard to the young, very old, or debilitated, and of course can create a carrier state in the individual. Shigellosis may be caused by a variety of shigella, including *Shigella dysentriae*, *S. sonnei*, *S. flexneri*, and others. Humans are the reservoir for the infectious spread of this disease for a period of up to 1 to 2 years.

Paratyphoid fever caused by *Salmonella paratyphi* and other salmonella comes from humans through urine or feces. Salmonellosis, which is caused by a variety of salmonella, and typhoid fever, which is caused by *S. typhi*, have already been described in Volume I, Chapter 3 on food.

Although the actinomycetes and the filamentous iron bacteria do not cause disease, they do cause odors that are annoying in water supplies. They may create accumulations, which cause taste, odor, and color problems, and also clog piping systems or wells.

Bacteria are commonly found in water when it either rains or snows. The rain washes the bacteria out of the air and into the water supply. The bacterial content varies with the amount of dust and the total bacteria count in the air. Large numbers of bacteria enter the water from runoff from various lands. The bacteria may act in unpredictable manners, depending on the quantity of nutrients present and the amount of bacteria available plus temperature, pH, speed of the stream, presence of predators, amount of sedimentation, antibiotics and toxins in the water, salinity, turbidity, and presence of industrial waste.

Water Quality and Shellfish

Shellfish receive their food by pumping water from their environment through their bodies. The food is strained out and absorbed into the body. Along with ordinary food, grains of sand, dirt particles, various bacteria, and viruses are found. As a result of this feeding characteristic, shellfish can and have become frequently contaminated with a variety of disease-causing bacteria. Two of the most recent problems have been infectious hepatitis and cholera. In addition, shellfish absorb pesticides, metals, radionuclides, and other hazardous chemicals. As a result, shellfish harvesting and growing areas must be strictly controlled. No pollution of any type must enter a certified shellfish growing area.

Viruses

A variety of potential viral diseases may also be transmitted through water from human feces. At times, small but very intense outbreaks of infectious hepatitis have been traced to drinking water from contaminated wells, ponds, or streams. Any of the enteric viruses may exist in the water, and therefore may create potential hazards to humans. It has been reported that apparent outbreaks of waterborne disease have caused diarrhea in tourists visiting summer resorts. The organisms causing these outbreaks are assumed to be viral in nature, but have not been tracked down adequately. Adenoviruses that produce acute respiratory diseases and eye infections have been associated with chlorinated swimming pools. It is known from laboratory studies that several viruses classified in the enteric group have survived storage in certain cold waters from 6 to 7 months. These apparently can then become a source of potential disease hazards.

Human enteric viruses may be present in wastewater and sludge. These viruses include the enteroviruses, such as polio virus, coxsackie virus A, coxsackie virus B, and echovirus. Also found in human waste are the hepatitis A virus, rotavirus (duovirus, reovirus-like agent), Norwalk-like agents (Norwalk, Hawaii, ditching), adenovirus, reovirus, astrovirus, calicivirus, and coronavirus-like particles.

When the viruses enter into the alimentary tract if they are not inactivated by the hydrochloric acid, bile acids, salts, and enzymes, they may multiply within the gut. Occasionally after continued multiplication in the lymphoid tissue of the pharynx and gut, viremia may occur. In viremia, the virus enters the bloodstream, leading to further virus proliferation in the cells of the reticuloendothelial system and finally the involvement of the major target organs.

Polioviruses cause poliomyelitis, an acute disease that may range from fever to aseptic meningitis, muscle weakness, or complete paralysis. The weakness and paralysis are caused by the destruction of motor neurons in the spinal cord. Although polio is rare in the United States today, it may be found in nonimmunized populations in the rest of the world. There does not appear to be any reliable evidence that the poliovirus is spread by wastewater at this time.

Coxsackie viruses may cause aseptic meningitis, myocarditis, pericarditis, pneumonia, rashes, common colds, fever, hepatitis, and infant diarrhea. The echoviruses may cause aseptic meningitis, paralysis, encephalitis, fever, rashes, colds, pericarditis, myocarditis, and diarrhea. The enteroviruses may cause pneumonia, bronchitis, acute hemorrhagic conjunctivitis, aseptic meningitis, encephalitis, and hand, foot, and mouth disease.

The prevalence of the diseases caused by the coxsackie viruses, echoviruses, and enteroviruses are not well known. These enteroviruses are found practically throughout the world and may spread rapidly in silent or overt epidemics, especially in the late summer and early fall in temperate regions. Children are the major target of enterovirus infections and are the main source of the spread of the diseases. Most of the infections are asymptomatic and natural immunity is acquired as the children get older. The poorer the sanitary conditions, the more rapidly immunity develops. About 90% of the children living under poor hygienic circumstances may be immune to the prevailing enteroviruses by the age of 5.

Hepatitis A virus causes infectious hepatitis that may range from an inapparent infection, especially in children to fulminating hepatitis and jaundice. Approximately 40,000 to 50,000 cases of hepatitis A are reported each year in the United States. About half of the U.S. population has antibodies to the hepatitis virus. Childhood infections are most common and tend to be asymptomatic.

Rotaviruses cause acute gastroenteritis with severe diarrhea, sometimes resulting in dehydration and death in infants. These viruses are the most important cause of acute gastroenteritis in infants and young children, especially during the winter, and can affect older children and adults.

Norwalk-like agents cause epidemic gastroenteritis, diarrhea, vomiting, abdominal pain, headache, and malaise. The disease is generally mild and self-limiting. Sporadic outbreaks of this disease occur in school children and in adults.

Adenoviruses are primarily the cause of respiratory and eye infections transmitted by the respiratory route. However, several of these may be enteric and be important causes of sporadic outbreaks of gastroenteritis in young children.

Reoviruses have been isolated from the feces of patients with a variety of diseases. However, no clear etiologic relationship has been established at this point. Reovirus infections may be common in people, but they may be mild or have clinical manifestations. Papovaviruses have been found in urine. Astroviruses, caliciviruses, and coronavirus-like particles may be associated with human gastroenteritis, producing diarrhea. These diseases are poorly understood.

Viruses are not normal inhabitants of the gastrointestinal tract or regular components, even in feces. Therefore, the use of coliforms and fecal streptococci as indicators of the potential viral contamination to the environment is useful. Viruses in wastewater do not behave the same as bacteria. Viruses are less easily removed by treatment processes and during passage through the soil. Removal of the polio virus varies from 0 to 69%; coxsackie virus, from 90 to 94%; and other viruses, from 53 to 71%. The concentration of viruses in unaffected people is normally zero. The concentration of viruses in an infected person is not well understood.

Estimates of the concentration of viruses in wastewater in the United States vary widely. Numbers tend to be higher in late summer and early fall than in other times of the year because of the increase in enteric viral infections at this time. The vaccine polio virus concentration tends to remain constant.

Viruses in aerosols are a concern as a potential route of transmission of disease, because the virus when inhaled may then be moved from the respiratory tract by the cilia into the oropharynx and then swallowed into the gastrointestinal tract. Viable viruses in aerosols have been found up to 500 ft from the source. Some enteroviruses may also multiply in the respiratory tract itself.

The survival time of viruses at a sludge application site is primarily of concern, because the longer the viruses survive at the surface, the greater the opportunity they have for moving through the soil toward the groundwater.

The factors affecting virus survival in soil are solar radiation, moisture, temperature, pH, and adsorption to soil particles. It appears that adsorption to inorganic surfaces prolongs the survival of viruses and that most virus inactivation occurs at the top few centimeters of soil where radiation forces are maximal. The persistence of virus particles that survive surface forces and enter the soil matrix is not well understood. Apparently viruses may penetrate up to 58 ft in sandy soil, but much less in loamy or clay soils. Enteric viruses can survive for about 100 days in soil unless subjected to very low temperatures, where the survival time increases.

The adsorption of enteric viruses by plants is theoretically possible. Some enteroviruses have been found to be adsorbed by tomato plant roots grown in hydroponic culture (the growing of plants in nutrient solutions with or without an inert medium to provide mechanical support). Although it may be possible to have viral uptake, the public health significance associated with the viral uptake through the root systems of plants appears to be minimal. Contamination of subsurface and low-growing crops in soil is possible where viruses have a long survival time of about 100 days.

Although viruses near the soil surface are rapidly inactivated, those that penetrate the aerobic zone can be expected to survive over a more prolonged period of time. The longer they survive, the greater is the chance that they may penetrate into the groundwater. Filtration plays a minor role in the removal of viruses in soils. Virus removal depends almost totally on adsorption. Because adsorption is a surface phenomenon, soils with high surface areas such as high clay content would be expected to have high virus removal capabilities, whereas soils such as sand would have low virus removal capabilities. The degree of adsorption of viruses onto soil is highly variable. Adsorption differs greatly among virus types, virus strains, and soils in which they are located. Heavy rainfall can move the viruses through the soil profile down toward the groundwater supply.

There seems to be little reliable information concerning viruses getting into groundwater beneath sludge application sites. Once enteric viruses do get into groundwater, however, they can survive for long periods of time ranging from 2 to 180 days. Enteroviruses have been found at depths of 54 to 201 ft from a septic tank leach field in a shallow sandy aquifer.

Human polio viruses, coxsackie viruses, echoviruses, and reoviruses have been recovered from, or found to produce, an infection in dogs, cats, swine, cattle, horses, and goats. Dogs and cats were found to be involved in the majority of these infections, probably because of their intimate association with people in the household. Information is insufficient at this point to determine the effect of virus transmission in animals and its relationship to transmission in people.

In contrast to bacteria where a large number of bacterial cells are usually needed to produce an infection, a few virus particles may be able to produce an infection under favorable conditions. This area needs continued study.

Because viruses do not regrow in foods or other environmental media as bacteria sometimes will, the risk of infection is totally dependent on exposure to an effective dose immediately transmitted to the material.

Fecally polluted vegetable garden irrigation water has been found to contain polio viruses and coxsackie viruses, and has been associated with epidemics among the people consuming the garden products. It is thought that epidemiological techniques are probably not sensitive enough to detect the low levels of viral disease transmission that might occur from a modern land application of sludge.

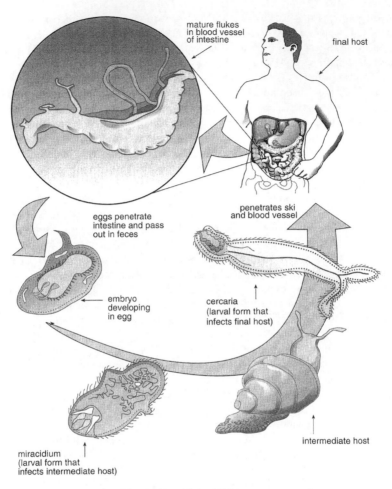

Figure 7.3   Protozoa infection cycle: blood fluke (*Shistosoma mansoni*).

Protozoa

Protozoa (Figure 7.3) and helminths (worms) are often grouped together as parasites. However, helminths are discussed in another subsection immediately following this one. Because of the large size of protozoan cysts and helminth eggs compared with bacteria and viruses, it is not likely that they are present in either aerosols or groundwater at the land application sites where sludge is applied to the ground. There has been an increasing recognition of parasitic infections in the United States, because military personnel and travelers from abroad have come back contaminated by them.

The most common protozoa found in wastewater and sludge are:

1. *Entamoeba histolytica* of the protozoan class Amoeba, with the nonhuman reservoir in domestic and wild mammals
2. *Giardia lamblia* of the protozoan class Flagellate, with the nonhuman reservoir in beavers, frogs, and sheep
3. *Balantidium coli* of the protozoan class Ciliate, with the nonhuman reservoir in pigs and other mammals
4. *Toxoplasma gondii* of the protozoan class sporozoan (Coccidia), with the nonhuman reservoir in cats
5. *Dientamoeba fragilis* of the protozoan class Amoeba
6. *Isopora belli* of the protozoan class sporozoan (Coccidia)
7. *I. hominis* of the protozoan class sporozoan (Coccidia)

Entamoeba histolytica causes amoebiasis or amoebic dysentery, an acute enteritis, with symptoms ranging from mild abdominal discomfort with diarrhea to fulminating dysentery with fever, chills, and bloody or mucoid diarrhea. Although most infections are asymptomatic, in severe cases abscesses appear in the liver, lungs, or brain, and death may result. The cysts are transmitted by contaminated water or food.

Giardia lamblia causes giardiasis, which is often an asymptomatic infection of the small intestine. This may be associated with chronic diarrhea, malabsorption of fats, abdominal cramps, bloating, fatigue, and weight loss. The disease is transmitted by cysts that have contaminated water or food, or may be spread person to person by contact.

Balantidium coli causes balantidiasis, which is a disease of the colon characterized by diarrhea or dysentery. Infections are often asymptomatic, and the incidence of disease in people is very low. The disease is transmitted by cysts that contaminate water. This is particularly true of water contaminated by pigs.

Toxoplasma gondii causes toxoplasmosis, which is an asymptomatic disease that rarely gives rise to critical illness. However, it can damage the fetus if a woman is infected and congenital transmission occurs during pregnancy. About 50% of the population of the United States is thought to be infected, but the infection is probably transmitted by oocysts in cat feces or the consumption of cyst contaminated inadequately cooked meat of infected animals.

The active stage of protozoans in the intestinal tract of infected individuals is the trophozolite. The trophozolites after a period of reproduction, may form precysts, which secrete tough membranes to become environmentally resistant cysts. The cysts are then excreted in the feces. The types and amounts of protozoan cysts found in wastewater depend on the levels of disease in the human population and on the degree of animal contribution to the system.

The protozoan cysts are sensitive to drying, and therefore should die off rapidly when exposed to the air or when deposited on the surfaces of plants. However, under poor management of the human food supply, cysts of *Entamoeba histolytica* and *Giardia lamblia* have been found on irrigated fruits and vegetables where wastewater has been used as the source of irrigation. A low infected dose of protozoan cysts may cause disease. It is, therefore, important that humans maintain minimum contact with any active land application of sewage sludge.

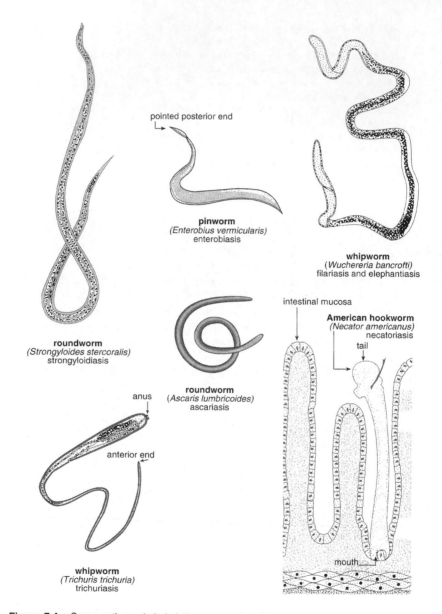

Figure 7.4   Some pathogenic helminths: nematodes. (Not to scale.)

Helminths

The pathogenic helminths (Figures 7.4 and 7.5) whose eggs are of major concern in wastewater and sludge are divided into two categories. They are the nematodes or roundworms, and cestodes or tapeworms. The trematodes or flukes are not

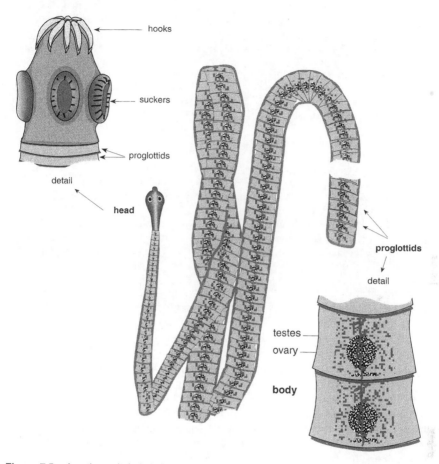

hooks

suckers

proglottids

detail

head

proglottids

detail

testes

ovary

body

Figure 7.5 A pathogenic helminth: typical tapeworm.

included as potential problems, although they do cause disease. The reason they are not included is because snails are necessary to complete their life cycles, and it is unlikely that snails will be found in sludge application sites.

The nematodes of concern are:

1. *Enterobius vermicularis* is the pinworm that causes enterobiasis.
2. *Ascaris lumbricoides* is the roundworm that causes ascariasis.
3. *A. suum* is the swine roundworm that causes ascariasis. The pig is the nonhuman reservoir.
4. *Trichuris trichiura* is the whipworm that causes trichuriasis.
5. *Necator americanus* is the hookworm that causes necatoriasis.
6. *Ancylostoma duodenale* is the hookworm that causes ancylostomiasis.
7. *A. braziliense* is the cat hookworm that causes cutaneous larva migrans. The nonhuman reservoir is cats and dogs.

8. *A. caninum* is the dog hookworm that causes cutaneous larva migrans. The non-human reservoir is the dog.
9. *Strongyloides stercoralis* is the threadworm that causes strongyloidiasis. The non-human reservoir is the dog.
10. *Toxocara canis* is the dog roundworm that causes visceral larva migrans. The nonhuman reservoir is the dog.
11. *T. cati* is the cat roundworm that causes visceral larva migrans. The nonhuman reservoir is the cat.

The cestodes or tapeworms are as follows:

1. *Taenia saginata* is the beef tapeworm that causes taeniasis.
2. *T. solium* is the pork tapeworm that causes taeniasis.
3. *Hymenolepis nana* is the dwarf tapeworm that causes an intestinal infection. The nonhuman reservoir is the rat and mouse.
4. *Echinococcus granulosus* is the dog tapeworm that causes unilocular hydatid disease. The nonhuman reservoir is the dog.
5. *E. multilocularis* causes alveolar hydatid disease. The nonhuman reservoir is the dog, fox, or cat.

Enterobius vermicularis, which is the pinworm, causes itching and discomfort in the perianal area. This occurs particularly at night when the female lays her eggs on the skin. Although this is the most common helminth infection in the United States, the eggs are not usually found in the feces but are spread by direct transfer from person to person. These pinworms live for only a few days.

Ascaris lumbricoides, the large roundworms, produce numerous eggs that require 1 to 3 weeks for embryonation. After the embryonated eggs are ingested, they hatch in the intestine, enter the intestinal wall, migrate through the circulatory system to the lungs, enter the alveoli, and migrate up to the pharynx. During the passage through the lungs, they may produce ascaris pneumonitis or Loeffler's syndrome. Loeffler's syndrome consists of coughing, chest pains, shortness of breath, fever, and eosinophilia, which can be especially severe in children.

The larval worms are then swallowed and they complete their maturation in the small intestine. Small numbers of worms usually do not produce symptoms. Large numbers of worms may cause digestive and nutritional disturbances, abdominal pain, vomiting, restlessness, disturbed sleep, and occasionally intestinal obstruction. Adult worms may cause death if they migrate into the liver, gallbladder, peritoneal cavity, or appendix. However, this is infrequent.

Trichuris trichiura, the human whipworm, lives in the large intestine with the anterior portion of its body threaded superficially through the mucosa. Eggs are passed in the feces and develop into the infected stage after about 4 weeks in the soil. Ingestion of the eggs causes an infection in the mucosa of the cecum and possibly in the proximal colon. Light infections are often asymptomatic. Heavy infections may cause intermittent abdominal pain, bloody stools, diarrhea, anemia, loss of weight, or rectal prolapse in the case of very heavy infections. Human infections have been reported with *T. suis*, the swine whipworm, and *T. vulpis*, the dog whipworm. *Ascaris*, *Trichuris*, and *Toxocara* are most frequently found in municipal wastewater sludge.

Necator americanus and *Ancylostoma duodenale*, the human hookworms, live in the small intestine by attaching themselves to the intestinal walls. The eggs are passed out in the feces and developed into the infective stage in 7 to 10 days in warm, moist soil. The larvae penetrate the bare skin of the foot and pass through the lymphatics and bloodstream to the lungs, enter the alveoli, migrate up the pharynx, are swallowed, and reach the small intestine. The larvae may also be swallowed in food or water. Light infections usually result with few clinical effects, but heavy infections can result in iron-deficiency anemia. This happens because the secreted anticoagulant causes bleeding at the site of attachment. The disease especially affects children and pregnant women.

The cat and dog hookworm do not live in the human intestinal tract. The larvae from the eggs in the cat and dog feces can penetrate the bare skin, especially the feet and legs on beaches, and burrow aimlessly intracutaneously, producing "cutaneous larvae migrans" or creeping eruption. After several weeks or months, the larvae die without completing the life cycle.

The threadworm lives in the mucosa of the upper small intestine. The eggs hatch within the intestine and reinfection may occur. Noninfective larvae can usually pass out in the feces. The larvae in the soil may develop into the infective stage or a free-living adult, which can produce infective larvae. Infective larvae penetrate the skin, usually the foot, and can complete their life cycle in a manner that is similar to hookworms. Intestinal symptoms include abdominal pains, nausea, weight loss, vomiting, diarrhea, weakness, and constipation. Massive infections and autoinfection can lead to wasting and death in patients who are receiving immunosuppressive medication. Dog feces is another source of threadworm larvae.

Dog and cat roundworms do not live in the human intestinal tract. When eggs from animal feces are ingested by people, especially children, the larvae hatch in the intestine and enter the intestinal wall similar to ascaris. However, because *Toxocara* cannot complete the life cycle, the larvae do not migrate, but instead wander aimlessly through the tissue, producing "visceral larvae migrans" until they die in several months to a year. The disease can cause fever, loss of appetite, cough, asthmatic episodes, abdominal discomfort, muscle aches, or neurological symptoms. This disease becomes particularly serious if the larvae invade the liver, lungs, eyes (often causing blindness), brain, heart, or kidneys. Approximately 50% of puppies and 50% of older dogs are affected with *T. canis*. *Toxocara* is one of the most common helminth eggs found in wastewater sludge.

The beef tapeworms live in the intestinal tract, where they cause nervousness, insomnia, anorexia, loss of weight, abdominal pain, and digestive disturbances, or the individual may be asymptomatic. The infection comes from eating incompletely cooked meat containing the larval stage of the tapeworm. People serve as a definitive host harboring the self-fertile adult. The eggs are passed in the feces, ingested by cattle and pigs, hatch into larvae, and migrate to tissues, where they develop into the cyst stage. The major direct hazard to people is in eating livestock grazing on land where sewage sludge has been applied. The dwarf tapeworm lives in the human intestinal tract and may be asymptomatic or produce the same symptoms as *Taenia*. Infected eggs are released and internal autoinfection may occur. Most of the problems appear in children under 15 years.

Helminth eggs and larvae live for long periods of time when applied to land, probably because soil is the transmission medium in which they grow. Under favorable conditions of moisture, temperature, and sunlight, *Ascaris*, *Trichuris*, and *Toxocara* can remain viable and infective for several years. Hookworms can survive up to 6 months. *Taenia* may survive a few days to 7 months. Because cattle consume considerable quantities of soil as they graze, it would be best not to have them graze on sludge-treated soil or any sludge treatment sites to avoid their becoming an intermediate host for the transmission of the helminths.

Single eggs of helminths are infectious to people. However, the infections are dose related and the larger the number of eggs, the greater the chance for infection.

Microorganisms in Wastewater Aerosols

Microorganisms, such as viruses, bacteria, protozoans, and fungi, may be present in aerosols coming from the wastewater treatment process. The aerosols may contaminate fomites, water, or food, or be inhaled or ingested through the breathing process. Aerosols are emitted by such wastewater treatment processes as activated sludge, trickling filters, and land application by spray irrigation. The process of activated sludge treatment generates small bubbles by diffused air aeration that rise through the sewage in the aeration tank to the surface boundary. The formation of sewage aerosols during aerobic wastewater treatment and their emission as mists or droplets have been shown. Rising bubbles can adsorb and concentrate suspended bacteria and viruses while moving toward a liquid surface. At the surface boundary, the film containing microorganisms is disrupted by the rising bubbles that burst, releasing tiny aerosol droplets containing the bubble-adsorbed as well as the surface film-associated microorganisms.

The characteristics of aerosols generated by processes of spray irrigation are affected by machinery design, as well as by the prevailing etiologic conditions. Spray irrigation equipment produces large quantities of respirable aerosols, whose number increases as a function of wind speed and the length of time exposed to the wind. During treatment with trickling filters, impacting sewage onto rock within the trickling filter unit causes a splashing effect, which results in sewage aerosol formation. The nearly instantaneous evaporation that may occur as these droplets become suspended in air leaves a dry residue, referred to as a *droplet nucleus*, which is subjected to downwind dispersion, depending on the size, density, and prevailing weather conditions.

Sewageborne aerosol formation is a potential contributor to air pollution problems and can cause risks to human health. Where liquid sludge is applied to the land by spray equipment, the aerosols that travel beyond the zone of application are suspensions of solid and liquid particles up to 50 μm in diameter.

The major health concern with aerosols is the possibility of direct human infection through the respiratory route by inhalation. The location where the aerosol particles are deposited is a function of the size, shape, and density of the particles. Particles above 2 μm in diameter are deposited primarily in the upper respiratory tract, including the nose, where they are carried by the cilia into the oropharynx. They then may be swallowed and enter the gastrointestinal tract. Because the smaller airways and alveoli do not contain cilia, pathogens deposited there have to be dealt with by local mechanisms. About 40% of 1-μm particles are removed, about half in the pulmonary region where the respiratory bronchiole, alveolar ducts, and alveoli are present. Any of the bacteria present in feces, urine, or wastewater may appear in aerosols coming from land application of sludge.

Chemical Hazards

A vast number of potentially hazardous chemicals are entering our receiving streams. As a result, great care must be taken to detect this spectrum of chemical compounds and to determine which of these chemicals are potentially hazardous to humans, through entering either the drinking water or the food chain. Definitive epidemiological studies have not been made in this area, although a statistical association seems to exist between cancer mortality rate and drinking water. Drinking water has also been implicated in reproductive failures in animals. Little information is available concerning the effect of inorganic chemicals on cardiovascular diseases, although epidemiological studies have shown that a consistently higher cardiovascular death rate occurs in soft water areas as compared with hard water areas. This raises many questions concerning the removal or type of removal process to be used on organic chemicals that may be cancerous.

Trihalomethanes

Trihalomethanes including chloroform, bromodichloromethane, chlorodibromomethane, and bromoform are often present in drinking water at concentrations between 10 and 100 μg/l. These compounds are chlorination by-products. Some of them are known to be carcinogenic in laboratory mammals at doses far greater than human intakes from drinking water. However, a variety of studies have shown that the ratio between the lowest dose level giving rise to a carcinogenic effect in animals from drinking water and likely human exposure is unlikely to provide a carcinogenic risk to humans. In addition, the effective disinfection of water supplies is clearly of great importance in the prevention of disease.

MTBE

MTBE and other airborne VOCs may enter water from the air. The U.S. Geological Survey National Water Quality Assessment Program is determining the frequency of occurrence and concentrations of a wide range of chemical compounds in the nation's surface and groundwater. The contaminants monitored include VOCs, pesticides, nutrients, and trace elements. Results of recent studies show that whereas chloroform and toluene may not be coming primarily from the atmosphere, MTBE may be coming from the atmosphere. A wide variety of possible sources exists for chloroform in surface waters. It is widely known that chloroform is a major disinfection by-product of the chlorination treatment process. Possible sources of chloroform to surface water include groundwater flow, incidental overland flow from the irrigation of lawns and parks, treatment plant effluents, and other sources of chlorinated water. Possible sources of toluene include effluents from industrial wastewater, sewage effluent, and storm water runoff from roadways and parking areas on which gasoline has accumulated.

Although MTBE sources can be associated with runoff from parking lots or streets, and industrial wastewater effluents, the main source appears to be the air. MTBE, which is an additive to gasoline to oxygenate it, can cause numerous symptoms including neurological and respiratory ones, extreme headaches, vomiting, nausea, diarrhea, fever, cough, excessive perspiration, muscle aches, sleepiness, disorientation, dizziness, and skin, nose, and eye irritation.

Polychlorinated Biphenyls

PCBs are mixtures of chlorinated biphenyls (209 individual compounds) with varying degrees of chlorination and isometric substitution. These compounds are manufactured by the reaction of a biphenyl with chlorine in the presence of a catalyst. During the formation of the PCBs, chlorine atoms are substituted for hydrogen atoms at one or more of the ten available positions on the biphenyl molecule. PCBs are used in the United States primarily as insulating fluids in electrical equipment. PCBs are used in small quantities in the investment casting industry and in hydraulic and heat-transfer fluids. PCBs are mobile oils or viscous fluids at room temperature, volatilizing slowly at normal ambient temperature and having a very low solubility in water. PCBs also are freely soluble in organic solvents; move easily from water into fats in biological systems, thereby creating a larger degree of accumulation in fatty tissue; and are strongly adsorbed onto solids, sediments, and airborne and waterborne particles. In the water environment, the adsorption process helps PCBs move along with sediment and they are found at the bottom of receiving streams. These compounds are released by the sediment and ingested by fish that do not even come into contact with the sediment. PCBs can affect bacteria, phytoplankton, aquatic invertebrates, and fish.

In birds, PCBs can cause death within a few weeks with rates as low as 30 to 40 ppm. In mammalian studies, monkeys receiving 1 to 300 ppm of substances containing PCBs die within 2 to 3 months of the initial feeding. Infant monkeys apparently contain levels of PCBs in their tissue because of the high levels of PCBs in their mothers' milk. Most of the PCBs are resistant to degradation by metabolism. PCBs have a long life in the environment. In one case they were identified in anaerobic marine sediments that could be dated back to the mid-1940s. One of the important properties of PCBs is that they are bioaccumulated or bioconcentrated by aquatic organisms into their tissues to levels that are much higher than the ambient water. This is because of the high solubility of PCBs in lipids and their low solubility in water.

The current EPA established criterion for ambient water for PCBs is 0.001 mg/l. The EPA is extremely concerned about the margin of safety provided for human beings in establishing any rules for PCB concentrations. It is felt that PCBs in humans may cause chronic effects at extremely low concentrations. It is difficult to determine with confidence if any PCB concentration above zero provides an ample margin of safety for humans. The reason is that humans are exposed for much longer periods including prenatal exposure. Humans also retain PCBs in their tissues more efficiently than experimental animals, with consequently greater exposure to sensitive organs. Human breast-fed infants probably ingest 30 to 40 times the quantity of PCBs as their mothers on a milligram per kilogram basis. Possibly a synergistic effect exists between drugs and other pollutants, such as PCBs, as they react on the human population. It has also been noted that PCBs can cause cancer in rats.

Pesticides

Since the advent of the commercially prepared synthetic pesticide era, there has been a concern over the amount of pesticides entering the environment. Within recent years, the concern has increased very sharply. The types of pesticides that are used include insecticides, herbicides, rodenticides, miticides, nematicides, and fungicides. They may be inorganic in nature and contain such substances as arsenic and fluorine; they may be botanical in nature, such as the pyrethrins and red squill; or they may be synthetic in nature and consist of the chlorinated hydrocarbons or the organic phosphorus compounds.

The first real awareness that synthetic pesticides were actually becoming a pollution problem occurred in 1950, when there were extensive fish kills in 14 of the streams that come from the Tennessee River in Alabama. These kills were caused by dieldrin that had run off from cotton fields. In 1959, a concerted project was undertaken to determine the cause of some of the fish kills that were occurring. By 1963, numerous additional fish kills occurred in lower Mississippi; and as the years have passed, the contamination of the water environment has increased sharply. Pesticides may be applied directly to the water supply to control aquatic insects or algae growth, or to poison certain fish that are causing undesirable kills of desirable fish in certain areas. Pesticides may also be discharged from industries or, as mentioned before, as runoff from agricultural lands and forests. The problems related to pesticides include solubility of the pesticide in water; persistence of the pesticide in the soil, depending on pH and temperature of the soil; quantity of pesticide applied to the soil; formulation of the pesticide; method of applying pesticides; slope of the land; soil characteristics; volume and intensity of rainfall; and type of soil conservation practices used.

Pesticides can cause acute illness and death, can result in chronic toxicity, or possibly can even cause cancer. Currently, although accidents continue to occur as a result of the release of large amounts of pesticides, the greatest concern is chronic toxicity over long periods of time to the animal as well as the human population. The EPA administrator banned the use of DDT, because DDT is a persistent pesticide that remains in the environment for long periods of time. Subsequently, other pesticides, including heptachlor, chlordane, aldrin, dieldrin, and others, have been banned. The pesticide mirex, which is chemically identical to kepone, except for one oxygen atom, is a possible carcinogen. The use of this chemical was temporarily halted until the USDA determined that it was necessary to utilize it for fire ant eradication.

One of the more spectacular pesticide accidents occurred in 1975 when kepone, a chlorinated hydrocarbon that has been used effectively for ant and roach control, was accidentally released into the James River. Kepone was also found in air and in shellfish as far away as 64 miles from the plant. Some 20 individuals within the plant were hospitalized with apparent brain and liver damage, sterility, slurred speech, loss of memory, and twitching of the eyes. Several lawsuits were filed on the kepone issue. The effect of kepone on the contamination of shellfish, crabs, and so forth will be felt for a considerable period of time.

Other Organic Compounds

The use of gas chromatography-mass spectrometry techniques for identifying organic materials in drinking water has greatly expanded our knowledge of the potential role of organic contaminants as etiologic agents in disease. Over 400 organic chemicals have been identified in drinking water. Of these, 300 have been identified in treated drinking water. Considerably improved techniques are still needed, both quantitative and qualitative, to get a better understanding of the potential risks involved in organic contamination of drinking water.

Chlorine is the most widely used method of disinfecting water that has been treated or is untreated. It is known that organic contaminants enter the receiving stream from agricultural and urban runoff, from industry, and from sewage treatment plants. Organic compounds, when combined with the chlorine in water treatment, form chlorinated hydrocarbons, which may create problems to the individuals who are consuming them. Some of the compounds that appear are the trihalogenated methanes. Further, a nationwide survey of drinking water quality showed a significant correlation between the nonvolatile total organic carbon content and the resulting concentration of the trihalogenated methanes.

Chloroform and carbon tetrachloride have been found in water supplies. Apparently chlorination of drinking water has contributed to the formation of these unwanted chlorinated hydrocarbons. A number of potentially toxic organic chemicals appear to be coming from other sources that are introduced into the receiving stream and therefore may become part of the drinking water.

An estimated 100 million Americans rely on underground sources of drinking water. Approximately one third of all public supplies and 95% of all rural domestic supplies use groundwater sources. Groundwater pollution is increasing as a result of contamination from domestic and industrial waste, sanitary landfills, wastewater sludges and effluents, feedlot wastes, fertilizers and agricultural chemicals, mine drainage, subsurface disposal of oil field brines, seepage from septic tanks and other on-site wastewater disposal systems, and USTs.

Potential Mutagens in Wastewater and Sludge

Most of the common organics found in domestic wastewater come from feces, urine, paper products, food waste, detergents, skin, excretions, and contaminants from bathing. In medium-strength sewage (700 ppm total solids contents), organics make up about 75% of the suspended solids and about 40% of the filterable solids that are colloidal and dissolved. Most of the organics found in municipal sludge of domestic origin are probably harmless in land application. However, it has been found that fecal material commonly contains mutagens, and mutagens form a large class of potential carcinogens. One of the causes of colorectal cancer is the presence of carcinogens or cocarcinogens produced by the bacterial degradation in the gut of bioacids or cholesterol. During anaerobic incubation in the presence of bile and bile acids, the mutagenicity of feces can be increased. High levels of chromosome-breaking mutagenic activity have been found in the feces of dogs, otters, gulls, cows, horses, sheep, chicken, and geese. The chemical nature of the fecal mutagens is unknown. It is not thoroughly understood whether any significance is related to these natural mutagens found in sludge coming from homes.

Toxic Organics from Homes

Toxic organics present in household wastewater come from the use of such products as cleaning agents and cosmetics containing solvents and heavy metals as their main ingredients. Deodorizers and disinfectants contain naphthalene, phenol, and chlorophenols. When pesticides, laundry products, paint products, polishes, and preservatives are put into the household wastewater stream, the level of organic pollutants increases sharply. These pollutants include benzene; phenol; 2,4,6-trichlorophenol; 2-chlorophenol; 1,2-dichlorobenzene; 1,4-dichlorobenzene; 1,1,1-trichloroethane; naphthalene; toluene; diethylphthalate; dimethylphthalate; trichloroethylene; aldrin; and dieldrin. It is very difficult to analyze these complex mixtures of organics found in wastewater. Advanced methods of extraction, gas and other chromatography, mass spectrometry, and computer analysis are used.

Priority Organics Found in Raw Municipal Wastewater

The most frequently detected priority organics as established by the EPA in raw municipal wastewater include phenol; 1,1,1-trichloroethane; trichloroethylene; tetrachloroethylene; ethylbenzene; trichloromethane (chloroform); di-*n*-butylphthalate; toluene; dichloromethane; *bis*-(2-ethylhexyl)-phthalate; naphthalene; 1,4-dichlorobenzene; phenanthrene; benzene; heptachlor; BHC-G (lindane); 1,2, dichlorobenzene; dimethylphthalate; dieldrin; 1,3-dichlorobenzene; DDT; di-*n*-octylphthalate; 1,1-dichloroethane; 1,2-dichloroethane; dichlorodiphenyldichloroethane (DDD); anthracene; aldrin; endolsulfan-D; and 1,2-*trans*-dichloroethylene. Many of these chemicals come from industrial waste discharges into municipal wastewater stream.

Priority Organics Found in Raw Municipal Sludge

The priority organics in raw municipal sludge include *bis*-(2-ethylhexyl)-phthalate; toluene; dichloromethane; ethylbenzene; benzene; 1,2-*trans*-dichloroethylene; trichloroethylene; pyrene; anthracene; di-*n*-butylphthalate; fluoranthene; butylbenzylphthalate; tetrachloroethylene; 1,1-dichloroethane; chrysene; 1,2-benzanthracene; trichloromethane (chloroform); 1,1,1-trichloroethane; 1,4-dichlorobenzene; 1,2 dichlorobenzne; 1,1,2,2 tetrachloroethane; pentachlorophenol; chlorobenzene; 1,2,4 trichlorobenzene; 3,4 benzofluoranthene; di-*n*-octylphthalate; and 1,2-dichloroethane. These organics come primarily from industrial discharges into the municipal wastewater stream.

Organic Chemicals in Soil and Groundwater

Organic compounds and sludge components may be volatilized, immobilized by adsorption, or transported through the soil possibly reaching the groundwater. At normal application rates and proper management, leaching, or soil migration of organics from municipal sludge, land application should be insignificant. Adsorbed organics may be degraded chemically or photochemically, or they may be desorbed. Iron and aluminum oxides adsorb organics. The adsorption of organic pesticides tends to increase with the concentration of the functional groups such as amine, amide, carboxyl, and phenol. Most pesticide residues remain in the surface soils during land treatment. PAH adsorption increases with increasing organic matter content of the soil, with increasing chlorine content, and with increasing hydrophobicity. The organics are immobilized by adsorption on the surfaces of soil particles. Microbial decomposition or biodegradation is probably the major mechanism of breakdown of these chemicals. If high levels of toxic organics were to be found in the sludge, they could inhibit the soil microflora and create a serious potential problem.

Inorganic Compounds

Lead and cadmium levels in Seattle, WA drinking water, and lead levels in the drinking water of Boston, MA, had been higher than recommended under the drinking water standards. Asbestos had also been found in drinking water supplies. Inorganic fluorides at levels above the recommended concentration of 1.0 ppm exist in certain bodies of water. Cyanides are usually involved in fish kills. Cyanides come from metal plating, case hardening, metal cleaning baths, silver and gold refining, gas scrubbers from pyrolytic processes, rubber acrylic fiber and plastic industries, and various other chemical processes. Unquestionably, cyanides may be a significant potential hazard in our water streams.

Mercury is naturally occurring. It may be found at levels ranging from 0.03 to 2.0 ppb. However, mercury has also been added to the various waters from a variety of industrial processes. Several large-scale outbreaks of mercury poisonings occurred in the 1950s and 1960s. In 1953, in Japan, 121 people suffered from severe nerve disorders, brain damage, birth defects, and death. Apparently shellfish were feeding in waters contaminated with methylmercury. The individuals who ate the seafood became ill with mercury poisoning. Mercury poisoning has been observed in many aquatic species throughout the entire food chain. When mercury combines with other substances, the results are more toxic than the original mercury alone. The inorganic mercury, when deposited in sediments, may be transformed into methylmercury by aquatic organisms. Methylmercury then becomes an extremely hazardous substance. Mercury has been used in a variety of industrial processes. Although the amount of mercury utilized has decreased, the levels of mercury in freshwater seem to be about the same as that in 1970. On a year-to-year basis, although direct discharge of mercury has been curtailed, a very complex cycling of mercury in the environment still seems not to be thoroughly understood; therefore, the mercury is still a source of great concern.

Other trace metals found in the environment include zinc, cadmium, arsenic, phosphorus, iron, molybdenum, manganese, beryllium, copper, silver, nickel, cobalt, lead, chromium, vanadium, barium, and strontium. The amount of these metals varies tremendously from one area to another. In certain cases, the amounts exceed the permissible levels established by the EPA.

The highest reported trace metal concentrations had been 1182 mg/l of zinc, 120 mg/l of cadmium, and 336 mg/l of arsenic in Lake Erie. The Colorado River had 1800 mg/l of boron, 38 mg/l of silver, and 300 mg/l of vanadium. The Missouri River had 5040 mg/l of phosphorus and 2760 mg/l of aluminum. The western Gulf had shown 952 mg/l of iron. The southwest lower Mississippi had shown 1100 mg/l of molybdenum and 5000 mg/l of strontium. It is obvious from these data that trace metals are found in various waters of the United States.

Heavy Metals in Sludge

Sewage sludge contains heavy metals, such as zinc, copper, nickel, cadium, chromium, mercury, lead, boron, molybdenum, cobalt, and selenium. The trace elements zinc, copper, and nickel are toxic to plants when they are available in significant quantities in the soil. Cadmium in high concentrations is toxic to plants. Normally, the cadmium content of most sludges is not high enough to cause plant injury. Chromium that is present in sludge applied to soil is converted to a form that is not taken up by, or harmful to, plants. Lead can be toxic to plants in acid soils that are also low in phosphates. However, lead in sludge appears to be nontoxic, because the large amount of phosphate in sludge ties up the lead and prevents injury to plants. Boron at the levels present in sludge is not harmful to plants, except in soils normally high in boron. The concentrations of molybdenum, selenium, and cobalt in domestic sludges are low, and very high rates of sludge would be required for soils to reach toxic levels of these metals.

The metals found in sewage sludge, with the exception of cadmium, copper, and zinc, are of no known hazard to the food chain through plant accumulation. However, surface contamination of vegetation of recently applied sludge containing these metals could be a special hazard to grazing animals. Excess zinc is toxic to animals when the diet contains between 500 and 1000 ppm of zinc. The potential hazard of cadmium to human and animal health is a major concern.

Factors affecting the accumulation of sludge bearing heavy metals in crops are pH, cation exchange capacity, organic matter, phosphorus, and crop species. Heavy metal redistribution in sludge-applied fields is affected by harvest removal, erosion, leaching, and tillage practices. The degree of downward movement or leaching of the heavy metals varies with the amount of sludge present, low soil pH, coarse-textured soils, amount of precipitation or irrigation, and long-term application of sludge. Usually, the availability of heavy metals in the soil decreases as the pH approaches 7. Little uptake of heavy metals occurs in plants when soil pH is greater than 6.5. Soils with a higher cation exchange capacity have greater retaining power for heavy metals. This reduces the amount of heavy metals available to the plants.

In studies that have been done with cattle that have been fed on soils and grass from sites that have had sludge applied to them, the cattle have shown an increase of cadmium and lead in specific tissues, such as the kidneys, liver, and bones. In Japan, rice that was contaminated by cadium produced a serious health problem.

Primary sewage treatment removes very little of the metals, while secondary treatment may remove a large portion. Some species of plants are capable of concentrating cadmium to levels much greater than those in the immediate environment. These plants include lettuce, wheat, potatoes, sweet corn, and oats. The acute effects of ingestion of ionized cadmium are nausea and vomiting. The chronic oral effect of cadmium intake is tied to the ultimate destruction of the kidneys and bones. Cadmium has a long biological half-life (16 to 33 years). Cadmium is severely toxic to human beings; can be absorbed in the body through the respiratory and intestinal tracts; and as has been mentioned accumulates in the liver, bones, and kidneys. Also plants may suffer from phytotoxicity, which is an intoxication of living plants by altering plant–water relationships, causing roots to become leaky, inhibiting photosynthesis, or affecting metabolic enzymes.

Detergents

Detergents and surfactants are a hazard to humans, fish, and birds. It is important that they, along with all of the toxic metals, be eliminated from the fish and wildlife system.

Insects

Vectors, such as insects (Figure 7.6) including mosquitoes and flies of varying types, and rodents, have become a serious nuisance in water areas. Mosquitoes are the most common problem. They cause not only disease but also considerable discomfort from their bites and swarming characteristics. The physical characteristics of bodies of water are such that mosquito breeding has increased sharply, and as a result of this the value of given bodies of water has decreased sharply. (See Volume I, Chapter 5, Insect Control.)

Swimmers itch exists in varying bodies of water because the organism *Cercaria stagnicolae* is shed from a variety of snails. Snails are more frequently present in polluted waters.

Effects on Fish and Wildlife

During the course of time, fish and wildlife have become acclimated to the various environmental factors associated with water. When humans entered the water pollution picture, they altered the environment in harmful ways through changes in temperature, dissolved oxygen, pH, carbonates, and so on. This section is concerned with the environmental, physiological, and accumulative effects on fish and wildlife. Substances that are either in suspension or in solution and may be solid, liquid, or gas determine the quality of the water in a major way. The aquatic organisms are affected not only directly by these materials but also indirectly by the effects that the materials have on other aquatic life that make up their food, are competitors, or are predators.

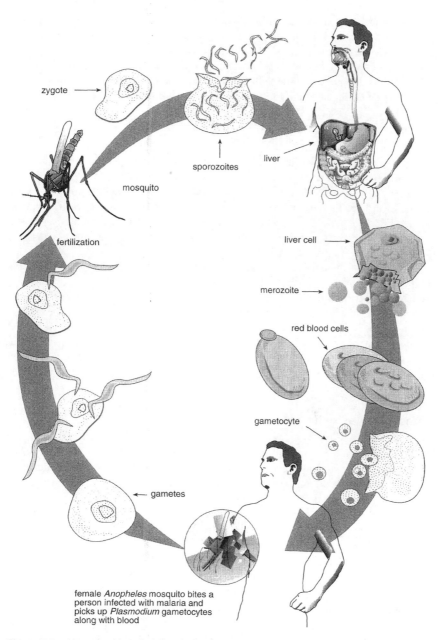

zygote

sporozoites

liver

mosquito

liver cell

merozoite

red blood cells

gametocyte

fertilization

gametes

female *Anopheles* mosquito bites a
person infected with malaria and
picks up *Plasmodium* gametocytes
along with blood

Figure 7.6 Mosquito (*Anopheles*) as a vector of infection (malaria).

Pollutants must not erect physical barriers to block spawning and migration of various fish species. It is also important that the natural tidal movement in estuaries is such that the plankton can move to where it ordinarily would become a source of food and the various parts of the ecological system in the water can function properly. The width of the zone and the volume of the flow within the zone of passage is an important determinant of the lifestyle of the fish within their environment.

Effects on Stream Life

Many streams in the United States are currently suffering from eutrophication, which occurs when the nutrient concentrations in the water increase and the number of algae cells increases. The nuisance conditions occurring include surface scums, algae-littered beaches, bad odors, and filter-clogging problems at the water treatment plants. Filamentous algae, especially *Caldophora*, grow profusely on any suitable subsurface. When these algae break loose and wash ashore, they form a mass of stinking vegetation.

The amount of nutrients depends on the amount of contaminants reaching the bodies of water. In reservoirs or lakes or any settling basins, the biological activity is affected by the amount of DO, pH, carbon dioxide, hardness, alkalinity, iron, manganese, phosphorus, and nitrogen. Also found in the polluted waters are blood-worms, other worms, nematodes, larvae, water fleas, water lice, sponges, clams, and snails. It is obvious that to make the stream or lake once more a productive center for humans, the sources of pollution must be eliminated, and removal of as much of the existing pollution as possible should be undertaken to allow the stream to purify.

Effects on Aesthetics and Recreation

Throughout the course of the discussion on water pollution, each of the problems that have been encountered have affected humans from a disease standpoint; have affected fish life, wildlife, and aquatic plant life; and also have affected the usage of the water for recreation or aesthetic purposes. Anything that can be done to remove the existing and potential pollutants can once again make available for humans one of the most vital and most used natural resources. It is unpleasant to see the river running through town turn into an open sewer or observe it catching fire because of pollutants. The river, instead of a stinking mass of algae, dead fish, and polluted wastes, should be a beautiful, aesthetically pleasing, restful setting for people.

Economic Impact of Water Pollution Control

The macroeconomic effect to the national economy of imposing pollution abatement standards is not severe. However, the microeconomic impact may have significance, depending on the industry and its status within the community.

Impact on Other Problems

Water pollution affects many other areas of the environment. Water pollution destroys recreation areas, harms our aesthetic values that we find within the environmental setting, causes disease and injury, kills fish and plants, and contaminates soil and groundwater.

POTENTIAL FOR INTERVENTION

The potential for intervention varies from poor to excellent depending on the type of water pollution problem and the kind of pollutant; the financial problem involved; and whether the pollutant is created by an individual source, a municipality, or a nonpoint source. Intervention consists of isolation, substitution, shielding, treatment, and prevention. Isolation is the removal of pollutants to areas where they cannot affect surface waters or underground waters. This technique is particularly used in the disposal of hazardous liquid wastes or radiological liquid wastes. Substitution is the changing of a process to use less hazardous materials. It is also the use of better techniques for treatment of wastes. Shielding is not appropriate for this area. Treatment is the major technique utilized. All the waste from point source pollution must be treated before it is released. Nonpoint source contamination needs modification and treatment where feasible. Prevention is the technique used to enforce the Water Pollution Control Act standards. It is also the development of appropriate standards and new techniques for the pretreatment of pollutants.

RESOURCES

The scientific and technical resources include the University of North Carolina School of Public Health, University of Michigan School of Public Health, National Environmental Health Association, American Public Health Association, American Chemical Society, American Institute of Chemical Engineers, and the American Society of Civil Engineers. The civic resources are the California Coastal Alliance, International Association for Pollution Control, and National Wildlife Federation. The governmental resources are the Maritime Administration, Department of Defense, The Department of Agriculture, NOAA, National Bureau of Standards, Energy Research and Development Administration, EPA, Department of Health and Human Services, Bureau of Mines, Geologic Survey, Bureau of Reclamation, National Science Foundation, National Aeronautics and Space Administration, and U.S. Coast Guard.

Other resources include the American Public Works Association, 1313 East 60th St., Chicago, IL; American Petroleum Institute, 1220 L Street N.W., Washington, D.C.; American Water Resources Association, 5410 Grosvenor Lane, Suite 220, Bethesda, MD; Interstate Conference on Water Problems, 2300 M Street, N.W., Suite 800, Washington, D.C.; Marine Technology Society, 1825 K St., N.W., Washington,

D.C.; National Council of the Paper Industry for Air and Stream Improvements, 260 Madison Ave., New York, NY; Water Pollution Control Federation, 601 Wythe St., Alexandria, VA; Water Quality Association, 4151 Naperville Road, Lisle, IL; and Water Resources Congress, 3800 North Fairfax Drive, Suite 7, Arlington, VA.

STANDARDS, PRACTICES, AND TECHNIQUES

Fish and Wildlife Areas

Standards of water for fish areas are as follows:

1. Dissolved materials have to be innocuous. Unnatural osmotic effects should not occur. The osmotic pressure should not be increased by more than one third of the natural concentration. The dissolved materials should have a maximum of the equivalent of 1500 mg/l of sodium chloride. Dissolved materials that are harmful in low concentrations should be excluded.
2. The pH should not be lower than 6 or raised above 9 by dissolved materials.
3. The total alkalinity must not be lowered under 20 mg/l.
4. The addition of weakly dissociated acids and alkalies must be regulated to prevent toxicity.
5. The temperature of the water is extremely important, and therefore the heat added to the stream in any given month should not exceed 5°F. In areas of lakes where important organisms may be affected, the temperature should not change by more than 3°F.
6. DO should be above 5 mg/l in normal warm water. In cold water, DO must be 7 mg/l or greater.
7. Free carbon dioxide should not exceed 25 mg/l.
8. Oil or petrochemicals should not be added to the receiving stream in such a way that a visible color film can be formed, an oily odor can be imparted, the banks and bottoms of the water can be coated, or the section can become toxic.
9. Turbidity in the receiving waters due to discharge of waste should not exceed 50 Jackson turbidity units in warm water streams or 10 Jackson units in cold water streams. A discharge should not cause the turbidity to exceed 25 Jackson units. A Jackson turbidity unit is a measure of the ratio of the light transmitted through the water in a straight line to the intensity of the incident light.
10. Settleable material in minor quantities inhibits the growth of the normal stream and lake flora; therefore, this material should be excluded from the receiving stream.
11. To provide effective photosynthesis, at least 10% of the light should reach to the bottom of any area where photosynthesis takes place.
12. All floating materials that are not part of the stream or lake should be excluded.
13. All materials that may give odor or taste to the fish or any of the edible invertebrates should be excluded from the receiving stream.
14. Radioactive materials should be excluded from the receiving stream. Radionuclides must meet the national primary drinking water regulations.
15. To avoid nuisance growths, these must be carefully controlled: the addition of organic wastes, such as sewage, food processing materials, cannery and industrial wastes that have nutrients, vitamins, trace elements, and stimulants. Anything that may increase the anaerobic decomposition of the lake must be controlled. The

concentration of total phosphorus should not exceed 100 mg/l in flowing streams or 50 mg/l where the streams enter lakes or reservoirs.

16. Pesticides, chlorinated hydrocarbons, and other hazardous substances should be excluded from the streams.

In marine areas and estuaries, the organisms must be protected in the same way as in freshwater. The following recommendations should be followed:

1. The basin and channels, as well as the freshwater influx, must not be altered to change the patterns of the water by more than 10%.
2. Currents are important for transporting nutrients, larvae, and sedimentary materials, and for flushing and purifying wastes. To protect these functions, the basic areas should not be altered.
3. Materials that may affect the pH by more than 0.1 must not be added to the water. At no point should the pH be under 6.7 or over 8.5.
4. The water that enters the stream must not raise the water temperature by more than 4°F during the fall, spring, and winter, and 1.5°F during the summer.
5. DO must be greater than 5.0 mg/l. In the estuaries and tidal tributaries, it should not be less than 4.0 mg/l.
6. Oil must be excluded.
7. Effluent that may cause any change in turbidity or color must not be added to the receiving waters.
8. Settleable floating substances must be kept out of the area.
9. Substances that cause off-flavors in other edible invertebrates must be kept out of the receiving streams.
10. The radionuclide recommendations are based on the national primary drinking water regulations.
11. The addition of plant nutrients that may cause growth of nuisance organisms must be limited so that the naturally occurring NO_3-N and PO_4-P is maintained. All toxic substances and industrial wastes must be kept out of the estuaries. The amount of fluorides must not exceed 1 ppm. Untreated sewage and other materials that may contain bacteria must not go into any seawater where shellfish is grown.

For wildlife to exist, the DO requirements must be such that the area stays aerobic to prevent the organisms causing botulism from producing botulism toxins. The pH should range between 7.0 and 9.2 for aquatic plants to thrive providing the greatest food value for waterfowl. The bicarbonate alkalinity should be between 30 and 130 mg/l, with fluctuations of less than 50 mg/l, from natural conditions. The salinity should be as close to normal conditions as possible. Light should penetrate at a rate of at least 10% at a depth of at least 6 ft. Settleable substances, oil, toxic substances, and disease organisms must be eliminated.

SWIMMING AREAS

The following recommendations are essential for the use of water for general recreation, bathing, and swimming (for more detail concerning swimming and bathing, see Chapter 4):

1. All recreational surface waters must be aesthetically pleasing. Objectionable odors and colors must not be present.
2. All microbiological hazards as defined in bathing and swimming areas must be eliminated or bathing and swimming cannot take place.
3. Recreational waters must not contain chemicals that are irritating to the skin or mucous membrane of the human body, or that may be toxic to humans if ingested in small quantities.
4. The thermal characteristics of bathing and swimming water should be such that the water does not exceed a temperature of 93 to 94°F, because this is hazardous; or drop to temperatures colder than 59°F, because this may be hazardous for individuals who are exposed for prolonged periods of time.
5. The pH of bathing and swimming waters must be a minimum of 6.5 to a maximum of 8.3. The preferred pH is 7.2 to 7.8. Where the pH is too high or too low, it can cause extreme eye and skin irritations.
6. The water must be clear enough that a bather can be seen at the bottom of the swimming area.
7. All potential subsurface or submerged hazards must be eliminated.
8. If shellfish are removed from the recreational area, the water must meet the standards of harvesting shellfish that have been discussed in other sections of this book and in Volume I, Chapter 3, Food Protection, and Chapter 4, Food Technology.
9. Where wild and scenic rivers exist, the areas must not be affected by any type of urban or industrial pollution.

Public Water Supplies

See Chapter 3 on public water supplies.

Sludge Disposal

Because ocean dumping of sludge was discontinued for sewage sludge and industrial waste December 31, 1991, other means of sludge disposal had to be found. Sludge conditioning is handled in several different ways. One of the most commonly used techniques is to add coagulants, such as ferric chloride, lime, or organic polymers. Ash from incinerated sludge can also be used as a conditioning agent. The chemical coagulants cause the solids to clump together and they are then more easily separated from the water. Chemical polymers are becoming increasingly more popular and easy to handle, require little storage space, and are very effective. The conditioning chemicals are injected into the sludge just before the thickening or dewatering process occurs. These chemicals are mixed with the sludge thoroughly. Another technique of conditioning is to heat the sludge between 350 and 450°F at pressures of 150 to 300 lb/in.[2]. Under these conditions, water bound in the solids is released and improves the dewatering characteristics of the sludge.

Another technique is the application of chlorine to the sludge under a pressure of 30 to 40 lb/in.[2]. This tends to condition the sludge and provides stabilization for organic sludges. After sludge conditioning, the sludge is further thickened. Thickening can be accomplished by floating the solids to the top of the liquid or by allowing the solids to settle to the bottom. As much water as possible should be removed from the sludge. The flotation thickening process utilizes air at a pressure of 40 to 80 lb/in.^{2}. The gravity thickening process simply allows sedimentation by gravity.

Sludge stabilization breaks down the organic solids biochemically so that they are more stable and therefore less odorous and less putrescent. The anaerobic digestion process takes place in closed tanks in the absence of oxygen. In the aerobic digestion process, air is injected into the closed tank. Sludge dewatering occurs or is carried out by spreading the sludge over drying sand beds, or by drawing the sludge onto a cylinder that is coated with a filtration material. The water is drawn through the cylinder and the sludge is left on the outside. The sludge then must be scraped off.

Centrifuges are used to dewater sludge. The centrifugal force causes the sludge particles to separate from the liquid. The unit operates at 1600 to 2000 r/min. The sludge is forced to the outside and the liquid to the center. Pressure filtration is another technique used for sludge dewatering. This is done by pumping sludge at pressures of 225 lb/in.^{2} through a filter medium that is attached to a series of plates. The water passes through the filter medium and the solids are retained; then the solids must be scraped off. After the dewatering process, sludge that contains essential plant nutrients and useful trace elements is used as a fertilizer or soil conditioner. Before this can be done, however, the sludge must be stabilized by digestion to kill off all pathogenic bacteria and viruses, and to reduce the potential for odors.

Sludge is also placed into an incinerator and burned at a temperature of 1500°F or higher. The ash escapes with the exhaust gases and can cause an air pollution control problem unless trapped. The final ash can then be placed into a landfill.

Incineration technologies are highly developed. The main types of equipment used are multiple hearths and fluidized-bed furnaces. Air pollution can be greatly reduced by use of wet systems, which are scrubbers; or dry systems, which are electrostatic precipitators, bag filters, or cyclones. Gaseous emissions can be controlled by scrubbers. Odors and organic compounds can be reduced by use of afterburners. PCB and pesticides can be removed at a temperature of 430°C for a 94% reduction, and 600°C for a 99.9% reduction. Mercury is not removed through sludge incineration, and even after water scrubbing, 97.6% of the mercury found in the sludge ends up in the exhaust gases.

Blast furnace dust can be used as an aid in the physical–chemical clarification of raw sewage. The sludge than can be used as a source of carbon to help in the smelting of the dust to form pig iron.

Sludge pasteurization can be accomplished in the following manner:

1. Temperature of 65°C for at least 30 min.
2. Temperature of 70°C for at least 25 min.
3. Temperature of 75°C for at least 20 min.
4. Temperature of 80°C for at least 10 min.

The sludge can be heated by gas-fired burners, by large continuous microwave ovens, or by the use of thermophilic aerobic bacteria (the most efficient way). Calcium oxide or unslaked lime can be added to dewatered sludge. This results in

a heating of the sludge to temperatures between 55°C and 70°C, caused by exothermic reactions of the calcium oxide and water. The pH must be at least 12.6

Electron-beam disinfection is used in conjunction with a highly efficient composting process. The technique relies on irradiating a thin layer of sludge with the
electron beam. The low penetrating capacity electron beam is overcome by using a
machine that produces a thin layer of sludge cake to be placed under the electron
beam. A radiation dose of 5 kGy can achieve complete disinfection. The subsequent
composting operations can be carried out successfully.

The concentration of heavy metals in sewage sludge can be reduced as follows:

1. Use of strict controls on industrial sources and other point sources of wastewater
   discharge
2. Control of diffuse sources, such as zinc-free roof gutters and copper-free water
   transport systems
3. Extraction of heavy metals from sewage sludge by use of inorganic acids

Wet air oxidation of high organic solids fluid occurs at temperatures >175°C at
elevated pressure, and in the presence of oxygen. The organic matter is oxidized to
carbon dioxide and water.

Land Application

Sewage sludge has been applied to land for many years in some parts of the
country. There is now a greater interest in land application of sludge for many parts
of the country resulting from higher cost of fertilizer, people needed for application
of fertilizers, and communities faced with the problem of how best to dispose of
sewage sludge produced by urban and industrial portions of the community. However, availability of suitable disposal sites is a concern.

Although farmers are concerned about the feasibility of applying sludge to their
land, they are interested in the potential economic benefits that may be available.
Sludge is a resource containing available nutrients. However, concerns exist about the
potential contaminants that are also found in the sludge. To determine one of the most
efficient means of application of sludge, it is necessary to understand the composition
of the sludge applied, physical and chemical properties of the soil, and crops to be
grown. Sludge needs to be analyzed, soils need to be tested, proper information needs
to be made available and appropriate application rates need to be determined.

Primary sludge is raw sludge obtained in the primary stage of the wastewater
treatment plant by collecting solids, both settled and floating. Activated sludge is
obtained in the secondary stage of the treatment plant by settling out the flocculated
bacteria-containing materials. Stabilized sludge contains primary and activated sludges that are treated, and the organic material has been decomposed into a relatively
stable material. Anaerobic and aerobic digestion are commonly used to stabilize the
solids in the digestive sludge. Sludges may also be stabilized by thermal conditioning
(wet oxidation) or lime stabilization. Chemical precipitation, gravity, and air flotation
processes are used to thicken the sludge. Vacuum filters, centrifuges, and filter presses

are used to dewater the sludge. Sand-drying beds or dehydrators may be used to produce a dried sludge.

The three basic types of sludge solids are:

1. Liquids have a solid content of 1 to 10% and may be handled by gravity flow, pumping, or transporting by tanker truck.
2. Semisolids (wet solids) contain 20 to 30% solids and can be handled through conveyors, buckets, and water-type truck transports.
3. Solids (dry solids) contain 25 to 80% solids and can be handled through conveyors, augers, and closed truck transports. Typically, a sludge with over 10% solids cannot flow by gravity.

The chemical characteristics of sludge include organic matter, organic chemicals, fertilizer, nutrients, and heavy metals plus other organics. The sludge composition varies depending on the waste entering the treatment plant and the type of waste treatment. The sludge may typically contain the following: boron, 10 to 800 ppm dry weight; cadmium, 3 to 3000 ppm dry weight; chromium, 20 to 30,000 ppm dry weight; cobalt, 2 to 20 ppm dry weight; copper, 85 to 11,000 ppm dry weight; nickel, 10 to 4000 ppm dry weight; mercury, 0.5 to 10,000 ppm dry rate; molybdenum, 20 to 30 ppm dry weight; lead, 50 to 20,000 ppm dry weight; and zinc, 100 to 28,000 ppm dry weight.

When sludge is analyzed, a determination is made for the pH, percentage of solids, and nutrients listed as ammonia nitrogen, total nitrogen, total phosphorus, and total potassium. A determination is also made for the amount of metals that are present including copper, cadmium, zinc, nickel, and lead. Other types of analysis for inorganic or organic constituents may be needed if an annually large input of industrial waste into the sewage occurs.

Nuisance odors must be considered in the use of sludge. This occurs when one or more of the following take place: mechanical failure of the digester, poor waste treatment plant management, or sludge has been removed during the periodic cleaning of the digester. Most anaerobic, aerobic, and lime stabilized sewage sludge do not have offensive odors.

The land on which sludge is deposited should not have slopes greater than 12%. If the land has a slope of 6 to 12%, the sludge should be spread only when at least 80% of the soil is covered with vegetation, when the immediate incorporation of the sludge into the ground occurs, or when special erosion control practices are followed. When the application rate of the sludge is less than 2 dry tons/acre, it may be spread on land up to 25 ft away from ponds, lakes, streams, or drainage ditches. If it is spread at rates less than 5 tons/acre and not incorporated into level ground, a vegetative barrier or land barrier is needed between the field and the pond, lake, stream, or ditch. At rates greater than 5 tons/acre, the sludge should not be spread within 300 ft of ponds, lakes, streams, or drainage ditches unless it is immediately incorporated into the ground. There should be at least a 25 ft distance from swales and small surface ditches in any area, providing that drainage does not go into ponds, lakes, or drainage ditches.

Sludge should not be applied to any soils where more than a 10% chance of flooding per year exists. Sludge application rates of 5 dry tons/acre or less may be used on soil where less than 5 ft of soil overlies fractured bedrock or permeable sand or gravels. Application of sludge in excess of 2 dry tons/acre is not recommended where the water table is within 1 ft of the surface. Where high rates of sludge application and potential for seepage occur, the sludge should not be applied to the land.

The soil must be tested for pH; cation exchange capacity (CEC); lime test index; available phosphorus; and exchangeable potassium, calcium, and magnesium. The pH of the plow layer, which is 0 to 8 in. below the soil, should be 6.5 or greater. Plants can accumulate more heavy metals from the soil with a pH less than 6.5. The CEC is important because soils with higher cation exchange capacities have greater ability to hold and immobilize heavy metals. Coarse soil can immobilize 5 to 10 meq/100 g of soil and organic soil can immobilize more than 50 meq/100 g of soil. The ability of mineral soils to inactivate the heavy metals and sludge increases as the organic matter content increases. Organic matter increases the CEC and also immobilizes some of the metals.

Phosphorus that is present in sludge in large amounts becomes available over time. At low sludge application rates, the phosphate is used by the crop and buildup in the soil is not a problem. At rates in excess of 2 dry tons/acre/year, the available phosphate levels increase in soils with a low phosphate retention capacity, such as sandy and organic soils. Excessive phosphate buildup in the soils could result in a movement of phosphates into groundwater.

The texture of the soil is part of the most important physical characteristic. It affects many of the other soil physical and chemical properties. In sandy and loamy sandy soil, the leaching of nitrates and other soluble sludge components is the major hazard. This should not be a problem if the sludge application rates do not exceed the nitrogen requirements of the crop, which indicates that all sludge must be tested for its various components. Sands have a low phosphate retention capacity, low CEC, and low buffer capacity, which does not help control changes in pH. Loamy soil and sandy loam have few limitations to sludge application. Silt loams have major limitations including soil crusting, erodibility, and potential for compaction. Silty clays, clay loams, and silty clay loams have major limitations due to poor drainage, poor aeration, slow permeability, and serious potential problems due to compaction. These limitations are less restrictive if the sludge application rate is less than 5 dry tons/acre and the application is made when the soils are not excessively wet.

Soil structure is important. Soils with massive subsurface structure restrict water movement, resulting in impaired drainage and poor aeration. High rates of sewage sludge should not be applied to these soils. Soil erodibility is very important. The ability of the soil to erode depends on many factors. These factors include slope, soil texture, and vegetative cover. The greatest hazard is with fine-texture soils. Sludge application of sloping fine-texture soils should not be done unless vegetative cover is maintained on these soils to increase infiltration. If the unincorporated sludge is put on bare slopes of greater than 6% grade, the sludge moves during runoff.

Liquid sewage sludge contains between 95 and 98% water. Therefore, soil permeability is very important. Soils with either very high or very low permeability should not be used. Highly permeable soils are susceptible to leaching and potential contamination of the groundwater by the sludge. Low permeable soil has internal drainage problems that restrict the decomposition of the sludge. These problems are minimal if the sludge application rates are about 2 dry tons/acre/year.

The drainage characteristics of the soil are important because the decomposition of the organic matter in sludge in the soil is aided by good aeration. Sludge application on soils of poor internal drainage may produce odors during sludge decomposition.

Proper transportation and application techniques must be used in the handling and disposal of sludge. These techniques are based on the type of sludge, and whether it is liquid, semisolid, or solid. This then would indicate the levels of water that would be present. The quantity of sludge produced, distance to the spreading site, site characteristics, time of application, and weather conditions are very important.

The two most common methods of applying sludge are in the liquid form using a tank truck or a major pipeline from barges, and the "cake" in which a spreader truck is used. In any case, these trucks and other pieces of equipment must not contaminate surrounding areas, streams, or highways. The commercial tank truck can spread the sludge on the ground by first blowing it into the air and allowing it to fall as rain onto the ground. The pipeline can temporarily flood an area and then allow the sludge to seep in.

On the average, a ton of dry sludge contains approximately 65 lb of nitrogen, 50 lb of phosphorus, and 5 lb of potassium. All the ammonia and nitrates are immediately available to the plants. About 30% of the organic nitrogen is available in the year of application and about 5% of the remaining organic nitrogen is available annually to the crop in future years.

Wastewater from Specific Industries

The organic chemical industry is made up of thousands of companies that manufacture huge numbers of products from such organic chemical sources as petroleum, coal, and natural gas. They typically produce synthetic detergents, fuel additives, solvents, plastics, resins, and synthetic fibers. The wastewater produced during the manufacture of these products is usually heavily contaminated with toxic substances. Both activated carbon treatment and wet-air oxidation treatment are utilized in trying to reduce the levels of toxic materials.

The activated carbon process is used because the material is highly adsorbent, and therefore the chemicals cling to the surface of the material. The chemicals can be released and possibly recycled by heating the activated carbon.

The wet-air oxidation process involves exposing the toxic organic wastewaters to high temperatures and pressures while they are confined in a reactor. These conditions cause the oxidation and conversion of toxic organic substances into nontoxic forms. The resulting effluent should be suitable for discharge directly to wastewater or nearby municipal treatment facility.

The petrochemical industry produces large amounts of wastewater containing a huge amount of chemicals that may be extremely hazardous. Many of these chemicals are potentially carcinogenic or mutagenic. Steam stripping, as previously discussed, in wet-air oxidation appears to be the major technique available for removing the toxic chemicals from the wastewater.

The petroleum refining industry is one of the five largest industrial wastewater dischargers in the country. It requires large volumes of water to produce its products. Sour water that comes from some of its refining processes contains high concentrations of sulfur and cyanide compounds as well as ammonia. The treatment techniques used to control sour water is a form of steam stripping known as sour water stripping. This helps recover ammonia and sulfur, which can later be sold. However, where high concentrations of ammonia are present, the system is not particularly efficient.

The petroleum refining industry is attempting to improve the catalytic cracking process, which is a major source of sour water. Catalytic cracking involves the splitting of large crude oil molecules into smaller molecules.

The pesticides manufacturing industry provides substantial quantities of toxic chemicals for the water waste stream. Currently, efforts are underway to determine the best way to treat these pesticides in the wastewater.

The inorganic chemical industry, which is made up of about 1600 plants, produces 110 million metric tons of products each year and generates 40 million metric tons of waste each year. Large quantities of water are used for cooling, processing, product washing, waste transport, and other production purposes. The wastewaters contain heavy metals, cyanides, suspended solids, fluorides, iron, and ammonia, and have a high COD. Despite the fact that the industry does a lot to treat the wastewater, significant water problems continue.

The battery-manufacturing industry produces low concentrations of the toxic metals, lead, arsenic, cadmium, and antimony in its wastewater discharges. Because most battery-manufacturing operations are inside cities, a problem arises with trying to treat the waste.

The metal-finishing industry produces more than 1 billion gal of wastewater containing toxic heavy metals and cyanide. Metal-finishing processes add a protective metal coating or plating to metal surfaces or to nonmetal surfaces. The substance plated is submerged first in the plating tanks and then in a series of rinse tanks. The concentration of the heavy metals declines in each of the consecutive rinse tanks until the product is essentially clean. As the concentration of heavy metals decreases on the object, it is increased in the rinse tanks. The most promising methods for reducing this waste is the use of membrane and electrochemical techniques, as well as a centralized waste treatment process. In the membrane technique, it is possible to concentrate the rinse water pollutants on the membrane through which the rinse water is passing, while at the same time generating an effluent stream that is relatively pollutant free. Major benefits of the membrane technologies are that the agents are recovered during the process, additional treatments are not needed, sludge is not generated, and energy consumption is low. The iron and steel industries produce approximately 150 million metric tons of waste. Wastewaters from steel plants contain dissolved solids, oils and greases, phenols, cyanides, ammonia, and sulfide, and have a high BOD.

The steam electric power industry uses large quantities of water for steam generation or cooling, and then returns the water to the source. The steam is a heat problem in the stream, and if boiler cleaning operations take place, the wastewater also contains toxic pollutants. During the operation of the plant, corrosion products accumulate in the boiler tubing, and if uncontrolled cause the power plant to become less efficient. These corrosion products contain heavy metals, such as iron, copper, zinc, and nickel. Strong chemicals are used to dissolve the corrosion products in the cleaning process. As a result, the heavy metals contained in the boiler tubing enter the solution and become part of the toxic industrial discharge. Lime treatment is utilized in precipitating and removing heavy metals in boiler cleaning solutions.

The textile industry produces wastewater in several parts of the country that contains a variety of dyes and chemicals such as acids, bases, salts, detergents, and finishes. The major types of wastewater treatment used are primary treatment, to remove pollutants that settle or float; biological treatment, to speed up the breakdown of degradable organic pollutants; and advanced treatment, for the further removal of pollutants. The biological treatment is inadequate.

The leather processing industry uses water extensively in cleaning, tanning, and dying operations. Cleaning is required to remove flesh and fat from the inside of animal skins. Tanning involves soaking the skins in chemicals to produce a flexible long-lasting product, and dyeing involves numerous chemicals to color the hides. The consequence of this is a wastewater stream high in BOD, COD, and suspended solids, and high in process chemicals and colors. The wastewater from leather processing is treated to settle suspended solids and then is passed into an oxidation ditch, in which many of the pollutants can be biologically degraded by bacteria and other microorganisms. Finally, powdered activated carbon is added to the oxidation ditch system to further reduce the pollutants that remain in the effluent in very low concentrations.

The paper industry produces more than 7 billion gal of wastewater daily. Soap-like resins and fatty acids of pulp and paper manufacturing effluents contribute to foam problems in rivers, streams, and lakes receiving effluents. Treatment technologies involve the elimination of foam through the recovery of resins and fatty acids. Foam separation and chemically assisted treatment techniques are also utilized.

WASTEWATER TREATMENT

Wastewater from domestic municipal and industrial sources are treated to remove pollutants that are harmful to human health and the environment. Of the 300 billion gal of water used in the United States each day, about 90% is used by industry. The variety of wastewaters are treated in several different ways. The water pollutants controlled by these treatment processes include organic substances, suspended solids, phosphorus-containing compounds (both suspended and dissolved), and ammonia.

Biological Waste

Biological waste concentrations are commonly expressed as a 5-day BOD_5 or COD, and total organic carbon (TOC) in milligrams per liter (mg/l). The BOD_5 has been used to express the biodegradable waste concentration in the municipal as well as industrial wastewater. The COD is a measure of the presence of refractory organic materials not amenable to biological treatment. The TOC measures total organic materials. All these measures of organic waste concentrations include both dissolved materials and suspended solids.

Suspended solids include both organic and inorganic, biologically inert materials, such as fine particles of silt. Organic suspended solids contribute to a portion of the total BOD_5 of the wastewater. Phosphorus (P) is found as dissolved phosphorus compounds and also found in some of the suspended solids. Ammonia (NH_3) is present both as dissolved ammonia gas and insoluble compounds. Ammonia can form from the degradation of nitrogeneous compounds in the waste.

The three technologies used as biological treatment processes are primary treatment, secondary treatment, and tertiary or advanced wastewater treatment.

Wastewater treatment processes that have an effluent level of 30 mg/l or less of 5-day BOD and 30 mg/l or less of suspended solids are considered to be conventional secondary treatment processes. The systems that achieve effluent levels of 10 mg/l or less of BOD_5 and 10 mg/l or less of suspended solids are referred to as advanced wastewater treatment.

The typical conventional secondary treatment system consists of preliminary treatment, influent pumping, primary clarification, activated sludge, secondary treatment, secondary clarification, effluent disinfection by chlorination, and sludge treatment. The typical advanced wastewater system contains everything in the secondary system plus primary chemical addition prior to primary clarification, secondary chemical addition, prior to secondary clarification, and granular media filtration of secondary clarifier effluent. For additional information on primary and advanced wastewater treatment processes, see Chapter 6.

Stabilization ponds and aerated lagoons are earthwork structures that can be either below grade or at grade, or can be built by damming a natural area. The ponds may be unlined or lined with clay, rubber, or plastic. They can be subdivided by earthwork partitions into several compartments or cells. The ponds treat wastewater by providing detention time for biological oxidation of BOD_5 and settling of suspended solids. These settled solids undergo anaerobic decomposition at the bottom of the pond. The detention time in the pond depends on the individual wastewater characteristics, pond waste load, and temperature. The ponds may be either absorption or flow-through ponds. The absorption pond is based on the percolation and evaporation of wastewater from the pond. From time to time, discharges may be needed when peak flows exceed the capacity of the pond.

The flow-through ponds are of four basic types: aerobic algae ponds; facultative ponds, where there is an aerobic upper layer and an anaerobic lower layer; anaerobic ponds; and aerator ponds or aerated lagoons. In aerobic ponds, algae growth provides the oxygen needed to satisfy the requirement for reduction of the wastewater BOD_5. The depth of these ponds is restricted to less than 5 ft to permit sunlight penetration. Some mixing is required to make sure that good oxygen distribution occurs. These shallow ponds rely on natural circulation. The algae must be removed from the treated water before discharge. This can be done by a special overflow design or a clarifier.

The facultative pond (stabilization pond) contains an upper water layer that behaves in the same manner as an aerobic pond. The bottom water layer and sludge on the pond bottom is anaerobic and the organic materials decompose to produce methane and other gases. The depths of these ponds are 3 to 6 ft.

The anaerobic ponds are anaerobic throughout the entire pond. They are relatively deep to minimize any generation of odors to the surface area and to retain heat so that the process can continue.

The aerated lagoons rely on mechanically promoted oxygen transfer to the wastewater. Diffused aeration or mechanical aeration systems can be used. These lagoons range in depth from 6 to 20 ft and can be subdivided into cells by making partitions of the earth. Although the upper layers of the pond are well aerated, anaerobic decomposition of solids occurs on the pond bottom. The surface mechanical aeration either can be a floating type or can be fixed. As the solids begin to accumulate at the bottom of the lagoon cells, a portion undergoes anaerobic decomposition. Suspended solids removal can be enhanced if several smaller aerated polishing cells follow the last aerated cell. Periodically solids must be removed from the ponds and taken to a landfill. These ponds can be used for both municipal and industrial wastewater where biological treatment is effective. Removal of BOD_5 ranges from about 60% to over 90%, depending on the character of the wastewater and the way the system is designed. These ponds are commonly used where the land is inexpensive and treatment costs can be minimized.

Phosphate removal and nitrate removal are discussed in Chapter 6 on sewage.

Chemical Wastes

Chemical wastes are treated in a variety of manners. These modes of treatment may be individual or combined. They include equalization, neutralization, flocculation, emulsion breaking, gravity-assisted separation, skimming, plate and tube separation, dissolved air flotation, chromium reduction, cyanide destruction, chemical precipitation, filtration, sand filtration, multimedia filtration, ultrafiltration, reverse osmosis, carbon adsorption, ion exchange, air stripping, biological treatment, activated sludge, sequencing batch reactors, vacuum filtration, and pressure filtration. To reduce the chemical wastes, an attempt is made to maximize the amount of materials recycled and minimize the amount of material going into waste waters.

Neutralization is a technique used to adjust high or low pH values by means of acids or alkalis. Flocculation is the agitation of chemically treated water to induce coagulation. Coagulation is a reduction of the net electrical repulse of forces at particle surfaces to cause an agglomeration of the destabilized particles by chemical joining and bridging. Inorganic electrolytes, natural organic polymers, and synthetic polyelectrolytes are used for this purpose. Emulsion breaking is achieved through the addition of chemicals, heat, or both to the emulsified oil and water mixture. This is accomplished by adding surfactants and coagulants as well as acid-cracking chemical treatment. Gravity separation is used to remove oil and grease. The wastewater is in a quiet state long enough for the oil droplets to rise to the top. Clarification utilizes gravity to provide continuous slow-moving effluents that allow removal of particulates. In dissolved air flotation, fine bubbles are introduced into the effluent to cause them to attach themselves to the particulates and to allow them to float free. Reduction is a chemical reaction in which electrons are transferred from one chemical to another.

Hexavalent chromium is reduced to trivalent chromium at a pH of 2 to 3, and then precipitated from the wastewater in conjunction with other metallic salts. Cyanide-bearing wastes from electroplating and metal finishing operations are destroyed by use of alkaline chlorination with gaseous chlorine, or with sodium hydrochlorite.

Chemical precipitation occurs when hydroxides, sulfides, or carbonates are added to metals. The metals precipitate and then are removed by filtration and clarification. Filtration is a method for separating solid particles from a fluid through the use of the porous medium. Sand filtration occurs through either a fixed or a moving bed that traps and removes suspended solids from water passing through the media. Multimedia filtration is a combination of two or more layers of granular materials used after chemical and biological treatment processes.

Ultrafiltration, used for the treatment of metal-finishing wastewater and oily wastes, is a process that can remove substances with molecular weights greater than 500, including suspended solids, oil and grease, large organic molecules, and complex heavy metals. It is a semipermeable microporous membrane process.

Reverse osmosis is a process in which the pressure applied to the more concentrated solution is greater than the normal osmotic pressure, thereby forcing the purified water through the membrane into the less concentrated stream. Activated carbon adsorption is a combination of physical, chemical, and electrostatic interactions between the activated carbon and the organic compounds.

Stripping is a method for removing dissolved VOCs from wastewater. The removal is done by passing air through the agitated waste stream. The process results in a contaminated gas stream that usually requires air pollution control equipment to remove the chemicals. Air stripping is accomplished by mass transfer of air that results in a significant difference between the concentrations in the air and water stream.

Liquid CO_2 extraction is a process used to extract and recover organic contaminants from aqueous waste streams. The waste stream is fed into the top of a pressurized extraction tower containing perforated plates, where it is in contact with a stream of liquid carbon dioxide going in a different direction. The organic contaminants in the waste stream are dissolved in the carbon dioxide. The extract is then sent to a separator where the carbon dioxide is redistilled, and the organics are separated out and removed.

Biological treatment is a method of using organic chemicals as a food source of carbon and energy for microbes. Certain organic pollutants including oils are amenable to biological degradation. Depending on the organic chemicals and processes, aerobic, anaerobic, or facultative organisms may be used. The success of biological treatment is dependent on many factors, such as pH and temperature of the wastewater, nature of the pollutants, nutrient requirements of the microbes, presence of inhibiting pollutants, and variations in the feed stream loading. Heavy metals may be toxic to the microorganisms and have to be removed before biological treatment. A sequencing batch reactor is a suspended growth system in which wastewater is mixed with existing biological floc in an aeration basin for a specific time period.

The process sludge is removed periodically. See Chapter 6 on sewage for information on trickling filters, oxidation ditches, and sludge treatment and disposal.

MODES OF SURVEILLANCE AND EVALUATION

Water Pollution Surveys

The water pollution survey is a combination of environmental studies conducted on bodies of water and on potential sources of pollution entering these bodies of water, as well as various microbiological and chemical analyses. The survey must take into account the general or nonpoint source pollution coming from a variety of areas where pollution may enter the stream. This survey must also account for the industrial waste point source pollution that enters the receiving stream.

Industrial Waste Surveys

Industrial waste surveys are a combination of the environmental survey, sampling, and laboratory analysis of the samples. The survey should include the production processes, types of waste generated, quantities of wastes, means of treatment of the wastes, and their effects on the receiving stream. The study should be conducted as a cooperative venture between the industrial plant and the environmental health practitioner from the health department or EPA.

The objectives of such a study are to determine the total pollution load of industry and its source, quantity, concentration, and characteristics. It also allows the industry to evaluate its own waste-creating processes and to make necessary process modifications or revisions to eliminate or reduce the pollutants. There have been cases where the reduction of pollutants have financially helped the industry by saving useful by-products. When the problem is defined jointly, then the industry and the regulatory agency can attempt to work together to resolve the problem. The study should be carried out using the following techniques:

1. Conduct the survey during normal operation of the industrial plant.
2. Assign professional staff to work with industrial technical staff.
3. Perform analytic work in agency laboratories.
4. Have industry install necessary flow-measuring devices.
5. Industry should provide personnel for sample collecting.
6. All information from surveys are to be held confidential.
7. Set a due date for the survey that best meets the needs of all concerned.
8. Have a conference with the industrial staff prior to the survey.
9. Conduct the survey with an individual who is knowledgeable about the plant and the processes under evaluation. The individual should have adequate authority to do what is necessary within the plant.
10. Examine an up-to-date plant sewer plan.
11. Establish the inspection and sampling stations.
12. Decide on the types of flow-measuring devices to be utilized and where.
13. Develop a cost estimate for the project.

14. Involve the engineer, environmental health practitioner, chemist, and biologist in
    the evaluation and the actual conducting of the study when needed.
15. Collect representative deposit samples over a 24-hour period.
16. Sample the major individual wastes and check against the outfall.
17. Review all the analytic data.
18. Prepare the necessary reports.

The liquid wastes should be evaluated with great care. The problems should be defined by studying the literature related to the process within the plant; by evaluating information on industries involved in this type of process; by evaluating the physical layout of the plant; and by determining raw materials used, intermediates, and final products. A schematic flow sheet should be prepared of the plant operation, including the alternate processes utilized by the industry, by-products and salvage operations, and liquid flow diagram. The final effluent discharge is a valuable adjunct to the study. It is important to determine the housekeeping procedures used and how they affect the liquid waste that leaves the plant and enters the receiving stream.

Samples may be taken either manually or automatically in the industrial waste survey. Manual sampling requires an adequate volume of material. Samples may also consist of grab samples, composite samples, proportional samples to the flow of effluent, and samples taken at the point where the effluent is most homogeneous. In this sample collecting procedure, it is necessary to take precautions against settleable solids, oil, or volatile materials. In automatic sampling, it is important to install the proper type of equipment, to calibrate it accurately, to maintain it accurately, and to understand its limitations.

In industrial plants, some problems occur in trying to determine the types of effluents due to the existence of submerged outfalls, deep manholes, steam-filled sewers, coated sewers, sedimentation, toxic or explosive liquids or vapors, solvents, and floating material; and an inability to get an ideal cross section of the effluent. When the samples have been taken and analyzed and the environmental survey completed, it is necessary to correlate all the data with the following plant operations: raw materials; finished products; intermediate products; by-products or salvaged materials; production yield; the processes used; and water use, reuse, and recycling.

In summation, it is important to recognize that the industrial waste survey is conducted by the official agency in cooperation with the industry involved. It is an attempt on the part of industry and government to determine actual hazards created in water by given industries and types of techniques that can be most effectively used in the most inexpensive and expedient manner.

General or Nonpoint Source Stream Surveys

Nonpoint and diffuse sources of pollution appear to be one of the primary causes of water quality problems. It is difficult to get a total understanding of the nonpoint source problem. However, measurements have been made (in areas where information is available) in a variety of water basins as noted in federal publications. The nonpoint source pollution appears to be an accumulation of pollutants from agricultural activities, erosion, acid mine drainage, and urban runoff. The agricultural nonpoint sources of pollution include cultivated crop fields, orchards, vineyards, pasturelands, and animal feedlots. The activities associated with harvesting, logging, and forestry contribute much of the nonpoint source pollution in the western states.

Mining also is a nonpoint source that contributes material from active and abandoned subsurface mines, spoils, tailing deposits, washing process areas, primary acid treatment, oil shale, and holding ponds. The problems created by materials from construction include devegetated slopes, waste building materials, and runoff from concrete and asphalt surfaces. Dam construction, dredging, and other channel activities add to nonpoint source pollution. The urban nonpoint sources include extensive impervious surfaces such as paved and roofed areas. Oil, street litter, salt, other ice-control chemicals, animal droppings, insecticides, dusts, industrial wastes, suspended solids, phosphates, nitrates, and heavy metals have come from urban runoff.

Sediment Monitoring

Sediment monitoring is an important tool for detecting loadings of pollutants in streams and lakes, because potential contaminants accumulate at far higher concentrations in the sediment than in the water above it. Nutrients, organic compounds, and heavy metals are found in the sediment and can be released to the water and made available for the biological community in the water. To determine levels of contamination, it is necessary to compare and contrast the amounts of a given contaminant in a noncontaminated area to a contaminated area and determine to what degree the contamination exceeds the maximum background level of the chemical that is acceptable.

Survey Procedures

The type of water quality survey is based on the limitations of personnel, facilities, time, and money. It is therefore necessary to develop a survey plan that gives the maximum amount of information for a reasonable amount of workforce and finances. The basic objectives behind the water quality surveys are to determine the natural quality of the stream; to measure in a selected and limited time period the existing effects of waste on water quality and on water use; to obtain data on waste loads, and how the waste loads affect water quality and stream characteristics; and to determine the corrective measures needed to protect the water quality of the receiving stream.

When planning the study, it is necessary to review all readily obtainable maps, information, reports, and other materials that may be available concerning the stream and its problems. A list of objectives of the study should be drawn up so that the actual study conducted may satisfy the planned objectives. The plan should include a map of the stream, including locations and strengths of known sources of waste; locations of areas of water use in the stream; various sections of the major stream and the important tributaries, including dams and diversions; possible sampling stations; sampling frequency and number of samples to be taken at each station; type of laboratory determinations to be made; existing stations, along the stream with gauges indicating the stream flow; any other hydrologic data available; location of potential laboratory facilities; special supplies and equipment that may be needed; types and numbers of personnel and how they are to be supervised; approximate cost of the field operation; environmental health survey techniques that are to be utilized; and arrangements to have experienced field people assist in the beginning of the actual field operation.

The environmental health practitioner should observe the entire reach of the stream by wading into it, walking along the bank, or using a boat. The areas where waste may be discharging into the stream should be specifically noted. Environmental health practitioners should evaluate and mark any visual evidence of pollution, sources of water entering the stream, pipes, or projections, either under or above the water that may be carrying effluent into the stream. They should also determine where sources of sewage may be coming from by evaluating the location of the sewered populations and the population areas where public sewerage is not available and where discharge may be carrying sewage into the entering stream.

During the course of the study, all industrial waste should be investigated and a diagram developed containing information concerning the operating plant; type of products, by-products, and waste that are produced or discharged; and location of the industrial plant waste disposal systems as well as waste disposal processes. During the course of this survey, sampling stations should be decided and samples taken. The samples should be taken from the bottom deposits wherever algae may be growing and also at least 50 ft above and below the active point of an entering pipe into the stream. Preliminary samples then should be utilized to determine where best to set up the general sampling. All information not only should be mapped but also recorded and then taken back to the office for further evaluation. Where appropriate, cameras should be used to take color photographs of areas where effluent of one type or another may be entering the stream or where the stream seems to have pollution problems.

Sampling Procedures

Water quality characteristics are not uniform from one body to another or from one place to another within any given body of water. The quality also varies by time and location, and by time at a given location in the body of water. The sampling program must be devised so that these variations are recognized. The purpose of collecting samples is to accumulate data that can be used to interpret the quality of the water. The surveys are undertaken for a variety of reasons. However, the objective generally is to determine quality of the water, varying problems influencing the quality, and eventually methods of eliminating the contaminating influences. Sample collecting, handling, and testing must be carried out in a careful well-planned manner. The location of the sampling stations and the number of samples taken may well affect the ultimate results. Therefore, there must be sufficient samples and properly located sampling stations. True variations in water sampling occur from the relationship of sunlight and darkness of the water, variation of waste discharges from the community, seasonal variations, or tidal influences. Samples may also be affected by rainfall, runoff, intermittent discharges of wastewater, and the addition of a certain amount of organic contaminants to a body of water. The survey helps to determine the actual source of effluent that may be a contaminant to the receiving stream. The flow rate and direction of flow determines not only how rapidly the contaminant enters the receiving stream but also the ability of the stream to recharge itself.

The actual laboratory sampling procedure is carried out in accordance with accepted water sampling tests for biological, chemical, physical, and radiological contaminants. Standard methods are utilized in these determinations. Automatic sampling devices consist of a continuous sampler or various types of samples that are taken intermittently. When samples are taken, it is necessary to record the weather and other conditions that affect the actual sample. Decisions should not be based on a single sample or a single series of samples, but instead on a large number of samples and the varying environmental conditions that may be affecting them.

Flow Measurement

The measurement of the flow of water is essential when collecting water samples and determining water quality. The analytic results are only as good as the determination of water flow. Water flow is determined by the use of a variety of gauges that are established in gauging stations at different points along the stream. These gauges include current meters, weirs, and Parshall flumes.

The current meter is a device used to measure the velocity of a flowing body of water. The cross section of the stream is divided into a number of sections, and the average velocity is taken in each of the sections. The discharge is then found by adding the products of area and velocity for each section.

Weirs are bulkheads that are placed across the channel (Figure 7.7). The notch is placed on the top of the weir through which liquid may flow. The liquid passing through the notch and flowing over the weir is measured. The partial flume is an open constricted channel in which differences in the elevation of the liquid flowing through the flume are translated into rates of flow (Figure 7.8). Head measuring devices include the staff gauge, plumb line, and water-level recorder. The measurement of water level is important to determine discharge. This can be accomplished by the use of some of the previously mentioned gauges.

Orifice Meters

Orifice meters are good for measuring flow rates from pipes discharging to the atmosphere. Orifice meters are usually circular orifices placed at the end of a horizontal discharge pipe. Flow rates are calculated from the orifice characteristics and are a measure of the pressure behind the orifice. The head on the orifice is measured with a water manometer. Orifice meters or easily constructed and economical flow-measuring devices for irrigation pump discharges. They are accurate within 2% of actual flow rates.

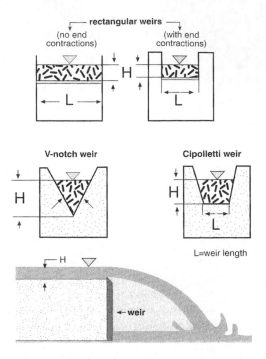

Figure 7.7   Cross sections of common weirs.

Tracing Natural Water Flow

To determine the pollutants entering a receiving stream, it is important to trace the natural flow of the water. The blending of the physical and chemical properties of the pollutants in the receiving water as it mixes and flows in the receiving stream cannot be thoroughly understood unless flow rates and sources of flow of the freshwater can be determined. This determination is made by utilizing wooden or plastic floats; by calculating the amount of salt found at a given point after it has been introduced into the water; or by adding certain dyes, such as fluorescein, the rhodamine series, pomtacyl, brilliant pink B (which is the most stable of the fluorescent dyes), or uranine (which fluoresces near the stream background level). The flow is also measured by the use of radioactive substances such as rubidium-86; iodine-131; tritium, which is best used for subsurface tracing; and krypton-85, which is used in gas transfer measurements in stream studies.

Determination of natural flow can also be made by checking the waste return characteristics, such as silt, loam, acid, temperature, and dyes, that are most prevalent in textile wastes. Also, measurement may be made by evaluating the mammals, fish, and shellfish present. The ideal tracer should be very low in toxicity; should be short term; should not have any carcinogenic or genetic effects; and should be stable in the stream despite the chemicals present, bacteria, sunlight, absorption, temperature,

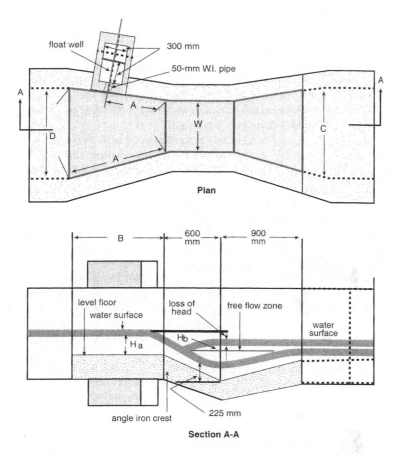

Figure 7.8 Parshall flume.

wind action, inherent decay, channel obstructions, or stratification. The tracer should be readily detectable by visual means or by means of instruments.

Effects of Weather

Climatological conditions have a very definite effect on the effluents discharged into the receiving stream and on the ultimate ability of the stream to rejuvenate itself. These climatological conditions include amount and frequency of rainfall, amount and pattern of runoff, stages and patterns of stream flow, temperature patterns of the receiving stream and the air, sunlight patterns, and wind patterns.

Preparation of Reports

The water quality survey report should provide a summary of findings and basic data; a listing of existing causes and effects, together with an explanation of how and why they occur; a projection of conditions that may occur due to natural variations in stream flow and temperature; a prediction of the effects of population growth and industrial changes; and an estimate of the water needs and water protection techniques. The report itself should have an introduction stating the problem; a description of the area, pertinent history, relationship of the study to other current water resource studies, water use, and economic data within the area; and a map explaining the various features of the stream and sources of pollution. The introduction should also include a statement of purpose.

A second part of the report includes the methodology used. A complete description should be given of the time period of the survey; sampling stations, including identification of locations; sampling and analytic methods utilized; frequency of sampling; and description of techniques used to determine the time of water travel, stream flow data, and wastewater measurements.

The next section contains the survey results. These results include sources of waste, based on the environmental and sampling surveys, strength of the waste, and amount of material flowing. It should also include a summary of stream data, hydrologic data, and actual results of the samples. In the survey results, reference should be made to the collected data that can be found in an appendix. The next section contains an analysis and interpretation of data; and should include a comparison of the survey results to appropriate water quality criteria, a projection of the survey data to provide a way of comparing stream conditions under varying adverse conditions, and an estimate of the amount of waste that may be put into the stream in the present and future.

The next section or conclusion, should clearly and concisely restate in a positive manner the critical parts of the report. Recommendations follow the conclusions section and should stipulate the major recommendations including cost estimates if possible, and alternate recommendations including cost estimates. Minor recommendations should be stated toward the end of this section. A bibliography is useful and enables individuals to properly research any phase of the report.

Interpretations of Analytic Procedures

The analytic laboratory provides qualitative and quantitative data that are helpful in making decisions. The data must accurately describe the characteristics or concentrations of the various materials found in the sample. Because the data may result in far-reaching decisions concerning water use, reuse, and discharge of effluents, it is essential for the laboratory to accurately utilize the latest standard procedures available. It is important for the laboratory staff to meet periodically with the field staff for the two groups to integrate their portion of the water quality study. At this time, the various problems encountered in water sampling or analysis can be discussed and the proper administrative techniques can be developed to handle any problems that may have arisen.

Obviously, all equipment must be carefully calibrated and all results should be carefully determined and recorded. In all cases, the specific method used for determination of various constituents of the water should be stated, along with the results of the water sample.

Bioassay and Biomonitoring

Bioassay and biomonitoring are evaluations in which living organisms provide the scale. The scale or degree of response is determined by rate of growth or decrease of a population, colony, or individual. Bioassays are important in evaluation of water quality, because they determine the effects of liquid waste on the aquatic environments in which experimental organisms, such as fish, may be subjected to a series of concentrations of known or suspected toxicants. The BOD test is a bioassay of the organic content of water subject to biodegradation.

Biomonitoring is a means of surveillance of water quality by observing the biota from the field and laboratory standpoint. It differs from the bioassay, particularly in its objective. The bioassay is an attempt to determine the specific defined value or threshold, whereas the biomonitoring program is an attempt to use living organisms to determine whether aquatic life is in danger. Bioassays are determined by initially introducing certain concentrations of potential toxicants into water where a captive fish population exists and then determining movement patterns, fish-breathing ratios, and number of fish that die within the 24-hour period. Bioassays have been utilized in areas where large fish kills have occurred and determinations had to be made concerning whether the industrial plants had caused the problem by leakage, spillage, accident, or purposeful discharge into the receiving stream.

Biomonitoring techniques are so reliable that they are now built into the regulatory process. The Clean Water Act of 1987 specifically refers to biological testing for assessing environmental hazards, especially where the mix of potential pollutants is complex. Such techniques as atomic absorption spectroscopy can detect concentrations as small as parts per trillion. However, these techniques cannot demonstrate effectively the interaction of chemicals with each other and their effect on living organisms. These tests are also not sensitive to other variables, such as acidity, hardness, solubility, exposure time, or effects on these living organisms.

Biomonitoring permits an assessment of the cumulative effects on aquatic life of multiple sources of pollution. Biomonitoring may range from exposure to a specific chemical, to a complex mixture of chemicals and their effects on the living organisms. The aquatic biologist continues to look for abnormalities in the fish and aquatic insects, as well as population diversity, size and structure of the organisms, and physical characteristics of the population that are under stress from the pollutant in the ecosystem. Biomonitoring allows for the screening of waters for signs of stress and confirms the results of chemical monitoring data.

The Department of the Interior, Fish and Wildlife Service has revised and expanded its 35-year-old National Contaminant Biomonitoring Program that assesses contaminant impact on ecosystems. The program is coordinated with various state and federal programs.

Microbiological Determinations

The coliform group has become the standard for determining the presence of fecal microorganisms in water. The coliform group includes all the aerobic and facultative anaerobic Gram-negative, non-spore-forming, rod-shaped bacteria that ferment lactose with formation of gas within a 48-hour period at 35°C (see Figure 7.9). The coliform group includes a variety of bacteria, such as *Escherichia coli*, *E. aurescens*, *E. freundii*, *E. intermedia*, *Aerobactor aerogenes*, *A. aloacae*, and the biochemical intermediates between *Escherichia* and *Aerobactor*. It is important to separate the coliform group into species of fecal origin and of nonfecal origin. It has been assumed that *E. coli* and its closely related strains are of fecal origin, whereas *A. aerogenes* and its close relatives are not of direct fecal origin. However, EPA studies have not fully borne out this distinction. To differentiate between *E. coli* and *A. aerogenes*, the following items may be utilized:

1. *Escherichia coli* organisms produce hydrogen and carbon dioxide during gas production in equal amounts from fermentation of glucose. *Aerobactor aerogenes* produce twice as much carbon dioxide as hydrogen.
2. In a methyl red test, during glucose fermentation, *E. coli* can drop the pH to a range of 4.2 to 4.6, which indicates a red or positive test. *Aerobactor aerogenes* produce a medium with a pH of 5.6 or greater, which is a yellow color or negative test.
3. Where tryptophan, which is an amino acid, is incorporated into the nutrient broth, the *E. coli* strains produce endol, which is a positive test, whereas *A. aerogenes* cannot, which is a negative test. During glucose fermentation, *A. aerogenes* strains produce acetylmethylcarbinol, whereas *E. coli* cannot produce this chemical. This is known as the Voges–Proskauer test.
4. *Escherichia coli* cannot use the carbon present in citrates, which is a negative test, whereas *A. aerogenes* can use the carbon in their metabolism, which is a positive test.

Coliform comes from a vast variety of sources, including the soil and materials of fecal origin. The values of using the coliform group as a pollution indicator are that the absence of coliform bacteria indicates bacteriologically safe water; density of coliforms present is roughly equivalent to the amount of excreta present; if pathogenic bacteria of intestinal origin are present in the source of pollution, coliform bacteria are also present in greater numbers; and coliforms are always present in the intestines of humans and other warm-blooded animals in large numbers and are deposited in fecal waste. Also, coliforms are more persistent and last longer in the aquatic environment than the pathogenic bacteria that come from intestinal sources; and coliforms are generally harmless to humans, with the exception of one highly virulent strain that causes severe infant diarrhea and other virulent strains that cause diarrhea in children and adults.

The limitations in the use of coliforms are that the coliform group may come from a vast number of sources, and the organisms occur in the intestines of all warm-blooded animals; some of the strains of coliform multiply in polluted waters and give a false reading; the age of the pollution may not be determined properly because of this multiplication; and the coliform tests may give false negatives if certain species of *Pseudomonas* are present.

The fecal coliform group are important as indicators of human fecal pollution, because the majority of them can grow at elevated test temperatures. These organisms infrequently occur, except when fecal pollution occurs. Those in the fecal coliform group do not survive as well in the water environment as the rest of the coliforms.

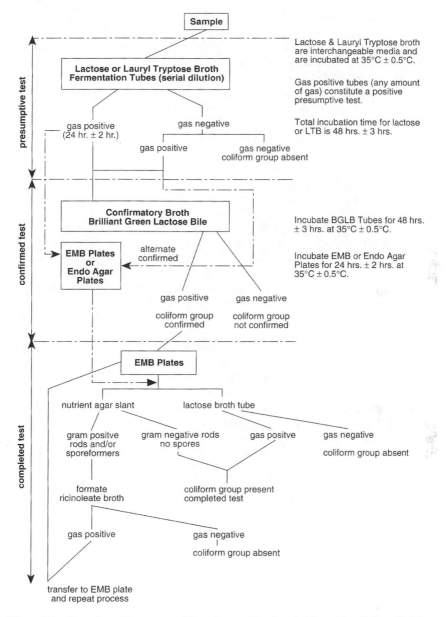

Figure 7.9 Tests for coliform group. (From Current Practices in Water Microbiology, Training Manual, U.S. Environmental Protection Agency, Water Programs Operations, Training Program, U.S. EPA, Washington, D.C., January 1973, pp. 9–21.)

Fecal coliforms generally do not multiply outside the intestines of warm-blooded animals. The limitations of the increased temperature fecal coliform group test is that the feces may come from any of the warm-blooded animals. No established correlation exists between total coliforms and fecal coliforms.

Fecal streptococci are an important indicator of fecal pollution, because they are consistently present in both the feces of all warm-blooded animals and in the environment into which the animals have discharged their feces. The fecal strepto-cocci are also known as enterococci, group D streptococci, and *Streptococcus fecalis*. By using serological procedures, group D streptococci can be separated from other groups. Under present conditions, a rigid definition of the fecal streptococci group is not available. The major human fecal strep is *S. salibarius*. This strep is usually produced in the human group and survives in the human fecal material. The non-human fecal streptococci include *S. bovis* and *S. equinus*. A substantial quantity of enterococcus types are found from wild birds. By utilizing specific techniques, such as the membrane filter or pour agar plates in the laboratory, it is possible to establish the sources of pollution due to feces in certain types of aquatic environments.

Fecal strep is found in storm waters, combined sewers, runoff into water systems, and runoff of irrigation waters. In some studies conducted by the EPA, the amount of fecal streptococci per 100 ml in domestic sewage varied from as low as 64,000 to as high as 4.5 million. In storm waters, the fecal streptococci per 100 ml have varied from a low of 5100 in business districts, to 58,000 in rural areas, to 15,000 in residential areas. Generally, the fecal streptococci are found in surface waters, and therefore, fecal pollution may be indicated. Where little or no fecal streptococci are found, warm-blooded fecal contamination probably has not occurred.

The advantages of the fecal strep test is that generally the fecal streptococci have more limited survival rates in the aquatic environment than those of the coliform group. Therefore, if these streptococci are found, it would be an indication of fecal pollution. In addition, they do not multiply in polluted waters. Another advantage to using fecal strep tests is that *S. salavarius* is a more reliable indicator when detected along with other fecal strep in bathing water. This organism has a greater resistance to chlorination than other fecal organisms. The disadvantages of using fecal streptococci tests are in setting up adequate laboratory procedures to determine fecal strep when ubiquitous strains make up the majority of the fecal strep present.

Another bacterial indicator of pollution utilized is the total bacterial count present. In effect, however, there is no such thing as a total bacterial count, because the microscopic count does not distinguish between dead and living bacteria, and the plate count only measures those bacteria that grow on certain types of media. However, the test still is of value, because it can indicate changes in the bacterial composition of the water source. It is used to measure process control procedures in treatment plants, and it can be used to determine the conditions of plant equipment and distributional systems.

Bacterial survival in the water environment is based on the numbers and types of bacteria entering the water from rain, land runoff, and surface waters entering the receiving stream. The amounts found in the tests are also determined by bacteria density and final bacteria load based on self-purification of the stream, predators present that consume microorganisms, amount of sunlight adsorption and sedimentation, amount of antibiotics and toxins present in the water, salinity of the water, temperature, pH, and turbidity.

The most probable number (MPN) is one of the major tests used for determining coliform presence. This involves the presumptive test, confirmed test, completed test, and fecal coliform test (Table 7.1).

The membrane filter is a technique utilized in many areas, to evaluate not only water samples but also other sampling procedures. The membrane filter is a flat, highly porous flexible plastic disk about 0.15 mm in thickness and usually 47 to 50 mm in diameter. The filter paper has an average pore diameter of 5 to 10 μm and thickness ranges from 70 to 150 μm. The standard water flow rate is 70 cc/min/cm². In testing, the sample is filtered through the membrane filter. The filter is then placed in a culture container, on an agar medium, or on a paper pad impregnated with moist culture medium. The inoculated filter is incubated, and after incubation the culture is examined and interpreted. In the coliform test, one or more special media are used for the growth of the bacteria. At least one of the media is selective for coliform bacteria growth. The filter may also be used for specific bacteria growth, such as salmonella and various shigella. The advantages of the tests are that the results can be obtained in 24 hours as compared with the usual 48 to 96 hours for the tube method. Larger, and therefore more representative, samples of water can be tested routinely. The numerical results from the membrane filter are more precise than the most probable number. The equipment and supplies are not bulky, are set up quickly, and need a minimum of space. The limitations of the membrane filter technique include the growth of large numbers of noncoliform bacteria that may interfere with coliform production. Also, where small quantities of coliform actually exist, when they are trapped with a large amount of suspended solids, bacterial growth can continue on the membrane surface. When 1 mg/l of copper or zinc is present, a regular coliform bacterial growth results and occasional strains of bacteria produce a sheen on the membrane filter that may be acid producing but not gas producing from lactose. This would give a false indication of coliform density. However, these limitations do not occur frequently, and the membrane filter is an acceptable way of determining the levels of coliform in water.

Mine Drainage

To determine the type of mine drainage problems that may be encountered, it is necessary to take water samples and to analyze the water for alkalinity, total acidity, conductivity, pH, turbidity, calcium, magnesium sulfate, total iron, ferrous iron, total solids, suspended solids, dissolved solids, settleable solids, aluminum, and manganese.

Air Deposition Monitoring

Air deposition monitoring is needed because air pollutants cause water pollution. Stream water sulfates may be due to air pollution. Increased calcium plus magnesium may also be due to air pollution. Over 200 air deposition monitoring sites are part of the National Atmospheric Deposition Program National Trends Network. They measure weekly wet deposition as rainfall of nitrates, ammonia, sulfates, chlorides, and pH.

Table 7.1   Most Probable Number Index and 95% Confidence Limits
for Various Combinations of Positive and Negative Results

Number of Tubes Giving Positive Reaction			MPN Index per 100 ml	95% Confidence Limits	
5 of 10 ml each	5 of 1 ml each	5 of 0.1 ml each		Lower	Upper
0	0	0	<2		
0	0	1	2	<0.5	7
0	1	0	2	<0.5	7
0	2	0	4	<0.5	11
1	0	0	2	<0.5	7
1	0	1	4	<0.5	11
1	1	0	4	<0.5	11
1	1	1	6	<0.5	15
1	2	0	6	<0.5	15
2	0	0	5	<0.5	13
2	0	1	7	1	17
2	1	0	7	1	17
2	1	1	9	2	21
2	2	0	9	2	21
2	3	0	12	3	28
3	0	0	8	1	19
3	0	1	11	2	25
3	1	0	11	2	25
3	1	1	14	4	34
3	2	0	14	4	34
3	2	1	17	5	46
3	3	0	17	5	46
4	0	0	13	3	31
4	0	1	17	5	46
4	1	0	17	5	46
4	1	1	21	7	63
4	1	2	26	9	78
4	2	0	22	7	67
4	2	1	26	9	78
4	3	0	27	9	80
4	3	1	33	11	93
4	4	0	34	12	93
5	0	0	23	7	70
5	0	1	31	11	89
5	0	2	43	15	110
5	1	0	33	11	93
5	1	1	46	16	120
5	1	2	63	21	150
5	2	0	49	17	130
5	2	1	70	23	170
5	2	2	94	28	220
5	3	0	79	25	190
5	3	1	110	31	250

Table 7.1 (continued)   Most Probable Number Index and 95% Confidence Limits for Various Combinations of Positive and Negative Results

Number of Tubes Giving Positive Reaction			MPN Index per 100 ml	95% Confidence Limits	
5 of 10 ml each	5 of 1 ml each	5 of 0.1 ml each		Lower	Upper
5	3	2	140	37	340
5	3	3	180	44	500
5	4	0	130	35	300
5	4	1	170	43	490
5	4	2	220	57	700
5	4	3	280	90	850
5	4	4	350	120	1000
5	5	0	240	68	750
5	5	1	350	120	1000
5	5	2	540	180	1400
5	5	3	920	300	3200
5	5	4	1600	640	5800
5	5	5	>2500		

Note: Five 10-ml portions, five 1-ml portions, and five 0.1-ml portions are used.

From *Current Practices in Water Microbiology,* Training Manual, Environmental Protection Agency, water Programs Operations, Training Program, U.S. EPA, Washington, D.C. January 1973, p. 11–16.

Biomonitoring of Environmental Status and Trends Program

The Biomonitoring of Environmental Status and Trends (BEST) Program is designed to identify and understand the effects of environmental contaminants on biological resources. The primary goals of the program are to determine the status and trends of environmental contaminants and their effects on biological resources; identify, assess, and predict the effects of contaminants on the ecosystems and biological populations; and provide summary information in a timely manner to managers and the public for guiding conservation efforts. At the national level, there is a network of sites for monitoring selected contaminants and effects on specific organisms. On the regional level, the focus is on selected, high-priority ecosystems to determine the overall impacts the contaminants are having on the habitats and biota within an ecosystem. At the local level, a site-specific contaminant assessment process is utilized.

Index of Watershed Indicators

The index of watershed indicators includes the following:

1. Assessed rivers
2. Fish and wildlife consumption advisories

3. Indicators of source water condition for drinking water systems
4. Contaminated sediments
5. Ambient water quality data for four toxic pollutants
6. Ambient water quality data for four conventional pollutants
7. Wetlands loss index
8. Pollutant loads discharged above permitted limits for toxic pollutants
9. Pollutant loads discharged above permitted limits for conventional pollutants
10. Aquatic and wetlands species at risk
11. Urban runoff potential
12. Index of agricultural runoff potential
13. Population change
14. Hydrologic modification caused by dams
15. Estuarine pollution susceptibility index

Beach Watch Survey

The beach watch program surveys a variety of ongoing local and state beach sampling programs. Its most recent survey indicated that over 4100 individual closings and advisories were issued for beaches at the ocean, bay, Great Lakes, and freshwater beaches. Major causes of beach closings and advisories were elevated bacteria in 69% of the cases, known pollution in 13%, and precautionary closing due to rain known to carry pollution to swimming waters. Major pollution sources included polluted runoff and storm water, sewage spills and overflows, sanitary sewer overflows, and rain-related problems.

National Shellfish Register of Classified Growing Waters

The National Shellfish Register of Classified Growing Waters is published every 5 years as a cooperative effort among the shellfish-producing states, federal agencies, and Interstate Shellfish Sanitation Conference. It includes information on the status of estuarine and nonestuarine growing waters. Of the classified waters, 69% were approved for harvest, whereas 31% were declared to be harvest limited. The top five pollution sources were urban runoff, upstream sources, wildlife, individual waste-water treatment systems, and wastewater treatment plants.

Marina and Coastal Protected Area Database

The geographic information system (GIS) database of the Federal Marine and Coastal Protected Areas in the United States and its territories includes information from the following sources:

- National Estuary Program
- National Park Service
- National Estuarine Research Services
- National Marine Sanctuaries
- National Marine Fisheries Service

Wetlands Water Quality Monitoring and Assessment

Wetlands water quality monitoring and assessment are carried out by the use of an index of biological integrity, which measures effects of pollutants on vascular plants, amphibians, birds, algae, snails, insects, clams, crayfish, etc. It also uses appropriate water quality criteria along with the index of watershed indicators. The index of watershed indicators is a compilation of information on the condition of aquatic resources in the United States. The indicator uses data on a watershed basis from a number of national assessment programs operated by the EPA, U.S. Department of Agriculture, NOAA, U.S. Geological Survey, and Army Corps of Engineers, Nature Conservancy; and from the various states, tribes, and other sources.

CONTROLS

The Water Quality Act has increased substantially the quantity of municipal sewage that is treated to improve the nation's water. It is necessary to take a very close look at the potential growth in population and the increasing amount of sewage that has to be treated and eventually put back into the receiving stream. Present research may have to be strengthened to find more efficient and cheaper methods of treating municipal waste, so that the water may be reutilized and so that the effluent cannot contribute to the overall pollution problem.

Municipal Sewage Waste

In Chapter 6 on public sewage disposal, a considerable discussion is given in the area of municipal waste disposal.

Sludge conditioning is discussed in the sections on the activated sludge process and on sludge digestors. Other types of sludge disposal processes consist of conditioning, thickening, dewatering, stabilization, and reduction. The conditioning process occurs when the sludge is treated with chemical or heat so that the water may be readily separated. Thickening is the separation of as much water as possible by gravity or flotation processes. Dewatering is the further separation of water by subjecting the sludge to vacuum pressure or drying processes. Stabilization is the stabilizing of organic solids so that they may be handled or used as soil conditioners without causing a nuisance or health hazard. Stabilization basically is the digestion process, which has already been discussed. Reduction is the turning of solids into a stable form by wet oxidation processes or incineration.

In the wastewater treatment process, the quantities of sludge are significant. In primary treatment there may be 2500 to 3500 gal of sludge per million gallons of wastewater. When the activated sludge process is used, 15,000 to 20,000 gal of sludge may be formed per million gallons of wastewater. When chemicals are used to remove phosphorus, an additional 10,000 gal worth of sludge may be formed. The more complex the treatment process of the effluent, the more sludge can be collected and the less solids can go into the receiving stream. Sludge consists of approximately 97% water. It is essential, to have an economical and effective means of disposal of sludge, to remove this water from the actual solids.

Mine Waste

Techniques utilized to control mine water pollution are water infiltration, treatment of ferrous-acid mine drainage with activated carbon, dewatering of mine drainage sludge, and acid mine drainage treatment by ion exchange. In water infiltration control, it is necessary to use reclamation techniques on the surface and to backfill the unreclaimed strip mines, with emphasis on those areas where surface water runoff goes into the deep mine areas. The elimination of excess water from the mine reduces the amount of mine drainage going into the receiving streams. Wherever possible, the mines should be sealed if they are no longer in use and the runoff waters from the mines should be directed into areas where they can be treated before they enter the normal water system.

Ferrous iron can be removed from acid mine drainage at low pHs by using activated carbon. This may result in almost complete removal of the ferrous iron. The acid mine drainage is passed through an aerated column containing bituminous coal-based activated carbon. The advantage of the process is that once the ferrous iron has been removed, the acid mine drainage can be more readily neutralized by using an inexpensive agent, such as limestone. The major disadvantage of the process is that the activated carbon soon loses its activity because of the adsorption of the iron on the carbon. The adsorbed iron is not easily removed and therefore the cost is quite high. The pH is limited to 2.5 to 4.0. Above 4.0, the precipitated iron physically clogs the column. Below 2.5, a significant reduction of iron back from the ferric to the ferrous state occurs, whereas the opposite reaction is desired.

Sludges can be produced with acid mine drainage by adding lime. This helps dewater the acid mine drainage. The rate at which this reaction occurs depends on the solids content of the sludge and the kind of neutralizing agent used. Flocculating agents can help improve the dewatering process by as much as 74%. Filter aid can help improve dewatering by as much as 11%. It is technically feasible to dewater all sludges. Conventional vacuum filtration is the least expensive method of dewatering sludge. Freezing the acid mine drainage sludge and then thawing it can significantly reduce the settled volume of the sludge.

Acid mine drainage has been treated in the laboratory by using an ion exchange process. The three recommended ion exchange processes include a two-resin system, which uses a strong acid cation exchanger followed by a weak base anion exchanger followed by posttreatment. A second process uses a weak base anion exchanger followed by aeration, lime treatment, etc. The third technique uses a strong base anion exchanger in its sulfate form.

Feedlot

The primary agricultural nonpoint source pollutants are nutrients, sediment, animal wastes, salts, and pesticides. The agricultural activities also have the potential to directly impact the habitat of aquatic species through physical disturbances caused by livestock, equipment, or management of the water. Nutrients, nitrogen and phosphorus, may be found in commercial fertilizer, manure, municipal and industrial treatment plant sludge, legumes, crop residues, irrigation water, and through the atmosphere deposition of the chemicals. Surface water runoff from agricultural land may transport the following pollutants:

- Particulate-bound nutrients, chemicals, and metals
- Soluble nutrients
- Sediment, particulate organic solids, and oxygen-demanding material
- Salts
- Bacteria, viruses, and other organisms

Nutrients may be reduced by proper animal diets and feed that is modified to reduce the amount of these chemicals in manure. Feed management may include the use of low phosphorus corn and an enzyme such as phytase, that can be added to nonruminant animal diets to increase the utilization of phosphorus.

Manure handling and storage if properly done prevent water pollution. Clean water needs to be diverted from feedlots, holding pens, animal manure, or manure storage systems. The construction and maintenance of buildings, collection and conveyance systems, and permanent and temporary storage facilities should be carried out in a manner that prevents leakage into the ground or into surface waters.

Manure should be handled and treated to reduce the loss of nutrients to the atmosphere during storage, to make the material a more stable fertilizer when applied to land, and to reduce pathogens, vectors, and odors. Manure can be used as a source of energy or for high-value, low-volume fertilizers.

Dead animals should be disposed of in a manner that cannot affect ground or surface water. Composting, rendering, and other practices are appropriate means of disposing of dead animals.

Today, runoff is prevented from entering the water bodies by using interception ditches and aerated or nonaerated lagoons. The feedlot surface is modified so that the runoff is stored right on the feedlot. It can then be pumped onto the cropland, evaporate, or infiltrate right on the site.

The major technique utilized today to improve the effluent from feedlots is the use of oxidation ditches. All treatable wastes must be treated before the effluent can be allowed to enter the receiving stream. The feedlot effluent is introduced into an oxidation ditch, where the wastes are continuously mixed and oxygenated by the use of a mechanical surface aerator. When the system remains completely aerobic it is odorless, and the solids and nutrients can be removed to a considerable degree. The system operates in the same way as a sewage treatment plant activated sludge process. The final effluent should have at least 95% of the contaminants removed. Final chlorination should take place before the effluent is permitted to enter the receiving stream. The EPA or the health department within a given state determines whether the effluent can be released.

Oil

Because of the sharp increase in oil spills, it is essential for existing legislation to be enforced by the U.S. Coast Guard and, if necessary, new legislation to be passed to prevent tankers that are unsafe and unable to properly haul oil from entering the U.S. coastal waters or U.S. ports. New technology is needed to find oil-herding agents that can force spilt oil masses to come together. These herding agents are quite useful, because the oil could then be vacuumed into other ships. In addition, further research is necessary to determine what microorganisms would destroy the oil and whether other techniques can be utilized to take care of oil spillages.

Polychlorinated Biphenyls

On February 2, 1977, the EPA printed rules and regulations for PCB standards in the *Federal Register*, that prohibit manufacturers of PCBs from discharging any of the PCBs into the water. Further stringent limitations have been placed on discharges from manufacturers of transformers and capacitors. However, PCBs were still found in stream sediment in 2001.

Waste from Watercraft

To have better water pollution control, watercraft must have adequate storage space for sewage and sewage must be placed in an onshore sewage disposal facility. The discharge of untreated or partially treated sewage into water can contribute to high bacterial counts and potential health problems. The sewage also disrupts coral reef communities and other ecosystems. In addition, ships should not be permitted to discharge bilge water, oils, or any other contaminants into the water. Where ships are within the territorial waters of the country, they are not permitted to discharge any solid waste or any other kinds of waste. Vessel sewage discharge is regulated under Section 312 of the Clean Water Act.

Combined Sewer Overflows

Combined sewer overflows occur because years ago communities combined their storm sewers with their sanitary sewers to save money. When large volumes of storm water need to be removed, they overwhelm the sewage treatment plant and are therefore diverted untreated to the stream. The overflows contain storm water or snowmelt, untreated human and industrial waste, toxic materials, and debris. Combined sewers serve about 40 million people in 950 communities. This is a major source of water pollution. EPA has a control policy for these combined sewer overflows.

New and Innovative Technologies

The new and innovative technologies for water pollution control are discussed next.

Wetlands-Based Treatment — This technology, developed by the Colorado School of Mines, is a process that uses natural geochemical and biological processes inherent in an artificial wetland ecosystem to accumulate and remove metals from influent waters. The treatment system incorporates principal ecosystem components found in wetlands, including organic soils, microbial fauna, algae, and vascular plants. Influent waters, with high metal concentrations and low pH, flow through the aerobic and anaerobic zones of the wetland ecosystem. Metals are removed by filtration, ion exchange, adsorption, absorption, and precipitation through geochemical and microbial oxidation and reduction. It is used to treat acid mine drainage from metal or coal mining activities.

In Situ **Mitigation of Acid Water** — This technology, developed by the University of South Carolina, is a process that deals with exposed sulfide-bearing minerals such as mine waste, rock, and abandoned metallic mines. The *in situ* mitigation strategy modifies the hydrology and geochemical conditions of the site through surface reconstruction and selective placement of limestone. It is used to mitigate acid drainage from abandoned waste dumps and mines.

Chemical Treatment — This technology, developed by Davy Research and Development Ltd., employs resin-in-pulp (RIP) or carbon-in-pump (CIP) technologies to treat soils, sediments, dredgings, and solid residues contaminated with organic and inorganic material. These technologies are based on resin ion exchange and resin or carbon adsorption of contaminants from a leached soil–slurry mixture. It is used to treat soils and other materials contaminated with inorganic and organic wastes. Inorganics include heavy metals such as copper, chromium, zinc, mercury, and arsenic. Potential applications include treatment of materials containing organics such as chlorinated solvents, pesticides, and polychlorinated biphenyls by selecting appropriate extractant reagents and sorbent materials.

Anaerobic Thermal Processor — This process, developed by Soiltech ATP Systems, Inc., is a process in which a rotary kiln unit desorbs, collects, and recondenses contaminants from the fed material. The unit also can be used in conjunction with a dehalogenation process to destroy halogenated hydrocarbons through a chemical process conducted at elevated temperatures. It is used to dechlorinate PCBs and chlorinated pesticides in soils and sludges; to separate oils and water from refinery wastes and spills; and, in general, to remove hazardous VOCs from soils and sludges.

Liquid and Solids Biological Treatment — This technology, developed by Remediation Technologies, Inc., is a process that remediates soils and sludges contaminated with biodegradable organics. The process is similar to activated sludge treatment of municipal and industrial wastewaters, but it occurs at substantially higher suspended solids concentrations (greater than 20%). First, an aqueous slurry of the waste material is prepared, and environmental conditions (nutrient concentrations, temperature, and pH) are optimized for biodegradation. The slurry is then mixed and aerated for a sufficient time to degrade the target waste constituents. It is used to treat sludges, sediments, and soils containing biodegradable organic materials.

Metals Immobilization and Decontamination of Aggregate Solids — This process, developed by PSI Technology Company, involves a modified incineration process in which high temperatures destroy organic contaminants and concentrate metals into fly ash. The bulk of the soil ends up as bottom ash and is rendered nonleachable. The fly ash is then treated with a sorbent to immobilize the metals. It is used to treat organics and heavy metals in soils, sediments, and sludges. The process has been effective in treating arsenic, cadmium, chromium, lead, nickel, and zinc.

Below-Grade Bioremediation Cell — This process, developed by Groundwater Technology Services, Inc., utilizes bioremediation to treat soils contaminated with cyclodiene insecticides, such as chlordane and heptachlor. Bioremediation is a proven technique for the remediation of soils containing a variety of organic compounds. The process involves stimulating the indigenous microbial population to degrade organic wastes into biomass and harmless by-products of microbial metabolism such as carbon dioxide, water, and inorganic salts. The process relies on aerobic metabolism of microorganisms present at the site. It is used to treat soil, sludge, and sediment and is applicable to all biodegradable organic compounds.

High-Energy Electron Beam Irradiation — This process, developed by High Voltage Environmental Applications, Inc., uses a high-energy electron beam irradiation technology that is a low temperature method for destroying complex mixtures and hazardous organic chemicals from solutions containing solids, such as slurried soils, river or harbor sediments, and sludges. The technology can also treat soils and groundwater containing these chemicals. It is used to treat a variety of organic compounds such as pesticides, petroleum residues, and PCBs.

The Engineering Approach to Stream Reaeration

Streams can be reaerated by using engineering techniques if a minimum level of 5 ppm of air is not met. The water systems using molecular oxygen are relatively inefficient. This is particularly true during the summer months, when the saturation concentration for DO is at its lowest value. The most efficient location for mechanical surface aerators and diffusors using air is at the point of maximum oxygen deficiency. To maintain a proper dissolved oxygen level, it is necessary to locate the units where the low points in the stream occur. At present, mechanical surface aerators and diffusors using air cannot efficiently operate in maintaining the dissolved levels above 4 ppm. Artificial aeration of rivers and streams should not be used as a substitute for waste treatment at the source. This is only a polishing technique used during periods of high temperature and low flow. The cost estimates of artificial aeration are not adequately calculated to determine the efficacy of this process. To fully understand what happens within a stream, it is necessary to first understand what the oxygen balance is all about and how it occurs. The four main processes that control oxygen concentrations in naturally aerated streams include (1) the consumption of oxygen during respiration of benthic, planktonic organisms, as well as chemical oxidation; (2) exchange of oxygen as a result of atmospheric reaeration; (3) production of oxygen during the day by photosynthesis, by benthic organisms and photoplankton; and (4) oxygen contributions, groundwater, surface drainage, and storage.

The consumption of oxygen is measured by BOD and COD. The amount of oxygen present in the water is measured as DO. The aeration systems are technically feasible. However, the cost must be evaluated before this becomes a definite technique that can be used on a practical basis in streams and other bodies of water. For further information, contact the EPA, because the agency is involved in conducting research in the area of engineering methodology for river and stream reaeration.

Water Pollution Control Programs of the 1970s and Early 1980s

The water pollution control programs in the United States are based on the Water Pollution Control Act of 1972 (PL92-500). This law improved the water pollution control programs initiated by Congress in 1948 and amended in 1956, 1961, 1965, 1966, and 1970. The 1972 law established a truly national program for the first time. Congress extended the water pollution program to all navigable water bodies in the United States. Previously, interstate but not intrastate waters were covered by legislation. The 1972 law also, for the first time, created a system of national effluent limitations and national performance standards for industries and publicly owned waste treatment plants. The 1972 law proclaimed two goals for the United States. On July 1, 1983, wherever possible, water was to be clean enough for swimming and other recreational use, and clean enough to protect fish, shellfish, and wildlife. By 1985, there were to be no more discharges of any pollutants into the nation's waters. Obviously, this did not happen and the National Municipal Policy of 1984 was developed and the Water Quality Act of 1987 was passed.

Development of National Effluent Limitations and Water Quality Standards

An effluent limitation is the maximum amount of a pollutant that a polluter may discharge into a water body. Effluent limits may permit either some discharge or no discharge of pollutants based on the type of pollutant. The 1970 law prohibited the discharge of any radiological, chemical, or biological warfare materials, or high-level radioactive wastes into the receiving streams. Other toxic pollutants have been controlled by effluent standards that have been issued in the last several years. These pollutants cause disease, death, behavioral abnormalities, cancer, genetic mutations, physiological malfunctions, or physical deformities directly or indirectly in people and other organisms. When setting a standard, there must be an adequate margin of safety. Previous legislation had already set strict standards for lead and mercury. The 1972 law strengthened these standards.

The EPA defined the most practical and best available water pollution control technology. This was based on the cost of pollution control, age of the industrial facility, process used, and environmental impact on other areas if controls applied. It was also the function of the EPA to provide control measures that would eliminate industrial discharges eventually. As of July 1, 1977, industries had to meet the effluent limit that reflected the best practical technology. By July 1, 1983, industries had to meet effluent limits that reflected the use of the best available technology. Industrial discharges into sewage treatment plants owned by municipalities were also subject to the national effluent limitations.

Effluent limits applied to publicly owned sewage treatment plants. After June 30, 1974, federal grants were made only where sewage treatment plants used the best practical treatment process. On July 1, 1977, all sewage treatment plants utilizing federal funds had to provide a minimum of secondary treatment. In addition, after July 1, 1977, all sewage treatment plants had to apply the additional more stringent effluent limitations that EPA or the various states had established to meet water quality standards, treatment standards, or compliance schedules. By July 1, 1983, all sewage treatment plants were to meet the best practical treatment process. Obviously, this did not occur. In fact, in 2001, there were 40,000 contaminated overflows into streams.

The EPA was responsible for establishing the national performance standards for new industrial sources of water pollution. The standards reflected the greatest degree of effluent reduction. They were carried out by applying the best demonstrated control technology, processes, operating methods, or alternatives available. Where it was practical, zero discharge of pollutants became the standard. Industries that had been covered by performance standards for new facilities included pulp, paper, paperboard, builders paper, board mills, end product, and rendering processing; dairy product processing; grain mills; can and preserved fruits, vegetables, and seafood processing; sugar processing; textile mills; cement manufacturing; feedlots; electroplating; manufacturing of organic and inorganic chemicals; manufacturing of plastic and synthetic materials; manufacturing of soaps and detergents; manufacturing of fertilizers; refining of petroleum; manufacturing of iron and steel; manufacturing of nonferrous materials; manufacturing of phosphates; steam electric power plants; ferro alloy manufacturing; leather tanning and finishing; manufacturing of glass and asbestos; and processing of rubber and processing of timber products.

Many of the final rules on these specific point sources were issued in 1998 and 1999. An example of this would be the final cluster rule for pulp, paper, and paperboard, which would cut air pollution by 60% and virtually eliminate all dioxin discharges into rivers and other surface waters. A 96% reduction in dioxins and furans, 99% reduction in chloroform, 59% reduction in air toxics, 47% reduction in sulfur and odors, 49% reduction in VOCs, and 37% reduction in particulate matter would occur. A series of new technologies are tested under grants from the EPA. One of them can test the feasibility of converting paper mill sludge or filter cake into pelletized, composite sorbent of activated charcoal, and highly porous clay.

The 1972 law continued to expand the water quality standards initiated in 1965. The federal government first issued guidelines and then criteria that helped the states set water standards for interstate use. The criteria contained all available scientific findings on the physical, chemical, temperature, and biological requirements for recreational and aesthetic waters, public water supplies, freshwater, marine water, agricultural water, and industrial water supply. The qualities and quantities had to be identified according to scientific determinations and then had to be controlled.

The identification was carried out by the use of analytic methods by chemists, biologists, engineers, recreational specialists, and environmental health practitioners. Monitoring was done by supplying measuring instruments to provide information for assessment and control. The standards were then written based on what was acceptable quality related to the local situation, including political, economic, and social factors, and also including the planned use of the water and its management. The criteria were an attempt to qualify water quality in terms of a variety of factors that had to be measured, and limits that had to be established based on the types of techniques utilized to carry out accurate measurements.

The difference between criteria and standards was that the criteria were evaluations of scientific data that were used to develop recommendations for uses of water, whereas standards utilized scientifically based recommendations for each of the water uses and then established a practical method for detecting and measuring a specific physical, chemical, biological, or aesthetic characteristic. In some cases, the techniques available were not adequate to do the proper job. It was recognized in setting up standards that allowances had to be made for a thorough understanding of local conditions.

1984 Water Pollution Study

In 1984, the Association of State and Interstate Water Pollution Control administrators prepared a study that showed that a marked improvement in 448,000 miles of streams and rivers had occurred because the Water Pollution Control Act of 1972 had gone into effect. However, the association found that there were 35,000 miles of streams and rivers that had degraded water quality and 400,000 acres of lake area that had degraded water quality. It also stated that significant stretches of streams and lakes only partially supported their designated areas. A study conducted in 1985 indicated that nonpoint sources were significant contributors of pollution in the degraded areas.

Water Quality Act of 1987

As a result of the previous studies and a major concern for water quality, Congress passed the Water Quality Act of 1987. This is commonly known as the Clean Water Act. Major provisions authorized $18 billion to be spent over 9 years for the construction of municipal sewage treatment systems, including $9.6 billion in direct federal grants and $8.4 billion in capitalization grants to help states establish permanent financing mechanisms for assisting their municipalities. It further authorized $400 million for a new state–federal program to control nonpoint source pollution that helped determine a variety of nonpoint sources instead of specific point sources. Examples of nonpoint source pollution include oil and grease runoff from city streets, pesticide runoff from farmlands, and polluted runoff from construction sites in mining areas. The amendments also restricted toxic water pollution in the country's waterways. The EPA 1984 National Water Quality Inventory showed that 37 states reported elevated levels of toxics in their waters. The law provides $60 million to be spent over 5 years for trying to solve pollution problems in estuaries.

The law also provides an additional effort to comply with the 1978 Great Lakes Water Quality Agreement. Under this agreement, the United States and Canada both committed to eliminating or reducing to the maximum extent practicable the discharge of pollutants into the Great Lakes. The agreement included needed construction of sewage treatment facilities, control of pesticides runoff, and control of toxic industrial discharges entering the Great Lakes area. The new law also provided for the states to carry out many new responsibilities in water quality control, whereas the 1972 law made water pollution a major federal government responsibility. It became a shared responsibility.

As a result of the Clean Water Act, the states had to adopt water quality standards for every stream within their borders. These standards included designated use, such as fishing or swimming, and prescribed criteria to protect the use. The criteria were pollutant specific and represented the permissible levels of substances in the water that would allow the use to be achieved. Water quality standards became the basis for nearly all water quality management decisions. Depending on the standards adopted for a particular stream, controls could be needed to reduce the pollutant level. The water quality standards were to be reviewed every 3 years and revised as needed. This act is currently enforced.

Natural lakes and reservoirs are often sources of drinking water. They are therefore covered under the Clean Water Act, Section 319, Section 303 (d); and the Safe Drinking Water Act of 1996. Lake protection and restoration activities are eligible for funding under Section 319 (h). Nonpoint sources are one of the primary concerns in lake management programs. In addition, such processes as dredging and aeration must be part of the management practices. The Clean Water Action Plan is the development of unified watershed assessments. This also includes lakes. Restoration activities are necessary where lakes have become severely contaminated and need assistance in going back to the original state. The Safe Drinking Water Act amendments of 1996 include new provisions for protecting source waters including lakes.

National Pollutant Discharge Elimination System and Guidelines

The EPA developed uniform nationally consistent effluent limitations that were pollutant-specific discharge limitations for industrial categories and sewage treatment plants. The purpose of the program is to protect human health and the environment. These limitations are based on the consideration of the best available technology that is economically feasible. A pollutant is any type of industrial, municipal, and agricultural waste discharged into water. The EPA and the states use these guidelines to establish NPDES permit limitations. The effluent guidelines are minimal or baseline limitations. Additional controls are required, in certain instances, to achieve the water quality standards.

All industrial and municipal facilities that discharge wastewater directly into the rivers and streams of the country must have an NPDES permit and are responsible for monitoring and reporting discharge levels. The states or the EPA inspect dischargers to determine if they are in compliance with their permit limitations.

Approximately 48,400 industrial and 15,300 municipal facilities have NPDES permits. The permit contains effluent limitations, monitoring, and reporting requirements. Effluent limitations are restrictions on the amount of specific pollutants that the facility can discharge into a stream, river, or harbor. The monitoring and reporting requirements are specific instructions on how the sampling of the effluent should be done to check to see whether the effluent limitations are met. Sampling may be done on a daily, weekly, or monthly basis, depending on the situation. The monitoring results must be reported to the EPA and state.

Because it would be an overwhelming job for the EPA to establish effluent limitations for each individual industrial and municipal discharger, Congress authorized the EPA to develop a uniform effluent limitations system for each category of point source pollutants, such as steel mills, paper mills, and pesticide manufacturers. The limitations were based on such factors as the efficiency of treatment technologies and the ability to recycle materials that would have gone out into the waterway stream. After the EPA proposed an effluent limit, it was open for public comment and then the order was issued for all point sources within that industry category to meet the limitations.

Limitations that are more stringent than those based on technology are sometimes needed to ensure that the state-developed water-quality standards are met. If several different facilities are discharging into one stream, creating pollutant levels harmful to fish, then more stringent treatment requirements, known as water-quality-based limitations, must be set forth. These limits are based on the water quality of the stream, instead of the amount of pollutants coming out of each of the polluter waste pipes. Therefore, the NPDES permits control of the individual polluters and the specific pollutants they put into the stream, whereas the water-quality-based limits control of the amount of pollutants in the stream by industry. In 1999, the NPDES Comprehensive Storm Water Phase II Regulations were finalized. Storm water needed to be regulated to protect surface water.

National Water Quality Assessment Program

In 1991, the U.S. Congress appropriated funds for the U.S. Geological Survey to begin the National Water Quality Assessment Program. As part of this program, the U.S. Geological Service works with other federal, state, and local agencies to understand the various problems related to water quality, how the water changes over time, effect of human activities, and how natural factors become a concern. Currently and into the future this program will continue to identify water quality needs. The focus currently is on how, when, and why nutrients and pesticides vary across the country. The information discovered is useful to help anticipate, prioritize, and manage water quality conditions for different land uses and environmental settings.

Concentrations of nitrogen and phosphorus have been found to commonly exceed levels that can contribute to excessive plant growth in streams. High levels of nitrate have been found in shallow groundwater, serving as an early warning of possible future contamination of underlying groundwater. Potential risks to humans and aquatic life implied by the studies can only be partially based on established standards and guidelines. Many pesticides and their breakdown products do not have standards or guidelines. Mixtures of chemicals are additional problems.

Land and chemical use are not the sole predictors of water quality. Concentrations of nutrients and pesticides vary considerably from season to season, and from watershed to watershed. Natural features, such as geology and soils, land management practices, tile drainage, and irrigation, can affect the movement of chemicals over land to aquifers or surface bodies of water. Water quality is constantly changing, from season to season and from year to year. The year 2001 studies are a snapshot of current events. These studies added to previous ones are part of the extremely valuable National Water Quality Assessment Program.

The U.S. Geological Survey issues a report titled, The Quality of Our Nation's Waters. This report presents insights on a variety of subjects that are a compilation of findings in numerous study units. The reports include such areas as nutrients and pesticides in water, pesticides in bed sediment and fish tissue, radon, arsenic, other trace elements, industrial chemicals, and physical and chemical effects on aquatic ecosystems. These reports should lead to the development of total maximum daily loads allowed in various streams to effectively utilize source water protection programs.

Ecoregion Approach

The ecoregion approach is based on the hypothesis that streams reflect the characteristics of the watersheds or lands that they drain. Streams within a particular ecoregion are more similar to one another in terms of their physical habitat, hydrology, water chemistry, and types of fish and aquatic insects than they are in the streams in other ecoregions. It is appropriate for the EPA to develop one set of water quality criteria for all similar ecoregions. By using this approach, it becomes possible to keep from having to develop unique water quality criteria for thousands of small segments for rivers or streams, which would not be cost effective. The concept can be used to reexamine state water quality criteria and standards to determine whether the standards need to be tightened or can be loosened.

Funding and Technical Assistance — Early Years

The federal government has assisted in the funding of sewage treatment plants for the last 45 years. In 1957, only 50 plants were authorized, and one actually had expenditures made on it. By 1975, as many as 7000 plants a year were authorized and 4000 a year had funds appropriated.

Under public law 92-500, the Water Pollution Control Act of 1972 the federal government was obligated for the expenditure of $18 billion for sewage treatment plants. The federal government provided 75% of the cost of these projects; the other 25% was divided between the state and local governments, and industrial users who hooked up to the municipal sewage treatment system. Municipalities could also get grants for demonstration projects that used new methods for treating sewage, for developing joint systems for treatment of municipal and industrial waste discharges, and for perfecting new water purification techniques. By 1979, about 1500 municipal wastewater treatment plants that served 2.3 million people were still discharging untreated sewage. Some 2700 additional plants were providing only primary treatment, which removed about 30% of some of the pollutants. By 1983, all these plants were to be in compliance. Obviously, this did not occur.

1987 Water Quality Act Financing

The Water Quality Act of 1987 reauthorized and extensively amended the Clean Water Act. It authorized $18 billion to be spent over 9 years for the construction of local sewage treatment systems while gradually phasing out the program. Including the Nonpoint Source Program and the Toxic Pollutant Program, the total amount approved by Congress was $20 billion. The law provided seed money for state revolving loan funds to help construct the necessary sewage treatment systems and to provide a source of self-sustaining funds to the states to finance local construction. About $2.4 billion was allotted for each of the years 1986, 1987 and 1988; $1.2 billion was allotted for 1989 and 1990. The federal capitalization funds for state revolving funds authorized $1.2 billion for each of the years 1989 to 1993. Finally, $600 million was allotted for 1994.

Federal construction grants covered 55% of a project's cost. Where a state has a rural population of at least 25%, between 4 and 7.5% of the state's construction grants were allocated to towns with populations of 3500 or less. Each state had to reserve between 4 and 5.5% of its allotment to help pay for sewage treatment projects that used innovative processes and techniques. This requirement applied to fiscal 1990. Two thirds of 1% of the money appropriated for construction grants were used to deal with pollution problems in marine-based estuaries where overflows from sewers carrying the storm water and domestic sewage occurred.

Planning and Basic Studies

Good planning is the first major step in the provision of good quality water. As part of the planning program, industrial users are encouraged to incorporate industrial waste where practical into municipal treatment plants, because this provides more effective pollution control by regionalizing waste treatment. The community can get part of its funding from the industrial user. Even if the community were able to eliminate all its problems, the industrial user still might contribute large quantities of pollution to the public receiving stream. Joint treatment must be thought out very carefully, and the kinds of industrial waste must be thoroughly understood to avoid problems within the municipal sewage treatment plant. When the data have been gathered concerning the current and projected needs of the area — population affected, severity of pollution problems, need for the preservation of high-quality water, priority of the treatment plant, kinds and nature of the waste to be treated — then a plan may be prepared by the group seeking the wastewater treatment plant and by consultants. This preliminary plan should include all the previous explanations, the set of design plans and specifications, the method of construction of the waste treatment facilities, and an environmental impact statement. All these documents are reviewed by the federal government, also by the state government, and by all others affected. It is important to determine at this point whether the design and the cost of the project is proper; project design is acceptable from environmental, social, and institutional standpoints; wastewater discharge permit has been issued; alternate wastewater management approaches are under evaluation; and plans for disposal of the sludge are sufficient.

After approval, specific plans and specifications meeting requirements of all governmental agencies and specific permits, etc. are to be granted. Construction must be carried out in an acceptable manner and inspected at its completion by both the EPA and the proper state agency.

Research Development and Demonstration Programs

To obtain better techniques for disposing of waste material, a variety of research development and demonstration programs are not only encouraged but also funded by the federal government. Two major categories of these programs include (1) how to define the water quality goals, and (2) how to reach a maximum effectiveness with minimum cost. In addition, research is carried out to determine the quality goals that are achievable, as well as the techniques to achieve these goals for water pollutants; in other words, how to best use the minimum amount of money and how to develop the best project to eliminate the greatest number of pollutants are determined.

Educational Programs

Because we are a democratic society, it is necessary to involve the citizens in all activities. National water pollution control will either fail or be successful depending on how people react to the needs and how willing they are to do something about resolving the problems of water pollution. It is necessary not only to train technicians but more importantly to advise the American public in as many ways as possible on the need for water pollution control and techniques utilized to achieve this.

It is important to work with the young people in the school systems and with a series of civic-minded groups who are eager to improve water quality. The news media, including newspapers, magazines, radio, and television, can carry the message of clean water. Further, literature can be supplied to any interested individuals or any interested group who may push for cleaner water. Only when individuals understand the necessity for clean water will they be willing to contribute their efforts toward a drive to eliminate pollutants from the water supply.

Humans have the capacity to utilize one of the greatest natural resources in an effective manner. Water can be used and reused numerous times. The reuse of the water is based on the pollutants that are present. Only the public, in the last analysis, can resolve the water pollution problem by supporting proper legislation, by not contaminating the water sources, and by supporting necessary ballot issues to build adequate numbers of quality wastewater treatment plants.

Enforcement Actions

The final technique used in water pollution control when other techniques do not work is strong, effective, equitable regulatory action. It is essential not only that permits are granted under the 1972 law for the selective discharge of pollutants into waters, as updated by subsequent laws, but also that these permits are reviewed carefully and subjected to scrutiny by the federal government, state government, and citizens groups who may seek legal action to stop the discharge of hazardous pollutants. The permit system is a means of establishing a control technique for pollutants. It not only grants an individual a certain amount of time to make necessary corrections but also stipulates what these corrections must be and how to go about achieving them. If the source is such that it discharges radiological, chemical, or biological materials or high-level radioactive wastes or materials that can impair health or navigation, a permit cannot be granted and the individual cannot be permitted to allow any of the effluent to enter the receiving stream. After a conference has been called with the individual polluter, either the regulatory agency may proceed directly to court enforcement, which requires a fine of up to $10,000 a day for each violation of water quality standards, or it may attempt to negotiate a settlement that eliminates the hazards. In any case, laws may eventually be enforced to eliminate specific hazards.

In addition to the specific hazard, a series of federal, state, and local laws controlling nonpoint source pollutants exist. Complex legal problems arise from defining pollution and distinguishing between contamination and pollution from nonpoint sources.

1987 Clean Water Act Penalties

The Clean Water Act stiffens the penalties for civil and criminal violations of clean water laws and gives the EPA new authority to assess civil penalties. Civil penalties for negligent violation of effluent standards may range between $2500 and $25,000 per day and up to a year in prison. The same penalties apply to anyone who negligently discharges into a sewer system a hazardous substance that the person knows can cause personal injury or property damage. In the case where these discharges are knowingly carried out by the individual, the penalty ranges from $5000 to $50,000 per day or up to 3 years in prison, or both. Any person who knowingly violates effluent standards, pretreatment requirements, or obtaining of permits and may put another person in imminent danger of death or serious bodily injury is subjected to a fine of up to $25,000 and up to 15 years in jail. An organization that commits this offense can be subjected up to a fine of $1 million.

Clean Lakes

The Clean Water Act of 1987 strengthens the previous program on improving the water quality in lakes. Every state is required to submit to the EPA every 2 years a report on lake quality. These reports include information on water quality of publicly owned lakes and methods of controlling pollution, including means for improving the acidified lakes. Reports also list and describe the lakes with impaired uses, including those that cannot meet water quality standards. The reports also include an assessment of the status and trends of water quality in the state lakes, the extent of pollution from point and nonpoint sources, and the extent to which the pollution impairs the use of the lakes. Funds have been provided to help the states improve their lake water quality and to help mitigate the harmful effects of high acidity in lakes.

Nonpoint Source Pollution

The Clean Water Act provides for the control of nonpoint source pollution, such as polluted runoff from city streets, farmland, mining sites, and other sources. It also encourages the state to pursue groundwater protection as part of the nonpoint source pollution control efforts. The program requires the states to provide the EPA with a report identifying the state waters that are not expected to meet water quality standards because of nonpoint source pollution. The types of nonpoint sources and where they are located are part of the report. The states must also prepare a management program for controlling nonpoint source pollution. The program describes the methods that are to be used to reduce nonpoint source pollution and explains how the program is to be conducted. If the states are unable to meet the water quality standards because of nonpoint source pollution coming from another state, the EPA may be asked to convene a conference to determine how best to deal with this problem. Funds have been authorized to help abate nonpoint source pollution.

Managing Nonpoint Source Pollution

Agriculture

Agricultural activities that cause nonpoint source pollution include confined animal facilities, grazing, plowing, pesticide spraying, irrigation, fertilization, planting, and harvesting. The major nonpoint source pollutants that result from these activities are sediment, nutrients, pathogens, pesticides, and salts. Over 40% of Section 319 of the Clean Water Act is related to grants used to control agricultural nonpoint source pollution. Several U.S. Department of Agriculture and state-funded programs provide cost-share programs, technical assistance, and economic incentives to implement nonpoint source pollution management practices. Farmers can implement nutrient management plans that help maintain high yields and save money in the use of fertilizers while reducing nonpoint source pollution. Discharges can be limited from confined animal facilities to storage areas. Use of appropriate waste management systems described in this book can sharply reduce nonpoint source pollution from animal facilities. Water use efficiency in irrigation improves crops, reduces costs, and reduces nonpoint source pollution by controlling excessive runoff. People can utilize integrated pest management techniques based on specific soils, climate, past history, and specific crops for particular fields. This avoids waste of chemicals and reduces nonpoint pollution. Overgrazing exposes soils, increases erosion, encourages invasion by undesirable plants, destroys fish habitat, and reduces filtration of sediment necessary for building stream banks, wet meadows, and floodplains. This can be corrected by altering grazing areas and grazing intensity, keeping livestock out of sensitive areas, providing different feed sources, and providing alternative sources of water and shade.

Bays

Bays can be improved in the following manner:

1. Limit the amount of impenetrable surfaces by using permeable paving materials such as wood, bricks, and concrete lattice to let water soak into the ground around homes.
2. Allow thick vegetation or buffer strips to grow along waterways to slow runoff and soak up pollutants. Trees, shrubs, and ground cover adsorb up to 14 times more rainwater than the grass lawn and do not require fertilizer.
3. Use natural alternatives to chemical fertilizers and pesticides.
4. Gutters and downspouts should drain onto vegetative or gravel-filled seepage areas and not directly onto paved surfaces. Use splash boxes.

5. Avoid hosing down driveways, sidewalks, or washing cars on the driveways.
6. Divert runoff from paved areas to grass or wooded areas.
7. Properly dispose of household hazardous waste.
8. Recycle all used motor oil.
9. Pick up animal wastes from pets.

All the preceding result in reducing nonpoint source pollutants from entering streams, bays, oceans, etc.

Boating and Marinas

To reduce pollution from these sources the following should be executed:

1. Select nontoxic cleaning products.
2. Use a drop cloth when cleaning and maintaining boats away from the water.
3. Do not dump any kind of sewage or solid waste into water.
4. Place the marinas where they cannot impact water quality.
5. Design marinas in such manner as to avoid contamination of water.

Forestry

Nonpoint source pollution in forests can be improved by preharvest planning, streamside management areas, appropriate road construction and reconstruction, better road management, improved timber harvesting techniques, appropriate site preparation and forest regeneration programs, forest chemical management, and wetlands forest management.

Hydromodification

Hydromodification is the process of modifying channels, constructing dams, changing stream banks, and shorelines. During this process it is necessary to consider the following to avoid or reduce nonpoint pollution:

1. Management measures related to the physical and chemical characteristics of the surface waters
2. Management measures for in-stream and riparian habitat management
3. Management measures for erosion and sediment control
4. Management measures for chemical and pollutant control
5. Management measures for eroding stream banks and shorelines

Roads and Highways

The Coastal Zone Act Reauthorization amendments of 1990 specifically charged coastal states and territories with upgrading their runoff pollution control programs to protect coastal waters. The EPA and the NOAA jointly oversee the development and implementation of these programs. The EPA published a document for the states titled Guidance Specifying Management Measures for Sources of Nonpoint Pollution in Coastal Waters. This document includes best management practices; technologies;

processes; siting criteria; and operating methods for roads, highways, and bridges that states can use to implement the management measures. Prior to construction a comprehensive erosion and sediment control plan is needed. Runoff from the land must be kept at a minimum and the velocity must be kept as low as possible. Straw bale barriers, filter fabrics, silt fences, sediment basins, and stabilized entrances to the construction site are useful to control runoff. Maintenance must be performed regularly to correct temporary erosion and runoff. Seeding with grass, fertilizing, and overlaying with mulch or mats help stabilize the ground. A wildflower cover or sod may be used. Permanent control can be achieved through the use of grassed swales, which are shallow, channeled grass depressions through which runoff is moved; filter strips or wide strips of vegetation; terracing that breaks a long slope into many flat surfaces; check dams, which are small temporary dams made of rock, logs, or other durable material; detention ponds or basins; infiltration trenches or basins; and constructed wetlands.

Other laws and rules affecting construction sites are the NPDES; Intermodal Surface Transportation Efficiency Act; and various parts of the Clean Water Act.

Sediment

To deal with the ecological and human health risks related to contaminated sediment in U.S. watersheds, the EPA published Contaminated Sediment Management Strategy. The strategy is an EPA working plan describing the actions that are needed to bring about consideration and reduction of risks caused by contaminated sediments. EPA has established four goals as follows:

1. To control sources of sediment contamination and prevent the volume of contaminated sediment from increasing
2. To reduce the volume of existing contaminated sediment
3. To ensure that sediment dredging and dredged material disposal are managed in an environmentally sound matter
4. To develop a range of scientifically sound sediment management tools for pollution prevention, source control, remediation, and dredged material management

On-Site Wastewater Systems

The exclusion of storm water runoff away from septic tanks and sanitary systems reduces the potential for water quality problems. Means of diverting the storm water runoff include tying the rain gutters into pipes removing the water to vegetative and rock-lined swales; using water retention ponds; and using perimeter drains to keep the water from higher areas from running across existing tile systems.

Urban Runoff

Urban runoff as shown in the National Water Quality Inventory is a leading source of impairment to estuaries and the third largest source of water quality problems in lakes. By 2010, more than half of the country will live in coastal cities and towns and thereby further degrade coastal waters. To avoid these problems the following measures need to be taken: promote comprehensive plans for new developments; promote comprehensive plans for controlling runoff from existing developments; control nutrients and pathogens from on-site disposal systems; and develop a comprehensive public education program on water quality.

Toxic Pollution Program

The 1987 amendments to the Clean Water Act created a new program to identify waterways that were contaminated by toxic pollutants and then the correction of these problems through control of point sources and nonpoint sources. A 1985 study of Boston Harbor by Greenpeace found that more than 1 ton of toxic pollutants entered the harbor each day from point sources and 100 to 1000 lb entered the harbor from nonpoint sources each day. By April 1, 1988, the states developed a list of point source discharges; and by February 4, 1989, they determined how to correct these problems. Nonpoint sources have followed.

National Estuary Program

The Clean Water Act provides for comprehensive conservation and management plans to control pollution problems in estuaries. Many of these have become badly contaminated by waste that has been generated from heavily populated areas. The EPA can convene management conferences to solve pollution problems in estuaries and then develop a comprehensive conservation and management plan for the estuary.

Currently, 28 National Estuary Programs exist. They work to resolve environmental problems through various combinations of regulation, innovative initiatives, balanced and inclusive planning and management, scientific research and monitoring, and public outreach in education. Each of the states has implemented its own programs with federal assistance. However, many of the problems they face are similar.

Nutrient overloading is a concern in at least 11 geographic management zones. The standards for dissolved oxygen in the water are either 5.0 or 6.0 mg/l. This is accomplished by setting standards for discharges of animal waste at confined animal facilities requiring that the facilities be designed and constructed to retain all wastewater generated together with all precipitation and drainage through areas where manure is produced. Further, in the state of Florida, the state legislature restricted the discharge of treated wastewater into the surface waters of Tampa and Sarasota Bays. In San Francisco, the Regional Water Quality Control Board adopted individual waste discharge requirements for the approximate 25 dairies within the watershed.

Pathogens are a concern in all areas. In Long Island Sound, attention was directed to the correction of problems resulting from combined sewer overflows, nonpoint source runoff, sewage treatment plant malfunctions, and vessel discharges. Septic tank systems were evaluated and standard tank design and performance criteria were set by various states. State shellfish monitoring and standards programs were implemented in accordance with the National Shellfish Sanitation program. The Corpus Christi Bay program consists of a modification of existing monitoring programs, assessment of additional public health measures, and development of a coordinated database and networking system for public health professionals and business. In Santa Monica Bay, an epidemiological study was conducted as a result of a partnership between the state of California, local cities and agencies, business, private foundations, local environmental groups, and U.S. EPA. As a result of this study, a beach regulatory protocol was established and beach closings and public advisories were issued. The Indian River Lagoon program recommended the development and implementation of the program to inspect septic tanks on a periodic basis to make sure they were properly operating. In North Carolina, nonpoint source teams have been established in each of the state's 17 major river basins to determine problems from agriculture, storm runoff, construction, on-site wastewater disposal, and solid waste areas

Toxic chemicals are found in virtually all estuary areas. The Delaware River Basin Commission is working with various states, federal agencies, and citizens to develop total maximum daily loads and waste load allocations for many of the toxic substances. In Morro Bay, pollution from boatyards and other urban sources have been controlled by use of best management practices and education. Corpus Christi officials have reviewed water quality standards, criteria, designated uses for all areas, establishment of biological criteria in water and sediment, establishment of waste-water discharge permits on a watershed basis, and development of new permit processes. The Washington Department of Ecology has adopted sediment standards for Puget Sound as part of state water quality standards. In Tampa Bay, sediment quality is determined by sediment chemistry, toxicity effects on standard test organisms, and benthic community structure. This program determines localized hot spots of contamination in Tampa Bay and helps reduce sources of toxics within the watersheds that drain into these hot spots.

Habitat loss and degradation is a concern in estuaries and upland areas. Trained people are studying the nature of the loss of the land and the wetlands and attempting to make necessary corrections. The Commonwealth of Massachusetts has established a complex structure of laws and regulations to combat wetlands loss in the Massachusetts Bay region.

Introduced species can destroy the native vegetation and create a variety of problems in the estuary. The U.S. Army Corps of Engineers is a leader in research and control of introduced aquatic plants. Alteration of natural flow may destroy existing water tables and delicate ecosystems. Optimum salinity may be affected by this process and may cause serious problems in the fish and shellfish industries.

Antibacksliding

The Water Pollution Control Act provides for the preservation of pollution control levels that have already been achieved by those discharging into the water. It prohibits the relaxation of treatment requirements when a discharged permit is renewed or rewritten, except in very special circumstances. If in the best professional judgment of the EPA a facility is undergoing substantial operations that justifies less stringent requirements, these may be allowed. If more information is made available that justifies less stringent requirements or if the EPA determines that technical mistakes or mistaken interpretations of the law were made in issuing the original permit, it can allow for less stringent requirements. These less stringent requirements are very tightly controlled.

Storm Water Discharges

The Clean Water Act specifies when and where permits are required for the discharge of storm water into a broader body. Industrial facilities were required to file applications for their storm water discharge by 1990. The permits were issued by the EPA and the states by 1991 and complied with by 1994.

City storm sewer systems are divided into different categories with different deadlines. Larger cities have to apply for and obtain permits for their storm water systems sooner than small cities and towns. By 1990, cities with storm sewer systems serving a population of 250,000 or more had to file for an application for storm water discharge permits. By 1991, the EPA and the states were required to issue the permits for discharges for these systems. The cities had to comply with the permit requirements by 1994. In 1995, the storm water regulation was finalized. In 1999, new rules were implemented for small municipal separate storm systems located in urbanized areas, and construction activities of between 1 and 5 acres. This rule affects an additional 19,000 municipalities and millions of individuals.

Cities with storm water systems serving a population of between 100,000 and 250,000 had until 1992 to file applications for storm water discharge permits. The EPA and state had to issue them by 1993, and the cities had to comply by 1996. Municipal storm sewer systems that serve populations of 100,000 or less did not fall under the act until after October of 1992. Municipalities have to improve requirements that prohibit any discharges other than storm waters in the storm sewers. They also require controls to reduce the discharge of pollutants to the maximum extent practical.

Combined Sewer Overflow

Combined sewer overflows occur when sanitary and storm sewers are connected. During rainstorms, the systems become overloaded and the contaminated water bypasses the treatment works and discharges as much as 90% of the pollutants they contain into streams. These wastes often contain high levels of suspended solids, floatables, heavy metals, nutrients, microorganisms, and other wastes. In the United States, about 1200 combined sewers serve about 43 million people. In 1991, the EPA required communities to obtain permits for these systems and to arrive at a means of correcting the situation.

Underground Injection

The EPA has negotiated under the Safe Drinking Water Act and issued orders to ten major oil companies to discontinue underground injection of waste into 1800 wells connected to service stations to avoid contaminating groundwater or soil.

Large capacity septic tanks are controlled under the Safe Drinking Water Act. These tanks serve 20 or more people per day. The septic tank is in many cases followed by a disposal well for the effluent disposal. The well becomes a form of underground injection and can cause contamination of the groundwater supply.

National Irrigation Water Quality Program

The Department of the Interior started this program in 1986 to identify, evaluate, and respond to any significant irrigation drainage contamination in the western states. Numerous studies had been made to determine whether the environment was contaminated, and in fact, in some areas, it was.

Sewage Sludge

The Clean Water Act required the identification of toxic pollutants that may be present in sewage sludge in concentrations that can adversely affect health or environment. Regulations were written specifying acceptable management practices for sludge containing these pollutants and establishing numerical limits on the pollutants for each use of sludge. The law required that clean water permits be issued to wastewater treatment plants that contained requirements for the use or disposal of sludge. The EPA was authorized to establish demonstration programs on the safe disposal of sludge.

Water Reclamation and Sludge

Treated municipal wastewater effluent can be used to irrigate agricultural lands, thereby conserving water resources. The effluent may contain some nitrogen and phosphorus. The sludge that results from the municipal wastewater treatment process contains organic matter and nutrients, which when properly treated and applied to farmland, can improve the physical properties of the soil and become a good means of disposal of the sludge. Currently, only a small amount of the reclaimed water is in use, whereas 36% of the sewage sludge is applied to the land for agriculture, turfgrass production, and reclamation of surface mining areas.

The Midwest has a long history of using treated sludge on croplands. In February 1993, the U.S. EPA promulgated standards for the use or disposal of sewage sludge. This rule builds on a number of federal and state regulations that are used to reduce pollutants entering the municipal waste stream through source controls and industrial pretreatment programs, which reduce the levels of contaminants in the sludge as well as in the final effluent. Water reclamation and sludge disposal, when managed properly, can be useful to the community. The EPA provides guidelines for reclaimed water quality and its use for crop irrigation in its 1992 guidelines for water reuse. Currently, 19 states regulate the practice of reclaimed water use and set standards for microbiological limits, crop restrictions, and waiting periods for human or grazing animal access to the crop.

Potentially harmful trace elements such as arsenic, cadmium, cobalt, copper, lead, mercury, molybdenum, nickel, selenium, and zinc are found in treated municipal wastewater effluents. The National Academy of Sciences has issued a report on water quality criteria that recommends limits on the concentrations of these trace elements.

Water Resources Development Act of 1986

From 1970 to 1986, no major water resources project authorization acts were passed by Congress. The United States was changing and many of the biggest water projects had either been built or were well under way toward completion. The urbanized society had many other interests to compete with, beside irrigation, navigation, and flood control. Recreation and environmental preservation of water quality became important issues. The need for rehabilitating or replacing the aging water resources infrastructure could not be pushed aside any longer. By 1980, a number of the locks on the Ohio, the upper Mississippi, and the Columbia Rivers were found to be too old (about 40 years), deteriorated, and too small to serve modern shipping. This waterway situation became particularly urgent during the energy crisis. New locks and deeper ports were needed to handle the transportation and exportation of coal and other energy supplies. In 1978, the traditional approach to navigation improvement that made the federal government totally responsible for these matters was changed, and the levying of waterway user fees for commercial traffic was started.

In 1986, the Water Resources Development Act (WRDA) was passed. This act was a major shift in the country's attitude toward water resources planning. Nonfederal interests could and should handle more of the financial burdens of water resources planning.

The 1986 WRDA authorized $16.23 billion in spending for water projects, of which the federal government would pay about $12 billion, with the remainder coming from nonfederal interests, such as states, port authorities, and local communities. The act authorized 377 new water projects for construction or study. This included 43 port projects, 24 shoreline protection projects, and 61 water resources conservation and development projects. Further, the act authorized 38 special studies, 63 project modifications, and 26 other miscellaneous projects and programs. The act also modified the approaches to planning and financing of the Corps of Engineers projects that had evolved over the last 50 years. The revision and cost-sharing requirements imposed cargo taxes to maintain harbors and inland waterways.

The Institute for Water Resources of the Army Corp of Engineers, in a report for the National Council on Public Works improvements identified 13 water resources functional service areas. They were ports, inland waterways, flood control, urban drainage, dam safety, shoreline protection, irrigation, hydropower, agricultural drainage, stream bank protection, fish and wildlife and recreation, municipal water supply, and wastewater treatment.

These service areas are responsible for the following:

1. Ports provide facilities including deep channels and associated dredging for supporting domestic and foreign waterborne commerce. The United States has about 200 ports.
2. Inland waterways facilities are provided for the transportation of goods to ports on the inland waterways and the coastal ports for export. In the United States, over 12,000 miles of inland waterways are commercially navigable, with over 200 locks and dams.

3. Flood control services are needed to prevent natural disasters. The country has invested billions of dollars in flood control.
4. Urban drainage removes excess water from the areas to storm drains and into streams, lakes and rivers. Currently the investment in urban drainage exceeds $50 billion.
5. Dams provide the means for controlled water for prevention of flooding, provision of adequate water at special times, and hydroelectric power. Currently, 3000 federal and more than 80,000 nonfederal dams are in service.
6. Shoreline protection is needed for the coastline of the United States. Shoreline protection is important for businesses, homes, various land uses, and preservation of lives. About 24% of the country's 85,240 miles of shoreline are significantly eroding. This includes about 33% of the Great Lakes shorelines and 43% of the oceanic coastline.
7. Irrigation is necessary to enhance crop production. About 13% of the nation's cropland is irrigated.
8. Hydropower facilities provide clean, efficient, and reliable sources of energy, which makes up about 12% of the energy needs of our society today.
9. Agricultural drainage removes excess water from soils to improve agriculture, but may add numerous contaminants to the water.
10. Stream bank protection is used to keep stream banks from failing, which would endanger buildings, highways, and property. It also helps reduce the deposition of sediment.
11. Fish, wildlife, and recreation water resources are needed to provide environmental and social benefits, as well as economic benefits for society.
12. Municipal water supply is necessary to provide potable water in adequate quantity to meet the needs of the population.
13. Wastewater treatment plants deal with billions of gallons of wastewater per day. They also treat waste from commercial and industrial establishments.
14. Dredged materials in excess of 300 million yd^3 have to be removed each year. Further, the local or state authorities remove an additional 100 to 150 million yd^3 of dredged material each year. About half of the dredged material is disposed of along the coasts or in the open ocean. About 30 to 40% is disposed of in upland sites. Approximately 3 to 5% do not meet water quality criteria for water disposal. Dredged material is often used for renourishment of beaches. This material may also be used for restoration or creation of wetland and for the construction of island habitats.

Effluent Trading in Watersheds

The EPA actively supports and promotes effluent trading within watersheds to achieve water quality objectives, including meeting water quality standards, with the extent authorized by the Clean Water Act and other regulations. Trading is a method to attain or maintain water quality standards, but allowing sources of pollution to achieve pollutant reductions through substituting a cost-effective and enforceable mix of controls on other sources of discharge. Under trading, the source can sell or barter its excess reduction in pollutants to another source unable to reduce its own pollutants as cheaply, thereby still reducing the overall pollutants in the waterway, while creating reduced costs and more efficient economic and environmental returns.

Earth Resources Observation Systems Data Center

The Earth Resources Observation Systems (EROS) Data Center is a data management, systems development, and research field center for the U.S. Geological Survey National Mapping Division. The U.S. Geological Survey is a bureau of the U.S. Department of the Interior. The Data Center holds the world's largest collection of civilian remotely sensed data covering Earth's land masses, including millions of satellite images and aerial photographs. This information is extremely valuable when evaluating a vast number of water control problems, as well as other situations.

The Department of the Interior is also responsible for the National Mapping Program, National Geologic Mapping Program, and Public Lands Survey System of United States.

Environmental Protection Agency Fish Contamination Program

The EPA Fish Contamination Program provides technical assistance related to persistent bioaccumulations of toxics in fish and wild fish and associated potential health problems. The EPA has issued a four-volume set of guidance documents titled Guidance for Assessing Chemical Contaminant Data for Use in Fish Advisories. Mercury, dioxin, and DDT are the most common contaminants. Special studies are conducted and recommendations are made for improving the water quality and correcting the chemical accumulation in fish.

Environmental Protection Agency National Geographic Information System Program

The EPA is one of the largest consumers of spatial data in the U.S. government. The Geographic Information System (GIS) program was established to provide leadership and support to the decision makers and program offices.

Mixing Zone

In 1999, the administrator of the EPA announced the elimination of mixing zones for toxic chemicals in the Great Lakes. New discharges of chemicals of concern including mercury, PCBs, dioxin, chlordane, DDT, and mirex were prohibited from entering the Great Lakes. The old concept that mixing these chemicals in water would dilute them enough to avoid health problems has been proved to be erroneous. These chemicals build up and become more concentrated as they move through the food chain from plants to fish and animals to humans. Pollution prevention is the technique in use to avoid these discharges.

National Coastal Water Program

The National Coastal Water Program includes the Chesapeake Bay Program, Great Lakes Program, and Gulf of Mexico Program. The mission of the program is to lead and empower others to protect and restore the ecosystem for future generations;

and to maintain and restore the chemical, biological, and physical integrity of the waters.

National Monitoring Program

The National Monitoring Program was established by the EPA under Section 319 of the Clean Water Act. The objectives of the program are:

1. To scientifically evaluate the effectiveness of watershed technologies designed to control nonpoint source pollution
2. To improve our understanding of nonpoint source pollution

As a result of this act, monitoring programs have been established on a multiyear basis to evaluate how improved land management reduces water pollution. A watershed approach is now in use to monitor and control nonpoint source pollution.

National Oceanic and Atmospheric Administration Coastal Ocean Program

The NOAA Coastal Ocean Program is involved in numerous projects including harmful algal blooms and toxic microbes. A number of new and innovative projects including models, methodologies, and technologies are carried out to reduce the risk to coastal waters.

Shore Protection Act Regulations

The EPA in conjunction with the Department of Transportation has developed regulations establishing waste handling practices for vessels and waste transfer stations for the hauling and handling of municipal and commercial waste. The rule ensures that wastes cannot be deposited during loading, off-loading, and transport.

Total Maximum Daily Load

The total maximum daily load (TMDL) is a calculation of the maximum amount of a pollutant that a water body can receive and still meet water quality standards. TMDL is a sum of the allowable loads of a single pollutant from all contributing point and nonpoint sources. The Clean Water Act, Section 303, establishes the water quality standards and TMDL programs. These programs include planning on a river-by-river, bay-by-bay, and lake-by-lake basis.

Wet Weather Flows Research Plan

The Risk Management Research Plan for Wet Weather Flows was prepared by the National Risk Management Research Laboratory, U.S. EPA, Office of Research and Development. Based on risk assessment and risk management techniques, the plan helps support better watershed management and controls nonpoint source pollution.

Wastewater Treatment Wetlands

Wetlands are known as biological filters, providing protection for water resources. Current research has been conducted concerning the use of wetlands for water and wastewater treatment. Wetlands provide highly efficient physical removal of contaminants. Sedimentation of suspended solids is promoted by the low flow velocity. Biological removal is most important. Contaminants such as nitrates, ammonium, and phosphates are readily taken up by wetland plants. A wide range of chemical processes are involved in removal of contaminants in wetlands. Research in these areas is currently moving forward with federal government assistance.

SUMMARY

Water pollution became a problem when the first major city was built. It has been increasing over the centuries and became a serious problem after the industrial revolution started. The sources of pollution include point sources from a variety of industrial plants and municipal plants, as well as nonpoint sources and runoff from land. Water pollutants include infectious agents; plant nutrients; vast variety of organic chemicals, including pesticides, salts, heavy metals, sediment, and silt; heat, or thermal pollution; mine waste; and oil pollution. The pollutants are destroying our water, our land, and our food supply. They also are getting into the food chain and causing a serious potential problem of direct poisoning, cancer, teratogenicity, and mutagenicity.

RESEARCH NEEDS

Research is needed in basic wastewater treatment technology. A better understanding is needed of sludges and their behavior within the environment. Studies must be made to determine the changes in form and where contaminants such as heavy metals, organics, pathogenic organisms, and radiological material finally locate within the soil in the land disposal process. It is necessary to understand the mechanisms that control the conditioning of solids taken from wastewater. Continued research is needed for dewatering equipment and for more efficient means of dewatering sludge at a low cost. Improved engineering design and construction practices are needed for a variety of treatment processes. New techniques must be developed to resolve the disposal of toxic materials. The ongoing work of upgrading the methods of removing nutrients must be continued. Further studies are needed for the disposal of materials in deep wells. Considerable research is needed in the study of the removal of enteric viruses from wastewater, as well as from finished water. Soil treatment systems must be devised to more efficiently handle the solids removed from water pollutants. Studies are necessary to determine the best techniques for managing urban runoff pollution. Studies are also needed to develop the techniques necessary for controlling agricultural land runoff pollution.

Research and development are needed in the planning, construction, and operation of dams, hydraulic structures, and power generation and transmission facilities. Better understanding of the physical processes involved in the interaction of water and shorelines to develop ways to predict natural occurrences and how people-type actions affect these occurrences need to be studied. The Dredged Material Research Program needs to be expanded. A better understanding of wetland ecosystems and how to preserve them is essential.

Additional research is needed on coastal waters where small as well as large estuaries are found. A determination needs to be made of the balance of contaminants, natural biological communities residing in these estuaries, and effects of contaminants on these communities. Finally, additional studies need to be conducted in coastal areas where large waste loads have been placed for many years.

CHAPTER 8

Terrorism and Environmental Health Emergencies

INTRODUCTION

On September 11, 2001, four jet airliners loaded with fuel were hijacked by terrorists and were used as missiles to destroy the World Trade Center in New York City and to damage the Pentagon in Virginia. Three of the four planes achieved their goal, whereas one was apparently downed by passengers, who knew from cell phone conversations, that the other three were used to severely disrupt the United States. These actions resulted in the deaths of 3063 U.S. citizens, and citizens of over 80 other countries. An additional 5000 casualties occurred. This single act of terrorism was meant to destroy the financial network of the United States, to kill as many innocent people as possible, and to change foreign policy.

The United States has faced many disasters since its inception. They have been caused by acts of nature, or by people. When an act of terrorism occurs, the environmental health practitioner becomes part of an overall team concerned with the problems created by the act of terrorism or a disaster, and with types of emergency services needed to protect people against disease and injury. It is necessary to understand the services provided for the homeless, injured, and overall affected area. This section of the chapter briefly describes various types of acts of terrorism, plans for preparedness and response, various types of disasters and disaster plans, and finally specific environmental health measures utilized during the environmental health emergency.

TERRORISM

Terrorism is violence, or the threat of violence, used to create an atmosphere of fear, coercing people into actions they otherwise would not undertake or into refraining from actions that they desire. All terrorist acts are crimes, and many of them violate the rules of war, because innocent civilians are involved. The motives of all terrorists are political, and the actions are carried out in a way to achieve maximum publicity.

A terrorist group is a collection of individuals belonging to a nonstate or anti-government movement dedicated to the use of violence to achieve their objectives. At least some structure of command and control apparatus provides an overall organizational framework and general strategic direction.

State-sponsored terrorism is the active involvement of a foreign government in training, arming, providing logistical and intelligence assistance and sanctuary for a period of time to a terrorist group for the purpose of carrying out violent acts on behalf of that government against its enemies. It is a form of surrogate warfare. In 1995, Iraq confirmed that it had produced, filled, and deployed bombs, rockets, and airplane spray tanks containing *Bacillus anthracis* and botulism toxin.

Terrorists attempt to inflict mass casualties and mass destruction by any means possible, even if it is a suicidal mission. They use airliners, crop dusters, trucks carrying hazardous materials or hazardous waste, car bombs, suicide bombers, chemicals, biological agents, nonexplosive radiological agents, nuclear weapons, or anything else available. The consequences of these types of terrorists attacks not only are immediate, but also may be long-term with the contamination of the air, soil, water, and food. A potential exists for release of toxic or hazardous materials from the sites attacked, as well as the effects of the agents used by the terrorists. The intended results are to disrupt society.

TYPES OF TERRORIST ACTS

Explosive Materials

Terrorists have used and continue to use explosives to kill and injure as many people as possible, especially civilians. They have used in the past, car bombs, truck bombs, explosives in luggage aboard planes, individuals covered with explosives (such as suicide bombers), roadside devices set off remotely, boats carrying explosives, etc. The acts of turning fully fueled planes into flying bombs has been the most devastating of all. More than 200,000 private aircraft are kept at regional and rural airports, where security is often nonexistent.

Biological and Chemical Terrorism

The terrorists attacks on the World Trade Center and Pentagon clearly demonstrate that there are individuals who are currently in the United States and in many other countries; these are sleepers, and willing to do anything, including committing suicide, to kill, injure, and destroy as much life and property as possible to achieve a political victory. These horrendous crimes may in fact be the prelude to a new era of chemical and biological terrorism. The U.S. national civilian vulnerability to the use of biological and chemical agents is considerable because of weapons development programs and existing arsenals in foreign countries, as well as chemical weapons in the United States waiting to be destroyed. Further, hazardous chemical wastes and hazardous chemicals travel daily on trucks and trains throughout the country, and in many parts of the industrialized world. Militants have attempted to acquire or possess chemical and biological agents to carry out high-profile terrorists attacks.

In 1995, the Japanese cult, Aum Shinrikyo, released the nerve gas Sarin in the Tokyo subway. The cult also had botulism toxin, anthrax cultures, and drone aircraft equipped with spray tanks to distribute the material. In 1992, members of the group traveled to Zaire to obtain samples of Ebola virus for weapons development. Russia has an extremely large and sophisticated former bioweapons facility, called Vector, in Koltsovo, Novosibirsk. This was a sophisticated 4000-person, 30-building facility with biosafety level 4 laboratories, used for the isolation of specimens and human cases. It had been protected by an elite guard. Now, because of financial problems, the facility is half empty and is protected by a few guards who have not been paid in months. The facility housed the smallpox virus as well as work on Ebola, Marburg, and hemorrhagic fever viruses. No one knows where the scientists have gone or whether material is missing from the site.

Biological terrorism is more likely than ever, and more effective than explosives or chemicals. Only marginal funding and minimal support exist for providing assistance against the threat to the civilian population. Preventing or countering bioterrorism is extremely difficult. Recipes for making biological weapons are available on the Internet. Detection of biological weapons is next to impossible.

The potential of aerosolized smallpox to spread over a considerable distance and to infect at low doses was demonstrated in an outbreak in Germany in 1970. A German electrician returning from Pakistan became ill with high fever and diarrhea. On January 11, he was admitted to a local hospital and was isolated in a separate room on the ground floor. He had contact with only two nurses over the next 3 days. On January 14, he developed a rash; on January 16, the diagnosis of smallpox was confirmed. He was immediately transported to one of Germany's special isolation hospitals, and more than 100,000 people were promptly vaccinated. However, the smallpox patient had a cough, a symptom seldom seen with smallpox, which produced a large volume of small particle aerosols. Subsequently, 19 cases occurred in the hospital, 4 of which were on his floor, 8 of which were on the floor above, and 9 of which were on the third floor. Smallpox in aerosol form can spread very rapidly.

Anthrax is readily produced in large quantities. In its dried form, it is extremely stable. Inhalation of anthrax is highly lethal. An accidental release from a bioweapons facility for a few minutes caused over 100 cases of anthrax, with 66 people dying.

From October 30 through December 23, 1998 the Centers for Disease Control and Prevention received reports of a series of bioterrorist threats of anthrax exposure. Letters alleged to contain anthrax were sent to health clinics in Indiana, Kentucky, Tennessee, and California. Although all the threats were hoaxes, they show how vulnerable the public is to bioterrorism.

Anthrax was used as a biological weapon in the United States in the fall of 2001. By the end of November 2001, 23 cases of anthrax had made the CDC case definition, including 11 cases of confirmed inhalation anthrax (see *Bacillus anthracis*). Apparently letters containing tiny forms of the spores, and material used to keep them separated, were mailed to various media organizations, government offices, and certain U.S. senators. The spores had been spread through mail rooms and through various buildings. A 61-year-old female in New York City, who died from the inhalation form of the disease, apparently had no known contact with government offices or the U.S. Postal Service. As of December 1, 2001, the 94-year-old female from Connecticut, who died from the inhalation form of anthrax, was thought possibly to have come in contact with a very lightly cross-contaminated letter that had passed through the contaminated postal facility in New Jersey. Apparently her immune system had been severely compromised.

In cutaneous anthrax, the incubation period ranges from 1 to 12 days. The case fatality rate without antibiotic treatment is 20%, whereas with antibiotic treatment it is less than 1%.

In gastrointestinal anthrax, which usually follows eating raw or undercooked contaminated meat, the incubation period is 1 to 7 days. The case fatality rate is estimated to be 25 to 60%. The effect of early antibiotic treatment has not been established.

In inhalation anthrax, the most lethal form of the disease, the incubation period is 1 to 7 days, but may be as long as 60 days. The disease results from the inspiration of 8000 to 50,000 spores of *Bacillus anthracis*. Host factors, dose of exposure, and chemoprophylaxis may affect the duration of the incubation period. The disease is not spread from person to person.

Prior to the October and November 2001 outbreak of anthrax in the United States, only 18 cases of inhalation anthrax occurred in the 20th century, with the last one reported in 1976. This compares with the 11 known cases in the 2-month period.

The United States as well as other countries have had recent large and complex outbreaks of disease. These include the epidemic of over 400,000 cases of waterborne cryptosporidiosis in Milwaukee, the severe unexplained acute respiratory decease determined to be hantavirus pulmonary syndrome; and the nationwide salmonellosis outbreak caused by contaminated ice cream, which resulted in 250,000 cases. These outbreaks are a good indication of how rapidly a bioterrorist attack might spread disease to large numbers of people.

In 1999, in New England, a woman was treated for brucellosis, when she was apparently exposed to cultures and laboratory flasks kept in the apartment of her boyfriend. There has been no good explanation for why this hazardous biological material was kept here and what its ultimate use would be. The boyfriend disappeared from this country.

An act of biological or chemical terrorism might range from the dispersion of aerosolized spores, viruses, or bacteria to a variety of chemicals. This would be equivalent to a massive outbreak of an emerging infectious disease.

Biological Terrorism

Terrorists can obtain biological warfare materials at least four ways, including:

1. Purchasing a biological agent from one of the world's 1500 germ banks
2. Stealing biological agents from a research laboratory, hospital, or public health service laboratory, where agents are cultivated for diagnostic or research purposes
3. Isolating and culturing of an agent from natural sources
4. Obtaining biological agents from a state that supports terrorism, or a disgruntled government or scientist

In 1984, in the Dallas, OR area, members of a religious cult led by the Bhagwan Shree Rajneesh contaminated the salad bars of ten restaurants and a water tank with *Salmonella typhimurium*. They were trying to reduce voting by causing the local population to become sick, thereby throwing the municipal election in their favor. Reportedly 751 people became ill with salmonella gastroenteritis as a result of the attack. Subsequently, five new cases appeared each year. This was low-level contamination.

In high-level contamination, the principal problem is the development of a lethal strain in sufficient quantities to cause mass casualties. To make the biological agent into a weapon, the process has to have stability and predictability. The cost of equipping a facility for the production of biological agents may be between $200,000 and $2 million, an amount that is easily raised by certain terrorist organizations. The agent would have to be put into an aerosol cloud to disseminate it. Environmental conditions, such as sunlight, smog, humidity, and temperature changes, reduce the ability of pathogens to survive and multiply, although biological agents disbursed into closed areas may not be subjected to these conditions.

The dissemination of the agent in a public place would not have an immediate impact because of the delay between exposure and onset of illness. Patients could experience a variety of symptoms, such as fever, back pain, headache, and nausea, similar to the flu or other viral infection; and their doctors would not necessarily recognize that the individuals might be in the early stage of smallpox and potentially could die. This highly communicable disease could spread very rapidly. Many other diseases could be spread by means of air, insects and rodents, water, or food. They might include anthrax, pneumonic and bubonic plague, salmonellosis, shigellosis, typhoid fever, viral hemorrhagic fevers, common drug-resistant or genetically engineered pathogens, or many other existing or emerging diseases, used in a solitary or in a combined manner.

Chemical Agents

Chemical agents come in five major categories as follows:

1. Choking agents, such as phosgene and chlorine
2. Blood agents, such as hydrogen cyanide and cyanogen chloride
3. Blister agents, such as mustard gas
4. G-series nerve agents, such as tabun (GA), sarin (GB), and soman (GD)
5. V-series nerve agents, such as VX

Although many of the chemical agents could be used for the purpose of causing mass casualty attacks, sarin may be most likely to be used. Sarin is highly toxic, volatile, and relatively easy to manufacture. It is a liquid at ambient temperatures, which turns into a vapor that is heavier than air and clings to floors, sinks into basements, and moves toward low terrain. Sarin interferes with the mechanisms through which nerves communicate with body organs causing the organs to become highly overstimulated. If a person inhales a small amount of the vapor, it can cause tightness in the chest, shortness of breath, and coughing. People who inhale larger amounts lose consciousness, go into convulsions, and then stop breathing.

For a large-scale attack with a chemical weapon, temperature, wind speed, common inversion conditions, and other meteorologic factors would likely determine effectiveness. Airplanes equipped with industrial or crop sprayers or trucks are likely means of transportation and transmission of the chemicals. Because of the cost of such an operation, terrorists may consider trying to engineer a chemical disaster using conventional means instead of developing a chemical weapon. It is believed that the Japanese cult, Aum Shinrikyo, had an 80-person program, housed in state-of-the-art facilities, led by a Ph.D. level scientist, to produce the sarin used in the subway attack. It took at least a year and cost as much as $30 million, for the research and development needed to synthesize and deliver the sarin.

Common industrial and agricultural chemicals can be as highly toxic as chemical weapons, as shown in 1984, in Bhopal, India. In that incident, a disgruntled employee at a pesticide plant precipitated an explosion in one of the storage tanks by simply adding water to it. The massive release of methyl isocynate that followed produced poisonous fumes affecting thousands of people living near the plant, eventually resulting in 3800 deaths and 11,000 disabled.

In the case of chemical agents, the terrorists attacks may be overt or covert. In an overt attack, such as a chemical bombing, chemical spraying, or hijacking of trains or trucks and dispersal of hazardous waste or hazardous chemicals, an immediate effect would result because the chemicals are absorbed through inhalation, or through the skin or mucous membranes. A long-term or chronic effect may also result, including potential growth of cancer, depending on the nature of the chemical and time and concentration of exposure. In the case of nerve gases such as sarin and VX, the response can be immediate or long term. In the case of dioxin contamination, the onset of symptoms are typically long term. This would be an example of a covert action.

Protecting the Armed Forces from a biological or chemical attack, although complex and difficult, is relatively easy compared with trying to protect the civilian population. Terrorists may be difficult to find and may not be deterred in releasing an aerosol of a virulent bacterium, virus, or toxin in such places as airports, stadiums, subways, large train stations, or large facilities of other types, including hospitals.

It is essential to develop the necessary organization to deal with the recognition of unusual outbreaks of disease. Need exists for tremendous increased knowledge of healthcare providers and understanding of the potential severity of the disease outbreak; reliable and knowledgeable public health laboratories; coordination of federal, state, and military public health personnel; appropriate organization and management of the care of patients; and provision for all necessary medications, treatment procedures, personnel, and facilities. The coordination of this federal, state, and military system, including all agencies, is of greatest concern and necessity.

Nuclear Devices

Nuclear devices are costly, complex, and difficult to transport. However, these weapons can be planted or deployed by ship or aircraft. With the end of the Cold War, the security around bases where nuclear weapons have been stored has decreased substantially. In addition, new nuclear powers are now leading to a proliferation of nuclear material that can be used for weapons. It is known that terrorists have been attempting to purchase or steal weapons grade nuclear material.

Radiological Material

Terrorists may consider stealing radioisotopes from hospitals, laboratories, or industry to use to intimidate or kill people. They may also try to disrupt radioactive waste facilities or nuclear power plants to create mass deaths among the civilian population.

Electromagnetic Pulse Device

A nuclear electromagnetic pulse device if detonated at high altitude could burn out and destabilize electrical systems on a huge scale. The military, since the end of the Cold War, has relaxed efforts to harden its systems against this type of attack. The civilian community has not taken any action, although an electromagnetic pulse device attack would be devastating to the electrical grid, the economy, and the modern way of life of the country.

Agricultural Terrorism

It is relatively easy to attack crops and livestock. Livestock are particularly vulnerable because of the increased use of antibiotics and steroids, which has increased their stress level and decreased their resistance. Many more agents are highly infectious to animals than to humans. All major food crops can be infected with pathogens that can decrease their yield or destroy them. The areas where farms are located are so vast that it becomes almost impossible to protect them. The financial and health costs due to agricultural terrorism are potentially huge.

Cyberterrorism

In recent years, it has been shown, that a single person can infect a large number of networks in business, industry, universities, governmental agencies, and even defense communication systems. Since 1999, the number of security breaches has more than doubled. A bored teenager can create a virus or worm that can travel around the world in a very rapid manner. A group of terrorists could potentially erase enormous quantities of data, transfer funds to their accounts, and create general chaos.

STRATEGIC PLAN FOR PREPAREDNESS AND RESPONSE

In June 1995, as a result of the Oklahoma City bombing, the President of the United States issued Presidential Decision Directive/NSC-39, an executive order seeking to reduce the nation's vulnerability to terrorists attacks, especially those involving mass casualties. This was supplemented by Presidential Decision Directive/NSC-62. These documents directed federal agencies to improve domestic response capabilities to manage the consequences of attacks using unconventional weapons and develop and publish the U.S. Government Interagency Domestic Terrorism Concept of Operations Plan. This was completed in January 2001.

In 1996, Congress passed the Defense against Weapons of Mass Destruction Act. The legislation came from a series of hearings conducted in 1995 and 1996 showing the growing dangers posed by potential terrorists' use of weapons of mass destruction and the potential for materials to be stolen from the former Soviet Union.

In 1996, the Congress passed the Domestic Preparedness Act and the administration provided funds for federal, state, and local antiterrorism programs and planning, which has led to a substantial expenditure of funds. More than 50 cities have held drills simulating chemical, biological, and bombing attacks.

In fiscal year 1996, the principal federal agencies involved in fighting terrorists spend $5.7 billion. For fiscal year 2000, the budget request was $10 billion, to be devoted to counterterrorism programs. For fiscal year 2002, the budget request was much higher.

In 1998, the Congress passed the National Defense Authorization Act for fiscal year 1999, which in section 1405 established an advisory panel known as the Gilmore Commission. The Gilmore Commission, which is chaired by James S. Gilmore, governor of Virginia, is also known as the Advisory Panel to Assess Domestic Response Capabilities for Terrorism Involving Weapons of Mass Destruction. The function of this panel is to evaluate all areas of potential terrorism, executive branch response, congressional branch response, state and local capabilities, and coordination of efforts; and to issue three reports to be used by Congress and the administration in revising and upgrading efforts to protect the country from terrorism. On December 15, 1999, it issued, First Annual Report to the President and the Congress of the Advisory Panel to Assess Domestic Response Capabilities for Terrorism Involving Weapons of Mass Destruction–Assessing the Threat. On December 15, 2000, it issued, Second Annual Report to the President and the Congress of the Advisory Panel to Assess Domestic Response Capabilities for Terrorism Involving Weapons of Mass Destruction — Toward a National Strategy for Combating Terrorism. The panel continued in 2001, and submitted its third and final annual report to the President and the Congress on December 15, 2001. The panel reviewed and analyzed existing federal programs designed to support or enhance preparedness for terrorist incidents, with emphasis on enabling legislation, training, communications, equipment, planning requirements, maritime regions, and coordination among various levels of government.

Toward a National Strategy for Combating Terrorism — December 15, 2000

The advisory panel in 1999 in its first report produced a comprehensive assessment of the terrorists threat to the United States. Unfortunately, no one ever anticipated using airplanes as missiles. The second report summary looked at specific programs as well as a national strategy and federal organization.

The major findings of the Gilmore Commission in the second report are:

1. The United States has no coherent, functional national strategy for combating terrorism.
2. The organization of the federal government programs for combating terrorism is fragmented, uncoordinated, and politically unaccountable.
3. The Congress shares responsibility for the inadequate coordination of programs to combat terrorism.
4. The executive branch and the Congress have not paid sufficient attention to state and local capabilities for combating terrorism and have not devoted sufficient resources to augment these capabilities to enhance preparedness of the nation as a whole.
5. Federal programs for domestic preparedness to combat terrorism lack clear priorities and are not efficient in numerous specific areas.

The major recommendations of the Gilmore Commission as of December 15, 2001 include:

1. The President should develop and present to the Congress a national strategy for combating terrorism within 1 year of assuming office.
2. The President should establish a National Office for Combating Terrorism located in the Executive Office of the President, and should seek a statutory basis for this office.
3. The Congress should consolidate its authority over programs for combating terrorism into a special committee, either a joint committee between the House and Senate, or separate committees; and congressional leadership should instruct all other committees to respect the authority of the new committee and conform strictly to authorizing legislation (this is in place of the 11 full committees in the Senate and 14 full committees in the house, as well as numerous subcommittees who claim oversight over various U.S. programs for combating terrorism).
4. The Executive branch should establish a strong institutional mechanism for ensuring the participation of high-level state and local officials in the development and implementation of a national strategy for terrorism preparedness.
5. The focus should continue to be on the needs of local and state response entities, because they always are the first response to an act of terrorism, and may be the only response for a period of time. However, any program established must be national in scope. It can no longer be a series of federal, state, and local entities attempting to fight through the morass to achieve a single goal.

The National Office for Combating Terrorism, should have the following five major sections, each headed by an assistant director:

1. Domestic Preparedness Programs
2. Intelligence
3. Health and Medical Programs
4. Research, Development, Test and Evaluation, and National Standards
5. Management and Budget

The reorganization of the congressional committees under the auspices of one committee, and the administrative functions under a single Director of Homeland

Security would strengthen our capabilities to prevent terrorists acts, and respond to them rapidly and rationally in a very coherent matter. Specifically, this would enable various governmental agencies at local, state, and federal levels to:

1. Collect intelligence, assess threats, and share information
2. Coordinate all operations in an effective manner
3. Train and equip people and perform exercises to protect against and respond to terrorism
4. Provide appropriate and timely health and medical services
5. Perform necessary research and development for combating terrorism and establish national standards
6. Provide cyber security against terrorism

CDC Strategic Plan

In the year 2000, the various sections of the Department of Health And Human Services came up with a strategic plan for terrorism. A work group made up of representatives of the National Center for Infectious Diseases, National Center for Environmental Health, Public Health Practice Program Office, Epidemiology Program Office, National Institute for Occupational Safety and Health, National Immunization Program, and National Center for Injury Prevention and Control developed the report. The Agency for Toxic Substances and Disease Registry also provided expertise in the area of industrial chemical terrorism. The document is entitled, Preparedness and Response to Biological and Chemical Terrorism, A Strategic Plan (CDC, unpublished report, 2000). The key focus areas of the report are preparedness and prevention, detection and surveillance, diagnosis and characterization of biological and chemical agents, response, and communication. Implementation of the plan requires cooperation among state and local public health agencies, medical research centers, healthcare providers and their networks, professional societies, medical examiners, emergency response units and responder organizations, safety and medical equipment manufacturers, other federal agencies, and international organizations.

U.S. Government Interagency Domestic Terrorism Concept of Operations Plan — January 2001

This interagency domestic terrorism concept of operations plan is in accordance with the appropriate presidential decision directives, ensures that policy is implemented in a coordinated manner, and provides guidance to federal, state, and local agencies.

The purpose of this plan is to facilitate an effective federal response to all threats or acts of terrorism within the United States. It does this by:

1. Establishing a structure for a systematic, coordinated, and effective national response to threats or acts of terrorism in the United States
2. Defining procedures for the use of federal resources to augment and support local and state governments
3. Encompassing both crisis and consequence management responsibilities and articulation and coordination relationships between these missions

Crisis management is predominantly a law enforcement function and includes identifying, acquiring, and planning the use of resources needed to anticipate, prevent, or resolve a threat or act of terrorism. Consequence management is predominantly an emergency management function, including measures to protect public health and safety; to restore central government services; and to provide emergency relief to governments, businesses, and individuals affected by the consequences of terrorism.

The primary federal agencies who respond to a terrorists threat or incident within the United States are to be coordinated at the federal level and have appropriate rapid communications with state and local levels. These agencies are the Department of Justice and Federal Bureau of Investigation (FBI, lead agency for crisis management), Federal Emergency Management Agency ((FEMA, lead agency for consequence management), Department of Defense (DOD), Department of Energy, Environmental Protection Agency (EPA), and Department of Health and Human Services.

The Attorney General is responsible for developing and implementing policies directed at preventing terrorist attacks domestically and undertaking the criminal prosecution of the acts of terrorism that violate U.S. law. The FBI designates a federal on-scene coordinator for overall U.S. government response until the Attorney General transfers the process for consequence management to the Federal Emergency Management Agency.

The Federal Emergency Management Agency (FEMA) manages federal response to the emergency effort related to people, companies, communities, and facilities. The agency works closely with the FBI.

The Department of Defense is a support agency to the FBI in technical operations, threat assessment, support for civil disturbances, transportation, and disposal of weapons of mass destruction. The department also assists FEMA in rescue efforts.

The Department of Energy is a support agency to the FBI for technical operations and to FEMA for consequence management. This department works in all areas of nuclear and radiological weapons of mass destruction terrorist incidents.

The EPA is a support agency for the FBI for technical operations and a support agency to FEMA for consequence management. The EPA is involved in threat assessment, emergency response team deployment, and technical advice and operational support for chemical, biological, and radiological releases. The agency is also involved in agent identification, hazard detection and reduction, environmental monitoring, sampling for forensic evidence collections and analysis, identification of contaminants, assessment of cleanup, on-site safety, protection, prevention, decontamination, and restoration activities. The EPA shares responsibility with the U.S. Coast Guard to respond to oil pollutant, or hazardous substances discharges into navigable waters.

The Department of Health and Human Services is a support agency for the FBI in technical operations and a support agency to FEMA for consequence management. The department provides personnel and supporting equipment during all aspects of the terrorists incident, regulatory follow-up when an incident involves food or drugs, epidemiologists, environmental health, and other technical specialists. It also provides mass immunizations, mass fatality management, pharmaceutical support, contingency medical records, patient tracking, inpatient evaluation, and definitive medical care provided through the National Disaster Medical System.

The various agencies working together utilize the following priorities:

1. Preserving life or minimizing the risk to health and safety
2. Preventing a threatened act from implementation or an existing terrorist act from expansion or exaggeration
3. Locating, accessing, rendering safe, controlling, containing, recovering, and disposing of a weapon of mass destruction that has not yet functioned
4. Rescuing, decontaminating, transporting, and treating victims while preventing secondary casualties
5. Releasing emergency public information that ensures adequate and accurate communications with the public
6. Restoring the essential services and mitigating suffering
7. Apprehending and successfully prosecuting terrorists
8. Conducting site restoration

According to the operations plan, the four distinct threat levels are:

1. Level 4 — minimal threat–threats that do not warrant actions beyond normal liaisons notifications
2. Level 3 — potential threat — intelligence or an articulated threat indicates a potential for terrorists incident
3. Level 2 — credible threat — a threat assessment indicating that the potential threat is credible and confirms the involvement of a weapon of mass destruction in the developing terrorist incident
4. Level 1 — weapon of mass destruction incident– a terrorist incident has occurred that requires the immediate process to identify, acquire, and plan the use of federal resources to augment state and local parties in response to limited or major consequences of a terrorist use or employment of a weapon of mass destruction

SOME CURRENT PROGRAMS

The Division of Laboratory Sciences of the National Center for Environmental Health is developing a rapid toxics screen, that will measure 150 chemical agents in blood and urine. The rapid toxics screen uses advanced analytic techniques, including tandem mass spectrometry and a high-resolution mass spectrometry, to quickly and accurately measure these chemical agents. A Laboratory Response Team also is on call 24 hr/day, 7 days/week to respond to known or potential chemical terrorist attacks. A chemical terrorism laboratory network is now in place in California, Michigan, New York, and Virginia.

The Division of Laboratory Sciences of the National Center for Environmental Health has been involved in a biomonitoring program for more than 25 years. They do a direct measurement of environmental chemicals, their primary metabolites, or their reaction products in people, through blood or urine tests. The goals of the biomonitoring program are to:

1. Determine which environmental chemicals actually get into people
2. Measure how much exposure each person has

3. Assess exposure for health studies of exposed populations
4. Determine which population groups are at high risk for exposure and adverse health effects
5. Assess the effectiveness of public health interventions to reduce exposures
6. Monitor trends in exposure levels over time

The U.S. Coast Guard Atlantic Strike Team conducted air-monitoring samples in the New York financial district after the World Trade Center bombings. This was done under the direction of the EPA.

CDC officials are working on a variety of programs including reinforcing systems of public health surveillance to detect unusual or covert events, building epidemiological capacity to investigate and control health threats from these events, enhancing public health laboratory capability to diagnose the illness and identify the causes, and developing and coordinating communications systems with other government agencies and the general public to disseminate critical information and reduce unnecessary fear. It also is involved in creating a national pharmaceutical stockpile to respond to a terrorist use of potential biological or chemical agents, as well as many other programs.

The EPA, because of its inherent role in protecting human health and the environment from possible harmful effects of certain chemical, biological, and nuclear materials, is actively involved in counterterrorism planning and response efforts. The EPA helps state and local responders to plan for emergencies, coordinates with federal planners, trains first responders, and provides resources in the event that a terrorist incident occurs.

Many people in programs from agencies already mentioned and other agencies that have not been mentioned are currently working in the area of counterterrorism and reacting to terrorist attacks. Individual environmental, occupational, and safety managers are in a position to develop contingency plans to deal with terrorism.This is a perfect time to update or improve existing plans or create new ones if they do not exist. Many organizations are already required to have a contingency plan in place because of the EPA Risk Management Planning Rule, or the Occupational Safety and Health Administration (OSHA) Process Safety Management Standard.

In the society in which we live today, a potential always exists for disasters caused by terrorists. This is a unique time and a unique opportunity for all environmental professionals to help develop plans and techniques; and also to carry out activities to protect the health of the public and the environment through comprehensive involvement in daily activities in all areas, as the individual goes about the normal work schedule. Teaching the public how to assess potentially dangerous situations, who to contact, and how to protect oneself, family, and community is an essential part of the work of the environmental professional.

TYPES OF DISASTERS

Disasters may include bioterrorism, chemical spills, earthquakes, floods, forest fires, hurricanes, landslides, radiological spills, tornadoes, and other windstorms.

Earthquakes

The continents are on large plates that are in motion. As these plates move, stresses form and accumulate until a fracture or abrupt slippage occurs. The release of a stress is called an earthquake. Although the focus of the earthquake is at the point of the release of stress, the mechanical energy flows forward in waves that radiate from a focus in all directions through the Earth. When the energy arrives at the surface of the Earth, it causes secondary surface waves. The severity of the earthquake depends on the amount of mechanical energy released at the focus, distance and depth of the focus, and structural properties of the rock or soil. Most earthquake tremors last only a few seconds. An earthquake of the magnitude of 8.5 on the Richter scale is equivalent to the energy released from 12,000 nuclear bombs.

The large earthquakes start with a deep rumbling, followed by a series of violent motions in the ground. Objects on top of the ground seem to disintegrate and cracks occur in the surface. Large earthquakes can cause buildings and bridges to collapse, dams to burst, and other rigid structures to be cut in half. Frequently, water is thrown from its confines. The effect of the earthquake may be felt for hundreds or thousands of miles from its center. Secondary effects include landslides, fires, tidal waves, and floods. Obviously, when an earthquake occurs, normal electrical and gas service may be disrupted, highways may be blocked, explosions and fires may readily occur, and various secondary problems may occur ranging from sudden rat migrations, to disruption of water and sewage systems, to a population vulnerable to numerous diseases, or to a disruption of all emergency services. Considerable research is under way in China, Russia, and the United States concerning preliminary signs that should be evaluated prior to the onset of an actual earthquake. After an earthquake has occurred, individuals should, wherever possible, shut off all gas and electricity, utilize water from the hot water heater where possible, and not consume any food or water that may have become contaminated. Emergency response can be enhanced by use of a geographic information system (GIS) program.

Annually throughout the world, the following types of earthquakes typically occur:

1. Great — 8+, one a year
2. Major — 7.0 to 7.9, 18 a year
3. Strong — 6.0 to 6.9, 120 a year
4. Moderate — 5.0 to 5.9, 800 a year
5. Light — 4.0 to 4.9, 6200 estimated a year
6. Minor — 3.0 to 3.9, 49,000 estimated a year

In 1994, a 6.8-magnitude earthquake struck a densely populated area of Los Angeles near Northridge. In the tremor 30 people were killed, and a total 61 deaths were attributed to direct and indirect causes. In the following 3 weeks, 2500 aftershocks occurred; 65,000 residential buildings had sustained damage; and the cost of the earthquake was between $18 and $20 billion. In 1995, several thousand people died, and hundreds of thousands of people were left homeless in Japan.

Earthquake preparation needs to include practicing drills, planning evacuations, establishing priorities, gathering and storing important documents in a fireproof safe, obtaining emergency supplies, providing emergency water storage and purification, providing emergency food and cooking equipment, and obtaining and using an emergency communications system. All facilities need to be checked very carefully before reentry and reuse.

Floods

Flooding may occur as the result of hurricanes, storms, tidal waves, extremely heavy rainfall, runoff from large amounts of snow, and changes brought about in the watershed due to forest fires or destruction of trees. Flooding not only may inundate homes, buildings, roads, railroads, and bridges, but also may destroy sanitary sewer systems, power sources, water sources, telephone installations, and croplands.

Over the years, extremely serious flood conditions have occurred in several parts of the United States, including Pennsylvania. In the Harrisburg and Wilkes-Barre, PA area, a flood developed when a summer storm moved from the ocean across the mountains and proceeded to hang over these areas for several days. The storm then came back from the mountains out toward the ocean and once again deposited huge quantities of rainwater. The water accumulated in the upper portions of the Susquehanna River and changed a pretty and quiet river into a raging monster that destroyed homes, killed people, overturned cemeteries, and caused enormous destruction. In some parts of Wilkes-Barre, especially the downtown area, the floodwaters were higher than the first-floor level.

California frequently has severe rainfalls causing flooding and mud slides. These severe flooding conditions are always a potential hazard to communities. Also in numerous areas, houses are situated in natural floodplains. These areas are flooded simply as a result of the normal increase in water that occurs in the spring. The individuals in these homes are particularly vulnerable to loss, and are also vulnerable to the health and safety hazards that exist.

Warning systems may be provided by the weather service concerning potential flooding conditions. Flash flood warnings may be issued by the weather bureau in advance by using meteorologic and radar data. However, flash floods may occur and strike without warning and cause enormous numbers of deaths and injury. Flood prevention may be carried out by means of flood abatement programs and by development of floodplain regulations for the usage of the land. Specifically, the U.S. Army Corps of Engineers is deeply involved in this particular flood prevention program. When flood warnings occur, all electrical and other services should be immediately shut off. Then the individuals should leave the areas and proceed to high ground, where they cannot be affected by the floods. All water, food, and other objects that have been inundated by the flood must be considered to be contaminated and should not be utilized. Special procedures should be carried out within homes after the floodwaters have gone down to properly clean and decontaminate the houses before people live in them again. Safety hazards related to flood cleanup include electrical hazards, carbon monoxide, musculoskeletal hazards, thermal stresses, heavy equipment, structural instability, hazardous materials, fire, drowning, confined spaces, power line hazards, agricultural hazards, stress, and fatigue.

Forest Fires

Forest fires are caused by people and nature. People, however, are responsible for roughly 65% of all the forest and grass fires. These fires, as well as the potential for fire, are influenced by weather conditions, topography, use of the land, and careless actions that may be carried out by people within the areas. Forest fires not only destroy huge quantities of usable timber but also create conditions where landslides may occur and where people and animals may die. Whenever weather conditions become such that an area becomes extremely dry, the danger of fires increases sharply. Proper warning systems, including the use of forest rangers, are essential to stop the fire before it has a chance to create a devastating effect.

Hurricanes

Conditions needed to produce hurricanes, which are cyclones, exist in the atmosphere over warm ocean areas. Winds reach speeds of 75 mi/hr or more, and blow in a large spiral around a relatively calm center that is the eye of the hurricane. The circulation of the winds is counterclockwise in the Northern Hemisphere and clockwise in the Southern Hemisphere. Hurricanes are spotted by the weather bureau and their path is determined long before they can cause hazardous situations to people. Today, because of the use of sophisticated equipment and because of mass communications systems, it is unnecessary for individuals to die as a result of hurricanes. However, a hurricane that heads inland may cause excessive damage due to winds and floods. The monetary damage can be astronomical. The best means of protection for people is to avoid the hurricane. Once the hurricane has moved on, a concerted effort must be made to bring the area back into an environmentally safe situation before people can reenter. Prior to leaving the hurricane area, all utilities should be shut off and wall appliances disconnected. All food, water, and drugs, as well as inanimate objects that may be utilized, have to be considered contaminated.

Landslides

Landslides may be due to a slow erosion of soil due to freezing, thawing, water, or wind. The slide may cause rock falls or move large masses of loose debris downward. Heavy rains can also create landslides or mud slides where the soil is no longer stabilized by forest or brush, or where large piles of mine wastes exist. Landslides may also be caused by the movement of glaciers, earthquakes, water construction projects, and the building of houses over areas where mining has occurred. The danger of landslides exists everywhere. Although some landslides may be anticipated because of the nature of the soil, others may occur without warning.

Radiological Spills

The potential for radiological spills exists because of the increasing use of radioisotopes and other radioactive materials. The major concern would be the possibility of a spill from a nuclear power plant or during the transportation and disposal of radiological waste. This area is highly specialized and therefore in the event such a problem did occur, it would be important to immediately call on the proper teams of specialists from local, state, and federal agencies to deal with the problem. Various state or local health officials may also be trained in this specific hazard.

Tornadoes and Windstorms

Tornadoes are locally occurring storms of brief duration formed by winds rotating at very high speeds in a counterclockwise direction. Windstorms may refer to any storm where wind becomes a problem or hazard. Tornadoes and windstorms each year cause extensive damage to property and kill and injure people. Tornadoes occur mostly in the spring, but may occur every month of the year; although most tornadoes frequently occur in an extended Midwest area, they may occur in any state. In 1990, over 100 tornadoes hit Indiana in 1 day. Most of the tornadoes occur between noon and midnight, with the greatest single time period between 4 and 6 P.M.

Tornado conditions and severe thunderstorms, as well as severe windstorm conditions, may be picked up by the weather service, and warnings may be broadcast in advance of the actual hazard. In some areas, local siren systems are utilized to advise individuals of tornado conditions. It is important that when these conditions do occur, electrical services as well as other services be shut off and individuals seek protection in steel-framed or reinforced concrete buildings or in areas away from windows. The potential for disrupted utilities, as well as water and sewage service, must be taken into account if a tornado occurs.

DISASTER PLANS

In the event of a disaster, the community should be prepared to take care of the emergency conditions that exist. Disaster plans are usually developed for the community by a joint task force of governmental and private agencies, including police, fire, health, hospitals, Red Cross, Civil Defense, and the Emergency Management Agency. Further, hospitals usually have external and internal disaster plans ready for use. Emergency vehicles, as well as emergency communication systems, must be provided as an integral part of disaster planning.

Emergency Planning and Community Right to Know

The reauthorization of Comprehensive Environmental Response, Compensation, and Liability Act (CERCLA), known as the Superfund Amendments and Reauthorization Act (SARA) of 1986, provided an amendment that included Title III, better known as the Emergency Planning and Community Right-To-Know Act. Title III promotes the public's awareness of the hazardous or toxic chemicals used or produced by industry. It also mandates that each community be prepared to respond to emergencies from the release or explosion of chemicals. Industrial and commercial facilities are required to report annually on the quantities of substances present in their facilities and released to the environment on a routine basis.

Title III builds on the EPA Chemical Emergency Preparedness Program (CEPP) and numerous state and local programs that are used to help communities meet their responsibilities in regard to potential chemical emergencies. Title III has these four major sections, emergency planning, emergency notification, community right-to-know reporting requirements, and toxic chemical release reporting emissions inventory. Together they are unified into the emergency planning and community right-to-know provisions of the act.

The emergency planning sections are designed to help state and local government emergency preparedness and response capabilities improve through better coordination and planning, especially at the local level. Title III requires that the governor of each state designates a State Emergency Response Commission (SERC), which has broad-based representation.

Public agencies and departments concerned with issues related to the environment, natural resources, emergency management, public health, occupational safety, and transportation all have important roles in Title III activities. Public and private groups and associations have interest and experience in Title III issues and can also be included on the SERC. The SERC designates local emergency planning districts and appoints local emergency planning committees (LEPC).

The SERC is responsible for supervising and coordinating the activities of the LEPCs for establishing procedures for receiving and processing public requests for information collected under other sections of Title III and for reviewing local emergency plans. The LEPC must include elected state and local officials, police, fire, civil defense, public health professionals, environmental, hospital, and transportation officials, as well as representatives of facilities, community groups, and the media. Interested persons may also petition to be on the LEPC. Facilities that are subjected to the emergency planning requirements must notify the LEPC of a representative who will be the facility emergency coordinator. This person can provide technical assistance and understanding of facility response procedures, information about chemicals and their potential effects on nearby persons in the environment, and opportunities they have for training. The LEPC establishes rules for giving the public notice of its activities and establishes procedures for handling public requests for information.

The primary responsibility of the LEPC is the development and update of an emergency response plan. The local committee evaluates available resources for preparing for, and responding to, a potential chemical accident. The plan must include:

1. Identification of hazardous materials facilities and extremely hazardous substances transportation routes
2. Emergency response procedures on-site and off-site
3. Designation of a community coordinator and facility coordinator to implement the plans
4. Emergency notification procedure
5. Methods for determining the occurrence of a release from the probable affected area and population

6. Description of community and industry emergency equipment and facilities and the identities of persons responsible for them
7. Evacuation plans
8. Description and schedules of training program for emergency response to chemical emergencies
9. Methods and schedules for carrying out emergency response plans

The LEPC, in preparing and reviewing plans, receives help from the National Response Team (NRT). This team is made up of individuals from 14 federal agencies with responsibilities for emergency preparedness and response. The team publishes guidelines for emergency planning.

The emergency plan must be reviewed by the SERC on completion and reviewed annually by the LEPC. The Regional Response Teams (RRTs) are made up of federal regional officials and state representatives, who review the plans and provide assistance at the LEPC request. Emergency planning activities of the LEPC and facilities focus on, but are not limited to, the extremely hazardous substances published in the *Federal Register*. The list includes the threshold planning quantity (TPQ) for each substance. The TPQs take into account the toxicity, reactivity, volatility, dispersability, combustibility, or flammability of a substance.

Any facility that produces, uses, or stores any of the listed chemicals in quantities greater than the TPQ must meet all emergency planning requirements. If a facility produces, uses, or stores one or more hazardous chemicals, it must immediately notify the LEPC and the SERC if a release of a listed hazardous substance exceeds the reportable quantity for the substance. The initial notification of release can be by telephone, by radio, or in person. Emergency requirements involve transportation incidence. This may be satisfied by dialing 911 if it is present in your community. In the absence of 911, call the operator.

The emergency notification needs to include chemical name and indication of whether the substance is an extremely hazardous substance; estimate of the quantity released into the environment; time and duration of the release; medium into which the release occurred; any known or anticipated acute or chronic health risks associated with the emergency; where appropriate, advice concerning medical attention necessary for exposed individuals; proper precautions, such as evacuation; and the name and telephone number of a contact person. This notification must be followed up in writing.

A facility that must prepare or have available Material Safety Data Sheets (MSDSs) under the OSHA hazard communications regulations has to submit copies of the MSDSs or a list of MSDS chemicals to the LEPC, SERC, and local fire department. The MSDSs must contain chemical name and basic characteristics, such as

- Toxicity, corrosivity, and reactivity
- Known health effects, including chronic effects from exposure
- Basic precautions in handling, storage, and use
- Basic countermeasures to take in the event of fire, explosion, or leak
- Basic protective equipment to minimize exposure

The inventory form must list, for each applicable OSHA category of health and physical hazard, the following:

1. Estimate in ranges of the maximum amount of chemicals for each category present at the facility at any time during the preceding calendar year
2. Estimate in ranges of the average daily amount of chemicals in each category
3. General location of hazardous chemicals in each category
4. Chemical name or common name as indicated on the MSDS
5. Estimate in ranges of the maximum amount of the chemical present at any time during the preceding calendar year
6. Brief description of the manner of storage of the chemical
7. Location of the chemical at the facility
8. Indication of whether the owner elects to withhold information from disclosure to the public

Title III requires the EPA to establish an inventory of toxic chemical emissions from certain facilities. These facilities must complete a toxic chemical release form for specified chemicals. The purpose of this recording requirement is to inform government officials and the public about releases of toxic chemicals into the environment. The requirement applies to owners and operators of facilities with ten or more full-time employees that manufacture, process, or otherwise use listed toxic chemicals in excess of specified threshold quantities.

Facilities using listed toxic chemicals in quantities over 10,000 lb in a calendar year are required to submit toxic chemical release forms by July 1 of the following year. Over 300 chemicals and categories are on these lists. The EPA can modify these combined lists. The EPA considers the following factors in putting chemicals on the list:

1. Does the substance cause cancer or serious reproductive or neurological disorders, genetic mutations, or other chronic health effects?
2. Can the substance cause significant adverse acute health effects as a result of continuous or frequent recurring releases?
3. Can the substance cause an adverse effect on the environment because of its toxicity, persistence, or tendency to bioaccumulate?

Hazardous materials referred to in any of the planning generally mean hazardous substances, such as petroleum, natural gas, synthetic gas, acutely toxic chemicals, and other toxic chemicals. Extremely hazardous substances, as defined by Title III of SARA, refer to those chemicals that can cause serious health effects following short-term exposure from accidental releases. The plan needs to specifically cover the response to hazardous materials, both at fixed facilities that may be involved in manufacturing, processing, storage, and disposal; or during transportation, which may occur on highways, waterways, or by rail and air.

The major federal agencies involved in emergency planning include:

1. Federal Emergency Management Agency (FEMA), Technological Hazards Division, Federal Center Plaza, 500 C Street, S.W., Washington, D.C.; (202) 646-2861
2. FEMA National Emergency Training Center, Emittsburg, MD; (301) 447-6771

3. U.S. Environmental Protection Agency, OSWER Preparedness Staff, 401 M Street, S.W., Washington, D.C.; (202) 475-8600, hot line (800) 535-0202
4. U.S. EPA OERR Emergency Response Division, 401 M Street, S.W., Washington, D.C.; (202) 475-8720
5. Agency for Toxic Substances and Disease Registry, Department of Health and Human Services, Chamblee Building, 30S, Atlanta, GA; (404) 452-4100
6. U.S. Department of Energy, 1000 Independence Ave., S.W., Washington, D.C.; (202) 252-5000
7. Department of Agriculture Forest Services, P.O. Box 96090, Washington, D.C.; (703) 235-8019
8. Department of Labor Occupational Safety and Health Administration, Directorate of Field Operations, 200 Constitution Ave., N.W., Washington, D.C.; (202) 523-7741
9. U.S. Coast Guard (G-MER) Marine Environmental Response Division, 2100 2nd St., S.W., Washington, D.C.; (202) 267-2010 for information
10. National Response Center; (800) 424-8802
11. U.S. Department of Transportation, Research and Special Programs Administration, Office of Hazardous Materials Transportation (Attention: DHM-50), 400 7th St., S.W., Washington, D.C.; (202) 366-4000
12. Department of Justice, Environmental Enforcement Section, Room 7313, 10th and Constitution, N.W., Washington, D.C.; (202) 633-3646
13. Department of Interior, 18th and C Street, N.W., Washington, D.C.; (202) 343-3891
14. Department of Commerce NOAA-Superfund Program Coordinator, 11400 Rockville Pike, MD; (301) 443-4865
15. Department of Defense OASD (A & L) E Room 3D, 833 The Pentagon, Washington, D.C.; (202) 695-7820
16. Department of State, Office of Oceans and Polar Affairs, Room 5801, 2001 C St., N.W., Washington, D.C.; (202) 647-3263
17. Nuclear Regulatory Commission, Washington, D.C.; (301) 492-7000

Federal Emergency Management Agency

FEMA and the U.S. Geological Survey (USGS) signed a memorandum of understanding on December 13, 2000 to form a partnership in a national disaster prevention initiative titled Project Impact: Building Disaster Resistant Communities. The partnership promotes federal efforts to improve disaster recovery and mitigation in communities throughout the country. The USGS provides FEMA with critical earth science information on natural hazards including earthquakes, floods, volcanoes, wildland fires, landslides, and other geological and hydrological hazards, needed to reduce vulnerability to natural disasters. Nearly 250 communities and 2500 business partners are now involved in Project Impact.

FEMA is trying to work with communities to help them become better prepared for the next disaster. Since 1995, the Preparedness Directorate has been producing a compendium of exemplary practices in emergency management. The objective of the compendium is to share information concerning innovative emergency management programs that have worked and may be adopted elsewhere. The four volumes of the compendium are available from FEMA. Individuals can find information concerning specific types of emergency situations and available programs that had been successful.

Emergency Management at the State Level

Emergency management at the state level has been addressed by the standing committees of the National Governors' Association, top-level state officials, and governors' offices. The governors recognize that emergencies can strike at any time with either minor or major human economic and political consequences. The range of natural disasters, as well as disasters due to technological emergencies, continues to increase. The state–federal emergency relationship that began over 35 years ago, was primarily responsible for preparing for nuclear attacks and natural disasters. Even at that time, conflicts existed among federal, state, and local authorities over their respective roles in emergencies. States found it difficult to deal with the many federal agencies involved in emergency preparedness and response. States also did not agree with the federal agencies on the use of civil defense funds for natural disasters. FEMA now works with the states to coordinate all available resources in an integrated program of hazard reduction, preparedness, and response.

The primary economic change currently is the tremendous increase in property damage caused by disasters. This is due to an increase in industrial, commercial, and residential development. Synthetic chemicals not only have resulted in a sizable increase in the types of emergencies but also have resulted in a need for new strategies and laws. Title III of SARA requires governors to establish chemical emergency planning programs. State emergency management is now a complex multilayered undertaking that is implemented by a combination of state and local programs, volunteer organizations, and private individuals.

Emergencies and disasters are used interchangeably. They refer to any sudden or unforeseen situation that requires a nonroutine response. Emergency and disaster also refer specifically to the role of the governor and aides and to any situations that threaten the lives and property of the citizens and requires a governor's coordination beyond the powers of any combination of state, local, and federal agencies.

Emergencies may consist of the following:

1. Technological and people-made hazards, including nuclear waste disposal spills; radiological, toxic substance, or hazardous materials; accidents; utilities failures; pollution; epidemics; crashes; explosions and urban fires
2. Natural disasters, such as earthquakes, floods, hurricanes, tornadoes, sea surges, extreme cold, blizzards, forest fires, drought, and infestation of large areas by insects
3. Internal disturbances, such as civil disorders, including riots, large-scale prison breaks, and strikes leading to violence, as well as acts of terrorism
4. Energy and material shortages due to strikes, price wars, labor problems, and scarcity of the resources
5. Nuclear, conventional, chemical, or biological warfare

Governors are placed in an extremely challenging and potentially serious position. The governor has to make a series of difficult decisions that often may be based on incomplete or premature data. The governor's office and the agencies must provide the governor the best available data and insight for use in making the decisions. Emergencies also are a public relations problem, because the public's perception of the governor may be based on the kinds of decisions or lack of decisions taken by the governor. Emergencies are often troublesome because they present many opportunities for criticism for a long period of time.

Emergency response is only part of emergency management. The media may stop reporting the emergency after a few days. However, reexaminations and debriefing of people may go on for months or years. The governor may respond quickly and professionally to a hazardous material spill that may be caused by a train wreck, yet be criticized later for allowing the train with a hazardous material to travel through residential areas in the first place. Governors may also be involved in interpersonal problems in dealing with local and federal officials, who may feel slighted if they do not receive the attention that the governor should give to them.

To make sure that the governor works effectively with others, this person can use the following emergency management approaches:

1. Understanding state emergency-related law, particularly as it refers to the governor's powers during the emergency
2. Ensuring that the state emergency office director and other key senior-level staff involved in emergency management have the political and technical skills necessary to do the job
3. Ensuring that the various state agencies, programs, and organizations involved in emergency management work together in a coordinated manner; designating a staff person to oversee this coordination
4. Preparing to make changes in state government and to accommodate changes during emergencies and federal programs

Emergency management requires a great deal of coordination among different levels of government and various private organizations, as well as governmental agencies. The state, especially the state governor, is the focal point in the Emergency Management System. The governor has the power to:

- Suspend state statutes
- Seize personal property where needed
- Procure materials and facilities without regard to limitations of any existing law
- Direct evacuations
- Control entrance and exit to and from the disaster area
- Authorize emergency funds

National Disaster Medical System

The National Disaster Medical System (NDMS) was established at the Park-Lawn Building, Room 4-81, 5600 Fishers Lane, Rockville, MD; (301) 443-4893. The NDMS is based on the concept of the Civilian Military Contingency Hospital System in which civilian hospitals voluntarily commit a portion of their beds for military casualties. In the NDMS, the hospital beds can be supplemented by medical teams and logistic support to enable the system to serve a large civilian disaster. The system is a cooperative effort of the Department of Health and Human Services, DOD, FEMA, state and local governments, and private sector. The NDMS comprises 150 disaster medical response teams, an evacuation system, and 100,000 precommitted beds in hospitals throughout the country. The system serves national needs in the event of a massive peacetime disaster or an overseas conventional military conflict.

The system was created because earthquakes, tidal waves, volcanic eruptions, storms, fires, industrial accidents, and many other disasters have struck and will continue to strike the United States. Although huge number of casualties have not occurred to this point in any one of these disasters, the potential for casualties compared with other incidences in other parts of the world is substantial.

A single city or state certainly cannot be expected to be prepared for major catastrophes. Therefore, all cities of the nation have health resources to assist each other in the event of extreme need.

Military planners are familiar with a massive number of casualties. Because the cost of adequate medical care has risen so rapidly and because adequate mass casualty capability is impossible to maintain in the high-risk areas of the country, the DOD established the Civilian-Military Contingency Hospital System (CMCHS). CMCHS is a military support program operated by the DOD in cooperation with the Veteran's Administration (VA) and the civilian hospitals of the nation. It is coordinated nationally by the Office of the Assistant Secretary of Defense for Health Affairs and is coordinated locally by major federal hospitals in urban areas. Each federal coordinating hospital recruits local general hospitals to participate in the system. With activation, the coordinating hospitals are also responsible for patient reception and sorting, local patient assignment and transportation, or patient administration. Participating hospitals agree to accept patients in proportion to their licensed bed capacity in the event of a military emergency and participate in exercise programs and mass casualty care.

The CMCHS system is of limited use for a civilian disaster. It contains no deployable medical elements capable of on-scene response. For a civilian emergency it is, therefore, a resource only for acute-care hospital beds. The CMCHS has, however, developed strong interest in a number of people in emergency preparedness in the American healthcare community. The interest led to the establishment of the NDMS and has provided a model for the design of the NDMS hospital program.

The Emergency Mobilization Preparation Board is charged with developing national policy and programs to improve emergency preparedness. Health programs development is delegated to the board's principal working group on health, which is chaired by the assistant secretary for Health of the Department of Health and Human Services. Major members of the board include individuals from the Public Health Service, DOD, Health Care Financing Administration, VA, and FEMA. All other federal agencies concerned with health services participate in the work of the board and its task force.

The NDMS is supposed to fulfill three main objectives:

1. To provide medical assistance to a disaster area in the form of medical assistance teams and medical supplies and equipment
2. To evacuate patients who cannot be cared for in the affected area to designated locations elsewhere in the nation
3. To provide hospitalization in a national network of hospitals that have agreed to accept patients in the event of a national emergency

The system is designed to accept up to 100,000 seriously injured patients requiring hospitalization. The NDMS recognizes the following five medical care functions involved in mass casualty care:

1. Field rescue and first aid
2. Casualty clearing, that is, triage medical stabilization and temporary care at the first point of medical care
3. Emergency surgical care, which may have to be done in the disaster area to save lives or to render critically injured patients fit for evacuation
4. Medical staging, that is, sorting and temporary care of stabilized casualties at transfer points in the evacuation system
5. Definitive care, which encompasses all remaining medical care required for proper treatment of the victim during the acute phase of injury; including a logistic support function, such as finance, supply, and transportation

Disaster Response Guidelines for Ambulance Providers

A medical disaster is any situation involving injury or illness of a number of persons under circumstances that are clearly beyond the capability of the usual emergency medical services (EMS) system to handle in a reasonable period of time. Information is provided to the EMS systems when weather patterns, flooding potential, and geographic characteristics may cause problems and allow providers to assess their operation, as well as be able to prepare for any situation. The responder must know the capabilities of the system and the backup services that are available. The EMS coordinators work with Civil Defense in disaster situations.

Community Disaster Plans

For localized disasters, the county commissioners or the mayor of the city are generally responsible by law for the community emergency operations. When a disaster starts to assume greater magnitude, the governor becomes the principal officer responsible for the disaster. Ultimately, if the governor believes that the situation is such that it is beyond the ability of the state to handle, the governor may call on the President of the United States to declare a *disaster area* or a state of emergency and send in federal aid. The governor has the power to utilize the National Guard to assist in any emergency situations. The LEPC and the SERC are immediately involved in the disaster.

In several states, Civil Defense acts as an arm of administrative officials to help provide adequate coordination of disaster relief efforts. Specialized emergency functions are carried out by various groups during the emergency. The Office of Civil Defense provides emergency staffing, support communications, emergency public information, and disaster warning system. The police and sheriff departments maintain law and order, control traffic flow, control access to the disaster scene, provide emergency public information, and provide necessary emergency services in a relief effort where needed. The fire department extinguishes fires, acts to prevent fires where gas may be leaking or other hazards are present, and also in many communities provides specialized emergency assistance through their paramedic teams and ambulances.

The streets department, or public works department, helps to clear debris, removes hazardous materials, and furnishes water for drinking and firefighting. The welfare department gives food, clothing, and emergency assistance; assists in registration of victims; and helps to coordinate the efforts of private welfare groups. The health department coordinates all emergency health and medical operations, provides specialized environmental services in food, insect and rodent control, housing, hazardous waste removal, solid waste removal, water and sewage treatment, testing, and accident hazards. This department also may provide emergency treatment and assist in the distribution of food and safe water; and is very much involved in intense public health education efforts that advise individuals how to avoid hazards due to a disaster.

The voluntary health agencies, such as the American Red Cross, provide emergency food, clothing, medical aid, bedding, shelter, and other essentials. These agencies may also provide assistance in the building and repair of homes, household furnishings, medical and nursing care, etc. A long list of additional groups are essential in emergencies. They include security services of various universities, Salvation Army, new car dealers, cab companies, funeral homes, community ambulance services, Army and Air Force National Guard, and Army and Naval Reserve units.

Six major phases of a community disaster plan include (1) security and traffic control, which is the basic function of the police, Army Reserve or National Guard units; (2) rescue operations, which is the basic function of the fire department, police, Army Reserve, and National Guard units, as well as volunteers; (3) transportation of casualties, which is the basic function of the ambulance services, police department, fire department, community ambulance associations, and other community groups; (4) communications, which is the basic function of the police department, as assisted by appropriate Army Reserve and National Guard units, as well as Civil Defense (citizens band [CB] operators may also be of major assistance in an emergency); (5) people with cellular phones, computers with modems, and fax machines; and people who come from all the preceding groups plus all voluntary organizations, which are of extreme importance; and (6) public information, which is generally the basic function of the mayor's, county commissioner's, or governor's offices. This function may in turn be given over to the Red Cross or other such agencies.

Additional people, as well as communications equipment, may be made available through the various city or county departments and through such groups as the telephone company, trucking companies, cab companies, and utility companies that have radio communication hookups.

Hospital Disaster Plans

All hospitals should have a disaster plan for external disasters and for internal disasters. A plan developed by the Terre Haute Regional Hospital of Terre Haute, IN, is recommended as a model for other hospitals. Although the plan is not discussed in its entirety, sufficient detail is now presented to give the practitioner a view of how internal and external disaster plans should operate.

The intent of an external disaster plan is to coordinate the activities of a community hospital with all available community emergency services and to provide assistance to other agencies during mass casualties. The plan provides for the coordination of available beds and supplies, backup medical support services, coordination with communication and transportation systems, and other logistical activities.

A disaster is any accident or series of accidents, usually catastrophic in nature, where large numbers of people need medical attention in a relatively short time period. Because most disasters occur with little or no warning, a hospital must be prepared, through proper prior planning and organization, to quickly modify its schedule to receive the sudden influx of patients, many of whom may be severely injured. The hospital function is to receive the injured person, provide immediate treatment, provide continuing care, and at the same time to continue the care for patients already in the hospital. When hospital staff members are informed of a disaster, they should determine the location, number of casualties, type and severity of injuries, and estimated arrival time of the injured at the hospital. The administrator, director of nursing, and medical director should be notified as quickly as possible. The administrators or their alternates or night nursing supervisors, if they are unable to reach the other personnel, should put the plan into action.

Notification of personnel to report to the hospital is extremely important. The hospital administrator or person in charge should notify the switchboard operator to make a general announcement that the hospital disaster plan is to be put into effect. The announcements should be made every 10 sec for three distinct times. A special disaster call list should be available to the switchboard to start contacting special individuals who should then report to the hospital. The first available individuals should start notifying doctors listed on a special disaster call list for doctors. Department heads should then be notified to start calling in personnel who will be needed. The main control center and communications center should be located in the administrator's office, with the ranking administrative person coordinating and directing the plan with the cooperation of the highest ranking medical staff officer.

Special locations should be set up within the hospital. These include:

1. The triage or sorting area should include the triage team consisting of an emergency room doctor, emergency room nursing supervisor, director of medical records, and other personnel as needed. The patients should remain within the triage area until moved by the triage physician.
2. The first aid station should be established to give first aid where needed; the patients should then be discharged and routed out of the hospital through an exit away from the incoming patients.
3. Special shock, burn, and trauma stations should be established for these types of patients.
4. Surgery stations should be established to receive patients who will be sent on for surgical techniques.
5. A mass-casualty holding area should be established for the overflow of patients.
6. A morgue should be established for those who are brought in dead or those who die in the hospital.
7. A special family and visitor holding area should be established where these individuals can wait for their family members and also receive information concerning them.

Communication systems in an emergency are extremely important. Several types of communications should be utilized, including:

1. Switchboard, which can be keyed into the entire hospital and also to the outside
2. Public telephones
3. Messenger service composed of individuals who are not directly involved in medical care
4. Hospital emergency radio network, which in the state of Indiana is known as IHERN; unit can be monitored and utilized in the emergency room, with a transmitting radius of 70 miles
5. Public information center to help provide necessary information, solicit blood donations, and provide information for the news media
6. Use of cellular phones owned by private individuals

Each of the various groups within the hospital has specialized assignments. The medical staff assigns the triage team directed by a trauma surgeon. The staff also assigns physicians to various treatment areas and physicians to discharge previously hospitalized patients who can be released.

The nursing staff, under the director of nursing, is responsible for overall direction of nursing personnel. One licensed nurse must be kept in each of the existing hospital units as a minimum, whereas the others are sent to a central place to be utilized as needed. In the specialized units, such as intensive care and intensive coronary care, additional nurses have to remain on duty.

Medical records personnel and business office personnel are responsible for completing the appropriate disaster tags, establishing and updating casualty lists, and maintaining proper records. The personnel director coordinates all personnel at a central location. The admitting supervisor maintains a complete and current available bed status and moves patients as necessary. Engineering and maintenance personnel are responsible for maintaining power, emergency water, heating or cooling, waste disposal, and basic sanitation.

The executive housekeeper should provide people to distribute and set up all disaster equipment; assist in the transportation of disaster victims; maintain adequate stretchers, wheelchairs, and cots in the triage area; assist in keeping rooms clean; and carry out necessary cleaning and disinfecting procedures where necessary.

Security guards are responsible for securing entrances; providing traffic control, including the triage area; routing relatives to the main visitor and family areas; and keeping other individuals away from the hospital.

The pharmacy is responsible for setting up anticipated drug and medication supplies in areas where casualties may be routed. The various other departments of the hospital, including central supply, laboratory, radiology, respiratory therapy, and physical therapy, should be ready to operate under emergency conditions. Food service must provide food for the existing patients, existing medical and nursing staff, additional medical and nursing staff, and disaster victims.

The laundry is responsible for providing adequate linens and materials needed in the temporary areas. Social service carries out public relations activities and assists in the public information center. During a disaster, when patients can be removed from the hospital and sent home, they should be, providing room for other patients.

The previously mentioned plan is a sample *external* disaster plan. The plan that is discussed now is a sample *internal* disaster plan. The internal disaster plan consists of establishing fire safety hospital floor plans and evacuation routes, disaster notification systems, evacuation procedures, guidelines for radiation disasters, and bomb threat procedures.

Fire safety is the job of every individual within the hospital. Many fires are easily extinguished if discovered early enough. It is essential that all individuals within the institution have a thorough understanding of fire safety, the fire alarm bell system, the various assignments in the event of fire, and how to immediately notify the proper individuals in the event of fire. The first responsibility is to remove the patients from immediate danger. It is not necessary to have permission to move such a patient. When a fire is discovered, the door should be closed; the individual should go to the nearest telephone and dial a predesignated number, such as 7; and then calmly give the operator the location, by zone, department, or room number, of the fire. The hospital fire alarm system automatically closes smoke safety doors, sounds an alarm signal, transmits a signal to the fire department, and shows the operator the general location of the fire. Beside using the hospital fire alarm system, the backup emergency call system to the operator must be utilized to confirm the fire.

It is essential to confine the fire as soon as possible. Specific instructions are available to those who are responsible for fire safety in the use of firefighting equipment. In the event of fire, the elevators should not be used without the authority of the fire department. Close as many doors as possible to contain the smoke and instruct visitors to remain with the patients or assemble them in the nearest lounge areas. The hospital has a fire brigade that responds. This group is supplemented quickly by the local fire department.

Evacuation plans should be drawn in such a way that each individual in each area of the hospital knows precisely how to remove patients from a fire area. Again it is recommended that a plan similar to the plan of Terre Haute Regional Hospital be utilized. This particular plan shows detailed diagrams of each of the floors and wings of the hospital and where to move patients.

When evacuation is needed, hospital administrators or their alternates, in conjunction with the chiefs of the medical staffs or their alternates, follow the evacuation plan established by the hospital. They determine which patients can be discharged to their home, which patients remaining should be removed to other units within the building or to a temporary hospital set up outside, secure litters from the triage closet, and determine the evacuation procedure down stairwells if elevators cannot be used. The radiation procedure that has been established for the hospital is discussed in the area under specific environmental health measures.

When bomb threats are made, as soon as the warning has been received, the switchboard operator or other person receiving the threat should try to keep the individual talking as long as possible, determine if the person is male or female, young or old; note distinguishing voice characteristics; be alert for distinguishing background noises; ask where the bomb is, what type it is, and when it will explode; and note if the caller seems to have knowledge of the hospital. Key personnel should be notified immediately. However, the hospital paging system should not be used to announce a bomb threat. In addition, the police, fire department, FBI, and bomb squads should be notified. The person in charge should assist proper authorities in making a complete check of the hospital. If necessary, this person should make determinations of where and how to move patients.

Personnel should always be alert for unusual packages, boxes, and other materials. All public areas such as lobbies, solariums, cafeterias, stairwells, and restrooms, should be thoroughly examined. Storerooms should be checked thoroughly. If areas are locked and unavailable to the public, these areas should be eliminated from the search unless the individual indicates the bomb is located there. Elevators should be kept available for use by the authorities. If an object appearing to be a bomb is discovered, it should not be touched. The area should be cleared and the object should be isolated as much as possible by closing all doors. Only properly trained individuals should decide where or how to move the object. When evacuation is necessary, it should follow the proper evacuation techniques established for fire disasters.

Emergency Vehicles and Emergency Communication Systems

As has already been mentioned, various emergency vehicles and emergency communication systems exist, such as police and fire vehicles, and vehicles from utilities and cab companies. In addition, a considerable number of CB enthusiasts are available who could be extremely valuable in the event of a disaster. The CB radio and cellular phone become instant forms of communication that can be utilized to bring in necessary professional personnel to a hospital or other types of assistance into areas of disasters. Automobile dealers who have vans available can provide the vehicles for additional transport in serious emergencies. Communication systems also exist at various National Guard and Reserve units, as well as Civil Defense units.

ENVIRONMENTAL RESPONSE TEAM

The environmental response team's major functions are:

1. Maintain a 24 hour a day activation system
2. Dispatch team members, when requested, to emergency sites to assist where needed
3. Provide consultation on water and air quality, toxicology, interpretation and evaluation of data, engineering, and scientific studies
4. Develop and conduct site-specific safety programs
5. Provide specialized equipment
6. Assist in developing new technologies
7. Train federal, state, and local government officials and industry in the latest technologies

Yearly the EPA responds to hundreds of requests for emergency assistance and cleanup. In the field, the team conducts studies for cleanup, containment, or disposal of hazardous waste; develops alternative systems; and implements new technical approaches as necessary. Environmental decisions are made after assessing toxicity of substances and determining what the priorities should be for preventing problems with people and the environment.

MENTAL HEALTH NEEDS IN DISASTERS

Mental health needs during major disasters comprise an area recognized by the federal government as very important. Emergency workers must be trained to deal with problems including depression, grief, anger, guilt, apathy, fear, burnout, bizarre behavior, and suicide. Individuals typically go through these four phases: (1) The heroic phase, which lasts for several days, is where people are altruistic and do things beyond their normal ability. (2) The honeymoon phase lasts from 1 week to 3 to 6 months after the disaster. For the survivors, a strong sense of shared experience exists during a dangerous and catastrophic experience. People work hard to clear the wreckage and resume life. (3) The disillusionment phase lasts from about 2 months to 1 to 2 years. People have strong feelings of disappointment, anger, resentment, and bitterness when delays occur and promises are not kept. The sense of shared disaster is leaving and individual problems are growing. (4) The reconstruction phase lasts for several years. People realize that they need to solve the problems of rebuilding their own lives, homes, and businesses. Fortunately the population is primarily normal. However, certain individuals may need further assistance.

SPECIFIC ENVIRONMENTAL HEALTH MEASURES

Housing

In emergency situations, it is necessary not only to provide temporary housing for individuals who cannot stay within their own homes but also to evaluate the housing that has been affected by the emergency. Temporary housing may consist of emergency fallout shelters, schools, arenas, warehouses, churches, or other sizable areas. The emergency housing must be clean, of sound construction, and accessible to the homeless and to the various emergency workers and suppliers. There is a need for beds, cots, clean blankets, clean sheets, proper water supplies, toilet facilities, heat, ventilation, light, food, and medical services. Each emergency unit should have nursing personnel and other workers who can supervise the operation of the unit, provide for immediate care, help maintain a sense of good morale, and provide whatever services, equipment, or supplies may be needed. It is essential that a proper communications system be located within the emergency housing units. It is important to maintain good personal hygiene and also to supervise the proper use and disposal of water, food, and various materials. Individual homes may also be utilized to take people who cannot go back to their own living quarters.

Where emergency housing is utilized, 50 ft² of space for each bed, 500 ft³ of air to each person, and 6 ft of distance between beds should be provided. If tents are used, the minimum size should be 16×16 ft and they should house no more than 6 people each.

Once the immediate emergency has passed, the Department of Housing and Urban Development can provide trailers for individuals whose houses have been destroyed or whose houses have been inundated and need considerable work before they are in livable condition. The cleanup of the houses should be carried out under the supervision of environmental health practitioners. These individuals, in consultation with building inspectors, fire inspectors, and plumbing inspectors, should decide whether houses are either unsafe or unfit for human habitation. The houses that cannot be reinstated to a livable condition or that constitute an immediate hazard should be destroyed. Assistance must be given to the population physically and also in building morale, to help them through the crisis. All portions of the house that have been inundated by floodwaters should be carefully scrubbed and disinfected. The rugs or other materials that cannot be disinfected should go to solid waste disposal sites.

Food

Frozen foods that have defrosted should be discarded. Refrigerated foods that have been kept below 41°F and that have not been contaminated by river water or other known or potential hazardous materials can be utilized. If perishables have been kept out of refrigeration, they should also be condemned. Foods that are in boxes, plastic bags, or any other containers other than glassware and cans; and that have been contaminated with water, by smoke, or by chemicals, should be condemned.

Where water from firefighting equipment or rainwater has come in contact with canned or bottled food, and where the cans or bottles are tightly sealed, they should be thoroughly washed with hot soapy water, rinsed in clear water, and then submerged in a chlorine solution for at least 5 min at a level of at least 100 ppm. If cans have lost their labels they should also be discarded. Utensils, pans, and other equipment, as well as storage, preparation, or food-serving areas, should be thoroughly washed and decontaminated before use. Plastic utensils and containers should be thrown away. Dishes, metal containers, and equipment utilized for food preparation, storage, or service should be thoroughly washed and decontaminated before they can be utilized. It is recommended that the equipment and utensils go through at least two complete washing and rinsing cycles in a dishwasher or be washed at least two complete times and then sanitized in chlorine at 200 ppm for at least 10 min before use. All foods that have been condemned should be removed to a sanitary landfill. The types and quantities of food should be determined and listed on appropriate sheets. The environmental health practitioner should accompany the food to the landfill site to make sure that it is properly discarded.

Water

Water becomes a major concern during an emergency situation. Where the water is even suspected of contamination by microorganisms, it should be boiled for a minimum of 10 min before use. The water can then be cooled and poured from one container to another to reaerate it and give it a better taste. Other emergency water-disinfecting techniques include the addition of 2 drops of household laundry bleach to a quart of clear water and 4 drops of household laundry bleach to cloudy water. The mixture should be thoroughly stirred or shaken, and then allowed to stand for 30 min. If a slight chlorine odor is not present at the end of 30 min, repeat the procedure for another 15 min and then again if necessary. A 2% tincture of iodine, which may be found in medicine chests or first aid kits, can be added at a rate of 5 drops to a quart of clean water or 10 drops to a quart of cloudy water. Mix the water thoroughly by stirring or shaking and allow it to stand for 30 min before drinking. It is possible to obtain iodine or chlorine tablets from either drugstores or sporting goods stores. Follow the directions on the label.

Where water has been contaminated with chemicals, boiling or chlorination cannot result in a pure water supply. The same problem occurs if the water has been contaminated with either oil or gasoline. Therefore, if chemical or oil pollution occurs, the water supply should not be utilized under any condition. Other sources of water should be provided. Emergency sources of water that are acceptable include bottled water, water that has been trapped in hot water heaters if the heater has not been inundated by floodwater, and other water-holding containers that have contained potable water that has not been subjected to flooding conditions. Usually, in major emergencies, because the water systems are disrupted, water is brought in by tank truck to the community. The tankers that are permissible are milk tank trucks that have been thoroughly washed and sanitized before using for hauling water. Other types of tank trucks can create serious potential hazards. If a well has become contaminated, water should be drawn from the well, disinfected with chlorine, and then poured back into the well. A slurry of chlorinated lime, calcium hypochlorite powder, or a mixture of sodium hypochlorite liquid in water may be utilized. The supernatant liquid should be poured down into the well but not the inert materials. After the chlorination has taken place, the water should run for a period of 24 hours, and then the well should be checked for chlorine odor. If a known contaminant exists, then a sample should be taken. Individuals should not consider the well to be safe until the sample comes back negative.

Insect and Rodent Control

During emergency situations and for a period of time thereafter, insects and rodents may proliferate at an alarming speed. The chance of the spread of disease increases sharply. As sewage systems are disrupted and as riverbanks are disturbed, rodents leave these areas and head for other sources of food and harborage. Obviously, after a disaster, considerable solid waste accumulates including materials that can serve as a food supply for the rodents. It is necessary not only to remove the source of rodent food and harborage by removing solid waste as quickly as possible, but also to use an extensive baiting program in the area that has been affected by the emergency situation.

Floods and heavy rains result in innumerable situations where standing water occurs. This leads to a vast amount of mosquito breeding, which in turn can cause considerable disease outbreaks. It is important that these situations be corrected as rapidly as possible. Because it is almost an impossible task to eliminate standing water, it is necessary to carry out an extensive spraying program. It may be necessary to use aircraft spraying in areas that are not readily accessible or if the area of mosquito breeding is so extensive that it cannot be covered by crews on the ground. Obviously, fly problems multiply rapidly in the aftermath of an emergency. Therefore, not only must solid waste be removed quickly and efficiently, and disposed of efficiently, but also all decaying food and other organic material must be removed quickly and efficiently. It may be important to conduct spraying programs to eliminate potential hazards and diseases from flies.

Sewage

During many emergencies, sewage systems are either disrupted, destroyed, or inundated by water. Because the normal removal of sewage from homes cannot occur, contamination of various areas may happen and the strong potential for an outbreak of disease is created. Two types of techniques are needed in the emergency situation. One is to put back into operation as quickly as possible the sewage treatment system and to disinfect all areas with chlorine where sewage may have come in contact with materials and structures that humans associate with. The second technique is to provide temporary privies, portable toilets, and holding tanks for the use of individuals during and in the aftermath of a disaster. The situation may be so bad that sewage may have to flow into the receiving stream while sewage treatment plants are put back into operation.

Solid and Hazardous Waste

Solid waste has already been referred to in several previous sections. During the course of disasters, solid waste disposal areas may become inundated and therefore may be unusable for the disposal of additional solid waste. It may be necessary to haul the waste to other areas where temporary landfill sites can be established or where incinerators may be available. In any case, solid waste disposal has to follow all the normal disposal procedures set forth in Chapter 2. Where hazardous substances are involved, they must be handled in the special manner discussed in other parts of this book.

It is essential that great care be given to the prevention of leaching of materials from the landfill sites into bodies of water, as well as the prevention of insect and rodent problems. Appropriate special pest control procedures should be set up to eliminate this potential hazard.

Radiation

In the event of a radiation disaster, the director of radiology or the chief radiological technologist of the hospital and the director of laboratories of the hospital should work with the appropriate public health officials and other radiation experts to determine the extent to which individuals have been exposed or injured by radiation and the existing potential hazards. Casualties from radioactive areas should be kept in a special area and should be closely monitored prior to admitting to special units of a hospital. Before leaving the contaminated room, the individuals should be hosed down thoroughly with water; should undress and leave all of their clothing, including underclothes and shoes in the radioactive area; should redress in clean linens; and should be remonitored before they are removed. If possible, showers should be taken at the site of the radiation exposure. The individual can then be removed by special ambulance to the hospital and then to the special hospital rooms for treatment. The ambulance has to then be thoroughly decontaminated by appropriate washing techniques and monitored before it can be put back into service. Wherever these affected individuals go, the area must be completely monitored to make sure that they are not spreading radiological material.

If individuals have been exposed to 100 R, they can experience nausea and vomiting in about 24 hours in about 5 to 10% of those exposed. Blood cell changes can also occur. Where individuals have been exposed to 150 R, nausea and vomiting can occur in 24% of the patients in about 24 hours. Where exposure has been to 200 R, nausea and vomiting occur in about 50% of the patients in 24 hours. In an exposure of 300 R, nausea and vomiting can occur for all exposed on the first day, with 20% of the individuals expected to die within a period of 2 to 6 weeks after exposure. At 450 R, nausea and vomiting can occur in all exposed, and in the absence of treatment the death rate can be expected to be 50%. At 600 R, nausea and vomiting can occur in all exposed in about 4 hours. The death rate may go to 100%. All patients admitted to the hospital should receive the necessary medical care, as stipulated by the medical staff.

SPECIAL ENVIRONMENTAL HEALTH PROBLEMS

Nuisance complaints are one of the most troublesome and yet one of the more important functions of environmental health personnel. Generally, boards of health or health departments have the authority to obtain corrective action in cases of nuisances caused by filth, hazardous substances, or potential health hazards. The specific authority may be present within the health code, or the health department may utilize common law related to nuisances. A nuisance complaint may be anything from stagnant ponds to lots full of weeds, to removal of dead animals, accumulation of trash, overflowing sewage, rats, mosquitoes, fly infestations, dilapidated property, presence of dog feces, and keeping of animals within city limits.

The environmental health practitioner is faced with an enormously touchy prob lem, because nuisance complaints may be real or they may be simply grudge fights between neighbors. It takes tact, diplomacy, and the ability of a King Solomon at times to resolve these problems. The environmental health practitioner must be careful in evaluating the problem; be accurate, calm, and attempt to mollify the individuals involved; and yet bring about the necessary correction. This particular area is especially good for students to work in, because it gives them excellent exposure to nontextbook cases, where they must deal with on-site problems and with citizens who may be extremely upset. The worst approach in handling the problem is to use legal action. The best approach is to make a good determination of what the problem is and to attempt to get individuals to take corrective action on a voluntary basis. Obviously, if this does not work, some form of legal action, varying from fines to court procedures, may be necessary.

It is well for the health department to develop a file, which should be indexed by name and by location of each of the complaints that come in during the course of a given period. The complaints should be put on a map of the community. Colored pins should be utilized to indicate the types of complaints. The complaint system should be set up in such a way that the person receiving the complaint, usually the secretary, should record all information concerning the location of the complaint, nature of the complaint, and name of the complainant, along with address and phone number.

It is important for public relations reasons, even though a definite health hazard may not exist, that the complaint be handled as rapidly as possible. It is usually recommended that the complaint be investigated within a 24-hour period and that the complainant be notified of the actions that are to be taken. However, the environmental health practitioner should not proceed from the area where the complaint is occurring to the complainant's home to notify the person of what is occurring, because this may lead to serious problems between neighbors. The complainant's name should never be released by the health department to anybody, because this may also create unpleasant difficulties. If the complaint involves emission of toxic fumes, toxic substances, or areas that are dangerous or unfit for human habitation, then immediate action may be necessary. In this case, the practitioner should advise the Public Health Commissioner through appropriate superiors or inform the board of health through the appropriate superiors to get immediate action to either suspend operations, clean up the problem, or carry out what other steps are necessary.

In most nonhazardous cases, if the complaint has been evaluated properly and if the individuals responsible for causing the nuisance have been properly advised of the nuisance and told how to correct it, some time should be given to them to make the necessary correction. The amount of time depends on the type of complaint. It is recommended that after the form has been completed by the secretary, a copy be placed in a tickler folder, and a copy be given to the appropriate environmental health practitioner. The individual, after completing the study, which includes determining all of the conditions existing, making a map on the back of the necessary study form, and issuing orders, should then return a copy of the form to the environmental office. These practitioners should determine when to make the restudy. If the restudy is not made within a reasonable amount of time, the secretary should advise the individual practitioner that the time has elapsed and that the restudy should be conducted. On completion of the restudy, if the situation has still not been corrected, the necessary appropriate legal action should start. It must be emphasized that the individual should be handled as carefully as possible.

The following specific procedures should be followed whenever a complaint is called in: record on the complaint form the complainee's name, address, and phone number; obtain specific details of complaint; environmentalists should identify themselves to all individuals when arriving on the scene of the complaint; list specific details of the study, including maps and pictures where needed; list all recommendations and when the specific recommendations must be completed; specify if the complaint is justified; make necessary referral to other governmental agencies where needed; set up special conferences or meetings with individuals when needed; make necessary restudies; use legal action, if necessary but only as a last resort; and file by address and cross-file by name of complainee (Figure 8.1).

Environmental Investigation		County

Location of Complaint	Name of Complainant Address of Complainant	Phone No.

Owner of Property	Address of Owner	Phone No.

Date of Complaint Registered Time	pm am	District or Township

Problem	Date of Referral

☐ Rodent ____ ☐ Insect _____ ☐ Air _____ ☐ Housing Dept. _____ ☐ State _____
☐ Sewage ____ ☐ Water _____ ☐ Food ____ ☐ Pub. Works Dept. ____ ☐ Wector
Program ____
☐ Housing ____ ☐ Hazardous ☐ Solid ☐ Water and
Substance ____ WWaste ____ Sewage Dept. _____ ☐ Other_____
☐ Other _____

Description of Complaint _____

_____

Conditions Observed _____

_____

_____

_____

Responsibility
☐ Owner   ☐ Occupant   ☐ City or County   ☐ Other _____

Name	Telephone	Address

Remarks and/or instructions given (include persons contacted) _____

_____

Recommendations _____

_____

Signature of Env. Health Practitioner	Date Surveyed	Time	am pm

(drawing on next page)

Figure 8.1   Typical complaint form.

Vast varieties of unusual situations are handled by the health department, depending on specific codes and ordinances. In all these situations, environmental health practitioners are not law enforcement officers. They carry out the routine functions as they would under any other environmental health program. In the case of turkish baths and massage parlors, it is their responsibility to evaluate water samples, equipment, or other types of environmental problems. It is not their function to become police officers and determine whether these operations are legal. If these facilities are licensed and meet environmental health standards, then they are legally operating. If the facilities are carrying out other types of functions that relate to moral problems, the environmental health officials should not become involved. This is a police action and not a health matter.

Barbershops and beauty shops are licensed by the state. The environmental health practitioner may be involved in determining the general housekeeping, equipment cleaning, and sanitizing; health of the operator; and other similar concerns. Again, the practitioners are not beauticians or barber inspectors. It is their function to be concerned with the environmental health aspects of these areas and not other types of problems.

In some health departments, the practitioners are still responsible for the evaluation of bedding, mattresses, and upholstery. Again, if this is required by law, the individual should carry out this function. However, wherever possible this function should be removed from the environmentalist's duties. Also, in the event weeds are found on lots that may contain narcotic substances, the weeds should be destroyed by the health department or police department in cooperation with the fire department.

CHAPTER 9

Major Instrumentation for Environmental Evaluation of Ambient Air, Water, and Soil

BACKGROUND AND STATUS

Major sampling and analytic instruments used to evaluate ambient soil, water, and air matrices are listed in Tables 9.1 and 9.2, respectively. Numerous physical, chemical, and biological parameters are evaluated for the three environmental matrices. Soil and water are commonly evaluated for various inorganic and organic chemical agents and biological organisms. Soil and water also are sampled to evaluate common physical parameters including pH, conductivity, turbidity, and percentage of solids. Air is commonly evaluated for inorganic and organic chemical particulates, gases, and vapors.

This chapter focuses on the commonly evaluated parameters and the related instrumentation required for sample collection and analysis. Summaries of major instruments used for characterizing the ambient soil, water, and air matrices are presented. Note that many instruments used to evaluate the air matrix, especially but not exclusively for indoor settings, are covered in Volume I, Chapter 12, which included major instruments for toxic and flammable chemicals, pathogenic microorganisms (i.e., bioaerosols), sound, and ionizing and nonionizing radiation.

SAMPLE COLLECTION

Collection of Soil Samples

Soil samples can be collected via both simplistic and relatively more complex methods. A simple method involves use of devices such as scoops or shovels to collect the soil sample by scraping or digging soil from upper surfaces. More complex methods involve use of manual devices called augers, corers, and triers (Figure 9.1). Powered augers also are available for sampling the denser and deeper strata of soil. Augers, corers, and triers are pushed, hammered, or twisted, or both

Table 9.1   Sample Collection Devices for Soil, Water, and Air Matrices

Environmental Matrix	Sample Collection Device
Soil	Shovel
	Scoop
	Auger
	Corer
	Trier
	Rigid glass or plastic jar
	Nonrigid plastic or paper bag
Water	Rigid glass or plastic jar or bottle
	Nonrigid resealable plastic bag
	Thief tube
	Coliwasa tube
	Bacon bomb
	Bailer
	Vacuum or peristaltic pump
Air	High-volume (Hi-Vol) samplers with filters for suspended particulates
	Dust fall bucket for settleable particulates
	Air sampling pumps with solid adsorbents
	Air sampling pumps with liquid absorbents
	Evacuated rigid glass or stainless canisters for gases and vapors
	Air sampling pumps with nonrigid plastic bags for gases and vapors
	Direct-reading instantaneous sampling and analysis meters for gases, vapors, and particulates

Table 9.2   Sample Analysis Devices for Common Parameters Measured in Soil, Water, and Air Matrices

Environmental Matrix	Parameter	Sample Collection Device
Soil and water	pH	pH meter
Soil and water	Conductivity	Conductivity meter
Soil	Solids (texture)	Stacked sieves and electrobalance
Soil and water	Total solids	Porcelain crucible, oven, and electrobalance
Soil	Volatile solids	Porcelain crucible, oven, furnace, and electrobalance
Water	Suspended and dissolved solids	Porcelain crucible, membrane filter assembly, oven, and electrobalance
Soil and water	Microorganisms	Membrane filter apparatus, nutrient growth media, thermal incubator, magnifier, and microscope
Water	Dissolved oxygen	Dissolved oxygen meter
Water	BOD	BOD bottle, dissolved oxygen meter, refrigerated incubator
Soil, water, and air	Organic chemicals	Gas chromatograph; UV/Vis spectrophotometer
Soil, water, and air	Inorganic chemicals (nonmetals)	UV/Vis spectrophotometer
Soil, water, and air	Inorganic chemicals (metals)	Atomic absorption spectrophotometer; inductively coupled plasma emission spectrophotometer
Air	Total and respirable particulates	Electrobalance

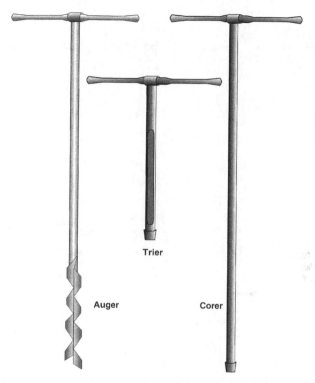

Trier

Auger   **Corer**

Figure 9.1   Auger, corer, and trier devices for collecting soil samples.

through the top surface of soil down into the lower strata and then removed to collect a sample representing vertical layers of soil. Collected samples are typically deposited into a glass or plastic sample jar, resealable plastic bag, paper bag, or other suitable containers.

Collection of Water Samples

Water samples also can be collected relatively simply by immersing a container, such as a rigid plastic or glass sample jar or bottle, or a nonrigid resealable plastic bag, into the water source and removing a sample (Figure 9.2). A thief tube and a Coliwasa tube are similar water sampling devices (Figure 9.3). Indeed, the devices are also used for sampling other liquids such as potentially hazardous chemicals contained in drums and tanks. Both devices are hollow, open-ended, rigid glass or plastic tubes approximately 1 meter long typically with a 1- to 2-cm internal diameter; the Coliwasa tube is tapered at one end. One end of each device is inserted into a source of water or other liquid material. For the thief, the other open end of the tube that is not immersed may have an aspirator bulb attached to create suction for pulling the sample up into the tube. Alternatively, the fluid can be contained in the tube by placing the thumb over the open end prior to withdrawing the thief from

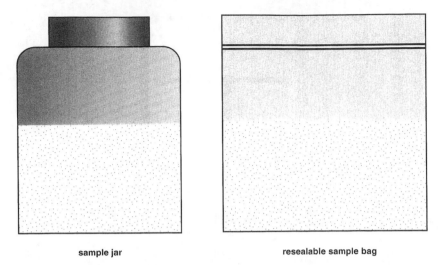

sample jar   resealable sample bag

Figure 9.2 Rigid plastic or glass sample jar and nonrigid plastic bag for collecting water samples.

the liquid. The tapered Coliwasa tubes may have an interior rod with a bulb or washer at the end. The rod is pushed into the Coliwasa tube to occlude the open end that is immersed in the liquid to contain the sample in the tube on withdrawal. Following withdrawal of either the thief or Coliwasa from the liquid, the sample collected is discharged into a glass or plastic sample collection jar, bottle, or plastic bag.

Other devices useful for manually collecting water from increased and varied depths are the bacon bomb and bailer (Figure 9.4). Each device is attached to a wire or rope, submerged into the water or containers such as drums or tanks of other liquids to a given depth where the sample is collected and then retrieved. This can be repeated to collect samples from various depths.

Water and other liquids also can be sampled by using flexible tubing and peristaltic or vacuum pumps. For example, groundwater can be sampled by inserting a flexible tube or hose into a drill well and using a pump to transport the subsurface water from the aquifer to the surface for sample collection and containment. Automated devices are used to collect surface water or groundwater samples at specific times during a defined period to generate a profile of water quality vs. time.

Samples collected at different times and different locations can be combined to generate what is called a composite sample. The composite sample represents an integrated sample averaging fluctuations in concentrations due to time or variations in concentrations due to location.

Collection of Air Samples for Particulates

Summarized in Volume I, Chapter 12 are direct reading aerosol meters for instantaneous sampling and analysis of airborne dusts and fibers, as well as active sampling powered air sampling pumps and filter media used to collect low volumes

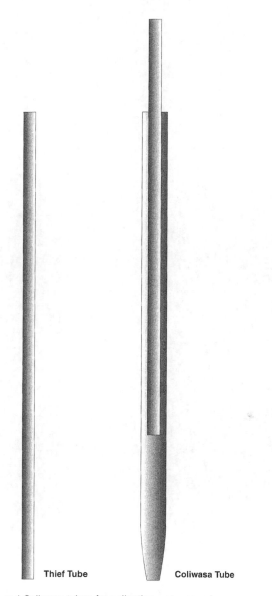

Figure 9.3 Thief and Coliwasa tubes for collecting water samples.

of air for particulates, especially in occupational environments. For larger sample volumes of ambient air, however, a major electronic particulate sampling device is the high-volume (Hi-Vol) sampler.

This device consists of a programmable ultrahigh volume air sampling pump or blower and a preweighed cellulose or glass fiber filter medium contained within a small houselike structure or shelter (Figure 9.5). The pump pulls air in under the

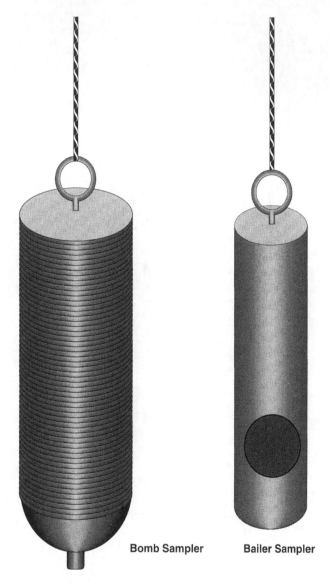

Bomb Sampler   **Bailer Sampler**

Figure 9.4   Bomb sampler and bailer sampler for collecting water samples.

eaves of the shelter's roof and through the filter. Air is sampled at high flow rates in excess of 1 m³/min for unattended periods typically not less than 24 hours. This results in a high volume of air passing through the approximately 20- × 25.4-cm filter. Particulates with diameters <100 μm, including respirable particles <10 μm, that were suspended within the airstream are collected by the filter. Following sampling, the filter is removed, desiccated to remove moisture, and weighed. The

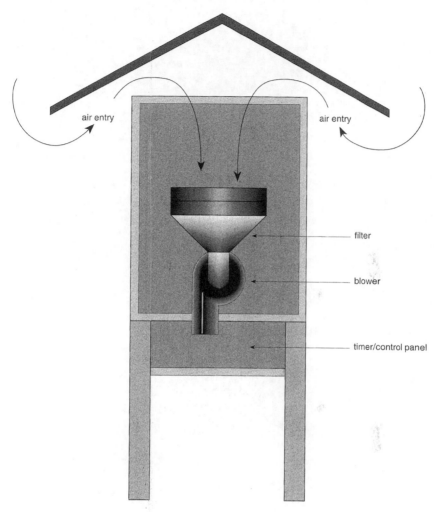

air entry

air entry

filter

blower

timer/control panel

Figure 9.5   High-volume (Hi-Vol) air monitoring device for collecting samples of suspended particulates.

postsampling weight of the filter actually represents the weight of the clean filter plus the deposited particulate. The amount of particulate collected by the filter is determined by subtracting the presampling weight from the postsampling weight of the filter. By dividing the amount of particulate collected by the volume of air sampled, the concentration of airborne particulate for the period can be calculated and expressed as milligrams particulate per cubic meter of air.

A dust fall bucket is a passive monitoring device for collecting larger settleable particulates from the ambient atmosphere. Sample collection involves a relatively simple device consisting of an open container (bucket) with known dimensions positioned on a 1.2 meters high stand (Figure 9.6). Both water soluble and insoluble

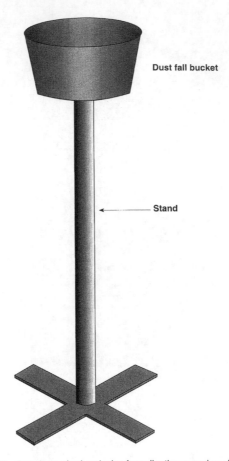

Dust fall bucket

Stand

Figure 9.6   Dust fall bucket air monitoring device for collecting samples of settleable particulates.

particles with diameters >10 μm can be collected. To collect soluble particulates, a known volume of reagent grade water treated with an algicide and fungicide to inhibit growth of unwanted organisms is placed in the container at the start of sampling. The sampler is placed in a selected acceptable location for a collection period of approximately 1 month (i.e., 30 days). Following sampling, the bucket is rinsed with a known volume of water that in turn is filtered using a preweighed filter. The filtered liquid is evaporated to measure the weight of soluble particulate. The filter is dried, and based on the difference between the presampling weight and postsampling weight of the filter, the weight of collected insoluble particulates can be measured. Total particulate equals the sum of insoluble plus soluble particulates. Based on the weight of particulate collected, duration of sampling, and sample collection area (i.e., area of the dust fall bucket), the insoluble, soluble, and total particulate fallout rates can be calculated. Fallout rates are commonly expressed as grams particulate per square meter per month (g/m²/mo) or converted to tons per square mile per month (tons/mi²/mo).

Collection of Air Samples for Gases and Vapors

Active sampling involving a powered air sampling pump and an adsorbent or absorbent medium can be used to sample for airborne gases and vapors. As discussed in Volume I, Chapter 12, common adsorbents are silica gel and activated carbon, and absorbents consist of various polar or nonpolar liquid solvents. Silica gel is a common solid sorbent media for adsorbing polar gases and vapors from the air. Activated carbon is another common solid sorbent media, but is used for adsorbing nonpolar organic gases and vapors from the air. An organic polymer, commercially known as Tenax®, is another common adsorbent for collecting nonpolar gaseous airborne organic compounds. The solid adsorbents are typically contained in sealed tubes that are opened at both ends prior to conducting sampling to permit air to pass through and across the sorptive material for collection and subsequent analysis (Figure 9.7).

Liquid absorbents are used to collect less volatile airborne inorganic and organic gases and vapors. The type of liquid used to absorb the gaseous air contaminants varies depending on the polarity of the compound. For example, polar solvents are used to collect polar vapors and nonpolar solvents are used to collect the less polar or more nonpolar vapors. During monitoring, the liquid sorbents are contained in glass devices called impingers or bubblers (Figure 9.8). The devices allow air to be pumped in through a hollow glass tube that is immersed in the absorbent. The air and gaseous contaminants mix with the liquid and the sampled contaminant is trapped or absorbed from the airstream and collected in the liquid for subsequent analysis.

Solid sorbents and liquid absorbents are commonly used for compound specific monitoring. Other rigid and nonrigid devices are available, however, to collect bulk samples of air (Figure 9.9). A special rigid stainless steel canister is evacuated to create negative pressure and placed in an area for collecting air samples. When the valve is opened, the ambient air purges into the canister to collect air for later detection and measurement of gaseous contaminants. A special nonrigid plastic bag with inlet and outlet valves also can be used with an active flow air sampling pump to collect a bulk sample of air. Air is pumped into the bag to collect the bulk sample that is subsequently transported in the sealed bag to a laboratory for analysis.

Several types of portable direct-reading instruments also are used to evaluate the atmosphere for gaseous chemical contaminants. These instruments are commonly used to instantaneously sample, detect, and measure airborne volatile organic vapors

air outlet   back section   front section   air inlet

Figure 9.7 Glass tube containing solid adsorbent for collecting samples of airborne gases and vapors.

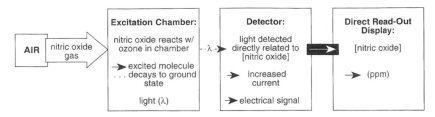

Figure 9.12   Summary of instantaneous monitoring and analysis of airborne nitric oxide gas using a chemiluminescence instrument.

Figure 9.13   Remote sensing FT-IR technology for instantaneous monitoring of airborne inorganic and organic gases and vapors over long distances.

Remote Monitoring of Gases, Vapors, and Particulates

Remote sensing of particulate and gaseous air contaminants is rapidly becoming a more common technology for instantaneous air sampling and analysis. Typical devices involve the generation of an electromagnetic signal through and across relatively long distances of atmosphere to a receptor or reflector (Figure 9.13). The presence of air contaminants in the atmosphere will interfere with the transmission of the electromagnetic energy. The remote optical sensing of emissions technology uses long, high-resolution IR or Fourier transfer infrared (FT-IR) wavelengths of electromagnetic energy to monitor inorganic and organic gaseous air contaminants. Gaseous contaminants each absorb a different wavelength of IR energy depending on the chemistry of each compound. Accordingly, certain wavelengths of IR energy are attenuated and absorbed by gaseous air contaminants that fall within the path of the IR beam. The remote sensor with FT-IR spectrophotometric capabilities detects and measures the transmitted IR energy and, in turn, indicates the degree of IR absorption by the atmospheric contaminants and the related concentration in parts per million. To date, over a hundred inorganic and organic gases and vapors can be detected and measured using this technology.

Laser Doppler velocimetry (LDV) is another type remote sensing technology. The LDV involves electromagnetic laser energy to detect and measure atmospheric particulates. Atmospheric particulates in the path of the projected laser beam cause light to reflect and scatter. The scattering of laser energy increases with higher atmospheric opacity due to the presence of suspended particulates.

Remote monitoring technology is very beneficial because it permits evaluation of plumes of contaminants discharged from stacks. In addition, it permits evaluation of air quality over a distance without need for using multiple sampling devices in various locations. In addition, the data represent "real-time" detection and measurement of contaminants because sampling and analysis are virtually instantaneous.

Stack Sampling for Gases, Vapors, and Particulates

Stack sampling refers to the collection of both particulate and gaseous agents emitted into the ambient atmosphere through discharge stacks typically associated with industrial sources and incinerators. Several configurations of equipment are used. In general, however, the equipment consists of a sampling probe connected to a filter, a series of sorbents, and a pump (Figure 9.14). Flow devices such as a dry gas meter and a Pitot tube with inclined manometer also are used to measure airflow rate through the sample probe and to measure air flow rate through the stack, respectively. This permits adjustments in the sampling train flow rate relative to the stack emission flow rate necessary for maintaining relatively equal flows or isokinetic conditions. Isokinetic conditions must be maintained to increase sampling efficiency and to collect representative samples.

The sampling probe is inserted through a hole drilled into the stack and collects samples at various traverse points across the diameter of the stack lumen. A pump pulls air emissions from inside the stack into the probe and through various sampling media. Various stack sampling train configurations are used. A common sampling media consists of a filter to collect particulate and sorbents (liquid absorbents or solid adsorbents) to collect gases and vapors.

SAMPLE ANALYSIS

Analysis of Soil and Water Samples for pH

The measurement of hydrogen ions in solution is referred to as pH. The pH scale ranges from 0 to 14 and is an indicator of acidic (pH 0 to <pH 7), neutral (pH 7), or alkaline conditions (>pH 7 to pH 14). Hydrogen ions (H^+) are released in aqueous solutions from disassociated acid (HA) materials (HA $\rightarrow$ H^+ + A^-). The pH of a solution or suspension equals the negative log of the measured concentration of hydrogen ions (pH = $-\log [H^+]$).

The pH of matrices such as soil and water can be determined using titration, chemical impregnated paper or strips, and most commonly a pH meter. The pH meter consists of a fluid-filled electrode probe attached to a volt meter with direct readout display (Figure 9.15). The probe or pH electrode responds to the difference in concentration of dissolved hydrogen ions in the medium under evaluation and the concentration of ions inside and indigenous to the solution inside the pH electrode. The difference in concentrations of ions inside vs. outside the electrode is called a membrane potential. The volt meter measures and displays the membrane potential or voltage typically in pH units. Because the actual potential across the pH electrode is measured, the analytic method is referred to as potentiometry. For soils, a measured quantity is mixed and suspended in deionized water to dissolve the available hydrogen ions. The pH probe is immersed in the soil suspension and the pH is recorded. For water, the pH probe is simply immersed in the water and the pH is measured directly.

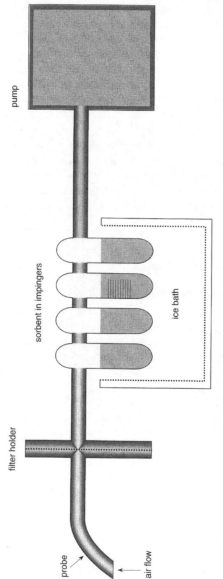

Figure 9.14   Stack sampling train for collecting samples of airborne particulates, gases, and vapors from within a stack.

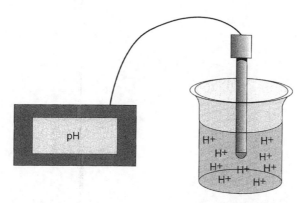

Figure 9.15 pH meter for measuring the concentration of hydrogen ions (H^+) in solution.

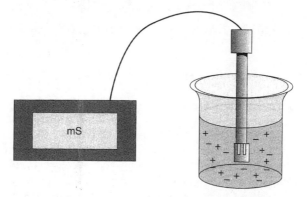

Figure 9.16 Conductivity meter for measuring the conduction of current relative to dissolved cations (+) and anions (−) in solution.

Analysis of Soil and Water Samples for Conductivity

The electrical conductivity of water or a soil suspension is directly related to the concentration of dissolved salts as cations and anions. The dissolved ions increase the ability of water and aqueous solution or suspension to transfer electrons and, as a result, conduct electricity. Accordingly, conductivity meters are used to measure the electrical conductivity of water and soil matrices as an indicator of dissolved solids. A conductivity meter consists of an electrode probe connected to a meter with direct readout display (Figure 9.16).

The probe consists of two electrodes (anode and cathode) spaced apart. The operation is commonly based on a Wheatstone Bridge or similar type of circuitry in which the presence of ions in solution conducts current between the two electrodes contained within the immersed probe. The current completes the circuit. The current generated increases as the concentration of ions increases and the conductivity is displayed on the meter in units of micro- or millimhos (mmho) or millisiemens (mS) per centimeter.

soil sample placed here ⌐

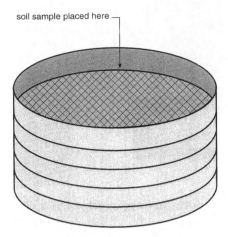

Figure 9.17   Stack of sieves for separating soil particles by size. Mesh sizes of sieves decrease from top to bottom sieve.

Analysis of Soil and Water Samples for Solids

Soil solids are composed of organic humic and fulvic acid colloids plus inorganic mineral fractions classified as sand, silt, and clay. The texture of soil is based on the ratio of sand (diameter 50 to 2000 μm), silt (diameter 2 to 50 μm), and clay (diameter <2 μm). Clays represent inorganic colloids in soil. The percentage of sand, silt, and clay can be determined by separating the inorganic fractions based on particle sizes using a series of sieves stacked with progressively smaller screen mesh sizes from top to bottom (Figure 9.17). Soil is weighed and placed in the top sieve, and the stack of sieves is manually or mechanically shaken. The separated fractions based on mesh sizes of the respective sieves are weighed and percentages of sand, silt, and clay are determined.

Soils are commonly evaluated for content of total solids and volatile or organic solids. Total solids are determined by placing a known weight of wet soil in a preweighed porcelain crucible or other suitable heat stable container and drying at approximately 105°C (Figure 9.18). The weight of the dry soil divided by the weight of the wet soil multiplied by 100 is the percentage of total solids. The organic or volatile solids in a soil sample are determined by placing the crucible containing the known weight of dry soil into a furnace and ashing at approximately 550°C. The residual material or ash represents the nonvolatile or inorganic solids. The percentage of volatile or organic solids equals the weight of the dry soil minus the weight of the ash divided by the weight of dry soil multiplied by 100.

Water samples are commonly collected to measure the content of total solids, dissolved solids, and suspended solids. Total solids can be measured directly by placing a known volume of water in a preweighed porcelain crucible and drying at 105°C. The weight of the dry solid residual divided by the weight of the water sample multiplied by 100 gives percentage of total solids. Alternatively, the weight

Figure 9.18 Porcelain crucible for drying water and soil samples and ashing residuals for measurement of solids. Porcelain withstands elevated temperatures required for drying and especially for ashing samples.

of the total solids divided by the volume of the water sample can be calculated and converted to express the concentration of total solids as milligrams per liter.

The total solids content of water can be determined indirectly by measuring both the suspended and dissolved solids. The sum of dissolved plus suspended solids equals total solids. Suspended solids are determined by filtering a known volume of water through a preweighed filter using a membrane filter apparatus (Figure 9.19). Following filtration and drying of the filter, the weight of the dry filtered solids divided by the volume of water sampled gives the concentration of suspended solids typically expressed in units of milligram per liter. The filtered water is dried and the dry residuals from that process represent the dissolved solids. The weight of the dissolved solids divided by the volume of water dried or evaporated yields the concentration of dissolved solids in milligram per liter. The concentration of dissolved solids is directly related to conductivity.

The presence of solids in water can increase turbidity. Turbidity is inversely related to the ability of light to pass through water. Thus, the presence of excessive concentrations of suspended solids increases the attenuation of light. Analytic devices are available that project a known intensity of light on a water sample and use a photodetector to measure the intensity of light that is able to pass unattenuated through the sample (Figure 9.20). Turbidity of surface water can be measured *in situ* using a Secchi disk (Figure 9.21). The black and white disk is attached to a line and gradually lowered into the water until it cannot be seen. Turbidity is measured based on the length of line needed to submerge the disk to the point where it is not visible.

Analysis of Water Samples for Dissolved Oxygen and Biological Oxygen Demand

The concentration of oxygen dissolved in water is important to determine whether conditions favor aerobic biological activity. The biological transformation and decomposition of organic matter in water occurs more rapidly and efficiently under aerobic conditions due to consumption and oxidation of the organic substrate by aerobic microorganisms. Dissolved oxygen is replenished naturally via diffusion of atmospheric oxygen into the water, yield from photosynthesis conducted by

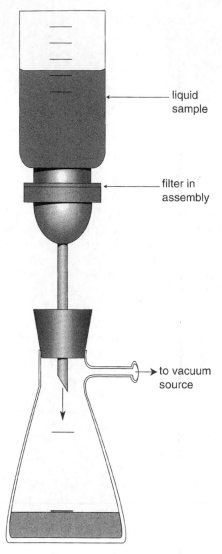

Figure 9.19 Glass or plastic membrane filter apparatus for filtering samples of water and soil suspensions for measurement of suspended and dissolved solids and for preparation of samples for microbiological analysis.

aquatic plants and algae, and turbulence caused by movement and agitation of water with simultaneous oxygenation due to wind and gravitational water flow.

Measurements of dissolved oxygen are commonly conducted in natural waters and water at treatment facilities. Levels of dissolved oxygen decrease with increased water temperature and increased aerobic microbial activity.

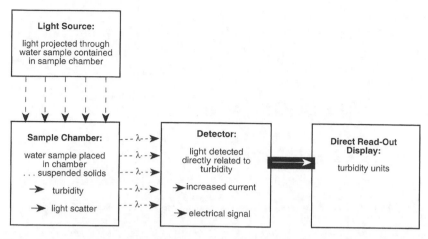

Figure 9.20 Summary of instantaneous monitoring and analysis of water samples for turbidity using a photometric instrument.

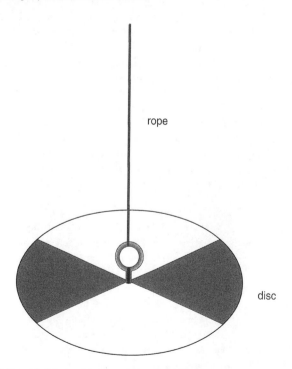

Figure 9.21 A Secchi disk used for measuring turbidity of water *in situ*.

Dissolved oxygen can be measured using a chemical assay (Winkler test) in which chemicals are mixed with a water sample and following titration ultimately yield a colorimetric change to a colorless end point that reflects the concentration

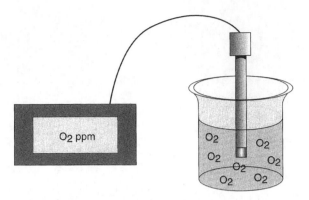

Figure 9.22   Dissolved oxygen meter for measuring the concentration of oxygen (O₂) in solution.

of dissolved oxygen in parts per million or milligrams per liter. Special electronic dissolved oxygen meters consisting of an electrode probe attached to a meter have been developed to instantaneously measure the concentration of dissolved oxygen in water (Figure 9.22). Oxygen diffuses through a semipermeable membrane at the end of the electrode. The presence of oxygen permits transport of electrons and, accordingly, increased current within a circuit. The meter converts the current to concentration of oxygen in units of parts per million.

In relation to dissolved oxygen, natural and wastewaters are commonly evaluated for the biological oxygen demand (BOD). When water contains biodegradable natural or anthropogenic organic matter, aerobic microorganisms can increase metabolic activity. The increased aerobic microbial activity may result in depletion of, or demand for, dissolved oxygen in excess of the rate of replenishment of oxygen in the water. Accordingly, the biological oxygen demand (BOD) is directly related to the concentration of labile organic matter in water. As a result, measurement of the BOD can be used as an indirect measure of the organic content or "organic load" in water.

In short, the assay involves collection of water in a BOD bottle (Figure 9.23). The concentration of dissolved oxygen is measured initially preceding incubation of the BOD sample in an incubator at 20°C under dark conditions typically for 5 days (BOD₅). Dark conditions are necessary to prevent possible photosynthetic activity and related generation of oxygen by algae that may be contained in the water sample. After 5 days, the dissolved oxygen is measured again. The BOD equals the initial concentration of dissolved oxygen minus the final concentration of dissolved oxygen. Commonly, the water sample suspected of contamination with excessive labile organic matter is seeded with aerobic microbes and trace nutrient minerals contained within dilution water. This assures that if an organic substrate is present, enough microbes may be available plus essential nutrients to accelerate biotransformational and decompositional activity. In addition, to assure that a measurable concentration of dissolved oxygen is available after 5 days or whatever other period that the water sample is incubated, the original sample is often diluted prior to incubation and analysis.

Figure 9.23 BOD bottle for collection and incubation of water samples for analysis of biological oxygen demand. Following collection of water sample (~300 ml), a glass stopper is inserted in the BOD bottle and water collects in the flared neck around the exterior of the stopper preventing entry or exit of air.

Analysis of Soil and Water Samples for Microorganisms

Sampling and analysis of airborne microorganisms (bioaerosols) are summarized in Volume I, Chapter 12. Soils can be analyzed for microorganisms such as bacteria, actinomycetes, and fungi by suspending a known weight of sample in a buffer and either spreading an aliquot (inoculum) over the surface of a nutrient growth agar or filtering the soil suspension using a membrane filter apparatus (shown earlier in Figure 9.19) and placing the filter on the surface of the agar. A known volume of a water sample also can be filtered using the membrane filter apparatus. The filter is removed from the apparatus and placed on the agar. For each analysis, the semisolid agar is contained in a culture dish to support growth of the microorganisms during incubation under conditions of controlled temperature, time, and light. Following incubation, the culture plates are viewed, usually with a magnifying device, to detect and count microbial colonies (Figure 9.24). Staining and microscopy can be used to assist in characterizing the type of microbes. Other alternatives are available. For example, a liquid nutrient broth contained within a culture tube can be inoculated with a small volume of water sample and observed for formation of gas bubbles

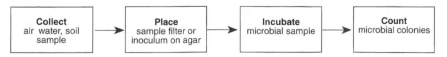

| Collect
air water, soil
sample | → | Place
sample filter or
inoculum on agar | → | Incubate
microbial sample | → | Count
microbial colonies |

Figure 9.24 Summary of steps involved for microbiological analysis of soil, water, and air samples.

following a defined period of incubation. This is described in more detail in this volume, Chapter 7.

Analysis of Soil, Water, and Air Samples for Chemicals

Some monitoring devices instantaneously sample, detect, and measure a chemical contaminant or other parameter. Most soil, water, and air samples collected for evaluation of chemical agents, however, must be submitted to a laboratory for subsequent analysis. These major analytic instruments are discussed in Volume I, Chapter 12. Summaries of their applications for analysis of soil, water, and air samples, however, follow:

Gas chromatography (GC) is the major analytic method for analyzing samples for organic analytes. Air samples collected on adsorbents are desorbed or extracted using a compatible organic solvent. The solvent extract containing dissolved contaminants or analytes are subsequently injected into a gas chromatograph for separation, detection, and measurement. Bulk samples of air collected using rigid canisters and nonrigid plastic bags can be injected directly into the instruments for analysis. Soil and water samples are extracted using solvents to remove the organic analytes from the solid soil and liquid water matrices. The solvent extract, in turn, is injected into the instrument for analysis.

Ultraviolet and visible (UV/Vis) light spectrophotometry is used to detect and measure organic compounds, but also has applications for numerous inorganic analytes. Air samples collected using liquid absorbents are mixed with chemical reagents to develop a color change associated with the presence of the contaminant or analyte. The liquid preparation is then inserted in the sample chamber of the spectrophotometer and absorbance is measured. Liquid extracts of soil and water samples also can be similarly prepared for UV/Vis spectrophotometric analyses.

Metals are commonly analyzed utilizing atomic absorption (AA) spectroscopy or inductively coupled plasma emission (ICPE) spectroscopy. Air samples of particulates collected on filters are analyzed for metals by ashing and resuspending or dissolving the filter in acid. The solid metal dusts and fumes dissolve in the acid and the resulting acid–metal solution is aspirated into the atomic absorption or inductively coupled plasma emission instrument for analysis. Soil and water samples also are prepared for analysis of metals by preparing with acid to dissolve the metal contaminants and yield a solution for analysis.

Gravimetry involving an electrobalance is used for evaluations based on change in presampling vs. postsampling weight of sample media for soil and water solids and air particulates. Commonly, filters are the media that are weighed presampling and postsampling to detect and measure soil and water solids or air particulates. Solid sorbents used to collect moisture during stack sampling, for example, also may be subject to gravimetric analysis.

Bibliography

CHAPTER 1

Ackerman, A.S., Toon, O.B., Stevens, D.E., Heymsfield, A.J. Ramanathan, V., Welton, E.J., Reduction of tropical cloudiness by soot, *Science*, 288:1042–1047, 2000.

Ackerman, F., Moomaw, W., Does emissions trading work?, *Electr. J.,* 10(7):August/September 1997.

Adam, K., Atmospheric elemental carbon, *Atmos. Environ.,* 24:597, 1990.

Agency for Toxic Substances and Disease Registry, Public Health Assessment Guidance Manual, Atlanta, GA, 1992.

Agency for Toxic Substances and Disease Registry, Toxicological Profile for Chlorodibenzofurans, Atlanta, GA, 1994.

Ahlborh, U.G., Brouwer, A., Fingerhut, M.A., et al., Impact of polychlorinated dibenzofurans, and biphenyls on human and environmental health, with special emphasis on application of the toxic equivalency factor concept, *Eur. J. Pharmacol.,* 228(4):179–199, 1992.

Ainsworth, S., Union carbide ethylene oxide plant explodes, *Chem. Eng. News,* 69:6, March 18, 1991.

Ambient air pollution: respiratory hazards to children, *Pediatrics,* 91(6):1210, 199.

Anto, J.M., Sunyer, J., Reed, E.E., et al., Preventing asthma epidemics due to soybeans by dust-control measures [see comments], *New Engl. J. Med.,* 329:1760–1763, 1993.

Atkeson, E., Joint implementation: lessons from Title IV's voluntary compliance programs, (WP-97003) Center for Energy and Environmental Policy Research, MIT, May 1997.

ATSDR, Toxicological profile for 1,3 butadiene, draft for public comment, ATSDR/U.S., Public Health Service, October 1993.

ATSDR, Toxicological profile for polycyclic aromatic hydrocarbons (PAHs), draft for public comment, ATSDR/U.S. Public Health Service, October 1993.

Bacon, S., Decadal variability in the outflow from the Nordic seas to the deep Atlantic Ocean, *Nature (London),* 394:871–874, 1998.

Baek et al., A review of atmospheric polycyclic aromatic hydrocarbons: sources, fate, and behavior, *Water Air Soil Pollut.,* 60:279–300, 1991.

Bailey, E.M., Allowance trading activity and state regulatory rulings: evidence from the U.S. Acid Rain Program, (WO-96002) Center for Energy and Environmental Policy Research, MIT, March 1996.

Bailey, W. et al., Cancer incidence and community exposure to air emissions from petroleum and chemical plants in Contra Costa County, California: a critical epidemiological assessment, *J. Environ. Health,* December 1993.

Ball, B.A., Folinsbee, L.J., Peden, D.B., Kehrl, H.R., Allergen bronchoprovocation of patients with mild allergic asthma after ozone exposure, *J. Allergy Clin. Immunol.*, 98:563–572, 1996.

Bauer, U., Berg, D., Kohn, M.A., Merriwether, R.A., Nickle, R.A., Acute effects of nitrogen dioxide after accidental release, *Public Health Rep.*, 113:62–70, 1998.

Benedick, R.E., *Ozone Diplomacy: New Directions in Safeguarding the Planet*, Harvard University Press, Cambridge, MA, 1991.

Bennett, M., Do you need on-site incineration?, *Process Eng.* 72:38–39, 1991.

Bhatia, R., Lopipero, P., Smith, A.H., Diesel exhaust exposure and lung cancer [see comments], *Epidemiology*, 9:84–91, 1994.

Bohi, D.R., Burtraw, D., SO$_2$ allowance trading: how experience and expectations measure up, *Resour. Future*, February 1997.

Bohi, D.R., Burtraw, D., Trading expectations and experience, *Electr. J.*, 10(7):August/September 1997.

Broecker, W.S., Sutherland, S., Peng, T.-H., A possible 20th-century slowdown of Southern Ocean deep water formation, *Science*, 286:1132–1135, 1999.

Brunekreef, B., Dockery, D.W., Krzynzanowki, M., Epidemiologic studies on short-term effects of low levels of major ambient air pollution components, *Environ. Health Perspect.*, 103:3–13, 1995.

Burnett, R.T., Cakmak, B.S., Jr., Krewski, D., The role of particulate size and chemistry in the association between summertime ambient air pollution and hospitalization for cardiorespiratory diseases, *Environ. Health Perspect.*, 105:614–620, 1997.

Burnett, R.T., Dales, R.E., Brook, J.R., Raizenne, M.C., Krewski, D., Association between ambient carbon monoxide levels and hospitalizations for congestive heart failure in the elderly in 10 Canadian cities, *Epidemiology*, 8:162–167, 1997.

Burtraw, D., Cost savings, market performance and economic benefits of the U.S. Acid Rain Program, in *Pollution for Sale: Emissions Trading and Joint Implementation*, Sorrell, S., Skea, J., Edgar Publishing, Northampton, MA, 1999.

Burtraw, D., Swift, B., A new standard of performance: an analysis of the Clean Air Act's acid rain program, *Environ. Law Rep.*, 26:10411, August 1996.

Butler, J., Comments on the Fossil Fuel Combustion Phase 2 Risk Assessment Final Reports, In peer review of *Fossil Fuel Combustion Risk Assessment*, Draft final report, prepared by Eastern Research Group, Lexington, MA, September 4, 1998.

Cagin, S., Dray, P., *Between Earth and Sky*, Pantheon Books, New York, 1993.

Calabrese, E.J., Kenyon, E.M., Air toxics and risk assessment, Lewis Publishers, Chelsea, MI, 1991.

Calderon-Garciduenas, L., Osnaya-Brizuela, N., Ramirez-Martinez., L., Villarreal-Calderon, A., DNA strand breaks in human nasal respiratory epithelium are induced upon exposure to urban pollution, *Environ. Health Perspect.*, 104:160–168, 1996.

Capaldo, K., Corbett, J.J., Kasibhatla, P., Fischbeck, P., Pandis, S.N., Effects of ship emissions on sulphur cycling and radiative climate forcing over the ocean, *Nature (London)*, 400:743–746, 1999.

Carpi, A., Lindberg, S.E., Sunlight-mediated emission of elemental mercury form soil amended with municipal sewage sludge, *Envir. Sci. Technol.*, 31(7):2085–2091, 1997.

CDC, Air pollution information activities at state and local agencies, *U.S. MMWR*, 41:967–969, 1992.

CDC, Populations at risk from air pollution — United States, *JAMA*, 269(19):2493, 1993.

Cerveny, R.S., Balling, R.C., Jr., Weekly cycles of air pollutants, precipitation and tropical cyclones in the coastal NW Atlantic region, *Nature (London)*, 394:561–563, 1998.

Chambers, R., Hydromatic coal, *Fuel*, 65:895–898, 1990.

Chestnut, L.G., Dennis, R., Economic benefits of improvements in visibility acid rain provisions of the 1990 Clean Air Act Amendments, *J. Air Waste Manage. Assoc.*, 47:395–402, March 1997.

Cioccale, M.A., Climatic fluctuations in the Central Region of Argentina in the last 1000 years, *Q. Int.*, 62:35–47, 1999.

Commission on Risk Assessment and Risk Management, Risk assessment and risk management in regulatory decision-making, draft report, June 13, 1996.

Committee of the Environmental and Occupational Health Assembly of the American Thoracic Society, Health effects of outdoor air pollution. *Am. J. Respir. Crit. Care Med.*, 153:3–50, 1996.

Conrad, K., Kohn, R.E., The U.S. market for SO_2 permits, *Energy Policy*, 24(12):1051–1059, 1996.

Cook, E., Marking a milestone in ozone protection: learning from the CFC phaseout., World Resources Institute, Washington, D.C., 1996.

Cook, E. et al., Ozone protection in the United States: elements of success, World Resources Institute, Washington, D.C., 1996.

Cordtz, T., Modifications of health behaviour in response to air pollution notifications, *Soc. Sci. Med.*, 33:621–626, 1991.

Cotton, P., Best data yet say air pollution kills below levels currently considered safe, *JAMA*, 269(24):3087–3088, 1993.

Crews, J.T. et al., Liming acid forest soils with flue gas desulfurization by-product: growth of Northern red oak and leachate water quality, *Environ. Pollut.*, 103:55–61, 1998.

Darcey, S., EPA includes recycling in new incineration regs, *Manage. World Wastes*, 33:28–29, January 1990.

Dempsey, C.R., Oppelt, E.T., Incineration of hazardous waste, a critical review update, *Air Waste*, 43:25–73, 1993.

Denissenko, M.F., Pao, A., Tang, M., Pfeifer, G.P., Preferential formation of buoyancy[a]pyrene adducts at lung cancer mutational hotspots in P53, *Science*, 274:430–432, 1996.

Dillon, H.K., Heinsohn, P.A., Miller, J.D., Eds., Field guide for the determination of biological contaminants in environmental samples, Biosafety Committee, American Industrial Hygiene Association, Fairfax, VA, 1996.

Dockery, D.W., Pope, C.A. III, Acute respiratory effects of particulate air pollution, *Annu. Rev. Public Health*, 15:107–132., 1994.

Dockery, D.W., Pope, A.C., Xu, X. et al., An association between air pollution and mortality in six U.S. cities [see comments], *N. Engl. J. Med.*, 329:1753–1759, 1993.

Dockery, D.W., Speizer, F.E., Stram, D.O., Ware, J.H., Spengler, J.D., Ferris, B.G., Jr., Effects of inhalable particles on respiratory health of children, *Am. Rev. Respir. Dis.*, 139(3):587–594.

Driscoll, N.W., Haug, A short circuit in thermohaline circulation: a cause for Northern Hemisphere glaciation?, *Science*, 282:436–438, 1998.

Dudek, D.J., Goffman, D.S., Wade, S., More clean air for the buck. Lessons from the U.S. Acid Rain Emissions Trading Program, Environmental Defense Fund, September 1997.

Durenberger, D., Air toxics: the problem, *JAMA*, January/February:30–31, 1991.

Dunford, R., *Your Health and the Indoor Environment*, New Dawn, Dallas, TX, 1991.

DuRose, R.A., New rules to prevent the accidental release of highly hazardous chemicals, *Plating Surface Finish.*, 78:16, 1991.

Easterling, D.R., Evans, J.L., Groisman, P. Y., Karl, T.R., Kunkel, K.E., Ambenje, P., Observed variability and trends in extreme climate events: a brief review, *Bull. Am. Meteorol. Soc.*, 81:417–425, 2000.

Elinder, C.G., Cadmium as an environmental hazard., *Iarc. Sci. Publ.*, 118:123–132, 1992.

Ellerman, A.D., Joskow, P.L., Schmalensee, R., Montero, J.-P., Bailey, E.M. *Markets for Clean Air: the U.S. Acid Rain Program*, Cambridge University Press, New York, 2000.

Ellerman, A., Joskow, P., Schmalensee, R., Working Paper 97-005: 1996 update on the compliance and emissions trading under the U.S. Acid Rain Program, MIT Center for Energy and Environmental Policy Research, Cambridge, MA, November 1997.

Ellerman, A.D., Schmalensee, R., Joskow, P., Montero, J.P., Bailey, E., Emissions trading under the U.S. Acid Rain Program: evaluation of compliance costs and allowance market performance, MIT Center for Energy and Environmental Policy Research, Cambridge, MA, October 1997.

Energy Information Administration, The effects of Title IV of the Clean Air Act Amendments on electric utilities: an update, March 1997.

Environmental Law Institute, Implementing an emissions cap and allowance trading system for greenhouse gases: lessons from the Acid Rain Program, *Res. Rep.,* September 1997.

Environmental Systems Research Institute, Inc. (ESR), Understanding GIS — The Arc/Info Method, 1991.

EPA, targets air quality no non-attainment areas, *Oil Gas J.,* 89:25, 1991.

Epstein, P.R., Diaz, H.F., Elias, S., Grabherr, G., Graham, N.E., Martens, W.J.M. et al., Biological and physical signs of climate change: focus on mosquito-borne disease, *Bull. Am. Meteorol. Soc.,* 78:409–417, 1998.

Executive Office of the President, The climate change action plan, EOP, Washington, D.C., 1993.

Executive Office of the President, Reinventing environmental regulation, EOP, Washington, D.C., 1995.

Executive Office of the President, Reinventing environmental regulation progress report, EOP, Washington, D.C., 1996.

Executive Office of the President, Office of Science and Technology ·Policy, climate change, July 1996: key findings of the second assessment report of the Intergovernmental Panel on Climate Change, EOP, Washington, D.C., 1996.

Executive Office of the President, Office of Science and Technology Policy, climate change July 1996: state of knowledge, EOP, Washington, D.C., 1996.

Facchini, M.C., Mircea, M., Fuzzi, S., Carlson, R.J., Cloud albedo enhancement by surface-active organic solutes in growing droplets, *Nature (London),* 401:257–259, 1999.

Fedorov, A.V., Philander, S.G., Is El Niño changing?, *Science,* 288, 1997–2002.

Ferek, R.J., Hegg, D.A., Hobbs, P.V., Durkee, P., Nielsen, K., Measurements of ship-induced tracks in clouds off the Washington coast, *J. Geophys.`Res.,* 103:23, 199–206, 1998.

Finlayson-Pitts, B.J., Pitts, J.N., Jr., Tropospheric air pollution: ozone, airborne toxics, polycyclic aromatic hydrocarbons, and particles, *Science,* 276:1045–1052, 1997.

Fishbein, L., Exposure from occupational versus other sources, *Scand. J. Work, Environ. Health,* 18:5–16, 1992.

Flannigan, B., Miller, J.D., Health implications of fungi in *Indoor Environments — An Overview, in Air Quality Monographs:* Volume 2: *Health Implications of Fungi in Indoor Environments,* Elsevier, Amsterdam, 1994, pp. 3–28.

Freeman, D.J., Cattell, C.R., Woodburning as a source of atmospheric polycyclic aromatic hydrocarbons, *Environ. Sci. Technol.,* 24(10):1581–1585, 1990.

Fullerton, D., McDermott, S.P., Caulkins, J.P., Sulfur dioxide compliance of a regulated utility, *J. Environ. Econ. Manage.,* 34:32–53 (Article No. EE971004), 1997.

Ghosh, A., Subbarao, C., Hydraulic conductivity and leachate characteristics of stabilized fly ash, *J. Environ. Eng.,* 124(9):812–820, 1997.

Gibbons, B., CEQ revisited: the role of the Council on Environmental Quality, Henry M. Jackson Foundation, 1995.

Golub, M.S., Donald, J.M., Reyes, J.A., Reproductive toxicity of commercial PCB mixtures: LOAELs and NOAELs from animal studies, *Environ. Health Perspect.*, 94:245–253, 1991.

Graveson, S., Frisvad, J.C., Samson, R.A., *Microfungi*, Munksgaard, Copenhagen, 1994.

Grosjean, D., Gas phase reaction of ozone, *Environ. Sci. Technol.*, 24:1428, 1990.

Hansen, J.E., Sato, M., Lacis, A., Ruedy, R., Tegan, I., Matthews, E., Climate forcings in the industrial era, *Proc. Natl. Acad. Sci. USA*, 95(12):753–712, 758, 1998.

Harvard Center for Risk Analysis, Annual report 1992, Harvard School of Public Health, Boston, MA, pp. 1–13, 1992.

Haug, G.H., Tiedemann, R., Effect of the formation of the Isthmus of Panama on Atlantic Ocean thermohaline circulation, *Nature (London)*, 393:673–676, 1998.

Hazardous Waste Identification Rule (HWIR), Hazardous waste management system: identification and listing of hazardous waste, Office of Solid Waste, 1995.

Hildemann, L. et al., Chemical composition of emissions from urban sources of fine organic aerosol, *Environ. Sci. Technol.*, 25(4):744–759, 1991.

Hopke, P.K., Xie, Y and Paatero, P., Mixed multiway analysis of airborne particle composition data, *J. Chemometrics*, 13:343–352, 1999.

Hoppel, W., Sub-micron aerosol size, *Atmos. Environ.*, 24:645, 1990.

Horiuchi, S., Greater lifetime expectations, *Nature (London)*, 405:744–745, 2000.

Hornung, C.A., Wel, M., Feigley, C.E., Macera, C.A., Draheim, L.A., Oldendick, R.W., Respiratory symptoms among the elderly in a community exposed to hazardous waste incineration and a comparison community, *J. Am. Geriatr. Soc.* (abstract), 41:5A65, 1993.

Hornung, C.A., Wel, M., Feigley, C.E., Macera, C.A., Olsen, G.N., Oldendick, R.W., Eleazer, G.P., Cardiac and respiratory symptoms associated with exposure to hazardous waste incineration, *I. Gen. Mt. Med.* (abstract), 9(2):31, 1994.

Houghton, J.T., Miera Filho, L.G., Callander, B.A., Harris, N., Kattenberg, A., Maskell, K. Eds., *Climate Change 1995: The Science, of Climate Change*, Cambridge University Press, Cambridge, UK, 1996.

Huebert, B.J., Sulphur emissions from ships, *Nature (London)*, 400:713–714, 1999.

Idso, S.B., CO_2 and the biosphere: the incredible legacy of the industrial revolution, Department of Soil, Water and Climate, University of Minnesota, St. Paul, MN, 1995.

Idso, S.B., CO_2-induced global warming: a skeptic's view of potential climate change, *Climate Res.*, 10:69–82, 1998.

Idso, S.B., Kimball, B.A., Pettit, G.R., III, Garner, L.C., Pettit G.R., Backhaus, R.A., Effects of atmospheric CO_2 enrichment on the growth and development of *Hymenocallis littoralis* (Amaryllidaceae) and the concentrations of several antineoplastic and antiviral constituents of its bulbs, *Am. J. Bot.*, 87:769–773, 2000,

Incineration, a special issue of waste management, *Waste Manage.*, 11(3):7–170, 1991.

Ish, Inc., Synthesis of available information on the management of coal combustion products in mines. Draft report for Ash Committee Utility Solid Waste Activities Group and EPRI, Cupertino, CA, docket number FF2P-50255, 1999.

Islam, K.K., Mulchi, C.L., Ali, A.A., Interactions of tropospheric CO_2 O_3 enrichments and moisture variations on microbial biomass and respiration in soil, *Global Change Biol.*, 6:255–265, 2000.

Ito, K., Thurston, G.D., Hayes, C., Lippmann, J., Associations of London, England, daily mortality with particulate matter, sulfur dioxide, and acidic aerosol pollution, *Arch. Environ. Health*, 48(4):213–220, 1993.

Jensen, A., Batterman, S., An evaluation of commercial hazardous waste thermal destruction capacity, *J. Air Waste Manage. Assoc.,* 44(8):995–1003, 1994.

Joos, E., Plume from a coal fired plant, *Atmos. Environ.,* 24:703, 1990.

Jorres, R., Nowak, C., Magnussen H., The effect of ozone exposure on allergen responsiveness in subjects with asthma or rhinitis, *Am. J. Crit. Care Med.,* 153:56–64, 1996.

Joskow, P.L., Schmalensee, R., The political economy of market-based environmental policy: the U.S. Acid Rain Program, (WP-96003), Center for Energy and Environmental Policy Research, MIT, March 1996.

Joskow, P.L., Schmalensee, R., Bailey, E.M., Auction design and the market for sulfur dioxide emissions, (WP96007), Center for Energy and Environmental Policy Research, MIT, August 1996.

Kaiser, J., Panel scores EPA on clean air science, *Science,* 280:193–194, 1998.

Karl, T.R., Knight, R.W., Plummer, N., Trends in high-frequency climate variability in the twentieth century, *Nature (London),* 377:217–220, 1995.

Kataoka, H., Methods for the determination of mutagenic heterocyclic amines and their applications in environmental analysis, *J. Chromatogr.,* 774:121–142, 1997.

Kavouras, I.G., Mihalopoulos, N., Stephanou, E.G., Formation of atmospheric particles from organic acids produced by forests, *Nature (London),* 395:683–686, 1998.

Keifer, M.E., Health effects of pesticides, *Occupational Medicine: State of the Art Reviews,* Vol. 12, Hanley & Belfus, Philadelphia, 1992, p. 2.

Kelly, R., Isotopic atmospheric tracers, *Atmos. Environ.,* 24:467, 1990.

Kepplinger, E.W., Cement clinker: an environmental sink for residues from hazardous waste treatment in cement kilns, *Waste Manage.,* 13(8):553–572, 1993.

Key, J.R., Chan, A.C.K., Multidecadal global and regional trends in 1000 mb and 500 mb cyclone frequencies, *Geophys. Res. Lett.,* 26:2053–2056, 1999.

Kim, C.S., Kang, T.C., Comparative measurement of lung deposition of inhaled fine particles in normal subjects and patients with obstructive airway disease, *Am. J. Respir. Crit. Care Med.,* 155:899–905, 1997.

Kruger, J., Dean, M., Looking back on SO_2 trading: what's good for the environment is good for the market, *Public Util. Fortnightly,* August(15):30–37, 1997.

Kyrklund, T., The use of experimental studies to reveal suspected neurotoxic chemicals as occupational hazards: acute and chronic exposures to organic solvents, *Am. J. Med.,* 21(1):15–24, 1992.

LaBar, G., Hazardous air and monitoring common gases, *Occup. Hazards,* 53:67–70, 1991.

Ladd, E.C., Bowman, K.H., Attitudes toward the environment: twenty-five years after Earth Day, American Enterprise Institute, Washington, D.C., 1995.

Latif, M., Roeckner, E., Mikolajewica, U., Voss, R., Tropical stabilization of the thermohaline circulation in a greenhouse warming simulation, *J. Climate,* 13:1809–1813, 2000.

Lave, L., Controlling emissions from motor vehicles, *Environ. Sci. Technol.,* 24:1128, 1990.

Lawson, D., Emissions from vehicles in L.A., *Air Waste Manage.,* 40:1096, 1990.

Leikauf, G.D., Kline, S., Albert, R.E., Baxter, C.S., Bernstein, K.I., Buncher, C.R., Evaluation of a possible association of urban air toxics and asthma, *Environ. Health Perspect.,* 103:253–271, 1995.

Lemly, A.D., Environmental implications of excessive selenium, *Biomed. Environ. Sci.,* 10:415–435, 1997.

Lemly, A.D., Guidelines for evaluating selenium data from aquatic monitoring and assessment studies, *Environ. Monitoring Assessment,* 28:83–100, 1993.

Leonard, T.L., Taylor, G.E., Gustin, M.E., Fernandez, G.C.J., Mercury and plants in contaminated soils: uptake, partitioning, and emission to the atmosphere, *Environ. Toxicol. Chem.,* 17(10):2063–2079, 1998.

Lettenmaier, D.P., Sheer, D.P., Climatic sensitivity of California water resources, *J. Water Resour. Plann. Manage.,* 117:108–125, 1991.

Lewtas, J., Complex mixtures of air pollutants: characterizing the cancer risk of polycyclic organic matter, *Environ. Health Perspect.,* 100:211, 1993.

Lile, R.D., Bohi, D.R., Burtraw, D., An assessment of the EPA's SO_2 emission allowance trading system, *Resour. Future,* February, 1997.

Lipperman, M., *Environmental Toxicants: Human Exposures and Their Health Effects,* Van Nostrand Reinhold, New York, 1992.

Lippmann, M., Ozone, in *Environmental and Occupational Medicine,* Rom, W., Ed., Lippincott-Raven, Philadelphia, 1998, pp. 601–615.

Lisk, D.J., Environmental implications of incineration of municipal solid waste and ash disposal, *Sci. Total Environ.,* 74:39–66, 1988.

Lobert, J., Biomass burning in the atmosphere, *Nature (London),* 346:552, 1990.

Lofroth, G., Zebuhr, Y., Polychlorinated dibenzo-*p*-dioxins (PCDDs) and dibenzofurans (PCDFs) in mainstream and sidestream cigarette smoke, *Bull. Environ. Contam. Toxicol.,* 48:789–794, 1992.

Lynch, J.A., Bowersox, V.C., Grimm, J.W., Trends in precipitation chemistry in the United States, 1983–1994: an analysis of the effects in 1995 of Phase 1 of the Clean Air Act amendments of 1990, Title IV, U.S. Geological Survey, USGS 96-0346, Washington, D.C., June 1996.

Makhijani, A., Gurney, K., *Mending the Ozone Hole,* MIT Press, Cambridge, MA, 1995.

Manegold, C.S., We didn't start the fire, *Newsweek,* 108:27, March 26, 1990.

Mansfield, G., Air quality monitors, *Safety Health,* 141:50–51, 1990.

Marty, M.C., Hazardous combustion products from municipal waste incineration, *Occup. Med.,* 8:603–620, 1993.

McConnell, M.A., Hallberg, L.M., Legator, M.S., Evaluation of the use of effects screening levels to ensure public health: a case study in Texas, *J. Clean Technol. Environ. Toxicol. Occup. Med.,* 6(1):23–58, 1997.

McGowan, T., Ross, R., Hazardous waste incineration is going mobile, *Chem. Eng.,* 9(10), October 1991.

McIlvaine, R.W., The 1991 global air pollution control industry, *J. Air Waste Manage. Assoc.,* 41:272–275, March 1991.

Mauderly, J.L., Toxicological and epidemiological evidence for health risks from inhaled engine emissions., *Environ. Health Perspect.,* 102:165–171, 1994.

Meerkotter, R., Schumann, U., Doelling, D.R., Minnis, P., Nakajima, T., Tsushima, Y., Radiative forcing by contrails, *Ann. Geophys.,* 17:1080–1094, 1999.

Mehlman, M.A., *Advances in Modern Environmental Technology,* Princeton Scientific Publishing, Vol. 2929, 1991, pp. 1–237.

Molfino, N.C., Wright, S.C., Katz, I. et al., Effect of low concentrations of ozone on inhaled allergen responses in asthmatic subjects [see comments], *Lancet,* 338:199–203, 1991.

Montero, J.P., Optimal design of a phase-in emissions trading program with voluntary compliance options, (WP-97004), Center for Energy and Environmental Policy Research, MIT, July 1997.

Montero, J.P., Volunteering for market-based environmental regulation: the substitution provision of the SO_2 emissions trading program, (WP-97001), Center for Energy and Environmental Policy Research, MIT, January 1997.

Montero, J.P., Why are allowance prices so low? An analysis of the SO_2 emissions trading program, (WP96001), Center for Energy and Environmental Policy Research, MIT, February 1996.

Montzka, S. et al., Decline in the tropospheric abundance of halogen from halocarbons: implications for stratospheric ozone depletion, *Science,* 272:1318–1322, May 31, 1996.

Morris, R.D., Naumova, E.N., Munasinghe, R.L., Ambient air pollution and hospitalization for congestive heart failure among elderly people in seven large U.S. cities [see comments], *Am. J. Public Health*, 85:1361–1365, 1995.

Morrison, S., Two-stage scrubbing system helps G-P Mill maintain air quality, *Pulp Paper*, 65:123, 1991.

Muraka, I.P., Fax and ash analysis of 14 EPRI sites, Batelle Pacific Northwest Laboratory, August 13, 1997, Docket no. FF29-50262, 1997.

Muto, H., Takizawa, Y., Potential health risk via inhalation/ingestion exposure to polychlorinated dibenzo-*p*-dioxins and dibenzofurans, *Bull. Environ. Contam. Toxicol.*, 49:701–707, 1992.

National Academy of Public Administration (NAPA), Setting priorities, getting results: a new direction for EPA, NAPA, Washington, D.C., 1995.

National Acid Precipitation Assessment Program (NAPAP), Biennial report to Congress: an integrated assessment, May 1998.

NCHS, Current estimates from the national health interview survey, 1990, U.S. Department of Health and Human Services, Public Health Services, CDC, Hyattsville, MD, 1991, DHHS no.(PHS)92–1509.

Newman, R.P., Air pollution controls for infectious hazardous waste incineration, *Pollut. Eng.*, 23:68–71, October 1991.

New report shows progress in air quality, *EPA J.*, 17:53–54, 1991.

Nilsson, U., Ostman, C., Chlorinated polycyclic aromatic hydrocarbons: method of analysis and their occurrence in urban air, *Environ. Sci. Technol.*, 27(9):1826–1831, 1993.

Oehme, M., Larssen, S., Brevik, E.M., Emission Factors of PCDD and PCDF for road vehicles obtained by tunnel experiment, *Chemosphere*, 23(11–12):1699–1708, 1991.

Observed variability and trends in extreme climate events. A brief review, *Bull. Am. Meteorol. Soc.*, 81:417–425, 2000.

Okey, A.B., Riddick, D.S., Harper, P.A., The Ah receptor: mediator of the toxicity of 2,3,7,8-tetrachlorodibenzo-*p*-dioxin (TCDD) and related compounds, *Toxicol. Lett.*, 70(1):1–22, 1994.

Olsen, J.R., Stedinger, J.R., Matalas, N.C., Stakhiv, E.Z., Climate variability and floodfrequency estimation for the Upper Mississippi and Lower Missouri Rivers, *J. Am. Water Resour. Assoc.*, 35:1509–1523, 1999.

On, W.R., Roberts, J.W., Everyday exposure to toxic pollutants, *Sci. Am.*, 278:86–91, 1998.

Oppelt, E.T., Air emission from the incineration of hazardous waste, *Toxicol. Ind. Health*, 6(5):23–51, 1990.

Organisation for Economic Co-operation and Development (OECD), Environmental performance reviews: United States, OECD, Paris, 1996.

Ostro, B., The association of air pollution and mortality: examining the case for inference, *Arch. Environ. Health*, 48:336–342, 1993.

Ott, W.R., Roberts, J.W., Everyday exposure to toxic pollutants, *Sci. Am.*, 278:1045–1052, 1997.

Peake, E., Ozone concentrations, *Atmos. Environ.*, 24:475, 1990.

Pearlman, M.E., Finklea, J.F., Creason, J.P., Shy, C.M., Young, M.M., Horton, J.M., Nitrogen dioxide and lower respiratory illness, *Pediatrics*, 47(2):391–398, 1971.

Pope, C.A., Particulate pollution and health: a review of the Utah valley experience, *J. Exposure Anal. Environ. Epidemiol.*, 6:23–34, 1996.

Pope, C.A., Bates, D.V., Raizenne, M.E., Health effects of particulate air pollution: time for reassessment?, *Environ. Health Perspect.*, 103:472–480, 1995.

Pope, C.A., III, Dockery, D.W., Spengler, J.I., Raizenne, M.E., Respiratory health and PM_{10} pollution, *Am. Rev. Respir. Dis.* (Part 1), 144(3):668–674, 1991.

Pope, C.A., Thun, M.J., Namboodiri, M.C. et al., Particulate air pollution as a predictor of mortality in a prospective study of U.S. adults, *Am. J. Respir. Crit. Care Med.,* 151:669–674, 1995.

Porteous, A., Barratt, R.S., An appraisal of incineration standards: when is a standard not standard, *J. Power Energy,* 204(A3):193–200, 1991.

President's Council on Sustainable Development, Sustainable America: a new consensus for prosperity, opportunity, and a healthy environment for the future, Government Printing Office, Washington, D.C., 1996.

Public Health Service, Healthy people 2000: national health promotion and disease prevention objectives, U.S. Department of Health and Human Services, Public Health Service, Washington, D.C., 1991, DHHS no.(PHS)91-50213.

Puckett, L., Ion sources in deciduous materials, *Atmos. Environ.,* 24:519, 1990.

Rams, L., Eljarrat, E., Hernandez, L.M., Alonso, L., Rivera, J., Gonzales, M.J., Levels of PCDDs and PCDFs in farm cow's milk located near potential contaminant sources in Asturias (Spain). Comparison with levels found in control, rural farms and commercial pasteurized cow's milk, *Chemosphere,* 35:2167–2179, 1997.

Rapiti, E., Sperati, A., Fano, V., Dell'Orco, V., Forastiere, F., Mortality among workers at municipal waste incinerators in Rome: a retrospective cohort study, *Am. J. Ind. Med.,* 31:659–661, 1997.

Raucy, J.L., Kraner, J.C., Lasker, J.M., Bioactivation of halogenated hydrocarbons by cytochroms P4502E1, *Crit. Rev. Toxicol.,* 23(1);1–20, 1993.

Reilly, W.K., The new clean air act: an environmental milestone, *EPA J.,* Jan./Feb.:2–4, 1991.

Remet, M., Castren, K., Jarvinen, K. et al., P53 protein expression is correlated with buoyancy[*a*]pyrene-DNA adducts in carcinoma cell lines, *Carcinogenesis,* 16:2117–2124, 1995.

Rensburg, L., Van, R.I., DeSousa-Correia, J., Booysen, Ginster, M., Revegetation on a coal fine ash disposal site in South Africa, *J. Environ. Q.,* 27:1479–1486, 1998.

Roan, S.L., *Ozone Crisis: the 15-Year Evolution of a Sudden Global Emergency,* John Wiley & Sons, New York, 1989.

Rogan, W.J., Gladen, B.C., Neurotoxicology of PCBs and related compounds, *Neurotoxicology,* 13(1):27–35, 1992.

Ruhlemann, C., Mulitza, S., Muller, P.J., Wefer, G., Zahn, R., Warming of the tropical Atlantic Ocean and slowdown of thermohaline circulation during the last deglaciation, *Nature (London),* 402:511–514, 1999.

Ruprecht, H., Interaction of aerosols and gases, *Atmos. Environ.,* 24:573, 1990.

Safe, S., Toxicology, structure-function relationship, and human and environmental health impacts of polychlorinated biphenyls: progress and problems, *Environ. Health Perspect.,* 100:259–268, 1993.

Sahai, A.K., Climate change: a case study over India, *Theoret. Appl. Climatol.,* 61:9–18, 1998.

Salo, L.F., Artiola, J.F., Goodrich-Mahoney, J.W., Effects of revegetation techniques of a saline flue gas desulfurization sludge pond, *J. Environ. Q.,* 28:218–225, 1999.

Satheesh, S.K., Ramanathan, V., Large differences in tropical aerosol forcing at the top of the atmosphere and Earth's surface, *Nature (London),* 405:60–63, 2000.

Sato, A., The effect of environmental factors on the pharmacokinetic behavior of organic solvent vapours, *Ann. Occup. Hyg.,* 35(5):525–541, 1991.

Sato, M., Bremmer, I., Oxygen free radicals and metallothionein, *Free Radic. Biol. Med.,* 14(3):325–337, 1993.

Schecter, A.J. et al., Dioxin levels in blood of municipal incinerator workers, *Med. Sci. Res.,* pp. 331–332, 1991.

Schwartz, J., Air pollution and daily mortality: a review and meta analysis, *Environ. Res.,* 64:36–52, 1994.

Schwartz, J., Short-term fluctuations in air pollution and hospital admissions of the elderly for respiratory disease, *Thorax,* 50:531–538, 1995.

Schwartz, J., What are people dying of on high air pollution days? *Environ. Res.,* 64:26–35, 1994.

Schwartz, J., Dockery, D.W., Neas, L.M., Is daily mortality associated specifically with fine particles? *J. Air Waste Manage. Assoc.,* 46:927–939,1996.

Schwartz, S.E., Buseck, P.R., Absorbing phenomena, *Science,* 288:989–990, 2000.

Science Advisory Board (SAB), Waste leachability: the need for review of current agency procedures, Environmental Engineering Committee, February 26, 1999.

Science Applications International Corporation (SAIC), Fossil fuel combustion waste risk assessment: revised groundwater analysis and sensitivity results, prepared for U.S. EPA Office of Solid Waste, October 9, 1998.

Sedman, R.M., Polisini, J.M., Esparza, J.R., The evaluation of stack metal emissions from hazardous waste incinerators: assessing human exposure through noninhalation pathways, *Environ. Health Perspect.,* 102:1105–1112, 1994.

Seitz, J., Urban air quality: the strategy, *EPA J.,* January 27–29, 1991.

Sevim, H., Unal, A., Promoting Illinois coal utilization through underground disposal of combustion products, *Mining Eng.,* August:68–72, 1998.

Shane et al., Organic toxicants and mutagens in ashes from eighteen municipal refuse incinerators, *Arch. Environ. Contam. Toxicol.,* 19(5):665–673, 1990.

Shy, C.M., Degnan, P., Fox, D.L., Mukerjee, S., Hazucha, M.J., Boehlecke, B.A., Rothenbacher, D., Briggs, P.M., Devlin, R.B., Wallace, D.D., Stevens, R.K., Bronmerg, P.A., Do waste incinerators induce adverse respiratory effects? An air quality and epidemiological study of six communities, *Environ. Health Perspect.,* 103(7–8):714–724, 1995.

Simo, R., Pedros-Alio, C., Role of vertical mixing in controlling the oceanic production of dimethyl sulphide, *Nature (London),* 402:386–399, 1999.

Snyder, R., Witz, G., Goldstein, B.D., The toxicology of benzene, *Environ. Health Perspect.,* 100:293–306, 1993.

Stahle, D.W., Cook, E.R., Cleaveland, M.K., Therrell, M.D., Meko, D.M., Grissino-Mayer, H.D., Watson, E., Luckman, B.H., Tree-ring data document 16th century megadrought over North America, *EOS, Trans. Am. Geophys. Union,* 81:121–125, 2000.

Steuhower, R., Sutton, P., Warren, D., Transport and plant uptake of soil-applied dry flue gas desulfurization byproducts, *Soil Sci.,* 161(9):562–574, 1996.

Stewart, R., Atmospheric soluble species, *Atmos. Environ.,* 24:519, 1990.

Stiling, P., Rossi, A.M., Hungate, B., Dijkstra, P., Hinkle, C.R., Knott, W.M., III, Drake, B., Decreased leaf-miner abundance in elevated CO_2: reduced leaf quality and increased parasitoid attack, *Ecol. Appl.,* 9:240–244, 1999.

Swift, B., Allowance trading and potential hot spots — good news from the Acid Rain Program, *Environ. Rep.,* 31:954–959, May 12, 2000.

Swift, B., *Environ. Forum,* 14(3):17–25, May/June 1997.

Tattum, L., BP Chemicals lead the way in emissions inventory, *Chem. Week.,* 148:95, July 17, 1991.

Thornton, J., McCally, M., Oris, P., Weinberg, J., Hospitals and plastics. Dioxin prevention and medical waste incinerators [see comments], *Public Health Rep.,* 111:299–313, 1996.

Thurston, B.D., D'Souza, N., Lippman, M., Bartoszek, M., Fine J. Associations between summertime haze air pollution and asthma exacerbations: a pilot camp study (abstract), *Am. Resp. Dis.,* 145:A429, 1992.

Thurston, G.C., A critical review of PM_{10}-mortality time-series studies, *J. Exposure Anal. Environ. Epidemiol.,* 6:3–21, 1996.

Thurston, G.D., Ito, K., Lippmann, M., Bates, D.V., Respiratory hospital admissions and summertime haze air pollution in Toronto, Ontario: consideration of the role of acid aerosols, *Environ. Res.,* 65:271–290, 1994.

Thurston, G.D., Kinney P.L., Air pollution epidemiology considerations in time-series modeling, *Inhal. Toxicol.,* 7:71–83, 1995.

Titus, J.G., Naraynan, V.K., The probability of sea level rise, U.S. Environmental Protection Agency, Washington, D.C., 1995.

Titus, J.C., Park, R.A., Leatherman, S.P. et al., Greenhouse effect and sea level rise: the cost of holding back the sea, *Coastal Manage.,* 19:171–204, 1991.

Tortora, G.J., Funke, B.R., Case, C.L., *Microbiology,* Benjamin/Cummings, New York, 1992.

Travis, C.C., Hattemer-Frey, H.A., Human exposure to dioxin, *Sci. Total Environ.,* 104(1–2):97–127, 1991.

Travis, C.C., Hester, S.T., Global chemical pollution, *ES&T,* 25:814–818, 1991.

Traynor, G., Space heaters organic pollutants, *Environ. Sci. Technol.,* 24:1265, 1990.

Truckers and HazMat regulations, *Traffic World,* 222:38, April 30, 1990.

Tuljapurkar, S., Li, N., Boe, C., A universal pattern of mortality decline in the G7 countries, *Nature (London),* 405:789–792, 2000.

Two, S.P., Derailments trigger review of HazMat handling practices, *Railway Age,* 192:24–25, September 1991.

U.S. Army Environmental Hygiene Agency, Inhalation risk from incinerator combustion products-Johnson Atoll chemical agent disposal system, Health Risk Assessment Number 42-21-MQ49-92, Aberdeen Proving Ground, MD, 1991.

U.S. Army Environmental Hygiene Agency, Inhalation risk from incinerator combustion products-Johnson Atoll chemical agent disposal system, Health Risk Assessment Number 42-21-MIBE-93, Aberdeen Proving Ground, MD, 1993.

U.S. Department of Agriculture, Soil Conservation Service, State Soil Geographic Data Base (STATSGO) Date Users Guide, Miscellaneous Publication Number 1492, 88 p. 1993.

U.S. Department of Energy, Energy Information Administration, Annual energy review 1995, DOE/EIA-0384(95), Government Printing Office, Washington, D.C., 1996.

U.S. Department of Energy Office of Fossil Energy, Coal combustion waste management study, draft report, prepared by ICF Resources Inc., Contract DE-AC01-91FE62017 Task 8, 1993.

U.S. Department of State, U.S. Climate Action Report, Government Printing Office, Washington, D.C., 1994.

U.S. Energy Information Agency, The effects of Title IV of the Clean Air Act Amendments of 1990 on electric utilities: an update, (DOE/EIA 0582-97), March 1997.

U.S. EPA, Acid deposition standard feasibility study: report to congress, EPA 430-R-95-010a, October 1995.

U.S. EPA, Assessment of potential health risks of gasoline oxygenated with methyl tertiary butyl ether (MTBE), EPA/6001R-93/206, November 1993,

U.S. EPA, The benefits and costs of the Clean Air Act, 1970–1990, draft report, May 30, 1996.

U.S. EPA, EPA Fact Sheets: Ozone depletion: the facts behind the phaseout; benefits of the CFC phaseout; fact sheet on the accelerated phaseout of ozone-depleting substances, EPA, Washington, D.C., 1993–1996.

U.S. EPA, Guidance for risk characterization, Science Policy Council, Washington, D.C., 1995.

U.S. EPA, HAP PRO: software program for control technologies for HAP, Control Technology Center, Research Triangle Park, NC, June 1991.

U.S. EPA, Human health benefits from sulfate reductions under Title IV of the 1990 Clean Air Act amendments, EPA 430-R-95-010, November 1995.

U.S. EPA, Intent to list chloroform as a hazardous air pollutant, *Fed. Regist.*, 50:39626, 1985.

U.S. EPA, Integrated Risk Information System (IRIS) on acetaldehyde, Office of Health and Environmental Assessment, Environmental Criteria and Assessment Office, Office of Research and Development, Cincinnati, OH, 1993.

U.S. EPA, Methodology for assessing health risks associated with indirect exposure to combustor emissions-interim final, Washington, D.C., 1990.

U.S. EPA, National priorities list sites: New York, EPA/540/4-90/032, Environmental Protection Agency, Washington, D.C., 1990.

U.S. EPA, Office of Air and Radiation, Air quality trends, EPA, Washington, D.C., 1995.

U.S. EPA, Office of Policy Planning and Evaluation, Inventory of U.S. greenhouse gas emissions and sinks: 1990–1994, EPA-230-R-96-006, U.D. EPA, Washington, D.C., 1995.

U.S. EPA, Office of Pollution Prevention and Toxics, 1994 toxics release inventory, EPA, Washington, D.C.,1996.

U.S. EPA, Office of Wastewater Management, A plain English guide to the EPA Part 305 Biosolids Rule, EPA832/R-93/0003, 1994.

U.S. EPA, Ozone and carbon monoxide areas designated nonattainment, U.S. Environmental Protection Agency, Office of Air Quality Planning and Standards, Research Triangle Park, NC, 1991.

U.S. EPA, Technical background document for the report to Congress on remaining wastes from fossil fuel combustion: waste characterization, March 15, 1999.

U.S. EPA, Technical Background document to support rulemaking pursuant to the clean air act- section 112(g). Ranking of pollutants with respect to hazards to human health, EPA/450/3-92/010, Emissions Standards Division, Office of Air Quality Planning and Standards, Research Triangle Park, NC, 1994.

U.S. Government Accounting Office, Acid rain: emissions trends and effects in the eastern United States, (GAO/RCED-00-47), March 2000.

U.S. General Accounting Office, Air pollution: state planning requirements will continue to challenge EPA and the states, GAO/RCED-93-113, June 1993.

U.S. General Accounting Office, Air pollution overview and issues on emissions allowance trading programs, statement by Peter Guerrero, testimony before the Joint Economic Committee, U.S. Congress (GAO/T-RCED-97-183), July 9, 1997.

U.S. General Accounting Office, Air pollution: reduction in EPA's 1994 air quality program budget, GAO/RCED-95-31BR, November 1994.

Van Wormer, M.B., Use air quality auditing as an environmental management tool, *Chem. Eng., Prog.* 82:62–67, November 1991.

Waalkes, M.P., Coogan, T.P., Barter, R.A., Toxicological principles of metal carcinogenesis with special emphasis on cadmium. *Crit. Rev. Toxicol.,* 22(3–4):175–201, 1992.

Walsh, M., Motor vehicles and fuels: the problem, *EPA J.,* January, 12–14, 1991.

Wild, M., Discrepancies between model-calculated and observed shortwave atmospheric absorption in areas with high aerosol loadings, *J. Geophys. Res.,* 104:27, 36127, 371, 1999.

Wild, M. et al., Wasted municipal solid waste incinerator fly ash as a source of polynuclear aromatic hydrocarbons to the environment, *Waste Manage. Res.,* 10(1):99–111, 1992.

Witwer, S.H., *Food, Climate, and Carbon Dioxide: the Global Environment and World Food Production,* CRC Press, Boca Raton, FL, 1995.

Woodhouse, C.A., Overpeck, J.T., 2000 years of drought variability in the central United States, *Bull. Am. Meteorol. Soc.,* 79:2693–2714, 1998.

Working Group I, Intergovernmental Panel on Climate Change (IPCC), *Climate Change 1995: The Science, of Climate Change,* Cambridge University Press, Cambridge, 1996.

Working Group II, Intergovernmental Panel on Climate Change (IPCC), *Climate Change 1995: Impacts, Adaptations, and Mitigation of Climate Change: Scientific-Technical Analyses,* Cambridge University Press, Cambridge, 1996.

Working Group III, Intergovernmental Panel on Climate Change (IPCC), *Climate Change 1995: The Economic and Social Dimensions of Climate Change,* Cambridge University Press, Cambridge, 1996.

World Meteorological Organization, Scientific assessment of ozone depletion: 1994, WMO Global Ozone Research and Monitoring Project, Report No. 37, WMO, Geneva, 1995.

Xiong, F.S., Meuller, E.C., Day, T.A., Photosynthetic and respiratory acclimation and growth response of Antarctic vascular plants to contrasting temperature regimes, *J. Bot.,* 87:700–710, 2000.

Yu, Z., Ito, E., Possible solar forcing of century-scale drought frequency in the northern Great Plains, *Geology,* 27:263–266, 1999.

Zile, M.H., Vitamin A homeostasis endangered by environmental pollutants [published erratum appears in *Proc. Soc. Exp. Biol. Med.,* 20(3):319, December 1992], *Proc. Soc. Exp. Biol. Med.,* 201(2):141–153, 1992.

CHAPTER 2

Aident, M., Foster, M., Stolte,W., MOTCO superfund site cleanup and restoration, *Waste Manage.,* 11(3):135–146, 1991.

Ainsworth, S., Lepkowski, W., Methan-sodium spill shows tank car safety flaws, *Chem. Eng. News,* 69:4, August 5, 1991.

Alexander, R., Innovations in compost marketing, *BioCycle,* October:36–39, 1996.

Allen, P.C., Foye, P., Henderson, T.M., Recycling and incineration: not mutually exclusive in Broward County, Florida, *Gov. Finan. Rev.,* 6(5):7–11, October 1990.

Anderton, D.L. et al., Environmental equity: the demographics of dumping, study by the Social and Demographic Research Institute, University of Massachusetts, Amherst, *Demography,* 31:2, May 1994.

Antler, S., Composting comes of age: highlights from a new Canada — wide study, *Solid Waste Recycling,* October/November:12–17, 1997.

Apotheker, S., *Resour. Recycling,* April 1993.

Aquino, J.T., Landfill reclamation attracts attention and questions, *Waste Age,* December:63–65, 68, 1994.

Asia buying more wastepaper from Japan, Europe; economic woes continue, *Paper Recycler,* June 1998.

Auger, J. et al., Decline in semen quality among fertile men in Paris during the past 20 years, *N. Engl. J. Med.,* 332:281–285, 1995.

Austin, T., Bio bonanza, *Civil Eng.,* 60(4):49–51, April 1990.

Avallone, E., *"Marks" Standard Handbook for Mechanical Engineers,* 10th ed., Avallone, E., Baumeister, T., Eds., McGraw-Hill, New York, 1996.

Bader, C.D., Beauty in landfill mining: more than skin deep, *MSW Manage.,* March/April:54–63, 1994.

Bailey, T.C., Gatrell, A.C., Empirical Bayes estimation, in *Interactive Spatial Data Analysis,* Longman Scientific & Technical, Longman Group, Essex, England, 1995, pp. 303–398.

Barr, L., Biotechnology unlocks the benefits of bacteria, *Plant Eng. Maint.,* 14(7):8, September 1991.

Barr, L., The role or aeration in waste water management, *Plant Eng. Maint.,* 14(3):8, March 1991.

Barton, J.R., Dalley, C., Patel V.S., Life cycle assessment for waste management, *Waste Manage.,* 16(1):35–50, 1996.

Belluck, D.A., Benjamin, S.L., Pesticides and human health, *J. Environ. Health,* July/August 1990.

Biosafety in Microbiological and Medical Laboratories, 4th ed., CDC/NIH/USDHHS, HHS Publication No. (CDC) 99-8395.

Biosafety in the Laboratory, National Research Council, National Academy Press, Washington, D.C., 1989.

Brauksieck, R., Abandoning commercial underground tanks, *ASHI Tech. J.,* 3(1):40–41, Spring 1993.

Brody, D.J. et al., Blood lead levels in the U.S. population: Phase 1 of the third national health and nutrition examination Survey (NHANES III, 1988 to 1991), *JAMA,* 272(4):277–283, July 27, 1994.

Bryant, B., Issues, Policies, and Solutions for Environmental Justice, University of Michigan, Ann Arbor, MI, 1994.

Bullard, R.D., Overcoming racism in environmental decision-making, *Environment,* 36:4, May 1994.

Bullard, R.D., Unequal protection: environmental justice and communities of color, Sierra Club, San Francisco, 1994.

Calaminus, B., Stahlberg, R., Thermal waste treatment: a better approach, *Chemtech,* October:40–60, 1998.

Caliper Corporation, Maptitude, v 4.02, Caliper Corporation, Newton, MA, www.caliper.com., 1998.

Canterbury, J.L., Pay-as-you-throw: a growing MSW management success story, *Resour. Recycling,* October 1997.

Canterbury, J.L., Pay-as-you-throw lessons learned about unit pricing, U.S. Environmental Protection Agency, EPA-R-94-004, April 1994.

Carlsen, E. et al., Evidence for decreasing quality of semen during the past fifty years, *Br. Med. J.,* 304:609–613, 1992.

Center for the Study of Environmental Endocrine Effects, Environmental endocrine effect: an overview of the state of scientific knowledge and uncertainties, [discussion draft], Center for the Study of Environmental Endocrine Effects, Washington, D.C., September 1955.

Centers for Disease Control and Prevention, Preventing lead poisoning in young children, CDC, Atlanta, 1991.

Chadzynki, L., Medical waste act requires physician compliance, *Mich. Med.,* 90(7):41, 43–47, July 1991.

Chaney, R.L. et al., Phytoremediation potential of *Thiaspi caerulescens* and bladder campion for zinc-and-cadmium contaminated soil, *J. Environ. Qual.,* 23:1151–1157.

Charles, D., Quicklime could quickly dispose of PCB's, *New Sci.,* 132:127–129, 1991.

Chen, Y., Inbar, Y., Chemical and spectroscopical analyses of organic matter transformations during composting in relation to; compost maturity, *Science, and Engineering of Composting,* Renaissance Publications, Worthington, OH, 1993.

CHP's hazardous waste investigative unit, *FBI Law Enforce. Bull.,* 60(4):12–13, 1991.

Ciminello, P.H., A primer on petroleum bulk storage tanks and petroleum contamination of property, *ASHI Tech. J.,* 3(1):35–39, Spring 1993.

Claritas Corporation, Annual census data 1990–1997, Claritas Corporation, Arlington, VA.

Cole, M.A., Liu, X., Zhang, L., Effect of compost addition on pesticide degradation in planted soils, in *Bioremediation of Recalcitrant Organics,* Hinchee, R.E., Anderson, D.B., Hoeppel, R.E., Eds., Batelle Press, Columbus, OH, 1995.

Congress of the United States, Office of Technology Assessment, *Finding the Reason for Managing Medical Wastes,* U.S. Government Printing Office, Washington, D.C., 1990.

Congress of the United States, Office of Technology Assessment, Green products by design: choices for a cleaner environment, OTA-E-541, October 1992.

Council on Environmental Quality et al., Improving federal facilities cleanup, Report of the Federal Facilities Policy Group, October 1995.

Council on Environmental Quality and the U.S. Postal Service, Recycling, Looking toward the next century, Final, Workshop Summary, White House Conference Center, Washington, D.C., May 19–21, 1998.

Cunningham, W.P., Saigo, B.W., *Environmental Science,: A Global Concern,* William C. Brown, Dubuque, IA, 1990.

Cynoweth, D.P., Bosch, G., Earle, J.F., Legrand, R., Liu, K., A novel process for anaerobic composting of municipal solid waste, *Appl. Biochem. Biotechnol.,* 28:421–432, Spring, 1991.

Daniel, D.E., Compacted clay and geosynthetic clay liners, American Society of Civil Engineers National Chapter Section: Geotechnical aspects of landfill design, National Academy of Sciences, Washington, D.C., January 1992.

Daniel, D.E., Gilbert, R.B., Geosynthetic clay liners for waste containment and pollution prevention, University of Texas at Austin, Austin, TX, February 1994.

Daniel, D.E., Koerner, R.M., Geotechnical aspects of waste disposal, in *Geotechnical Practice for Waste Disposal,* Daniel, D.E., Ed., Chapman & Hall, London, 1993, chap. 18.

Davis, P.A., Senate acts to force cleanup by government polluters, *Congr. Q. Wkly. Rep.,* 49(43):3121, 1991.

Delmar, D., Commercial market for synthetic marsh builds, *Environ. Today.,* 2(1):41, January/February, 1991.

Department of Housing and Urban Development and the U.S. Environmental Protection Agency, The effects of environmental hazards and regulation on urban redevelopment, August 1997.

Devine, O.J., Louis, T.A., A constrained empirical Bayes estimator for incidence rates in areas with small populations, *Stat. Med.,* 12(11):1119–1133, 1994.

Devine, O.J., Louis, T.A., Halloran, M.C., Empirical Bayes methods for stabilizing incidence rates before mapping, *Epidemiology,* 5(6):622–630, 1994.

Dickinson, W., Landfill mining comes of age, *Solid Waste Technol.,* March/April:46, 1995. Dioxins and dioxin-like chemicals, *Environ. Health Mon.,* 7(6), March 1955.

Dockery, D.W. et al., An association between air pollution and mortality in six U.S. cities, *N. Engl. J. Med.,* 329(24):1753–1759, December 1993.

Domenico, A., Guidelines for the definition of environmental action alert thresholds for polychlorodibenzodioxins and polychlorodibenzofurans, *Regul. Toxicol. Pharmacol.,* 11(1):8–23, February 1990.

Donegan, T.A., Landfill mining: an award-winning solution to an environmental problem, *Westchester Eng.,* April:56(8), 1992.

Edelstein, J., Who Should Pay?, *EPA J.,* 17(3):33, 1991.

Elan, K., Glass recycling rate drops seven percent in 1997, *Waste Age's Recycling Times,* June 1, 1998.

Evans, W.D., Stark, T.D., The Rumpke landslide, *Waste Age,* September:91–105, 1997.

Ewel, D., Solid waste flow control update, *BioCycle,* 36(5):38–39, May 1995.

Executive Office of the President, Reinventing environmental regulation, EOP, Washington, D.C., 1995.

Executive Office of the President, Reinventing environmental regulation, progress report, EOP, Washington, D.C., 1996.

Fang, H.Y., Pamukcu, S., Chaney, R.C., Soil-pollution effects on geotextile composite walls, American Society for Testing and Materials, Special Technical Publication 1129:103–116, 1992.

Farland, S.H., EPA's scientific reassessment of dioxin, *Health Environ. Dig.,* 5(12), March 1992.

Fisch, H. et al., Semen analyses in 1,238 men from the United States over a 25-year period: no decline in fertility, *Fertil. Sterility,* 65:1009–1014, 1996.

Fishbein, B.K., Gelb, C., Making less garbage: a planning guide for communities, *Inform,* New York, 1992.

Fogerty, A.M., Tuovinen, O.H., Microbiological degradation of pesticides in yard waste composting, *Microbiol. Rev.,* 55(2):225–233, June 1991.

Folinsbee, L.J., Human Health effects of air pollution, *Environ. Health Perspect.,* 100:45–56, 1992.

Food and Drug Administration (FDA), Pesticide program residue monitoring 1994, FDA, Washington, D.C., 1994.

Food and Drug Administration, Pesticide program residues monitoring, 1997, FDA, Washington, D.C., 1998.

Fordham, W., Yard trimmings composting in the Air Force, *BioCycle,* 36:44, 1995.

Forster, G., Assessment of landfill mining and the effects of age on combustion of recovered municipal solid waste, Landfill Reclamation Conference, Lancaster, PA, 1994.

Forster, G., Assessment of landfill reclamation and the effects of age on combustion of recovered municipal solid waste, National Renewable Energy Laboratory, Golden, CO, January 1995.

Friedman, D.J., Leaky oil tanks, *ASHI Tech. J.,* January:42–43, 1992.

Gagliardo, P.F., Steele, T.L., Taking steps to extend the life of San Diego's landfill, *Solid Waste Power,* June:34–40, 1991.

Garland, G.S., Grist, T.A., Green, R.E., The compost story: from soil enrichment to pollution remediation, *BioCyle,* 36:53–56, 1995.

Garner, J.W., Treatment technologies emerging to meet organochlorine removal needs, *Pulp Paper,* 65(11):137–143, November 1991.

Geiser, K., The greening of industry: making the transition to a sustainable economy, *Tech. Rev.,* 94(6):64–72, August/September 1991.

GeoLytics, Inc., Census CD + maps, U.S. 1990 Census (STF3A, C, and D), GeoLytics, Inc. East Brunswick, NJ, <www.Geolytics.com>, 1998.

Glass, D.J., Waste management: biological treatment of hazardous wastes, *Environment,* 33(9):43, 1991.

Glass Container Recycling, *Container Recycling Rep.,* November 1998.

Glenn, J., MSW composting in the United States, *BioCycle,* November 1997.

Glenn, J., The state of garbage in America, *BioCycle,* April 1998.

Goldman, B.A., Fitton, L.J., Toxic wastes and race revisited: An update of the 1987 report on the racial and socioeconomic characteristics of communities with hazardous waste sites, Center for the Policy Alternatives, Washington, D.C., 1994.

Goldstein, N., National trends in food residuals composting, Part I, *BioCycle,* July 1997.

Goldstein, N., Block, D., Nationwide inventory of food residuals composting, Part II, *BioCycle,* August 1997.

Goldstein, N., Glenn, J., Gray, K., Nationwide overview of food residuals composting, *Bio-Cycle*, August 1998.

Gould, M. et al., Source separation and composting of organic municipal solid waste, *Resour. Recycling*, July 1992.

Grehan, D.M., Dodd, V.A., Dennison, G.J., An experimental assessment of greenwaste compost for horticultural applications, *J. Solid Waste Technol. Manage.*, 23(1):28–33, 1996.

Grittner, N., Kaminsky, W., Obst, G., Fluid bed pyrolysis of anhydride-hardened epoxy resins and polyether-polyurethane by the Hamburg process, *J. Anal. Appl. Pyrolysis*, 25:293–299, 1993.

Hammer, S., Garbage in/garbage out: a hard look at mixed MSW composting, *Resour. Recycling*, February 1992.

Hanson, D., Hazardous wastes: EPA adds 25 organics to RCRA list, *Chem. Eng. News*, 68(11):4, 1990.

Hanson, D., Hazardous waste rules: EPA sent back to the drawing board, *Chem. Eng. News*, 69(50):4–5, 1991.

Hanson, D., Tighter rules urged for chemical rail shipments, *Chem. Eng. News*, 69:5, May 20, 1991.

Harris, P., Companies, cities square off over landfill cleanups, *Environ. Today*, 2(6):1, 11, 62, July/August 1991.

Hawkes, G., Thirteen reasons not to sink nuclear subs, *Oceans*, 16(5):70, 1983.

Hearne, S.A., Tracking toxics: chemical use and the public's right-to-know, *Environment*, July/August:5–9, 28–34, 1996.

Hegberg, B.A., Hallenbeck, W.H., Brenniman, G.R., Wadden, R.A., Setting Standards for yard waste compost, *BioCycle*, February 1991.

Hermes, L.H., Syme, J.D., Industrial pretreatment: cooperation to a point, *Water Eng. Manage.*, 137(8):40–41, August 1990.

Hickman, H.L., Jr., A broken promise reversing 35 years of progress, *MSW Manage.*, 8(4):78, July/August, 1994.

Hodges-Copple, J., The economic advantages of preventing pollution, *Bus. Econ. Rev.*, 36(4):38–41, July–September 1990.

Hollister, C., Subseabed disposal of nuclear wastes, *Science*, 214(4514):1432–1325, 1981.

Holmes, W.H., Converting waste to energy, *Plant Eng.*, 45(6):120–122, March 21, 1991.

Honshu, J., Brawn, I., Targeting commercial businesses for recycling, *Resour. Recycling*, November 1991.

How one college helps firms recycle industrial wastes, *Environ. Today*, 2(8):18–19, October 1991.

Interagency Working Group on Environmental Justice, Second annual report to the President by the Interagency Working Group on Environmental Justice on Implementation of Executive Order 12898, draft, Interagency Working Group on Environmental Justice, Washington, D.C., 1996.

International City/County Management Association, Brownfields redevelopment: a guidebook for local government and communities, 1997.

Jewell, W.J., Cummings, R.J., Richards, B.K., Methane fermentation of energy crops: maximum conversion kinetics and in situ biogas purification, *Biomass Bioenergy*, 5(4), 1993.

Kaminsky, W., Franck, J., Monomer recovery by pyrolysis of poly(methyl methacrylate) (PMMA), *J. Anal. Appl. Pyrolysis*, 19:311–318, 1991.

Kaminsky, W, Kastner, H., Recycle plastics into feedstocks, *Hydrocarbon Process.*, 1995(5):109–112, 1995.

Kaminsky, W., Kock, O., Sinn, H., Pyrolysis of "Fusen," *J. Anal. Appl. Pyrolysis,* 25:285–291, 1993.

Kaminsky, W., Röbler, H., Sinn, H, Pyrolyse von Elastomeren und Gummi zur Werstoffrück-gewinnung, *Kautsch. Gummi Kunstst.,* 44(9):846–851, 1991.

Kaminsky, W., Roch, J., Pyrolysis of a refinery sewage sludge, *Energ. Kohle-Erdgas-Petrochem.,* 46(0):323–325, 1993.

Kaminsky, W., Schlesselman, B., Simon, C.M., Thermal degradation of mixed plastic waste to aromatics and gas, *Polym. Degrad. Stability,* 53:189–197, 1996.

Kaminsky, W., Sinn, H., Pyrolase, in *Recycling von Kunststoffen, Munchen,* Menges, G., Michaeli, W., Bittner, M., Eds., Carl Hanser, Verlag, 1992, pp. 243–252.

Kavlock, R.J. et al., Research needs for the risk assessment of health and environmental effects of endocrine disruptors: A report of the U.S. EPA-sponsored workshop, *Environ. Health Perspect.,* 104:Suppl. 4, 1996.

Kelly, W.R., Buried treasure, *Civil Eng.,* April:52–54, 1990.

Kemezis, P., Truckers and chemical firms seek a common voice, *Chem. Week,* 45, February 19, 1992.

Kemezis, P., Hazardous waste disposal: three proposals in Mississippi, *Chem. Week,* 128:22, 1991.

Kemezis, P., Nuclear mercenaries name their price, *New Sci.,* 132:12, 1991.

Kenworth, L.A., *A Citizen's Guide to Promoting Toxic Waste Reduction,* Inform, New York, 1990.

Kenworthy, W.E., Lessons from the California spill, *Traffic World,* 227:47, August 5 1991.

Kilback, D., Barrett, G., Closing a landfill: a case study, *Pulp Paper Can.,* 98(6):T205-T208, 1997.

Kiser, J.V.L., The future role of municipal waste combustion, *Waste Age,* November 1991.

Kiser, J.V.L., Municipal waste combustion in North America: 1992 update, *Waste Age,* November 1992.

Knight, J.A., Planning pollution response to trucking spill accidents, *Risk Manage.,* 37:84–86, April 1990.

Koerner, R.M., *Designing with Geosynthetics,* 3rd ed., Prentice Hall, New York, 1994.

Koerner, R.M., Narejo, D., Bearing capacity of hydrated geosynthetic clay liners, *J. Geotech. Eng.,* January:82–85, 1995.

Korzun, E.A., Heck, H.H., Sources and fates of lead and cadmium in municipal solid waste, *J. Air Waste Manage. Assoc.,* 40(9):1220–1226, September 1990.

Kovalick, W.W., Cummings, J.B., Technological innovation in hazardous waste remediation, *J. Air Waste Manage. Assoc.,* 41(3):347–349, March 1991.

Kreis, I.A., Cadmium contamination of the countryside, a case study on health effects, *Toxicol. Ind. Health,* 6(5):181–188, October 1990.

Kreith, F., *Handbook of Solid Waste Management,* McGraw-Hill, New York, 1994.

Kunzler, C., Farrell, M., Food service composting projects update, *BioCycle,* May 1996.

Kunzler, C., Roe, R., Food service composting projects on the rise, *BioCycle,* April 1995.

La Bar, G., Du Pont: watching its waste, *Occup. Hazards,* 52(7):51–55, July 1990.

La Bar, G., Reducing the Flow, *Occup. Hazards,* 52(11):32–36, November 1990.

LaCarruba, J., Passing a medical waste inspection, *N. Engl. J. Med.,* 88(4):269–227, April 1991.

LaGasse, R.C., Marketing organic soil products, *BioCycle,* March 1992.

Laine, D., Darilek, G., Detecting leaks in geomembranes, *Civil Eng.,* 63:50–53, 1993.

Lee, G.F., Jones-Lee, A., Advantages and limitations of leachate recycle in MSW landfills, *World Waste,* 73(8):16, 19, August 1994.

Lee, G.F., Jones-Lee, A., Landfill post-closure care: can owners guarantee the money will be there?, *Solid Waste Power,* 7:35–39, 1993.

Lee, G.F., Jones, R.A., Managed fermentation and leaching: an alternative to MSW landfills, *Bicycle*, 31(5):78–80, 83, 1990.

Lewis, J., Superfund, RCRA, and UST: the cleanup threesome, *EPA J.*, 17(3):7, 1991.

Liss, G.M., Crimi, C., Jaczek, K.H., Anderson, A., Slattery, B., D'Cunha, C., Improper office disposal of needles and other sharps: an occupational hazard outside of health care institutions, *Can. J. Public Health*, 81(6):417–420, November/December 1990.

Lober, D.J., Informing the process and outcomes of recycling in the United Stated: the National Municipal Sold Waste Recycling Symposium, *J. Solid Waste Technol. Manage.*, 23(4):181–195, 1996.

Long, J., Chemical rail transport: another tank car spill hits California, *Chem. Eng. News*, 69:4, August 5, 1991.

Lucks, J.O., Dispose hazardous wastes safely, *Chem. Eng.*, 97(3):141–144, March 1990.

Lueck, G.W., Landfill mining yields buried treasure, *Waste Age*, March:118–120, 1990.

Lustig, T., Reducing ink disposal lost, *Graphic Arts Mon.*, 63:114–115, 1991.

Magnuson, A., Cap repair leads to landfill reclamation, *Waste Age*, September:121–124, 1990.

Magnuson, A., Landfill reclamation at Edinburg, *Waste Age*, November:75–78, 1991.

Malveaux, F.J., Fletcher-Vincent, S.A., Environmental risk factors of childhood asthma in urban centers, *Environ. Health Perspect.*, 103(Suppl. 6):59–62, 1995.

Manning, S., Waste exchanges: why dump when you can deal?, *Plant Eng. Manage.*, 13(6):32–41, June 1990.

Markets, *Plastic Recycling Update*, November 1998.

Marzulla, R.J., Superfund '91 — Congress' chance to clean up its act, *Risk Manage.*, 37(4):32–40, April 1990.

McCormick, R.C., Liability of property owners for pollution damage caused by former owners or tenants, *Rough Notes*, 134(10):40, October 1991.

McGrath, L.T., Creamer, P.D., Geosynthetic clay liner application, *Waste Age*, 99:104, May 1995.

Merchant, V.A., Molinari, J.A., Evacuation system lines and solid waste filter traps: associated flora and infection control, *Gen. Dent.*, 38(3):189–193, May/June 1990.

Mertinkat, J., Kirsten, A, Predel, M., Kaminsky, W., Cracking catalysts used as fluidized bed material in the Hamburg pyrolysis process, *J. Anal. Appl. Pyrolysis*, 49:87–95, 1999.

Metals markets take a tumble, *Waste Age's Recycling Times*, November 16, 1998.

Michaels, A., Solid waste forum: landfill recycling, *Public Works*, May:6668, 1993.

Migden, J.L., State policies on waste-to-energy facilities, *Public Util. Fortnightly*, 126:26, September 13, 1990.

Mohan, R.K., Herbich, J.B., Hossner, L.R., Williams, F.S., Reclamation of solid waste landfills by capping with dredged material., *J. Hazardous Mater.*, 53:141–164, 1997.

Morelli, J., Landfill reuse strategies, *BioCycle*, March:40–43, 1990.

Morelli, J., Town of Edinburg landfill reclamation demonstration project, New York State Energy Research and Development Authority, Albany, NY, doc. 92–4, December 1992.

Morelli, J., Town of Edinburg landfill reclamation demonstration project: report supplement, New York State Energy Research and Development Authority, Albany, NY, doc.93-7, December 1993.

Mumma, R.O., Raupach, D.C., Sahadewan, K., Manos, C.O., Rutzke, M., Kuntz, H.J., Bache, C.A., Lisk, D.J., National survey of elements and radioactivity in municipal incinerator ashes, *Archeol. Environ. Contam. Toxicol.*, 19(3):399–404, May/June 1990.

National Association of Local Governmental Professionals, Brownfields revitalization: challenges and opportunities for local governments, *Munic. Lawyer*, March/April 1998.

National Association of Local Government Environmental Professionals, National Incentives for Smart Growth Communities, *Nat. Resour. Environ.*, Summer 1998.

National Research Council, *Environmental Epidemiology: Public Health and Hazardous Wastes,* National Academy Press, Washington, D.C., 1991.

New Hazmat regulations, *Railway Age,* 192:44–45, April 1991.

Newell, T., Markstahler, E., Synder, M., Commercial food waste from restaurants and grocery stores, *Resour. Recycling,* February 1993.

Nightingale, D., Hartz, K., Hazardous waste minimization: what works, what doesn't?, *Am. City County,* 105(8):RR10–RR11, 1990.

No imminent rebound in U.S. market as weakness pervades all grades, *Paper Recycler,* June 1998.

Norris, T.C., The outlook for solid waste recycling, *Graphic Arts Mon..* 63:90, July 1991.

Nyamwange, M., Public perception of strategies for increasing participation in recycling programs, *J. Environ. Educ.,* 27(4):19–22, 1996.

Ohnesorgen, F., Sharing solid waste solutions, *Public Manage.,* 73:6–8, June 1991.

O'Leary, P., Walsh, P., Disposal of hazardous and special waste, *Waste Age,* February:81–88, 1992.

Organisation for Economic Co-operation and Development (OECD), *Environ. Performance Rev.,* United States, OECD, Paris, 1996.

O'Sullivan, D., Soil redemption gains momentum, *Chem. Eng., News,* 69(47):24, 1991.

Parkinson, G., Reducing wastes can be cost-effective, *Chem. Eng.,* 97(7):30–33, July 1990.

Patrico, J., Looking into the fire, *Farm J.,* 115(11):36–37, 1991.

Predel, M., Kaminsky, W., Pyrolysis of rape-seed in a fluidised-bed reactor, *Bioresource Technol.,* 66:113–117, 1998.

Price, R.L., Stopping waste at the source, *Civil Eng.,* 60(4):67–69, April 1990.

Recycling more glass rests on assured quality, *Purchasing,* 110:114, March 7, 1991.

Remich, N.C., Jr., Goal: no-waste metal treatment, *Appliance Man.,* 39(10):95–96, October 1991.

Repa, E.W., Blakey, A., Municipal Solid Waste Disposal Trends: 1996 update, Environmental Industry Associations, Washington, D.C., 1996.

Repa, E., Blakey, A., Municipal solid waste disposal trends: 1996 update, *Waste Age,* May 1996.

Rettenberger, G., Urban-Kiss, S., Schneider, R., Goschl, R., German project reconverts a sanitary landfill, *BioCycle,* June:44–47, 1995.

Richard, T.L., The key to successful MSW compost marketing, *BioCycle,* April 1992.

Richard, T.L., Woodbury, P., Strategies for separating contaminants from municipal solid waste, MSW Composting Fact Sheet, No. 3, Cornell Waste Management Institute, 1993.

Richards, B.K., Cummings, R.J., Jewell, W.F., Herndon, F.G., Jewell, W.J., High solids Anaerobic methane fermentation of sorghum and cellulose, *Biomass Bioenergy,* 1(1):47–53, 1991.

Richards, B.K., Cummings, R.J., Jewell, W.J., High rate low solid anaerobic methane fermentation of sorghum, corn and cellulose, *Biomass Bioenergy,* 1(5):249–260, 1991.

Richards, B.K., Cummings, R.J., Jewell, W.J., White, T.E., Methods for kinetic analysis of methane fermentation in high solids biomass digesters, *Biomass Bioenergy,* 1(2):65–73, 1991.

Riggle, D., How to promote backyard composting, *BioCycle,* April:48–49, 1996.

Riggle, D., Tapping textile recycling, *BioCycle,* February 1992.

Robison, R., Detecting leaks electronically, *Civil Eng.,* 66:16A, 1996.

Roulac, J., Pedersen, M., Home composting heats up, *Resour. Recycling,* April 1993.

Rules and regulations, *Fed. Regist.,* 63(115):32743–32753, June 16, 1998.

Rumer, R.R., Mitchell, J.K., Assessment of barrier containment technologies, International Containment Technology Workshop, Baltimore, MD., 1995, pp. 355–394.

Rutledge, G., Pollution abatement and control expenditures, 1987–1989, *Surv. Curr. Bus.,* 71(11):46, 1991.

Savage, G.M., The history and utility of waste characterizations studies, *MSW Manage.,* May/June 1994.

Schaefer, M.E., Hazardous waste management, *Dental Clin. N. Am.,* 35(2):383–390, April 1991.

Schmitt, C.J., Bunck, C.M., Persistent environmental contamination in fish and wildlife, in U.S. Department of the Interior, National Biological Service, *Our Living Resources,* GPO, Washington, D.C., 1995.

Schubert, W.R., Bentonite matting in composite lining systems. *Geotechnical Practice for Waste Disposal,* American Society of Civil Engineers, New York, 1987.

Schulz, J.D., Key rail, truck executives urge realism in Hazmat laws, not more regulation, *Traffic World,* 226:15, June 24, 1991.

Scudder, K., Blehm, K.D., Household hazardous waste, *J. Environ. Health,* 53(6):18–20, 1991.

Sellers, C., Combined sample preparation and inductively coupled plasma emission spectroscopy method for determination of 23 elements in solid wastes: summary of collaborative study, *J. Assoc. Off. Anal. Chem.,* May/June:110–118, 1990.

SERI claims solid waste gives pipeline-quality gas, *Chem. Market. Rep..* 240:240, August 12, 1991.

Shan, H.Y., Daniel, D.E., Results of laboratory tests on a geotextile/bentonite liner material, in Proceedings Geosynthetics, Industrial Fabrics Association International, St. Paul, MN, 2:517–535, 1991.

Shanklin, C.W., Solid waste management: how will you respond to the challenge?, *J. Am. Diet. Assoc.,* 91(6):663–664, June 1991.

Simon, C.M., Kaminsky, W., Chemical recycling of polytetrafluoroethylene by pyrolsis, *Polym. Degradation Stability,* 62:1–7, 1998.

Simon, C.M., Kaminsky, W., Schlesselman, B., Pyrolysis of polyolefins with steam to yield olefins, *J. Anal. Appl. Pyrolysis,* 38:75–87, 1996.

Skaja, J., Anaerobic gasification advances, *BioCycle,* 32(10):74–77, October 1991.

Skeely, S., Health benefits predicted from hazardous waste, *J. Environ. Health,* 53:10, 1991.

Skumatz, L.A., Truitt, E., Green, J., The state of variable rates: economic signals more into the mainstream, *Resour. Recycling,* August 1997.

Solomon, M.B., Hazmat bill pits moral concerns against fear of economic debacle, *Traffic World,* 222:34–35, May 14, 1990.

Spencer, R., Landfill space reuse, *BioCycle,* February:30–32, 1990.

Spencer, R., Mining Landfills for recyclables, *BioCycle,* February:34, 1991.

State-of-the-art procedures and equipment for internal inspection and upgrading of USTs, EPA 600-R-97-085, September 1997.

Steel can recycling, *Container Recycling Rep.,* November 1998.

Steuteville, R, The state of garbage in America, Part 1, *BioCycle,* 36(4):54–63, 1995.

Steuteville, R., The state of garbage in America, Part 2, *BioCycle,* 36(5):30–37, 1995.

Steuteville, R., The state of garbage in America, Part 1, *BioCycle,* 37(4):54–61, 1996.

Steuteville, R., The state of garbage in America, Part 2, *BioCycle,* 37(5):35–41, 1996.

Stevenson, M.E., Provoking a fire storm: waste incineration, *Environ. Sci. Tech.* 25:1808–1813, 1991.

Stinson, M.C., Galanek, M.S., Ducatman, A.M., Masse, F.X., Kuritzkes, D.R., Model for inactivation and disposal of infectious human immunodeficiency virus and radioactive waste in a BL-3 facility, *Appl. Environ. Microbiol.,* 56(1):264–268, January 1990.

Stundza, T., Recycling resources: new raw material sources for the '90s, *Purchasing,* 109(7):64B26–64B28, November 8, 1990.

Suprenant, B.A., Lahrs, M.C., Smith, R.L., Oilcrete: facing tighter hazardous-waste legisla-
    tion, oil companies search for ways to detoxify oil-waste pits, *Civil Eng.,* 60(4):61–63,
    April 1990.
Takeshita, R., Akimoto, Y., Leaching of polychlorinated dibenzo-*p*-dioxins and dibenzofurans
    in fly ash from municipal solid waste incinerators to a water system, *Archeol. Environ.
    Contam. Toxicol.,* 21(2):245–252, August 1991.
Taylor, S.M., Elliot, S., Eyles, J., Frank, J., Haight, M., Steiner, D., Walter, S., White, N.,
    Williams, D., Psychosocial impacts in populations exposed to solid waste facilities,
    Soc. Sci. Med., 33(4):441–447, 1991.
Tedder, D.W., Pohland, F.G., Emerging technologies in hazardous waste management II, ACS
    Symposium Series 468, 1991.
Thayer, A.M., Bioremediation: innovation technology for cleaning up hazardous waste, *Chem.
    Eng., News,* 69(34):23–44, 1991.
Thayer, A.M., Hazardous waste management industry to grow, *Chem. Eng. News,* 70(1–12):12,
    1991.
Thibodeaux, L., Hazardous material management in the future, *Environ. Sci. Technol.,*
    April:456–459, 1990.
Thornloe, Landfill gas recovery/utilization–options and economics, Presented at the 16th
    Annual Conference by the Institute of Gas Technology on Energy from Biomass and
    Wastes, Orlando, FL, March 15, 1992.
Thorneloe, Cosulich, Pacey, Roqueta, Landfill gas utilization-Survey of United States projects.
    Presented at the Solid Waste Association of North America's 20th Annual Interna-
    tional Landfill Gas Symposium, Monterey, CA, March 25–27; published in Confer-
    ence Proceedings, Environmental Protection Agency Office of Research and
    Development, EPA-ORD, 1997.
Thornloe and Pacey, Database of North American landfill gas-to-energy projects, Presented
    at the 17th Annual International Landfill Gas Symposium by the Solid Waste Asso-
    ciation of North America, Long Beach, CA, March 22–24, 1944; published in Con-
    ference Proceedings, 1994a.
Thorneloe and Pacey, Database of North American landfill gas-to-energy projects. Presented
    at the 18th Annual International Landfill Gas Symposium by the Solid Waste Asso-
    ciation of North America, New Orleans, LA, March 27–30, 1995; published in
    Conference Proceedings, 1995.
Thorneloe and Pacey, Landfill gas utilization — technical and non-technical considerations.
    Presented at the 17th Annual International Landfill Gas Symposium by the Solid
    Waste Association of North America, Long Beach, CA, March 22–24, 1944; published
    in Conference Proceedings, 1994b.
Titalsky, S., Turning tires into electricity, *Public Util. Fortnightly,* 127:17, May 1, 1990.
Toxic waste is banks' latest hazard, *U.S. Banker,* 100(5):58, May 1990.
Travis, C.C., Hattemer-Frey, H.A., Human exposure to dioxin, *Sci. Total Environ.,*
    104(1–2):97–127, May 1991.
UCB market analysis, *Container Recycling Rep.,* November 1998.
Underground storage tanks: requirements and options, EPA 510-F- 97-005, June 1997.
U.S. Code of Federal Regulations, Protection of the environment, Title 40, Office of the
    Federal Register National Archives and Records Administration, Washington, D.C.,
    1991.
U.S. Conference of Mayors, Recycling America's land: a national report on brownfields
    redevelopment, January 1998, 46 pp.
U.S. Congress, Senate Committee on Environment and Public Works, Liability and Resource
    Issues. (Brownfields), 105th Congress 1st Session, March 4, 1997, pp. 105–142.

U.S. Congress, Subcommittee on Transportation and Hazardous Materials of the Committee on Energy and Commerce, Hazardous Wastes-United States, U.S. Government Printing Office, Washington, D.C., 1991.

U.S. Congressional Budget Office, The total costs of cleaning up nonfederal superfund sites, January 1994, 54 pp.

U.S. Department of Agriculture, Estimating and addressing America's food losses, Economic Research Service, www.econ.ag.gov/., July 1997.

U.S. Department of Agriculture, Food consumption, prices, and expenditures, 1996, Economic Research Service, Putnam, J.J., April 1996.

U.S. Department of Commerce, Bureau of the Census, Combined annual and revised monthly retail trade, *Curr. Bus. Rep.,* BR/95-RV.

U.S. Department of Commerce, Bureau of the Census, Monthly retail trade, *Curr. Bus. Rep.,* April 1997

U.S. Department of Commerce, Bureau of the Census, Statistical abstract of the United States, 1997.

U.S. Department of Commerce, U.S. Environmental Protection Agency, Landview III environmental mapping software, <www.census.gov/apsd/pp98/pp.html>, 1997.

U.S. Department of Energy, Energy Information Administration, Electric Power Annual, Volume I and Volume II, DOE/EIA-0348(97), 1997.

U.S. Department of Energy, Energy Information Administration, Instructions for form EIA-1605 (1998), voluntary reporting of greenhouse gases, OMB No. 1905-0194, 1998.

U.S. Department of Energy, Office of Environmental Management, Annual report of waste generation and pollution prevention progress 1997, DOE/Em-03635. DOE, Washington, D.C., 1998.

U.S. Department of Health and Human Services–NIEHS, Brownfields: Hazmat cleanup, but more, April 1998.

U.S. EPA, Administration and Resource Management, Access EPA Public Information Tools, U.S. Government Printing Office, Washington, D.C., 1991.

U.S. EPA, Administration and Resource Management, Hazardous Waste Superfund Collection, U.S. Government Printing Office, Washington, D.C., 1990.

U.S. EPA, Anthropogenic methane emissions in the United States: estimates for 1990, report to Congress, EPA 430-R-93-003, United States Environmental Protection Agency, 1993a.

U.S. EPA, Characterization of municipal solid waste in the United States: 1990 update, EPA/530-SW-90-042, June 1991.

U.S. EPA, Characterization of municipal solid waste in the United States: 1992 update, EPA/530-R-92-019, July 1992.

U.S. EPA, Characterization of municipal solid waste in the United States: 1994 update, EPA/530-R-94-042, November 1994.

U.S. EPA, Characterization of municipal solid waste in the United States, 1995 update, EPA/530-R-96-001, November 1995.

U.S. EPA, Characterization of municipal solid waste in the United States: 1995 update, EPA/530-R-945-001, March 1996.

U.S. EPA, Characterization of municipal solid waste in the United States: 1996 update, EPA/530-R-97-015, June 1997.

U.S. EPA, Characterization of municipal solid waste in the United States: 1997 update, EPA/530-R-98-007, May 1998.

U.S. EPA, Compilation of information on alternative barriers for liner and cover systems, Prepared by Daniel, D.E., Estornell, P.M. for Office of Research and Development, Washington, D.C., October 1990.

U.S. EPA, Construction quality management for remedial action and remedial design waste containment systems, OH, EPA540-8-92-073, Technical guidance document, Risk Reduction Engineering Laboratory, Cincinnati, OH, 1992.

U.S. EPA, The consumer's handbook for reducing solid waste, EPA/530-D-92-003, August 1992.

U.S. EPA, Does your business produce hazardous waste? Many small businesses do, EPA/530-SW-90-027,U.S. EPA Office of Solid Waste and Energy Response, Washington, D.C., 1990.

U.S. EPA, Effect of freeze/thaw on the hydraulic conductivity of barrier materials: laboratory and field evaluation, prepared by Kraus, J.D., Benson C.H. for the Risk Reduction Engineering Laboratory, Cincinnati, OH, EPA600-R-95-118, 1995.

U.S. EPA, Environmental justice 1994 report: focusing on environmental protection for all people, EPA/200-R-95-003, EPA, Office of Environmental Justice, Washington, D.C., April 1995.

U.S. EPA, Environmental justice fact sheet: interagency working group, EPA/200-A-95-009, EPA, Office of Environmental Justice, Washington, D.C., June 1995.

U.S. EPA, Environmental justice implementation plan, draft, EPA/300-R—96-004, EPA, Office of Environmental Justice, Washington, D.C., April 1996.

U.S. EPA, Environmental justice strategy: Executive Order 12898, EPA/200-R-95-002, EPA, Office of Environmental Justice, Washington, D.C., April 1955.

U.S. EPA, EPA's ongoing regulatory program, Dioxin Facts, EPA, Office of Communications, Education, and Public Affairs, Washington, D.C., September 1994.

U.S. EPA, Guide to accessing pollution prevention information electronically, EPA, Washington, D.C., 1997.

U.S. EPA, Hazardous Waste Land Treatment, Cincinatti, OH, 1991.

U.S. EPA, Helping landfill owners achieve effective, low-cost compliance with federal gas regulations, United States Environmental Protection Agency, 1998.

U.S. EPA, Opportunities to reduce anthropogenic methane emissions in the United States: report to Congress, United States Environmental Protection Agency, EPA 430-R-93-012, 1993.

U.S. EPA, Pollution Prevention Directory, EPA, Washington, D.C., 1994.

U.S. EPA, Quality assurance and quality control for waste containment facilities, technical guidance document, EPA600-R-93-182, Risk Reduction Engineering Laboratory, Cincinnati, OH, September 1992.

U.S. EPA, The RCRA Orientation Manual, U.S. Government Printing Office, Washington, D.C., 1991.

U.S. EPA, Report of 1995 workshop on geosynthetic clay liners, EPA600-R-96-149, Washington, D.C., June 1996.

U.S. EPA, Report of workshop on geosynthetic clay liners, EPA600-R-93-171, Office of Research and Development, Washington, D.C., August 1993.

U.S. EPA, Second forum on innovative hazardous waste treatment technologies: domestic and international, Office of Solid Waste and Emerging Response, Philadelphia, PA, 1990.

U.S. EPA, Superfund administrative reforms annual report, Fiscal Year 1996, December 1996, 56 pp.

U.S. EPA, Superfund: outlook for and experience with natural resource damage settlements, April 1996, 42 pp.

U.S. EPA, Turning a liability into an asset: a landfill gas-to-energy handbook for landfill owners and operators, United States Environmental Protection Agency, 1994.

U.S. EPA, Waste wise: second year progress report, EPA/530-R-96/016, September 1996.

U.S. EPA, Workshop report on toxicity equivalency factors for polychlorinated biphenyl congers, U.S. Government Printing Office, Washington, D.C., 1991.

U.S. EPA, Office of Emergency and Remedial Response, Superfund clean-up figures, EPA, Washington, D.C., 1998.

U.S. Office of the Federal Register, National Archives and Records Administration, *Code of Federal Regulations,* Title 40, Part 171, U.S. Government Printing Office, Washington, D.C., 1991.

U.S. Office of the Federal Register, National Archives and Records Administration, *Code of Federal Regulations,* Title 40, Part 260, U.S. Government Printing Office, Washington, D.C., 1991.

U.S. Office of the Federal Register, National Archives and Records Administration, *Code of Federal Regulations,* Title 40, Part 261, U.S. Government Printing Office, Washington, D.C., 1991.

U.S. Office of the Federal Register, National Archives and Records Administration, *Code of Federal Regulations,* Title 40, Part 263, U.S. Government Printing Office, Washington, D.C., 1991.

U.S. Office of the Federal Register, National Archives and Records Administration, *Code of Federal Regulations.* Title 40, Part 302, U.S. Government Printing Office, Washington, D.C., 1991.

U.S. Office of the Federal Register, National Archives and Records Administration, *Code of Federal Regulations,* Title 49, Part 127, U.S. Government Printing Office, Washington, D.C., 1991.

U.S. Office of the Federal Register, National Archives and Records Administration, *Code of Federal Regulations,* Title 49, Part 172, U.S. Government Printing Office, Washington, D.C., 1991.

U.S. Office of the Federal Register, National Archives and Records Administration, *Code of Federal Regulations,* Title 49, Part 173, U.S. Government Printing Office, Washington, D.C., 1991.

U.S. EPA, Office of Outreach and Special Projects, *Handbook of Tools for Managing Federal Superfund Liability Risks at Brownfields and Other Sites,* November 1998.

U.S. EPA, Office of Pollution Prevention and Toxics, 1994 toxics release inventory executive summary, EPA, Washington, D.C., 1996.

U.S. EPA, Office of Pollution Prevention and Toxics, 1996 toxics release inventory: public data release, EPA, Washington, D.C., 1998.

U.S. EPA, Office of Pollution Prevention and Toxics, fiscal year 1997 annual report, EPA, Washington, D.C., 1998.

U.S. EPA, Office of Pollution Prevention and Toxics, Pollution prevention 1997: a national progress report, EPA, Washington, D.C., 1997.

U.S. EPA, Office of Research and Development, Hazardous Waste Management, U.S. Government Printing Office, Washington, D.C., 1991.

U.S. EPA, Office of Research and Development, Remediation of Contaminated Sediments, U.S. Government Printing Office, Washington, D.C., 1991.

U.S. EPA, Office of Research and Development, Toxic treatments, in situ steam/hot air stripping technology, U.S. Government Printing Office, Washington, D.C., 1991.

U.S. EPA, Office of Solid Waste and Emergency Response, Characterization of municipal solid waste in the United States: 1995 update, EPA, Washington, D.C., 1996.

U.S. EPA, Office of Solid Waste and Emergency Response, Characterization of municipal solid waste in the United States: 1997, EPA, Washington, D.C., 1998.

Wagner, T.P., *The Complete Guide to Hazardous Waste Regulations,* Van Nostrand Reinhold, New York, 1991.

Walker, J.M., O'Donnell, M.J., Comparative assessment of MSW compost characteristics, *BioCycle,* August 1991.

Walsh, P., Pferdehirt, W., O'Leary, P., Collection of recyclables from multifamily housing and businesses, *Waste Age,* April 1993.

White, C.C., Barker, R.D., Electrical leak detection system for landfill liners, a case history, *Ground Water Monitor Remediation,* 17(3):153–159, 1997.

The World Bank, Pollution prevention and abatement handbook: toward cleaner production, The World Bank, Washington, D.C., 1998.

CHAPTER 3

Allen, H.E., *Metals in Groundwater,* Lewis Publishers, Boca Raton, FL, 1991.

American Water Works Association, *AWWA J.,* 88(9):53–136, *AWWA J.,* 87(9):83–121, 1996.

Appel, C.A., Reilly, T.C., Summary of selected computer programs produced by the U.S. Geological Survey for simulation of ground-water flow and quality, U.S. Geological Survey Circular 1104, 1994, 98 pp.

Aspelin, A.L., Pesticides industry sales and usage, 1992 and 1993 market estimates, U.S. EPA, Office of Pesticides Programs, Biological and Economic Analysis Div., Economic Analysis Branch Report 733-K-94-001, 1994, 33 pp.

Barbash, J.E., Pesticides in ground water: current understanding of distribution and major influences, U.S. Geological Survey Fact Sheet FS-244-95, 1995, 4 pp.

Barbash, J.E., Pesticides in ground water: an overview of current understanding [abstract]: American Geophysical Union Spring Meeting. The Rubin Symposium on Soil Science, Solute Transport, and National-Scale Water Quality Assessments, 1 June 1995, Baltimore, MD, *Eos,* 76(17):S143, 1995.

Barbash, J.E., Problems associated with the use of solute transport models and vulnerability assessments for predicting the behavior of pesticides in the subsurface [abstract], American Chemical Society Abstracts of Papers, 208th National Meeting, Part 1, 22 August 1994.

Barbash, J.E., Resek, E.A., Pesticides in ground water: distribution, trends, and governing factors, *Pesticides in the Hydraulic System,* Vol. 2, Chelsea Press, Ann Arbor, MI, 1996, 590 pp.

Barbash, J.E., Thelin, G.P., Kolpin, D.W., Gillliom, R.J., Distribution of major herbicides in ground water of the United States, U.S. Geological Survey Water-Resources Investigations Report 98-4245, 1999.

Bell, A., Guasparini, R., Meeds, D. et al., A swimming pool-associated outbreak of cryptosporidiosis in British Columbia, *Can. J. Public Health,* 84:334–337, 1993.

Bender, D.A., Selection procedure and salient information for volatile organic compounds emphasized in the National Water-Quality Assessment Program, U.S. Geological Survey Open-File Report, OFR 99-182, 1999, 32 pp.

Burkart, M.R., Kolpin, D.W., James, D.E., Assessing groundwater vulnerability to agrichemical contamination in the Midwest U.S., *Water Sci. Technol.,* 39(3):103–112, 1999.

Butler, B.J., Mayfield, C.I., *Cryptosporidium* spp. — a review of the organism, the disease, and implications for managing water resources, Waterloo Centre for Groundwater Research, Waterloo, Ontario, Canada, August 31, 1996.

Cantor, K.P., Lynch, C.F., Hildesheim, M., Chlorinated drinking water and risk of glioma: a case control study, Iowa, U.S.A, *Epidemiology,* 7(4):PS83, (T25), 1996.

Cantor, K.P., Lynch, C.F., Hildesheim, M.E., Dosemeci, M., Lubin, J., Alavanja, M., Craun, G., Drinking water source and chlorination byproducts. I. Risk of bladder cancer, *Epidemiology,* 9:21–18, 1998.

Capel, P.D., Organic chemical concepts, in *Regional Ground-Water Quality,* Alley, W.M., Ed., Van Nostrand Reinhold, New York, 1993, pp. 155–179.

Casemore, D.P., Epidemiological aspects of human cryptosporidiosis, *Epidemiol. Infect.,* 104:1–28, 1990.

CDC, Cercarial dermatitis outbreak at a state park — Delaware, *MMWR,* 41:225–228, 1992.

CDC, Methemoglobinemia in an infant — Wisconsin, *MMWR,* 42:217–219, 1992.

CDC, Primary amebic meningoencephalitis — North Carolina, 1991, *MMWR,* 41:437–440, 1991.

Commission on Physical Sciences, Mathematics, and Resources, National Research Council, *Groundwater Models: Scientific and Regulatory Applications,* National Academy Press, Washington, D.C., 1990.

Cory, D., Moy, M., Reauthorization of the Safe Drinking Water Act and the viability of rural public water systems, *Environ. Law Forum,* May:1–5, 18, 1995.

Cothern, R., Rebers, P., Eds., *Radon in Drinking Water,* Lewis Publishers, Boca Raton, FL, 1990.

Craun, G.F., Ed., *Safety of Water Disinfection: Balancing Chemical & Microbial Risks,* ILSI Press, Washington, D.C., 1993, 690 pp.

Craun, G.F., Waterborne disease outbreaks in the United States of America: causes and prevention, *World Health Stat. Q.,* 45:192–195, 1992.

Craun, G.F., McGoldrick, J.L., Workshop on methods for investigation of waterborne disease outbreaks, EPA publication no. 600/9-90/021, U.S. Environmental Protection Agency, Research Triangle Park, N.C., 1990.

Cryptosporidium capsule, *Gov. Epidemiol. Res. Bus. Technol. News,* Vols. 1, 2, 1996.

Cryptosporidium in drinking water, Arkansas Department of Health, educational packet, Little Rock, AR, 1996.

Current, W.L., Garcia, L.S., Cryptosporidiosis, *Clin. Microbiol. Rev.,* July:325–328, 1991.

Davis, W.F., A case study of lead in drinking water: protocol, methods, and investigative techniques, *Am. Ind. Hyg. Assoc. J.,* 51(12):620–624, 1990.

Delzer, G.C., Occurrence of the gasoline oxygenate MTBE and BTEX compounds in urban storm water in the United States, U.S. Geological Survey Water-Resources Investigation Report, 1996, 6 pp.

Delzer, G.C., Quality of methyl *tert*-butyl ether (MTBE) data for ground-water samples collected during 1003-95 as part of the National Water-Quality Assessment Program, U.S. Geological Survey Fact Sheet, FS-101-99, 1999, 4 pp.

Delzer, G.C., Setmire, J.G., Quality of methyl *tert*-butyl ether (MTBE) data for ground-water samples collected during 1993-95 as part of the National Water-Quality Assessment Program: U.S. Geological Survey Fact Sheet, FS-101-99, 1999, 4 pp.

Delzer, G.C., Zogorski, J.S., Lopes, T.J., Occurrence of the gasoline oxygenate MTNE and BTEX compounds in urban storm water in the United States, 1991-95, in American Chemical Society Division of Environmental Chemistry preprints of papers, 213th, San Francisco, CA, *Am. Chem. Soc.,* 37(1):374–377, 1997.

De Zuane, J., *Drinking Water Quality: Standards and Controls,* Van Nostrand Reinhold, New York, 1990.

Doyle, T.J., Zheng, W., Cerhan, J.R., Hong, C.P., Sellers, T.A., Kushi, L.H., Folsom, A.R., The association of drinking water source and chlorination by-products with cancer incidence among postmenopausal women in Iowa: a prospective cohort study, *Am. J. Public Health,* 87:1168–1176, 1997.

DuPont, H.L., Chappell, C.L., Sterling, C.R., Okhuysen, P.C., Rose, J.B., Jakubowski, W., The infectivity of *Crytpsporidium parvum* in healthy volunteers, *N. Engl. J. Med.,* 332:855–859, 1995.

Fagliano, J., Berry, M., Bove, F., Burke, T., Drinking water contamination and the incidence of leukemia: an ecologic study, *Am. J. Public Health,* 80:1209–1212, 1990.

Fennemore, C., Groundwater quality protection, in *Arizona Environmental Law*, Federal Publications, Washington, D.C., chap. 9.

Financial viability training for small drinking water systems, Arizona Department of Environmental Quality, 1995.

Ford, T.E., Keelhaul, R.R., A global decline in microbiological safety of water: a call for action, American Academy of Microbiology, 1325 Massachusetts Ave., N.W., Washington, D.C., 1996.

Freedman, D.M., Cantor, K.P., Lee, N.L., Chen, L.S., Lei, H.H., Ruhl, C.E., Wang, S.S., Bladder cancer and drinking water: a population-based case-control study in Washington County, MD (United States), *Cancer Causes Control*, 8:738–744, 1997.

Frey, M., Hancock, C., Logsdon, G., 10 common questions about *Cryptosporidium*, *Opflow*, 24(2):February 1998.

Geldreich E.E., Fox, K.R., Goodrich, J.A., Rice, E.W., Clark, R.M., Swerdlow, D.L., Searching for a water supply connection in the Cabool, Missouri–disease outbreak of *Escherichia coli* O157:H7, *Water Res.*, 26:1127–1137, 1992.

General Accounting Office (GAO Report), Drinking water: widening gap between needs and available resources threatens vital EPA Program, Washington, D.C., 1992.

Grady, S., Ground-water issues and recent findings on MTBE, Ground Water, Augusta, Maine, April 15, 1999.

Grady, S., A plan for assessing the occurrence and distribution of methyl-*tert*-butyl ether and other volatile organic compounds in drinking water and ambient ground water in the northeast and mid-Atlantic regions of the United States, U.S. Geological Survey Open-File Report 99-207, 1999, 36 pp.

Griffin, P.M., Tauxe, R.V., The epidemiology of infections caused by *Escherichia coli* O157:H7, other enterohemorrhagic *E. coli* and the associated hemolytic uremic syndrome, *Epidemiol. Rev.*, 13:60–68, 1991.

Gustafson, D.I., *Pesticides in Drinking Water*, Van Nostrand Reinhold, New York, 1993.

Haas, C.N., Rose, J.B., Reconciliation of microbial risk models and outbreak epidemiology: the case of the Milwaukee outbreak [abstract], in Proceedings of the American Water Works Association 1994 Annual Conference: Water Quality, Denver, CO, American Water Works Association, 1994, pp. 517–523.

Halde, M.J., Study design and analytical results used to evaluate a surface-water point sampler for volatile organic compounds, U.S. Geological Survey Open-File Report, OFR 98-651, 1998, 31 pp.

Hayes, E.B., Matte, T.D., O'Brien, T.R. et al., Large community outbreak of cryptosporidiosis due to contamination of a filtered public water supply, *N. Engl. J. Med.*, 320:1372–1376, 1989.

Herwaldt, B.L., Craun, G.F., Stokes, S.L., Juranek, D.D., Waterborne-disease outbreaks, 1989–1990, in CDC Surveillance Summaries, December 1991, *MMWR*, 40(SS-3):1–21, 1991.

Hildesheim, M.E., Cantor, K., Lynch, C.F., Dosemeci, M., Lubin, J., Alavanja, M., Craun, G., Drinking water and chlorination byproducts II. Risk of colon and rectal cancers, *Epidemiology*, 9:29–35, 1998.

Hoff, G., Moen, I.E., Mowinkel, P., Rosef, O., Nordbro, E., Sauar, J., Vatn, M.H., Torgrimsen, T., Drinking water and the prevalence of colorectal adenomas; an epidemiologic study in Telemark, Norway, *Eur. J. Cancer Prev.*, 1:423–428, 1992.

Ijsselmuiden, C.B., Gaydos, C., Feighner, B., Novakoski, W.L., Serwadda, D., Caris, L.H., Comstock, G.W., Cancer of the pancreas and drinking water: a population based case-control study in Washington County, MD, *Am. J. Epidemiol.*, 139:836–842, 1992.

King, W., Great Lakes water and your health. A summary of Great Lakes Basin cancer risk assessment: A case control study of cancers of the bladder, colon and rectum, Health Canada publication, December 1995.

King, W.D., Marrett, L.D., Case-control study of bladder cancer and chlorination by-products in treated water (Ontario, Canada), *Cancer Causes Control,* 7:596–604, 1996.

Koivusalo, M., Hakulinen, T., Vartiainen, T., Pukkala, E., Jaakkola, J.J.K., Tuomisto, J., Drinking water mutagenicity and urinary track cancers: a population-based case-control study in Finland, *Am. J. Epidemiol.,* 148:704–712, 1998.

Koivusalo, M., Jaakkola, J.J.K., Vartiainen, T., Hakulinen, T., Karjalainen, S., Pukkala, E., Tuomisto, J., Drinking water mutagenicity and gastrointestinal and urinary tract cancers: an ecological study in Finland, *Am. J. Public Health,* 84:1223–1228, 1994.

Koivusalo, M., Pukkala, E., Vartiainen, T., Jaakkola, J.J.K., Hakulinen, T., Drinking water chlorination and cancer–a historical cohort study in Finland, *Cancer Causes Control,* 8:192–200, 1997.

Koivusalo, M., Vartiainen, T., Hakulinen, T., Pukkala, E., Jaakkola, J., Drinking water mutagenicity and leukaemia, lymphomas and cancers of the liver, pancreas and soft tissue, *Arch. Environ. Health,* 50:269–276, 1995.

Kolpin, D.W., Barbash, J.E., Gilliom, R.J., Occurrence of pesticides in shallow groundwater of the United States, initial results from the National Water-Quality Assessment, 1998.

Kolpin, D.A., Hallberg, G.R., Libra, R.D., Temporal trends of selected agicultural chemicals in Iowa's groundwater, 1982–1995: are things getting better?, *J. Environ. Qual.,* 26(4):1007–1017, 1997.

Kolpin, D.W., Kalkhoff, S.J., Goolsby, D.A., Sneck-Faher, D.A., Thurman, E.M., Occurrence of selected herbicides and herbicide degradation products in Iowa's ground water, *Ground Water,* 35(4):679–688, 1995.

Kolpin, D.W., Squillace, P.J., Zogorski, J.S., Barbash, J.E., Pesticides and volatile organic compounds in shallow urban groundwater of the United States, in Chilton, J. et al., Eds., Congress on Groundwater in the Urban Environment [Proceedings], A.A. Balkema, Netherlands, 1997, pp. 469–474.

Kolpin, D.W., Thurman, E.M., Linhart, S.M., The environmental occurrence of herbicides: the importance of degradates in ground water, *Arch. Environ. Contamin. Toxicol.,* 35:385–390, 1998.

Kool, J.L., Effect of monochloramine disinfection of municipal drinking water on risk of nosocomial Legionnaires' disease, *Lancet Interactive,* January 23, 1999.

Kramer, M.H., Herwaldt, B.L., Craun, G.F., Calderon, R.L., Juranek, D.D., Surveillance for waterborne-disease outbreaks — United States, 1993–1994, in CDC Surveillance Summaries, April 12, 1996, *MMWR,* 45(SS-l):1–33, 1996.

Kukkula, M., Lofroth, G., Chlorinated drinking water and pancreatic cancer. A population based case-control study, *Eur. J. Public Health,* 7:297–301, 1997.

Lapham, W.W., Enhancements of nonpoint-source monitoring programs to assess volatile organic compounds in the nation's ground water, in National Monitoring Conference, July 7–9, 1998.

Lapham, W.W., Ground-water data-collection protocols and procedures for the National Water-Quality Assessment Program — selection, installation, and documentation of wells and collection of related data, U.S. Geological Survey Open-File Report, OFR 95-398, 1995, 69 pp.

Lapham, W.W., Plan for assessment of the occurrence, status, and distribution of volatile organic compounds in aquifers of the United States, U.S. Geological Survey Open-File Report, OFR 96-199, 1996, 44 pp.

Lapham, W.W., USGS compiles data set for national assessment of VOCs in ground water, *Ground Water Monitoring Remediation,* Fall:147–157, 1997.

Lapham, W.W., Neitzert, K.M., Moran, M.J., Zogorski, J.S., USGS compiles data set for national assessment of VOC's in ground water, *Ground Water Monitoring Remediation,* Fall:147–157, 1997.

Lapham, W.W., Tadayon, S., Plan for assessment of the occurrence, status, and distribution of volatile organic compounds in aquifers of the United States, U.S. Geological Survey Open-File Report, OFR 96–199, 1996, 44 pp.

Lapham, W.W., Wilde, F.D., Koterba, M.T., Ground-water data-collection protocols and procedures for the National Water-Quality Assessment Program — selection, installation, and documentation of wells, and collection of related data, U.S. Geological Survey Open-File Report, OFR 95–398, 1995, 69 pp.

Lawhorn, B., Human *Cryptosporidium* and Cryptosporidosis, Texas Agricultural Extension Service, publication L-5162, College Station, TX, 1996, 4 pp.

LeChevallier, M.W., Norton, E.D., Lee, R.G., *Giardia* and *Cryptosporidium* spp. in filtered drinking water supplies, *Appl. Environ. Microbiol.,* 57:2617–2621, 1991.

Leland, C., McAnulty, J., Keene, W., Stevens, G., A cryptosporidiosis outbreak in a filtered-water supply, *J. Am. Water Works Assoc.,* 85:34–42, 2993.

Levy, D.A., Bens, M.S., Craun, G.F., Calderon, R.L., Herwaldt, B.L., Surveillance for waterborne-disease outbreak — United States, 1995–1996, in Surveillance Summaries, December 11, 1998, *MMWR,* 47(SS-5):1–33, 1998.

Liu, S., Lu, J., Kolpin, D.W., Meeker, W.Q., Analysis of environmental data with censored observations, *Environ. Sci. Technol.,* 32(23):3358–3362, 1997.

Lopes, T.J., Bender, D.A., Nonpoint sources of volatile organic compounds in urban areas–relative importance of urban land surfaces and air, *Environ. Pollut.,* 101:221–230, 1998.

Lopes, T.J., Price, C.W., Study plan for urban stream indicator sites of the National Water-Quality Assessment Program, U.S. Geological Survey Open-File Report, OFR 997-25, 1997, 15 pp.

Lorenzo-Lorenzo, J.J., Ares-Mazas, M.E., Villacorta-Martinez de Matron, I., Duran-Oreiro,D., Effect of ultraviolet disinfection of drinking water on the viability of *Cryptosporidium parvum* oocysts, *J. Parasitol.,* 79:67–70, 1993.

Lynch, C.F., Woolson, R.F., O'Gorman, T., Cantor, K.P., Chlorinated drinking-water and bladder cancer: effect of misclassification on risk estimates, *Arch. Environ. Health,* 44:252–259, 1989.

Marcus, P.M., Savitz, D.A., Millikan, R.C., Morgenstern, H., Female breast cancer and trihalomethane levels in drinking water in North Carolina, *Epidemiology,* 9:156–160, 1998.

McGeehin, M.A., Reif, J.S., Becher, J.C., Mangione, E.J., Case- control study of bladder cancer and water disinfection methods in Colorado, *Am. J. Epidemiol.,* 138:492–501, 1993.

Montgomery, J.H., Welkom, L.M., *Ground Waters Chemicals Desk Reference,* Lewis Publishers, Boca Raton, FL, 1990.

Moore, A.C., Herwaldt, B.L., Craun, G.F., Calderon, R.L., Highsmith, A.K., Juranek, D.D., Surveillance for waterborne-disease outbreaks — United States, 1991–1992, in CDC Surveillance Summaries, *MMWR,* 42(SS-5):1–22, 1993.

Moran, M.J., Clawges, R.M., Zogorski, J.S., Identifying the usage patterns of methyl-*tert*-butyl ether (MTBE) and other oxygenates in gasoline using gasoline surveys, to be presented at the American Chemical Society 219th National Meeting, March 26–31, 1999, 5 pp.

Moran, M.J., Davis, A.D., Occurrence of selected volatile organic compounds in ground water of the United States, 1985–1995 — relations with hydrogeologic and anthropogenic variables, in Hydrology Days, 18th, Fort Collins, CO, 1998 [proceedings], American Geophysical Union, Hydrology Days Publications, 1998, pp. 201–210.

Moran, J.J., Halde, M.J., Clawges, R.M., Zogorski, J.S., Relations between the detection of methyl-*tert*-butyl ether (MTBE) in surface and ground water and its content in gasoline. To be presented at the American Chemical Society 219th National Meeting March 26–31, 2000, 1999, 7 pp.

Moran, M.J., Zogorski, J.S., Squillace, P.J., MTBE in ground water of the United States — occurrence, potential sources, and long-range transport, in Water Resources Conference, American Water Works Association, Norfolk, VA, September 26–29, [proceedings], American Water Works Association, 1999.

Morris, R.D., Audet, A.M., Angelillo, I.F., Chalmers, T.C., Mosteller, F., Chlorination, chlorination by-products and cancer: a meta-analysis, *Am. J. Public Health*, 82:955–963, 1992.

National water quality inventory: 1996 report to Congress, EPA841-F-97-003, 12 pp.

National water quality inventory: 1996 report to Congress, EPA841-R-97-008, 588 pp.

Nielson, D.M., Ed., *Practical Handbook of Ground Water Monitoring*, Lewis Publishers, Boca Raton, FL, co-published with the National Water Well Association, 1990.

Noonan, D.C., Curtis, J.T., *Groundwater Remediation and Petroleum: A Guide for Underground Storage Tanks*, Lewis Publishers, Boca Raton, FL, 1990.

Olson, E.D. et al., Trouble on tap: arsenic, radioactive radon, and trihalomethanes in our drinking water, Natural Resources Defense Council with Clean Water Action and U.S. Public Interest Research Group, October 1995, 62 pp.

Pankow, J.F., Rathbun, R.E., Zogorski, J.S., Calculated volatilization rates of fuel oxygenate compounds and other gasoline-related compounds from rivers and streams, *Chemosphere*, 33(5):921–937, 1996.

Pankow, J.F., Thompson, N.R., Johnson, R.L., *Modeling the Atmospheric Input of MTBE to Groundwater Systems*, 11th ed., Society of Environmental Toxicology and Chemistry, Washington, D.C., 1996, p. 115.

Pankow, J.F., Thomson, N.R., Johnson, R.L., Baehr, A.L., Zogorski, J.S., The urban atmosphere as a non-point source for the transport of MTBE and other volatile organic compounds (VOCs) to shallow groundwater, *Environ. Sci. Technol.*, 31(10):2821–2828, 1997.

Pankow, J.F., Wentai, L., Isabelle, L.M., Bender, D.A., Baker, R.J., Determination of a wide range of volatile organic compounds (VOCs) in ambient air using multisorbent absorption/thermal desorption (ADY) and gas chromatography/mass spectrometry (GC/MS), *Anal. Chem.*, 70(24):5213–5221, 1998.

Pontius, F.W., Roberson, J.A., The current regulatory agenda: an update, *J. Am. Water Works Assoc.*, 86:54–63, 1994.

Pontius, F.W., Implementing the 1996 SDWA amendments, *J. Am. Water Works Assoc.*, 89:18–36, 1997.

Price, C.V., Characterization of the urban landscape using Landsat-derived land cover, Census, GIRAS and DLG data: Pecora Thirteen–human interactions with the environment, perspective from space, U.S. Geological Survey, Eros Data Center, Sioux Falls, SD, 1996.

Price, C.V., Mapping residential growth using Landsat Thematic Mapper data in conjunction with ancillary index.html data: American Congress of Surveying and Mapping, American Society for Photogrammetry and Remote Sensing, Seattle, WA, 1997, p. 726.

The quality of our nation's water, EPA841-S-97-001, 1996, 197 pp.

Ram, N.M., Christman, R.F., Cantor, K.P., Eds., *Significance and Treatment of Volatile Organic Compounds in Water Supplies,* Lewis Publishers, Boca Raton, FL, 1990.

Raucher, R.S. et al., Cost-effectiveness of SDWA regulations, *J. Am. Water Works Assoc.,* August:28–36, 1994.

Rice, R.G., *Ozone Drinking Water Treatment Book,* Lewis Publishers, Boca Raton, FL, 1990.

Richardson, A.J., Frankenberg, R.A., Buck, A.C. et al., An outbreak of waterborn cryptosporidosis in Swindon and Oxfordshire, *Epidemol. Infect.,* 107:485–495, 1991.

Robertson, J.B., Edberg, S.C., Natural protection of spring and well drinking water against surface microbial contamination, in hydrogeological parameters, *Crit. Rev. Microbiol.,* 23:143–178, 1997.

Rose, D.L., Schroeder, M.P., Methods of analysis by the U.S. Geological Survey National Water-Quality Laboratory, Determination of volatile organic compounds in water by purge and trap capillary gas chromatography/mass spectrometry, U.S. Geological Survey Open-File Report 94-708, 1995, 26 pp.

Savrin, J.E., Cohn, P.D., Comparison of bladder and rectal cancer incidence with trihalomethanes in drinking water [abstract], *Epidemiology,* 7(4 Suppl.):S63, 1996.

Shanaghan, P., Small systems and SDWA reauthorization, *J. Am. Water Works Assoc.,* May:52–61, 1994.

Sorvillo, F.J., Fujioka, K., Nahlen, B. et al., Swimming-associated cryptosporidosis, *Am. J. Public Health,* 82:742–744, 1992.

Squillace, P.J., Occurrence of the gasoline additive MTBE in shallow ground water in urban and agricultural areas, U.S. Geological Survey Fact Sheet, 1995, 4 pp.

Squillace, P.J., Preliminary assessment of the occurrence and possible sources of MTBE in groundwater of the United States, 1993–1994, in Congress on Groundwater in the Urban Environment [Proceedings], Chilton, J. et al., Eds., A.A. Bakema, Netherlands, 1997, pp. 537–542.

Squillace, P.J., A preliminary assessment of the occurrence and possible sources of MTBE in ground water of the United States, 1993–1994, U.S. Geological Survey Open-File Report, OFR 95-456, 1995, 16 pp.

Squillace, P.J., Preserving ground-water samples with hydrochloric acid does not result in the formation of chloroform, *Ground Water Monitoring Remediation,* Winter:67–74, 1999.

Squillace, P.J., Moran, J.J., Lapham, W.W., Price, C.V., Clawges, R.M., Zogorski, J.S., Volatile organic compounds in untreated ambient groundwater of the United States, 1985–1995, *Environ. Sci. Technol.,* 33(23):4176–4187, 1999.

Squillace, P.J., Moran, M.J., Lapham, W.W., Price, C.V., Clawges, R.M., Zogorski, J.S., Volcanic organic compounds in untreated ambient groundwater of the United States, 1985–1995, in Geological Society of America 1999 Annual Meeting, Denver, CO, October 25–28, 1999 [abstract], Geological Society of America, 1999, p. A157.

Squillace, P.J., Moran, M.J. Zogorski, J.S., Occurrence of MTBE in ground water of the United States, 1993–1998, and a preliminary analysis of explanatory variables, in American Water Works Association Annual Conference, Chicago, IL, June 20–24, 1999 [proceedings], American Water Works Association, CD-ROM.

Squillace, P.J., Pankow, J.F., Zogorski, J.S., Environmental behavior and fate of methyl tertiary butyl ether (MTBE), in the Southwest Focused Ground Water Conference — discussing the issue of MTBE and perchlorate in ground water, National Ground Water Association, 1998, p. 4–9.

Squillace, P.J., Zogorski, J.S., Price, C.V., Urban land-use study plan for the National Water-Quality Assessment Program, U.S. Geological Survey, in Chilton, J. et al., Eds., Congress on Groundwater in the Urban Environment [proceedings], A.A. Balkema, Netherlands, 1997, pp. 665–670.

Squillace, P.J., Zogorski, J.S., Wilber, W.G., Price, C.V., Preliminary assessment of the occurrence and possible sources of MTBE in groundwater in the United States, 1993–1994, in South Dakota Department of Environmental and Natural Resources Annual Ground-Water Conference, 9th, Pierre, SD, March 18–20, 1997, South Dakota Department of Environmental and Natural Resources, 1997, p. 27.

Squillace, P.J., Zogorski, J.S.,Wilber, W.G., Price, C.V., A preliminary assessment of the occurrence and possible sources of MTBE in ground water of the United States, 1993–94, U.S. Geological Survey Open-File Report, 1995, 16 pp.

Squillace, P.J., Zogorski, J.S., Wilier, W.G., Price, C.V., Preliminary assessment of the occurrence and possible sources of MTBE in groundwater in the United States, 1993–1994, *Environ. Sci. Technol.*, 30(5):1721–1730, 1996.

Squillace, P.J., Zogorski, J.S., Wilber, W.G., Price, C.V., Preliminary assessment of the occurrence and possible sources of MTBE in groundwater in the United States, 1993–1994, U.S. Geological Survey, in Chilton, J. et al., Eds., Congress on Groundwater in the Urban Environment [proceedings], A.A. Balkema, Netherlands, 1997, p. 537–542.

Squillace, P.J., Zogorski, J.S., Wilber, W.G., Price, C.V., A preliminary assessment of the occurrence and possible sources of MTBE in groundwater of the United States, 1993–1994, in American Chemical Society Division of Environmental Chemistry preprints of papers, 213rd, San Francisco, CA, *Am. Chem. Soc.*, 37(l):372–374, 1997.

Suarez-Varela, M.M.M., Gonzales, A.L., Perez, M.L.T., Caraco, E.J., Chlorination of drinking water and cancer incidence, *J. Environ. Pathol. Toxicol. Oncol.*, 13:39–41, 1994.

Sun, R.J., Johnston, R.H., Regional aquifer-system analysis program of the U.S. Geological Survey, 1978–1992, U.S. Geological Survey Circular 1099, 1994, 126 pp.

Thomas, D.L., Mundy, L.M., Tucker, P.C., An outbreak of hot-tub legionellosis, Abstracts of the 1991 ICAAC Abstr. No. 310, Chicago, IL, September–October 1991.

Trager, S.M. et al., Safe Drinking Water Act reauthorization: in the eye of the storm, *Nat. Resour. Environ.*, 9:17–19, 54–55, summer 1994.

U.S. Congress, House Committee on Commerce, Safe Drinking Water Act amendments of 1996, report on H.R. 3604, 104th Congress, 2nd session, June 24, 1996, U.S. Government Printing Office, Washington, D.C., H. Rept. 104–632, 1996, 134 pp.

U.S. Congress, Senate Committee on Environment and Public Works, Safe Drinking Water Act amendments of 1995, report on S. 1316, 104th Congress, 1st session, November 7, 1995, S. Rept. 104-169, U.S. Government Printing Office, Washington, D.C., 230 pp.

U.S. Congress, Senate Committee on Environment and Public Works, Safe Drinking Water Act amendments of 1995, Hearing on S. 1316, 104th Congress, 1st session, October 19, 1995, S. Rept. 104-354, U.S. Government Printing Office, Washington, D.C., 1995, 532 pp.

U.S. Congress, House Committee on Transportation and Infrastructure, Subcommittee on Water Resources and Environment, Hearing on H.R. 2747, The Water Supply Infrastructure Assistance Act of 1995, 104th Congress, 2nd session, Jan. 31, 1996, U.S. Government Printing Office, H. Rept 104-45, 1996, 218 pp.

U.S. Congress, Congressional Budget Office, The Safe Drinking Water Act: a case study of an unfunded Federal mandate. September 1995, 46 pp.

U.S. Department of Agriculture (USDA), Water Quality, NRCS/RCA Issue Brief 9, USDA, Natural Resources Conservation Service, Washington, D.C., March 1996.

U.S. Department of Commerce, We asked…you told us: source of water and sewage disposal, census questionnaire content, 1990 CQC-28, Department of Commerce, Washington, D.C., February 1995.

U.S. Department of Health and Human Services, Healthy people 2000 review, Department of Health and Human Services, Washington, D.C., 1994.

U.S. EPA, Announcement of the drinking water contaminant candidate list, [notice], *Fed. Regist.*, 63:10274–10287, 1997.

U.S. EPA, Best management practices for protecting groundwater — automotive service stations using shallow waste disposal wells (Class V BMP Fact Sheet number 1), EPA Document EPA 570/91-036A, U.S. Environmental Protection Agency, 1992.

U.S. EPA, Best management practices for protecting groundwater — facilities using shallow industrial waste disposal wells, (Class V BPA Fact Sheet number 2), EPA Document EPA 570/9/91-036B, U.S. Environmental Protection Agency, 1992.

U.S. EPA, Best management practices for protecting groundwater — additional BMPs for dry cleaners using shallow industrial waste disposal wells, (Class V BMP Fact Sheet number 2A), EPA Document EPA 570/9/91-036A, U.S. Environmental Protection Agency, 1992.

U.S. EPA, Best management practices for protecting groundwater — additional BMPs for photographic processing establishments using shallow industrial waste disposal wells, (Class V BMP Face Sheet number 2B), EPA Document EPA 570/9/91-036D, U.S. Environmental Protection Agency, 1992.

U.S. EPA, Best management practices for protecting groundwater — additional BMPs for furniture strippers using shallow industrial waste disposal wells, (Class V BMP Fact Sheet number 2C), EPA Document EPA 570/9/91-036E, U.S. Environmental Protection Agency, 1992.

U.S. EPA, Best management practices for protecting groundwater — additional BMPs for electroplaters using shallow industrial waste disposal wells, (Class V BMP Face Sheet number 2D), EPA Document EPA 570/9-91-036F, U.S. Environmental Protection Agency, 1992.

U.S. EPA, Best management practices for protecting groundwater — additional BMPs for medical service facilities using shallow industrial waste disposal wells, (Class V BMP Fact Sheet number 2H), EPA Document EPA 570/9-91-036J, U.S. Environmental Protection Agency, 1992.

U.S. EPA, Best management practices for protecting groundwater — additional BMPs for lawn care establishments using shallow industrial waste disposal wells, (Class V BMP Face Sheet number 2I), EPA Document EPA 570/9-91-036K, U.S. Environmental Protection Agency, 1992.

U.S. EPA, Drinking water, national primary drinking water regulations: filtration, disinfection, turbidity, *Giardia lamblia*, viruses, Legionella, and heterotrophic bacteria, [final rule], 40 CFR Parts 141 and 142, *Fed. Regist.*, 54:27–27486–27541, 1989.

U.S. EPA, Drinking water regulations: maximum contaminant level goals and national primary drinking water regulations for lead and copper, [final rule], 40 CFR Parts 141 an 142, *Fed. Regist.*, 56:26460–26464, 1991.

U.S. EPA, Drinking water, national primary drinking water regulations, total coliforms, corrections and technical amendments [final rule], 40 CFR Parts 141 and 142, *Fed. Regist.*, 55:25064–25065, 1990.

U.S. EPA, Drinking water, national primary drinking water regulations, total coliforms (including fecal coliforms and *E. coli*), [final rule], 40 CFR Parts 141 and 142, *Fed. Regist.*, 54:27544–27568, 1989.

U.S. EPA, Drinking water regulations and health advisories, Office of Water, Washington, D.C., EPA 822-R-96-001m, 1996, 16 pp.

U.S. EPA, Environmental pollution control alternatives: drinking water treatment for small communities, 1990.

U.S. EPA, *Fed. Regist.*, 40 CFR Parts 152 and 156, June 26, 1966.

U.S. EPA, National drinking water program redirection strategy, EPA, Washington, D.C., June 1996.

U.S. EPA, National primary drinking water regulations: interim enhanced surface water treatment, [final rule], 40 CFR Parts 141 and 142, *Fed. Regist.*, 63:69477–69521, 1998.

U.S. EPA, National primary drinking water regulations: monitoring requirements for public drinking water supplies, [final rule], 40 CFR Part 141, *Fed. Regist.*, 61:24353–24388, 1996.

U.S. EPA, The Safe Drinking Water Act amendments of 1996: strengthening protection for America's drinking water, [draft paper], EPA, Office of Water, Washington, D.C., August 1996.

U.S. EPA, The Safe Drinking Water Act: a pocket guide to the requirements for the operators of small water systems, booklet prepared by the Environmental Protection Agency, Region 9, San Francisco, CA, 1993.

U.S. EPA, Science Advisory Board report: safe drinking water, future trends and challenges, EOASAAB-DWC-95002, EPA, Washington, D.C., 1995.

U.S. EPA, Strengthening the safety of our drinking water: a report on progress and challenges and an agenda for action, EPA 810-R-95-001, March 1995, 22 pp.

U.S. EPA, Technical and economic capacity of states and public water systems to implement drinking water regulations, report to Congress, EPA 810-R-93-001, September 1993, 125 pp.

U.S. EPA, Underground injection control regulations for class V injection wells, revision, [final rule], 40 CFR Parts 9, 144, 145, and 146, *Fed. Regist.*, 64:68545–68573, 1999.

U.S. EPA and American Water Works Association, *Point of Use/Entry Treatment of Drinking Water*, Noyes Data Corp, Park Ridge, NJ, 1990.

U.S. EPA, Office of Drinking Water and Office of Pesticides, *Drinking Water Advisory: Pesticides*, Lewis Publishers, Boca Raton, FL, 1990.

U.S. EPA, Office of Water. Office of Water Environmental and Program Information Systems Compendium FY 1990, Washington, D.C., 1990.

U.S. General Accounting Office, Drinking water: stronger efforts essential for small communities to comply with standards, report to the chairman, Subcommittee on Environment, Energy, and Natural Resources, Committee on Government Operations, House of Representatives, GAO/RCED-94-40, March 1994, 60 pp.

Van Der Leeden, F., Troise, F.L., Todd, D.K., Eds., *The Water Encyclopedia*, 2nd ed., Lewis Publishers, Boca Raton, FL, 1990.

Varady, R.G., Mack, M., Transboundary water resources and public health in the U.S.–Mexico border region, *J. Environ. Health*, 57(8):8–14, 1995.

von Guerard, P., Weiss, W.B., Water quality of storm runoff and comparison of procedures for estimating storm-runoff loads, volume, event-mean concentrations, and the mean load for a storm for selected properties and constituents for Colorado Springs, southeastern Colorado, 1992, U.S. Geological Survey Water-Resources Investigations Report 94-4194, 1995, 68 pp.

Waggoner, P.E. Ed., *Climate Change and U.S. Water Resources*, John Wiley & Sons, New York, 1990.

Water quality conditions in the United States, EPA841-F-97-001, 2 pp.

Zogorski, J.S., MTBE — Summary of findings and research by the U.S. Geological Survey, in Annual Conference of the American Water Works Association — Water Quality, June 21–25, 1998, Dallas, TX [proceedings], AWWA, Denver, CO, 1998, p. 287–309.

Zogorski, J.S., Fuel oxygenates and water quality — current understanding of sources, occurrence in natural waters, environmental behavior, fate, and significance — final report, Office of Science and Technology Policy, Executive Office of the President, 37 pp. with attachments. [Also published by National Science and Technology Council, 1997, Interagency assessment of oxygenated fuels, Fuel Oxygenates and Water Quality, Washington, D.C., Office of Science and Technology Policy, the Executive Office of the President, chap. 2.

Zogorski, J.S., Baehr, A.L., Bauman, B.J., Conrad, D.L., Drew, R.T., Korte, N.E., Lapham, W.W., Morduchowitz, A., Pankow, J.F., Washington, E.R., Significant findings and water-quality recommendations of the Interagency Oxygenated Fuel Assessment, in *Contaminated Soils,* Vol. 2, Kostecki, P.T., Calabrese, E.J., Bonazountas, M., Eds., Amherst Scientific, Amherst, MA, 1997, pp. 661–679.

Zogorski, J.S., Deizer, G.C., Bender, D.A., Squillace, P.J., Lopes, T.J., Baehr, A.L., Stackelbert, P.A., Landmeyer, J.E., Boughton, C.J., Lico, M.S., Pankow, J.S., Johnson, R.L., Thomson, N.R., MTBE — Summary of findings and research by the U.S. Geological Survey, in Annual Conference of the American Water Works Association — Water Quality, June 21–25, Dallas, TX [proceedings], AWWA, Denver CO, pp. 287–309, 1998.

Zogorski, J.S., Morduchowitz, A., Baehr, A.L., Bauman, B.J., Conrad, D.C., Drew, R.T., Korte, N.W., Lapham, W.W., Pankow, J.F., Washington, E.R., Fuel oxygenates and water quality. Current understanding of sources, occurrence in natural waters, environmental behavior, fate, and significance, Office of Science and Technology Policy, Washington, D.C., 1996, 91 pp.

CHAPTER 4

Acheson, D.K., Bennish, M.L., *Shigella* and enteroinvasive *Escherichia coli,* in *Infections of the Gastrointestinal Tract,* Blazer, M.J., Smith, P.D., Ravdin, J.I., Greenberg, H.B., Guerrant, R.L., Eds., Raven Press, New York, 1995.

Bell, A., Guasparini, R., Meeds, D., Mathias, R.G., Farley, J.D., A swimming pool-associated outbreak of cryptosporidiosis in British Columbia, *Can. J. Public Health,* 84:334–337, 1993.

Boating statistics 1992, CMDTPUB P1674.8, U.S. Department of Transportation, Washington, D.C., 1992.

Boyce, T.G., Pemberton, A.G., Addiss, D.G., *Cryptosporidium* testing practices among clinical laboratories in the United States, *Pediatr. Infect. Dis. J.,* 15:87–88, 1996.

Calderon, R.L., Mood, E.W., Dufour, A.P., Health effects of swimmers and nonpoint sources of contaminated water, *Int. J. Environ. Health Res.,* 1:21–31, 1991.

California Department of Health Services, Primary amebic meningoencephalitis associated with a natural hot spring in San Bernardino County, *Calif. Morbidity,* 13/14, 1992.

Carpenter, C., Fayer, T., Trout, J., Beach, M.J., Chlorine disinfection of recreational water for *Cryptosporidium parvum, Emerg. Infect. Dis.,* 5:579–584, 1999.

CDC, Assessing the public health threat associated with waterborne cryptosporidiosis: report of a workshop, *MMWR,* 44(RR-6):1995.

CDC, *Cryptosporidium* infections associated with swimming pools — Dane County, Wisconsin 1993, *MMWR,* 43:561–563, 1994.

CDC, Engineering and administrative recommendations for water fluoridation 1995, *MMWR,* 44(RR-13), 1995.

CDC, Lead-contaminated drinking water in bulk-water storage tanks — Arizona and California 1993, *MMWR,* 43:751–758, 1994.

CDC, Outbreak of acute febrile illness among athletes participating in triathlons — Wisconsin and Illinois 1998, *MMWR,* 47:585–588, 1998.

CDC, Preventing lead poisoning in young children: a statement by the Centers for Disease Control, Atlanta, U.S. Department of Health and Human Services, Public Health Service, CDC, October 21, 1991.

CDC, Surveillance for waterborne-disease outbreaks — United States, 1989–1990, in CDC Surveillance Summaries (December), *MMWR,* 40(SS-3), 1991.

CDC, Surveillance for waterborne-disease outbreaks — United States, 1991–1992, in CDC Surveillance Summaries (November), *MMWR,* 42(SS-5), 1993.

CDC, Surveillance for waterborne-disease outbreaks — United States, 1993–1994, in CDC Surveillance Summaries (April), *MMWR,* 45(SS-1), 1996.

CDC, Surveillance for waterborne-disease outbreaks — United States, 1995–1996, in CDC Surveillance Summaries (December), *MMWR,* 47(SS-5), 1998.

CDC, Surveillance for waterborne-disease outbreaks — United States, 1997–1998, in CDC Surveillance Summaries (May), *MMWR,* 49(SS-4), 2000.

CDC, USPHS/IDSA guidelines for the prevention of opportunistic infections in persons infected with human immunodeficiency virus: a summary, *MMWR,* 44(RR-8), 1995.

CDC, Waterborne disease outbreaks, 1986–1988, *MMWR,* 39(SS-1):1–13, 1990.

CDC, Water fluoridation: a manual for engineers and technicians, U.S. Department of Health and Human Services, Public Health Service, Atlanta, GA, 1986.

Craun, G.F., Ed., Methods for the investigation and prevention of waterborne disease outbreaks, EPA600/1-90/005a, Cincinnati, OH, U.S. Environmental Protection Agency, Health Effects Research Laboratory, 1990.

Dupont, H.L., Chappell, C.L., Sterling, C.R., Okhuysen, P.C., Rose, J.B., Jakubowski, W., The infectivity of *Cryptosporidium parvum* in healthy volunteers, *N. Engl. J. Med.,* 332:855–859, 1995.

Effect of last summer's outbreak on Wild Water Adventure's theme park, *Crypto. Capsule,* 2:1–3, 1997.

Fayer, R., Effect of sodium hypochlorite exposure on infectivity of *Cryptosporidium parvum* oocysts for neonatal BALB/c mice, *Appl. Environ. Microbiol.,* 61:844–846, 1995.

Finch, G.R., Black, E.K., Gyurek, L.L., Ozone and chlorine inactivation of *Cryptosporidium,* Proceedings of the AWWA Water Quality Technical Conference, San Francisco, CA, 1994, pp. 1303–1308.

Fobbs, M., Skala, M., Waterborne hepatitis A associated with a church and school, *MO Epidemiol.,* 14:6–8, 1992.

Fuortes, L., Nettleman, M., Leptospirosis: a consequence of the Iowa flood, *Iowa Med.,* 84:449–450, 1994.

Gilbert, L., Blake, P., Outbreak of *Escherichia coli* O157:H7 infections associated with a water park, *GA Epidemiol. Rep.,* 14:1–2, 1998.

Goldstein, S.T., Juranek, D.D., Ravenholt, O. et al., Cryptosporidiosis: an outbreak associated with drinking water despite state-of-the-art water treatment, *Ann. Intern. Med.,* 124:459–468, 1996.

Griffin, P.M., Tauxe, R.V., The epidemiology of infections caused by *Escherichia coli* O157:H7, other enterohemorrhagic *E. coli,* and the associated hemolytic uremic syndrome, *Epidemiol. Rev.,* 13:60–98, 1991.

Haas, C.N., Rose, J.B., Reconciliation of microbial risk models and outbreak epidemiology: the case of the Milwaukee outbreak, in Proceedings of the American Water Works Association 1994 Annual Conference — Water Quality, American Water Works Association, Denver, CO, 1994, pp. 517–523.

Hayes, E.B., Matte, T.D., O'Brien, T.R. et al., Large community outbreak of cryptosporidiosis due to contamination of a filtered public water supply, *N. Engl. J. Med.,* 320:1372–1376, 1989.

Howland, J., Hingson, R., Alcohol as a risk factor for drownings: a review of the literature 1950–1985, *Accid. Anal. Prev.,* 20:19–25, 1988.

Howland, J. et al., A pilot survey of aquatic and related consumption of alcohol, with implications for drowning, *Public Health Rep.*, 105:415–419, 1990.

Hunt, D.A., Sebugwawo, S., Edmonson, S.G., Casemore, D.P., Cryptosporidiosis associated with a swimming pool complex, *Commun. Dis. Rep. CDR Rev.*, 4:R20–22, 1994.

Jackson, L.A., Kaufmann, A.F., Adams, W.G. et al., Outbreak of leptospirosis associated with swimming, *Pediatr. Infect. Dis. J.*, 12:48–54, 1993.

John, D.T., Howard, M.J., Seasonal distribution of pathogenic free-living amebae in Oklahoma waters, *Parasitol. Res.*, 81:193–201, 1995.

Juranek, D.D., Cryptosporidiosis: sources of infection and guidelines for prevention, *Clin. Infect. Dis.*, 21(Suppl. 1):S57–S61, 1995.

Juranek, D.D., Addiss, D.G., Bartlett, M.E. et al., Cryptosporidiosis and public health: workshop report, *J. Am. Water Works Assoc.*, 87:69–80, 1995.

Kaminski, J.C., *Cryptosporidium* and the public water supply [letter], *N. Engl. J. Med.*, 331:1529–1530, 1994.

Keene, W.E., McAnulty, J.M., Hoesly, F.C. et al., A swimming-associated outbreak of hemorrhagic colitis caused by *Escherichia coli* O157:H7 and *Shigella sonnei*, *N. Engl. J. Med.*, 331:579–584, 1994.

Knobeloch, L., Ziarnick, M., Howard, J. et al., Gastrointestinal upsets associated with ingestion of copper-contaminated water, *Environ. Health Perspect.*, 102:958–961, 1994.

Korich, D.G., Mead, J.R., Madore, M.S., Sinclair, N.A., Sterling, C.R., Effects of ozone, chlorine dioxide, chlorine, and monochloramine on *Cryptosporidium parvum* oocyst viability, *Appl. Environ. Microbiol.* 56:1423–1428, 1990.

Kramer, M.H., Herwaldt, B.L., Craun, G.F., Calderon, R.L., Juranek, D.D., Waterborne disease: 1993 and 1994, *J. Am. Water Works Assoc.*, 88:66–80, 1996.

Kramer, M.H., Sorhage, F., Goldstein, S., Dalley, E., Wahlquist, S., Herwaldt, B., First reported outbreak in the United States of cryptosporidiosis associated with a recreational lake, *Clin. Infec. Dis.*, 26:27–33, 1996.

LeChevallier, M.W., Norton, W.D., Lee, R.G., *Giardia* and *Cryptosporidium* spp. in filtered drinking water supplies, *Appl. Environ. Microbiol.*, 57:2617–2621, 1991.

LeChevallier, M.W., Norton, W.D., Lee, R.G., *Giardia* and *Cryptosporidium* in raw and finished water, *J. Am. Water Works Assoc.*, 87:54–68, 1995.

LeChevallier, M.W., Norton, W.D., Lee, R.G., Occurrence of *Giardia* and *Cryptosporidium* spp. in surface water supplies, *Appl. Environ. Microbiol.*, 57:2610–2616, 1991.

Lemmon, J.M., McAnulty, J.M., Bawden-Smith, J., Outbreak of cryptosporidiosis linked to an indoor swimming pool, *Med. J. Aust.*, 165:613–616, 1996.

Levy, D.A., Bens, M.S., Craun, G.F., Calderon, R.L., Herwaldt, B.L., Surveillance forwaterborne disease outbreaks — United States, 1995–1996, *MMWR*, 47(SS-5):1–34, 1998.

Litovitz, T.L., Clark, L.R., Soloway, R.A., 1993 annual report of the American Association of Poison Control Centers Toxic Exposure Surveillance System, *Am. J. Emerg. Med.*, 12:546–584, 1994.

MacKenzie, W.R., Hoxie, N.J., Proctor, M.E. et al., A massive outbreak in Milwaukee of *Cryptosporidium* infection transmitted through the public water supply, *N. Engl. J. Med.*, 331:161–167, 1994.

MacKenzie, W.R., Kazmierczak, J.J., Davis, J.P., An outbreak of cryptosporidiosis associated with a resort swimming pool, *Epidemiol. Infect.*, 115:545–553, 1995.

Mandell, G.M., Bennett, J.E., Dolin, R., *Principles and Practice of Infectious Diseases*, 4th ed., Churchill Livingstone, New York, 1995, pp. 2033–2035.

McAnulty, J.M., Fleming, D.W., Gonzalez, A.H., A community-wide outbreak of cryptosporidiosis associated with swimming at a wave pool, *JAMA*, 272:1597–1600, 1994.

Meinhardt, P.L., Casemore, D.P., Miller, K.B., Epidemiologic aspects of human cryptosporidiosis and the role of waterborne transmission, *Epidemiol. Rev.,* 18:118–136, 1996.

Moore, A.C., Herwaldt, B.L., Craun, G.F., Calderon, R.L., Highsmith, A.K., Juranek, D.D., Surveillance for waterborne disease outbreaks — United States 1991–1992, *MMWR,* 42(SS-5):L1–22, 1993.

Most cases in Australia linked to swimming pool exposure, *Crypto. Capsule,* 3(7):5–6, 1998.

National Center for Health Statistics (NCHS), National mortality data 1995, NCHS, Hyattsville, MD, 1997.

New Zealand's pool related outbreak is over, *Crypto. Capsule,* 3(9):5–6, 1998.

Number of cryptosporidiosis cases increase in Australia, *Crypto. Capsule,* 3:1–2, 1998.

Outbreak of cryptosporidiosis associated with a water sprinkler fountain — Minnesota 1997, *MMWR,* 47:856–860, 1998.

Outbreak of pharyngoconjunctival fever at a summer camp — North Carolina 1991 (news), *Infect. Control Hosp. Epidemiol.,* 13:499–500, 1992.

Outbreaks in England and Wales: first half of 1997, *Crypto. Capsule,* 3:4, 1998.

Past water-related outbreaks in Florida, *Crypto. Capsule,* 3:3, 1998.

Recreational outbreak in Oregon, *Crypto. Capsule,* 4(1):1, 1998.

Rose, J.B., Gerba, C.P., Jakubowski, W., Survey of potable water supplies for *Cryptosporidium* and *Giardia, Environ. Sci. Technol.,* 25:1393–1400, 1991.

Rosenberg, M.L., Hazlet, K.K., Schaefer, J., Wells, J.G., Pruneda, R.C., Shigellosis from swimming, *J. Am. Med. Assoc.,* 236:1849–1852, 1996.

Samonis, G., Elting, L., Skoulika, E. et al., An outbreak of diarrhoeal disease attributed to *Shigella sonnei, Epidemiol. Infect.,* 112:235–245, 1994.

Schwartz, J., Low-level lead exposure and children's IQ: a meta-analysis and search for a threshold, *Environ. Res.,* 65:42–55, 1994.

Sorvillo, F.J., Fujioka, K., Nahlen, B., Tormey, M.P., Kebabjian, R., Mascola, L., Swimming-associated cryptosporidiosis, *Am. J. Public Health,* 82:742–744, 1992.

Sterling, C.B., Christian, B.B., Cope, J.O., A public health laboratory: handling a parasitic outbreak, *Tenn. Department of Health (THD) Laboratory Services Newsl.,* 4:1–3, 1995.

Sundkvist, T., Dryden, M., Gabb, R., Soltanpor, N., Casemore, D., Stuart, J. et al., Outbreaks of cryptosporidiosis associated with a swimming pool in Andover, *Commun. Dis. Rep.,* 7:R190–192, 1997.

Swimming pools implicated in Australia's largest cryptosporidiosis outbreak ever reported, *Crypto. Capsule,* 3:1–2, 1998.

U.S. EPA, 40CFR Parts 141 and 142, Drinking water: national primary drinking water regulations: filtration, disinfection, turbidity, *Giardia lamblia,* viruses, *Legionella,* and heterotrophic bacteria, final rule, *Fed. Regist.,* 54:27486–27541, 1989.

U.S. EPA, 40CFR Parts 141 and 142, Drinking water: national primary drinking water regulations: total coliforms (including fecal coliforms and *E. coli*), final rule, *Fed. Regist.,* 54:27544–27568, 1989.

U.S. EPA, 40CFR Parts 141 and 142, Drinking water: national primary drinking water regulations: total coliforms, corrections and technical amendments, final rule, *Fed. Regist.,* 54:25064–25065, 1990.

U.S. EPA, 40 CFR Parts 141 and 142, Drinking water regulations: maximum contaminant level goals and national primary drinking water regulations for lead and copper, final rule, *Fed. Regist.,* 56:26460–26464, 1991.

U.S. EPA, 40 CFR Parts 141 and 142, National primary drinking water regulations: enhanced surface water treatment requirements, proposed rule, *Fed. Regist.,* 59:38832–38858, 1994.

U.S. EPA, 40 CFR Part 141, National primary drinking water regulations: monitoring require-
    ments for public drinking water supplies: *Cryptosporidium, Giardia,* viruses, disin-
    fection byproducts, water treatment plant data and other information requirements,
    proposed rule, *Fed. Regist.,* 59:6332–6444, 1994.
Unintentional injuries: the problems and some preventive strategies, in *Handbook of Black
    American Health: The Mosaic of Condition, Issues, Policies and Prospects,* Living-
    stone, E.L., Ed., Greenwood, Westport, CT, 1994.
Visvesara, G.S., Stehr-Green, J.K., Epidemiology of free-living ameba infections, *J. Proto-
    zool.,* 37(Suppl.):25S–33S, 1990.
Wilberschied, L., A swimming pool-associated outbreak of cryptosporidiosis, *Kan. Med.,*
    96:67–68, 1995.

CHAPTER 5

Boyce, T.G., Pemberton, A.G., Addiss, D.G., *Cryptosporidium* testing practices among clinical
    laboratories in the United States, *Pediatr. Infect. Dis. J.,* 15:87–88, 1996.
CDC, Outbreak of acute febrile illness among athletes participating in triathlons — Wisconsin
    and Illinois 1998, *MMWR,* 47:585–588, 1998.
CDC, Outbreak of cryptosporidiosis associated with a water sprinkler fountain — Minnesota
    1997, *MMWR,* 47:856–860.
CDC, Surveillance for waterborne disease outbreaks — United States 1991–1992, *MMWR,*
    42(SS-5):1–22, 1993.
Craun, G.F., Ed., Methods for the investigation and prevention of waterborne disease out-
    breaks, EPA600/1-90/005a, U.S. Environmental Protection Agency, Health Effects
    Research Laboratory, Cincinnati, OH, 1990.
Craun, G.F., Ed., *Waterborne Diseases in the United States,* CRC Press, Boca Raton, FL, 1986.
Cross-Connection Control Manual, United States Environmental Protection Agency, Wash-
    ington, D.C., 1985.
Gilbert, L., Blake, P., Outbreak of *Escherichia coli* O157:H7 infections associated with a
    water park, *GA Epidemiol. Rep.,* 14:1–2, 1998.
Herwaldt, B.L., Craun, G.F., Stokes, S.L., Juranek, D.D., Waterborne disease outbreaks
    1989–1990, in CDC Surveillance Summaries, December 1991, *MMWR,* 40(SS-3):1–21,
    1991.
International Association of Plumbing and Mechanical Officials, Uniform plumbing code,
    2000 edition, Walnut Grove, CA, September 1999, 380 pp.
International Association of Plumbing and Mechanical Officials, Uniform plumbing code,
    illustrated training manual, 2000 edition, Walnut Grove, CA, September 1999, 560 pp.
Kramer, M.H., Herwaldt, B.L., Craun, G.F., Calderon, R.L., Juranek, D.D., Surveillance for
    waterborne-disease outbreaks — United States 1993–1994, in CDC Surveillance
    Summaries, April 12, 1996, *MMWR,* 45(SS-1):1–33, 1996.
Levy, D.A., Bens, M.S., Craun, G.F., Calderon, R.L., Herwaldt, B.L., Surveillance for water-
    borne-disease outbreaks — United States 1995–1996, in CDC Surveillance Summa-
    ries, December 11, 1998, *MMWR,* 47(SS-5):1–33, 1998.
MacKenzie, W.R., Hoxie, N.J., Proctor, M.E. et al., A massive outbreak in Milwaukee of
    Cryptosporidium infection transmitted through the public water supply, *N. Engl. J.
    Med.,* 331:161–167, 1994.
Moore, A.C., Herwaldt, B.L., Craun, G.F., Calderon, R.L., Highsmith, A.K., Juranek, D.D.,
    Surveillance for waterborne-disease outbreaks–United States 1991–1992, in CDC
    Surveillance Summaries, November 19, 1993, *MMWR,* 42(SS-5):1–22, 1993.

Pontius, F.W., Roberson, J.A., The current regulatory agenda: an update, *J. Am. Water Works Assoc.,* 86:54–63, 1994.

Pontius, F.W., Implementing the 1996 SDWA amendments, *J. Am. Water Works Assoc.,* 89:18–36, 1997.

U.S. EPA, Announcement of the drinking water contaminant candidate list [notice], *Fed. Regist.,* 63:10274–10287, 1998.

CHAPTER 6

Alloway, B.C., Jackson, A.P., The behavior of heavy metals in sewage sludge-amended soils, *Sci. Total Environ.,* March:151–176, 1991.

Aravena, R., Evans, M.L., Cherry, J.A., Stable isotopes of oxygen and nitrogen in source identification of nitrate from septic systems, *Ground Water,* 31(2):180–186, 1993.

Arnold, J.A., Osmond, D.L., Hoover, M.T., Rubin, A.R., Line, D.E., Coffey, S.W., Spooner, J., On-site wastewater management — guidance manual, North Carolina Cooperative Extension Service and North Carolina Department of the Environment, Health, and Natural Resources, Raleigh, NC, North Carolina Administrative Code, Title 15A, Subchapter 18A, Rules 1901-1968, Rules effective July 1, 1982, Rules amended effective January 20, 1997.

Baleux, B., On-site sewage disposal: the importance of the wet season water table, *J. Environ. Health,* 52:277–279, 1990.

Ball, H.I., Nitrogen reduction in an on-site trickling filter/upflow filter wastewater treatment system, Proceedings of the 7th International Symposium on Individual and Small Community Sewage Systems, December 1994.

Barbier, D., Parasitic hazard with sewage sludge applied to land, *Appl. Environ. Microbiol.,* 56:1420–1422, 1990.

Beck, M.B., Distribution and survival of motile *Aeromanas* spp. in brackish water receiving sewage treatment effluent, *Appl. Environ. Microbiol.,* 57(A6):2459–2467, 1991.

Bell, P.F., James, B.R., and Chaney, R.L., Heavy metal extractability in long-term sewage sludge and metal salt-amended soils, *J. Environ. Qual.,* 20:481–486, 1991.

Brady, H.C., *The Nature and Properties of Soils,* MacMillan, New York, 1984.

Chang, A.C., Granato, T.C., Page, A.L., A methodology for establishing phytotoxicity criteria for chromium, copper, nickel, and zinc in agricultural land application of municipal sewage sludges, *J. Environ. Qual.,* 21:521–536, 1992.

Charles, G.M., Waller, D.H., Mooers, J.D., Long-term performance of contour trench systems for subsurface disposal of septic tank effluent from multiple dwellings, submitted to Canada Mortgage and Housing Corporation Research Division, 1990.

Check, G., Waller, D.H., Lee, S.A., Pask, D., Mooers, J., The lateral flow sand filter system for septic tank effluent treatment, Laboratory Modeling and Field Investigations, *Water Environ. Fed.,* 66(7):919–928, 1994.

Chemical fixation treatment cleans toxic metals from soil, *Pollut. Eng.,* 23:120–122, March 1991.

Coghlan, A., Toxic sludge that just won't die, *New Sci.,* July 15, 1993.

Considering the alternatives, a guide to onsite wastewater systems in North Carolina, North Carolina Rural Communities Assistance Project, February 1994.

Cunningham, W.P., Saigo, B.W., *Environmental Science, a Global Concern,* W.C. Brown, Dubuque, IA, 1992.

DelPorto, D.A., Steinfeld, C.J., *The Composting Toilet Book,* Chelsea Green, Whiteriver Junction, VT, 1998.

Duncan, C.S., Reneau, R.B., Jr., Hagedorn, D., Impact of effluent quality and soil depth on renovation of domestic wastewater, On-Site Wastewater Treatment Proceedings of the 7th International Symposium on Individual and Small Community Sewage Systems, Atlanta, GA, 1994.

Eiceman, G.A., Urquhart, N.S., O'Connor, G.A., Logistic and economic principals in gas chromatography–mass spectrometry use for plant uptake investigations, *J. Environ. Qual.*, 22:167–173, 1993.

Fent, K., Europe plans for cleaner water, *New Sci.*, 129(N52):38–42, 1991.

Fernandez, P., Alternative ways to treat sewage cheaply, *New Sci.*, 127(N53):44, 1990.

Frost, P., Camenzind, R., Magert, A., Bonjour, R., Karlaganis, G., Organic micropollutants in Swiss sewage sludge, *J. Chromatogr.*, 643:379–388, 1993.

Geldreich, E.E., Goodrich, K.R., Rice, E.W., Clark, R.M., Swerdlow, D.L., Searching for a water supply connection in the Cabool, Missouri disease outbreak of *Escherichia coli* O157:H7, *Water Resour.*, 26:1127–1137, 1992.

Gerritse, R.B., Adeney, J.A., Hosking, J., Nitrogen losses from a domestic septic tank system on the Darling plateau in western Australia, *Water Res.*, 29:2055–2058, 1995.

Goldstein, N., Sludge management practices in the U.S., *BioCycle*, 32(3):46, 1991.

Grimalt, J.K., Plant upgrade tackles odors, *ENR*, 226:17, 1991.

Hinson, T.H., Hoover, M.T., Evans, R.O., Sand lined trench septic system performance on wet, clayey soil, On-Site Wastewater Treatment, Proceedings of the 7th International Symposium on Individual and Small Community Sewage Systems, ASAE publication, 1994, pp. 245–255.

Hoffman, M.R., Catalytic autoxidation of hydrogen sulfide in wastewater, *Environ. Sci. Technol.*, 25:1153–1160, 1991.

Hoover, M.T., Evans, R.O., Hinson, T.H., Heath, R.C., Performance of sand lined trench septic systems on wet, clayey soils in northeastern North Carolina, College of Agriculture and Life Sciences, North Carolina State University, Raleigh, 1993.

Ijzerman, M.M., Hagedorn, C., Reneau, R.B., Jr., Microbial tracers to evaluate an on-site shallow-placed low pressure distribution system, *Water Res.*, 27:3443–3447, 1992.

Lamb, B.E., Gold, A.J., Loomis, G.W., McKiel, C.G., Nitrogen removal for on-site sewage disposal: field evaluation of buried sand filter/greywater systems, Kingston, RI, *Trans. ASAE*, 34(3):883–889, 1991.

McIntyre, C., D'Amico, C., Willenbrock, J.H., Residential wastewater treatment and disposal: on-site spray irrigation systems, Proceedings of the 7th International Symposium on Individual and Small Community Sewage Systems, December 1994.

McKee, J.A., Brooks, J.L., Peat filters for on-site wastewater treatment, Proceedings of the 7th International Symposium on Individual and Small Community Sewage Systems, December 1994.

McNeillie, J.I., Anderson, D.L., Belanger, T.V., Investigation of the surface water contamination potential from on-site wastewater treatment systems (OSTS) in the Indian River lagoon basin, St. Joseph, MI, *ASAE Pub. Proc.*, 7(18–94):156–163, 1994.

Mooers, J.D., Contribution of toxic chemicals to groundwater from domestic on-site sewage disposal systems, Centre for Water Resources Studies, Internal Report 92-12, 1992.

Mooers, J.D., Measurement of hydraulic conductivity of soils in-situ, Centre for Water Resources Studies, Internal Report 92-4, 1992.

Mooers, J.D., Microbiological field analysis techniques, Centre for Water Resources Studies, Internal Report 92-10, 1992.

Mooers, J.D., Numerical modeling of the interaction between an on-site overburden well and disposal field, Centre for Water Resources Studies, Internal Report 92-9, 1992.

Mooers, J.D., Numerical modeling of the interaction between on-site bedrock wells and disposal fields, Proceedings of the 8th Atlantic Region CWRA/CSCE Hydrotechnical Conference, November 1992, Halifax, N.S., 1993.

Mooers, J.D., Septic tank additive pamphlet, Centre for Water Resources Studies, Internal Report 92-11, 1992.

Mooers, J.D., Verification of a mathematical model of a septic tank distribution system, Centre for Water Resources Studies, Internal Report 92-8, 1992.

Mooers, J.D., Waller, D.H., Innovative on-site wastewater disposal, Proceedings of Annual Conference of the Canadian Society for Civil Engineering, 1998, pp.11–20.

Mooers, J.D., Waller, D.H., On-site wastewater research program phase II, Final Report, Province of Nova Scotia, 1993.

Mooers, J.D., Waller, D.H., On-site wastewater research program phase III, Final Report, Province of Nova Scotia, 1994.

Mooers, J.D., Waller, D.H., Potential for groundwater degradation from seasonal septic systems, in Proceedings CSCE Annual Conference, June 1993, Frederickton, N.B., Hydrotechnical, 1:239–248, 1993.

Mooers, J.D., Waller, D.H., Remediation of septic fields through flow reduction techniques, published by Canada Mortgage and Housing Corporation Research Division, 1992.

Mooers, J.D., Waller, D.H., Wastewater management districts: the Nova Scotia experience, in Proceedings of the Wastewater Nutrient Removal Technologies and Onsite Management Districts, Waterloo Centre for Groundwater Research, Waterloo, Ontario, 1994.

Mote, C.R., Ruiz, E.E., Design and operating criteria for nitrogen removal in a recirculating sand filter, Proceedings of the 7th International Symposium on Individual and Small Community Sewage Systems, December 1994.

National Research Council, Use of reclaimed water and sludge in food crop production, National Academy Press, Washington, D.C., 1996.

North Carolina On-Site Wastewater Management, Guidance manual, March 1996, pp. 1.1–1.3, 2.1–2.8, 4.1–4.3.

Osesek, B., Shaw, B., Graham, J., Design and optimization of two recirculating sand filters, Proceedings of the 7th International Symposium on Individual and Small Community Sewage Systems, Atlanta, GA, December 1994.

Otis, R.J., Anderson, D.L., Meeting public health and environmental goals: performance standards for on-site wastewater systems, On-Site Wastewater Treatment Proceedings of the 7th International Symposium on Individual and Small Community Sewage Systems, Atlanta, GA, December 1994.

Piluk, R.J., Peters, E.C., Performance and cost of on-site recirculating sand filters, Proceedings of the 7th International Symposium on Individual and Small Community Sewage Systems, Atlanta, GA, December 1994.

Piluk, R.J., Peters, E.C., Small recirculating sand filters for individual homes, Proceedings of the 7th International Symposium on Individual and Small Community Sewage Systems, Atlanta, GA, December 1994.

Powelson, D.K., Simpson, J.R., Gerba, C.P., Effects of organic matter on virus transport in unsaturated flow, Appl. Environ. Microbiol. August:2192–2196, 1991.

Prescott, L.M., Harley, J.P., Klien, D.A., Microbiology, 2nd ed. W.C. Brown, Dubuque, IA, 1993.

Reed, S.C., Middlebrooks, E.J., Crites, R.W., Natural Systems for Waste Management and Treatment, McGraw-Hill, New York, 1995.

Regulations governing individual onsite wastewater disposal systems, Mississippi State Department of Health, draft regulation, 1995.

Robertson, W.D., Cherry, Sudicky, E.A. Ground-water contamination from two small septic systems on sand aquifers, *Ground Water*, 29(1):82–92, 1991.

Ross, A.D., Lawrie, R.A., Keneally, J.P., Whatmuff, M.S., Risk characterization of sewage sludge on agricultural land: implications for the environment and the food chain, *Aust. Vet. J.*, 69(8):177–181, August 1992.

Rubin, A.R., Slow rate spray irrigation and drip disposal systems for treatment and renovation of domestic wastewater from individual homes, Proceedings of the 7th On-Site Wastewater Treatment Short Course and Equipment Exhibition, Seattle, WA, September 1992.

Rubin, A.R., Greene, S., Sinclair, T., Jantrania, A., Performance evaluation of drip disposal system for residential treatment, Proceedings of the 7th International Symposium on Individual and Small Community Sewage Systems, December 1994.

Scandura, J.E., Sobsey, M.D., Viral and bacterial contamination of groundwater from on-site sewage treatment systems, *Water Sci. Technol.*, 35(11):141–146, 1997.

Scott, R.S., Mooers, J.D., Waller, D.H., Rain water cistern systems — a regional approach to cistern sizing in Nova Scotia, Proceedings of the 7th International Rain Water Cistern Systems Conference, Beijing, China, 1995, pp. 10–48.

Septage Handling Task Force, Septage handling, Water Environment Federation (WEF) Manual of Practice No. 24, Alexandria, VA, 1997.

Sherman, K., What we have learned from the research — bridging the technology gap: taking on-site systems from concept to implementation, 12th Annual North Carolina On-Site Wastewater Treatment Conference, Raleigh, NC, 1997.

Simpson, C., Sludge regulations finalized, *Pollut. Eng.*, February:20–21, 1993.

Sims, J.T., Kline, J.S., Chemical fractionation and plant uptake of heavy metals in soils amended with co-composted sewage sludge, *J. Environ. Qual.*, 20:387–395, 1991.

Small Flows Newsletter, 8(2), Spring 1994, National Small Flows Clearinghouse, West Virginia University, Morgantown, WV.

Soares, A.C., Occurrence of Enteroviruses and Giardia Cysts in Sludge before and after Anaerobic Digestion, Department of Microbiology and Immunology/University of Arizona, Tuscon, AZ, 1990.

Straub, T.M., Pepper, I.L., Gerba, C.P., Hazards from pathogenic microorganisms in land-disposed sewage sludge, *Rev. Environ. Contam. Toxicol.*, 132:55–191, 1993.

Strauch, R., Survival of pathogenic microorganisms and parasites in excreta, manure, and sewage sludge, *Rev. Sci. Technol.*, 10(3):813–846, September 1991.

Sunarko, B., Degradation of vinyl acetate by soil, sewage, sludge, and the newly isolated aerobic bacterium C2, *Appl. Environ. Microbiol.*, 56:3032–3038, 1990.

Texas Department of Health policy statement — On-site surface application of treated wastewater, November 20, 1990.

U.S. Congress, Senate Committee on Environment and Public Works, Espanola Valley and Projaaque Valley wastewater master plan, Washington, D.C., 1990.

U.S. Department of Agriculture, Sewage Sludge in Agriculture, Beltsville, MD, 1991.

U.S. EPA, Cleaner water through conservation, USEPA 841-B-95-002, 1995.

U.S. EPA, Guidance specifying management measures for sources on nonpoint pollution in coastal waters, USEPA 840-B-92-002, 1993.

U.S. EPA, A guide to the biosolids risk assessments for the EPA part 503 rule, EPA/832/B-93-005, September 1995.

U.S. EPA, Guide to septage treatment and disposal, EPA/625/R-94/002, EPA Office of Research and Development, Washington, D.C., 1994.

U.S. EPA, Permits for municipal sewage sludge, Washington, D.C., 1991.

U.S. EPA, A plain English guide to the EPA Part 503 biosolids rule, EPA/832/R-93-003, September 1994.

U.S. EPA, State sludge management program guidance manual, Washington, D.C., 1990.

U.S. EPA, Surface disposal of sewage sludge, a guide for owners/operators of surface disposal facilities on the monitoring, record-keeping, and notification requirements of the Federal Standards for the Use or Disposal of Sewage Sludge, 40 CFR Part 503, 1994.

U.S. EPA, Region 8, Biosolids management handbook for small publicly owned treatment works (POTWs), September 1995.

Wakakuwa, J.R., Toxic sludge, *Nature (London)*, 346(2):617–618, 1991.

Waller, D.H., Mooers, J.D., Options for wastewater management in non-urban areas in Nova Scotia, Proceedings of Rural Resources Development Conference, Truro, 1997.

Weiskel, P.K., Howes, B.L., Differential transport of sewage-derived nitrogen and phosphorus through a coastal watershed, *Environ. Sci. Technol.*, 26: 352–360, 1992.

Wild, S.R., Jones, K.C., Organic chemicals entering agricultural soils in sewage sludges: screening for their potential to transfer to crop plants and livestock, *Sci. Total Environ.*, 85–119, June 1, 1992.

Wilhelm, S.R., Schiff, S.L., Robertson, W.D., Chemical fate and transport in a domestic septic system: unsaturated and saturated zone geochemistry, *Environ. Toxicol. Chem.*, 13(2):193–203, 1994.

Wilson, A., Ocean dumpers seek new options on land, *ENR*, 226:65, 1991.

Yingming, L., Corey, R.B., Heavy metals in the environment, redistribution of sludge-borne cadmium, copper, and zinc in a cultivated plot, *J. Environ. Qual.*, 22:1–8, 1993.

CHAPTER 7

Abbruzzese, B.S., Leibowitz, S.G., Sumner, R., Application of the synoptic approach to wetland designation: a case study in Louisiana, final report, submitted to U.S. Environmental Protection Agency, Region 10, Seattle, WA, 1990b.

Abelson, P., Desalination of brackish and marine waters, *Science*, 251:189, March 15, 1991.

Alexander, R.B., Slack, J.R., Fitzgerald, A.S., Schertz, T.L., Data from selected U.S. Geological Survey national stream water quality monitoring networks, *Water Resour. Res.*, 34(9):2401–2405, 1997.

Alexander, R.B., Smith, R.A., Schwarz, G.E., Effect of stream channel size on the delivery of nitrogen to the Gulf of Mexico, *Nature (London)*, 403:758–761, February 17, 2000.

Alton, H.E., Fu, G., Deng, B., Analysis of acid-volatile sulfide (AVS) and simultaneously extracted metals (SEM) for the estimation of potential toxicity in aquatic sediments, *Environ. Toxicol. Chem.*, 12:1441–1453, 1993.

Ambrose, R.B., Martin, J.L., Jirka, G.H., Technical guidance manual for performing waste load allocations, Book III: *Estuaries*, Part 3, Use of mixing zone models in estuarine waste load allocations, EPA-823/R-92/004, U.S. Environmental Protection Agency, Washington, D.C., 1992.

American Water Works Association, *Water Fluoridation Principles and Practices*, American Water Works Association, Denver, CO.

Anderson, R., Rockel, M., Economic valuation of wetlands, American Petroleum Institute, Washington, D.C., 1991.

Andrews, W.J., Few volatile organic compounds detected in rivers and groundwater in the Upper Mississippi River Basin, Minnesota and Wisconsin, U.S. Geological Survey Fact Sheet, FS-095-96, 1996, 2 pp.

Andrews, W.J., Fallon, J.D., Kroening, S.E., Water-quality assessment of part of the Upper Mississippi River Basin, Minnesota and Wisconsin — volatile organic compounds in surface and groundwater, 1978–94, U.S. Geological Survey Water-Resources Investigation Report, WRIR 95-4216, 1995, 39 pp.

API, Users guide and technical resource document: evaluation of sediment toxicity tests for biomonitoring programs, API pub. no. 4607, Prepared for American Petroleum Institute, Health and Environmental Sciences Department, Washington, D.C., by PTI Environmental Services, Bellevue, WA, 1994.

Arar, E.J., Pfaff, J.D., Determination of dissolved hexavalent chromium in industrial wastewater effluents by ion chromatography and post-column derivation with diphenylcarbazide, *J. Chromatogr.* 546(1–2):335–340, June 1991.

Aspelin, A.L., Pesticides industry sales and usage, 1994 and 1995 market estimates, U.S. Environmental Protection Agency, Washington, D.C., 733-R-97-002, 1997, p. 2.

Bacher, A.L., Stackelberg, P.E., Baker, R.J., Kauffman, L.J., Hopple, J.A., Ayers, M.A., Design of a sampling network to determine the occurrence and movement of methyl *tert*-butyl ether and other organic compounds through the urban hydrologic cycle, Symposium on Fate and Transport of Fuel Oxygenates, Division of Environmental Chemistry *Am. Chem. Soc.,* San Francisco, CA, April 13–17, Preprints of Extended Abstracts, 37(1):400-401, 1997.

Bachman, L.J., Lindsey, B.D., Brakebill, J., Powars, D.S., Groundwater discharge and base flow nitrate loads of nontidal streams, and their relation to a hydrogeomorphic classification of the Chesapeake Bay watershed, Middle Atlantic coast, U.S. Geological Survey Water-Resources Investigations Report 98-4059, 1998, 71 pp.

Baker, J.E., Church, T.M., Eisenreich, S.J., Fitzgerald, W.F., Scudlark, S.R., Relative atmospheric loadings of toxic contaminants and nitrogen to the great waters, Prepared for U.S. Environmental Protection Agency, Office of Air Quality Planning and Standards, Pollution Assessment Branch, Durham, NC, 1993. Cited in USEPA, 1994a.

Barbash, J.E., Resek, E.A., Pesticides in ground water — distribution, trends, and governing factors, in *Pesticides in the Hydrologic System,* Vol. 2, Gilliom, R.J., Ed., Ann Arbor Press, Chelsea, MI, 1996, 588 pp.

Baudo, R., Muntau, H., Lesser known in-place pollutants and diffuse source problems, in *Sediments: Chemistry and Toxicity of In-Place Pollutants,* Baudo, R., Giesy, J., Muntau, H., Eds., Lewis Publishers, Chelsea, MI, 1990, chap. 1.

Becher, K.D., Schnoebelen, D.J., Akers, K.B., Nutrients discharged to the Mississippi River from eastern Iowa watersheds, *J. Am. Water Resour. Assoc.,* 36(1):161–173, February 2000.

Bender, C.A., Zogorski, J.S., Halde, M.J., Rowe, B.L., Selection procedure and salient information for volatile organic compounds emphasized in the National Water-Quality Assessment Program, U.S. Geological Survey Open-File Report 99-182, 1999.

Bender, D.A., Zogorski, J.S., Luo, W., Pankow, J.F., Mejewski, M., Baker, R., Atmosphere-water interaction of chloroform, toluene, and MTBE in small perennial urban streams, in Proceedings of the Air and Waste Management 93rd Annual Conference and Exposition, June 18–22, 2000, Salt Lake City, UT, 2000.

Bennion, H., A diatom-phosphorus transfer function for shallow, eutrophic ponds in southeast England, *Hydrobiologia,* 275(6):391–410, 1994.

Bleier, A., Waste management/marine sanitation, in *Proceedings of the 1991 National Applied Marina Research Conference,* Ross, N., Ed., International Research Institute, Wickford, RI, 1991.

Bohlen, C., Protecting the coast, *Bioscience,* 40(4):243, April 1990.

Bopp, R.F., Gross, M.L., Tong, H., Simpson, H.J., Monson, S.J., Deck, B.L., Moser, F.C., A major incident of dioxin contamination: sediments of New Jersey estuaries, *Environ. Sci. Technol.*, 25(5):951–956, 1991.

Bopp, R.F., Simpson, H.J., Chillrud, S.N., Robinson, D.W., Sediment derived chronologies of persistent contaminants in Jamaica Bay, New York, *Estuaries*, 16(3B):608–616, 1993.

Braam, G.A., Jansen, W.A., North Point Marina A case study, in World Marina '91: Proceedings of the First International Conference, September 4–8, 1991, American Society of Civil Engineers, Long Beach, CA, 1991.

Bricker, S.B., Historical trends in contamination of estuarine and coastal sediments: the history of Cu, Pb, and Zn inputs to Narragansett Bay, Rhode Island, as recorded by salt marsh sediments, *Estuaries*, 16(3B):589–607, 1993.

Brigham, M.E., Pesticides detected in surface waters and fish of the Red River of the north drainage basin, North Dakota Water Quality Symposium Proceedings, March 30–31, 1994, Fargo, ND, 1994, pp. 256–269.

Brigham, M.E., Tornes, L.H., trace elements and organic contaminants in stream sediments from the Red River of the north basin, 1992–1995, North Dakota Water Quality Symposium Proceedings, March 20–21, 1996, Bismarck, North Dakota State University Extension Service, 1996, pp. 135–144.

British Columbia Research Corporation, Urban runoff quality and treatments: a comprehensive review, *GVRD*, 1991.

British Waterways Board, Waterway ecology and the design of recreational craft, Inland Waterways Amenity Advisory Council, London, England, 1983.

Buell, G.R., Transport of trace elements and semi-volatile organic compounds in fluvial sediments of the lower Flint River (Georgia) and Apalachicola River (Georgia, Florida, Alabama) basins during the tropical storm Alberto, July 1994, in U.S. Geological Survey reprints from *Proceedings of the 1997 Georgia Water Resources Conference, March 20–22, 1997*, Hatcher, K.J., Ed., University of Georgia, Athens, GA, 1997, pp. 24–32.

Buell, G.R., Couch, C.A., Environmental distribution of organochlorine compounds in the Apalachicola–Chattahoochee–Flint River basin, in U.S. Geological Survey reprints from *Proceedings of the 1995 Georgia Water Resources Conference April 11–12, 1995*, Hatcher, K.J., Ed., University of Georgia, Athens, GA, 1995, pp. 13–19.

Buell, G.R., Couch, C.A., National Water Quality Assessment Program — Environmental and distribution of organochlorine compounds in the Appalachicola–Chattahoochee–Flint River Basin, in *Proceedings of the 1995 Georgia Water Resources Conference, Athens, GA,* Hatcher, K.J., Ed., Vinson Institute of Government, University of Georgia, 1995, pp. 46–53.

Burke, W., Restoring water naturally, *Tech. Rev.,* 94(1):16–17, January 1991.

Cahill Associates, Limiting NPS pollution from new development in the New Jersey coastal zone, New Jersey Department of Environmental Protection, 1991.

Callender, E., Rice, K.C., The urban environmental gradient: anthropogenic influences on the spatial and temporal distributions of lead and zinc in sediments, *Environ. Sci. Technol.,* 34(2):232–238, 2000.

Canfield, T.J., Kemble, N.E., Brumbaugh, W.J., Dwyer, F.J., Ingersoll, C.G., Fairchild, F.J., Use of benthic invertebrate community structure and the sediment quality triad to evaluate metal-contaminated sediment in the Upper Clark Fork River, Montana, *Environ. Toxicol. Chem.,* 13:1999–2012, 1994.

Capel, P.D., Organic chemical concepts, in *Regional Ground-Water Quality,* Van Nostrand Reinhold, New York, 1993, pp. 155–179.

Capel, P.D., Larson, S.J., A chemodynamic approach for estimating losses of target organic chemicals from water during sample holding time, *Chemosphere*, 30(6):1097–1107, 1995.

Capel, P.D., Ma, L., Schroyer, B.R., Larson, S.J., Gilchrist, T.A., Analysis and detection of the new corn herbicide acetochlor in river water and rain, *Environ. Sci. Technol.*, 29(6):1702–1705, 1995.

Chapelle, F.H., McMahon, P.B., Dubrovsky, N.M., Fujii, R.F., Oaksford, E.T., Vroblesky, D.A., Deducing the distribution of terminal electron-accepting processes in hydrologically diverse groundwater systems, *Water Resour. Res.*, 31(2),359–371, 1995.

Cherfas, J., The fringe of the ocean — under siege from land, *Science*, 248:163–165, April 13, 1990.

Clarke, J.U., McFarland, V.A., Assessing bioaccumulation in aquatic organisms exposed to contaminated sediments, Long-Term Effects of Dredging Operations Program, misc. pap. D-91-2, U.S. Army Corps of Engineers, Waterways Experiment Station, Vicksburg, MS, 1991.

Chow, D.W., Mast, M.A., Long-term trends in stream water and precipitation chemistry at five headwater basins in the northeastern United States, *Water Resour. Res.*, 35(2):541–554, 1999.

Church, C.D., Isabelle, L.M., Pankow, J.F., Rose, D.L., Tratnyek, P.G., Method for determination of methyl-*tert*-butyl ether and its degradation products in water, *Environ. Sci. Technol.*, 31(12):3723–3726, 1997.

Clawges, R.M., Stackelberg, P.E., Ayers, M.A., Vowinkel, E.F., Nitrate, volatile organic compounds, and pesticides in ground water — a summary of selected studies from New Jersey and Long Island, New York, U.S. Geological Survey Water-Resources Investigation Report 99-4027, 1999, 32 pp.

Cochran, K., Lessons from the deep sea, *Nature (London)*, 346:219–220, July 19, 1990.

Cohn-Lee, R., Cameron, D., Poison runoff in the Atlanta region, Natural Resources Defense Council, Washington, D.C., 1991.

Coler, R.A., Rockwood, J.P., *Water Pollution Biology: A Laboratory/Field Handbook*, Technomic Publishing, Lancaster, PA, 1989.

Colman, J.A., Sanzolone, R.F., Geochemical characterization of streambed sediment in the upper Illinois River basin, *Water Resour. Bull.*, 28(5):933, October 1992.

Comis, D., Reviving the Chesapeake Bay, *Agric. Res.*, 38(9):4–11, September 1990.

Commission on Physical Science, Mathematics, and Resources, National Academy of Sciences, *Irrigation Induced Water Quality Problems*, National Academy Press, Washington, D.C., 1989.

Connor, B.F., Rose, D.L., Noriega, M.C., Murtagh, L.K., Abney, S.R., Methods of analysis by the National Water-Quality Laboratory — determination of 86 volatile organic compounds in water by gas chromatography/mass spectrometry, including detections less than reporting limits, U.S. Geological Survey Open-File Report 97-829, 1998.

Cooper, A.B., Nitrate depletion in the riparian zone and stream channel of a small headwater catchment, *Hydobiologica*, 202:13–26, 1990.

Corbitt, R., *Standard Handbook of Environmental Engineering*, McGraw-Hill, New York, 1990.

Couch, C.A., Environmental distribution of mercury related to land use and physiochemical setting in watersheds of the Apalachicola–Chattahoochee–Flint River basin, in U.S. Geological Survey reprints from *Proceedings of the 1997 Georgia Water Resources Conference, March 20–22, 1997*, Hatcher, K.J., Ed., University of Georgia, Athens, GA, 1997, pp. 33–39.

Couch, C.A., National Water-Quality Assessment Program: environmental setting of the Apalachicola–Chattahoochee–Flint River basin in Proceedings of the 1993 Georgia Water Resources Conference, April 20–21, 1993, Athens, GA, 1993, 3 pp.

Cowdery, T.K., Goff, K.L, Nitrogen concentrations near the water table of the Cheyenne Delta aquifer beneath cropland areas, Ransom and Richland counties, North Dakota, North Dakota Water Quality Symposium Proceedings, March 30–31, 1994, Fargo, ND, 1994, pp. 89–102.

Crandall, C.A., Katz, B.G., Hirten, J.J., Hydrochemical evidence for mixing of river water and groundwater during high-flow conditions, lower Suwannee River basin, Florida, U.S.A., *Hydrology J.,* 7:454–467, 1999.

Crawford, D.W., Bonnevie, N.L., Wenning, J., Sources of pollution and sediment contamination in Newark Bay, New Jersey, *Ecotoxicol. Environ. Saf.,* 30(1):85–100, 1995.

Crawford, J.K., Luoma, S.N., Guidelines for studies of contaminants in biological tissues for the National Water-Quality Assessment Program, U.S. Geological Survey Open-File Report 92-494, 1994, 69 pp.

Dahl, T.E., Johnson, C.E., Wetlands: status and trends of wetlands in the coterminous United States, mid-1970s to mid-1980s, U.S. Department of the Interior, Fish and Wildlife Service, Washington, D.C., 1991.

Degen, M.B., Renbeau, R.B., Jr., Hegedorn, C., Martens, D.C., Denitrification in onsite wastewater treatment and disposal systems, Virginia Polytechnic Institute, Blacksburg, VA, 1991.

Delzer, G.C., Zogorski, J.S., Lopes, T.J., Bosshart, R.L., Occurrence of the gasoline oxygenate MTBE and BTEX compounds in urban stormwater in the United States, 1991–1995, U.S. Geological Survey Water-Resources Investigation Report 96-4145, 1996.

DeVivo, J.C., Impact of introduced shiner, *Cyprinella lutrensis,* on stream fishes near Atlanta, GA, in U.S. Geological Survey reprints from Proceedings of the 1997 Georgia Water Resources Conference, March 20–22, 1997, Hatcher, K.J., Ed., University of Georgia, Athens, GA, 1997, pp. 95–98.

DeVivo, J.C., Couch, C.A., Freeman, B.J., Use of a preliminary index of biotic integrity in urban streams around Atlanta, GA, in U.S. Geological Survey reprints from *Proceedings of the 1997 Georgia Water Resources Conference, March 20–22, 1997,* Hatcher, K.J., Ed., University of Georgia, Athens, GA, 1997, pp. 40–43.

DeVivo, J.C., Frick, E.A., Hippe, D.J., Buell, G.R., Effect of restricted phosphate detergent use and mandated upgrades at two wastewater-treatment facilities on water quality, metropolitan Atlanta, GA, 1983–1993, in in U.S. Geological Survey reprints from *Proceedings of the 1995 Georgia Water Resources Conference, April 11–12, 1995,* Hatcher, K.J., Ed., University of Georgia, Athens, GA, 1997, pp. 20–22.

DiToro, D.M., Mahony, J.D., Hansen, D.J., Scott, K.J., Hicks, M.B., Mays, S.M., Redmond, M.S.,Toxicity of cadmium in sediments: the role of acid volatile sulfide, *Environ. Toxicol. Chem.,* 9:1487–1502, 1990.

Dreher, D.W., Price, T.H., Best management practice handbook for urban development, Northeastern Illinois, Planning Commission, Chicago, IL, 1992.

Driscoll, C.T., Van Dressen, R., Seasonal and long-term temporal patterns in the chemistry of Adirondack lakes, *Water, Air, Soil Pollut.,* 67:319–344, 1993.

Driscoll, E., Shelley, P., Strecker, E., *Pollutant Loadings and Impacts from Highway Stormwater Runoff,* Vol. 1, Federal Highway Administration, April 1990.

Duedall, I.W., Global inputs, characteristics, and fates of ocean dumped sewage and industrial wastes, *Water Pollut. Control Fed. J.,* 57(5):358–365, May 1990.

Dunkle, S.A., Plummer, L.N., Busenberg, E., Phillips, P.J., Denver, J.M., Hamilton, P.A., Michel, R.L., Coplen, T.B., Chlorofluorocarbons (CC13F and CC12F2) as dating tools and hydrologic tracers in shallow groundwater of the Delmarva Peninsula, Atlantic Coastal Plain, United States, *Water Resour. Res.,* 29(12)3837–3860, 1993.

Eisler, R., Electroplating wastes in marine environments. A case history at Quonset Point, Rhode Island, in *Handbook of Ecotoxicology,* Hoffman, D.J., Rattner, B.A., Burton, G.A., Jr., Cairns, J., Jr., Eds., Lewis Publishers, Boca Raton, FL, 1995, pp. 609–630.

Engle, B.W., Jarrett, A.R., Improved sediment retention efficiencies of sedimentation basins, American Society of Agricultural Engineers, Chicago, IL, Paper No. 90-2629, 1990.

Enserink, E.L. et al., Combined effects of metals: an ecotoxicological evaluation, *Water Res.,* 25(6):679–687, 1991.

Executive Office of the President, Reinventing environmental regulation, EOP, Washington, D.C., 1995.

Executive Office of the President, Reinventing environmental regulation, progress report, EOP, Washington, D.C., 1996.

Exner, M.E., Burbach, M.E., Watts, D.G., Shearman, R.C., Spalding, R.F., Deep nitrate movement in the unsaturated zone of a simulated urban lawn, *J. Environ. Qual.,* 20:658–662, 1991.

FDEP, Approach to the assessment of sediment quality in Florida coastal water, Vol 1, Development and evaluation of sediment quality assessment guidelines, Prepared for Florida Department of Environmental Protection, Office of Water Policy, Tallahassee, FL, by MacDonald Environmental Sciences, Ladysmith, British Columbia, 1994.

Fisher, D.C., Oppenheimer, M., Atmospheric nitrogen deposition and the Chesapeake Bay estuary, *AMBIO,* 20(3-4):102–108, 1991.

Fitzpatrick, F.A., Use of a geographic information system in the upper Illinois River basin pilot project of the National Water-Quality Assessment Program, in Remote Sensing and GIS Application to Nonpoint Source Planning: USEPA Northeastern Illinois Planning Commission Workshop, October 1–3, 1990, Chicago, Illinois, 1990, pp. 55–66.

Fitz-Simons, T., Freas, W., Guinnup, D., Hemby, J., Mintz, D., Sansevero, C., Schmidt, M., Thompson, R., Wayland, M., Damberg, R., National air quality and emissions trends report, 1995: Research Triangle Park, NC, Environmental Protection Agency Report, EPA-454/R-96-005, 1995, 168 pp.

Fleisher, J.M., Conducting recreational water quality surveys: some problems and suggested remedies, *Marine Pollut. Bull.,* 21(12):562–567, 1990.

Fleisher, J.M., A reanalysis of data supporting U.S. federal bacteriological water quality criteria governing marine recreational waters, *Res. J. WPCF,* 63(3):259–265, 1991.

Fleisher, J.M., Jones, F., Kay, D., Stanwell-Smith, R., Wyer, M., Morano, R., Water and non-water-related risk factors for gastroenteritis among bathers exposed to sewage-contaminated marine waters, *Int. J. Epidemiol.,* 22(4):698–708, 1993.

Fleischer, S., Stibe, L., Leonardson, L., Restoration of wetlands as a means of reducing nitrogen transport to coastal waters, *Ambio,* 20(6):271–272, 1991.

Focazio, M.J., Welch, A.H., Watkins, S.A., Helsel, D.R., Horn, M.A., A retrospective analysis on the occurrence of arsenic in ground-water resources of the United States and limitations in drinking water supply characterizations, U.S. Geological Survey Water Resources Investigations Report WRIR 99-4279, 2000, 21 pp.

Foster, B., Alternative technologies for deicing highways, National Conference of State Legislatures, *State Legislative Rep.,* 15(10), April 1990.

Fox, R.G., Tuchman, M., The assessment and remediation of contaminated sediments (ARCS) program, *J. Great Lakes Res.,* 22:493–494, 1996.

Frahm, A., Cleanup of Puget Sound, *EPA J.,* 16(6):29–33, November 1990.

Francy, D.S., Helsel, D.H., Nally, R.A., Occurrence and distribution of microbiological indicators in ground water and stream water, *Water Environ. Res.,* 72(2):152–161, 2000.

Freeman, H.M., *Standard Handbook of Hazardous Waster Treatment and Disposal,* U.S. EPA, McGraw-Hill, New York, 1989.

Freeman, W.O., Ed., *NAWQA Notes*–Newsletter of the National Water-Quality Assessment: Hudson River basin, 1(1), May 1994, U.S. Geological Survey NAWQA Hudson River Basin Study Unit, 1994, 4 pp.

Frick, E.A., Suface-water and shallow ground-water quality in the vicinity of Metropolitan Atlanta, upper Chattahoochee River basin, GA, 1992–1995, in U.S. Geological Survey reprints from *Proceedings of the 1997 Georgia Water Resources Conference, March 20–22, 1997,* Hatcher, K.J., Ed., University of Georgia, Athens, GA, 1997, pp. 44–48.

Frick, E.A., Buell, G.R., Relation of land use to nutrient and suspended-sediment concentrations, loads, and yields in the upper Chattahoochee River basin, GA, 1993–1998, in U.S. Geological Survey reprints from *Proceedings of the 1999 Georgia Water Resources Conference, March 30–31, 1999,* Hatcher, K.J., Ed., University of Georgia, Athens, GA, 1997, pp. 170–179.

Frick, E.A., Crandall, C.A., Water quality in superficial aquifers in two agricultural areas in Georgia, Alabama, and Florida, in U.S. Geological Survey reprints from *Proceedings of the 1995 Georgia Water Resources Conference, April 11–12, 1995,* Hatcher, K.J., Ed., University of Georgia, Athens, GA, 1995, p. 9–13.

Furness, R.W., Rainbow, P.S., *Heavy Metals in the Marine Environment,* CRC Press, Boca Raton, FL, 1990.

Galli, J., Dubose, R., Water temperature and freshwater stream biota: an overview, Maryland Department of the Environment, Sediment and Stormwater Administration, Baltimore, 1990.

Garton, L.S., Bonner, J.S., Ernest, A.N., Autenrieth, R.L., Fate and transport of PCBs at the New Bedford Harbor Superfund site, *Environ. Toxicol. Chem.,* 15(5):736–745, 1996.

GESAMP, The state of the marine environment, United Nations Environment Program (UNEP) Regional Seas Reports and Studies no. 115, IMO/FAO/UNESCO/WMO/WHO/IAEA/UN/UNEP Joint Group of Experts on the Scientific Aspects of Marine Pollution, New York, 1990.

Gilliom, R.J., Alley, W.M., Gurtz, M.E., Design of the National Water Quality Assessment Program — occurrence and distribution of water quality conditions, U.S. Geological Survey Circular 1112, 1995, 33 pp.

Gilliom, R.J., Barbash, J.E., Kolpin, D.W., Larson, S.J., Testing water quality for pesticide pollution, *Environ. Sci. Technol.,* April:164A–169A, 1999.

Gilliom, R.J., Clifton, D.G., Organochlorine pesticide residues in bed sediments of the San Joaquin River, California, *Water Res. Bull.,* 26(1):11–24, 1990.

Gilliom, R.J., Mueller, D.K., Nowell, L.H., Methods for comparing water quality conditions among assessment study units, 1992–1995, U.S. Geological Survey Open File Report 97-589, 1997, 54 pp.

Glass, G.E., Rapp, G.R., Jr., Schmidt, K.W., Sorenson, J.A., New source identification of mercury contamination in the Great Lakes, *Environ. Sci. Technol.,* 24:1059–1069, 1991.

Glick, R., Wolfe, M.L., Thurow, T.L., Urban runoff quality as affected by native vegetation, presented at the 1991 International Summer Meeting sponsored by American Society of Engineers, Albuquerque, NM, ASAE Paper No. 91-2067, 1991.

Gobas, F.A., McNeil, E.J., Lovett-Doust, L., Bioconcentration of chlorinated aromatic hydrocarbons in aquatic macrophytes, *Environ. Sci. Technol.,* 25:924–929, 1991.

Goldstein, R.M., Brigham, M.E., Stauffer, J.C., Comparison of mercury concentrations in liver, muscle, whole bodies, and composites of fish bodies, and composites of fish from the Red River of the north, *Can. J. Fish. Aquatic Sci.,* 53(2):244–252, 1996.

Goldstein, R.M., Simon, T.P., Bailey, P.A., Ell, M., Pearson, E., Schmidt, K., Enblom, J.W., Concepts for an index of biotic integrity for streams of the Red River of the north basin, North Dakota Water Quality Symposium Proceedings, March 30–31, 1994, Fargo, North Dakota, 1994, pp. 169–180.

Goplerud, C.P., Water pollution law: milestones from the past and anticipation of the future, *Nat. Resour. Environ.*, 10(2):7–12, Fall 1995.

Graffy, E.A., Low-level detection of pesticides — so what?, *J. Soil Water Conserv.*, 53(1):11–12, First Quarter 1998.

Grimalt, J., Assessment of fecal sterols and ketones as indicators of urban sewage inputs to the coastal waters, *Environ. Sci. Technol.*, March:357–363, 1990.

Hakapaa, K., Marine pollution, *Am. J. Int. Law*, 84(4):972–973, October 1990.

Halde, M.J., Delzer, G.C., Zogorski, J.S., Study design and analytical results used to evaluate a surface-water point sampler for volatile organic compounds, U.S. Geological Survey Open-File Report OFR 98-651, 1998, 31 pp.

Hall, J.V., Frayer, W.E., Bill, O., Status of the Alaska wetlands, U.S. Department of the Interior, Fish and Wildlife Service, Alaska Region, Anchorage, AK, 1994.

Hamilton, P.A., Welch, A.H., Christenson, S.C., Alley, W.M., Uses and limitations of existing ground-water-quality data, in *Regional Ground-Water Quality*, Van Nostrand Reinhold, New York, 1993, pp. 613–622.

Hammer, D.A., Designing constructed wetlands systems to treat agricultural nonpoint source pollution, *Ecol. Eng.*, 1(1992):49–82.

Hammer, D.A., Pullin, B.P., Watson, J.T., Constructed wetlands for livestock waste treatment, Tennessee Valley Authority, Knoxville, TN, 1989.

Hansen, D.J., Assessment tools that can be used for the National Sediment Inventory. Memorandum from D.J. Hansen, Environmental Research Laboratory, Narragansett, to C. Fox, USEPA Office of Water, February 28, 1995.

Hansen, D.J., Berry, W.J., Mahoney, J.D., Boothman, W.S., DiToro, D.M., Robson, D.L., Ankley, G.T., Ma, D., Yan, Q., Pesch, C.E., Predicting the toxicity of metal-contaminated field sediments using interstitial concentrations of metals and acid-volatile sulfide normalizations, *Environ. Toxicol. Chem.*, 15(12):2080–2094, 1996.

Hanson, J.S., Malanson, G.P., Armstrong, M.P., Landscape fragmentation and dispersal in a model of riparian forest dynamics, *Ecol. Modeling*, 49:277–296, 1990.

Hefner, J.M., Wilen, B.O., Dahl, T.E., Frayer, W.E., Southeast wetlands: status and trends, mid-1970s to mid-1980s, U.S. Department of the Interior, Fish and Wildlife Service, Atlanta, GA, 1994.

Henny, C.J., Blus, L.J., Hoffman, D.J., Grove, R.A., Lead in hawks, falcons and owls downstream from a mining site on the Coeur d'Alene River, Idaho, *Environ. Monit. Assess.*, 29(3):267–288, 1994.

Hetsel, D.R., Hirsch, R.M., *Statistical Methods in Water Resources*, Elsevier, Inc., New York, 1992.

Hippe, D.J., Hatzell, H.H., Ham, L.K., Hardy, P.S., Pesticide occurrence and temporal distribution in streams draining urban and agricultural basins in Georgia and Florida, 1993–1994, in U.S. Geological Survey reprints from *Proceedings of the 1995 Georgia Water Resources Conference, April 11–12, 1995*, Hatcher, K.J., Ed., University of Georgia, Athens, GA, 1995, pp. 1–8.

Hippe, D.J., Wangsness, D.J., Frick, E.A., Garrett, J.W., Do the pesticides I use contaminate the rivers everyone uses?, Based on U.S. Geological Survey Water-Resources Investigations Report 94-4183, 1994, 2 pp.

Hippe, D.J., Wangsness, D.J., Frick, E.A., Garrett, J.W., Suspended sediment and agricultural chemicals in floodwaters caused by tropical storm Alberto, Based on U.S. Geological Survey Water-Resources Investigations Report 94-4183, 1994, 2 pp.

Hitt, K.J., Refining 1970's land-use data with 1990 population data to indicate new residential development, U.S. Geological Survey Water-Resources Investigations Report 94-4250, 1994, 15 pp.

Horowitz, A.J., Elrick, K.A., Cook, R.B., Effect of mining and related activities on the sediment trace element geochemistry of Lake Coeur d'Alene, Idaho, USA, Part 1: surface sediments, *Hydrol. Processes,* 7(4):403–423, 1993.

Houck, O.A., TMDLs: the resurrection of water quality standards-based regulation under the Clean Water Act, *Environ. Law Rep. News Anal.,* 27(7):10329–10344, July 1997.

IEP, Inc., Vegetated buffer strip designation method guidance manual, Narragansett Bay Project, prepared for U.S. Environmental Protection Agency and the Rhode Island Department of Environmental Management, Providence, RI, 1991.

Ingersoll, C.G., Sediment toxicity and bioaccumulation testing, *ASTM Stand. News,* 19:28–33, 1991.

Ingersoll, C.G., Haverland, P.S., Brunson, E.L., Canfield, T.J., Dwyer, F.J., Henke, C.E., Kemble, N.E., Calculation and evaluation of sediment effect concentrations, *J. Great Lakes Res.* 22:602–623, 1996.

Jacobs, H.M., Planning the use of land for the 21st century, *J. Soil Water Conserv.,* 47(1):32–34, 1992.

Joint Task Force, Design of municipal wastewater treatment plants, MOP 8, Water Environment Federation, Alexandria, VA, 1991.

Kadlec, R.H., Knight, R.L., *Treatment Wetlands,* Lewis Publishers, Boca Raton, FL, 1996.

Kalkhoff, S.J., Kolpin, D.W., Thurman, E.M., Ferrer, I., Barcelo, D., Degradation of chloro-acetanilide herbicides: the prevalence of sulfonic and oxanilic acid metabolites in Iowa groundwaters and surface waters, *Environ. Sci. Technol.,* 32(11):1738–1740, 1998.

Kaplan, I., Sedimentary coprostanol as an index of sewage addition in Santa Monica Basin, *Environ. Sci. Technol.,* February:208–214, 1990.

Karp, C.A., Penniman, C.A., Boater waste disposal "briefing paper" and proceedings from Narragansett Bay Project Management Committee, Narragansett Project, Rhode Island, 1991.

Kay, D., Fleisher, J.M., Salmon, R.L., Jones, F., Wyer, M.D., Godfree, A.F., Zelenauch-Jacquotte, Z., Shore, R., Predicting likelihood of gastroenteritis from sea bathing: results from randomised exposure, *Lancet,* 344:905–909, 1994.

Keeler, G.J., Lake Michigan Urban Air Toxics Study, U.S. Environmental Protection Agency, Office of Research and Development, Atmospheric Research and Exposure Assessment Laboratory, Research Triangle Park, NC, Cited in USEPA, 1994a.

Keko, Inc., Letter dated April 13, 1992, to Geoffrey Grubbs, Director, Assessment and Watershed Protection Division, U.S. Environmental Protection Agency, from W. Kenton, President, Keko, Inc.

Kennicutt, M.C., Wade, T.L., Presly, B.J., Requejo, A.G., Brooks, J.M., Denoux, G.J., Sediment contaminants in Casco Bay, Maine: inventories, sources, and potential for biological impact, *Environ. Sci. Technol.,* 28(1):1–15, 1994.

Kiraly, S.J., Cross, F.A., Buffington, J.D., Federal coastal wetland mapping programs, U.S. Department of the Interior Fish and Wildlife Service, Washington, D.C., *Biol. Rep.* 90(18), 1880.

Klein, R.D., The effects of boating activity and related facilities upon small, tidal waterways in Maryland, Community and Environmental Defense Services, Maryland Line, MD, 1992.

Knopman, D.S., Smith, R.A., Twenty years of the Clean Water Act, *Environment,* 35(1):17–34, 1993.

Knopman, D.S., Smith, R.A., Twenty years of the Clean Water Act: has U.S. water quality improved?, *Environment*, 31(1):16–20, 34–41, January/February 1993.

Kolpin, D.W., Agricultural chemicals in ground water of the Midwestern United States: relations to land use, *J. Environ. Qual.*, 26(4):1025–1037, July–August 1997, (also listed under NS–Nutrients).

Kolpin, D.W., Barbash, J.E., Gilliom, R.J., Occurrence of pesticides in shallow ground water of the United States: initial results from the National Water-Quality Assessment Program, *Environ. Sci. Technol.*, 32(5):558–566, 1998.

Kolpin, D.W., Kalkhoff, S.J., Goolsby, D.A., Sneck-Fahrer, D.A., Thurman, E.M., Occurrence of selected herbicides and herbicide degradation products in Iowa's ground water, 1995, *Groundwater*, 35(4):670–688, July/August 1997.

Kolpin, D.W., Nations, B.K., Goolsby, D.A., Thurman, E.M., Acetochlor in the hydrologic system in the Midwestern United States, 1994, *Environ. Sci. Technol.*, 30(5):1459–1464, 1996.

Kolpin, D.W., Sneck-Fahrer, D.A., Hallberg, G.R., Libra, R.D., Temporal trends of selected agricultural chemicals in Iowa's ground water, 1982–1995, are things getting better?, *J. Environ. Qual.*, 26(4):1007–1017, July/August 1997.

Kolpin, D.W., Squillace, P.J., Barbash, J.E., Zogorski, J.S., Pesticides and volatile organic compounds in shallow urban ground water of the United States, in *Ground Water in the Urban Environment*, Chilton, J., Ed., A.A. Balkema, Netherlands, 1997, pp. 469–474 (also listed under NS–VOCs).

Kolpin, D.W., Thurman, E.M., Linhart, S.M., The environmental occurrence of degradates in groundwater, *Arch. Environ. Contamin. Toxicol.*, 35:385–390, 1998.

Kortmann, R.W., Rich, P.H., Lake ecosystem energetics: the missing management link, In *Lake Reserv. Manage.*, 8(2):77–97, 1994.

Koterba, M.T., Banks, W.S., Shedlock, R.J., Pesticides in ground water in the Delmarva Peninsula, *J. Environ. Qual.*, 22(3):500–518, 1993.

Koterba, M.T., Wilde, F.D., Lapham, W.W., Ground-water data collection protocols and procedures for the National Water-Quality Assessment Program — collection and documentation of water-quality samples and related data, U.S. Geological Survey Open-File Report 95-399, 1995, 113 pp.

Kusler, J.A., Kentula, M.E., Eds., *Wetland Creation and Restoration: The Status of the Science*, Island Press, Washington, D.C., 1990.

Lake, J.L., Pruell, R.J., Osterman, F.A., An examination of dechlorination processes and pathways in New Bedford Harbor sediments, *Mar. Environ. Res.*, 33(1):31–47, 1992.

Landis, W.G., Yu, M.-H., *Introduction to Environmental Toxicology: Impacts of Chemicals upon Ecological Systems*, Lewis Publishers, Boca Raton, FL, 1995.

Lanier, A.L., Database for evaluating the water quality effectiveness of best management practices, North Carolina State University, Department of Biological and Agricultural Engineering, Chapel Hill, NC, 1990.

Lapham, W.W., Moran, M.J., Zogorski, J.S., Enhancements of nonpoint-source monitoring programs to assess volatile organic compounds in the nation's groundwater, in National Monitoring Conference, July 7–9, 1998, Reno, NV [proceedings], National Water Quality Monitoring Council, 1998, p. III-371–III-381.

Lapham, W.W., Neitzert, K.M., Moran, M.J., Zogorski, J.S., USGB compiles data set for national assessment of VOCs in ground water, *Ground Water Monitoring Remediation*, Fall:147–157, 1997.

Lapham, W.W., Wilde, F.D. Koterba, M.T., Ground-water data collection protocols and procedures for the National Water-Quality Assessment Program — selection, installation, and documentation of wells, and collection of related data, U.S. Geological Survey Open-File Report 95-398, 1995, 69 pp.

Larson, S.J., Capel, P.D., Goolsby, D.A., Zaugg, S.D., Sandstrom, M.W., Relations between pesticide use and riverine flux in the Mississippi River basin, *Chemosphere*, 31(5):3305–3321, 1995.

Larson, S.J., Capel, P.D., Mejewski, M.S., *Pesticides in Surface Waters, Distribution, Trends, and Governing Factors,* Vol. 3 of *Pesticides in the Hydrologic System,* Gilliom, R.J., Ed., Ann Arbor Press, Chelsea, MI, 1997, 390 pp.

Larson, S.J., Gilliom, R.J., Capel, P.D., Pesticides in streams of the United States — initial results from the National Water-Quality Assessment Program, U.S. Geological Survey Water Resources Investigation Report 98-4222, 1999, 92 pp.

Lee, K.H., Wetlands detection methods investigation, prepared for U.S. Environmental Protection Agency, Environmental Monitoring Systems Laboratory, Las Vegas, NV, EPA/600/4-91/014, 1991.

Leigh, D.S., Mercury contamination and floodplain sedimentation from former gold mines in north Georgia, *Water Res. Bull.,* 30(4):739–748, 1994.

Lewis, W.M., Jr., Wetlands on the line, *Geotimes,* July 1996, p. 5.

Likens, G.E., Bormann, F.H., *Biogeochemistry of a Forested Ecosystem,* 2nd ed., Springer-Verlag, New York, 1995, 159 pp.

Likens, G.E., Driscoll, C.T., Buso, D.C., Long-term effects of acid rain–response and recovery of a forest ecosytem, *Science,* 272:244–246, 1996.

Liu, S., Yen, S.T., Kolpin, D.W., Pesticides in ground water: do atrazine metabolites matter?, *J. Am. Water Resour. Assoc.,* 32(4):845–853, August 1996.

Loeb, P., Very troubled waters, *U.S. News World Rep.,* 125(12):39, 41–42, September 28, 1998.

Long, E.R., MacDonald, D.D., Smith, S.L., Calder, F.D., Incidence of adverse biological effects within ranges of chemical concentrations in marine and estuarine sediments, *Environ. Manage.,* 19(1):81–97, 1995.

Long, E.R., Robertson, A., Wolfe, D.A., Hameedi, I., Sloane, G.M., Estimates of the spatial extent of sediment toxicity in major U.S. estuaries, *ES&T,* 30(12):3585–3592, 1996.

Long, R.P., Horsley, S.B., Lilja, P.R., Impact of forest liming on growth and crown vigor of sugar maple and associated hardwoods, *Can. J. For. Res.,* 27:1560–1573, 1997.

Lopes, T.J., Nonpoint sources of volatile organic compounds in urban areas — relative importance of urban land surfaces and air, *Environ. Pollut.,* 101:221–230, 1998.

Lopes, T.J., Occurrence and distribution of semivolatile organic compounds in stream bed sediments, United States, 1992–1995, in *Environ. Toxicol. Risk Assessment: ASTM STP 1333,* Little, E.E., Delonay, A.J., Greenberg, B.M., Eds., American Society for Testing Materials, 1998, pp. 105–119.

Lopes, T.J., Occurrence of the gasoline additives MTBE and BTEX compounds in urban stormwater in the United States, 1991–1995, in *Am. Geophys. Union, San Francisco, California: Fall Meeting,* Washington, D.C., 77(46):F283, 1996.

Lopes, T.J., Study plan for urban stream indicator sites of the National Water-Quality Assessment Program, U.S. Geological Survey Open-File Report OFR 98-25, 1998, 15 pp.

Lopes, T.J., Semivolatile organic compounds (SVOCs) in streambed sediment, United States, 1992–1995, in *Environmental Toxicology and Chemistry 20th Annual Meeting, Philadelphia, PA, Nov. 14–18, 1999* [abstract], Washington, D.C., SETAC, 1999, p. 49.

Lopes, T.J., Compositing water samples for analysis of volatile organic compounds, *J. Environ. Eng.,* 2000.

Lopes, T.J., Bender, D.A., Nonpoint sources of volatile organic compounds in urban areas–relative importance of land surfaces and air, *Environ. Pollut.,* 101:221–230, 1998.

Lopes, T.J., Dionne, S.G., A review of semivolatile and volatile organic compounds in highway runoff and urban stormwater, U.S. Geological Survey Open-File Report OFR 98-409, 1998, 67 pp.

Lopes, T.J., Furlong, E.T., Pritt, J.W., Occurrence and distribution of semivolatile organic compounds in stream bed sediments, United States, 1992–1995, in *Environmental Toxicology and Risk Assessment,* Little, E.E., DeLonay, A.J., Greenberg, B.M., Eds., American Testing and Materials, ASTM STP 1333, 7:105–119, 1998.

Lopes, T.J., Price, C.V., Study plan for urban stream indicator sites for the National Water-Quality Assessment Program, U.S. Geological Survey Open-File Report 97-25, 1997.

Love, J.T., Delzer, G.C., Abney, S.R., Zogorski, J.S., Study design and analytical results used to evaluate stability of volatile organic compounds in water matrices, U.S. Geological Survey Open-File Report OFR 98-637, 1998, 156 pp.

Lundgren, R.F., Lopes, T.J., Occurrence, distribution and trends of volatile organic compounds in the Ohio River and its major tributaries, 1987–1996, U.S. Geological Survey Water-Resources Investigations Report 99-4257, 1999, 99 pp.

Lynch, J.A., Bowersox, V.C., Grimm, J.W., Trends in precipitation chemistry in the United States, 1983–1944 — analysis of the effects in 1995 of phase I of the Clean Air Act Amendments of 1990, title IV, U.S. Geological Survey Open-File Report 96–346, 1996, 100 pp.

Majewski, M.S., Capel, P.D., Pesticides in the atmosphere–distribution, trends, and governing factors, Vol. 1 of the series: *Pesticides in the Hydrologic System,* Ann Arbor Press, Chelsea, MI, 1995, 228 pp.

Mancini, J.L., Plummer, A.H., Jr., A review of EPA sediment criteria, Proceedings of the Water Environment Federation, 6th Annual Conference and Exposition, October 15–19, Chicago, IL, in *Surface Qual. Ecol.,* 4:681–694, 1994.

Mancl, K., Magette, W., Maintaining your septic tank, *Water Resources 28,* Cooperative Extension Service, University of Maryland, College Park, MD, 1991.

Mantell, M.A., Harper, S.F., Propst, L., *Creating Successful Communities: A Guidebook to Growth Management Strategies,* Island Press, Washington, D.C., 1990.

Marble, A.D., A guide to wetland functional design, Federal Highway Administration, Washington, D.C., July 1990.

Mast, M.A., Turk, J.T., Environmental characteristics and water quality of hydrologic benchmark stations in the eastern United States, 1963–1995, U.S. Geological Survey Circular 1173-A, 1999, 158 pp.

Mastran, T.A., Dietrich, A.M., Gallagher, D.L., Grizzard, T.J., Distribution of polyaromatic hydrocarbons in the water column and sediments of a drinking water reservoir with respect to boating activity, *Water Res.,* 28(11):2353–2366, 1994.

McCain, B.B. et al., Chemical contamination and associated fish diseases in San Diego Bay, *Environ. Sci. Technol.,* 26:725, 1992.

McLean et al., Effects of three primary treatment sewage outfalls on metal concentrations in the fish *Cheilodactylus fuscus* collected along the coast of Sydney, Australia, *Ma. Pollut. Bull.,* (G.B.) 22:134, 1992.

McMahon, P.B., Bruce, B.W., Distribution of terminal electron-accepting processes in an aquifer having multiple contaminant sources, *Appl. Geochem.,* 12:507–516, 1997.

McMahon, P.B., Dennehy, K.F., N_2O emissions from a nitrogen-enriched river, *Environ. Sci. Technol.,* 33(1):21–25, 1999.

McMahon, P.B., Litke, D.W., Paschal, J.E., Dennehy, K.F., Ground water as a source of nutrients and atrazine to streams in the South Platte River basin, *Water Resour. Bull.,* 30(3):521–530, 1994.

McMahon, P.B., Tindall, J.A., Collins, J.A., Lull, K.J., Hydrologic and geochemical effects on oxygen uptake in bottom sediments of an effluent-dominated river, *Water Resour. Res.,* 31(10):2561–2569, October 1995.

MDDNR, A guidebook for marina owners and operators on the installation and operation of sewage pumpout stations, Maryland Department of Natural Resources, Boating Administration, Annapolis, MD, 1991.

Meade, R.H., Ed., Contaminants in the Mississippi River, 1987–1992, U.S. Geological Survey Circular 1133, 1997, 139 pp.

Metayer, C., Amiard-Triquet, C., Baud, J.P., Species-related variations of silver bioaccumulation and toxicity to three marine bivalves, *Water Res.,* 24:995–1001, 1990.

METRO, Boatyard wastewater treatment guidelines, Municipality of Metropolitan Seattle Water Pollution Control Department, Industrial Waste Section, Seattle, WA, 1992b.

METRO, Maritime Industrial Waste Project, reduction of toxicant pollution from the Maritime Industry in Puget Sound, Municipality of Metropolitan Seattle Water Pollution Control Department, Industrial Waste Section, Seattle, WA, 1992a.

Meyer, J.S., Davison, W., Sundby, B., Ores, J.T., Lauren, D.J., Forstner, U., Hong, J., Crosby, D.G., Bioavailability: physical, chemical, and biological interactions, Synopsis of discussion sessions: the effects of variable redox potentials, pH, and light on bioavailability in dynamic water-sediment environments, in *Proceedings of the Thirteenth Pellson Workshop,* Hamelink, J.L., Landrum, P.F., Bergman, H.L., Benson, W.H., Eds., Lewis Publishers, Boca Raton, FL, 1994, pp. 155–170.

Michigan Department of Environmental Quality, Office of the Great Lakes, Office of the Great Lakes activity report, Michigan Department of Environmental Quality, Lansing, MI, June 1996.

Michigan Department of Environmental Quality, Office of the Great Lakes, State of the Great Lakes: annual report for 1995, Michigan Department of Environmental Quality, Lansing, MI, 1996.

Minter, J.B., Comparison of macroinvertebrate communities from three substrates in the lower South Platte River, Degree of Master of Science thesis, Colorado State University, Fall 1996, 221 pp.

Mitsch, W.J., Landscape design and the role of created, restored, and natural riparian wetlands in controlling nonpoint source pollution, *Ecol. Eng.,* 1:27–47, 1992.

Mitsch, W.J., Wetlands for the control of nonpoint source pollution: preliminary feasibility study for Swan Creek watershed of Northwestern Ohio, Ohio Environmental Protection Agency, Columbus, OH, 1990.

Mitsch, W.J., Gosselink, J.G., *Wetlands,* Van Nostrand Reinhold, New York, 1993.

Moran, M.J., MTBE in ground water of the United States — occurrence, potential sources, and long range transport, in Water Resources Conference, American Water Works Association, Norfolk, VA, September 26–29, 1999 [proceedings], American Water Association, 1999.

Moran, M.J., Clawges, R.M., Zogorski, J.S., Indentifying the usage patterns of methyl-*tert*-butyl ether (MTBE) and other oxygenates in gasoline using gasoline surveys, in Proceedings of the American Chemical Society, 219th ACS National Meeting, 2000, San Francisco, CA, *Am. Chem. Soc.,* 40(1):209–218, 2000.

Moran, M.J., Halde, M.J., Clawges, R.M., Zogorski, J.S., Relations between the detection of methyl-*tert*-butyl ether (MTBE) in surface and ground water and its content in gasoline, in Proceedings of the American Chemical Society, 219th ACS National Meeting, 2000, San Francisco, CA, *Am. Chem. Soc.,* 40(1):195–198, 2000.

Moran, M.J., Zogorski, J.S., Squillace, P.J., MTBE in ground water of the United States — occurrence, potential sources, and long range transport in Proceedings of the 1999 Water Resources Conference, American Water Works Association, 1999.

Mueller, D.K., Hamilton, P.A., Helsel, D.R., Hitt, K.J., Ruddy, B.C., Nutrients in ground water and surface water of the United States — an analysis of data through 1992, U.S. Geological Survey Water Resources Investigations Report 95-4031, 1995, 74 pp.

Mueller, D.K., Helsel, D.R., Nutrients in the nation's waters — too much of a good thing?, U.S. Geological Survey Circular 1136, 1996, 24 pp.

Mueller, D.K., Ruddy, B.C., Battaglin, W.A., Logistic model of nitrate in streams of the upper Midwestern United States, *J. Environ. Qual.*, 26(5):1223–1230, 1997.

Mueller, D.K., Stoner, J.D., Identifying the potential for nitrate contamination of streams in agricultural areas of the United States, in Proceedings of the National Water Quality Monitoring Conference, July 7–9, Reno, NV, 1998, p. III163–III173.

Munson, T., A flume study examining silt fences, in Proceedings of the 5th Federal Interagency Sedimentation Conference, Las Vegas, NV, March 18, 1991.

Murdoch, P.S., Stoddard, J.L., The role of nitrate in the acidification of streams in the Catskill Mountains of New York, *Water Resour. Res.*, 28(10):2707–2720, 1992.

Murphy, T.P., Prepas, E.E., Lime treatment of hardwater lakes to reduce eutrophication, *Verh. Int. Verein. Limnol.*, 24:327–334, 1990.

Muzik, K., Coral grief, *Tech. Rev.*, 94(3):60–67, April 1991.

National Academy of Public Administration (NAPA), Setting priorities, getting results: a new direction for EPA, NAPA, Washington, D.C., 1995.

National Atmospheric Deposition Program, National atmospheric deposition program (NRSP-3)/national trends network, 1997, NADP/NTN Program Office, Illinois State Water Survey, Champaign, IL, 1997.

National Ocean Survey, The 1990 national shellfish register of classified estuarine waters, National Oceanic and Atmospheric Administration, National Ocean Survey, Rockville, MD, 1991.

National Research Council, (NRC) *Contaminated Marine Sediments — Assessment and Remediation*, National Academy Press, Washington, D.C., 1989.

National Research Council, *A Review of the U.S.G.S. National Water Quality Assessment Pilot Program*, National Academy Press, Washington, D.C., 1990.

National Research Council, *Restoration of Aquatic Ecosystems: Science, Technology, and Public Policy*, National Academy Press, Washington, D.C., 1991.

Natural Resources Defense Council, Testing the waters, volume IV, who knows what you're getting into, NRDC, New York, 1996.

NAWQA News, National Water-Quality Assessment Program — Western Lake Michigan Drainage basin, Vol. 1, no.1, March 1995, 6 pp.

NAWQA News, National Water-Quality Assessment Program — Western Lake Michigan Drainage basin, Vol. 2, September 1995, 6 pp.

NAWQA News, National Water-Quality Assessment Program — Western Lake Michigan Drainage basin, Vol. 3, June 1996, 2 pp.

NAWQA News, National Water-Quality Assessment Program — Western Lake Michigan Drainage basin, Vol. 4, April 1997, 4 pp.

Nichols, M.M., Sedimentologic rate and cycling of kepone in an estuarine system: example from the James River estuary, *Sci. Tot. Environ.*, 97–98:407–440, 1990.

Nimmo, D.R. et al., Three studies using ceriodaphnia to detect nonpoint sources of metals from mine drainage, *Res. J. WPCF*, 62(1):7–15, January/February 1990.

NOAA, Coastal nonpoint pollution control program: program development and approval guidance, U.S. Department of Commerce, National Oceanic and Atmospheric Administration, and U.S. Environmental Protection Agency, Office of Water, Washington, D.C., 1991.

NOAA, Inventory of chemical concentrations in coastal and estuarine sediments, NOAA tech. mem. NOS ORCA 76, National Oceanic and Atmospheric Administration, National Ocean Service, Silver Spring, MD, 1994.

Nolan, B.T., Nitrate behavior in ground waters of the southeastern U.S.A., *J. Environ. Qual.,* 28(5):1518–1527, September/October 1999.

Nolan, B.T., Nutrients in groundwaters of the coterminus United States, 1992-1995, *Environ. Sci. Technol.,* 34(7):1156–1165, 2000.

Nolan, B.T., Ruddy, B.C., Hitt, K.J., Helsel, D.R., A national look at nitrate contamination of ground water, *Water Cond. Purif.,* January:76–79, 1998.

Nolan, B.T., Ruddy, B.C., Hitt, K.J., Helsel, D.R., Risk of nitrate in ground waters of the United States — a national perspective, *Environ. Sci. Technol.,* 31(8):229–236, 1998.

Nowell, L.H., Capel, P.D., Dileanis, P.D., Pesticides in stream sediment and aquatic biota–distribution, trends, and governing factors, *Pesticides in the Hydrologic System,* Vol. 4, Ann Arbor Press, Chelsea, MI, 1995, 1001 pp.

Nriagu, J., Global metal pollutants, *Environment,* 32(7):6–11, September 1990.

North Carolina Department of Transportation, NCDOT erosion and sediment control manual, New Standards, 1991.

North Carolina State University, Evaluation of the North Carolina erosion and sedimentation control program, North Carolina Sedimentation Control Commission, Raleigh, 1990, V6–V13.

Novotny, V., Urban diffuse pollution: sources and abatement, *Water Environ. Technol.,* December 1991.

Nurnberg, G.K., Dillon, P.J., Iron budgets in temperate lakes, *Can. J. Fish. Aquatic. Sci.,* 50:1728–1737, 1993.

Nurnberg, G.K., Phosphorus release from anoxic sediments: what we know and how we can deal with it, *Limnetica,* 10(1):1–4, 1994.

Nurnberg, G.K., Quantifying anoxia in lakes, *Limnol. Oceanogr.,* 40(6):1100–1111, 1995.

Nurnberg, G.K., The anoxic factor, a quantitative measure of anoxia and fish species richness in central Ontario lakes, *Am. Fish. Soc.,* 124:677–686, 1995.

Oberts, G.L., Osgood, R.A., Water-quality effectiveness of a detention/wetland treatment system and its effect on an urban lake, *Environ. Manage.,* 15(1):131–138, 1991.

Olem, H., Flock, G., Eds., *Lake and Reservoir Restoration Guidance Manual,* 2nd ed., EPA 440/4-90-006, Prepared by N. Am. Lake Manage. Soc. for U.S. EPA, 1990, 326 pp.

Ontario Ministry of Environment and Energy, Science and Technology Branch, Aquatic Science Section, 1996, Barley straw for algae control in ponds, STB Tech. Bull. No. AqSS-1, 1996, 2 pp.

Ontario Ministry of Environment and Energy, Science and Technology Branch, Aquatic Science Section, 1996, Establishing the credibility of biological data, STB Tech. Bull. No. AqSS-2, 1996, 2 pp.

Ontario Ministry of Environment and Energy, Science and Technology Branch, Aquatic Science Section, 1996, Lake Simcoe–the need for major reductions in phosphorus loading, STB Tech. Bull. No. AqSS-3, 1996, 3 pp.

Ontario Ministry of Environment and Energy, Science and Technology Branch, Aquatic Science Section, 1996, Strategies and parameters for trophic status and water quality assessment, STB Tech. Bull. No. AqSS-4, 1996, 2 pp.

Ontario Ministry of Environment and Energy, Science and Technology Branch, Aquatic Science Section, 1996, Hypolimnetic oxygen: data collection strategies for use in predictive models, STB Tech. Bull. No. AqSS-5, 1996, 2 pp.

Ontario Ministry of Environment and Energy, Science and Technology Branch, Aquatic Science Section, 1996, Mercury in Ontario's environment, STB Tech. Bull. No. AqSS-6, 1996, 3 pp.

Ontario Ministry of Environment and Energy, Science and Technology Branch, Aquatic Science Section, 1996, Mercury in Ontario's environment, who is at risk?, STB Tech. Bull. No. AqSS-7, 1996, 3 pp.

Ontario Ministry of Environment and Energy, Science and Technology Branch, Aquatic Science Section, 1996, The limnology of Lakes Muskoka, Joseph and Rosseau: a collaborative study with MNR, STB Tech. Bull. No. AqSS-8, 1996, 3 pp.

Ontario Ministry of Environment and Energy, Science and Technology Branch, Aquatic Science Section, 1996, (Draft) Crayfish as vectors of transfer and bioconcentration of mercury, STB Tech. Bull. No. AqSS-9, 1996, 3 pp.

Ontario Ministry of Environment and Energy, Science and Technology Branch, Aquatic Science Section, 1996, The trouble with chlorophyll: cautions regarding the collection and use of chlorophyll data, STB Tech. Bull. No. AqSS-10, 1996, 3 pp.

Ontario Ministry of Environment and Energy, Science and Technology Branch, Aquatic Science Section, 1996, Western Lake Erie clean up: phosphorus control or Zebra Mussel effect? an example of the value of long-term data, STB Tech. Bull. No. AqSS-11, 1996, 3 pp.

Organisation for Economic Cooperation and Development, Environmental Performance Reviews: United States, OECD, Paris, 1996.

Overend, R.P., Rivard, C.J., Thermal and biological gasification, in Proceedings of the First Biomass Conference of the Americas: Energy, Environment, Agriculture, and Industry: August 30–September 2, 1993, Burlington, VT, NREL/CP-200-678, National Renewable Energy Laboratory, 1993, pp. 470-497.

Pankow, J.F., Calculated volatilization rates of fuel oxygenate compounds and other gasoline-related compounds from rivers and streams, *Chemosphere,* 33(5):921–937, 1996.

Pankow, J.F., Determination of a wide range of volatile organic compounds (VOCs) in ambient air using multisorbent adsorption/thermal desorption (ATD) and gas chromatography/mass spectrometry (GC/MS), *Anal. Chem.,* 70(24):5213–5221, 1998.

Pankow, J.F., Low detection methods for the determination of VOCs in air and water as applied in the USGS National Water Quality Assessment (NAWQA) Program, in 22nd Conference on Environmental Analysis, Philadelphia, PA, June 1–4, 1999 [proceedings], U.S. Environmental Protection Agency, 1999.

Pankow, J.F., Modeling the atmospheric input of MTBE to groundwater systems, 11th Society of Environmental Toxicology and Chemistry, Washington, D.C., 1996, p. 115.

Pankow, J.F., The urban atmosphere as a nonpoint source for the transport of MTBE and other VOCs to shallow groundwater, *Environ. Sci. Technol.,* 31(10):2821–2828, 1997.

Pankow, J.F., Thomson, N.R., Johnson, R.L., Baehr, A.L., Zogorski, J.S., The urban atmosphere as a nonpoint source for the transport of MTBE and other volatile organic compounds (VOCs) to shallow groundwater, *Environ. Sci. Technol.,* 31:2821–2828, 1997.

Pankow, J.F., Thomson, N.R., Johnson, R.L., Baehr, A.L., Zogorski, J.S., The urban atmosphere as a nonpoint source for the transport of MTBE and other volatile organic compounds (VOCs) to shallow groundwater, in American Chemical Society Division of Environmental Chemistry preprints of papers, 213th, San Francisco, CA, *Am. Chem. Soc.,* 37(1):385–387, 1997.

Pennsylvania Department of Environmental Resources, *Erosion and Sediment Pollution Control Program Manual,* 1990.

Pereira, W.E., Hostettler, F.D., Rapp, R.B., Bioaccumulation of hydrocarbons derived from terrestrial and anthropogenic sources in the Asian clam, *Potamocorbula amurensis*, in San Francisco Bay estuary, *Mar. Pollut. Bull.*, 24(2):103–109, 1994.

Peters, N.E., Kandell, S.J., Evaluation of streamwater quality in the Atlanta region, U.S. Geological Society reprints from proceedings of the 1997 Georgia Water Resources Conference, March 20–22, 1997, University of Georgia, Athens, GA, 1997, pp. 51–55.

Pfenning, K., McMahon, P.B., Effect of nitrate, organic carbon, and temperature on denitrification rates in an alluvial aquifer, *J. Hydrol.*, 187:283–295, 1996.

Phillips, P.J., Regional water-quality investigation and the ecology of a rare plant species on the Delmarva Peninsula, U.S. Geological Survey, *WRD Ecol. Newsl.*, 3:1–2, 1990.

Phillips, P.J., Riva-Murray, K., Hollister, H.M., Flanary, E.A., Distribution of DDT, chlordane, and total PCBs in bed sediments in the Hudson River basin, *New York Earth Sci. Environ.*, 3(1):27–47, Spring 1997.

Phillips, P.J., Shedlock, R.J., Hydrology and chemistry of ground water and seasonal ponds in the Atlantic coastal plain in Delaware, U.S.A., *J. Hydrol.*, 141:157–178, 1993.

Phillips, P.J., Denver, J.M., Shedlock, R.J., Hamilton, P.A., Effect of forested wetlands on nitrate concentrations in ground water and surface water on the Delmarva Peninsula, *Wetlands*, 13(2):75–83, 1993.

Phillips, P.J., Shedlock, R.J., Hamilton, P.A., National Water-Quality Assessment Program activities on the Delmarva Peninsula in parts of Delaware, Maryland, and Virginia, in Proceedings of the 1988 U.S. Geological Survey Workshop on the Geology and Geohydrology of the Atlantic Coastal Plain, U.S. Geological Survey Circular 1059, 1992, pp. 75–78.

Phillips, P.J., Wall, G.R., Eckhardt, D.A., Freehafer, D.A., Rosenmann, L.A., Pesticide concentrations in surface waters of New York state in relation to land use — 1997, U.S. Geological Survey Water Resources Investigations Report 98-4101, 1998.

Phillips, P.J., Wall, G.R., Thurman, E.M., Eckhardt, D.A., Vanhoesen, J., Metolachlor and its metabolites in tile drain and stream runoff in the Canajoharie Creek watershed, *Environ. Sci. Technol.*, 33(20):3531–3537, 1997.

Pitt, R.E., Effects of urban runoff on aquatic biota, in *Handbook of Ecotoxicology*, Hoffman, D.J., Rattner, B.A., Burton, G.A., Jr., Cairns, J., Jr., Eds., Lewis Publishers, Boca Raton, FL, 1995, pp. 609–630.

Pitt, R.E., McLean, J., Stormwater, baseflow, and snowmelt pollutant contributions from an industrial area, Water Environment Federation 65th Annual Conference & Exposition, Surface Water Quality & Ecology Symposia, Volume VII, September 20–24, New Orleans, LA 1992, Order No. C2007.

Pocernich, M., Litke, D.W., Nutrient concentrations in wastewater-treatment-plant effluents, South Platte River basin, *Water Resour. Bull.*, 33(1):205–214, 1997.

Power, E.A., Chapmann, P.M., Assessing sediment quality, in *Sediment Toxicity Assessment*, Burton, G.A., Jr., Ed., Lewis Publishers, Ann Arbor, MI, 1992.

Puckett, L.J., Estimation of nitrate contamination of an agro-ecosystem outwash aquifer using a nitrogen mass-balance budget, *J. Environ. Qual.*, 28(6):2015–2025, November/December 1999.

Puckett, L.J., Identifying the major sources of nutrient water pollution, *Environ. Sci. Technol.*, 29(9):408–414, September 1995.

Puckett, L.J., Nonpoint and point sources of nitrogen in major watersheds of the United States, U.S. Geological Survey Water-Resources Investigations Report 94-4001, 1994, 9 pp.

Puget Sound Water Quality Authority, Pesticides in Puget Sound, Puget Sound Water Quality Authority, Seattle, WA, 1990.

Puget Sound Water Quality Authority, Puget Sound water quality management plan, action plan, Household Hazardous Waste Program, Puget Sound Water Quality Authority, Seattle, WA 1991, pp. 134–139.

Pytte, A., New Jersey members push for coastal water laws, *Congr. Q. Wkly. Rep.,* 48(29):2295, July 21, 1990.

Rathbun, R.E., Transport, behavior, and fate of volatile organic compounds in streams, U.S. Geological Survey Professional Paper 1589, 1998, 151 pp.

Reddy, K.R., D'Angelo, E.M., Soil processes regulating water quality in wetlands, in *Global Wetlands of Old World and New,* Mitsch, W.J., Ed., Elsevier Science, Amsterdam, 1994.

Reef Relief, Brochure for public education on septic tanks, Key West, FL, 1992.

Reilly, R.E., Plummer, L.N., Phillips, P.J., Busenberg, E., The use of simulation and multiple environmental tracers to quantify groundwater flow in a shallow aquifer, *Water Resour. Res.,* 30(2):421–433, 1994.

Reinhelt, L.E., Horner, R.R., Characterization of the hydrology and water quality of Palustrine Wetlands affected by urban stormwater, Puget Sound wetlands and Stormwater Management Research Program, 1990.

Reiser, R.G., Relation of pesticide concentrations to season, streamflow, and land use in seven New Jersey streams, U.S. Geological Survey Water-Resources Investigation Report 99-4154, 1999, 20 pp.

Reiser, R.G., O'Brien, A.K., Occurrence and seasonal variability of volatile organic compounds in seven New Jersey streams, U.S. Geological Survey Water-Resources Investigation Report 98-4074, 1998, 11 pp.

Reiser, R.G., O'Brien, A.K., Pesticides in streams in New Jersey and Long Island, New York, and relation to land use, U.S. Geological Survey Water-Resources Investigation Report 98-4261, 1999, 12 pp.

Rice, D.W., Seltenrich, S.P., Spies, R.B., Keller, M.L., Seasonal and annual distribution of organic contaminants in marine sediments from Elkhorn Slough, Moss Landing Harbor and nearshore Monterey Bay, California, *Environ. Pollut.,* 82:79–91, 1993.

Richter, K.O., Azous, A., Cooke, S.S., Wisseman, R., Horner, R., Effects of stormwater runoff on wetland zoology and wetland soils characterization and analysis, King County Resource Planning Section, Washington State Department of Ecology, 1991.

Rivard, C.J., Anaerobic bioconversion of municipal solid wastes: effects of total solids levels on microbial numbers and hydrolytic enzyme activities, *Appl. Biochem. Biotechnol.,* 39/40:101–117, 1993.

Rivard, C.J., Anaerobic bioconversion of municipal solid wastes using a novel high-solids reactor design: maximum organic loading rate and comparison with low-solids reactor systems, *Appl. Biochem. Biotechnol.,* 39/40:71–82, 1993.

Rivard, C.J., Anaerobic digestion as a waste disposal option for American Samoa, NREL/TP-422-5043, National Renewable Energy Laboratory, Golden, CO, 1993.

Robertson, D.M., Roerish, E.D., Influence of various water quality sampling strategies on load estimates for small streams, *Water Resour. Res.,* 35(12):3747–3759, 1999.

Rogers, C.S., Responses of coral reefs and reef organisms to sedimentation, *Mar. Ecol. Prog. Ser.,* 62:185–202, 1990.

Rowe, B.L., Summary of published aquatic toxicity information and water-quality criteria for selected volatile organic compounds, U.S. Geological Survey, Open-File Report, OFR 97-563, 1997, 60 pp.

Rowe, B.L., Landrigan, S.J., Lopes, T.J., Summary of published aquatic toxicity information and water-quality criteria for selected volatile organic compounds, U.S. Geological Survey Open-File Report OFR 97-563, 1997.

Rupert, M.G., Probability of detecting atrazine/desethylatrazine and elevated concentrations of nitrate ($NO_2 + NO_3$–N) in groundwater in the Idaho part of the Upper Snake River Basin, U.S. Geological Survey Water Resources Investigation Report 98–4203, 1998, 32 pp.

Rushton, B.T., Dye, C., Hydrologic and water quality characteristics of a wet detention pond, in *The Science, of Water Resources: 1990 and Beyond, November 4–9, 1990,* Jennings, M., Ed., American Water Resources Association, Bethesda, MD, 1990.

Rushton, B.T., Dye, C.W., Tampa office wet detention stormwater treatment, in Annual Report for Stormwater Research Program Fiscal Year 1989–1990, Southwest Florida Water Management District, 1990, pp. 39–74.

Santa Clara Valley Water Control District, Best management practices for automotive-related industries, Santa Clara Valley, Nonpoint Source Pollution Control Program and the San Jose Office of Environmental Management, Santa Clara, CA, 1992.

Schertz, T.L., Alexander, R.B., Ohe, D.J., The computer program Estimate Trend (ESTREND), a system for the detection of trends in water-quality data, U.S. Geological Survey Water-Resources Investigations Report 91-4040, 1991, 63 pp.

Schiffer, D., Impact of stormwater management practices on groundwater, U.S. Geological Survey and the Florida Department of Transportation, Tallahassee, FL, 1990b.

Schiffer, D., Wetlands for stormwater treatment, U.S. Geological Survey and the Florida Department of Transportation, Tallahassee, FL, 1990a.

Shedlock, R.J., The Delmarva study, *Geotimes,* 38(12):12–14, December 1993.

Shedlock, R.J., Hamilton, P.A., Denver, J.M., Phillips, P.J., Multiscale approach to regional ground-water-quality assessment of the Delmarva Peninsula, in *Regional Ground-Water Quality,* Van Nostrand Reinhold, New York, 1993, pp. 563–587.

Shelton, L.R., Field Guide for collecting samples for analysis of volatile organic compounds in stream water for the National Water-Quality Assessment Program, U.S. Geological Survey Open-File Report 97-401, 1997.

Shortle, W.C., Smith, K.T., Minocha, R., Lawrence, G.B., David, M.B., Acidic deposition, cation mobilization, and stress in healthy red spruce trees, *J. Environ. Qual.,* 26:871–876, 1997.

Schroeder, P.R., Palermo, M.R., The automated dredging and disposal alternatives management system (ADDAMS), Environmental Effects of Dredging Technical Note EEDP-06-12, U.S. Army Engineer Waterways Experiment Station, Vicksburg, MS, 1990.

Schubel, J.R., Long Island Sound: facing tough choices, *EPA J.,* 16(6):26–28, November 1990.

Schueler, T., A current assessment of urban best management practices, Metropolitan Washington Council of Governments, Washington, D.C., 1992.

Schueler, T.R., Pollutant dynamics of pond muck, *Watershed Prot. Tech.,* 1(2):39–46, Summer 1994.

Schueler, T.R., Urban pesticides: from the lawn to the stream, *Watershed Prot. Tech.,* 2(1):247–253, Fall 1995.

Schueler, T.R., Galli, J., Herson, L., Kumble, P., Shepp, D., Developing effective BMP strategies for urban watersheds, in Nonpoint Source Watershed Workshop, September 1, 1991, Seminar Publication, U.S. Environmental Agency, Washington, D.C., EPA/625/4-91/027, 1991, pp. 69–83.

Schueler, T.R., Kumble, P.A., Heraty, M.A., A current assessment of urban best management practices: techniques for reducing nonpoint source pollution in the coastal zone, Department of Environmental Programs, Metropolitan Washington Council of Governments, Washington, D.C., 1992.

Schueler, T.R., Lugbill, J., Performance of current sediment control measures at Maryland construction sites, Metropolitan Washington Council of Governments, Washington, D.C., 1990.

Scott, M., Dilutions the solution, *EPA J.,* 17(3):28, November 1991.

Scribner, E.A., Goolsby, D.A., Thurman, E.M., Battaglia, W.A., Reconnaissance for selected herbicides, metabolites, and nutrients in streams of nine Midwestern states, U.S. Geological Survey Open File Report, 98-181, 1998, 36 pp.

Segarra-Garcia, R., Loganathan, V.G., Stormwater detention storage design under random pollutant loading, *J. Water Res. Plan. Manage.,* ASCE 118(5):475–491, 1992.

Semmler, J.A., PCB volatilization from dredged material, Indiana Harbor, Indiana, Environmental Effects of Dredging Technical Note EEDP-02-12, U.S. Army Engineer Waterways Experiment Station, Vicksburg, MS, 1990.

Shaver, E., Sand filter design for water quality treatment, presented at 1991 ASCE Stormwater Conference in Crested Butte, CO, 1991.

Shelton, L.R., Field guide for collecting and processing stream-water samples for the National Water-Quality Assessment Program, U.S. Geological Survey Open-File Report 94-458, 1994, 20 pp.

Small Flows Clearinghouse, West Virginia University, Morgantown, Ed., More states using contructed wetlands for onsite wastewater treatment, *Small Flows,* 6(1):1992.

Smith, R.A., Alexander, R.B., Tasker, G.D., Price, C.V., Robinson, K.W., White, D.A., Statistical modeling of water-quality in regional watersheds, *Watershed,* 1993, pp. 751–754.

Smith, R.A., Schwarz, G.E., Alexander, R.B., Regional interpretation of water-quality monitoring data, *Water Resour. Res.,* 33(12):2781–2798, December 1997.

Smith-Vargo, L., Difficult wastewaters, *Water Eng. Manage.,* 137(10):35–37, October 1990.

Sorensen, J.A., Glass, G.E., Schmidt, K.W., Huber, J.K., Rapp, G.R., Jr., Airborne mercury deposition and watershed characteristics in relation to mercury concentrations in water, sediments, plankton, and fish of eighty northern Minnesota lakes, *Environ. Sci. Technol.,* 24(11):1716–1727, 1990.

Speiran, G.K., Geohydrology and geochemistry near coastal ground-water-discharge areas of the Eastern Shore, Virginia, U.S. Geological Survey Water Supply Paper 2479, 1996, 73 pp.

Squillace, P.J., Environmental behavior and fate of methyl-*tert*-butyl ether (MTBE), U.S. Geological Survey Fact Sheet FS-203-96, 1996, 6 pp. (revised 2/98).

Squillace, P.J., Urban land-use study plan for the National Water Quality Assessment Program, U.S. Geological Survey Open-file Report, OFR 96-217, 1996, 19 pp.

Squillace, P.J., Moran, M.J., Lapham, W.W., Price, C.V., Clawges, R.M., Zigorski, J.S., Volatile organic compounds in untreated ambient groundwater of the United States, 1985-1995, *Environ. Sci. Technol.,* 33:4176–4187, 1999.

Squillace, P.J., Pankow, J.F., Korte, N.E., Zogorski, J.S., Review of the environmental behavior and fate of methyl *tert*-butyl ether, *Environ. Toxic. Chem.,* 16(9):1836–1844, September 1997.

Stackelberg, P.E., Hopple, J.A., Kauffman, L.J., Occurrence of nitrate, pesticides, and volatile compounds in the Kirkwood-Cohansey aquifer system, southern New Jersey, U.S. Geological Survey Water-Resources Investigation Report 97-4241, 1997, 8 pp.

Stamer, J.K., Wieczoreck, M.E., Pesticides in streams in central Nebraska, U.S. Geologicial Survey Fact Sheet FS-232-95, 1995, 4 pp.

Stauffer, J.C., Goldstein, R.M., Comparison of three qualitative habitat indices and their applicability to prairie streams, *N. Am. J. Fish. Manage.,* 17(2):348–361, 1997.

Stell, S.M., National Water-Quality Assessment Program: analysis of available information on pesticides for the Apalachicola–Chattahoochee–Flint River Basin, Proceedings of the 1993 Georgia Water Resources Conference, April 20–21, 1993, Athens, GA, p. 1.

Stewart, J.S., Assessment of alternative methods for stratifying Landsat TM data to improve land-cover classification accuracy across areas with physiographic variation, Master's thesis, University of Wisconsin–Madison, 1994.

Stewart, J.S., Stratification of Landsat thematic mapper data, based on regional landscape patterns, to improve land-cover classification accuracy of large study areas, American Congress on Surveying and Mapping and the American Society for Photogrammetry and Remote Sensing Proceedings, 1995, pp. 826–835.

Stoddard, J.L., Trends in Catskill stream water quality–evidence from historical data, *Water Resour. Res.*, 27(11):2855–2864, 1991.

Stoner, J.D., Lorenz, D.L., Wiche, G.J., Goldstein, R.M., Red River of the north basin, Minnesota, North Dakota, and South Dakota, *Water Resour. Bull.*, 29(4):575–615, July/August 1993. (Reprinted 1993, U.S. Geological Survey's National Water-Quality Assessment Program (NAWQA), American Water Resource Association Monograph Series No. 19.)

Sullivan, D.J., Richards, K.D., National water-quality assessment — western Lake Michigan drainage basin, Clean Water — Clean Environment 21st Century Conference Proceedings, USDA Working Group on Water Quality, 1995, Vol. 3, p. 271.

Svenson, B.-G., Exposure to dioxins and dibenzofurans through the consumption of fish, *N. Engl. J. Med.*, 324(1):8–12, January 3, 1991.

Swartz, R.C., Schults, D.W., Lamberson, J.O., Ozretich, R.J., Stull, J.K., Vertical profiles of toxicity, organic carbon, and chemical contaminants in sediment cores from the Palos Verdes Shelf and Santa Monica Bay, California, *Mar. Environ. Res.*, 31:215–225, 1991.

Swartz, R.C., Schults, D.W., Ozretich, R.J., Lamberson, J.O., Cole, F.A., DeWitt, T.H., Redmond, M.S., Ferraro, S.P., PAH: a model to predict the toxicity of polynuclear aromatic hydrocarbon mixtures in field-collected sediments, *Environ. Toxicol. Chem.*, 14(11):1977–1987, 1995.

Tate, C.M., Heiny, J.S., Organochlorine compounds in bed sediment and fish tissue in the South Platte River basin, U.S.A., 1992-1993, *Arch. Environ. Contamin. Toxicol.*, 30(1):62–78, 1996.

Tate, C.M., Heiny, J.S., The ordination of benthic invertebrate communities in the South Platte River basin in relation to environmental factors, *Freshwater Biol.*, 33(3):439–454, 1995.

Tesoriero, A.J., Voss, F.D., Predicting the probability of elevated nitrate concentrations in the Puget Sound basin, *Ground Water*, 35:1029–1039, 1997.

Transportation Research board, Highway deicing: comparing salt and calcium magnesium acetate, Transportation Research Board, Special Report No. 235, Washington, D.C., 1991.

USACE, Anacostia River Basin reconnaissance study, U.S. Army Corps of Engineers, Baltimore District, 1990.

U.S. Department of Commerce, National Marine Fisheries Service, Fisheries of the United States, 1995, Government Printing Office, Washington, D.C., 1996.

U.S. Department of Commerce, National Marine Fisheries Service, Our living oceans: 1995, NMFS, Silver Spring, MD, 1996.

U.S. Department of Commerce, National Marine Fisheries Service Northeast Region, Status of the fishery resources off the northeast United States for 1994, NMFS, Northeast Fisheries Science Center, Woods Hole, MA, 1995.

U.S. Department of Commerce, National Oceanic and Atmospheric Administration, National Ocean Service, Recent trends in coastal environmental quality: results from the Mussel Watch Project, NOAA, Silver Spring, MD, 1995.

USDOI-BLM, New Mexico State Office, New Mexico riparian-wetland 2000: a management strategy, U.S. Department of the Interior, Bureau of Land Management, 1990.

U.S. EPA, Assessing human health risks from chemically contaminated fish and shellfish: a guidance manual, EPA-503/8-89-002, U.S. Environmental Protection Agency, Office of Marine and Estuarine Protection and Office of Water Regulations and Standards, Washington, D.C., 1989a.

U.S. EPA, Clean Lakes Program, Terrene Institute, 1991, 9 pp.

U.S. EPA, Clean water agenda, remaking the laws that protect our water resources, *EPA J.,* 20(1-2):whole issue, Summer 1994.

U.S. EPA, Cleaning up the nation's waste sites: markets and technology trends, EPA 542-R-92-012 (PB93-140762), Rockville, MD, Government Institutes Inc, for EPA Office of Solid Waste and Emergency Response, 1994a.

U.S. EPA, Criteria for municipal solid waste landfills, U.S. Environmental Protection Agency, Code of Federal Regulations, 40 CFR 258, 1991b.

U.S. EPA, Deposition of air pollutants to the great waters, U.S. Environmental Protection Agency, Office of Air Quality, Research Triangle Park, NC, 1994a.

U.S. EPA, Derivation of EPA's sediment quality advisory levels (Draft), U.S. Environmental Protection Agency, Office of Water, Washington, D.C., 1996.

U.S. EPA, Drinking water regulations and health advisories, U.S. Environmental Protection Agency Report, EPA 822-B-96-001 [variously paged], 1996.

U.S. EPA, Environmental impacts of stormwater discharges, U.S. Environmental Protection Agency, Office of Water, Washington, D.C., 1992a.

U.S. EPA, Environmental impacts of stormwater discharges: a national profile, U.S. Environmental Protection Agency, Office of Water, Washington, D.C., 1992b.

U.S. EPA, EPA administered permit programs: the national pollutant discharge elimination system. Subpart B: permit application and special NPDES program requirements. Storm water discharges, U.S. Environmental Protection Agency, Code of Federal Regulations, 40 CFR 122.26, 1990b.

U.S. EPA, Facility pollution prevention guide, EPA/600/R-92/088, U.S. EPA, Office of Research and Development, Washington, D.C., May 1992.

U.S. EPA, Fact sheet for the multi-sector stormwater general permit. Part VIII: specific requirements for discharges associated with specific industrial activities, U.S. Environmental Protection Agency, *Fed. Regist.,* 60:R61146, November 19, 1993b.

U.S. EPA, Framework for the development of the National Sediment Inventory, U.S. Environmental Protection Agency, Office of Science and Technology, Washington, D.C., 1993a.

U.S. EPA, Great Lakes National Program Office, A report to Congress on the Great Lakes ecosystem, EPA, Chicago, IL, 1994.

U.S. EPA, Great Lakes Water Quality Initiative criteria documents for the protection of wildlife, EPA-820-B-95-008, U.S. Environmental Protection Agency, Office of Water, Washington, D.C., 1995a.

U.S. EPA, Guidance for assessing chemical contamination data for use in fish advisories, Vol. II, Risk assessment and fish consumption limits, EPA-823-B-94-004, U.S. Environmental Protection Agency, Office of Water, Washington, D.C., 1994f.

U.S. EPA, Guidance for assessing chemical contaminant data for use in fish advisories, Vol. 1: fish sampling and analysis (second edition), EPA-823-R-95-007, U.S. Environmental Protection Agency, Office of Water, Washington, D.C., 1995c.

U.S. EPA, Guidance specifying management measures for sources on nonpoint pollution in coastal waters, EPA 840-B-002, U.S. Environmental Protection Agency, Office of Water, January 1993.

U.S. EPA, Guidelines for deriving site-specific sediment quality criteria for the protection of benthic organisms, EPA 822-R-93-017, U.S. Environmental Protection Agency, Office of Science and Technology, Health and Ecological Criteria Division, Washington, D.C., 1993b.

U.S. EPA, Guides to pollution prevention: the fabricated metal products industry, EPA/625/7-90/006, U.S. EPS Office of Research and Development, Washington, D.C., July 1990.

U.S. EPA, Guides to pollution prevention, the automotive refinishing industry, U.S. Environmental Protection Agency, Office of Research and Development, Washington, D.C., EPA/625/7-91/016, October 1991, 1991a.

U.S. EPA, Hazardous waste management system: identification and listing of hazardous waste; recycled used oil management standards. Part IX: relationship to other programs–International Convention for Prevention of Pollution from Ships (MARPOL 73/78). U.S. Environmental Protection Agency, final rule, *Fed. Regist.*, 57FR 41605, September 10, 1992a.

U.S. EPA, Identification and listing of hazardous waste. Definition of hazardous waste, U.S. Environmental Protection Agency, Code of Federal Regulations, 40 CFR 261.3, 1992c.

U.S. EPA, Identification and listing of hazardous waste. Subpart C, characteristic of hazardous waste; toxicity characteristic, U.S. Environmental Protection Agency, Code of Federal Regulations, 40 CFR 261.24, 1990a.

U.S. EPA, Identification of sources contributing to the contamination of the great waters by toxic compounds, U.S. Environmental Protection Agency, Office of Planning and Standards, Durham, NC, 1993c.

U.S. EPA, The incidence and severity of sediment contamination in surface waters of the United States, v.1 of national sediment quality survey, U.S. Environmental Protection Agency, Office of Science and Technology, EPA 823-R-97-006, 1997.

U.S. EPA, Interim methods for development of inhalation reference doses, U.S. Environmental Protection Agency, Office of Research and Development, EPA 600/8-90-066A, Washington, D.C., 1990c.

U.S. EPA, A method for tracing on-site effluent from failing septic systems, in U.S. EPA Nonpoint Source News Notes, U.S. Environmental Protection Agency, Office of Water, Washington, D.C., 1991b.

U.S. EPA, Methods for assessing the toxicity of sediment-associated contaminants with estuarine and marine amphipods, EPA 600-R-94-025, U.S. Environmental Protection Agency, Office of Research and Development, Washington, D.C., 1994b.

U.S. EPA, Methods for measuring the toxicity and bioaccumulation of sediment-associated contaminants with fresh water invertebrates, EPA 600-R-94-024, U.S. Environmental Protection Agency, Duluth, MN, 1994c.

US. EPA, Monitoring guidance for the National Estuary Program, EPA842-B-92-004, U.S. Environmental Protection Agency, Office of Wetlands, Ocean and Watersheds, Oceans and Coastal Protection Division, Washington, D.C., 1992a.

U.S. EPA, National water quality inventory: 1998 report to Congress, EPA841-R-00-001, Washington, June 2000.

U.S. EPA, National water quality inventory: 1992 report to Congress, EPA-841-R-94-001, U.S. Environmental Protection Agency, Office of Water, Washington, D.C., 1994c.

U.S. EPA, Notes of riparian and forestry management, in U.S. EPA Nonpoint Source News Notes, U.S. Environmental Protection Agency, Office of Water, Washington, D.C., March 1992, pp. 10–11.

U.S. EPA, A Phase I inventory of current EPA efforts to protect ecosystems, EPA-841-S-95-001, U.S. Environmental Protection Agency, Office of Water, Washington, D.C., 1995b.

U.S. EPA, POTW sludge sampling and analysis guidance document, U.S. Environmental Protection Agency, Office of Water Enforcement and Permits, Washington, D.C., 1989b.

U.S. EPA, Proceedings of EPAs contaminated sediment management strategy forums, Chicago, IL, April 21–22, Washington, D.C., EPA 823-R-92-007, May 27–28/June 16, 1992 (1992a).

U.S. EPA, Proceedings of the National Sediment Inventory Workshop, April 26–27, 1994, Washington, D.C., U.S. Environmental Protection Agency, Office of Science and Technology, Washington, D.C., 1994d.

U.S. EPA, Proposed guidance specifying management measures for sources of nonpoint pollution in coastal waters, U.S. Environmental Protection Agency, Office of Water, Washington, D.C., 1991d.

U.S. EPA, Provisional guidance for quantitative risk assessment of polycyclic aromatic hydrocarbons, EPA/600/021, U.S. Environmental Protection Agency, Office of Health and Environmental Assessment, Cincinnati, OH, 1993c.

U.S. EPA, Recycled used oil management standards, U.S. Environmental Protection Agency, *Fed. Regist.*, 57 FR 41566, 1992b.

U.S. EPA, Report to Congress on hydrogen sulfide air emissions associated with the extraction of oil and natural gas, EPA–453/R-3-045, U.S. Environmental Protection Agency, Office of Air Quality Planning and Standards, Research Triangle Park, NC, 1993a.

U.S. EPA, Sediment classification methods compendium, EPA823-R-006, U.S. Environmental Protection Agency, Office of Water, Washington, D.C., 1992b.

U.S. EPA, Sequencing batch reactors for nitrification and nutrient removal, U.S. Environmental Protection Agency, Office of Water Enforcement and Compliance, Washington, D.C., 1992c.

U.S. EPA, Snowmelt literature review, Prepared by Tetra Tech for the U.S. Environmental Protection Agency, Washington, D.C., 1991b.

U.S. EPA, Standards for the management of specific hazardous wastes and specific types of hazardous waste management facilities, Subpart H: hazardous waste burned in boilers and industrial furnaces, U.S. Environmental Protection Agency, Code of Federal Regulations, 40 CFR 266m /Subpart HH, 1991a.

U.S. EPA, Standards for the management of used oil, U.S. Environmental Protection Agency, Code of Federal Regulations, 40CFR 279, 1992d.

U.S. EPA, Supplemental manual on the development and implementation of local discharge limitations under the pretreatment program. Residential and commercial toxic pollutant loadings and POTW removal efficiency estimation, U.S. Environmental Protection Agency, Office of Wastewater and Compliance, Washington, D.C., 1991a.

U.S. EPA, Suspended, cancelled, and restricted pesticides, U.S. Environmental Protection Agency, Office of Pesticides and Toxic Substances 20T-4002, 1990.

U.S. EPA, Technical basis for establishing sediment quality criteria for nonionic organic contaminants for the protection of benthic organisms by using equilibrium partitioning [Draft], U.S. Environmental Protection Agency, EPA 822-R-93-011, Office of Science and Technology, Health and Ecological Criteria Division, Washington, D.C., 1993d.

U.S. EPA, Technical support document for water quality-based toxics control, EPA-505/2-90-001, U.S. Environmental Protection Agency, Office of Water Regulations and Standards, Washington, D.C., 1991b.

U.S. EPA, Volunteer lake monitoring: a methods manual, EPA440/4-91-002, 1991, 122 pp.

U.S. EPA, The watershed protection approach: 1993/1994 activity report, EPA-840-S-94-001, U.S. Environmental Protection Agency, Office of Water, Washington, D.C., 1994g.

U.S. EPA et al., Clean Water Action Plan–restoring and protecting America's waters, U.S. Environmental Protection Agency Report EPA-840-R-98-001, 1998, 89 pp.

U.S. EPA, Office of Water, Environmental indicators of water quality in the United States, EPA841-R-96-002, Washington, D.C., 1996, 26 pp.

U.S. EPA, Office of Water, National water quality inventory: 1994 report to Congress, EPA, Washington, D.C., 1995.

U.S. EPA, Office of Water, Liquid assets: a summertime perspective on the importance of clean water to the American economy, EPA, Washington, D.C., 1996.

U.S. EPA, Office of Water, The clean water state revolving fund, financing America's environmental infrastructure–a report of progress, EPA, Washington, D.C., 1995.

U.S. General Accounting Office, Key EPA and state decisions limited by inconsistent and incomplete data, (GAO/RCED-00-54) March 2000, 73 pp.

U.S. General Accounting Office, Many [water pollution] violations have not received appropriate enforcement attention, GAO/RCED-96-23) March 1996, 23 pp.

U.S. General Accounting Office, State revolving fund loans to improve water quality, GAO/RCED-97-19) December 1996, 20 pp.

U.S. Department of Agriculture, Natural Resources Conservation Service, Wetlands: values and trends, NRCS/RCA Issue Brief 4, Washington, D.C., January 1995.

U.S. Department of Agriculture, Natural Resources Conservation Service, Wetlands: programs and partnerships, NRCS/RCA Issue Brief 8, Washington, D.C., January 1996.

U.S. EPA, Guidance for assessing chemical contaminant data for use in fish advisories, Fish Sampling and Analysis, 2nd ed., Vol. 1, U.S. Enviromental Protection Agency, Office of Water, EPA 823-R-95-007.

VanMetre, P.C., Wilson, J.T., Callender, E., Fuller, C.C., Similar rates of decrease of persistent, hydrophobic and particle-reactive contaminants in riverine systems, *Environ. Sci. Technol.*, 32(21):3312–3317, 1998.

Wangsness, D.J., Design and implementation of the National Water-Quality Assessment Program — a United States example, in *Protecting the Danube River Basin Resources*, Murphy, I.L., Ed., Kluwer Academic, Dordrecht, Netherlands, 1997, pp. 89–103.

Wangsness, D.J., National Water-Quality Assessment Program: overview of the Apalachicola-Chattahoochee-Flint River Basin: Proceedings of the 1993 Georgia Water Resources Conference, April 20–21, 1993, Athens, GA, p. 1.

Wangsness, D.J., Frick, E.A., Buell, G.R., DeVivo, J.C., Effect of the restricted use of phosphate detergent and upgraded wastewater-treatment facilities on water quality in the Chattahoochee River near Atlanta, GA, Based on U.S. Geological Survey Open-file Report 94-99, 1994, 4 pp.

Welch, A.H., Helsel, D.R., Focazio, M.J., Watkins, S.A., Arsenic in ground water supplies of the United States in *Arsenic Exposure and Health Effects*, Chappell, W.R., Abernathy, C.O., Calderon, R.L., Eds., Elsevier Science, Amsterdam, 1999.

Welch, A.H., Watkins, S.A., Helsel, D.R., Focazio, M.J., Arsenic in ground-water resources of the United States, U.S. Geological Fact Sheet FS-063-00, 2000, 4 pp.

Wenning, R.J., Bonnevie, N.L., Huntley, S.L., Accumulation of metals, polychlorinated biphenyls, and polycyclic aromatic hydrocarbons in sediments from the lower Passaic River, New Jersey, *Arch. Environ. Contam. Toxicol.*, 27(1):64–81, 1994.

Viessman, W., Hammer, M.J., *Water Supply and Pollution Control*, Harper Collins, New York, 1993.

Wilber, W.G., Davis, J.V., Assessing the quality of the nation's water–point vs nonpoint sources of phosphorus in rivers, *U.S. Geological Survey Yearbook, Fiscal Year 1993*, 1994, pp. 48-49.

Woodward-Clyde, The use of wetlands for controlling stormwater pollution, Prepared for
    U.S. Environmental Protection Agency, Region 5, Chicago, IL, 1991a.
Woodward-Clyde, Urban BMP cost and effectiveness summary data for 6217(g) guidance:
    erosion and sediment control during construction-draft, December 12, 1991, 1991b.
Woodward-Clyde, Urban nonpoint source pollution resource notebook, final draft report,
    1991c.
Woodward-Clyde, Urban management practices cost and effectiveness summary data for
    6217(g) guidance: onsite sanitary disposal systems, prepared for U.S. Environmental
    Protection Agency, Washington, D.C., 1992a.
Woodward-Clyde, Urban BMP cost and effectiveness summary data for 6217(g) guidance:
    erosion and sediment control during construction, prepared for U.S. Environmental
    Protection Agency, Washington, D.C., 1992b.
Yousef, Y.A., Lin, L., Sloat, J., Kay, K., Maintenance guidelines for accumulated sediments
    in retention/detention ponds receiving highway runoff, Florida Department of Trans-
    portation, Tallahassee, FL, 1991.
Zogorski, J.S., Baehr, A.L., Bauman, B.J., Conrad, D.L., Drew, R.T., Korte, N.E., Lapham,
    W.W., Morduchowitz, A., Pankow, J.F., Washington, E.R., Significant findings and
    water-quality recommendations of the Interagency Oxygenated Fuel Assessment, in
    Contaminated Soils, Vol. 2, Kostecki, P.T., Calabrese, E.J., Bonazountas, M., Eds.,
    Amherst Scientific, Amherst, MA, 1997, pp. 661–679.
Zynjuk, L.D., Majedi, B.F., January 1996 floods deliver large loads of nutrients and sediment
    to the Chesapeake Bay, U.S. Geological Survey Fact Sheet 140-96, 1996, 2 pp.

CHAPTER 8

Abramova, F., Grinberg, L., Yampolskaya, O., Walker, D., Pathology of inhalational anthrax
    in forty-two cases from the Sverdlovsk outbreak of 1979, *Proc. Natl. Acad. Sci. USA,*
    90:2291–2294, 1993.
Army Reserve National Guard, Stakeholders III Conference medical panel, National Guard
    Bureau, Arlington, VA, December 1998.
Ashford, D.A., Rotz, L.D., Perkins, B.A., Use of anthrax vaccine in the United States:
    recommendations of the Advisory Committee on Immunization Practice (ACIP),
    MMWR, 49(RR-15), 2000.
ATSDR, Recent evidence of illnesses linked to exposure to hazardous substances, ATSDR,
    Atlanta, GA, 1996.
Auf Der Heide, E., *Disaster Response: Principles of Preparation and Coordination,* Mosby-
    Yearbook, St. Louis, MO, 1990.
Balf, T., *Community Right to Know Compliance Handbook,* Business and Legal Reports,
    Madison, CT, 1988.
Bioterrorism alleging use of anthrax and interim guidelines for management — United States,
    1998, *MMWR,* 48(01):69–74, February 5, 1999.
Bioweapons and bioterrorism, *JAMA,* 278:351–370, 389–436, 1997.
Brachman, P.S., Friedlander, A.M., Anthrax, in *Vaccines,* Plotkin, S.A., Mortimer, E.A., Eds.,
    W.B. Saunders, Philadelphia, 1994.
Brachman, P.S., Kaufmann, A., Anthrax, in *Bacterial Infections of Humans,* Evans, A.S.,
    Brachman, P.S., Eds., Plenum Medical, New York, 1998, pp. 95–107.
Broad, W., Miller, J., The threat of germ weapons is rising. Fear, too, *New York Times,*
    December 27, 1998.

Burke, Robert, *Hazardous Materials Chemistry for Emergency Responders,* Lewis Publishers, Boca Raton, FL, 1997.

Cann, A.J., *Principles of Molecular Virology,* 2nd ed., Harcourt Brace, Academic Press, San Diego, 1997, p. 230.

Carus, S., Bioterrorism and biocrimes: the illicit use of biological agents in the 20th century, Center for Counterproliferation Research, National Defense University, Washington, D.C., 1998.

CDC, Beyond the flood: a prevention guide for personal health and safety, U.S. Department of Health and Human Services, Public Health Service, Atlanta, GA, 1993.

CDC, Bioterrorism Readiness Plan: A Template for Healthcare Facilities, 1998.

CDC, Congressional testimony addressing biological terrorism by Dr. James Hughes, June 2, 1998.

CDC, Congressional testimony addressing biological terrorism by Dr. James Hughes, April 20, 1999.

CDC, Congressional testimony addressing chemical terrorism by Dr. Richard Jackson, June 2, 1998.

CDC, Congressional testimony addressing chemical terrorism by Dr. Jeffrey P. Koplan, March 25, 1999.

CDC, Congressional testimony addressing chemical terrorism by Dr. Stephen M. Ostroff, May 20, 1999.

CDC, Preliminary report: medical examiner reports of deaths associated with Hurricane Andrew — Florida, August 1992, *MMWR,* 41:641–644, 1992.

CDC, Preventing emerging infectious diseases: a strategy for the 21st century, 1998.

Centers for Disease Control and Prevention, Emergency mosquito control associated with Hurricane Andrew — Florida and Louisiana 1992, *MMWR,* 42:240–242, 1993.

Centers for Disease Control and Prevention, Human anthrax associated with an epizootic among livestock — North Dakota, 2000, *MMWR,* 50:677–680, 2001.

Centers for Disease Control and Prevention, Notice to readers: ongoing investigation of anthrax — Florida, 2001, *MMWR,* 50:877, 2001.

Centers for Disease Control and Prevention, Rapid assessment of vectorborne diseases during the Midwest flood — United States, 1993, *MMWR,* 42:240–242, 1994.

Centers for Disease Control and Prevention, Update: investigation of anthrax associated with intentional exposure and interim public health guidelines, October 2001, *MMWR,* 50:889–903, 2001.

Centers for Disease Control and Prevention, Update: investigation of bioterrorism-related anthrax and interim guidelines for clinical evaluation of persons with possible anthrax, *MMWR,* 50:941–948, 2001.

Centers for Disease Control and Prevention, Update: investigation of bioterrorism-related anthrax and interim guidelines for exposure management and antimicrobial therapy, October 2001, *MMWR,* 50:909–919, 2001.

Commission on Engineering and Technical Systems, National Research Council, *Managing Troubled Waters: The Role of Marine Environmental Monitoring,* National Academy Press, Washington, D.C., 1990.

Cotton, P., Health threat from mosquitoes rises as flood of the century finally recedes, *JAMA,* 270:685–686, 1993.

Committee on R&D Needs for Improving Civilian Medical Response to Chemical and Biological Terrorism Incidents, *Chemical and Biological Terrorism: Research and Development to Improve Civilian Medical Response,* Institute of Medicine, National Research Council, National Academy Press, Washington, D.C., 1999, pp. 129–132.

Danzig, R., Berkowsky, P.B., Why should we be concerned about biological warfare?, *JAMA*, 285:431–432, 1997.

Daplan, E., Marchell, A., *The Cult at the End of the World*, Crown Publishing Group, New York, 1996.

Dixon, T., Meselson, M., Guillemin, J., Hanna, P., Anthrax, *N. Engl. J. Med.*, 341:815–826, 1999.

Ekeus, R., Iraq's biological weapons programme: UNSCOM's experience, Memorandum report to the United Nations Security Council, New York, November 20, 1996.

Emmons, R.W., Ascher, M.S., Dondero, D.V., Enge, B., Milby, M.M., Hui, L.T. et al., Surveillance for arthropod-borne viral activity and disease in California during 1991, in Proceedings of the California Mosquito and Vector Control Association, 1992.

Emmons, R.W., Ascher, M.S., Dondero, D.V., Enge, B., Reisen, W.K., Milby, M.M. et al., Surveillance for arthropod-borne viral activity and disease in California during 1992, in Proceedings of the California Mosquito and Vector Control Association, 1993.

Emmons, R.W., Ascher, M.S., Dondero, D.V., Enge, B., Reisen, W.K., Milby, M.M. et al., Surveillance for arthropod-borne viral activity and disease in California during 1993, in *Proceedings of the California Mosquito and Vector Control Association*, 1994.

Epidemiologic report, smallpox, Canada, *MMWR*, 11:258, 1962.

Federal Bureau of Investigation, Domestic terrorism in the United States, Washington, D.C., 1995.

Fenner, F., Henderson, D.A., Arita, I., Jezek, Z., Ladnyi, I., Smallpox and its eradication, World Health Organization, Geneva, 1988.

Fletcher, J., Human anthrax in the United States: a descriptive review of case reports, 1955–1999, Tollins School of Public Health, Emory University, Atlanta, GA, 2000.

Franz, D.R., Jahrling, P.B., Friedlander, A.M., McClain, D.J., Hoover, D.L., Bryne, W.R., Pavlin, J.A., Christopher, G.W., Eitzen, E.M., Jr., Clinical recognition and management of patients exposed to biological warfare agents, *JAMA*, 278:399–411, 1997.

Friedlander, A.M., Welkos, S.L., Pitt, M.L. et al., Postexposure prophylaxis against experimental inhalation anthrax, *J. Infect. Dis.*, 167:1239–1243, 1993.

Glass, R.I., Noji, E.K., Epidemiologic surveillance following disasters, in *Public Health Surveillance*, Halperin, W.E., Baker, E.L., Monson, R.R., Eds., Van Nostrand Reinhold, New York, 1992, pp. 195–205.

Henderson, D.A., Bioterrorism as a public health threat, *Emerging Infect. Dis.*, July/September 4(3), 1998.

Hughart, J.L., Chemicals and terrorism: human health threat analysis, mitigation and prevention information for Kanawha County and Nitro, West Virginia, ATSDR, Atlanta, GA, 1998.

Hughart, J.L., Common toxics and terrorism, Proceedings of the 1996 Southwest Counter-Terrorism Training Symposium, Nevada Department of Public Safety, Las Vegas, NV, 1996.

Hughart, J.L., Health risk assessment and Las Vegas case study, Proceedings of the 1997 Annual Conference of the Nevada Public Health Association, Las Vegas, NV, 1997.

Hughart, J.L., Terrorist use of expedient chemical agents, 15th Chemical Conference and NBC Symposium, U.S. Army Chemical Corps, Ft. McClellan, AL, 1997.

Ikle, F., Waiting for the next Lenin, *The National Interest*, Spring 1997.

Kolavic, S.A., Kimura, A., Simons, S.L., Slutsker, L., Barth, S., Haley, C.E., An outbreak of *Shigella dysenteriae* Type 2 among laboratory workers due to intentional food contamination, *JAMA*, Abstracts, August 6, 1997.

Laqueur, W., Postmodern terrorism, *Foreign Affairs*, September/Octtober 1996.

Lee, C., U.S. can't ignore implications of TB plaguing Russia, *USA Today*, February 8, 1999, pp. 16A, 232.

Lee, L.E., Fonseca, V., Brett, K.M. et al., Active morbidity surveillance after Hurricane Andrew — Florida, 1992, *JAMA*, 270:591–594, 1993.

McDade, J.E., Franz, D., Bioterrorism as a public health threat, *Emerging Infect. Dis.*, July/September 4(3), 1998.

Meselson, M., Guillemin, V., Hugh-Jones, M., Langmuir, A., Popova, I., Shelokov, A. et al., The Sverdlovsk anthrax outbreak of 1979, *Science*, 266:1202–1208, 1994.

Mitchel, A., Panel discussion at conference, Integrating Medical and Emergency Response, Office of State and Local Domestic Preparedness Support, U.S. Department of Justice, Washington, D.C., March 10, 1999.

Moore, C.G., McLean, R.G., Mitchell, C.J., Nasci, R.S., Tsai, T.F., Calisher, C.H. et al., Guidelines for arbovirus surveillance programs in the United States, U.S. Department of Health and Human Services, Centers for Disease Control and Prevention, 1993.

Nasci, R.S., Moore, C.G., Planning for emergency mosquito surveillance and control, *Wing Beats*, 4:4–7, 1993.

National Research Council, *Assessing the Nation's Earthquakes: The Health and Future of Regional Seismograph Networks*, National Academy Press, Washington, D.C., 1990.

National Response Team, Oil contingency planning: national status: a report to the President from Samuel K. Skinner, Secretary, Department of Transportation and William K. Reilly, Administrator, Environmental Protection Agency, Washington, D.C., 1990.

Noji, E.K., Disaster epidemiology: challenges for public health action, *J. Public Health Policy*, 13:332–340, 1992.

Patrick, P., Brenner, S.A., Noji, E.K., Lee, J., The American Red Cross — Centers for Disease Control natural disaster morbidity and mortality surveillance system [letter], *Am. J. Public Health*, 82:1690, 1992.

Philen, R., Combs, D.L., Miller, L. et al., Hurricane Hugo, 1989, *Disasters*, 15:177–179, 1992.

Pile, J.C., Malone, J.D., Eitzen, E.N., Friedlander, A.M., Anthrax as a potential biological warfare agent, *Arch. Intern. Med.*, 158:429–434, 1998.

Rall, D.P., Toxic agent and radiation control: progress toward objectives for the nation for the year 1990, *Public Health Rep.*, 103(4):342–347, July/August 1988.

Reeves, W.C., Epidemiology and control of mosquito-borne arboviruses in California, 1943–1987, Sacramento, CA, California Mosquito and Vector Control Association, 1990.

Reich, M.R., Spong, J.K., Kepone: a chemical disaster in Hopewell, Virginia, *Int. J. Health Serv.*, 13(2):227–246, 1983.

Roberts, B., New challenges and new policy priorities for the 1990s, in *Biologic Weapons: Weapons of the Future*, Center for Strategic and International Studies, Washington, D.C., 1993.

Seaman, J., Leivesley, S., Hogg, C., Epidemiology of natural disasters, in *Contributions to Epidemiology and Biostatistics*, Vol. 5, Klingberg, M.A., Papier, C., Eds., Karger, Basel, New York, 1984, pp. 49–70.

Sheridan, R., Brief review of smoke inhalation injury, Shriner Burns Institute, Boston, MA, 1998.

Stuempfle, A.K. et al., Final Report of Task Force 25: Hazard from industrial chemicals, U.S., Department of Defense, March 18, 1996.

Tachakra, S.S., The Bhopal disaster, *J. R. Soc. Health*, 107(1):1–2, February 1987.

Tonat, K., Panel discussion at conference, Integrating Medical and Emergency Response, Office of State and Local Domestic Preparedness Support, U.S. Department of Justice, Washington, D.C., March 10, 1999.

Török, T.J., Tauxe, R.V., Wise, R.P., Livengood, J.R., Sokolow, R., Mauvais, S. et al., A large
      community outbreak of salmonellosis caused by intentional contamination of restau-
      rant salad bars, *JAMA,* 278:389–395, 1997.
Tucker, J.B., National health and medical services response to incidents of chemical and
      biological terrorism, *JAMA,* 285:362–368, 1997.
U.S. Army Medical Research Institute for Infectious Diseases/CDC/Food and Drug Admin-
      istration, Medical response to biological warfare and terrorism [Satellite Broadcast],
      U.S. Department of Defense/U.S. Department of Health and Human Services, CDC,
      Atlanta, Georgia, September 22–24, 1998.
U.S. EPA, Hazardous substances in our environment: a citizen's guide to understanding health
      risks and reducing exposure, Washington, D.C., 1990.
Vorobyov, A., Criterion rating as a measure of probable use of bio agents as biological
      weapons, in papers presented to the Working Group on Biological Weapons Control
      of the Committee on International Security and Arms Control, National Academy of
      Sciences, Washington, D.C., April 1994.
Wehrle, P.F., Posch, J., Richter, K.H., Henderson, D.A., An airborne outbreak of smallpox in
      a German hospital and its significance with respect to other recent outbreaks in
      Europe, *Bull. WHO,* 4:669–679, 1970.
Zalinskas, R.A., Iraq's biological weapons: the past as future?, *JAMA,* 278:418–424, 1997.

CHAPTER 9

American Conference of Governmental Industrial Hygienists, *Air Sampling Instruments*, 9th
      ed. ACGIH, Cincinnati, OH, 2001.
Bisesi, M.S., Kohn, J.P., *Industrial Hygiene Evaluation Methods*, Lewis Publishers/CRC
      Press, Boca Raton, FL, 1995.
Bisesi, M.S., Bisesi, C.A., Hazardous Waste Management, in *The Occupational Environment
      — Its Evaluation and Control,* Dinardi, S., Ed., American Industrial Hygiene Asso-
      ciation Publications, Fairfax, VA, 1997.
Boulding, J.R., *Description and Sampling of Contaminated Soils*, 2nd ed., Lewis Publish-
      ers/CRC Press, Boca Raton, FL, 1994.
Carter, M.R., Ed., *Soil Sampling and Methods of Analysis*, Lewis Publishers, Boca Raton,
      FL, 1993.
Chou, J., *Hazardous Gas Monitors: A Practical Guide to Selection, Operation and Applica-
      tions*, McGraw-Hill Book, New York, 2000.
Clesceri, L.S., Greenberg, A.E., Eaton, A.D., Eds., *Standard Methods for the Examination of
      Water and Wastewater*, 20th ed., APHA/AWWA/WEF, Washington, D.C., 1998.
Keith, L.H., *Environmental Sampling and Analysis: A Practical Guide*, Lewis Publishers,
      Chelsea, MI, 1991.
Kenkel, J., *Analytical Chemistry Refresher Manual*, Lewis Publishers, Chelsea, MI, 1992.
Marshall, T.L., Chaffin, C.T., Hammaker, R.M., Fately, W.G., Introduction to open path FT-IR
      atmospheric monitoring, *Environ. Sci. Technol.,* 28(5):224A–232A, 1994.
Wight, G.D., *Fundamentals of Air Sampling*, Lewis Publishers, Chelsea, MI, 1994.

Index

VOLUME II

Index

VOLUME I